BICENTENNIAL

1807

WILEY

2007

BICENTENNIAL

THE WILEY BICENTENNIAL—KNOWLEDGE FOR GENERATIONS

*E*ach generation has its unique needs and aspirations. When Charles Wiley first opened his small printing shop in lower Manhattan in 1807, it was a generation of boundless potential searching for an identity. And we were there, helping to define a new American literary tradition. Over half a century later, in the midst of the Second Industrial Revolution, it was a generation focused on building the future. Once again, we were there, supplying the critical scientific, technical, and engineering knowledge that helped frame the world. Throughout the 20th Century, and into the new millennium, nations began to reach out beyond their own borders and a new international community was born. Wiley was there, expanding its operations around the world to enable a global exchange of ideas, opinions, and know-how.

For 200 years, Wiley has been an integral part of each generation's journey, enabling the flow of information and understanding necessary to meet their needs and fulfill their aspirations. Today, bold new technologies are changing the way we live and learn. Wiley will be there, providing you the must-have knowledge you need to imagine new worlds, new possibilities, and new opportunities.

Generations come and go, but you can always count on Wiley to provide you the knowledge you need, when and where you need it!

WILLIAM J. PESCE
PRESIDENT AND CHIEF EXECUTIVE OFFICER

PETER BOOTH WILEY
CHAIRMAN OF THE BOARD

MATTER & INTERACTIONS I

MODERN MECHANICS

SECOND EDITION

RUTH W. CHABAY

BRUCE A. SHERWOOD

North Carolina State University

JOHN WILEY & SONS, INC.

EXECUTIVE PUBLISHER	Kaye Pace
SENIOR ACQUISITIONS EDITOR	Stuart Johnson
PRODUCTION ASSISTANT	Andrea Juda
SENIOR MARKETING MANAGER	Amanda Wygal
EDITORIAL ASSISTANT	Aly Rentrop

This book was set in New Baskerville by the authors and printed and bound by Courier Companies. The cover was printed by Courier Companies.

This book is printed on acid free paper.

To order books or for customer service please, call 1-800-CALL WILEY (225-5945).

ISBN 978- 0-470-10830-7

Printed in the United States of America

10 9 8 7 6 5 4

Preface to Volume I

Matter & Interactions I: Modern Mechanics focuses on the atomic structure of matter and the interactions that matter undergoes. This two-volume textbook emphasizes that there are only a small number of fundamental principles that underlie the behavior of matter, and that using these powerful principles it is possible to construct models that can explain and predict a wide variety of physical phenomena.

Prerequisites

This book is intended for introductory calculus-based college physics courses taken by science and engineering students. You will need a basic knowledge of derivatives and integrals, which can be obtained by studying calculus concurrently. A brief review of derivatives is available in an appendix at the end of this volume.

Modeling

This textbook places a major emphasis on constructing and using physical "models." A central aspect of science is the modeling of complex real-world phenomena. A physical model is based on what we believe to be fundamental principles; its intent is to predict or explain the most important aspects of an actual situation. Modeling necessarily involves making approximations and simplifying assumptions in order that the model can be analyzed in detail. The principles of classical mechanics and thermal physics that are the focus of Volume I have wide applicability. We can use these principles to model systems as different as molecules and galaxies.

Computer modeling

Computer modeling has become important, because it makes it possible to analyze complex systems which would otherwise require very sophisticated mathematics or which could not be analyzed at all without a computer. Numerical calculations based on the Momentum Principle give us the opportunity to watch the dynamical evolution of the behavior of a system. Simple models frequently need to be refined and extended. This can be done straightforwardly with a computer model but is often impossible with a purely analytical (non-numerical) model.

Computer modeling is now as important as theory and experiment in contemporary science and engineering. We introduce you to serious computer modeling right away to help you build a strong foundation in the use of this important tool.

Experiments using simple equipment

Some end-of-chapter problems involve experiments using simple equipment such as weights, string, coffee filters, a stopwatch, a weak spring, a styrofoam cup, and a toy gyroscope. You will find that by experimenting with very simple equipment you can gain insight into rather deep scientific issues.

? Stop and Think

As you read the text, you will frequently come to a paragraph that asks you to stop and think by making a prediction, carrying out a step in a derivation or analysis, or applying a principle. Usually these questions are answered in the following paragraphs, but it is important that you make a serious effort to answer the questions on your own before reading further. Be honest in comparing your answers to those in the text. Paying attention to surprising or counterintuitive results can be a useful learning strategy.

Exercises

Small exercises that require you to apply new concepts are found at the end of many sections of the text. These may involve qualitative reasoning or simple calculations. You should work these exercises when you come to them, to consolidate your understanding of the material you have just read. Answers to the exercises are at the end of each chapter.

Conventions used in diagrams and equations

The conventions most commonly used to represent vectors and scalars in diagrams in this text are shown in the adjacent figure. In equations and text, a vector will be written with an arrow above it:

$$\vec{p}$$

The magnitude of $\vec{p}$ is written as $|\vec{p}|$ or simply as p.
The x component of $\vec{p}$ is written as p_x.
A vector may be written in terms of components or in terms of unit vectors along the x, y, and z axes:

$$\vec{p} = \langle p_x, p_y, p_z \rangle = p_x\hat{\imath} + p_y\hat{\jmath} + p_z\hat{k}$$

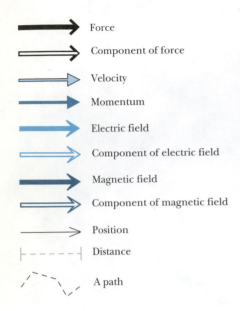

Force

Component of force

Velocity

Momentum

Electric field

Component of electric field

Magnetic field

Component of magnetic field

Position

Distance

A path

To the instructor

The approach to mechanics and thermal physics taken in this volume differs significantly from that in most introductory physics texts. Key emphases of the approach are these:
- Starting from fundamental principles rather than secondary formulas
- Atomic-level description and analysis
- Modeling the real world through idealizations and approximations
- Computer modeling of physical systems
- Qualitative reasoning that complements the quantitative reasoning
- Unification of mechanics and thermal physics
- Tight integration of theory and experiment

The following one-semester introductory course on electricity and magnetism is addressed by *Matter & Interactions II: Electric & Magnetic Interactions*, Ruth Chabay & Bruce Sherwood, John Wiley & Sons.

Instructor web site

For a wealth of useful information, go to the Matter & Interactions web site:

http://www4.ncsu.edu/~rwchabay/mi

This web site includes information of a general character, including useful educational software, and information on how to obtain access to additional resources for instructors, such as a daily log of what we do in own classrooms, sample course web sites, sample exams, etc. At the start of the daily log is procurement information for the simple experiments used with this volume.

Articles dealing with Matter & Interactions

Three key articles dealing with this curriculum are the following:

Chabay, R. W. & Sherwood, B. A. (1999). Bringing atoms into first-year physics. *American Journal of Physics* **67**, 1045-1050.

Chabay, R. W. & Sherwood, B. A. (2004). Modern mechanics. *American Journal of Physics* **72**, 439-445.

Chabay, R. W. & Sherwood, B. A. (2006). Restructuring Introductory E&M. *American Journal of Physics*, **74**, 329-336.

These articles are available on the Matter & Interactions web site, along with an article from an on-line book edited by Edward Redish and Patrick Cooney dealing with introductory physics curricula that are based on physics education research. The article includes implementation aspects, evaluations, etc.

Computer homework problems

Some critically important homework problems are designed for the student to write a computer program. We strongly recommend VPython as the best tool for student use. An adequate subset of the underlying Python programming language can be taught in an hour or two, even to students who have never written a program before, and instructional materials are available to instructors.

Real-time 3D animations are generated as a side effect of student computations, and these animations provide powerfully motivating and instructive visualizations of fields and motions. VPython supports true vector computations, which encourages students to begin thinking about vectors as powerful tools rather than as barriers to understanding. VPython can be obtained at no cost for Windows, Macintosh, and Linux at http://vpython.org.

Homework problems

In addition to the large homework problems provided at the end of a chapter, there are many small-scale problems in the form of exercises distributed throughout the chapter, and small review questions at the end of the chapter. You may wish to ask students to write out some exercises or review questions as part of the homework, or to have them work out exercises at intervals during a lecture. We have prepared an extensive and sophisticated set of exercises and problems for the WebAssign computer homework system (http://www.webassign.net); see the Matter & Interactions web site.

Choosing topics

In the big course for engineering and science students at North Carolina State University, with three 50-minute lectures and one 110-minute small-group studio lab every week, we are able to complete most but not all of this modern mechanics volume in a 15-week semester, up to the middle of Chapter 11 (not including the Boltzmann distribution), and you can read the organizational details in the daily log available in the instructor resources mentioned above. In an honors course, or a course for physics majors, it is possible to do almost the entire book.

What can be omitted if there is not enough time to do everything? Let us say that the one thing we feel should *not* be omitted is the introduction to entropy in terms of quantum statistical mechanics (Chapter 11). This is a climax of the theme of integrating mechanics and thermal physics. One way to decide what can be omitted is to be guided by what needs to be done to ensure dealing adequately with entropy.

Any starred section (*) can safely be omitted. Material in these sections is not referenced in later work. Next we offer chapter by chapter discussions of what other topics might be omitted, and what the ramifications are of such omissions.

Chapter 1 (Interactions and Motion): Unless your students have a particularly strong background in vectors, all of this chapter is important for later work (except for starred sections).

Chapter 2 (The Momentum Principle: Impulse and Momentum Change): The starred sections are reference material for those interested but not essential. Note the examples that engage students right away with concepts of modeling: making approximations, simplifying assumptions, and estimates. It is important to talk about these concepts, because they usually are unfamiliar even to students with strong physics backgrounds.

Chapter 3 (The Momentum Principle: Non-constant Forces): You can omit the section on determinism, as we do.

Chapter 4 (Contact Forces and the Momentum Principle): We omit buoyancy and pressure (you could return to these topics during Chapter 12 on gases).

Chapter 5 (The Energy Principle): This is the foundation for the energy concept, and nothing should be omitted (other than starred sections, of course). The starred section on "a puzzle" is an amusing teaser that hints at field energy. The starred derivations at the end of the chapter are probably of interest to few students but are provided for completeness.

Chapter 6 (Energy in Macroscopic Systems): If you are pressed for time, you might choose to omit the second half of the chapter on energy dissipation, beginning with Section 6.9. We ourselves are able to mention dissipative processes only briefly.

Chapter 7 (Energy Quantization): None of the unstarred sections of this short chapter should be omitted.

Chapter 8 (Multiparticle Systems): The starred sections on modeling friction are not mathematically difficult but are conceptually challenging. We skip over the formalism of finding the center of mass, because the important applications have obvious centers of mass locations.

Chapter 9 (Collisions): A good candidate for omission is the analysis of collisions in the center-of-mass frame. Note that there is a basic introduction to collisions in Chapter 2, but that was before energy had been introduced.

Chapter 10 (Angular Momentum): The starred sections on gyroscopes are interesting but students find them difficult, presumably because the analysis includes both a nonzero torque and three dimensions. We typically show the students a bicycle wheel precessing at the end of a rope and sketch briefly the analysis, but we do not have time to go into detail.

Chapter 11 (Entropy: Limits on the Possible): We strongly recommend doing at least the first half of this chapter, on the Einstein solid. To make this work well, exposition and student computing should be interwoven. It might seem odd to ask the students simply to reproduce the results shown in the textbook, but in our experience and that of our students it is enormously educational to be required to make the ideas concrete in a computer program. You find yourself asking "What exactly is q_1?" and answering very concretely by actually calculating with the initially unfamiliar quantities. It is the difference between writing an essay about entropy and calculating the actual value of the entropy.

Despite its intrinsic importance, you might omit the second half of the chapter, on the Boltzmann distribution, as we do (for lack of time). However, if you do Chapter 12 you will need a few key results from this analysis.

Chapter 12 (Gases and Engines): If you do this chapter, you should not omit the initial sections, including the relationship of pressure to the results from Chapter 11. We should point out that the important formula for particle flow ($nA\bar{v}$) and the concept of mean free path are both used in Volume II (though they are rederived there). You might decide to omit the sections on macroscopic energy transfers (isothermal and adiabatic processes). The sections on heat engines represent additional applications of the second law of thermodynamics and may be omitted. Another possibility is to omit just the sections on nonzero-power engines.

Acknowledgments for Volume I

We owe much to the unusual working environment provided by the Department of Physics and the former Center for Innovation in Learning at Carnegie Mellon, which made it possible during the 1990s to carry out the research and development leading to the first edition of this two-volume textbook. We are grateful for the open-minded attitude of our colleagues in the Carnegie Mellon physics department toward curriculum innovations.

We are grateful to the support of our colleagues Robert Beichner and John Risley in the Physics Education Research and Development group at North Carolina State University, and to other colleagues in the NCSU physics department.

We thank Fred Reif for emphasizing the role of the three fundamental principles of mechanics, and for his view on the reciprocity of electric and gravitational forces. We thank Robert Bauman, Gregg Franklin, and Curtis Meyer for helping us think deeply about energy. Vidhya Ramachandran contributed a useful take-home experiment on heat capacity.

Much of Chapter 11 on quantum statistical mechanics is based on an article by Thomas A. Moore and Daniel V. Schroeder, "A different approach to introducing statistical mechanics," *American Journal of Physics* vol. 65, pp. 26-36 (January 1997). We have benefited from many stimulating conversations with Thomas Moore, author of another introductory textbook that takes a contemporary view of physics, "Six Ideas that Shaped Physics." Michael Weissman and Robert Swendsen provided particularly helpful critiques on some aspects of our implementation of Chapter 11.

We thank David Andersen, David Scherer, and Jonathan Brandmeyer for the development of tools that enabled us and our students to write associated software.

We thank our colleagues David Brown, Krishna Chowdary, Laura Clarke, Thomas Foster, Chris Gould, Mark Haugan, Joe Heafner, Andrew Hirsch, Matthew Kohlmyer, Barry Luokkala, Jonathan Mitschele, Prabha Ramakrishnan, Michael Schatz, Robert Swendsen, Aaron Titus, Michael Weissman, and Hugh Young.

We thank a group of reviewers assembled by the publisher, who gave us useful critiques on this volume:

Kelvin Chu	University of Vermont
Michael Dubson	University of Colorado-Boulder
Tom Furtak	Colorado School of Mines
Dave Goldberg	Drexel University
Javed Iqbal	University of British Columbia
Shawn Jackson	University of Tulsa
Craig Ogilvie	Iowa State University
Michael Politano	Marquette University
Rex Ramsier	The University of Akron
Larry Weinstein	Old Dominion University
Michael Weissman	University of Illinois

This project was supported, in part, by the National Science Foundation (grants MDR-8953367, USE-9156105, DUE-9554843, DUE-9972420, DUE-0320608, and DUE-0237132). Opinions expressed are those of the authors, and not necessarily those of the Foundation.

Ruth W. Chabay
Bruce A. Sherwood
Department of Physics
North Carolina State University
November 2006

Image Credits

Chapter 1: Randall Feenstra (Figure 1.3). Chapter 4: Randall Feenstra (Figure 4.5). Chapter 6: Stacey Benson (Figure 6.36 and Figure 6.37). Judith Harrison (Figure 6.44).

All other figures were created by the authors, using Adobe Illustrator, Adobe Photoshop, VPython, POV-Ray, and Specular Infini-D.

The text was produced by the authors as print-ready copy, using Adobe FrameMaker. The font is 10-point New Baskerville.

CHAPTER 1

INTERACTIONS AND MOTION

This textbook deals with the nature of matter and its interactions. The variety of phenomena that we will be able to explain and understand is very wide, including the orbit of stars around a black hole, nuclear fusion, and the speed of sound in a solid.

> The main goal of this textbook is to have you engage in a process central to science: the attempt to explain in detail a broad range of phenomena using a small set of powerful fundamental principles.

> The specific focus is on learning how to model the nature of matter and its interactions in terms of a small set of physical laws that govern all mechanical interactions, and in terms of the atomic structure of matter.

Chapter 1 introduces the notion of interactions between material objects and the changes they produce.

KEY CONCEPTS
- Fundamental physics principles apply to all kinds of matter, from galaxies to subatomic particles
- Change is an indication of an interaction
- Position and motion in 3D space can be described precisely by vectors
- The momentum of an object depends on both mass and velocity

1.1 KINDS OF MATTER

We will deal with material objects of many sizes, from subatomic particles to galaxies. All of these objects have certain things in common.

Atoms and nuclei
Ordinary matter is made up of tiny atoms. An atom isn't the smallest type of matter, for it is composed of even smaller objects (electrons, protons, and neutrons), but many of the ordinary everyday properties of ordinary matter can be understood in terms of atomic properties and interactions. As you probably know from studying chemistry, atoms have a very small, very dense core, called the nucleus, around which is found a cloud of electrons. The nucleus contains protons and neutrons, collectively called nucleons. Electrons are kept close to the nucleus by electric attraction to the protons (the neutrons don't interact with the electrons).

> **?** Recall your previous studies of chemistry. How many protons and electrons are there in a hydrogen atom? In helium or carbon atoms?

If you don't remember the properties of these atoms, see the periodic table on the inside front cover of this textbook. Hydrogen is the simplest atom, with just one proton and one electron. A helium atom has two protons and two electrons. A carbon atom has six protons and six electrons. Near the other end of the chemical periodic table, a uranium atom has 92 protons and 92 electrons. Figure 1.1 shows the approximate cloud of electrons for several elements but cannot show the nucleus to the same scale; the tiny dot marking the nucleus in the figure is much larger than the actual nucleus.

The radius of the electron cloud for a typical atom is about 1×10^{-10} meter. The reason for this size can be understood using the principles of

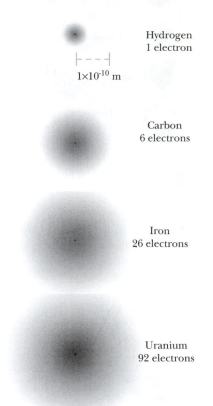

Hydrogen
1 electron

1×10^{-10} m

Carbon
6 electrons

Iron
26 electrons

Uranium
92 electrons

Figure 1.1 Atoms of hydrogen, carbon, iron, and uranium. The black dot shows the location of the nucleus. On this scale, however, the nucleus would be much too small to see.

? "Stop and think" questions

Throughout this text you will encounter "Stop and Think" questions, preceded by a large question mark. Try to answer each question before reading the next paragraph. Trying to answer a question by using what you already know, as well as material you have just read, helps you learn more than just reading passively.

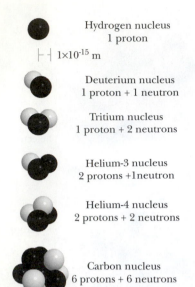

Hydrogen nucleus
1 proton

⊢⊣ 1×10⁻¹⁵ m

Deuterium nucleus
1 proton + 1 neutron

Tritium nucleus
1 proton + 2 neutrons

Helium-3 nucleus
2 protons +1neutron

Helium-4 nucleus
2 protons + 2 neutrons

Carbon nucleus
6 protons + 6 neutrons

Figure 1.2 Nuclei of hydrogen, helium, and carbon. Note the *very* much smaller scale than in Figure 1.1!

quantum mechanics, a major development in physics in the early 20th century. The radius of a proton is about 1×10^{-15} meter, very much smaller than the radius of the electron cloud.

Nuclei contain neutrons as well as protons (Figure 1.2). The most common form or "isotope" of hydrogen has no neutrons in the nucleus. However, there exist isotopes of hydrogen with one or two neutrons in the nucleus (in addition to the proton). Hydrogen atoms containing one or two neutrons are called deuterium or tritium. The most common isotope of helium has two neutrons (and two protons) in its nucleus, but a rare isotope has only one neutron; this is called helium-3.

The most common isotope of carbon has six neutrons together with the six protons in the nucleus (carbon-12), while carbon-14 with eight neutrons is an isotope that plays an important role in dating archeological objects.

Near the other end of the periodic table, uranium-235, which can undergo a fission chain reaction, has 92 protons and 143 neutrons, while uranium-238, which does not undergo a fission chain reaction, has 92 protons and 146 neutrons.

Molecules and solids

When atoms come in contact with each other, they may stick to each other ("bond" to each other). Several atoms bonded together can form a molecule—a substance whose physical and chemical properties differ from those of the constituent atoms. For example, water molecules (H_2O) have properties quite different from the properties of hydrogen atoms or oxygen atoms.

An ordinary-sized rigid object made of bound-together atoms and big enough to see and handle is called a solid, such as a bar of aluminum. A new kind of microscope, the scanning tunneling microscope (STM), is able to map the locations of atoms on the surface of a solid, which has provided new techniques for investigating matter at the atomic level. Two such images appear in Figure 1.3. You can see that atoms in a crystalline solid are arranged in a regular three-dimensional array. The arrangement of atoms on the surface depends on the direction along which the crystal is cut. The irregularities in the bottom image reflect "defects," such as missing atoms, in the crystal structure.

Liquids and gases

When a solid is heated to a higher temperature, the atoms in the solid vibrate more vigorously about their normal positions. If the temperature is raised high enough, this thermal agitation may destroy the rigid structure of the solid. The atoms may become able to slide over each other, in which case the substance is a liquid.

At even higher temperatures the thermal motion of the atoms or molecules may be so large as to break the interatomic or intermolecular bonds completely, and the liquid turns into a gas. In a gas the atoms or molecules are quite free to move around, only occasionally colliding with each other or the walls of their container.

We will learn how to analyze many aspects of the behavior of solids and gases. We won't have much to say about liquids, because their properties are much harder to analyze. Solids are simpler to analyze than liquids because the atoms stay in one place (though with thermal vibration about their usual positions). Gases are simpler to analyze than liquids because between collisions the gas molecules are approximately unaffected by the other molecules. Liquids are the awkward intermediate state, where the atoms move around rather freely, but always in contact with other atoms. This makes the analysis of liquids very complex.

Figure 1.3 Two different surfaces of a crystal of pure silicon. The images were made with a scanning tunneling microscope. Images courtesy of Randall Feenstra, Carnegie Mellon University.

Planets, stars, solar systems, and galaxies
In our brief survey of the kinds of matter that we will study, we make a giant leap in scale from atoms all the way up to planets and stars, such as our Earth and Sun. We will see that many of the same principles that apply to atoms apply to planets and stars. By making this leap we bypass an important physical science, geology, whose domain of interest includes the formation of mountains and continents. We will study objects that are much bigger than mountains, and we will study objects that are much smaller than mountains, but we don't have time to apply the principles of physics to every important kind of matter.

Our Sun and its accompanying planets constitute our Solar System (Figure 1.4). It is located in the Milky Way galaxy, a giant rotating disk-shaped system of stars. On a clear dark night you can see a band of light (the Milky Way) coming from the huge number of stars lying in this disk, which you are looking at from a position in the disk, about two-thirds of the way out from the center of the disk. Our galaxy is a member of a cluster of galaxies that move around each other much as the planets of our Solar System move around the Sun. The Universe contains many such clusters of galaxies.

1.2 DETECTING INTERACTIONS

Objects made of different kinds of matter interact with each other in various ways: gravitationally, electrically, magnetically, and through nuclear interactions. How can we detect that an interaction has occurred? In this section we consider various kinds of observations that indicate the presence of interactions.

? Before you read further, take a moment to think about your own ideas of interactions. How can you tell that two objects are interacting with each other?

Change of direction
Suppose you observe a proton moving through a region of outer space, far from almost all other objects. The proton moves along a path like the one shown in Figure 1.5. The arrow indicates the initial direction of the proton's motion, and the "x's" in the diagram indicate the position of the proton at equal time intervals.

? Do you see evidence that the proton is interacting with another object?

Evidently a change in direction is a vivid indicator of interactions. If you observe a change in direction of the motion of a proton, you will find another object somewhere that has interacted with this proton.

? Suppose that the only other object nearby was another proton. What was the approximate initial location of this second proton?

Since two protons repel each other electrically, the second proton must have been located to the right of the bend in the first proton's path.

Change of speed
Suppose that you observe an electron traveling in a straight line through outer space far from almost all other objects (Figure 1.6). The path of the electron is shown as though a camera had taken multiple exposures at equal time intervals.

? Where is the electron's speed largest? Where is the electron's speed smallest?

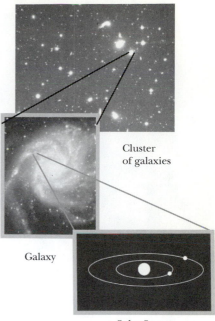

Figure 1.4 Our Solar System exists inside a galaxy, which itself is a member of a cluster of galaxies.

Figure 1.5 A proton moves through space, far from almost all other objects. The initial direction of the proton's motion is upward, as indicated by the arrow. The x's represent the position of the proton at equal time intervals.

Figure 1.6 An electron moves through space, far from almost all other objects. The initial direction of the electron's motion is upward and to the left, as indicated by the arrow. The x's represent the position of the electron at equal time intervals.

The speed is largest at the top, where the dots are farther apart, which means that the electron has moved farthest during the time interval between exposures. The speed is smallest at the bottom, where the dots are closer together, which means that the electron has moved the least distance during the time interval between exposures.

? Suppose that the only other object nearby was another electron. What was the approximate initial location of this other electron?

The other electron must have been located directly below the starting point of the path, since electrons repel each other electrically.

Evidently a change in speed is an indicator of interactions. If you observe a change in speed of an electron, you will find another object somewhere that has interacted with the electron.

Change of velocity: change of speed or direction

In physics, the word "velocity" has a special technical meaning which is different from its meaning in everyday speech. In physics, the quantity called "velocity" indicates a combination of speed and direction. (In contrast, in everyday speech, "speed" and "velocity" are often used as synonyms. In physics, however, all words have precise meanings and there are no synonyms.)

For example, consider an airplane that is flying with a speed of 1000 kilometers/hour in a direction that is due east. We say the velocity is 1000 km/hr, east, where we specify both speed and direction. An airplane flying west with a speed of 1000 km/hr would have the same speed, but a different velocity.

We have seen that a change in an object's speed, or a change in the direction of its motion, indicates that the object has interacted with at least one other object. The two indicators of interaction, change of speed and change of direction, can be combined into one compact statement:

> A change of velocity (speed or direction or both) indicates the existence of an interaction.

1   2

Figure 1.7 Two successive positions of a particle (indicated by a dot), with arrows indicating the velocity of the particle at each location. The shorter arrow indicates that the speed of the particle at location 2 is less than its speed at location 1.

Diagrams showing changes in velocity

In physics diagrams, the velocity of an object is represented by an arrow: a line with an arrowhead. The tail of the arrow is placed at the location of the object, and the arrow points in the direction of the motion of the object. The length of the arrow is proportional to the speed of the object. Figure 1.7 shows two successive positions of a particle at two different times, with velocity arrows indicating a change in speed of the particle (it's slowing down). Figure 1.8 shows three successive positions of a different particle at three different times, with velocity arrows indicating a change in direction but no change in speed.

We will see a little later in the chapter that velocity is only one example of a physical quantity that has a "magnitude" (an amount or a size) and a direction. Other examples of such quantities are position relative to an origin in 3D space, force, and magnetic field. Quantities having magnitude and direction can be usefully described as "vectors". Vectors are mathematical quantities which have their own special rules of algebra, similar (but not identical) to the rules of ordinary algebra. Arrows are commonly used in diagrams to denote vector quantities. We will use vectors extensively.

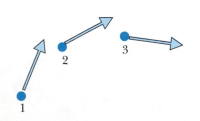

Figure 1.8 Three successive positions of a particle (indicated by a dot), with arrows indicating the velocity of the particle at each location. The arrows are the same length, indicating the same speed, but they point in different directions, indicating a change in direction and therefore a change in velocity.

Uniform motion

Suppose you observe a rock moving along in outer space far from all other objects. We don't know what made it start moving in the first place; presumably a long time ago an interaction gave it some velocity and it has been coasting through the vacuum of space ever since.

It is an observational fact that such an isolated object moves at constant, unchanging speed, in a straight line. Its velocity does not change (neither its direction nor its speed changes). We call such motion with unchanging velocity "uniform motion" (Figure 1.9).

An object at rest

A special case of uniform motion is the case in which an object's speed is zero and remains zero—the object remains at rest. In this case the object's speed is constant (zero) and the direction of motion, while undefined, is not changing.

Uniform motion implies no net interaction

When we observe an object in uniform motion, we conclude that since its velocity is not changing, either it is not interacting significantly with any other object, or else it is undergoing multiple interactions that cancel each other out. In either case, we can say that there is no "net" (total) interaction.

1.3 NEWTON'S FIRST LAW OF MOTION

The basic relationship between change of velocity and interaction is summarized qualitatively by Newton's "first law of motion":

NEWTON'S FIRST LAW OF MOTION

An object moves in a straight line and at constant speed except to the extent that it interacts with other objects.

Figure 1.9 "Uniform motion"—no change in speed or direction.

The words "to the extent" imply that the stronger the interaction, the more change there will be in direction and/or speed. The weaker the interaction, the less change. If there is no net interaction at all, the direction doesn't change and the speed doesn't change (uniform motion). This case can also be called "uniform velocity" or "constant velocity," since velocity refers to both speed and direction. It is important to remember that if an object is not moving at all, its velocity is not changing, so it too may be considered to be in uniform motion.

Newton's first law of motion is only qualitative, because it doesn't give us a way to calculate quantitatively how much change in speed or direction will be produced by a certain amount of interaction, a subject we will take up in the next chapter. Nevertheless, Newton's first law of motion is important in providing a conceptual framework for thinking about the relationship between interaction and motion.

The English physicist Isaac Newton was the first person to state this law clearly. Newton's first law of motion represented a major break with ancient tradition, which assumed that constant pushing was required to keep something moving. This law says something radically different: no interactions at all are needed to keep something moving!

Does Newton's first law apply in everyday life?

Superficially, Newton's first law of motion may at first seem not to apply to many everyday situations. To push a chair across the floor at constant speed, you have to keep pushing all the time.

? Doesn't Newton's first law of motion say that the chair should keep moving at constant speed without anyone pushing it? In fact, shouldn't the speed or direction of motion of the chair change due to the interaction with your hands? Does this everyday situation violate Newton's first law of motion? Try to answer these questions before reading farther.

The complicating factor here is that your hands aren't the only objects that are interacting with the chair. The floor also interacts with the chair, in a way that we call friction. If you push just hard enough to compensate exactly for the floor friction, the sum of all the interactions is zero, and the chair moves at constant speed as predicted by Newton's first law. (If you push harder than the floor does, the chair's speed does increase.)

Motion without friction

It is difficult to observe motion without friction in everyday life, because objects almost always interact with many other objects, including air, flat surfaces, etc. This explains why it took people such a long time (Newton was born in 1642) to understand clearly the relationship between interaction and change.

You may be able to think of situations in which you have seen an object keep moving at constant (or nearly constant) velocity, without being pushed or pulled. One example of a nearly friction-free situation is a hockey puck sliding on ice. The puck slides a long way at nearly constant speed in a straight line (constant velocity) because there is little friction with the ice. An even better example is the uniform motion of an object in outer space, far from all other objects.

EXERCISES

At the end of a section you will usually find exercises such as these. It is important to work through exercises as you come to them, to make sure that you can apply what you have just read. Simply reading about concepts in physics is not enough—you must be able to use the concepts in answering questions and solving problems.

Make a serious attempt to do an exercise before checking the answer at the end of the chapter. This will help you assess your own understanding.

1.X.1 Which of the following objects are moving with constant velocity?
(a) A ship sailing northeast at a speed of 5 meters per second
(b) The moon orbiting the Earth
(c) A tennis ball traveling across the court after having been hit by a tennis racket
(d) A can of soda sitting on a table
(e) A person riding on a Ferris wheel which is turning at a constant rate

1.X.2 Apply Newton's first law to each of the following situations. In which situations can you conclude that the object is undergoing a net interaction with one or more other objects?
(a) A book slides across the table and comes to a stop
(b) A proton in a particle accelerator moves faster and faster
(c) A car travels at constant speed around a circular race track
(d) A spacecraft travels at a constant speed toward a distant star
(e) A hydrogen atom remains at rest in outer space

1.X.3 A spaceship far from all other objects uses its thrusters to attain a speed of 10^4 m/s. The crew then shuts off the power. According to Newton's first law, what will happen to the motion of the spaceship from then on?

1.4 OTHER INDICATORS OF INTERACTION

Change of identity

Change of velocity (change of speed and/or direction) is not the only indicator of interactions. Another is change of identity, such as the formation of water (H_2O) from the burning of hydrogen in oxygen. A water molecule behaves very differently from the hydrogen and oxygen atoms of which it is made.

Change of shape or configuration

Another indicator of interaction is change of shape or configuration (arrangement of the parts). For example, slowly bend a pen or pencil, then hold it in the bent position. The speed hasn't changed, nor is there a change in the direction of motion (it's not moving!). The pencil has not changed identity. Evidently a change of shape can be evidence for interactions, in this case with your hand.

Other changes in configuration include "phase changes" such as the freezing or boiling of a liquid, brought about by interactions with the surroundings. In different phases (solid, liquid, gas), atoms or molecules are arranged differently. Changes in configuration at the atomic level are another indication of interactions.

Change of temperature

Another indication of interaction is change of temperature. Place a pot of cold water on a hot stove. As time goes by, a thermometer will indicate a change in the water due to interaction with the hot stove.

Other indications of interactions

Is a change of position an indicator of an interaction? That depends. If the change of position occurs simply because a particle is moving at constant speed and direction, then a mere change of position is not an indicator of an interaction, since uniform motion is an indicator of no net interaction.

? If however you observe an object at rest in one location, and later you observe it again at rest but in a different location, did an interaction take place?

Yes. You can infer that there must have been an interaction to give the object some velocity to move the object toward the new position, and another interaction to slow the object to a stop in its new position.

In later chapters we will consider interactions involving change of identity, change of shape, and change of temperature, but for now we'll concentrate on interactions that cause a change of velocity (speed and/or direction).

Indirect evidence for an additional interaction

Sometimes there is indirect evidence for an additional interaction. When something doesn't change although you would normally expect a change due to a known interaction, this indicates that another interaction is present. Consider a balloon that hovers motionless in the air despite the downward gravitational pull of the Earth. Evidently there is some other kind of interaction that opposes the gravitational interaction. In this case, interactions with air molecules have the net effect of pushing up on the balloon ("buoyancy"). The lack of change implies that the effect of the air molecules exactly compensates for the gravitational interaction with the Earth.

When you push a chair across the floor and it moves with constant velocity despite your pushing on it (which ought to change its speed), that means that something else must also be interacting with it (the floor).

The stability of the nucleus of an atom is another example of indirect evidence for an additional interaction. The nucleus contains positively charged protons that repel each other electrically, yet the nucleus remains intact. We conclude that there must be some other kind of interaction present, a nonelectric attractive interaction that overcomes the electric repulsion. This is evidence for an interaction called the "strong interaction" that acts between protons and neutrons in the nucleus.

Summary: changes as indicators of interactions

Here then are the most common indicators of interactions:
- change of velocity (change of direction and/or change of speed)
- change of identity
- change of shape
- change of temperature
- lack of change when change is expected (indirect evidence)

The important point is this: **Interactions cause change.**

In the absence of interactions, there is no change, which is usually uninteresting. The exception is the surprise when nothing changes despite our expectations that something should change. This is indirect evidence for some interaction that we hadn't recognized was present, that more than one interaction is present and the interactions cancel each others' effects.

For the next few chapters we'll concentrate on change of velocity as evidence for an interaction (or lack of change of velocity, which can give indirect evidence for additional interactions).

1.5 DESCRIBING THE 3D WORLD: VECTORS

Physical phenomena take place in the 3D world around us. In order to be able to make quantitative predictions and give detailed, quantitative explanations, we need tools for describing precisely the positions and velocities of objects in 3D, and the changes in position and velocity due to interactions. These tools are mathematical entities called 3D "vectors."

3D coordinates

We will use a 3D coordinate system to specify positions in space and other vector quantities. Usually we will orient the axes of the coordinate system as shown in Figure 1.10: $+x$ axis to the right, $+y$ axis upward, and $+z$ axis coming out of the page, toward you. This is a "right-handed" coordinate system: if you hold the thumb, first, and second fingers of your right hand perpendicular to each other, and align your thumb with the x axis and your first finger with the y axis, your second finger points along the z axis. (In some math and physics textbook discussions of 3D coordinate systems, the y axis points out and the z axis points up, but we will also use a 2D coordinate system with y up, so it makes sense always to have the y axis point up.)

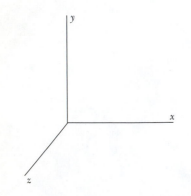

Figure 1.10 Right-handed 3D coordinate system.

Basic properties of vectors: magnitude and direction

A vector is a quantity that has a magnitude and a direction. For example, the velocity of a baseball is a vector quantity. The magnitude of the baseball's velocity is the speed of the baseball, for example 20 meters/second. The direction of the baseball's velocity is the direction of its motion at a particular instant, for example "up" or "to the right" or "west" or "in the $+y$ direction." A symbol denoting a vector is written with an arrow over it:

$$\vec{v} \text{ is a vector.}$$

A position in space can also be considered to be a vector, called a position vector, pointing from an origin to that location. Figure 1.11 shows a position vector that might represent your final position if you started at the origin and walked 4 meters along the x axis, then 2 meters parallel to the z axis, then climbed a ladder so you were 3 meters above the ground. Your new position relative to the origin is a vector that can be written like this:

$$\vec{r} = \langle 4, 3, 2 \rangle \text{ m}$$

$$x \text{ component } r_x = 4 \text{ m}$$

$$y \text{ component } r_y = 3 \text{ m}$$

$$z \text{ component } r_z = 2 \text{ m}$$

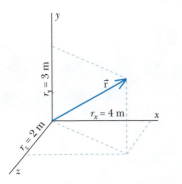

Figure 1.11 A position vector $\vec{r} = \langle 4, 3, 2 \rangle$ m and its x, y, and z components.

In three dimensions a vector is a triple of numbers $\langle x, y, z \rangle$. Quantities like the position of an object and the velocity of an object can be represented as vectors:

$$\vec{r} = \langle x, y, z \rangle \text{ (a position vector)}$$

$$\vec{r}_1 = \langle 3.2, -9.2, 66.3 \rangle \text{ m (a position vector)}$$

$$\vec{v} = \langle v_x, v_y, v_z \rangle \text{ (a velocity vector)}$$

$$\vec{v}_1 = \langle -22.3, 0.4, -19.5 \rangle \text{ m/s (a velocity vector)}$$

Components of a vector

Each of the numbers in the triple is referred to as a "component" of the vector. The x component of the vector $\vec{v}$ is the number v_x. The z component of the vector $\vec{v}_1 = \langle -22.3, 0.4, -19.5 \rangle$ m/s is -19.5 m/s. A component such as v_x is not a vector, since it is only one number.

It is important to note that the x component of a vector specifies the difference between the x coordinate of the tail of the vector and the x coordinate of the tip of the vector. It does not give any information about the location of the tail of the vector (compare Figure 1.11 and Figure 1.12).

Equality of vectors

A vector is equal to another vector if and only if all the components of the vectors are equal. If $\vec{r} = \langle 4, 3, 2 \rangle$ m,

$$\vec{w} = \vec{r} \text{ means that}$$

$$w_x = r_x \text{ and } w_y = r_y \text{ and } w_z = r_z, \text{ so } \vec{w} = \langle 4, 3, 2 \rangle \text{ m}$$

If two vectors are equal, their magnitudes and directions are the same.

Drawing vectors

In Figure 1.11 we represented your position vector relative to the origin graphically by an arrow whose tail is at the origin and whose arrowhead is at your position. The length of the arrow represents the distance from the origin, and the direction of the arrow represents the direction of the vector, which is the direction of a direct path from the initial position to the final position (the "displacement"; by walking and climbing you "displaced" yourself from the origin to your final position).

Since it is difficult to draw a 3D diagram on paper, when working on paper you will usually be asked to draw vectors which all lie in a single plane. Figure 1.13 shows an arrow in the xy plane representing the vector $\langle -3, -1, 0 \rangle$.

Vectors and scalars

A quantity which is represented by a single number is called a *scalar*. A scalar quantity does not have a direction. Examples include the mass of an object, such as 5 kg, or the temperature, such as -20 C. Vectors and scalars are very different entities; a vector can never be equal to a scalar, and a scalar cannot be added to a vector. Scalars can be positive or negative:

$$m = 5 \text{ kg}$$
$$T = -20 \text{ C}$$

Although a component of a vector such as v_x is not a vector, it's not a scalar either, despite being only one number. An important property of a true scalar is that its value doesn't change if we orient the xyz coordinate axes differently. Rotating the axes doesn't change an object's mass, or the temperature, but it does change what we mean by the x component of the velocity since the x axis now points in a different direction.

1.X.4 How many numbers are needed to specify a 3D position vector?

1.X.5 How many numbers are needed to specify a scalar?

1.X.6 Does the symbol $\vec{a}$ represent a vector or a scalar?

1.X.7 Which of the following are vectors?
(a) 5 m/s   (b) $\langle -11, 5.4, -33 \rangle$ m   (c) $\vec{r}$ (d) v_z

1.X.8 $\vec{a} = \langle -3, 7, 0.5 \rangle$. If $\vec{b} = \vec{a}$, what is the y component of $\vec{b}$?

Magnitude of a vector

Consider again the vector in Figure 1.14, showing your displacement from the origin. Using a 3D extension of the Pythagorean theorem for right tri-

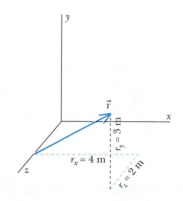

Figure 1.12 The arrow represents the vector $\vec{r} = \langle 4, 3, 2 \rangle$ m, drawn with its tail at location $\langle 0, 0, 2 \rangle$.

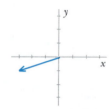

Figure 1.13 The position vector $\langle -3, -1, 0 \rangle$, drawn at the origin, in the xy plane. The components of the vector specify the displacement from the tail to the tip. The z axis, which is not shown, comes out of the page, toward you.

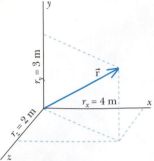

Figure 1.14 A vector representing a displacement from the origin.

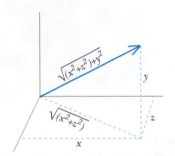

Figure 1.15 The magnitude of a vector is the square root of the sum of the squares of its components (3D version of the Pythagorean theorem).

Figure 1.16 Two vectors (Exercise 1.X.10).

angles (Figure 1.15), the net distance you have moved from the starting point is

$$\sqrt{(4\text{ m})^2 + (3\text{ m})^2 + (2\text{m})^2} = \sqrt{29}\text{ m} = 5.39\text{ m}$$

We say that the *magnitude* $|\vec{r}|$ of the position vector $\vec{r}$ is

$$|\vec{r}| = 5.39\text{ m}$$

The magnitude of a vector is written either with absolute-value bars around the vector as $|\vec{r}|$, or simply by writing the symbol for the vector without the little arrow above it, r.

The magnitude of a vector can be calculated by taking the square root of the sum of the squares of its components (see Figure 1.15).

MAGNITUDE OF A VECTOR

If the vector $\vec{r} = \langle r_x, r_y, r_z \rangle$ then $|\vec{r}| = \sqrt{r_x^2 + r_y^2 + r_z^2}$ (a scalar).

The magnitude of a vector is always a positive number. The magnitude of a vector is a single number, not a triple of numbers, and it is a scalar, not a vector.

The magnitude of a vector is a true scalar, because its value doesn't change if you rotate the coordinate axes. Rotating the axes changes the individual components, but the length of the arrow representing the vector doesn't change.

Can a *vector* be positive or negative?

? Consider the vector $\vec{v} = \langle 8\times10^6, 0, -2\times10^7 \rangle$ m/s. Is this vector positive? Negative? Zero?

None of these descriptions is appropriate. The x component of this vector is positive, the y component is zero, and the z component is negative. Vectors aren't positive, or negative, or zero. Their components can be positive or negative or zero, but these words just don't mean anything when used with the vector as a whole.

On the other hand, the *magnitude* of a vector such as $|\vec{v}|$ is always positive.

1.X.9 Does the symbol $|\vec{v}|$ represent a vector or a scalar?

1.X.10 Consider the vectors $\vec{r}_1$ and $\vec{r}_2$ represented by arrows in Figure 1.16. Are these two vectors equal?

1.X.11 If $\vec{r} = \langle -3, -4, 1 \rangle$ m, find $|\vec{r}|$.

1.X.12 Can the magnitude of a vector be a negative number?

1.X.13 What is the magnitude of the vector $\vec{v}$, where $\vec{v} = \langle 8\times10^6, 0, -2\times10^7 \rangle$ m/s?

Mathematical operations involving vectors

Although the algebra of vectors is similar to the scalar algebra with which you are very familiar, it is not identical. There are some algebraic operations that cannot be performed on vectors.

Algebraic operations that *are* legal for vectors include the following operations, which we will discuss in this chapter:

- adding one vector to another vector: $\vec{a} + \vec{w}$
- subtracting one vector from another vector: $\vec{b} - \vec{d}$
- finding the magnitude of a vector: $|\vec{r}|$
- finding a unit vector (a vector of magnitude 1): $\hat{r}$
- multiplying (or dividing) a vector by a scalar: $3\vec{v}$ or $\vec{w}/2$

• finding the rate of change of a vector: $\dfrac{\Delta \vec{r}}{\Delta t}$ or $\dfrac{d\vec{r}}{dt}$

In later chapters we will also see that there are two more ways of combining two vectors:

the vector dot product, whose result is a scalar
the vector cross product, whose result is a vector

Operations that are *not* legal for vectors

Although vector algebra is similar to the ordinary scalar algebra you have used up to now, there are certain operations that are not legal (and not meaningful) for vectors:

A vector cannot be set equal to a scalar.
A vector cannot be added to or subtracted from a scalar.
A vector cannot occur in the denominator of an expression. (Although you can't divide by a vector, note that you can legally divide by the *magnitude* of a vector, which is a scalar.)

Multiplying a vector by a scalar

A vector can be multiplied (or divided) by a scalar. If a vector is multiplied by a scalar, each of the components of the vector is multiplied by the scalar:

$$\text{If } \vec{r} = \langle x, y, z \rangle \text{ then } a\vec{r} = \langle ax, ay, az \rangle$$

$$\text{If } \vec{v} = \langle v_x, v_y, v_z \rangle \text{ then } \frac{\vec{v}}{b} = \langle \frac{v_x}{b}, \frac{v_y}{b}, \frac{v_z}{b} \rangle$$

$$\left(\frac{1}{2}\right)\langle 6, -20, 9 \rangle = \langle 3, -10, 4.5 \rangle$$

Multiplication by a scalar "scales" a vector, keeping its direction the same but making its magnitude larger or smaller (Figure 1.17). Multiplying by a negative scalar reverses the direction of a vector.

Magnitude of a scalar

You may wonder how to find the magnitude of a quantity like $-3\vec{r}$, which involves the product of a scalar and a vector. This expression can be factored:

$$|-3\vec{r}| = |-3| \cdot |\vec{r}|$$

The magnitude of a scalar is its absolute value, so:

$$|-3\vec{r}| = |-3| \cdot |\vec{r}| = 3\sqrt{r_x^2 + r_y^2 + r_z^2}$$

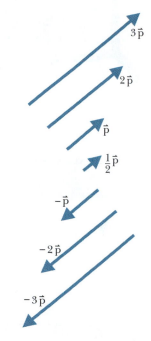

Figure 1.17 Multiplying a vector by a scalar changes the magnitude of the vector. Multiplying by a negative scalar reverses the direction of the vector.

1.X.14 If $\vec{v} = \langle 2, -3, 5 \rangle$ m/s , what is $3\vec{v}$?

1.X.15 If $\vec{r} = \langle 2, -3, 5 \rangle$ m , what is $\dfrac{\vec{r}}{2}$?

1.X.16 What is the result of multiplying the vector $\vec{a}$ by the scalar f, where $\vec{a} = \langle 0.02, -1.7, 30.0 \rangle$ and $f = 2.0$?

1.X.17 How does the direction of the vector $-\vec{a}$ compare to the direction of the vector $\vec{a}$?

1.X.18 Is $3 + \langle 2, -3, 5 \rangle$ a meaningful expression? If so, what is its value?

1.X.19 Is $\dfrac{4}{\langle 6, -7, 4 \rangle}$ a meaningful expression? If so, what is its value?

Direction of a vector: Unit vectors

One way to describe the direction of a vector is by specifying a *unit vector*. A unit vector is a vector of magnitude 1, pointing in some direction. A unit vector is written with a "hat" (caret) over it instead of an arrow. The unit vector $\hat{a}$ is called "a-hat".

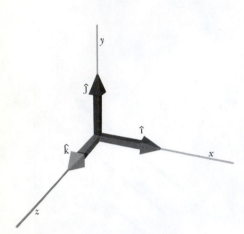

Figure 1.18 The unit vectors $\hat{i}$, $\hat{j}$, $\hat{k}$.

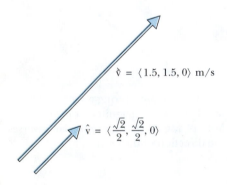

Figure 1.19 The unit vector $\hat{v}$ has the same direction as the vector $\vec{v}$, but its magnitude is 1, and it has no physical units.

? Is the vector $\langle 1, 1, 1 \rangle$ a unit vector?

The magnitude of $\langle 1, 1, 1 \rangle$ is $\sqrt{1^2 + 1^2 + 1^2} = 1.73$. so this is not a unit vector.

The vector $\langle 1/\sqrt{3}, 1/\sqrt{3}, 1/\sqrt{3} \rangle$ is a unit vector, since its magnitude is 1:

$$\sqrt{\left(\frac{1}{\sqrt{3}}\right)^2 + \left(\frac{1}{\sqrt{3}}\right)^2 + \left(\frac{1}{\sqrt{3}}\right)^2} = 1$$

Note that every component of a unit vector must be less than or equal to 1.

In our 3D Cartesian coordinate system, there are three special unit vectors, oriented along the three axes. They are called i-hat, j-hat, and k-hat, and they point along the x, y, and z axes, respectively (Figure 1.18):

$$\hat{i} = \langle 1, 0, 0 \rangle$$
$$\hat{j} = \langle 0, 1, 0 \rangle$$
$$\hat{k} = \langle 0, 0, 1 \rangle$$

One way to express a vector is in terms of these special unit vectors:

$$\langle 0.02, -1.7, 30.0 \rangle = 0.02\hat{i} + (-1.7)\hat{j} + 30.0\hat{k}$$

We will usually use the $\langle x, y, z \rangle$ form rather than the ijk form in this book, because the familiar $\langle x, y, z \rangle$ notation, used in many calculus textbooks, emphasizes that a vector is a single entity.

Not all unit vectors point along an axis, as shown in Figure 1.19. For example, the vectors

$$\hat{g} = \langle 0.5774, 0.5774, 0.5774 \rangle \text{ and } \hat{F} = \langle 0.424, 0.566, 0.707 \rangle$$

are both unit vectors, since the magnitude of each is equal to 1. Note that every component of a unit vector is less than or equal to 1.

Calculating unit vectors

Any vector may be factored into the product of a unit vector in the direction of the vector, multiplied by a scalar equal to the magnitude of the vector.

$$\vec{w} = |\vec{w}| \cdot \hat{w}$$

For example, a vector of magnitude 5, aligned with the y axis, could be written as:

$$\langle 0, 5, 0 \rangle = 5\langle 0, 1, 0 \rangle$$

Therefore, to find a unit vector in the direction of a particular vector, we just divide the vector by its magnitude:

CALCULATING A UNIT VECTOR

$$\hat{r} = \frac{\vec{r}}{|\vec{r}|} = \frac{\langle x, y, z \rangle}{\sqrt{(x^2 + y^2 + z^2)}}$$

$$\hat{r} = \langle \frac{x}{\sqrt{(x^2 + y^2 + z^2)}}, \frac{y}{\sqrt{(x^2 + y^2 + z^2)}}, \frac{z}{\sqrt{(x^2 + y^2 + z^2)}} \rangle$$

For example, if $\vec{v} = \langle -22.3, 0.4, -19.5 \rangle \text{m/s}$, then

$$\hat{v} = \frac{\vec{v}}{|\vec{v}|} = \frac{\langle -22.3, 0.4, -19.5 \rangle \text{ m/s}}{\sqrt{(-22.3)^2 + (0.4)^2 + (-19.5)^2} \text{ m/s}} = \langle -0.753, 0.0135, -0.658 \rangle$$

Remember that to divide a vector by a scalar, you divide each component of the vector by the scalar. The result is a new vector. Note also that a unit vector has no physical units (such as meters per second), because the units in the numerator and denominator cancel.

1.X.20 What is the unit vector in the direction of $\langle 0, 6, 0 \rangle$?

1.X.21 What is the unit vector in the direction of $\langle -300, 0, 0 \rangle$?

1.X.22 What is the unit vector in the direction of $\langle 2, 2, 2 \rangle$? What is the unit vector in the direction of $\langle 3, 3, 3 \rangle$?

1.X.23 What is the unit vector $\hat{a}$ in the direction of $\vec{a}$, where $\vec{a} = \langle 400, 200, -100 \rangle$ m/s^2?

1.X.24 Write the vector $\vec{a} = \langle 400, 200, -100 \rangle$ m/s^2 as the product $|\vec{a}| \cdot \hat{a}$.

Vector addition

The sum of two vectors is another vector, obtained by adding the components of the vectors:

$$\vec{A} = \langle A_x, A_y, A_z \rangle$$

$$\vec{B} = \langle B_x, B_y, B_z \rangle$$

$$\vec{A} + \vec{B} = \langle (A_x + B_x), (A_y + B_y), (A_z + B_z) \rangle$$

For example,

$$\langle 1, 2, 3 \rangle + \langle -4, 5, 6 \rangle = \langle -3, 7, 9 \rangle$$

Don't add magnitudes!

The magnitude of a vector is *not* in general equal to the sum of the magnitudes of the two original vectors! For example, the magnitude of the vector $\langle 3, 0, 0 \rangle$ is 3, and the magnitude of the vector $\langle -2, 0, 0 \rangle$ is 2, but the magnitude of the vector $(\langle 3, 0, 0 \rangle + \langle -2, 0, 0 \rangle)$ is 1, not 5!

Adding vectors graphically: Tip to tail

The sum of two vectors has a geometric interpretation. In Figure 1.20 you first walk along displacement vector $\vec{A}$, followed by walking along displacement vector $\vec{B}$. What is your net displacement vector $\vec{C} = \vec{A} + \vec{B}$? The x component C_x of your net displacement is the sum of A_x and B_x. Similarly, the y component C_y of your net displacement is the sum of A_y and B_y.

To add two vectors $\vec{A}$ and $\vec{B}$ graphically (Figure 1.20):

- Draw the first vector $\vec{A}$
- Move the second vector $\vec{B}$ (without rotating it) so its tail is located at the *tip* of the first vector
- Draw a new vector from the tail of vector $\vec{A}$ to the tip of vector $\vec{B}$

Vector subtraction

The difference of two vectors will be very important in this and subsequent chapters. To subtract one vector from another, we subtract the components of the second from the components of the first:

$$\vec{A} - \vec{B} = \langle (A_x - B_x), (A_y - B_y), (A_z - B_z) \rangle$$

$$\langle 1, 2, 3 \rangle - \langle -4, 5, 6 \rangle = \langle 5, -3, -3 \rangle$$

Subtracting vectors graphically: Tail to tail

To subtract one vector $\vec{B}$ from another vector $\vec{A}$ graphically:

- Draw the first vector $\vec{A}$
- Move the second vector $\vec{B}$ (without rotating it) so its tail is located at the *tail* of the first vector
- Draw a new vector from the tip of vector $\vec{B}$ to the tip of vector $\vec{A}$

Note that you can check this algebraically and graphically. As shown in Figure 1.21, since the tail of $\vec{A} - \vec{B}$ is located at the tip of $\vec{B}$, then the vector $\vec{A}$ should be the sum of $\vec{B}$ and $\vec{A} - \vec{B}$, as indeed it is:

$$\vec{B} + (\vec{A} - \vec{B}) = \vec{A}$$

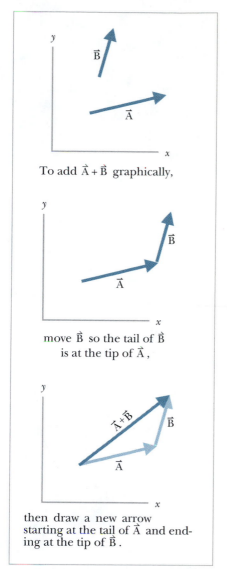

To add $\vec{A} + \vec{B}$ graphically,

move $\vec{B}$ so the tail of $\vec{B}$ is at the tip of $\vec{A}$,

then draw a new arrow starting at the tail of $\vec{A}$ and ending at the tip of $\vec{B}$.

Figure 1.20   The procedure for adding two vectors graphically: draw vectors tip to tail.

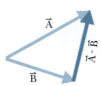

Figure 1.21 The procedure for subtracting vectors graphically: draw vectors tail to tail; draw new vector from tip of second vector to tip of first vector.

The zero vector

It is convenient to have a compact notation for a vector whose components are all zero. We will use the symbol $\vec{0}$ to denote a zero vector, in order to distinguish it from a scalar quantity that has the value 0.

$$\vec{0} = \langle 0, 0, 0 \rangle$$

For example, the sum of two vectors $\vec{B} + (-\vec{B}) = \vec{0}$.

Change in a quantity: The Greek letter Δ

Frequently we will want to calculate the change in a quantity. For example, we may want to know the change in an moving object's position or the change in its velocity during some time interval. The Greek letter Δ (capital delta suggesting "d for difference") is used to denote the change in a quantity (either a scalar or a vector).

We use the subscript i to denote an *initial* value of a quantity, and the subscript f to denote the *final* value of a quantity. If a vector $\vec{r}_i$ denotes the initial position of an object relative to the origin (its position at the beginning of a time interval), and $\vec{r}_f$ denotes the final position of the object, then

$$\Delta\vec{r} = \vec{r}_f - \vec{r}_i$$

$\Delta\vec{r}$ means "change of $\vec{r}$" or $\vec{r}_f - \vec{r}_i$ (displacement)

Δt means "change of t" or $t_f - t_i$ (time interval)

The symbol Δ (delta) always mean "final minus initial", not "initial minus final". For example, when a child's height changes from 1.1 m to 1.2 m, the change is $\Delta y = +0.1$ m, a positive number. If your bank account dropped from \$150 to \$130, what was the change in your balance? Δ(bank account) = −20 dollars.

Relative position vectors

Vector subtraction is used to calculate relative position vectors, vectors which represent the position of an object relative to another object. In Figure 1.22 object 1 is at location $\vec{r}_1$ and object 2 is at location $\vec{r}_2$. We want the components of a vector that points from object 1 to object 2. This is the vector obtained by subtraction: $\vec{r}_{2\ \text{relative to 1}} = \vec{r}_2 - \vec{r}_1$. Note that the form is always "final" minus "initial" in these calculations.

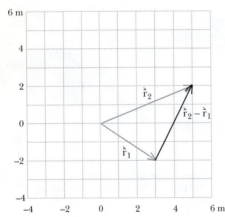

Figure 1.22 Relative position vector.

1.X.25 If $\vec{F}_1 = \langle 300, 0, -200 \rangle$ and $\vec{F}_2 = \langle 150, -300, 0 \rangle$, what is the sum $\vec{F}_1 + \vec{F}_2$?

1.X.26 What is the magnitude of $\vec{F}_1$ (see Exercise 1.X.25)? What is the magnitude of $\vec{F}_2$? What is the magnitude of $\vec{F}_1 + \vec{F}_2$?

1.X.27 What is the magnitude of $\vec{F}_1$ (see Exercise 1.X.25) plus the magnitude of $\vec{F}_2$? Is $|\vec{F}_1 + \vec{F}_2| = |\vec{F}_1| + |\vec{F}_2|$?

1.X.28 What is the difference $\vec{F}_1 - \vec{F}_2$? What is $\vec{F}_2 - \vec{F}_1$?

1.X.29 A snail is initially at location $\vec{r}_1 = \langle 3, 0, -7 \rangle$ m. At a later time the snail has crawled to location $\vec{r}_2 = \langle 2, 0, -8 \rangle$ m. What is $\Delta\vec{r}$, the change in the snail's position?

1.X.30 In Figure 1.22, $\vec{r}_1 = \langle 3, -2, 0 \rangle$ m and $\vec{r}_2 = \langle 5, 2, 0 \rangle$ m. Calculate the position of object 2 relative to object 1, as a relative position vector. Before checking the answer at the back of this chapter, see whether your answer is consistent with the appearance of the vector $\vec{r}_{2\ \text{relative to 1}} = \vec{r}_2 - \vec{r}_1$ shown in Figure 1.22.

1.X.31 What is the position of object 1 relative to object 2, as a vector?

1.X.32 At 10:00 AM you are at location $\langle -3, 2, 5 \rangle$ m. By 10:02 AM you have walked to location $\langle 6, 4, 25 \rangle$ m. (a) What is $\Delta\vec{r}$, the change in your position? (b) What is Δt, the time interval during which your position changed?

Unit vectors and angles

Suppose a taut string is at an angle θ_x to the $+x$ axis, and we need a unit vector in the direction of the string. Figure 1.23 shows a unit vector $\hat{A}$ pointing along the string. What is the x component of this unit vector? Consider the triangle whose base is A_x and whose hypotenuse is $|\hat{A}| = 1$. From the definition of the cosine of an angle we have this:

$$\cos\theta = \frac{\text{adjacent}}{\text{hypotenuse}} = \frac{A_x}{1}, \text{ so } A_x = \cos\theta_x$$

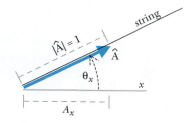

Figure 1.23 A unit vector whose direction is at a known angle from the $+x$ axis.

In Figure 1.23 the angle θ_x is shown in the first quadrant (θ_x less than $90°$), but this works for larger angles as well. For example, in Figure 1.24 the angle from the $+x$ axis to a unit vector $\hat{B}$ is in the second quadrant (θ_x greater than $90°$) and $\cos\theta_x$ is negative, which corresponds to B_x being negative.

What is true for x is also true for y and z. Figure 1.25 shows a 3D unit vector $\hat{r}$ and indicates the angles between the unit vector and the x, y, and z axes. Evidently we can write

$$\hat{r} = \langle \cos\theta_x, \cos\theta_y, \cos\theta_z \rangle$$

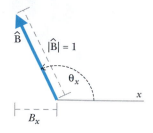

Figure 1.24 A unit vector in the second quadrant from the $+x$ axis.

These three cosines of the angles between a vector (or unit vector) and the coordinate axes are called the "direction cosines" of the vector. The cosine function is never greater than 1, just as no component of a unit vector can be greater than 1.

A common special case is that of a unit vector lying in the xy plane, with zero z component (Figure 1.26). In this case $\theta_x + \theta_y = 90°$, so that $\cos\theta_y = \cos(90° - \theta_x) = \sin\theta_x$, so that you can express the cosine of θ_y as the sine of θ_x, which is often convenient. However, in the general 3D case shown in Figure 1.25 there is no such simple relationship among the direction angles, nor among their cosines.

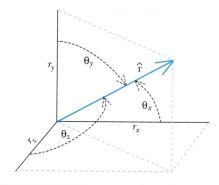

Figure 1.25 A 3D unit vector and its angles to the x, y, and z axes.

Example: From angle to unit vector

A rope lying in the xy plane, pointing up and to the right, supports a climber at an angle of $20°$ to the vertical. What is the unit vector pointing up along the rope?

Solution: $\langle \cos(90° - 20°), \cos(20°), 0 \rangle = \langle 0.342, 0.940, 0 \rangle$

Example: From unit vector to angles

A vector $\vec{v}$ points from the origin to the location $\langle -600, 0, 300 \rangle$ m. What is the angle that this vector makes to the x axis? To the y axis? To the z axis?

$$\hat{v} = \frac{\langle -600, 0, 300 \rangle \text{ m}}{\sqrt{(-600)^2 + (0)^2 + (300)^2} \text{ m}} = \langle -0.894, 0, 0.447 \rangle$$

But we also know that $\hat{v} = \langle \cos\theta_x, \cos\theta_y, \cos\theta_z \rangle$, so

$\cos\theta_x = -0.894$, and the arccosine gives $\theta_x = 153.4°$.

Similarly,

$\cos\theta_y = 0$, $\theta_y = 90°$ (which checks; no y component)

$\cos\theta_z = 0.447$, $\theta_z = 63.4°$

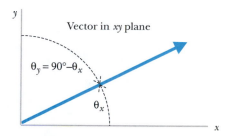

Figure 1.26 If a vector lies in the xy plane, $\cos\theta_y = \sin\theta_x$.

1.X.33 A unit vector lies in the xy plane, at an angle of $160°$ from the $+x$ axis, with a positive y component. What is the unit vector? (It helps to draw a diagram.)

1.X.34 A string runs up and to the left in the xy plane, making an angle of $40°$ to the vertical. Determine the unit vector that points along the string.

1.X.35 A cube is 3 cm on a side, with one corner at the origin. What is the unit vector pointing from the origin to the diagonally opposite corner at location $\langle 3, 3, 3 \rangle$ cm? What is the angle from this diagonal to one of the adjacent edges of the cube?

1.6 SI UNITS

In this book we use the SI (Système Internationale) unit system, as is customary in technical work. The SI unit of mass is the kilogram (kg), the unit of distance is the meter (m), and the unit of time is the second (s). In later chapters we will encounter other SI units, such as the newton (N), which is a unit of force.

It is essential to use SI units in physics equations; this may require that you convert from some other unit system to SI units. If mass is known in grams, you need to divide by 1000 and use the mass in kilograms. If a distance is given in centimeters, you need to divide by 100 to convert the distance to meters. If the time is measured in minutes, you need to multiply by 60 to use a time in seconds. A convenient way to do such conversions is to multiply by factors which are equal to 1, such as $(1\text{ min})/(60\text{ s})$ or $(100\text{ cm})/(1\text{ m})$. As an example, consider converting 60 miles per hour to SI units, meters per second. Start with the 60 mi/hr and multiply by factors of 1:

$$\left(60\, \frac{\text{mi}}{\text{hr}}\right)\left(\frac{1\text{ hr}}{60\text{ min}}\right)\left(\frac{1\text{ min}}{60\text{ s}}\right)\left(\frac{5280\text{ ft}}{1\text{ mi}}\right)\left(\frac{12\text{ in}}{1\text{ ft}}\right)\left(\frac{2.54\text{ cm}}{1\text{ in}}\right)\left(\frac{1\text{ m}}{100\text{ cm}}\right) = 26.8\, \frac{\text{m}}{\text{s}}$$

Observe how most of the units cancel, leaving final units of m/s.

1.X.36 A snail moved 80 cm (80 centimeters) in 5 minutes. What was its average speed in SI units? Write out the factors as was done above.

1.7 VELOCITY

We use vectors not only to describe the position of an object but also to describe velocity (speed and direction). If we know a object's present speed in meters per second and the object's direction of motion, we can predict where it will be a short time into the future. As we have seen, *change* of velocity is an indication of interaction. We need to be able to work with velocities of objects in 3D, so we need to learn how to use 3D vectors to represent velocities. After learning how to describe velocity in 3D, we will also learn how to describe *change* of velocity, which is related to interactions.

Average speed

The concept of speed is a familiar one. Speed is a single number, so it is a scalar quantity (speed is the magnitude of velocity). A world class sprinter can run 100 meters in 10 seconds. We say the sprinter's average speed is $(100\text{ m})/(10\text{ s}) = 10$ m/s. In SI units speed is measured in meters per second, abbreviated "m/s".

A car that travels 100 miles in 2 hours has an average speed of (100 miles)/(2 hours) = 50 miles per hour (about 22 m/s). In symbols,

$$v_{avg} = \frac{d}{t}$$

where v_{avg} is the "average speed," d is the distance the car has traveled, and t is the elapsed time.

There are other useful versions of the basic relationship among average speed, distance, and time. For example,

$$d = v_{avg}t$$

expresses the fact that if you run 5 m/s for 7 seconds you go 35 meters. Or you can use

$$t = \frac{d}{v_{\text{avg}}}$$

to calculate that to go 3000 miles in an airplane that flies at 600 miles per hour will take 5 hours.

Units

While it is easy to make a mistake in one of the formulas relating speed, time interval, and change in position, it is also easy to catch such a mistake by looking at the units. If you had written $t = v_{\text{avg}}/d$, you would discover that the right hand side has units of (m/s)/m, or 1/s, not s. Always check units!

Instantaneous speed compared to average speed

If a car went 70 miles per hour for the first hour and 30 miles an hour for the second hour, it would still go 100 miles in 2 hours, with an average speed of 50 miles per hour. Note that during this two hour interval, the car was almost never actually traveling at its average speed of 50 miles per hour.

To find the "instantaneous" speed—the speed of the car at a particular instant—we should observe the short distance the car goes in a very short time, such as a hundredth of a second: If the car moves 0.3 meters in 0.01 s, its instantaneous speed is 30 meters per second.

Vector velocity

Earlier we calculated vector differences between two different objects. The vector difference $\vec{r}_2 - \vec{r}_1$ represented a *relative* position vector—the position of object 2 relative to object 1 at a particular time. Now we will be concerned with the change of position of *one* object during a time interval, and $\vec{r}_f - \vec{r}_i$ will represent the "displacement" of this single object during the time interval, where $\vec{r}_i$ is the initial 3D position and $\vec{r}_f$ is the final 3D position (note that as with relative position vectors, we always calculate "final minus initial"). Dividing the (vector) displacement by the (scalar) time interval $t_f - t_i$ (final time minus initial time) gives the average (vector) velocity of the object:

DEFINITION: AVERAGE VELOCITY

$$\vec{v}_{\text{avg}} = \frac{\vec{r}_f - \vec{r}_i}{t_f - t_i}$$

Another way of writing this expression, using the "Δ" symbol (Greek capital delta, defined on page 14) to represent a change in a quantity, is:

$$\vec{v}_{\text{avg}} = \frac{\Delta \vec{r}}{\Delta t}$$

Remember that this is a compact notation for:

$$\vec{v}_{\text{avg}} = \frac{\Delta \vec{r}}{\Delta t} = \langle \frac{\Delta x}{\Delta t}, \frac{\Delta y}{\Delta t}, \frac{\Delta z}{\Delta t} \rangle$$

The magnitude of the average velocity, $|\vec{v}_{\text{avg}}|$, is called the average speed.

Determining average velocity from change in position

Consider a bee in flight (Figure 1.27). At time $t_i = 15$ s after 9:00 AM, the bee's position vector was $\vec{r}_i = \langle 2, 4, 0 \rangle$ m . At time $t_f = 15.1$ s after 9:00 AM,

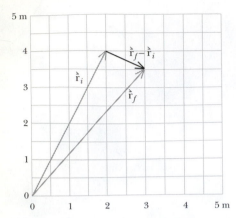

Figure 1.27 The displacement vector points from initial position to final position.

the bee's position vector was $\vec{r}_f = \langle 3, 3.5, 0 \rangle$ m. On the diagram, we draw and label the vectors $\vec{r}_i$ and $\vec{r}_f$.

Next, on the diagram, we draw and label the vector $\vec{r}_f - \vec{r}_i$, with the tail of the vector at the bee's initial position. One useful way to think about this graphically is to ask yourself what vector needs to be added to the initial vector $\vec{r}_1$ to make the final vector $\vec{r}_f$, since $\vec{r}_f$ can be written in the form $\vec{r}_f = \vec{r}_i + (\vec{r}_f - \vec{r}_i)$.

The vector we just drew, the change in position $\vec{r}_f - \vec{r}_i$, is called the "displacement" of the bee during this time interval. This displacement vector points from the initial position to the final position, and we always calculate displacement as "final minus initial".

Note that the displacement $\vec{r}_f - \vec{r}_i$ refers to the positions of one object (the bee) at two different times, not the position of one object relative to a second object at one particular time ("relative position vector"). However, the vector subtraction is the same kind of operation for either kind of situation.

We calculate the bee's displacement vector numerically by taking the difference of the two vectors, final minus initial:

$$\vec{r}_f - \vec{r}_i = \langle 3, 3.5, 0 \rangle \text{ m} - \langle 2, 4, 0 \rangle \text{ m} = \langle 1, -0.5, 0 \rangle \text{ m}$$

This numerical result should be consistent with our graphical construction. Look at the components of $\vec{r}_f - \vec{r}_i$ in Figure 1.27. Do you see that this vector has an *x* component of +1 and a *y* component of −0.5 m? Note that the (vector) displacement $\vec{r}_f - \vec{r}_i$ is in the direction of the bee's motion.

The average velocity of the bee, a vector quantity, is the (vector) displacement $\vec{r}_f - \vec{r}_i$ divided by the (scalar) time interval, $t_f - t_i$. Calculate the bee's average velocity:

$$\vec{v}_{avg} = \frac{\vec{r}_f - \vec{r}_i}{t_f - t_i} = \frac{\langle 1, -0.5, 0 \rangle \text{ m}}{(15.1 - 15) \text{ s}} = \frac{\langle 1, -0.5, 0 \rangle \text{ m}}{0.1 \text{ s}} = \langle 10, -5, 0 \rangle \text{ m/s}$$

Since we divided $\vec{r}_f - \vec{r}_i$ by a scalar ($t_f - t_i$), the average velocity $\vec{v}_{avg}$ points in the direction of the bee's motion, if the bee flew in a straight line.

What is the speed of the bee?

$$\text{speed of bee} = |\vec{v}_{avg}| = \sqrt{10^2 + (-5)^2 + 0^2} \text{ m/s} = 11.18 \text{ m/s}$$

What is the direction of the bee's motion, expressed as a unit vector?

$$\text{direction of bee: } \hat{v}_{avg} = \frac{\vec{v}_{avg}}{|\vec{v}_{avg}|} = \frac{\langle 10, -5, 0 \rangle \text{ m/s}}{11.18 \text{ m/s}} = \langle 0.894, -0.447, 0 \rangle$$

Note that the "m/s" units cancel; the result is dimensionless. We can check that this really is a unit vector:

$$\sqrt{0.894^2 + (-0.447)^2 + 0^2} = 0.9995$$

This is not quite 1.0 due to rounding the velocity coordinates and speed to three significant figures.

Put the pieces back together and see what we get. The original vector factors into the product of the magnitude times the unit vector:

$$|\vec{v}|\hat{v} = (11.18 \text{ m/s})\langle 0.894, -0.447, 0 \rangle = \langle 10, -5, 0 \rangle \text{ m/s}$$

This is the same as the original vector $\vec{v}$.

1.X.37 At a time 0.2 seconds after it has been hit by a tennis racket, a tennis ball is located at $\langle 5, 7, 2 \rangle$ m, relative to an origin in one corner of a tennis court. At a time 0.7 seconds after being hit, the ball is located at $\langle 9, 2, 8 \rangle$ m.

(a) What is the average velocity of the tennis ball?
(b) What is the average speed of the tennis ball?
(c) What is the unit vector in the direction of the ball's velocity?

1.X.38 A spacecraft is observed to be at a location $\langle 200, 300, -400 \rangle$ m relative to an origin located on a nearby asteroid, and 5 seconds later is observed at location $\langle 325, 25, -550 \rangle$ m .
(a) What is the average velocity of the spacecraft?
(b) What is the average speed of the spacecraft?
(c) What is the unit vector in the direction of the spacecraft's velocity?

Scaling a vector to fit on a graph

We can plot the average velocity vector on the same graph that we use for showing the vector positions of the bee (Figure 1.28). However, note that velocity has units of m/s while positions have units of m, so in a way we're mixing apples and oranges.

Moreover, the magnitude of the vector, 11.18 m/s, doesn't fit on a graph that is only 5 units wide (in meters). It is standard practice in such situations to scale the arrow representing the vector down to fit on the graph, preserving the correct direction. In Figure 1.28 we've scaled the velocity vector down by about a factor of 3 to make the arrow fit on the graph. Of course if there is more than one velocity vector we use the same scale factor for all the velocity vectors. The same kind of scaling is used with other physical quantities that are vectors, such as force and momentum, which we will encounter later.

Predicting a new position

We can rewrite the velocity relationship in the form

$$(\vec{r}_f - \vec{r}_i) = \vec{v}_{avg}(t_f - t_i)$$

That is, the (vector) displacement of an object is its average (vector) velocity times the time interval. This is just the vector version of the simple notion that if you run at a speed of 7 m/s for 5 s you move a distance of $(7 \text{ m/s})(5 \text{ s}) = 35$ m, or that a car going 50 miles per hour for 2 hours goes $(50 \text{ mi/hr})(2 \text{ hr}) = 100$ miles.

Is $(\vec{r}_f - \vec{r}_i) = \vec{v}_{avg}(t_f - t_i)$ a valid vector relation? Yes, multiplying a vector $\vec{v}_{avg}$ times a scalar $(t_f - t_i)$ yields a vector. We make a further rearrangement to obtain a relation for updating the position when we know the velocity:

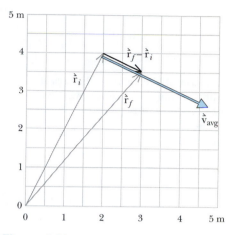

Figure 1.28 Average velocity vector: displacement divided by time interval.

THE POSITION UPDATE FORMULA

$$\vec{r}_f = \vec{r}_i + \vec{v}_{avg}(t_f - t_i)$$

or

$$\vec{r}_f = \vec{r}_i + \vec{v}_{avg}\Delta t$$

This equation says that if we know the starting position, the average velocity, and the time interval, we can predict the final position. This equation will be important throughout our work.

Using the position update formula

The position update formula $\vec{r}_f = \vec{r}_i + \vec{v}_{avg}\Delta t$ is a vector equation, so we can write out its full component form:

$$\langle x_f, y_f, z_f \rangle = \langle x_i, y_i, z_i \rangle + \langle v_{avg,x}, v_{avg,y}, v_{avg,z} \rangle \Delta t$$

Because the x component on the left of the equation must equal the x component on the right (and similarly for the y and z components), this compact vector equation represents three separate component equations:

$$x_f = x_i + v_{avg,x}\Delta t$$

$$y_f = y_i + v_{avg,y}\Delta t$$

$$z_f = z_i + v_{avg,z}\Delta t$$

Example: Updating the position of a ball

At time $t_i = 12.18$ s after 1:30 PM a ball's position vector is $\vec{r}_i = \langle 20, 8, -12 \rangle$ m. The ball's velocity at that moment is $\vec{v} = \langle 9, -4, 6 \rangle$ m/s. At time $t_f = 12.21$ s after 1:30 PM, where is the ball, assuming that its velocity hardly changes during this short time interval?

$$\vec{r}_f = \vec{r}_i + \vec{v}(t_f - t_i) = \langle 20, 8, -12 \rangle\, m + (\langle 9, -4, 6 \rangle\, m/s)(12.21 - 12.18)s$$

$$\vec{r}_f = \langle 20, 8, -12 \rangle\ m + \langle 0.27, -0.12, 0.18 \rangle\ m$$

$$\vec{r}_f = \langle 20.27, 7.88, -11.82 \rangle\ m$$

Note that if the velocity changes significantly during the time interval, in either magnitude or direction, our prediction for the new position may not be very accurate. In this case the velocity at the initial time could differ significantly from the average velocity during the time interval.

1.X.39 A proton traveling with a velocity of $\langle 3 \times 10^5, 2 \times 10^5, -4 \times 10^5 \rangle$ m/s passes the origin at a time 9.0 seconds after a proton detector is turned on. Assuming the velocity of the proton does not change, what will its position be at time 9.7 seconds?

1.X.40 How long does it take a baseball with velocity $\langle 30, 20, 25 \rangle$ m/s to travel from location $\vec{r}_1 = \langle 3, 7, -9 \rangle$ m to location $\vec{r}_2 = \langle 18, 17, 3.5 \rangle$ m ?

1.X.41 A "slow" neutron produced in a nuclear reactor travels from location $\langle 0.2, -0.05, 0.1 \rangle$ m to location $\langle -0.202, 0.054, 0.098 \rangle$ m in 2 microseconds ($1\mu s = 1 \times 10^{-6}$ s).
(a) What is the average velocity of the neutron?
(b) What is the average speed of the neutron?

Instantaneous velocity

The curved colored line in Figure 1.29 shows the path of a ball through the air. The colored dots mark the ball's position at time intervals of one second. While the ball is in the air, its velocity is constantly changing, due to interactions with the Earth (gravity) and with the air (air resistance).

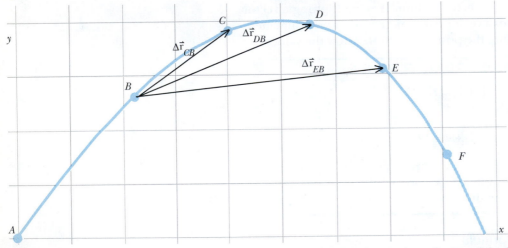

Figure 1.29 The trajectory of a ball through air. The axes represent x and y distance from the ball's initial location; each square on the grid corresponds to 10 meters. Three different displacements, corresponding to three different time intervals, are indicated by arrows on the diagram.

Suppose we ask: What is the velocity of the ball at the precise instant that it reaches location *B*? This quantity would be called the "instantaneous velocity" of the ball. We can start by approximating the instantaneous velocity of the ball by finding its average velocity over some larger time interval.

The table in Figure 1.30 shows the time and the position of the ball for each location marked by a colored dot in Figure 1.29. We can use these data to calculate the average velocity of the ball over three different intervals, by finding the ball's displacement during each interval, and dividing by the appropriate Δt for that interval:

$$\vec{v}_{EB} = \frac{\Delta \vec{r}_{EB}}{\Delta t} = \frac{\vec{r}_E - \vec{r}_B}{t_E - t_B} = \frac{(\langle 69.1, 31.0, 0 \rangle - \langle 22.3, 26.1, 0 \rangle)\,\text{m}}{(4.0 - 1.0)\,\text{s}}$$

$$= \langle 15.6, 1.6, 0 \rangle \frac{\text{m}}{\text{s}}$$

$$\vec{v}_{DB} = \frac{\Delta \vec{r}_{DB}}{\Delta t} = \frac{\vec{r}_D - \vec{r}_B}{t_D - t_B} = \frac{(\langle 55.5, 39.2, 0 \rangle - \langle 22.3, 26.1, 0 \rangle)\,\text{m}}{(3.0 - 1.0)\,\text{s}}$$

$$= \langle 16.6, 6.55, 0 \rangle \frac{\text{m}}{\text{s}}$$

$$\vec{v}_{CB} = \frac{\Delta \vec{r}_{CB}}{\Delta t} = \frac{\vec{r}_C - \vec{r}_B}{t_C - t_B} = \frac{(\langle 40.1, 38.1, 0 \rangle - \langle 22.3, 26.1, 0 \rangle)\,\text{m}}{(2.0 - 1.0)\,\text{s}}$$

$$= \langle 17.8, 12.0, 0 \rangle \frac{\text{m}}{\text{s}}$$

loc.	t (s)	position (m)
A	0.0	$\langle 0, 0, 0 \rangle$
B	1.0	$\langle 22.3, 26.1, 0 \rangle$
C	2.0	$\langle 40.1, 38.1, 0 \rangle$
D	3.0	$\langle 55.5, 39.2, 0 \rangle$
E	4.0	$\langle 69.1, 31.0, 0 \rangle$
F	5.0	$\langle 80.8, 14.8, 0 \rangle$

Figure 1.30 Table showing elapsed time and position of the ball at each location marked by a dot in Figure 1.29.

Not surprisingly, the average velocities over these different time intervals are not the same, because both the direction of the ball's motion and the speed of the ball were changing continuously during its flight. The three average velocity vectors that we calculated are shown in Figure 1.31.

Figure 1.31 The three different average velocity vectors calculated above are shown by three arrows, each with its tail at location B. Note that since the units of velocity are m/s, these arrows use a different scale from the distance scale used for the path of the ball. The three arrows representing average velocities are drawn with their tails at the location of interest. The dashed arrow represents the actual instantaneous velocity of the ball at location B.

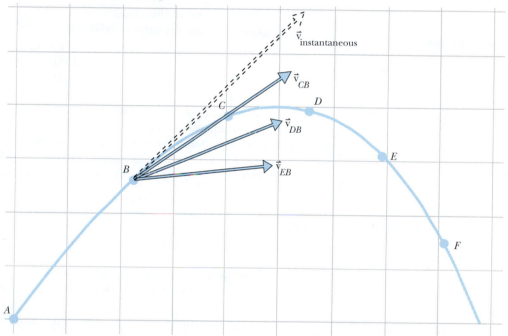

? Which of the three average velocity vectors depicted in Figure 1.31 best approximates the instantaneous velocity of the ball at location B?

Simply by looking at the diagram, we can tell that $\vec{v}_{CB}$ is closest to the actual instantaneous velocity of the ball at location B, because its direction is closest to the direction in which the ball is actually traveling. Because the direction of the instantaneous velocity is the direction the ball is moving at a particular instant, the instantaneous velocity is tangent to the ball's path. Of the three average velocity vectors we calculated, $\vec{v}_{CB}$ best approximates a tangent to the path of the ball. Evidently $\vec{v}_{CB}$, the velocity calculated with the shortest time interval, $t_C - t_B$, is the best approximation to the instantaneous velocity at location B. If we used even smaller values of Δt in our calculation of average velocity, such as 0.1 second, or 0.01 second, or 0.001 second, we would presumably have better and better estimates of the actual instantaneous velocity of the object at the instant when it passes location B.

Two important ideas have emerged from this discussion:

- The direction of the instantaneous velocity of an object is tangent to the path of the object's motion.
- Smaller time intervals yield more accurate estimates of instantaneous velocity.

Connection to calculus

You may already have learned about derivatives in calculus. The instantaneous velocity is a derivative, the limit of $\Delta \vec{r}/\Delta t$ as the time interval Δt used in the calculation gets closer and closer to zero:

$$\vec{v} = \lim_{\Delta t \to 0} \frac{\Delta \vec{r}}{\Delta t}, \text{ which is written as } \vec{v} = \frac{d\vec{r}}{dt}$$

In Figure 1.31, the process of taking the limit is illustrated graphically. As smaller values of Δt are used in the calculation, the average velocity vectors approach the limiting value: the actual instantaneous velocity.

A useful way to see the meaning of the derivative of a vector is to consider the components:

> The rate of change of a vector (the derivative) is itself a vector.

$$\vec{v} = \frac{d\vec{r}}{dt} = \frac{d}{dt}\langle x, y, z \rangle = \left\langle \frac{dx}{dt}, \frac{dy}{dt}, \frac{dz}{dt} \right\rangle = \langle v_x, v_y, v_z \rangle$$

The derivative of the position vector $\vec{r}$ gives components that are the components of the velocity, as we should expect.

Informally, you can think of $d\vec{r}$ as a very small ("infinitesimal") displacement, and dt as a very small ("infinitesimal") time interval. It is as though we had continued the process illustrated in Figure 1.29 to smaller and smaller time intervals, down to an extremely tiny time interval dt with a correspondingly tiny displacement $d\vec{r}$. The ratio of these tiny quantities is the instantaneous velocity.

The ratio of these two tiny quantities need not be small. For example, suppose an object moves in the x direction a tiny distance of 1×10^{-15} m, the radius of a proton, in a very short time interval of 1×10^{-23} s:

$$\vec{v} = \frac{\langle 1 \times 10^{-15}, 0, 0 \rangle \text{ m}}{1 \times 10^{-23} \text{ s}} = \langle 1 \times 10^{8}, 0, 0 \rangle \text{ m/s},$$

which is one-third the speed of light (3×10^{8} m/s)!

Change of magnitude and/or change in direction

Note that the time rate of change of a vector $\vec{r} = |\vec{r}|\hat{r}$ has two parts:

$$\text{rate of change of the magnitude of the vector } \frac{d|\vec{r}|}{dt}$$

$$\text{rate of change of the direction of the vector } \frac{d\hat{r}}{dt}$$

We will discuss this further in later sections.

1.8 MOMENTUM

Newton's first law of motion:

> **An object moves in a straight line and at constant speed except to the extent that it interacts with other objects.**

gives us a conceptual connection between interactions and their effects on the motion of objects. However, this law does not allow us to make quantitative (numerical) predictions or explanations—we could not have used this law to predict the exact trajectory of the ball shown in Figure 1.29, and we could not use this law alone to figure out how to send a rocket to the Moon.

In order to make quantitative predictions or explanations of physical phenomena, we need a quantitative measure of interactions and a quantitative measure of effects of those interactions.

Newton's first law of motion does contain the important idea that if there is no interaction, a moving object will continue to move in a straight line, with no change of direction or speed, and an object that is not moving will remain at rest. A quantitative version of this law would provide a means of predicting the motion of an object, or of deducing how it must have moved in the past, if we could list all of its interactions with other objects.

Changes in velocity

? What factors make it difficult or easy to change an object's velocity?

You have probably noticed that if two objects have the same velocity but one is much more massive than the other, it is more difficult to change the heavy object's speed or direction. It is easier to stop a baseball traveling at a hundred miles per hour than to stop a car traveling at a hundred miles per hour! It is easier to change the direction of a canoe than to change the direction of a large, massive ship such as the Titanic (which couldn't change course quickly enough after the iceberg was spotted).

Momentum involves both mass and velocity

To take into account both an object's mass and its velocity, we can define a vector quantity called "momentum" that involves the product of mass (a scalar) and velocity (a vector). Instead of saying "the stronger the interaction, the bigger the change in the velocity," we now say "the stronger the interaction, the bigger the change in the momentum."

Momentum, a vector quantity, is usually represented by the symbol $\vec{p}$. We might expect that the mathematical expression for momentum would be simply $\vec{p} = m\vec{v}$, and indeed this is almost, but not quite, correct.

Experiments on particles moving at very high speeds, close to the speed of light $c = 3 \times 10^8$ m/s, show that changes in $m\vec{v}$ are not really proportional to the strength of the interactions. As you keep applying a force to a particle near the speed of light, the speed of the particle barely increases, and it is not possible to increase a particle's speed beyond the speed of light.

Through experiments it has been found that changes in the following quantity are proportional to the amount of interaction:

DEFINITION OF MOMENTUM

$$\vec{p} = \gamma m\vec{v}$$

The proportionality factor γ (lower-case Greek gamma) is defined as

$$\gamma = \frac{1}{\sqrt{1 - \left(\frac{|\vec{v}|}{c}\right)^2}}$$

In these equations $\vec{p}$ represents momentum, $\vec{v}$ is the velocity of the object, m is the mass of the object, $|\vec{v}|$ is the magnitude of the object's velocity (the speed), and c is the speed of light (3×10^8 m/s). Momentum has units of $\text{kg} \cdot \text{m/s}$. To calculate momentum in these units, you must specify mass in kg and velocity in meters per second.

This is the "relativistic" definition of momentum. Albert Einstein in 1905 in his special theory of relativity predicted that this would be the appropriate definition for momentum at high speeds, a prediction that has been abundantly verified in a wide range of experiments.

Example: Momentum of a fast moving proton

Suppose that a proton (mass 1.7×10^{-27} kg) is traveling with a velocity of $\langle 2 \times 10^7, 1 \times 10^7, -3 \times 10^7 \rangle$ m/s. (a) What is the momentum of the proton? (b) What is the magnitude of the momentum of the proton?

(a)
$$|\vec{v}| = \sqrt{(2 \times 10^7)^2 + (1 \times 10^7)^2 + (-3 \times 10^7)^2} = 3.7 \times 10^7 \frac{m}{s}$$

$$\frac{|\vec{v}|}{c} = \frac{3.7 \times 10^7 \frac{m}{s}}{3 \times 10^8 \frac{m}{s}} = 0.12$$

$$\gamma = \frac{1}{\sqrt{1 - (0.12)^2}} = 1.007$$

$$\vec{p} = (1.007)(1.7 \times 10^{-27} kg)\langle 2 \times 10^7, 1 \times 10^7, -3 \times 10^7 \rangle \frac{m}{s}$$

$$\vec{p} = \langle 3.4 \times 10^{-20}, 1.7 \times 10^{-20}, -5.1 \times 10^{-20} \rangle \frac{kg \cdot m}{s}$$

$$|\vec{p}| = \sqrt{(3.4 \times 10^{-20})^2 + (1.7 \times 10^{-20})^2 + (-5.1 \times 10^{-20})^2} \frac{kg \cdot m}{s}$$

(b)
$$= 6.4 \times 10^{-20} \frac{kg \cdot m}{s}$$

Approximate expression for momentum

In the example above, we found that $\gamma = 1.007$. Since in that calculation we used only two significant figures, we could have used the approximation that $\gamma \approx 1.0$ without affecting our answer. Let's examine the expression for γ to see if we can come up with a guideline for when it is reasonable to use the approximate expression

$$\vec{p} \approx 1 \cdot m\vec{v}$$

Looking at the expression for γ

$$\gamma = \frac{1}{\sqrt{1 - \left(\frac{|\vec{v}|}{c}\right)^2}}$$

we see that it depends only on the ratio of the speed of the object to the speed of light (the object's mass doesn't appear in this expression).

If $\frac{|\vec{v}|}{c}$ is a very small number, then $1 - \left(\frac{|\vec{v}|}{c}\right)^2 \approx 1 - 0 \approx 1$, so $\gamma \approx 1$.

APPROXIMATION: MOMENTUM AT LOW SPEEDS

$$\vec{p} \approx m\vec{v} \text{ when } |\vec{v}| \ll c$$

Some values of $(|\vec{v}|/c)$ and γ are displayed in Figure 1.32. From this table you can see that even at the very high speed where $|\vec{v}|/c = 0.1$, which means that $|\vec{v}| = 3 \times 10^7$ m/s, the relativistic factor γ is only slightly different from 1. For large-scale objects such as a space rocket, whose speed is typically only about 1×10^4 m/s, we can ignore the factor γ, and momentum is to a good approximation $\vec{p} \approx m\vec{v}$. It is only for high-speed cosmic rays or particles produced in high-speed particle accelerators that we need to use the full relativistic definition for momentum, $\vec{p} = \gamma m\vec{v}$.

$\lvert\vec{v}\rvert$ m/s	$\lvert\vec{v}\rvert/c$	γ
0	0	1.0000
3	1×10^{-8}	1.0000
300	1×10^{-6}	1.0000
3×10^6	1×10^{-2}	1.0001
3×10^7	0.1	1.0050
1.5×10^8	0.5	1.1547
2.997×10^8	0.999	22.3663
2.9997×10^8	0.9999	70.7124
3×10^8	1	infinite! (and impossible)

Figure 1.32 Values of γ calculated for some speeds. γ is shown to four decimal places, which is more accuracy than we will usually need.

From this table you can also get a sense of why it is not possible to exceed the speed of light. As you make a particle go faster and faster, approaching the speed of light, additional increases in the speed become increasingly difficult, because a tiny increase in speed means a huge increase in momentum, requiring huge amounts of interaction. In fact, for the speed to equal the speed of light, the momentum would have to increase to be infinite! There is a cosmic speed limit in the Universe, 3×10^8 m/s.

We will repeatedly emphasize the role of momentum throughout this textbook because of its fundamental importance not only in classical (prequantum) mechanics but also in relativity and quantum mechanics. The use of momentum clarifies the physics analysis of certain complex processes such as collisions, including collisions at speeds approaching the speed of light.

Direction of momentum

Like velocity, momentum is a vector quantity, so it has a magnitude and a direction.

? A leaf is blown by a gust of wind, and at a particular instant is traveling straight upward, in the $+y$ direction. What is the direction of the leaf's momentum?

The mathematical expression for momentum can be looked at as the product of a scalar part times a vector part. Since the mass m must be a positive number, and the factor gamma (γ) must be a positive number, this scalar factor cannot change the direction of the vector (Figure 1.33). Therefore the direction of the leaf's momentum is the same as the direction of its velocity: straight up (the $+y$ direction).

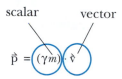

Figure 1.33 The expression for momentum is the product of a scalar times a vector. The scalar factor is positive, so the direction of an object's momentum is the same as the direction of its velocity.

1.X.42 A good sprinter can run 100 meters in 10 seconds. What is the magnitude of the momentum of a sprinter whose mass is 65 kg and who is running at a speed of 10 m/s?

1.X.43 A baseball has a mass of about 155 g. What is the magnitude of the momentum of a baseball thrown at a speed of 100 miles per hour? (Note that you need to convert mass to kilograms and speed to meters/second. See the inside back cover of the textbook for conversion factors.)

1.X.44 What is the magnitude (in $kg \cdot m/s$) of the momentum of a 1000 kg airplane traveling at a speed of 500 miles per hour? (Note that you need to convert speed to meters per second.)

1.X.45 If a particle has momentum $\vec{p} = \langle 4, -5, 2 \rangle$ kg·m/s, what is the magnitude $|\vec{p}|$ of its momentum?

1.X.46 What is the momentum of an electron traveling at a velocity of $\langle 0, 0, -2 \times 10^8 \rangle$ m/s? (Masses of particles are given on the inside back cover of this textbook.) What is the magnitude of the momentum of the electron?

1.X.47 A proton in an accelerator is traveling at a speed of $0.99c$. (Masses of particles are given on the inside back cover of this textbook.) (a) If you use the approximate nonrelativistic formula for the magnitude of momentum of the proton, what answer do you get? (b) What is the magnitude of the correct relativistic momentum of the proton? (c) The approximate value (the answer to part a) is significantly too low. What is the ratio of magnitudes you calculated (correct / approximate)? Such speeds are attained in particle accelerators.

Using momentum to update position

If you know the momentum of an object, you can calculate the change in position of the object over a given time interval. This is straightforward if the object is traveling at a speed low enough that the approximate expression for momentum can be used, since in this case $\vec{v} \approx \vec{p}/m$. For speeds near the speed of light, we must use the exact expression

$$\vec{v} = \frac{\vec{p}/m}{\sqrt{1 + \left(\frac{|\vec{p}|}{mc}\right)^2}}$$

This expression was derived simply by rearranging the equation $\vec{p} = \gamma m \vec{v}$. The detailed derivation is given at the end of the chapter.

Example: Displacement of a fast proton

A proton with constant momentum $\langle 0, 0, 2.72 \times 10^{-19} \rangle \, \text{kg} \cdot \text{m/s}$ leaves the origin 10.0 seconds after an accelerator experiment is started. What is the location of the proton 2 ns later? (ns = nanosecond = 1×10^{-9} s).

$$\vec{v} = \frac{\vec{p}/m}{\sqrt{1 + \frac{|\vec{p}|^2}{mc}}} = \frac{\langle 0, 0, 2.72 \times 10^{-19} \rangle \, \text{kg} \cdot \text{m/s}}{(1.7 \times 10^{-27} \, \text{kg}) \sqrt{1 + \left(\frac{2.72 \times 10^{-19} \, \text{kg} \cdot \text{m/s}}{(1.7 \times 10^{-27} \, \text{kg})(3 \times 10^8 \, \text{m/s})}\right)^2}}$$

$$= \langle 0, 0, 1.6 \times 10^8 \rangle \, \text{m/s}$$

$$\vec{r}_f = \vec{r}_i + \vec{v}_{\text{avg}} \Delta t$$

$$= \langle 0, 0, 0 \rangle \, \text{m} + \langle 0, 0, 1.6 \times 10^8 \text{m/s} \rangle \, (2 \times 10^{-9} \, \text{s})$$

$$= \langle 0, 0, 0.32 \rangle \, \text{m}$$

The proton traveled 32 cm in 2 ns.

Example: Displacement of an ice skater

An ice skater whose mass is 50 kg is moving with constant momentum $\langle 40, 0, 30 \rangle \, \text{kg} \cdot \text{m/s}$. At a particular instant in her skating program she passes location $\langle 0, 0, 3 \rangle$ m. What was her location at a time 3 seconds earlier?

$$\vec{v} \approx \frac{\vec{p}}{m} = \frac{\langle 400, 0, 300 \rangle \, \text{kg} \cdot \text{m/s}}{50 \, \text{kg}} = \langle 8, 0, 6 \rangle \, \text{m/s}$$

$$(\vec{r}_f - \vec{r}_i) = \vec{v} \Delta t$$

$$\vec{r}_i = \vec{r}_f - \vec{v} \Delta t$$

$$= \langle 0, 0, 3 \rangle \, \text{m} - (\langle 8, 0, 6 \rangle \, \text{m/s})(3 \, \text{s})$$

$$= \langle -24, 0, -15 \rangle \, \text{m}$$

1.X.48 A 4.5 kg bowling ball rolls down an alley with nearly constant momentum of $\langle 0.9, 0, 22.5 \rangle \, \text{kg} \cdot \text{m/s}$, starting from the origin. What will the location of the ball be after 2 seconds?

1.9 CHANGE OF MOMENTUM

In the next chapter we will introduce "the Momentum Principle" which quantitatively relates change in momentum $\Delta \vec{p}$ to the strength and duration of an interaction. In order to be able to use the Momentum Principle we need to know how to calculate changes in momentum.

Momentum is a vector quantity, so just as was the case with velocity, there are two aspects of momentum that can change: magnitude and direction. A mathematical description of change of momentum must include either a change in the magnitude of the momentum, or a change in the direction of the momentum, or both.

Change of the vector momentum

The change in the momentum $\Delta\vec{p}$ during a time interval is a vector: $\Delta\vec{p} = \vec{p}_f - \vec{p}_i$. This vector expression captures both changes in magnitude and changes in direction. Figure 1.34 is a graphical illustration of a change from an initial momentum $\vec{p}_i$ to a final momentum $\vec{p}_f$. Place the initial and final momentum vectors tail to tail, then draw a vector from initial to final. This is the vector representing $\Delta\vec{p} = \vec{p}_f - \vec{p}_i$. This is the same procedure you used to calculate relative position vectors by subtraction, or displacement vectors by subtraction. The rule for subtracting vectors is always the same: Place the vectors tail to tail, then draw from the tip of the initial vector to the tip of the final vector. This resultant vector is "final minus initial".

Change in magnitude of momentum

If an object's speed changes (that is, the magnitude of its velocity changes), the magnitude of the object's momentum also changes. However, note that the change of the magnitude of momentum is not in general equal to the magnitude of change of momentum.

Change of direction of momentum

There are various ways to specify a change in the direction of motion. For example, if you use compass directions, you could say that an airplane changed its direction from 30° east of North to 45° east of North: a 15° clockwise change. One can imagine various other schemes, involving other kinds of coordinate systems. The standard way to deal with this is to use vectors, as in Figure 1.35, which shows the changing momentum of an object traveling at constant speed along a circular path.

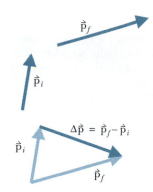

Figure 1.34 Calculation of $\Delta\vec{p}$.

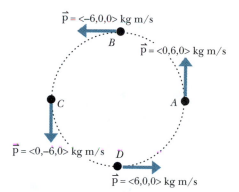

Figure 1.35 The momentum of an object traveling in a circle changes, even if the magnitude of momentum does not change.

Example: Change in momentum of a proton

Figure 1.36 shows a portion of the trajectory of a ball in air, subject to gravity and air resistance. When the ball is at location B, its momentum is $\vec{p}_B = \langle 3.03, 2.83, 0 \rangle$ kg·m/s. When it is at location C, its momentum is $\vec{p}_C = \langle 2.55, 0.97, 0 \rangle$ kg·m/s. (a) Find the change in the ball's momentum between these locations, and show it on the diagram. (b) Find the change in the magnitude of the ball's momentum.

(a)

$$\Delta\vec{p} = \vec{p}_C - \vec{p}_B = \langle 2.55, 0.97, 0 \rangle \text{ kg·m/s} - \langle 3.03, 2.83, 0 \rangle \text{ kg·m/s}$$

$$= \langle -0.48, -1.86, 0 \rangle \text{ kg·m/s}$$

Both the x and y components of the ball's momentum decreased, so $\Delta\vec{p}$ has negative x and y components. This is consistent with the graphical subtraction shown in Figure 1.37.

(b)

$$\left|\vec{p}_B\right| = \sqrt{(3.03)^2 + (2.83)^2 + (0)^2} = 4.15 \text{ kg·m/s}$$

$$\left|\vec{p}_C\right| = \sqrt{(2.55)^2 + (0.97)^2 + (0)^2} = 2.73 \text{ kg·m/s}$$

The magnitude of the proton's momentum decreased.

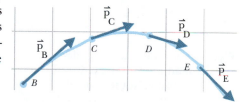

Figure 1.36 A portion of the trajectory of a ball moving through air, subject to gravity and air resistance. The arrows represent the momentum of the ball at the locations indicated by letters.

Example: Change in momentum and in magnitude of momentum

Suppose you are driving a 1000 kilogram car at 20 m/s in the +x direction. After making a 180 degree turn, you drive the car at 20 m/s in the −x direction. (20 m/s is about 45 miles per hour or 72 km per hour.) (a) What is the change of magnitude of the momentum of the car $\Delta\left|\vec{p}\right|$? (b) What is the magnitude of the change of momentum of the car $\left|\Delta\vec{p}\right|$?

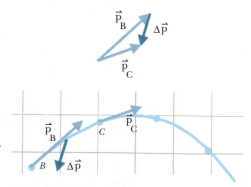

Figure 1.37 Graphical calculation of $\Delta\vec{p}$. The result is also shown superimposed on the ball's path, midway between the initial and final locations.

These speeds are very small compared to the speed of light, so we can use the approximate nonrelativistic formula for momentum.

(a)

$$\Delta |\vec{p}| = |\vec{p}_2| - |\vec{p}_1| \approx |m\vec{v}_2| - |m\vec{v}_1| = |m||\vec{v}_2| - |m||\vec{v}_1|$$

$$\Delta |\vec{p}| = (1000 \text{ kg})(20 \text{ m/s}) - (1000 \text{ kg})(20 \text{ m/s})$$

$$\Delta |\vec{p}| = 0 \text{ kg·m/s}$$

(b)

$$\Delta \vec{p} = \vec{p}_2 - \vec{p}_1 = 1000 \text{ kg} \langle 20, 0, 0 \rangle \text{ m/s} - 1000 \text{ kg} \langle -20, 0, 0 \rangle \text{ m/s}$$

$$= \langle 4 \times 10^4, 0, 0 \rangle \text{ kg·m/s}$$

$$|\Delta \vec{p}| = 4 \times 10^4 \text{ kg·m/s}$$

So $|\Delta \vec{p}| \neq \Delta |\vec{p}|$.

1.X.49 A tennis ball of mass 57 g travels with velocity $\langle 50, 0, 0 \rangle$ m/s toward a wall. After bouncing off the wall, the tennis ball is observed to be traveling with velocity $\langle -48, 0, 0 \rangle$ m/s.
(a) Draw a diagram showing the initial and final momentum of the tennis ball.
(b) What is the change in the momentum of the tennis ball?
(c) What is the change in the magnitude of the tennis ball's momentum?

1.X.50 The planet Mars has a mass of 6.4×10^{23} kg, and travels in a nearly circular orbit around the Sun, as shown in Figure 1.38. When it is at location A, the velocity of Mars is $\langle 0, 0, -2.5 \times 10^4 \rangle$ m/s. When it reaches location B, the planet's velocity is $\langle -2.5 \times 10^4, 0, 0 \rangle$ m/s.
(a) What is $\Delta \vec{p}$, the change in the momentum of Mars between locations A and B?
(b) On a copy of the diagram in Figure 1.38, draw two arrows representing the momentum of Mars at locations *C* and *D*, paying attention to both the length and direction of each arrow.
(c) What is the direction of the change in the momentum of Mars between locations *C* and *D*? Draw the vector $\Delta \vec{p}$ on your diagram.

1.X.51 A 50 kg child is riding on a carousel (merry-go-round) at a constant speed of 5 m/s. What is the magnitude of the change in the child's momentum $|\Delta \vec{p}|$ in going all the way around (360°)? In going halfway around (180°)? Draw a diagram showing the initial vector momentum and the final vector momentum, then subtract, then find the magnitude.

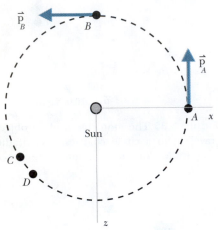

Figure 1.38 The nearly circular orbit of Mars around the Sun, viewed from above the orbital plane. Not to scale: the sizes of the Sun and Mars are exaggerated.

Average rate of change of momentum

The rate of change of the vector position is such an important quantity that it has a special name: "velocity". In Section 1.7 we discussed how to find both the average rate of change of position (average velocity) and the instantaneous rate of change of position (instantaneous velocity) of an object.

The average rate of change of momentum and the instantaneous rate of change of momentum are also extremely important quantities. In some situations, we will only be able to find an average rate of change of momentum:

AVERAGE RATE OF CHANGE OF MOMENTUM

$$\frac{\Delta \vec{p}}{\Delta t} = \frac{\vec{p}_f - \vec{p}_i}{t_f - t_i}$$

This quantity is a vector, and points in the direction of $\Delta \vec{p}$.

Example: **Average rate of change of momentum**

If the momentum of a ball changes from $\langle 1, 2, 0 \rangle$ kg $\cdot$ m/s to $\langle 0.5, 0, 0.5 \rangle$ kg $\cdot$ m/s in half a second, the average rate of change of momentum of the ball is

$$\frac{(\langle 0.5, 0, 0.5 \rangle - \langle 1, 2, 0 \rangle)\ \text{kg} \cdot \text{m/s}}{0.5\ \text{s}} = \langle -1, -4, 1 \rangle\ \frac{\text{kg} \cdot \text{m}}{\text{s}^2}$$

1.10 *THE PRINCIPLE OF RELATIVITY

Sections marked with a "*" are optional. They provide additional information and context, but later sections of the textbook don't depend critically on them. This optional section deals with some deep issues about the "reference frame" from which you observe motion. Newton's first law of motion only applies in an "inertial reference frame," which we will discuss here in the context of the principle of relativity.

A great variety of experimental observations has led to the establishment of the following principle:

THE PRINCIPLE OF RELATIVITY

**Physical laws work in the same way for observers
in uniform motion as for observers at rest.**

This principle is called "the principle of relativity." (Einstein's extensions of this principle are known as "special relativity" and "general relativity.") Phenomena observed in a room in uniform motion (for example, on a train moving with constant speed on a smooth straight track) obey the same physical laws in the same way as experiments done in a room that is not moving. According to this principle, Newton's first law of motion should be true both for an observer moving at constant velocity and for an observer at rest.

For example, suppose you're riding in a car moving with constant velocity, and you're looking at a map lying on the dashboard. As far as you're concerned, the map isn't moving, and no interactions are required to hold it still on the dashboard. Someone standing at the side of the road sees the car go by, sees the map moving at a high speed in a straight line, and can see that no interactions are required to hold the map still on the dashboard. Both you and the bystander agree that Newton's first law of motion is obeyed: the bystander sees the map moving with constant velocity in the absence of interactions, and you see the map not moving at all (a zero constant velocity) in the absence of interactions.

On the other hand, if the car suddenly speeds up, it moves out from under the map, which ends up in your lap. To you it looks like "the map sped up in the backwards direction" without any interactions to cause this to happen, which looks like a violation of Newton's first law of motion. The problem is that you're strapped to the car, which is an accelerated reference frame, and Newton's first law of motion applies only to nonaccelerated reference frames, called "inertial" reference frames. Similarly, if the car suddenly turns to the right, moving out from under the map, the map tends to keep going in its original direction, and to you it looks like "the map moved to the left" without any interactions. So a change of speed or a change of direction of the car (your reference frame) leads you to see the map behave in a strange way.

The bystander, who is in an inertial (non-accelerating) reference frame, doesn't see any violation of Newton's first law of motion. The bystander's reference frame is an inertial frame, and the map behaves in an understandable way, tending to keep moving with unchanged speed and direction when the car changes speed or direction.

The cosmic microwave background

The principle of relativity, and Newton's first law of motion, apply only to observers who have a constant speed and direction (or zero speed) relative to the "cosmic microwave background," which provides the only backdrop and frame of reference with an absolute, universal character. It used to be that the basic reference frame was loosely called "the fixed stars," but stars and galaxies have their own individual motions within the Universe and do not constitute an adequate reference frame with respect to which to measure motion.

The cosmic microwave background is low-intensity electromagnetic radiation with wavelengths in the microwave region, which pervades the Universe, radiating in all directions. Measurements show that our galaxy is moving through this microwave radiation with a large, essentially constant velocity, toward a cluster of a large number of other galaxies. The way we detect our motion relative to the microwave background is through the "Doppler shift" of the frequencies of the microwave radiation, toward higher frequencies in front of us and lower frequencies behind. This is essentially the same phenomenon as that responsible for a fire engine siren sounding at a higher frequency when it is approaching us and a lower frequency when it is moving away from us.

The discovery of the cosmic microwave background provided major support for the "Big Bang" theory of the formation of the Universe. According to the Big Bang, the early Universe must have been an extremely hot mixture of charged particles and high-energy, short-wavelength electromagnetic radiation (visible light, x-rays, gamma rays, etc.). Electromagnetic radiation interacts strongly with charged particles, so light could not travel very far without interacting, making the Universe essentially opaque. Also, the Universe was so hot that electrically neutral atoms could not form without the electrons immediately being stripped away again by collisions with other fast-moving particles.

As the Universe expanded, the temperature dropped. Eventually the temperature was low enough for neutral atoms to form. The interaction of electromagnetic radiation with neutral atoms is much weaker than with individual charged particles, so the radiation was now essentially free, dissociated from the matter, and the Universe became transparent. As the Universe continued to expand (the actual space between clumps of matter got bigger!), the wavelengths of the electromagnetic radiation got longer, until today this fossil radiation has wavelengths in the relatively low-energy, long-wavelength microwave portion of the electromagnetic spectrum.

Inertial frames of reference

It is an observational fact that in reference frames that are in uniform motion with respect to the cosmic microwave background, far from other objects (so that interactions are negligible), an object maintains uniform motion. Such frames are called "inertial frames" and are reference frames in which Newton's first law of motion is valid. All of these reference frames are equally valid; the cosmic microwave background simply provides a concrete example of such a reference frame.

? Is the surface of the Earth an inertial frame?

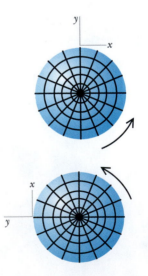

Figure 1.39 Axes tied to the Earth rotate through 90° in a quarter of a day (6 hours).

No! The Earth is rotating on its axis, so the velocity of an object sitting on the surface of the Earth is constantly changing direction, as is a coordinate frame tied to the Earth (Figure 1.39). Moreover, the Earth is orbiting the Sun, and the Solar System itself is orbiting the center of our Milky Way galaxy, and our galaxy is moving toward other galaxies. So the motion of an object sitting on the Earth is actually quite complicated and definitely not uniform with respect to the cosmic microwave background.

However, for many purposes the surface of the Earth can be considered to be (approximately) an inertial frame. For example, it takes 6 hours for the rotation of the Earth on its axis to make a 90° change in the direction of the velocity of a "fixed" point. If a process of interest takes only a few minutes, during these few minutes a "fixed" point moves in nearly a straight line at constant speed due to the Earth's rotation, and velocity changes in the process of interest are typically much larger than the very small velocity change of the approximate inertial frame of the Earth's surface.

Similarly, although the Earth is in orbit around the Sun, it takes 365 days to go around once, so for a period of a few days or even weeks the Earth's orbital motion is nearly in a straight line at constant speed. Hence for many purposes the Earth represents an approximately inertial frame despite its motion around the Sun.

The special theory of relativity

Einstein's special theory of relativity (published in 1905) built on the basic principle of relativity but added the conjecture that the speed of a beam of light must be the same as measured by observers in different frames of reference in uniform motion with respect to each other. In Figure 1.40, observers on each spaceship measure the speed of the light c emitted by the ship at the top to be the same ($c = 3\times10^8$ m/s), despite the fact that they are moving at different velocities.

This additional condition seems peculiar and has far-reaching consequences. After all, the map on the dashboard of your car has different speeds relative to different observers, depending on the motion of the observer. Yet a wide range of experiments has confirmed Einstein's conjecture: all observers measure the same speed for the same beam of light, $c = 3\times10^8$ m/s. (The color of the light is different for the different observers, but the speed is the same.)

On the other hand, if someone on the ship at the top throws a ball or a proton or some other piece of matter, the speed of the object will be different for observers on the three ships; it is only light whose speed is independent of the observer.

Einstein's theory has interesting consequences. For example, it predicts that time will run at different rates in different frames of reference. These predictions have been confirmed by many experiments. These unusual effects are large only at very high speeds (a sizable fraction of the speed of light), which is why we don't normally observe these effects in everyday life, and why we can use nonrelativistic calculations for low-speed phenomena.

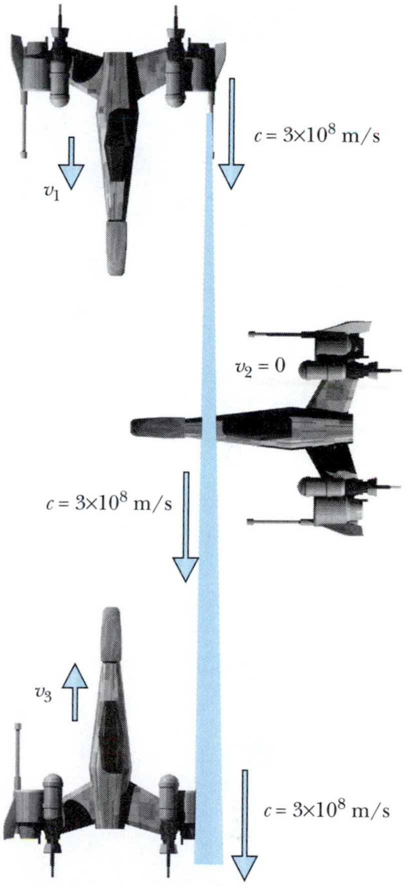

Figure 1.40 Light emitted by the top spaceship is measured to have the same speed by observers in all three ships.

1.11 *UPDATING POSITION AT HIGH SPEED

If $v << c$, $\vec{p} \approx m\vec{v}$ and $\vec{v} \approx \vec{p}/m$. But at high speed it is more complicated to determine the velocity from the (relativistic) momentum. Here is a way to solve for $\vec{v}$ in terms of $\vec{p}$:

$$|\vec{p}| = \frac{1}{\sqrt{1 - (|\vec{v}|/c)^2}} m|\vec{v}|$$

Divide by m and square:
$$\frac{|\vec{p}|^2}{m^2} = \frac{|\vec{v}|^2}{1 - (|\vec{v}|/c)^2}$$

Multiply by $(1 - (|\vec{v}|/c)^2)$:
$$\frac{|\vec{p}|^2}{m^2} - \left(\frac{|\vec{p}|^2}{m^2 c^2}\right)|\vec{v}|^2 = |\vec{v}|^2$$

Collect terms: $\left(1 + \dfrac{|\vec{p}|^2}{m^2 c^2}\right)|\vec{v}|^2 = \dfrac{|\vec{p}|^2}{m^2}$

$$|\vec{v}| = \dfrac{|\vec{p}|/m}{\sqrt{1 + \left(\dfrac{|\vec{p}|}{mc}\right)^2}}$$

The expression above gives the magnitude of $\vec{v}$, in terms of the magnitude of $\vec{p}$. To get an expression for the vector $\vec{v}$, recall that any vector can be factored into its magnitude times a unit vector in the direction of the vector, so

$$\vec{p} = |\vec{p}|\hat{p} \quad \text{and} \quad \vec{v} = |\vec{v}|\hat{v}$$

But since $\vec{p}$ and $\vec{v}$ are in the same direction, $\hat{v} = \hat{p}$, so

$$\vec{v} = |\vec{v}|\hat{p} = \dfrac{|\vec{p}|/m}{\sqrt{1 + \left(\dfrac{|\vec{p}|}{mc}\right)^2}}\hat{p} = \dfrac{(|\vec{p}|\hat{p})/m}{\sqrt{1 + \left(\dfrac{|\vec{p}|}{mc}\right)^2}}$$

$$\vec{v} = \dfrac{\vec{p}/m}{\sqrt{1 + \left(\dfrac{|\vec{p}|}{mc}\right)^2}}$$

THE RELATIVISTIC POSITION UPDATE FORMULA

$$\vec{r}_f = \vec{r}_i + \dfrac{1}{\sqrt{1 + \left(\dfrac{|\vec{p}|}{mc}\right)^2}}\left(\dfrac{\vec{p}}{m}\right)\Delta t \quad \text{(for small } \Delta t)$$

Note that at low speeds $|\vec{p}| \approx m|\vec{v}|$, and the denominator is

$$\sqrt{1 + \left(\dfrac{|\vec{v}|}{c}\right)^2} \approx 1$$

so the formula becomes the familiar $\vec{r}_f = \vec{r}_i + (\vec{p}/m)\Delta t$.

1.12 SUMMARY

Interactions (Section 1.2 and Section 1.4)

Interactions are indicated by
- change of velocity (change of direction and/or change of speed)
- change of identity
- change of shape of multiparticle system
- change of temperature of multiparticle system
- lack of change when change is expected

Newton's first law of motion (Section 1.3)

An object moves in a straight line and at constant speed
except to the extent that it interacts with other objects.

Vectors (Section 1.5)

A 3D *vector* is a quantity with magnitude and a direction, which can be expressed as a triple $\langle x, y, z \rangle$. A vector is indicated by an arrow: $\vec{r}$

A *scalar* is a single number.

Legal mathematical operations involving vectors include:
- adding one vector to another vector
- subtracting one vector from another vector
- multiplying or dividing a vector by a scalar
- finding the magnitude of a vector
- taking the derivative of a vector

Operations that are *not* legal with vectors include:
- A vector cannot be added to a scalar
- A vector cannot be set equal to a scalar
- A vector cannot appear in the denominator (you can't divide by a vector)

A unit vector $\hat{r} = \vec{r}/|\vec{r}|$ has magnitude 1.
A vector can be factored using a unit vector:
$\vec{F} = |\vec{F}|\hat{F}$

Direction cosines: $\hat{r} = \langle \cos\theta_x, \cos\theta_y, \cos\theta_z \rangle$

The symbol Δ

The symbol Δ (delta) means "change of": $\Delta t = t_f - t_i$, $\Delta\vec{r} = \vec{r}_f - \vec{r}_i$

Δ always means "final minus initial".

Velocity and change of position (Section 1.7)

Definition of average velocity

$$\vec{v}_{avg} = \frac{\Delta\vec{r}}{\Delta t} = \frac{\vec{r}_f - \vec{r}_i}{t_f - t_i}$$

Velocity is a vector. $\vec{r}$ is the position of an object (a vector). t is the time. Average velocity is equal to the change in position divided by the time elapsed. SI units of velocity are meters per second (m/s).

The position update formula

$$\vec{r}_f = \vec{r}_i + \vec{v}_{avg}\Delta t$$

The final position (vector) is the vector sum of the initial position plus the product of the average velocity and the elapsed time.

Definition of instantaneous velocity

$$\vec{v} = \lim_{\Delta t \to 0}\frac{\Delta\vec{r}}{\Delta t} = \frac{d\vec{r}}{dt}$$

The instantaneous velocity is the limiting value of the average velocity as the time elapsed becomes very small.

Velocity in terms of momentum

$$\vec{v} = \frac{\vec{p}/m}{\sqrt{1 + \left(\frac{|\vec{p}|}{mc}\right)^2}} \text{ or } \vec{v} \approx \vec{p}/m \text{ at low speeds}$$

Momentum (Section 1.8)

Definition of momentum

$$\vec{p} = \gamma m\vec{v}$$

where $\gamma = \dfrac{1}{\sqrt{1 - (|\vec{v}|/c)^2}}$ (lower-case Greek gamma)

Momentum (a vector) is the product of the relativistic factor "gamma" (a scalar), mass, and velocity.

Combined into one equation: $\vec{p} = \dfrac{1}{\sqrt{1 - (|\vec{v}|/c)^2}}m\vec{v}$

Approximation for momentum at low speeds

$\vec{p} \approx m\vec{v}$ at speeds such that $|\vec{v}| \ll c$

Useful numbers:

Radius of a typical atom: about 1×10^{-10} meter.
Radius of a proton or neutron: about 1×10^{-15} meter
Speed of light: 3×10^8 m/s

These and other useful data and conversion factors are given on the inside back cover of the textbook.

1.13 REVIEW QUESTIONS

The purpose of review questions is to help you re-flect on the most important concepts in the chap-ter. Try to answer the questions without flipping through the chapter looking for an answer, but think about what you know already from having done the exercises throughout the chapter. If you are stumped, look at the "Summary" on the pre-ceding page.

If you are still stumped after looking at the chapter summary, make a note in your notebook, as a reminder that you may need to spend some extra time studying this particular concept.

Detecting interactions

1.RQ.52 Give two examples (other than those discussed in the text) of interactions that may be detected by observing:
- (a) change in velocity
- (b) change in temperature
- (c) change in shape
- (d) change in identity
- (e) lack of change when change is expected

1.RQ.53 In which of the following situations is there observa-tional evidence for significant interaction between two objects? How can you tell?
- (a) a book rests on a table
- (b) a baseball that was hit by a batter flies toward the out-field
- (c) water freezes in an ice cube tray in the freezer
- (d) a communications satellite orbits the earth
- (e) a space probe leaves the solar system traveling at con-stant speed toward a distant star
- (f) a charged particle leaves a curving track in a particle detector

1.RQ.54 Which of the following observations give conclusive evidence of an interaction? (Choose all that apply.)
- (a) Change of velocity, either change of direction or change of speed.
- (b) Change of shape or configuration without change of ve-locity.
- (c) Change of position without change of velocity.
- (d) Change of identity without change of velocity.
- (e) Change of temperature without change of velocity.
Explain your choice.

1.RQ.55 Moving objects left the traces labelled *A - F* in the di-agram. The dots were deposited at equal time intervals (for ex-

ample, one dot each second). In each case the object starts from the square.

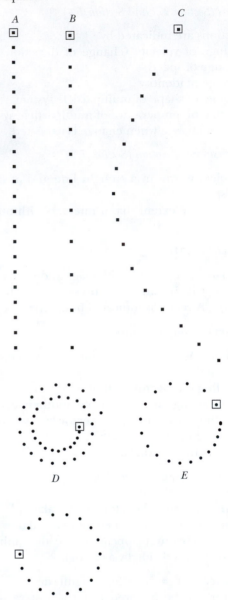

Traces left by moving objects. The dots mark the objects' positions at equal time intervals.

Which trajectories show evidence that the moving object was interacting with another object somewhere? If there is evi-dence for an interaction, what is the evidence?

1.RQ.56 A spaceship far from all other objects uses its rockets to attain a speed of 1×10^4 m/s. The crew then shuts off the power. According to Newton's first law, which of the following statements about the motion of the spaceship after the power is shut off are correct? (Choose all statements that are correct.)
- (a) The spaceship will move in a straight line.
- (b) The spaceship will travel on a curving path.
- (c) The spaceship will enter a circular orbit.
- (d) The speed of the spaceship will not change.
- (e) The spaceship will gradually slow down.
- (f) The spaceship will stop suddenly.

1.RQ.57 Why do we use a spaceship in outer space, far from other objects, to illustrate Newton's first law? Why not a car or a train? (More than one of the following statements may be correct.)

A car or train touches other objects, and interacts with them.

A car or train can't travel fast enough.

The spaceship has negligible interactions with other objects.

A car or train interacts gravitationally with the Earth.

A spaceship can never experience a gravitational force.

1.RQ.58 You slide a coin across the floor, and observe that it travels in a straight line, slowing down and eventually stopping. A sensitive thermometer shows that the coin's temperature increased. What can we conclude? (Choose all statements that are correct.)

(a) Because the coin traveled in a straight line, we conclude that it did not interact with anything.

(b) Because the coin did not change shape, we conclude that it did not interact with anything.

(c) Because the coin slowed down, we conclude that Newton's first law does not apply to objects in everyday life, such as coins.

(d) Because the coin's speed changed, we conclude that it interacted with one or more other objects.

(e) Because the coin got hot, we conclude that it interacted with one or more other objects.

1.RQ.59 Some science museums have an exhibit called a Bernoulli blower, in which a volleyball hangs suspended in a column of air blown upward by a strong fan. If you saw a ball suspended in the air but didn't know the blower was there, why would Newton's first law suggest that something must be holding the ball up?

1.RQ.60 Place a ball on a book and walk with the book in uniform motion. Note that you don't really have to do anything to the ball to keep the ball moving with constant velocity (relative to the ground) or to keep the ball at rest (relative to you). Then stop suddenly, or abruptly change your direction or speed. What does Newton's first law of motion predict for the motion of the ball (assuming the interaction between the ball and the book is small)? Does the ball behave as predicted? It may help to take the point of view of a friend who is standing still, watching you.

Velocity

1.RQ.61 How does average velocity differ from instantaneous velocity?

1.RQ.62 Start with the definition of average velocity and derive the position update formula from it. Show all steps in the derivation.

1.RQ.63 In the expression $\Delta \vec{r}/\Delta t$, what is the meaning of $\Delta \vec{r}$? What is the meaning of Δt?

Momentum

1.RQ.64 Which of the following statements about the velocity and momentum of an object are correct?

(a) The momentum of an object is always in the same direction as its velocity

(b) The momentum of an object can either be in the same direction as its velocity or in the opposite direction

(c) The momentum of an object is perpendicular to its velocity

(d) The direction of an object's momentum is not related to the direction of its velocity

(e) The direction of an object's momentum is tangent to its path

1.RQ.65 In which of these situations is it reasonable to use the approximate formula for the momentum of an object, instead of the full relativistically correct formula?

(a) A car traveling on an interstate highway

(b) A commercial jet airliner flying between New York and Seattle

(c) A neutron traveling at 2700 meters per second

(d) A proton in outer space traveling at 2×10^8 m/s

(e) An electron in a television tube traveling 3×10^6 m/s

1.RQ.66 Answer the following questions about the factor γ (gamma) in the full relativistic formula for momentum.

(a) Is γ a scalar or a vector quantity?

(b) What is the minimum possible value of γ?

(c) Does γ reach its minimum value when an object's speed is high or low?

(d) Is there a maximum possible value for γ?

(e) Does γ become large when an object's speed is high or low?

(f) Does the approximation $\gamma \approx 1$ apply when an object's speed is low or when it is high?

Change of momentum

1.RQ.67 A tennis ball of mass m traveling with velocity $\langle v_x, 0, 0 \rangle$ hits a wall and rebounds with velocity $\langle -v_x, 0, 0 \rangle$.

(a) What was the change in momentum of the tennis ball?

(b) What was the change in the magnitude of the momentum of the tennis ball?

Relativity

1.RQ.68 Which of the following observers might observe something that appears to violate Newton's first law of motion? Explain why.

(a) a person standing still on a street corner

(b) a person riding on a roller coaster

(c) a passenger on a starship travelling at $0.75c$ toward the nearby star Alpha Centauri

(d) an airplane pilot doing aerobatic loops

(e) a hockey player coasting across the ice

1.RQ.69 A spaceship at rest with respect to the cosmic microwave background emits a beam of red light. A different spaceship, moving at a speed of 2.5×10^8 m/s toward the first ship, detects the light. Which of the following statements are true for observers on the second ship? (More than one statement may be correct.)

(a) They observe that the light travels at 3×10^8 m/s.

(b) They see light that is not red.

(c) They observe that the light travels at 5.5×10^8 m/s

(d) They observe that the light travels at 2.5×10^8 m/s.

1.14 PROBLEMS

Vectors

1.HW.70 On a piece of graph paper, draw arrows representing the following vectors. Make sure the tip and tail of each arrow you draw are clearly distinguishable.

(a) Placing the tail of the vector at $\langle 5, 2, 0 \rangle$ draw an arrow representing the vector $\vec{p} = \langle -7, 3, 0 \rangle$. Label it $\vec{p}$.

(b) Placing the tail of the vector at $\langle -5, 8, 0 \rangle$ draw an arrow representing the vector $-\vec{p}$. Label it $-\vec{p}$.

1.HW.71 The following questions refer to the vectors depicted by arrows in the diagram.

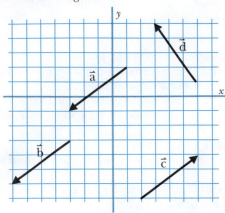

(a) What are the components of the vector $\vec{a}$? (Note that since the vector lies in the *xy* plane, its z component is zero.)

(b) What are the components of the vector $\vec{b}$?

(c) Is this statement true or false? $\vec{a} = \vec{b}$

(d) What are the components of the vector $\vec{c}$?

(e) Is this statement true or false: $\vec{c} = -\vec{a}$?

(f) What are the components of the vector $\vec{d}$?

(g) Is this statement true or false: $\vec{d} = -\vec{c}$?

1.P.72

(a) What are the components of the vector $\vec{d}$?

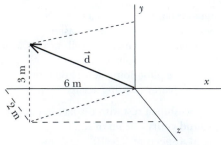

An arrow representing a position vector.

(b) If $\vec{e} = -\vec{d}$, what are the components of $\vec{e}$?

(c) If the tail of vector $\vec{d}$ were moved to location $\langle -5, -2, 4 \rangle$ m, where would the tip of the vector be located?

(d) If the tail of vector $-\vec{d}$ were placed at location $\langle -1, -1, -1 \rangle$ m, where would the tip of the vector be located?

1.HW.73 Here are several arrows representing vectors in the *xy* plane.

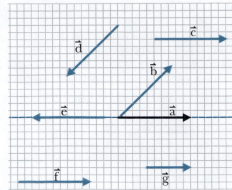

(a) Which vectors have magnitudes equal to the magnitude of $\vec{a}$?

(b) Which vectors are equal to $\vec{a}$?

1.HW.74 Consider a vector $\vec{u} = \langle u_x, u_y, u_z \rangle$, and another vector $\vec{p} = \langle p_x, p_y, p_z \rangle$. If $\vec{u} = \vec{p}$, then which of the following statements must be true? Some, all, or none of the following may be true:

(i) $u_x = p_x$

(ii) $u_y = p_y$

(iii) $u_z = p_z$

(iv) The direction of $\vec{u}$ is the same as the direction of $\vec{p}$.

Vector operations

1.HW.75 In the diagram three vectors are represented by arrows in the *xy* plane. Each square in the grid represents one meter.

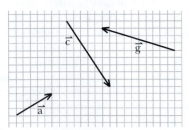

For each vector:

(a) Write out the components of the vector.

(b) Calculate the magnitude of the vector.

1.HW.76 Imagine that you have a baseball and a tennis ball at different locations. The center of the baseball is at $\langle 3, 5, 0 \rangle$ m, and the center of the tennis ball is at $\langle -3, -1, 0 \rangle$ m. On a piece of graph paper, do the following:

(a) Draw dots at the locations of the center of the baseball and the center of the tennis ball.

(b) Draw the position vector of the baseball, which is an arrow whose tail is at the origin and whose tip is at the location of the baseball. Label this position vector $\vec{B}$. Clearly show the tip and tail of each arrow.

(c) Complete this equation: $\vec{B} = \langle \underline{\quad}, \underline{\quad}, \underline{\quad} \rangle$ m.

(d) Draw the position vector of the tennis ball. Label it $\vec{T}$.

(e) Complete this equation: $\vec{T} = \langle \underline{\quad}, \underline{\quad}, \underline{\quad} \rangle$ m.

(f) Draw the relative position vector for the tennis ball relative to the baseball. The tail of this vector is at the center of the baseball, and the tip of the vector is at the center of the tennis ball. Label this relative position vector $\vec{r}$.

(g) Complete the following equation by reading the coordinates of $\vec{r}$ from your graph: $\vec{r}$ = < ____ , ____ , ____ > m

(h) Calculate the following difference: $\vec{T} - \vec{B}$ =< ____ , ____ , ____ > m.

(i) Is the following statement true? $\vec{r} = \vec{T} - \vec{B}$ _____

(j) Write two other equations relating the vectors $\vec{B}$, $\vec{T}$, and $\vec{r}$.

(k) Calculate the magnitudes of the vectors $\vec{B}$, $\vec{T}$, and $\vec{r}$.

(l) Calculate the difference of the magnitudes $|\vec{T}| - |\vec{B}|$.

(m) Does $|\vec{T}| - |\vec{B}| = |\vec{T} - \vec{B}|$?

1.HW.77 Which of the following are vectors?
(a) 3.5 (b) 0 (c) $\langle 0.7, 0.7, 0.7 \rangle$ (d) $\langle 0, 2.3, -1 \rangle$ (e) -3×10^6
(f) $3 \cdot \langle 14, 0, -22 \rangle$

1.HW.78 Which of the following are vectors?
(a) $\vec{r}/2$ (b) $|\vec{r}|/2$ (c) $\langle r_x, r_y, r_z \rangle$ (d) $5 \cdot \vec{r}$

1.HW.79 (a) What is the vector whose tail is at $\langle 9.5, 7, 0 \rangle$ m and whose head is at $\langle 4, -13, 0 \rangle$ m ? (b) What is the magnitude of this vector?

1.HW.80 A man is standing on the roof of a building with his head at the position $\langle 12, 30, 13 \rangle$ m . He sees the top of a tree, which is at the position $\langle -25, 35, 43 \rangle$ m .

(a) What is the relative position vector that points from the man's head to the top of the tree?

(b) What is the distance from the man's head to the top of the tree?

1.HW.81 (a) On a piece of graph paper, draw the vector $\vec{f} = \langle -2, 4, 0 \rangle$, putting the tail of the vector at $\langle -3, 0, 0 \rangle$. Label the vector $\vec{f}$.

(b) Calculate the vector $2\vec{f}$, and draw this vector on the graph, putting its tail at $\langle -3, -3, 0 \rangle$, so you can compare it to the original vector. Label the vector $2\vec{f}$.

(c) How does the magnitude of $2\vec{f}$ compare to the magnitude of $\vec{f}$?

(d) How does the direction of $2\vec{f}$ compare to the direction of $\vec{f}$?

(e) Calculate the vector $\vec{f}/2$, and draw this vector on the graph, putting its tail at $\langle -3, -6, 0 \rangle$, so you can compare it to the other vectors. Label the vector $\vec{f}/2$.

(f) How does the magnitude of $\vec{f}/2$ compare to the magnitude of $\vec{f}$?

(g) How does the direction of $\vec{f}/2$ compare to the direction of $\vec{f}$?

(h) Does multiplying a vector by a scalar change the magnitude of the vector?

(i) The vector $a(\vec{f})$ has a magnitude three times as great as that of $\vec{f}$, and its direction is opposite to the direction of $\vec{f}$. What is the value of the scalar factor a?

1.HW.82 (a) On a piece of graph paper, draw the vector $\vec{g} = \langle 4, 7, 0 \rangle$ m . Put the tail of the vector at the origin.

(b) Calculate the magnitude of $\vec{g}$.

(c) Calculate $\hat{g}$, the unit vector pointing in the direction of $\vec{g}$.

(d) On the graph draw $\hat{g}$. Put the tail of the vector at $\langle 1, 0, 0 \rangle$ m so you can compare $\hat{g}$ and $\vec{g}$.

(e) Calculate the product of the magnitude $|\vec{g}|$ times the unit vector $\hat{g}$, $(|\vec{g}|)(\hat{g})$.

1.HW.83 A proton is located at $\langle 3\times10^{-10}, -3\times10^{-10}, 8\times10^{-10} \rangle$ m .

(a) What is $\vec{r}$, the vector from the origin to the location of the proton?

(b) What is $|\vec{r}|$?

(c) What is $\hat{r}$, the unit vector in the direction of $\vec{r}$?

1.HW.84 Which of the following statements about the vectors depicted by these three arrows are correct?

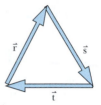

(a) $\vec{s} = \vec{t} - \vec{r}$ (b) $\vec{r} = \vec{t} - \vec{s}$ (c) $\vec{r} + \vec{t} = \vec{s}$ (d) $\vec{s} + \vec{t} = \vec{r}$
(e) $\vec{r} + \vec{s} = \vec{t}$

1.HW.85 Which of the following are unit vectors? (Numerical values are given only to 3 significant figures.)
(a) $\langle 0, 0, -1 \rangle$ (b) $\langle 0.5, 0.5, 0 \rangle$ (c) $\langle 0.333, 0.333, 0.333 \rangle$
(d) $\langle 0.9, 0, 0.1 \rangle$ (e) $\langle 0, 3, 0 \rangle$ (f) $\langle 1, -1, 1 \rangle$
(g) $\langle 0.577, 0.577, 0.577 \rangle$ (h) $\langle 0.949, 0, -0.316 \rangle$

1.HW.86 Two vectors, $\vec{f}$ and $\vec{g}$, are equal: $\vec{f} = \vec{g}$. Which of the following statements are true?
$\hat{f} = \hat{g}$ (b) $g_x = f_x$ (c) $f_z = g_z$ (d) the directions of $\vec{f}$ and $\vec{g}$ may be different (e) the magnitudes of $\vec{f}$ and $\vec{g}$ may be different

1.HW.87 A proton is located at $\vec{r}_p = \langle 2, 6, -3 \rangle$ m . An electron is located at $\vec{r}_e = \langle 4, 12, -6 \rangle$ m . Which of the following statements are true?
(a) $2\vec{r}_p = \vec{r}_e$ (b) $2\hat{r}_p = \hat{r}_e$ (c) $|2\vec{r}_p| = |\vec{r}_e|$

1.HW.88 A proton is located at $\langle x_p, y_p, z_p \rangle$. An electron is located at $\langle x_e, y_e, z_e \rangle$. What is the vector pointing from the electron to the proton? What is the vector pointing from the proton to the electron?

1.HW.89 The vector $\vec{a} = \langle 0.03, -1.4, 26.0 \rangle$ and the scalar $f = -3.0$. What is $f\vec{a}$?

1.HW.90 The vector $\vec{g} = \langle 2, -7, 3 \rangle$ and the scalar $h = -2$. What is $h + \vec{g}$?
(a) $\langle 0, -9, 1 \rangle$ (b) $\langle 4, -5, 5 \rangle$ (c) $\langle 4, 9, 5 \rangle$
(d) This is a meaningless expression.

1.HW.91 Write each of these vectors as the product of the magnitude of the vector and the appropriate unit vector:
  (a) $\langle 0, 0, 9.5 \rangle$ (b) $\langle 0, -679, 0 \rangle$
  (c) $\langle 3.5 \times 10^{-3}, 0, -3.5 \times 10^{-3} \rangle$ (d) $\langle 4 \times 10^6, -6 \times 10^6, 3 \times 10^6 \rangle$

1.HW.92 $\vec{A} = \langle 3 \times 10^3, -4 \times 10^3, -5 \times 10^3 \rangle$ and

$\vec{B} = \langle -3 \times 10^3, 4 \times 10^3, 5 \times 10^3 \rangle$. Calculate the following:
  (a) $\vec{A} + \vec{B}$ (b) $|\vec{A} + \vec{B}|$ (c) $|\vec{A}|$ (d) $|\vec{B}|$ (e) $|\vec{A}| + |\vec{B}|$

Problems on velocity and momentum

1.HW.93 A baseball has a mass of 0.155 kg. A professional pitcher throws a baseball 90 miles per hour, which is 40 m/s. What is the magnitude of the momentum of the pitched baseball?

1.HW.94 The position of a golf ball relative to the tee changes from $\langle 50, 20, 30 \rangle$ m to $\langle 53, 18, 31 \rangle$ m in 0.1 second. As a vector, write the velocity of the golf ball during this short time interval.

1.HW.95 A hockey puck with a mass of 0.4 kg has a velocity of $\langle 38, 0, -27 \rangle$ m/s. What is the magnitude of its momentum, $|\vec{p}|$?

1.HW.96 A proton in an accelerator attains a speed of $0.88c$. What is the magnitude of the momentum of the proton?

1.HW.97 The crew of a stationary spacecraft observe an asteroid whose mass is 4×10^{17} kg. Taking the location of the spacecraft as the origin, the asteroid is observed to be at location $\langle -3 \times 10^3, -4 \times 10^3, 8 \times 10^3 \rangle$ m at a time 18.4 seconds after lunchtime. At a time 21.4 seconds after lunchtime, the asteroid is observed to be at location $\langle -1.4 \times 10^3, -6.2 \times 10^3, 9.7 \times 10^3 \rangle$ m. Assuming the velocity of the asteroid does not change during this time interval, calculate the vector velocity $\vec{v}$ of the asteroid.

1.HW.98 An electron with a speed of $0.95c$ is emitted by a supernova, where c is the speed of light. What is the magnitude of the momentum of this electron?

1.HW.99 A "cosmic-ray" proton hits the upper atmosphere with a speed $0.9999c$, where c is the speed of light. What is the magnitude of the momentum of this proton?

1.HW.100 The position of a baseball relative to home plate changes from $\langle 15, 8, -3 \rangle$ m to $\langle 20, 6, -1 \rangle$ m in 0.1 second. As a vector, write the average velocity of the baseball during this time interval.

1.HW.101 The diagram below shows the trajectory of a ball traveling through the air, affected by both gravity and air resistance.

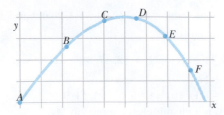

The table below gives the position of the ball at several successive times.

loc.	t (s)	position (m)
A	0.0	$\langle 0, 0, 0 \rangle$
B	1.0	$\langle 22.3, 26.1, 0 \rangle$
C	2.0	$\langle 40.1, 38.1, 0 \rangle$

  (a) What is the average velocity of the ball as it travels between location A and location B?
  (b) If the ball continued to travel at the same average velocity during the next second, where would it be at the end of that second? (That is, where would it be at time $t = 2$ seconds?)
  (c) How does your prediction from part (b) compare to the actual position of the ball at $t = 2$ seconds (location C)? If the predicted and observed locations of the ball are different, explain why.

1.HW.102 In a laboratory experiment, an electron passes location $\langle 0.02, 0.04, -0.06 \rangle$ m, and 2 μs later is detected at location $\langle 0.02, 1.84, -0.86 \rangle$ m (1 microsecond is 1×10^{-6} s).
  (a) What is the average velocity of the electron?
  (b) If the electron continues to travel at this average velocity, where will it be in another 5 μs?

1.HW.103 At 6 seconds after 3:00, a butterfly is observed leaving a flower whose location is $\langle 6, -3, 10 \rangle$ m relative to an origin on top of a nearby tree. The butterfly flies until 10 seconds after 3:00, when it alights on a different flower whose location is $\langle 6.8, -4.2, 11.2 \rangle$ m relative to the same origin. What was the location of the butterfly at a time 8.5 seconds after 3:00? What assumption did you have to make in calculating this location?

1.HW.104 The colored line shows a portion of the trajectory of a ball traveling through the air. Arrows indicate its momentum at several locations.

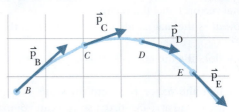

At various locations, the ball's momentum is:
  $\vec{p}_B = \langle 3.03, 2.83, 0 \rangle$ kg·m/s
  $\vec{p}_C = \langle 2.55, 0.97, 0 \rangle$ kg·m/s
  $\vec{p}_D = \langle 2.24, -0.57, 0 \rangle$ kg·m/s
  $\vec{p}_E = \langle 1.97, -1.93, 0 \rangle$ kg·m/s
  $\vec{p}_F = \langle 1.68, -3.04, 0 \rangle$ kg·m/s
  (a) Calculate the change in the ball's momentum between each pair of adjacent locations.
  (b) On a copy of the diagram, draw arrows representing each $\Delta\vec{p}$ you calculated in part (a)
  (c) Between which two locations is the magnitude of the change in momentum greatest?

1.HW.105 A spacecraft traveling at a velocity of $\langle -20, -90, 40 \rangle$ m/s is observed to be at a location $\langle 200, 300, -500 \rangle$ m relative to an origin located on a nearby asteroid. At a later time the spacecraft is observed to be at location $\langle -380, -2310, 660 \rangle$ m .

a) How long did it take the spacecraft to travel between these locations?

b) How far did the spacecraft travel?

c) What is the speed of the spacecraft?

d) What is the unit vector in the direction of the spacecraft's velocity?

Computational problems

These problems are intended to introduce you to using a computer to model matter, interactions, and motion. You will build on these small calculations to build models of physical systems in later chapters.

Some parts of these problems can be done with almost any tool (spreadsheet, math package, etc.). Other parts are most easily done with a programming language. We recommend the free 3D programming language VPython (http://vpython.org). Your instructor will introduce you to an available computational tool and assign problems, or parts of problems, that can be addressed using the chosen tool.

1.P.106 Move an object across a computer screen

(a) Write a program that makes an object move from left to right across the screen at speed v. Make v a variable, so you can change it later. Let the time interval for each step of the computation be a variable dt, so that the position x increases by an amount $v*dt$ each time.

(b) Modify a copy of your program to make the object run into a wall and reverse its direction.

(c) Make a modification so that the object's speed v is no longer a constant but changes smoothly with time. Is the speed change clearly visible to an observer? Try to make one version in which the speed change is clearly noticeable, and another in which it is not noticeable.

(d) Corresponding to part (c), make a computer graph of x vs. t, where t is the time.

(e) Corresponding to part (c), make a computer graph of v vs. t, where t is the time.

Turn in your programs for parts (c), (d), and (e).

1.P.107 Move an object at an angle

(a) Write a program that makes an object move at an angle.

(b) Change the component of velocity of the object in the x direction but not in the y direction, or vice versa. What do you observe?

(c) Start the object moving at an angle and make it bounce off at an appropriate angle when it hits a wall.

Turn in your answer to part (b), and the final version of your computation, part (c).

1.P.108 Move an object, leave a trail

Write a program that makes an object move smoothly from left to right across the screen at speed v, leaving a trail of dots on the screen at equal time intervals. If the dots are too close together, leave a dot every N steps, and adjust N to give a nice display.

1.15 ANSWERS TO EXERCISES

1.X.1 (page 6)   a, d

1.X.2 (page 6)   a, b, c

1.X.3 (page 6)   Continues to move in same direction at 1×10^4 m/s.

1.X.4 (page 9)   3

1.X.5 (page 9)   1

1.X.6 (page 9)   vector

1.X.7 (page 9)   b, c

1.X.8 (page 9)   7

1.X.9 (page 10)   scalar

1.X.10 (page 10)   no

1.X.11 (page 10)   5.10 m

1.X.12 (page 10)   no

1.X.13 (page 10) 2.15×10^7 m/s

1.X.14 (page 11) $\langle 6, -9, 15 \rangle$ m/s

1.X.15 (page 11) $\langle 1, -1.5, 2.5 \rangle$ m/s

1.X.16 (page 11)   0.04, −3.4, 60.0

1.X.17 (page 11)   in the opposite direction

1.X.18 (page 11)   no

1.X.19 (page 11)   no

1.X.20 (page 12) $\langle 0, 1, 0 \rangle$

1.X.21 (page 13) $\langle -1, 0, 0 \rangle$

1.X.22 (page 13) $\langle \frac{1}{\sqrt{3}}, \frac{1}{\sqrt{3}}, \frac{1}{\sqrt{3}} \rangle$, $\langle \frac{1}{\sqrt{3}}, \frac{1}{\sqrt{3}}, \frac{1}{\sqrt{3}} \rangle$

1.X.23 (page 13) $\langle 0.873, 0.436, -0.218 \rangle$

1.X.24 (page 13) $458 \langle 0.873, 0.436, -0.218 \rangle \frac{\text{m}}{\text{s}^2}$

1.X.25 (page 14) $\langle 450, -300, -200 \rangle$

1.X.26 (page 14)   361, 335, 577

1.X.27 (page 14)   696, no

1.X.28 (page 14) $\langle 150, 300, -200 \rangle$, $\langle -150, -300, 200 \rangle$

1.X.29 (page 14) $\langle -1, 0, -1 \rangle$ m

1.X.30 (page 14) $\langle 2, 4, 0 \rangle$ m

1.X.31 (page 14) $\langle -2, -4, 0 \rangle$ m

1.X.32 (page 14) $\langle 9, 2, 20 \rangle$ m, 120 s

1.X.33 (page 15) $\langle -0.940, 0.342, 0 \rangle$

1.X.34 (page 15) $\langle -0.643, 0.766, 0 \rangle$

1.X.35 (page 16) $\langle 0.577, 0.577, 0.577 \rangle$, 54.7°

1.X.36 (page 16) 2.67×10^{-3} m/s

1.X.37 (page 18) $\langle 8, -10, 12 \rangle$ m/s, 17.55 m/s, $\langle 0.456, -0.570, 0.684 \rangle$

1.X.38 (page 19) $\langle 25, -55, -30 \rangle$ m/s, 67.45 m/s, $\langle 0.371, -0.815, -0.445 \rangle$

1.X.39 (page 20) $\langle 2.1 \times 10^5, 1.4 \times 10^5, -2.8 \times 10^5 \rangle$ m

1.X.40 (page 20)   0.5 s

1.X.41 (page 20) $\langle -2.01 \times 10^5, 5.20 \times 10^4, -1.00 \times 10^3 \rangle$ m/s, 2.08×10^5 m/s

1.X.42 (page 25) $650 \text{ kg} \cdot \text{m/s}$

1.X.43 (page 25) $6.9 \text{ kg} \cdot \text{m/s}$

1.X.44 (page 25) $2.2 \times 10^5 \text{ kg} \cdot \text{m/s}$

1.X.45 (page 25) $6.71 \text{ kg} \cdot \text{m/s}$

1.X.46 (page 25) $2.415 \times 10^{-22} \text{ kg} \cdot \text{m/s}$

1.X.47 (page 25) $5.0 \times 10^{-19} \text{kg} \cdot \text{m/s}$, $3.6 \times 10^{-18} \text{kg} \cdot \text{m/s}$, $\gamma = 7.09$

1.X.48 (page 26) $\langle 0.4, 0, 10 \rangle$ m

1.X.49 (page 28) $\vec{p}_i$ to the right $(+x)$, $\vec{p}_f$ to the left $(-x)$, $\langle -5.59, 0, 0 \rangle$ kg·m/s, 0.114 kg·m/s

1.X.50 (page 28) $\langle -1.6 \times 10^{28}, 0, 1.6 \times 10^{28} \rangle$ kg·m/s, downward to the left

1.X.51 (page 28) $0, 500 \text{ kg} \cdot \text{m/s}$

CHAPTER 2

THE MOMENTUM PRINCIPLE: IMPULSE AND MOMENTUM CHANGE

Key concepts:
- One or more objects can be considered to be a "system."
  - Everything not in the system is part of the "surroundings."
- The momentum of a system can be changed only by interactions with the surroundings.
  - Force is a quantitative measure of interaction.
  - Force is a vector (it has magnitude and direction).
- The change in momentum of a system is equal to the net impulse applied, which involves:
  - The total force exerted on it by all objects not in the system
  - The time interval over which the interaction occurs.
- Momentum is *conserved*: The change in momentum of a system plus the change in momentum of its surroundings is zero.

The Momentum Principle is the first of three fundamental principles of mechanics which together make it possible to predict and explain a very broad range of real-world phenomena (the other two are the Energy Principle and the Angular Momentum Principle). The Momentum Principle makes a quantitative connection between amount of interaction and change of momentum.

A fundamental principle relates the change in a quantity to the interaction causing the change, and is special because it applies in absolutely every situation.

2.1 SYSTEM AND SURROUNDINGS

One or more objects can be considered to be a "system." Everything that is not included in the system is part of the "surroundings." The Momentum Principle relates the change in momentum of a system to the amount of interaction with its surroundings. No matter what system we choose, the Momentum Principle will correctly predict the behavior of the system.

As an example, consider the following situation: We are in a spacecraft in outer space, far from massive objects such as planets. In the spacecraft are two identical small, hard balls, each with a mass of 2 kg. Initially ball *B* is at rest, and ball *A* is moving toward ball *B* with velocity $\langle 3, 0, 0 \rangle$ m/s, as shown in Figure 2.1, so its momentum is

$$\vec{p}_A = (2 \text{ kg})\langle 3, 0, 0 \rangle \text{ m/s} = \langle 6, 0, 0 \rangle \text{ kg} \cdot \text{m/s}$$

If *A* collides exactly head-on with *B*, *A* stops dead, and *B* is now observed to move with velocity $\langle 3, 0, 0 \rangle$ m/s, and momentum $\langle 6, 0, 0 \rangle$ kg·m/s.

System: One object (ball *A*)
In thinking about the collision described above, we could choose ball *A* to be the system, and ball *B* to be part of the surroundings. In this case, we would say that the momentum of ball *A* changed because ball *A* interacted with its surroundings (ball *B*). The change of momentum of ball A was:

$$\Delta \vec{p}_A = \langle 0, 0, 0 \rangle \text{ kg} \cdot \text{m/s} - \langle 6, 0, 0 \rangle \text{ kg} \cdot \text{m/s} = \langle -6, 0, 0 \rangle \text{ kg} \cdot \text{m/s}$$

The momentum of the surroundings (ball B) also changed:

$$\Delta \vec{p}_B = \langle 6, 0, 0 \rangle \text{ kg} \cdot \text{m/s} - \langle 0, 0, 0 \rangle \text{ kg} \cdot \text{m/s} = \langle +6, 0, 0 \rangle \text{ kg} \cdot \text{m/s}$$

The sum of the change of momentum of the system (*A*) and the change of momentum of the surroundings (*B*) is zero.

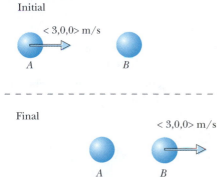

Figure 2.1 A head-on collision between two identical balls. Initially *B* is at rest. After the collision *A* is at rest.

In a sense, momentum is "sticky"—it's hard to get rid of. Once a system has momentum, the only way to change that momentum is through an interaction which transfers momentum to the surroundings.

System: Both objects (ball *A* and ball *B*)

Alternatively, we could choose to include both *A* and *B* in the system; in this case the surroundings include the air in the spacecraft, the spacecraft itself, etc. In this case, at every instant we need to consider the total momentum of the system, which is simply the sum of the momenta of balls *A* and *B*.

Initially, the momentum of the two-ball system was:

$$\vec{p}_i = \langle 6, 0, 0 \rangle \text{ kg} \cdot \text{m/s} + \langle 0, 0, 0 \rangle \text{ kg} \cdot \text{m/s} = \langle 6, 0, 0 \rangle \text{ kg} \cdot \text{m/s}$$

After the collision, the momentum of the two-ball system was:

$$\vec{p}_f = \langle 0, 0, 0 \rangle \text{ kg} \cdot \text{m/s} + \langle 6, 0, 0 \rangle \text{ kg} \cdot \text{m/s} = \langle 6, 0, 0 \rangle \text{ kg} \cdot \text{m/s}$$

The change in momentum of the two-ball system was $\vec{0}$, because the system did not interact significantly with its surroundings (momentum simply "flowed" from one object within the system to another). The change in momentum of the surroundings was also $\vec{0}$, so the sum of these changes was $\vec{0}$.

Because the sum of the change of momentum of the system and the change of momentum of the surroundings is always zero, we say that momentum is a "conserved" quantity. We will discuss this in more detail in Section 2.10.

2.2   THE MOMENTUM PRINCIPLE

The Momentum Principle is a fundamental principle that is sometimes called Newton's second law. It restates and extends Newton's first law of motion in a quantitative, causal form that can be used to predict the behavior of objects. The validity of the Momentum Principle has been verified through a very wide variety of observations and experiments, involving large and small objects, moving slowly or at speeds near the speed of light. It is a summary of the way interactions affect motion in the real world.

The Momentum Principle is a Fundamental Principle because:

It applies to every possible system, no matter how large or small (from clusters of galaxies to subatomic particles), or how fast it is moving.

It is true for any kind of interaction (electric, gravitational, etc.)

It relates an effect (change in momentum) to a cause (an interaction).

> ### THE MOMENTUM PRINCIPLE
>
> $$\Delta \vec{p} = \vec{F}_{net} \Delta t$$
>
> In words: change of momentum of a system (the effect) is equal to the net force acting on the system times the duration of the interaction (the cause).

As usual, the capital Greek letter delta (Δ) means "change of" (something), or "final minus initial." We will see that Δt must be small enough that the net force on the system does not change significantly during the interval.

$$\boxed{\Delta \vec{p}} = \vec{F}_{net} \Delta t$$

Change of momentum

As we saw in the previous chapter, the change in momentum of a system can involve

- a change in the magnitude of momentum
- a change in the direction of momentum
- change in both magnitude and direction

$$\Delta \vec{p} = \vec{p}_f - \vec{p}_i$$

You are already familiar with change of momentum $\Delta \vec{p}$ and with time interval Δt. The new element is the concept of "force." The "net" force $\vec{F}_{net}$ is the vector sum of all the forces acting on an object.

Net force

Scientists and engineers employ the concept of "force" to quantify interactions between two objects. Force is a vector quantity because a force has a magnitude and is exerted in a particular direction. Examples of forces include the following:

$$\Delta \vec{p} = \boxed{\vec{F}_{net}} \Delta t$$

- the repulsive electric force a proton exerts on another proton
- the attractive gravitational force the Earth exerts on you
- the force that a compressed spring exerts on your hand
- the force on a spacecraft of expanding gases in a rocket engine
- the force of the air on the propeller of an airplane or swamp boat

Measuring the magnitude of the velocity of an object (in other words, measuring its speed) is a familiar task, but how do we measure the magnitude of a force?

A simple way to measure force is to use the stretch or compression of a spring. In Figure 2.2 we hang a block from a spring, and note that the spring is stretched a distance *s*. Then we hang two such blocks from the spring, and we see that the spring is stretched twice as much. By experimentation, we find that any spring made of the same material and produced to the same specifications behaves in the same way. Similarly, we can observe how much the spring compresses when the same blocks are supported by it (Figure 2.3). We find that one block compresses the spring by the same distance $|s|$, and two blocks compress it by $2|s|$. (Compression can be considered negative stretch, because the length of the spring decreases.)

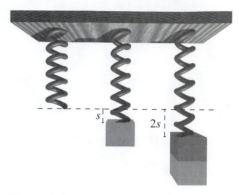

Figure 2.2 Stretching of a spring is a measure of force.

We can use a spring to make a scale for measuring forces, calibrating it in terms of what force is required to produce a given stretch. The SI unit of force is the "newton," abbreviated as "N." One newton is a rather small force. A newton is approximately the downward gravitational force of the Earth on a small apple, or about a quarter of a pound. If you hold a small apple at rest in your hand, you apply an upward force of about one newton, compensating for the downward pull of the Earth.

The net force, the vector sum of all the forces acting on a system, acting for some time Δt causes changes of momentum.

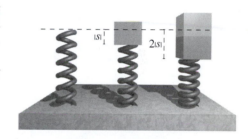

Figure 2.3 Compression of a spring is also a measure of force.

DEFINITION OF NET FORCE

$$\vec{F}_{net} \equiv \vec{F}_1 + \vec{F}_2 + \ldots$$

"Net" means "total". The net force acting on a system is the vector sum of all forces acting on the system.

The net force acting on a system is the vector sum of all of the forces exerted on the system by objects in the surroundings. Forces internal to the system may exist, but cannot change the momentum of the system. See the discussion and proof in Section 2.8.

We find experimentally that the magnitude of the net force acting on an object affects the magnitude of the change in its momentum. Many introductory physics laboratories have air tracks like the one illustrated in Figure 2.4. The long triangular base has many small holes in it, and air under pressure is blown out through these holes. The air forms a cushion under the glider, allowing it to coast smoothly with very little friction.

Suppose we place a block on a glider on a long air track, and attach a calibrated spring to it (Figure 2.4). We pull on the spring so that it is stretched a distance *s*, so that it exerts a force *F* on the block, and we pull for a short time. We observe that the momentum of the block increases from zero to an amount mv (since $v \ll c$). If instead we pull on the spring so that it stretches a distance $2s$, the spring exerts a force $2F$ on the block, and we observe that the block's momentum increases to $2mv$ in the same amount of time. Apparently the magnitude of the change in momentum is proportional to the magnitude of the net force applied to the object.

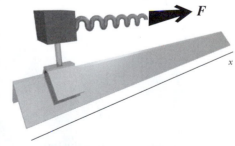

Figure 2.4 A block mounted on a nearly frictionless air track, pulled by a spring.

Duration of interaction

Another experiment will show us that not only the magnitude of the force, but the length of time during which it acts on an object, affects the change

$$\Delta \vec{p} = \vec{F}_{net} \left(\Delta t\right)$$

of momentum of the object. Suppose we again place a block on an air track and attach a spring to it. We pull on the spring for a time Δt with a force F, and we observe that the momentum of the block increases from zero to an amount mv. However, if we repeat the experiment but pull for a time twice as long with the same force F, the block's momentum increases to $2mv$; the change in its momentum is twice as great. We observe that the magnitude of the change in momentum is directly proportional to the length of time $\Delta t = t_2 - t_1$ during which the force acts on the object.

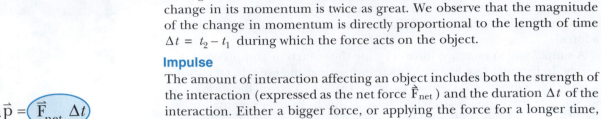

Impulse

The amount of interaction affecting an object includes both the strength of the interaction (expressed as the net force $\vec{F}_{net}$) and the duration Δt of the interaction. Either a bigger force, or applying the force for a longer time, will cause more change of momentum.

The product of a force and a time interval is called "impulse".

DEFINITION OF IMPULSE

Impulse $\equiv \vec{F}\Delta t$ (for small enough Δt)

Impulse has units of N·s (newton-seconds)

With this definition of impulse we can state the Momentum Principle in words like this:

**The change of momentum of a system
is equal to the net impulse applied to it.**

What if the force is not constant during the interval Δt?

In the air track experiment described above, we were able to keep the force constant during the time interval Δt, by keeping the stretch of the spring constant. In many cases (perhaps most real-world cases), the force applied to an object is not constant, but changes as the object moves. For example, the magnitude of the force exerted on an object by a spring attached to the object changes when the stretch of the spring changes. As a comet moves, both the direction and the magnitude of the force exerted on the comet by a star changes (Figure 2.5).

If the magnitude or the direction of a force changes during a time interval Δt, what value of the force should we use in calculating the impulse? There are two possibilities:

- Use an approximate average value of the force
- Divide the time interval into several time intervals small enough that the force remains approximately constant over each smaller time interval

Each of these possibilities requires that we make an approximation. This is fine, but it is important to be aware that we are making an approximation!

This is the same issue we met with the position update relation, $\vec{r}_f = \vec{r}_i + \vec{v}_{avg}\Delta t$, where we need to use a short enough time interval that the velocity isn't changing very much, or else we need to know the average velocity during the time interval.

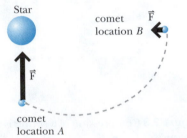

Figure 2.5 The direction and magnitude of the gravitational force on a comet by a star change as the comet's position changes. The size of the comet is exaggerated in this diagram.

Δt must be small enough that the force can be considered to be approximately constant over the time interval.

?   A constant net force $\langle 3, -5, 4\rangle$ N acts on an object for 10 s. What is the net impulse applied to the object? What was the change in momentum of the object?

$$\text{impulse} = \vec{F}_{net}\Delta t = \langle 3, -5, 4\rangle \text{ N} \cdot (10 \text{ s}) = \langle 30, -50, 40\rangle \text{ N} \cdot \text{s}$$

The change in the object's momentum is equal to the net impulse, so

$$\Delta\vec{p} = \langle 30, -50, 40\rangle \text{ kg} \cdot \text{m/s}$$

A newton can be expressed in terms of kilograms, meters, and seconds.

$1 \text{ N} = 1 \text{ kg·m/s}^2$

Evidently $1 \text{ N} \cdot \text{s} = 1 \text{ kg} \cdot \text{m/s}$, so $1 \text{ N} = 1 \text{ kg} \cdot \text{m/s}^2$.

2.X.1 A constant net force of $\langle -0.5, -0.2, 0.8 \rangle$ N acts on an object for 2 minutes. (a) What is the impulse applied to the object, in SI units? (b) What is the change in the momentum of the object?

2.X.2 A hockey puck initially has momentum $\langle 0, 2, 0 \rangle$ kg · m/s. It slides along the ice, gradually slowing down, until it comes to a stop. (a) What was the impulse applied by the ice to the hockey puck? (b) It took 3 seconds for the puck to come to a stop. During this time interval, what was the net force on the puck by the ice and the air (assuming this force was constant)?

Update form of the Momentum Principle

Since $\Delta \vec{p} = \vec{p}_f - \vec{p}_i = \vec{F}_{net}\Delta t$ ("final minus initial"), we can rearrange the Momentum Principle to the update form:

THE MOMENTUM PRINCIPLE (UPDATE FORM)

$$\vec{p}_f = \vec{p}_i + \vec{F}_{net}\Delta t$$

for a time interval Δt short enough that the net force is approximately constant over this time interval

This update version of the Momentum Principle emphasizes the fact that if you know the initial momentum, and you know the net force acting during a "short enough" time interval, you can predict the final momentum.

The "update form" of the Momentum Principle can be used to predict the new momentum of a system to which a known force has been applied for a known time interval.

Separation of components

A great deal of information is expressed compactly in the vector form of this equation, $\vec{p}_f = \vec{p}_i + \vec{F}_{net}\Delta t$. The Momentum Principle written in terms of vectors can be interpreted as three ordinary scalar equations, for components of the motion along the x, y, and z axes:

$$p_{fx} = p_{ix} + F_{net, x}\Delta t$$

$$p_{fy} = p_{iy} + F_{net, y}\Delta t$$

$$p_{fz} = p_{iz} + F_{net, z}\Delta t$$

An important aspect of this is the fact that the x component of an object's momentum cannot be affected by forces in the y or z directions.

This fact can be very useful in solving problems. In some situations, for example, if we know that the y and z components of an object's momentum are not changing, we may choose to work only with the x component of the momentum update equation.

Applying the Momentum Principle

We can use the Momentum Principle to predict the change in momentum of an object, in an experiment using the air track apparatus described above. We will choose the x axis to point in the direction of the motion. We stretch a calibrated spring, which exerts a force on the block of 20 N when stretched 4 cm. Suppose the block starts from rest, and the spring pulls the block for 1 second in the $+x$ direction with a a force of magnitude 20 N (you will have to move forward in order to keep the spring stretched).

? The block starts from rest, so $\vec{p}_i = \langle 0, 0, 0 \rangle$ kg·m/s. What would the Momentum Principle predict the new momentum of the block to be after 1 second?

Since the friction force on the glider is negligibly small, the net force on the glider is just the force exerted by the spring. The Momentum Principle (update form) applied to the glider is:

$$\vec{p}_f = \langle 0, 0, 0 \rangle \text{ kg·m/s} + \langle 20, 0, 0 \rangle \text{ N(1 s)}$$

Since the net force is in the x direction, we know that the y and z components of the glider's momentum will not change, and we can work with just the x component of the momentum.

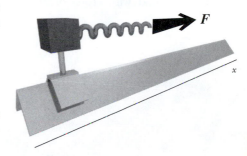

Figure 2.6 Apply a constant force to a block on a low-friction air track.

$$p_{fx} = p_{ix} + F_{net,x}\Delta t = 0 + (20\text{ N})(1\text{ s}) = 20\text{ kg·m/s}$$

If you do the experiment, this is what you will observe.

Suppose the spring keeps exerting a force on the block for another second, but now with the spring stretched half as much, 2 cm, so you know that the spring is exerting half the original force (10 N) on the block.

? What would the Momentum Principle predict the new x component of the momentum of the block to be now?

The Momentum Principle would predict the following, where we take the final momentum from the first pull and consider that to be the initial momentum for the second pull:

$$p_{fx} = p_{ix} + F_{net,x}\Delta t = (20\text{ kg·m/s}) + (10\text{ N})(1\text{ s}) = 30\text{ kg·m/s}$$

If you do the experiment, this is what you will observe (Figure 2.7). Note that the effects of the interactions in the two 1-second intervals add; we add the two momentum changes.

Instead of varying the net force, we could try varying the duration of the interaction. Start over with the block initially at rest. Pull for 2 seconds with the spring stretched 4 cm, so the force of the spring on the block is 20 N.

? The block starts from rest ($p_{ix} = 0$). What would the Momentum Principle predict the new x component of the momentum of the block to be?

$$p_{fx} = p_{ix} + F_{net,x}\Delta t = 0 + (20\text{ N})(2\text{ s}) = 40\text{ kg·m/s}$$

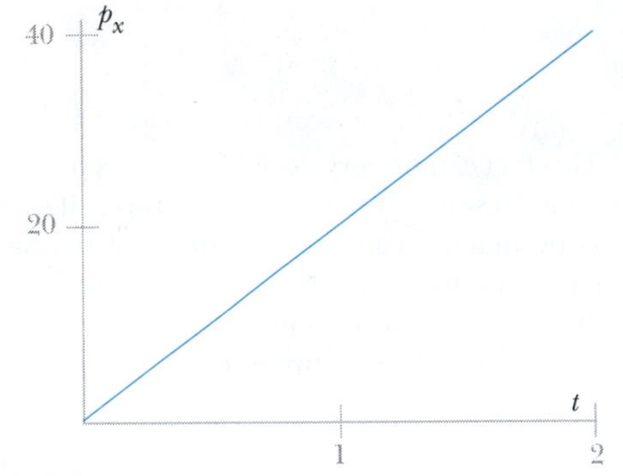

Here (Figure 2.8) the final x component of momentum was 40 kg·m/s after applying a force of 20 N for 2 s, whereas in the previous experiments we got 30 kg·m/s after the spring applied a force of 20 N for 1 s plus a force of 10 N for 1 s. In our calculations we can use big values of Δt as long as the force isn't changing much. When the force changed from 20 N for a second to 10 N for a second we had to treat the two time intervals separately.

Many different experiments have shown the validity of the Momentum Principle. If we use two springs to move the block, we find that it is indeed the vector sum of the two spring forces, the "net" force, that accounts for the change in momentum.

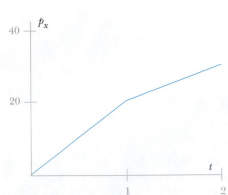

Figure 2.7 x component of momentum of he block (kg·m/s) *vs.* time (s), when pulled for 1 second with a force of 20 N, then for 1 second with a force of 10 N.

Figure 2.8 x component of momentum of the block (kg·m/s) *vs.* time (s), when pulled for 2 seconds with a force of 20 N.

2.3 APPLYING THE MOMENTUM PRINCIPLE

To apply the Momentum Principle to a real-world situation, several steps are required:

- Choose a system, consisting of a portion of the Universe. The rest of the Universe is called the surroundings.
- Make a list of the objects in the surroundings that exert significant forces on the chosen system, and make a labeled diagram showing the external forces exerted by the objects in the surroundings.
- Choose initial and final times (often these are obvious; in examples we will comment on this when necessary)
- Apply the Momentum Principle to the chosen system:

$$\vec{p}_f = \vec{p}_i + \vec{F}_{net}\Delta t$$

Substitute known values into the terms of the Momentum Principle.

- Check units; check the reasonableness of your answer (direction of momentum, change in magnitude of momentum), based on the physical situation.

Example: Ball and string (2D, single force)

Inside a spaceship in outer space there is a small steel ball of mass 0.25 kg. At a particular instant, the ball has momentum $\langle -8, 3, 0 \rangle$ kg·m/s. At this instant the ball is being pulled by a string, which exerts a net force $\langle 10, 25, 0 \rangle$ N on the ball. What is the ball's approximate momentum 0.5 seconds later?

Assumption: The force didn't change much during 0.5 second.

System: the steel ball
Surroundings: the string

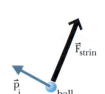

Momentum Principle:

$$\vec{p}_f = \vec{p}_i + \vec{F}_{net} \Delta t$$

$$\vec{p}_f = (\langle -8, 3, 0 \rangle \text{ kg·m/s}) + (\langle 10, 25, 0 \rangle \text{ N})(0.5 \text{ s})$$

$$\vec{p}_f = (\langle -8, 3, 0 \rangle \text{ kg·m/s}) + (\langle 5, 12.5, 0 \rangle \text{ N·s})$$

$$\vec{p}_f = \langle -3, 15.5, 0 \rangle \text{ kg·m/s}$$

Check: units correct (momentum: kg·m/s and position: m); direction of final momentum makes sense (Figure 2.9).

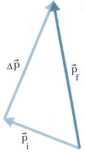

Figure 2.9 Initial momentum, change in momentum, and final momentum for a ball pulled by a string.

More than one force

An easy way to arrange to apply a nearly constant force is to mount a battery powered fan on a cart (Figure 2.10). If the fan directs air backwards, the interaction with the air pushes the cart forward with a nearly constant force. Swamp boats used in the very shallow Florida Everglades are built in a similar way, with large fans on top of the boats propelling them through the swamp.

Air resistance and friction

When an object gets going very fast, air resistance, sometimes called "drag", becomes important and at high speeds can be as big as the propelling force. At this point the momentum doesn't increase any more, and the object travels at constant speed. When the speed of an object is low, air resistance is small, so we can make the approximation that the net force is due solely to the fan and is nearly constant.

Friction between the object and the surface on which it moves can also be important. Friction is usually independent of speed, so this is a constant force. The friction between the cart and the track is relatively low, so it is reasonable to say it is approximately zero in this case.

Figure 2.10 A fan cart on a track.

Other interacting objects

If we choose the fan cart as the system, the air is not the only external object that interacts with the system. The Earth exerts a downward gravitational force on the cart, and the track exerts an upward force, as shown in Figure 2.11. We must consider these forces as well, because they contribute to the net external force on the fan cart.

If we see the cart rolling horizontally along the track, it is clear that the y component of its momentum remains zero and does not change:

$$\Delta p_y = 0$$

Figure 2.11 Forces on a moving fan cart.

? Using the Momentum Principle, what can we conclude about the relative magnitudes of the force on the cart by the Earth and the force on the cart by the track?

According to the Momentum Principle the y component of the net force on the cart is zero. Evidently these two forces must be exactly equal in magnitude, and opposite in direction.

You may wonder how the track "knows" exactly how much upward force to exert on the cart. Basically, the cart compresses the track slightly, and the compressed track pushes up, like a spring. Understanding this in detail requires a microscopic view, discussed in Chapter 4.

Example: A fan cart (1D, several forces, constant net force)

Suppose you have a fan cart whose mass is 400 grams (0.4 kg), and with the fan turned on, the force acting on the cart due to the air and friction with the track, is $\langle 0.2, 0, 0 \rangle$ N and constant. You give the cart a shove, and you release the cart at position $\langle 0.5, 0, 0 \rangle$ m with initial velocity $\langle 1.2, 0, 0 \rangle$ m/s. What is the momentum of the cart 3 seconds later? What is its velocity at this time?

We choose the initial time to be just after you release the cart, so your hand no longer exerts a force on the cart.

System: the cart (including the fan). System is indicated by a circle on the diagram.
Surroundings: the Earth, the track, the air
Initial time: see comment in margin

Momentum Principle

Since the y component of the cart's momentum does not change, we know that the y component of the net force must be zero, and $\left| \vec{F}_{track} \right| = \left| \vec{F}_{Earth} \right|$.

$$\vec{p}_f = \vec{p}_i + \vec{F}_{net}\Delta t = \vec{p}_i + (\vec{F}_{track} + \vec{F}_{Earth} + \vec{F}_{air})(\Delta t)$$

$$\vec{p}_f = \vec{p}_i + \langle 0.2, (\left| \vec{F}_{track} \right| - \left| \vec{F}_{Earth} \right|), 0 \rangle \text{N}(3 \text{ s})$$

$$\vec{p}_f = \langle 0.4 \text{kg} 1.2 \text{m/s}, 0, 0 \rangle + \langle 0.2, 0, 0 \rangle \text{N}(3 \text{ s})$$

We can use a large time interval Δt because the force isn't changing very much in either magnitude or direction.

$$\vec{p}_f = \langle 1.08, 0, 0 \rangle \text{ kg} \cdot \text{m/s}$$

$$\vec{v}_f \approx \frac{\vec{p}_f}{m} = \frac{\langle 1.08, 0, 0 \rangle \text{ kg} \cdot \text{m/s}}{0.4 \text{kg}} = \langle 2.7, 0, 0 \rangle \text{ m/s}$$

Since $v \ll c$, we can use the approximation that $\vec{p} \approx m\vec{v}$.

Check: Speed increased; reasonable since force was in same direction as momentum.

Example: Fast proton (1D, constant net force, relativistic)

A proton in a particle accelerator is moving with velocity $\langle 0.96c, 0, 0 \rangle$, so the speed is $0.96 \times 3\times10^8$ m/s $= 2.88\times10^8$ m/s. A constant electric force is applied to the proton to speed it up, $\vec{F}_{net} = \langle 5\times10^{-12}, 0, 0 \rangle$ N. What is the proton's speed as a fraction of the speed of light after 20 nanoseconds ($1 \text{ ns} = 1\times10^{-9}$ s)?

System: the proton
Surroundings: electric charges in the accelerator

$\langle 5\times10^{-12}, 0, 0 \rangle$ N

Momentum Principle:

$$\vec{p}_f = \vec{p}_i + \vec{F}_{net}\Delta t$$

$$\langle p_{fx}, 0, 0 \rangle = \langle \gamma_i m v_{ix}, 0, 0 \rangle + ((\langle 5\times10^{-12}, 0, 0 \rangle \text{ N})(20\times10^{-9} \text{ s})$$

$$p_{fx} = \frac{1}{\sqrt{1 - \left(\frac{0.96c}{c}\right)^2}}(1.7\times10^{-27} \text{ kg})(0.96 \times 3\times10^8 \text{ m/s}) + (1\times10^{-19} \text{ N·s})$$

$$p_{fx} = (1.75\times10^{-18} \text{ kg·m/s}) + (1\times10^{-19} \text{ N·s}) = 1.85\times10^{-18} \text{ kg·m/s}$$

$$\frac{v_{fx}}{c} = \frac{\dfrac{p_{fx}}{mc}}{\sqrt{1 + \left(\dfrac{p_{fx}}{mc}\right)^2}}$$

Evaluate $\dfrac{p_{fx}}{mc} = \dfrac{1.85\times10^{-18}\ \text{kg}\cdot\text{m/s}}{(1.7\times10^{-27}\ \text{kg})(3\times10^8\ \text{m/s})} = 3.62$ (no units)

$\dfrac{v_{fx}}{c} = \dfrac{3.62}{\sqrt{1 + 3.62^2}} = 0.964$ (no units)

2.X.3 A hockey puck is sliding along the ice with nearly constant momentum $\langle 10, 0, 5\rangle$ kg $\cdot$ m/s when it is suddenly struck by a hockey stick with a force $\langle 0, 0, 2000\rangle$ N that lasts for only 3 milliseconds $(3\times10^{-3}\ \text{s})$. What is the new momentum of the puck?

2.X.4 You were driving a car with velocity $\langle 25, 0, 15\rangle$ m/s. You quickly turned and braked, and your velocity became $\langle 10, 0, 18\rangle$ m/s. The mass of the car was 1000 kg. What was the (vector) change in momentum $\Delta \vec{p}$ during this maneuver? Pay attention to signs. What was the (vector) impulse applied to the car by the ground?

2.X.5 In the previous exercise, if the maneuver took 3 seconds, what was the average net (vector) force $\vec{F}_{\text{net}}$ that the ground exerted on the car?

2.X.6 A truck driver slams on the brakes and the momentum changes from $\langle 9\times10^4, 0, 0\rangle$ kg $\cdot$ m/s to $\langle 5\times10^4, 0, 0\rangle$ kg $\cdot$ m/s in 4 seconds due to a constant force of the road on the wheels of car. As a vector, write the force exerted by the road.

2.X.7 At a certain instant, a particle is moving in the $+x$ direction with momentum $+10$ kg·m/s. During the next 0.1 s, a constant force $\langle -6, 3, 0\rangle$ N acts on the particle. What is the momentum of the particle at the end of this 0.1 s interval?

(See Section 2.13, page 65; obtaining v from p when the speed is near the speed of light.)

Although the magnitude of the momentum increased from 1.75×10^{-18} kg·m/s to 1.85×10^{-18} kg·m/s, the speed didn't increase very much, because the proton's initial speed, $0.96c$, was already close to the cosmic speed limit, c. Because the speed hardly changed, the distance the proton moved during the 20 ns was approximately equal to:

$(0.96 \times 3\times10^8\ \text{m/s})(20\times10^{-9}\ \text{s}) = 5.8$ m

2.4 UPDATING POSITION IF MOMENTUM IS CHANGING

In the fan cart example (page 48) the momentum, and therefore the velocity, of the fan cart is changing with time. If we want to predict where the fan cart will be at the end of the three second time interval, we need to use an appropriate value for the average velocity of the cart.

We can't calculate the average velocity directly from the definition

$$\vec{v}_{\text{avg}} \equiv \frac{\Delta \vec{r}}{\Delta t} = \frac{\vec{r}_f - \vec{r}_i}{\Delta t}$$

because the final location $\vec{r}_f$ of the cart is unknown. Since in this case we do not have any other information that would allow us to determine the average velocity, we need to find an approximate value for the average velocity in this situation. There are two possible approaches:

- If the change in velocity is very small over the time interval of interest, we can use the final velocity (or the initial velocity) as an approximation to the average velocity
- We can try to estimate an average value for the velocity

In the fan cart example on page 48, the final velocity of the cart is $\langle 2.7, 0, 0\rangle$ m/s, which is more than twice the initial velocity of $\langle 1.2, 0, 0\rangle$ m/s. Neither the final nor the initial velocity is a good approximation to the average velocity.

? What might be a better approximation for $\vec{v}_{\text{avg}}$ in this case?

Let's consider using the arithmetic average:

$$v_{avg,x} \approx \frac{(v_{ix} + v_{fx})}{2} \text{ is an approximation for } v_{avg,x} = \frac{\Delta x}{\Delta t}$$

The arithmetic average lies between the two extremes. For example, the arithmetic average of 6 and 8 is $(6+8)/2 = 14/2 = 7$, halfway between 6 and 8.

The arithmetic average is often a good approximation, but it is not necessarily equal to the true average $v_{avg,x} = \Delta x/\Delta t$. The arithmetic average does not give the true average unless v_x is changing at a constant rate, which is the case only if the net force is constant, as it happens to be for a fan cart (See Figure 2.12). The proof that $v_{avg,x} = (v_{ix} + v_{fx})/2$ when v_x changes at a constant rate is given in optional Section 2.11 at the end of this chapter. (The proof is more complicated than one might expect.)

? When is the arithmetic average a poor approximation for $\vec{v}_{avg}$?

If you drive 50 mi/hr for four hours, and then 20 mi/hr for an hour, you go 220 miles, and your average speed is $(220 \text{ mi})/(5 \text{ hr}) = 44$ mi/hr, whereas the arithmetic average is $(50+20)/2 = 35$ mi/hr (see Figure 2.13 for another such example). In situations where the force is not constant, we have to choose short enough time intervals that the velocity is nearly constant during the brief Δt.

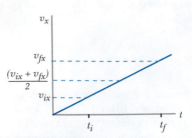

Figure 2.12 v_x is changing at a constant rate (linearly with time), so the arithmetic average is equal to $v_{avg,x}$

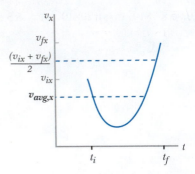

Figure 2.13 v_x is not changing linearly with time. In this case the arithmetic average is much higher than the true average value of $v_{avg,x}$

APPROXIMATE AVERAGE VELOCITY

$$v_{avg,x} \approx \frac{(v_{ix} + v_{fx})}{2} \text{ is an approximation for } v_{avg,x} = \frac{\Delta x}{\Delta t}$$

Exactly true only if the net force is constant (linear change in v).

May be a poor approximation if net force is not constant.

Now that we have a way to calculate the x component of the average velocity in this special situation (constant x component of force), we can predict the position of the fan cart at the end of the time interval:

The net force on the cart is constant, so this calculation of average velocity gives the correct value.

$$v_{avg,x} = \frac{(v_{ix} + v_{fx})}{2} = \frac{(1.2 + 2.7)}{2} \text{ m/s}$$

$$\vec{v}_{avg} = \langle 1.95, 0, 0 \rangle \text{ m/s}$$

$$\vec{r}_f = \vec{r}_i + \vec{v}_{avg}\Delta t = \langle 0.5, 0, 0 \rangle \text{m} + \langle 1.95, 0, 0 \rangle \frac{\text{m}}{\text{s}}(3\text{s})$$

$$\vec{r}_f = \langle 6.35, 0, 0 \rangle \text{m}$$

2.5 PREDICTING MOTION: CONSTANT FORCE

We can now combine the Momentum Principle with the position update formula to predict the future motion of an object, if the net force on the object is approximately constant over the time interval of interest. The basic approach is:

1. Apply the Momentum Principle (this requires specifying a system and interacting objects) to find the new momentum

2. Use the momentum to find the new position

3. Check reasonableness of results

In ordinary life, there are very few situations in which the net force on a moving system is truly constant, because objects interact with so many other objects at all times. However, it is sometimes useful to simplify a situation by considering the net force to be approximately constant.

Special case: An analytical solution

In the preceding examples we have obtained a numerical solution for the final momentum and position of an object. In a few very special cases it is possible to obtain an "analytical," or algebraic solution. The case of a constant net force is one of these special cases. The analytical solution is an equation that gives the position of an object as a function of time.

To find an analytical solution we follow the procedure stated above, but we use symbolic expressions instead of numbers. We will solve the problem generally, and then apply our solution to particular situations.

Example: 2D motion, constant net force

An object which is initially in motion is subject to a constant force. Its initial momentum has nonzero x and y components, but the force acts only in the y direction. (For example, this could be a ball thrown slowly enough that air resistance is negligible; it could be a helicopter taking off, or an electron moving through a television tube, or a ball thrown on the airless Moon.) The initial location of the object is $\langle x_i, y_i, 0 \rangle$, and its initial velocity is $\langle v_{ix}, v_{iy}, 0 \rangle$. The object moves slowly compared to the speed of light. Predict the velocity and position of the object after a time Δt. We'll assume the constant force acts in the $+y$ direction, but later we will be able to use our results for the case where there is a constant force that acts in the $-y$ direction, as is the case with the gravitational force near the Earth's surface.

1. Momentum principle:

System: the object
Surroundings: Constant net force $\langle 0, F_y, 0 \rangle$

$$\vec{p}_f = \vec{p}_i + \vec{F}_{net}\Delta t$$

$$\langle p_{fx}, p_{fy}, 0 \rangle = \langle p_{ix}, p_{iy}, 0 \rangle + \langle 0, F_y, 0 \rangle \Delta t$$

$$\langle p_{fx}, p_{fy}, 0 \rangle = \langle p_{ix}, (p_{iy} + F_y\Delta t), 0 \rangle$$

$$p_{fx} = p_{ix}$$

$$p_{fy} = (p_{iy} + F_y\Delta t)$$

$$p_{fz} = 0$$

A possible situation

Since the net force acts only in the y direction, the x and z components of momentum do not change. Only the y component of momentum (or velocity) is changed by this force.

2. Position update ($v \ll c$, so $\vec{p} \approx m\vec{v}$)

$$mv_{fx} = mv_{ix}$$

$$v_{fx} = v_{ix}$$

$$mv_{fy} = mv_{iy} + F_y\Delta t$$

$$v_{fy} = v_{iy} + \left(\frac{F_y}{m}\right)\Delta t$$

$$mv_{fz} = 0$$

$$v_{fz} = 0$$

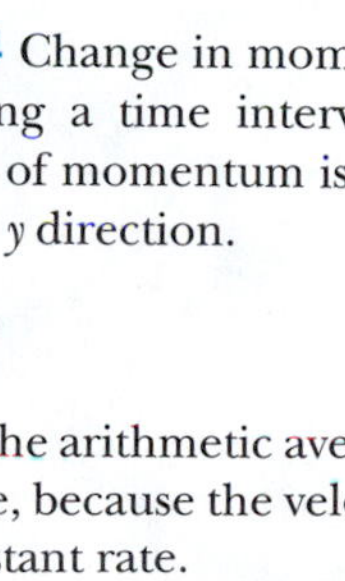

$$v_{avg,x} = \frac{v_{ix} + v_{fx}}{2} = v_{ix}$$

$$v_{avg,y} = \frac{\left(v_{iy} + \left(v_{iy} + \left(\frac{F_y}{m}\right)\Delta t\right)\right)}{2} = v_{iy} + \frac{1}{2}\left(\frac{F_y}{m}\right)\Delta t$$

$$v_{avg,z} = 0$$

$$\vec{r}_f = \vec{r}_i + \vec{v}_{avg}\Delta t$$

Figure 2.14 Change in momentum of the object during a time interval Δt. The y component of momentum is affected by a force in the y direction.

In this case the arithmetic average gives the correct value, because the velocity is changing at a constant rate.

$$x_f = x_i + v_{ix}\Delta t$$

$$y_f = y_i + v_{iy}\Delta t + \frac{1}{2}\left(\frac{F_y}{m}\right)(\Delta t)^2$$

$$z_f = 0$$

3. Units check: all terms have units of meters.

Component of momentum perpendicular to force

One of the most important things to note about this result is that the x component of momentum was completely unaffected by a force in the y direction. This is a consequence of the vector nature of the Momentum Principle.

Air resistance

There are very few real world situations in which the net force on a moving object is truly constant. The results of the previous solution are useful only if we can reasonably say that the net force on a moving object is approximately constant.

One significant factor in determining the motion of objects near the Earth is air resistance, sometimes called drag. You have probably felt the effects of air resistance yourself, for example, when coasting downhill on a bicycle. The air resistance force on a moving object depends on the speed of the object, so as the speed of an object changes, the air resistance force changes too, and the net force on the object also changes—it is not constant. Additionally, the direction of the air resistance force changes, since the direction of this force is always opposite to the direction of motion. Note that this is very different from the constant force used in the preceding analysis.

A low-density object such as a styrofoam ball experiences air resistance that is comparable to the small gravitational force on the ball, so air resistance is important unless the styrofoam ball is moving very slowly (air resistance is small at low speeds and big at high speeds, as you may have experienced if you put your hand out the window of a car). At low speeds a baseball, which has a fairly high density, moves with negligible air resistance. But at the speed that a professional pitcher can throw a baseball (about 100 mi/hr or 44 m/s), a baseball goes only about half as far in air as it would in a vacuum, because air resistance is large at this high speed (Figure 2.15).

We will discuss air resistance in more detail in Chapter 6.

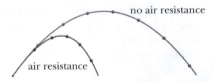

Figure 2.15 The trajectory of a baseball thrown at high speed (around 100 mph, or 44 m/s), ignoring air resistance (top curve) and including the effect of air resistance (bottom curve). The dots indicate the ball's position at equal time intervals. The ball travels about half as far in air as it would in a vacuum.

Magnitude of the gravitational force near the Earth's surface

The gravitational force on an object of mass m near the surface of the Earth is approximately mg, where g is a positive constant $g = +9.8$ N/kg. We will discuss this further in the next chapter.

Example: A ball with negligible air resistance

A ball of mass 500 g is initially on the ground, at location $\langle 0, 0, 0 \rangle$ m, and you kick it with initial velocity $\langle 3, 7, 0 \rangle$ m/s. (a) Where will the ball be half a second later? (b) At what time will the ball hit the ground? Make the approximation that air resistance is negligible, and use the previous analytical result for motion with a constant force.

1. Momentum principle

System: ball
Surroundings: Earth (neglecting air resistance)
Initial time: just after the kick
Final time: just before hitting the ground

Pick initial and final times so that during the time interval Δt only the Earth exerts a force on the ball.

2. Position update

(a) Using results from the analytical solution in the previous example:

$$\vec{F}_{net} = \langle 0, -mg, 0 \rangle, \text{ so } F_{net,y} = -mg$$

$$x_f \approx x_i + v_{ix}\Delta t$$

$$x_f \approx (0 + (3\text{m/s})(0.5 \text{ s}) = 1.5 \text{ m})$$

$$y_f \approx y_i + v_{iy}\Delta t + \frac{1}{2}\left(\frac{(-mg)}{m}\right)(\Delta t)^2$$

$$y_f = 0 + (7\text{m/s})(0.5 \text{ s}) - \frac{1}{2}(9.8\text{N/kg})(0.5 \text{ s})^2 = 2.275 \text{ m}$$

$$\vec{r}_f = \langle 1.5, 2.275, 0 \rangle \text{ m}$$

3. Check: correct units. Ball has moved in appropriate direction.

(b) At the instant the ball hits the ground, $y_f = 0$, so

$$0 = 0 + v_{iy}\Delta t + \frac{1}{2}(-g)(\Delta t)^2$$

Solving this quadratic equation for the unknown time Δt, we find two possible values:

$$\Delta t = 0 \text{ and } \Delta t = \frac{2v_{iy}}{g}$$

The first value, $\Delta t = 0$, corresponds to the initial situation, when the ball is near the ground, just after the kick. The second value is the time when the ball returns to the ground, just before hitting:

$$\Delta t = \frac{2(7\text{m/s})}{(9.8\text{N/kg})} = 1.43 \text{ s}$$

? Could we use these equations for x and y as a function of time to find the location of the ball 10 seconds after you kick it?

No. Our result would be that the ball was far underground, since

$$y_f = (7\text{m/s})(10 \text{ s}) - \frac{1}{2}(9.8\text{N/kg})(10 \text{ s})^2 = -420 \text{ m}$$

which is not physically reasonable! (The ball would have hit the ground and stopped before 10 seconds had passed, due to other interactions not included in our model.)

Graphs of motion

Figure 2.16 and Figure 2.17 show graphs of position and velocity components vs. time for the ball in the preceding example. In Figure 2.16 the first graph, v_x vs. t, is simply a horizontal line, because v_x doesn't change, since there is no x component of force. The graph of x is a straight line (second graph), rising if v_x is positive. Note that the slope of the x vs. t graph is equal to v_x.

In Figure 2.17 the graph of v_y is a falling straight line (third graph), because the y component of the force is $-mg$, which constantly makes the y component of momentum decrease. At some point the y component of momentum decreases to zero, at the top of the motion, after which the ball heads downward, with negative v_y. The graph of y vs. time t is an inverted parabola (fourth graph), since the equation for y is a quadratic function in the time.

Note that the slope of the y vs. t graph (Figure 2.17) at any time is equal to v_y at that time. In particular, when the slope is zero (at the maximum y),

In this case, the mass of the ball cancels, because the gravitational force is proportional to mass.

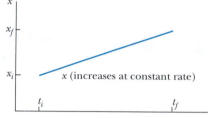

Figure 2.16 Motion graphs for the thrown ball. Top: v_x vs. t, bottom: x vs. t. Note that v_x does not change because the net force acted only in the y direction.

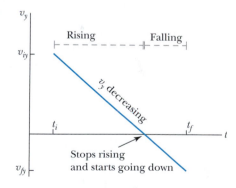

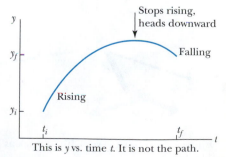

Figure 2.17 Motion graphs for the thrown ball. Top: v_y vs. t; bottom: y vs. t.

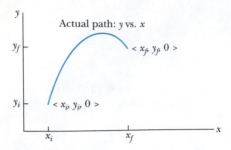

Figure 2.18 The actual trajectory of the thrown ball, with negligible air resistance (y vs. x).

v_y is momentarily equal to zero. Before that point the slope is positive, corresponding to $v_y > 0$, and after that point the slope is negative, corresponding to $v_y < 0$.

The actual path of the ball, the graph of y vs. x (Figure 2.18), is also an inverted parabola (the bottom graph). Since x increases linearly with t, whether we plot y vs. t or y vs. x we'll see a similar curve. The scale factor along the horizontal axis is different, of course (meters instead of seconds).

2.X.8 A ball is kicked from a location $\langle 9, 0, -5 \rangle$ m (on the ground) with initial velocity $\langle -10, 13, -5 \rangle$ m/s .
(a) What is the velocity of the ball 0.6 seconds after being kicked?
(b) What is the location of the ball 0.6 seconds after being kicked?
(c) At what time does the ball hit the ground?
(d) What is the location of the ball when it hits the ground?

2.6 PROBLEMS OF GREATER COMPLEXITY

So far all of the examples we have considered have involved finding a change in momentum (and position), given a known force acting over a known time interval. The following problems require you to find either the duration of an interaction (time interval), or the force exerted during an interaction. These large problems involve several steps in reasoning.

Example: Strike a hockey puck

In Figure 2.19 an 0.4 kg hockey puck is sliding along the ice with velocity $\langle 20, 0, 0 \rangle$ m/s . As the puck slides past location $\langle 1, 0, 2 \rangle$ m on the rink, a player strikes the puck with a sudden force in the +z direction, and the hockey stick breaks. Some time later, the puck's position on the rink is $\langle 13, 0, 21 \rangle$ m . When we pile weights on the side of a hockey stick we find that the stick breaks under a force of about 1000 N (this is roughly 250 pounds; a force of one newton is equivalent to a force of about a quarter of a pound, approximately the weight of a small apple).

(a) Make a sketch of the path of the puck before and after it is hit. (b) For approximately how much time $\Delta t_{\text{contact}}$ was the hockey stick in contact with the puck? Evidently the contact time is quite short, since you hear a short, sharp crack. State what approximations and/or simplifying assumptions you make in your analysis.

1. Momentum principle

System: the hockey puck
Surroundings: Earth, ice, hockey stick, air

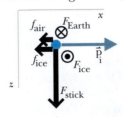

Top view, looking down on the ice. $\otimes$ means "into page" and $\odot$ means "out of page". By convention a lower case f usually denotes a frictional force.

Initial time: when stick first makes contact with puck
Final time: when stick breaks

$$\vec{p}_f = \vec{p}_i + \vec{F}_{\text{net}}\Delta t$$

$$\langle p_{fx}, 0, p_{fz} \rangle = \langle p_{ix}, 0, 0 \rangle + \langle -f_{\text{ice}} - f_{\text{air}}, (F_{\text{ice}} - mg), F_{\text{stick}} \rangle \Delta t_{\text{contact}}$$

x component: $p_{fx} = p_{ix} - (f_{\text{ice}} + f_{\text{air}})\Delta t_{\text{contact}}$

$$f_{\text{ice}} \approx 0 \text{ and } f_{\text{air}} \approx 0 \text{ , so}$$

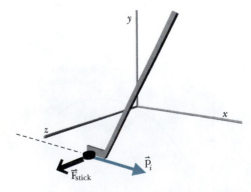

Figure 2.19 A hockey stick hits a puck as it slides by.

$$p_{fx} \approx p_{ix} + (0)\Delta t_{\text{contact}} \approx p_{ix} \text{ so no change in } p_x$$

y component: $0 = 0 + (F_{\text{ice}} - mg)\Delta t_{\text{contact}}$ therefore $F_{\text{ice}} = mg$

z component: $p_{fz} = F_{\text{stick}}\Delta t_{\text{contact}}$

$$p_{fz} = F_{\text{stick}}\Delta t_{\text{contact}} = (1000 \text{ N})\Delta t_{\text{contact}}$$

2. Position update

Initial time: when stick breaks
Final time: when puck is at location $\langle 13, 0, 21 \rangle$ m

(a) The path of the puck looks something like Figure 2.20.
$$\vec{r}_f = \vec{r}_i + \vec{v}_{\text{avg}}\Delta t_{\text{slide}}$$

$$\langle 13, 0, 21 \rangle \text{ m} = (\langle 1, 0, 2 \rangle \text{ m}) + \langle 20 \text{ m/s}, 0, v_z \rangle \Delta t_{\text{slide}}$$

x component: $(13 \text{ m}) = (1 \text{ m}) + (20 \text{ m/s})\Delta t_{\text{slide}}$

$$\Delta t_{\text{slide}} = (12 \text{ m})/(20 \text{ m/s}) = 0.6 \text{ s}$$

y component: $0 = 0 + 0\Delta t_{\text{slide}}$ so $0 = 0$

z component: Since $\Delta t_{\text{slide}} = 0.6$ s :

$$(21 \text{ m}) = (2 \text{ m}) + v_z(0.6 \text{ s})$$

$$v_z = (19 \text{ m})/(0.6 \text{ s}) = 31.7 \text{ m/s}$$

1*. Back to the Momentum Principle

$p_{fz} = (1000 \text{ N})\Delta t_{\text{contact}}$ where $p_{fz} \approx mv_{fz}$ (since $v \ll c$)

$$(0.4 \text{ kg})(31.7 \text{ m/s}) = (1000 \text{ N})\Delta t_{\text{contact}}$$

(b) $\Delta t_{\text{contact}} = (0.4 \text{ kg})(31.7 \text{ m/s})/(1000 \text{ N}) = 0.013$ s

3. Check: Units okay (contact time is in seconds). Time reasonable? The contact time is very short; this is consistent with the sound you hear (a sharp crack).

Further discussion

? How good were our assumptions?

The neglect of sliding friction and air resistance is probably pretty good, since a hockey puck slides for long distances on ice with nearly constant speed.

We know the hockey stick exerts a maximum force of $F_{\text{stick}} = 1000$ N, because we observe that the stick breaks. We approximate the force as nearly constant during contact. Actually, this force grows quickly from zero at first contact to 1000 N, then abruptly drops to zero when the stick breaks.

The final approximation is somewhat questionable. Although 0.013 s is a short time, the puck moves $(20 \text{ m/s})(0.013 \text{ s}) = 0.26$ m (a bit less than one foot) in the x direction during this time. Also during this time v_z increases from 0 to 31.7 m/s, with an average value of about 15.8 m/s, so the z displacement is about $(15.8 \text{ m/s})(0.013 \text{ s}) = 0.2$ m during contact. On the other hand, these displacements aren't very large compared to the displacement from $\langle 1, 0, 2 \rangle$ m to $\langle 13, 0, 21 \rangle$ m, so our result isn't terribly inaccurate due to this approximation. Nevertheless, a more accurate sketch of the path of the puck should show a bend as in Figure 2.21.

Even though our analysis of the stick contact time (0.013 s) isn't exact, it is adequate to get a reasonably good determination of this short time, something that we wouldn't know without using the Momentum Principle and

Approximations and simplifying assumptions:
Ice exerts little force in the x or z directions (low sliding friction); negligible air resistance.
Force of stick roughly constant during $\Delta t_{\text{contact}}$.
The puck doesn't move very far during the contact time.
After contact, velocity is nearly constant.
There are two unknown quantities in the z component equation: p_{fz} and $\Delta t_{\text{contact}}$. We need another equation to find p_{fz}. Let's try the position update equation.

Δt_{slide} is the time during which the puck slides after the impact.

The y component equation doesn't give any useful information in this case.

Now we know p_{fz}, and can use it to solve for $\Delta t_{\text{contact}}$

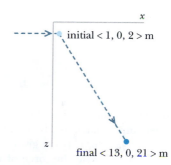

Figure 2.20 The x component of the momentum (and velocity) hardly changes, but the z component of momentum (and velocity) changes quickly from zero to some final value when the puck is hit.

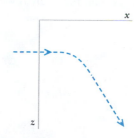

Figure 2.21 A more accurate overhead view of the path of the hockey puck, showing the bend during impact.

the position update formula. The very short duration of the impact explains why we hear a sharp, short crack.

Estimating times

Real-world problems often require the estimation of one or more quantities. Estimating masses or distances is usually not difficult, but most people find it more challenging to estimate time durations, especially if the times are very short. It is common to guess what seems a rather short time, such as a second, or half a second. However, we found in the previous example that the contact time between a puck and a hockey stick was significantly shorter than this—about a hundredth of a second.

A useful, systematic way to estimate a short time is to use the relationship between velocity and position (a.k.a. the position update formula).

> **?** Suppose we had guessed that the contact time between the stick and the puck was about one second. We know that the original speed of the puck was 20 m/s. At this speed, how far would the puck have traveled in 1 s?

$$(20 \text{ m/s})(1 \text{ s}) = 20 \text{ m}$$

It is clear that the puck did not move 20 meters during the impact. We must conclude that the impact took much less than 1 second.

Suppose we guess that the puck may have slid 10 cm along the stick during the impact:

$$\Delta t \approx \frac{(0.10 \text{ m})}{(20 \text{ m/s})} = 0.005 \text{ s}$$

This estimate of the contact time differs from our result by only a factor of around 2, instead of a factor of 100, so it is a much better estimate. Using an object's speed and estimating the distance traveled during an interval allows us to come up with a much better estimate of interaction times than we would otherwise get.

Example: Colliding students

This problem is rather ill-defined and doesn't seem much like a "textbook" problem. No numbers have been given, yet you're asked to estimate the force of the collision. This kind of problem is typical of the kinds of problems engineers and scientists encounter in their professional work.

Two students who are late for tests are running to classes in opposite directions as fast as they can. They turn a corner, run into each other head-on, and crumple into a heap on the ground. Using physics principles, estimate the force that one student exerts on the other during the collision. You will need to estimate some quantities; give reasons for your choices and provide checks showing that your estimates are physically reasonable.

1. Momentum Principle

We could have chosen the student on the right as the system; the analysis would be very similar.

System: the student on the left
Surroundings: Earth, ground, other student, air.
Initial time: just before impact
Final time: when speeds become zero

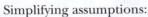

Simplifying assumptions:
- Students have same mass and same speed
- f_{ground} and f_{air} are negligible compared to F

$$\langle 0, 0, 0 \rangle = \langle p_{ix}, 0, 0 \rangle + \langle -F - f_{\text{ground}} - f_{\text{air}}, (F_{\text{ground}} - mg), 0 \rangle \Delta t$$

x component: $0 = p_{ix} - (F + f_{\text{ground}} + f_{\text{air}})\Delta t$

$$0 \approx p_{ix} - (F)\Delta t$$

y component: $0 = (F_{ground} - mg)\Delta t$, so $F_{ground} = mg$

z component: $0 = 0$

Since $0 = p_{ix} - F\Delta t$, $mv = F\Delta t$ (since $v \ll c$)

Estimates: See comments in margin. Mass of the student is around 60 kg. The student's average speed during the impact is 3 m/s. Perhaps the student's body is compressed by about 5 cm during the collision, which will be painful.

2. Position update (during collision)

$$\Delta x = v_{avg,x}\Delta t$$

$$\Delta t = \Delta x / v_{avg,x} = (0.05 \text{ m})/(3 \text{ m/s}) = 0.017 \text{ s}$$

Since $\Delta v_x = (0 - 6)$ m/s and $|\Delta p_x| = |F_x \Delta t|$,

$$|F_x| = \left|\frac{\Delta p_x}{\Delta t}\right| \approx \left|\frac{(60 \text{ kg})(-6 \text{ m/s})}{(0.01 \text{ s})}\right| = 21000 \text{ N}$$

3. Check
• Units check (force is in newtons, collision time is in seconds)
• Is the result reasonable? The contact time is very short, as expected. Is the force reasonable or not? See discussion below.

The student's heels are pushed to the left along the ground due to the collision, and the resisting ground pushes to the right. We'll assume that this force is much smaller than the force exerted by the other student, because the student's shoes can slip.

The y and z component equations don't yield any useful information in this situation.

Notes on estimates

A 1 kg mass weighs 2.2 lb.

An Olympic sprinter can run the 100 meter dash in less than 10 seconds, so 10 m/s is an upper limit. A brisk walking speed is around 2 m/s, so 6 m/s is an intermediate value.

During the collision the student's speed decreases from 6 m/s to 0 m/s, so his or her average speed is about 3 m/s.

Force has units (kg)(m/s)/s, which has the units of momentum (kg·m/s) divided by s, which is correct for a force (change of momentum divided by time).

Further discussion
21000 N is a very large force. For example, the gravitational force on a 60 kg student (the "weight") is only about $(60 \text{ kg})(9.8 \text{ N/kg}) \approx 600 \text{ N}$. The force of the impact is about 35 times the weight of the student! It's like having a stack of 35 students sitting on you. If the students hit heads instead of stomachs, the squeeze might be less than 1 cm, and the force would be over 5 times as large! This is why heads can break in such a collision.

Our result of 21000 N shows why collisions are so dangerous. Collisions involve very large forces acting for very short times, giving impulses of ordinary magnitude.

How good were our approximations?
We made the following approximations and simplifying assumptions:
• We estimated the running speed from known 100 m dash records.
• We estimated the masses of the students.
• We assumed that the horizontal component of the force of the ground on the bottom of the student's shoe was small compared to the force exerted by the other student. Now that we find that the impact force is huge, this assumption seems quite good.
• We made the approximation that the impact force was nearly constant during the impact, so what we've really determined is an average force.
• We assumed that the students had similar masses and similar running speeds, to simplify the analysis. If this is not the case, the analysis is significantly more complicated, but we would still find that the impact force is huge.

You might object that with all these estimates and simplifying assumptions the final result of 21000 N for the impact force is not useful. It is certainly the case that we don't have a very accurate result. But nevertheless we gained valuable information, that the impact force is *very* large. Before doing this analysis based on the Momentum Principle, we had no idea of whether the force was small compared to the student's weight, comparable, or much bigger. Now we have a quantitative result that the force is about 35 times the weight of a student, and we can appreciate why collisions are so dangerous.

2.7 PHYSICAL MODELS

Our model of the colliding student situation is good enough for many purposes. However, we left out some aspects of the actual motion. For example, we mostly ignored the flexible structure of the students, how much their shoes slip on the ground during the collision, etc. We have also quite sensibly neglected the gravitational force of Mars on the students, because it is so tiny compared to the force of one student on the other.

Making and using models is an activity central to physics, and the criteria for a "good" physical model will depend on how we intend to use the model. In fact, one of the most important problems a scientist or engineer faces is deciding what interactions must be included in a model of a real physical, chemical, or biological system, and what interactions can reasonably be ignored.

Order of magnitude estimates

If we neglect some effects, we say that we are constructing a simplified model of the situation. A useful model should omit extraneous detail but retain the most important features of the real-world situation. When we make many estimates, our goal is often to determine the order of magnitude of the answer: is the force on a colliding student closer to 0.0001 newton or 10,000 Newtons? Knowing the order of magnitude to expect in an answer can be critically important in solving a real problem, designing an experiment, or designing a crash helmet. The designer of a crash helmet needs an approximate value for the maximum force it must withstand. Actual collisions will vary, so there is no "right" or simple answer.

Idealized models

Some models are "idealized," by which we mean they involve simple, clean, stripped-down situations, free of messy complexities (Figure 2.22). "Ideally," a ball will roll forever on a level floor, but a real ball rolling on a real floor eventually comes to a stop. An "ideal" gas is a fictitious gas in which the molecules don't interact at all with each other, as opposed to a real gas whose molecules do interact, but only when they come close to each other. A model of a single Earth orbiting a Sun is an idealized model because it leaves out the effects of the other bodies in the Solar system.

The behavior of idealized models allows us to investigate simple patterns of motion, and learn what factors are important in determining these patterns. Once we understand these factors, we can revise and extend our models, including more interactions and complexities, to see what effects these have.

An important aspect of physical modeling is that we will engage in making appropriate approximations to simplify the messy, real-world situation enough to permit (approximate) analysis using Newton's laws. Actually, using Newton's laws is itself an example of modeling and making approximations, because we are neglecting the effects of quantum mechanics and of general relativity (Einstein's treatment of gravitation). Newton's laws are only an approximation to the way the world works, though frequently an extremely good one.

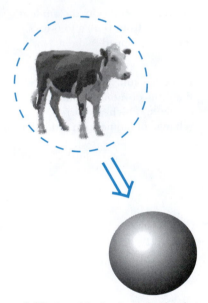

Figure 2.22 An old physics joke begins, "Consider a spherical cow..." Sometimes this degree of idealization is actually appropriate.

2.8 SYSTEMS CONSISTING OF SEVERAL OBJECTS

Reciprocity

When you push on a spring, compressing it, you exert a force on the spring. The compressed spring also exerts a force on your hand. It turns out that the force exerted by your hand on the spring is equal in magnitude (though opposite in direction) to the force exerted by the spring on your hand. This "reciprocity" of forces is a fundamental property of the electric interaction between the electrons and protons in the atoms of your hand and the electrons and protons in the atoms making up the spring. Similarly, the gravitational force that a falling apple exerts on the Earth is just as big as the gravitational force the Earth exerts on the apple. We will say more about the reciprocity of electric and gravitational forces in the next chapter. (Interestingly, reciprocity does not always apply to magnetic forces.)

External forces

In our applications of the Momentum Principle up to this point, we have usually chosen a single object as the system of interest. In some situations, however, it can be very useful to choose a system that consists of two or more interacting objects. This is a legitimate choice of system; we will show that the Momentum Principle applied to a system of two or more objects says that the change in the *total* momentum of the system (Figure 2.23) can be determined by finding the net *external* impulse applied to the system:

MOMENTUM PRINCIPLE FOR MULTIPARTICLE SYSTEMS

$$\Delta \vec{p}_{total} = (\vec{p}_{total,f} - \vec{p}_{total,i}) = \vec{F}_{net}\Delta t$$

Where:

$\vec{p}_{total} \equiv \vec{p}_1 + \vec{p}_2 + \vec{p}_3 + ...$ sum of momenta of all objects in the system

$\vec{F}_{net} \equiv \vec{F}_1 + \vec{F}_2 + \vec{F}_3 + ...$ sum of all external forces on the system

---

This is an interesting statement of the Momentum Principle; it implies that if we know the external forces acting on a multi-object system, we can draw conclusions about the change of momentum of the system over some time interval without worrying about any of the details of the interactions of the objects with each other. This can greatly simplify the analysis of the motion of some very complex systems. It should not be surprising that internal forces alone cannot change a system's momentum. If they could, you could lift yourself off the ground by pulling up on your own feet!

We have implicitly been using the multiparticle version of the Momentum Principle when we have treated macroscopic objects like humans, spacecraft, and planets as if they were single pointlike objects.

Proof of The Momentum Principle for multiparticle systems

In a three-particle system (Figure 2.24) we show all of the forces acting on each particle, where the lower case $\vec{f}$'s are forces the particles exert on each other (so-called "internal" forces), and the upper case $\vec{F}$'s are forces exerted by objects in the surroundings that are not shown and are not part of our chosen system (these are so-called "external" forces, such as the gravitational attraction of the Earth, or a force that you exert by pulling on one of the particles).

We will use the following shorthand notation: $\vec{f}_{1,3}$ will denote the force exerted on particle 1 by particle 3; $\vec{F}_{2,\,surr}$ will denote the force on particle 2 exerted by objects in the surroundings, and so on. We start by applying the Momentum Principle to each of the three particles separately:

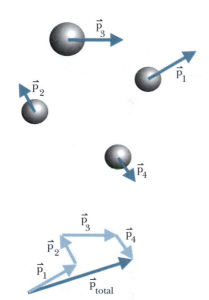

Figure 2.23 The total momentum of the system of four objects is the sum of the individual momenta of each object.

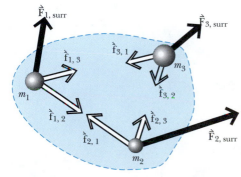

Figure 2.24 External and internal forces acting on a system of three particles.

$$\Delta \vec{p}_1 = (\vec{F}_{1,\text{surr}} + \vec{f}_{1,2} + \vec{f}_{1,3})\Delta t$$

$$\Delta \vec{p}_2 = (\vec{F}_{2,\text{surr}} + \vec{f}_{2,1} + \vec{f}_{2,3})\Delta t$$

$$\Delta \vec{p}_3 = (\vec{F}_{3,\text{surr}} + \vec{f}_{3,1} + \vec{f}_{3,2})\Delta t$$

Nothing new so far. But now we add up these three equations. That is, we create a new equation by adding up all the terms on the left sides of the three equations, and adding up all the terms on the right sides, and setting them equal to each other:

$$\Delta \vec{p}_1 + \Delta \vec{p}_2 + \Delta \vec{p}_3 = (\vec{F}_{1,\text{surr}} + \vec{f}_{1,2} + \vec{f}_{1,3} +$$
$$\vec{F}_{2,\text{surr}} + \vec{f}_{2,1} + \vec{f}_{2,3} +$$
$$\vec{F}_{3,\text{surr}} + \vec{f}_{3,1} + \vec{f}_{3,2})\Delta t$$

Many of these terms cancel. By the principle of reciprocity (see page 59), which is obeyed by gravitational and electric interactions, we have this:

$$\vec{f}_{1,2} = -\vec{f}_{2,1}$$

$$\vec{f}_{1,3} = -\vec{f}_{3,1}$$

$$\vec{f}_{2,3} = -\vec{f}_{3,2}$$

Thanks to reciprocity, all that remains after the cancellations is this:

$$\Delta \vec{p}_1 + \Delta \vec{p}_2 + \Delta \vec{p}_3 = (\vec{F}_{1,\text{surr}} + \vec{F}_{2,\text{surr}} + \vec{F}_{3,\text{surr}})\Delta t$$

The total momentum of the system is $\vec{P}_{\text{tot}} = \vec{p}_1 + \vec{p}_2 + \vec{p}_3$, so we have:

$$\Delta \vec{P}_{\text{tot}} = \vec{F}_{\text{net,surr}}\Delta t$$

The importance of this equation is that reciprocity has eliminated all of the internal forces (the forces that the particles in the system exert on each other); internal forces cannot affect the motion of the system as a whole. All that matters in determining the rate of change of (total) momentum is the net external force. The equation has exactly the same form as the Momentum Principle for a single particle.

In a later chapter we will see that the total momentum can be expressed as $\vec{P}_{\text{tot}} = M_{\text{total}}\vec{v}_{\text{center of mass}}$, where the center of mass is a mathematical point obtained from a weighted average of the masses in the system.

Moreover, if an object is a sphere whose density is only a function of radius, the object exerts a gravitational force on other objects as though the sphere were a point particle. We can predict the motion of a star or a planet or an asteroid as though it were a single point particle of large mass.

2.9 COLLISIONS: NEGLIGIBLE EXTERNAL FORCES

An event is called a "collision" if it involves an interaction that takes place in a relatively short time and has a large effect on the momenta of the objects compared to the effects of other interactions during that short time. A collision does not necessarily involve actual physical contact between objects, which may be interacting via long distance forces like the gravitational force or the electric force. For example, a spacecraft is deflected as it passes close to Mars on its way to Jupiter, and Mars exerts a large gravitational force for a relatively short time. During that short time the effects of the other planets are negligible.

Often it can be useful to analyze collisions by choosing a system that includes all of the colliding objects, as we will see in the following example.

Example:

Two lumps of clay travel through the air toward each other, at speeds much less than the speed of light (Figure 2.25), rotating as they move. When the lumps collide they stick together. The mass of lump 1 is 0.2 kg and its initial velocity is $\langle 6, 0, 0 \rangle$ m/s, and the mass of lump 2 is 0.5 kg and its initial velocity is $\langle -5, 4, 0 \rangle$ m/s. What is the final velocity of the stuck-together lumps?

1. Momentum Principle

System: both lumps
Surroundings: air, Earth

$$\vec{p}_{total,f} = \vec{p}_{total,i} + \vec{F}_{net}\Delta t$$

$$\vec{p}_{total,f} \approx m_1\vec{v}_{1i} + m_2\vec{v}_{2i} + \langle 0, 0, 0 \rangle \Delta t$$

$$(m_1 + m_2)\vec{v}_f = m_1\vec{v}_{1i} + m_2\vec{v}_{2i}$$

$$\vec{v}_f = \frac{m_1\vec{v}_{1i} + m_2\vec{v}_{2i}}{m_1 + m_2}$$

$$\vec{v}_f = \frac{(0.2 \text{ kg})\langle 6, 0, 0 \rangle + (0.5 \text{ kg})\langle -5, 4, 0 \rangle}{(0.7 \text{ kg})} \text{ m/s}$$

$$\vec{v}_f = \langle -1.86, 2.86, 0 \rangle \text{ m/s}$$

2. Not necessary to update position

3. Check: units correct.

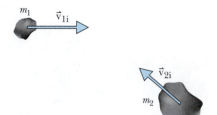

Figure 2.25 Two lumps of clay traveling through the air, just before colliding.

$v \ll c$; neglect air resistance and gravitational force (which is negligible compared to the contact force during the very short time of the collision).

The rotation of the lumps doesn't affect their momentum, and the Momentum Principle still applies. See Figure 2.26.

Figure 2.26 The momentum of the stuck-together balls just after the collision is equal to the sum of the initial momenta of the two balls.

Further discussion

By choosing both lumps as the system, we were able to find the final momentum of the system without needing to know anything about the details of the complex forces that the lumps exerted on each other during the collision.

? Why doesn't it matter that the lumps are rotating?

The rotation of an object doesn't affect its momentum; the total momentum of the system is still the sum of the individual momenta of the objects within the system. However, the Momentum Principle does not tell us anything about how fast the stuck-together objects will be rotating; for this, we will have to apply the Angular Momentum Principle, which is discussed in Chapter 10.

The velocities involved are actually the "center-of-mass" velocities of each lump. You can think of this as the velocity of a point at the center of each lump. This concept will be made more quantitative in Chapter 8.

Some problems require more than one principle

Because the balls in the previous example had the same final velocities (Figure 2.26), we had enough information to solve for the final velocity of the stuck-together balls—we had one equation with only one unknown (vector) quantity. In more complex situations, we will find that we sometimes do not have enough information to solve for all of the unknown quantities by applying only the Momentum Principle.

For example, consider a collision between two balls that bounce off each other, as shown in Figure 2.27. The Momentum Principle tells us that

$$\vec{p}_{total,f} = \vec{p}_{total,i}$$

$$\vec{p}_{1f} + \vec{p}_{2f} = \vec{p}_{1i} + \vec{p}_{2i}$$

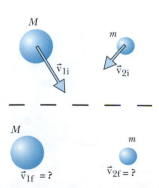

Figure 2.27 The initial momenta of the colliding objects are known, but the final momenta after the collision are unknown.

but we are left with two unknown (vector) quantities, $\vec{p}_{1f}$ and $\vec{p}_{2f}$, and only one equation. We will not be able to analyze problems of this kind fully until we can invoke the Energy Principle (Chapter 5) and the Angular Momentum Principle (Chapter 10), along with the Momentum Principle.

2.X.9 You and a friend each hold a lump of wet clay. Each lump has a mass of 20 grams. You each toss your lump of clay into the air, where the lumps collide and stick together. Just before the impact, the velocity of one lump was $\langle 5, 2, -3 \rangle$ m/s, and the velocity of the other lump was $\langle -3, 0, -2 \rangle$ m/s. What was the total momentum of the lumps just before impact? What is the momentum of the stuck-together lump just after the impact? What is its velocity?

2.X.10 In outer space, far from other objects, two rocks collide and stick together. Before the collision their momenta were $\langle 10, 20, -5 \rangle$ kg·m/s and $\langle 8, -6, 12 \rangle$ kg·m/s. What was their total momentum before the collision? What must be the momentum of the combined object after the collision?

2.X.11 At a certain instant, the momentum of a proton is $\langle 3.4 \times 10^{-21}, 0, 0 \rangle$ kg·m/s as it approaches another proton which is initially at rest. The two protons repel each other electrically, without coming close enough to touch. When they are once again far apart, one of the protons now has momentum $\langle 2.4 \times 10^{-21}, 1.6 \times 10^{-21}, 0 \rangle$ kg·m/s. At this instant, what is the momentum of the other proton?

2.10 CONSERVATION OF MOMENTUM

The choice of system affects the detailed form of the Momentum Principle. Consider the case of two stars orbiting each other, a "binary star." If you choose as the system of interest just one of the stars (Figure 2.28), the other star is in the "surroundings" and exerts an external force which changes the first star's momentum.

If on the other hand you choose both stars as your system (Figure 2.29), the surroundings consist of other stars, which may be so far away as to have negligible effects on the binary star. In that case the net external force acting on the system is nearly zero, which makes the analysis simpler.

? What happens to the total momentum of the isolated binary star as time goes by?

The Momentum Principle $\vec{p}_f = \vec{p}_i + \vec{F}_{net}\Delta t$ reduces in this case to $\vec{p}_f = \vec{p}_i$, which predicts that the total momentum of the system $\vec{p} = \vec{p}_1 + \vec{p}_2$ remains constant in magnitude and direction. This is an important special case: the total momentum of an isolated system, a system with negligible interactions with the surroundings, doesn't change but stays constant.

In a later chapter we will see that the total momentum can be expressed as $M_{total}\vec{v}_{center\ of\ mass}$, where the center of mass is a mathematical point between the two stars, closer to the more massive star. The velocity of the center of mass does not change but is constant as the binary star drifts through space (or is zero if the binary star's total momentum is zero).

One way to think about this result is to say that momentum gained by one star is lost by the other, because the gravitational forces (and impulses) are equal in magnitude but opposite in direction (see the short discussion of reciprocity on page 59). The effect is that the total momentum doesn't change. Let $\vec{F}$ be the force exerted on star 1 by star 2, so $-\vec{F}$ is the force exerted on star 2 by star 1. After a short time interval Δt the new total momentum is this:

$$(\vec{p}_1 + \vec{F}\Delta t) + (\vec{p}_2 - \vec{F}\Delta t) = \vec{p}_1 + \vec{p}_2$$

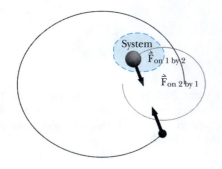

Figure 2.28 A binary star. The gray lines show the trajectories of the individual stars. Choose just one of the stars as the system. The momentum of the system changes due to the external force.

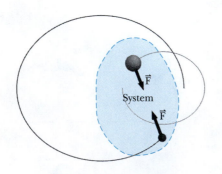

Figure 2.29 A binary star: choose both stars as the system. The momentum of the combined system doesn't change.

Momentum lost by one star is gained by the other star. This is a simple example of an important restatement of the Momentum Principle, "conservation of momentum," which says that the change of momentum in a system plus the change of momentum in the surroundings adds up to zero.

CONSERVATION OF MOMENTUM

$$\Delta \vec{p}_{\text{system}} + \Delta \vec{p}_{\text{surroundings}} = \vec{0}$$

In the case of the two stars there are no objects in the surroundings, and no external forces, so the momentum of the system doesn't change. Zero net external impulse ($\vec{F}_{\text{net}}\Delta t$), zero net momentum change. One of the stars can gain momentum (due to a force acting on it), but only if the other star loses the same amount.

Relativistic momentum conservation

Particle accelerators produce beams of particles such as electrons, protons, and pions at speeds very close to the speed of light. When these high-speed particles interact with other particles, experiments show that the total momentum is conserved, but only if the momentum of each particle is defined in the way Einstein proposed, $\vec{p} \equiv \gamma m\vec{v}$. When the speed v approaches the speed of light c, the low-speed approximation $\gamma \approx 1$ ($\vec{p} \approx m\vec{v}$) is not valid (that is, the quantity $m\vec{v}$ is not conserved when the speed v approaches the speed of light c).

What is not conserved?

In later chapters we will see that energy and angular momentum are also conserved quantities. However, most quantities are not conserved quantities. For example, velocity is not conserved in an interaction, as we saw clearly in the collision between the two lumps of clay (Figure 2.25 and Figure 2.26). Temperature is another example of a quantity that is not conserved.

2.X.12 Consider the head-on collision of two identical bowling balls (Figure 2.30).

(a) Choose a system consisting only of ball A. What is the momentum change of the system during the collision? What is the momentum change of the surroundings?

(b) Choose a system consisting only of ball B. What is the momentum change of the system during the collision? What is the momentum change of the surroundings?

(c) Choose a system consisting of both balls. What is the momentum change of the system during the collision? What is the momentum change of the surroundings?

2.X.13 You hang from a tree branch, then let go and fall toward the Earth. As you fall, the y component of your momentum, which was originally zero, becomes large and negative.

(a) Choose yourself as the system. There must be an object in the surroundings whose y momentum must become equally large, and positive. What object is this?

(b) Choose yourself and the Earth as the system. The y component of your momentum is changing. Does the total momentum of the system change? Why or why not?

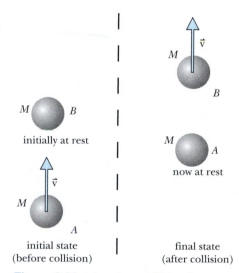

Figure 2.30 A head-on collision between two identical bowling balls, each of mass M.

2.11 *DERIVATION: SPECIAL CASE AVERAGE VELOCITY

Here are two proofs, one geometric and one algebraic (using calculus), for the following special-case result concerning average velocity:

$$v_{avg,x} = \frac{(v_{ix} + v_{fx})}{2} \; only \; if \; v_x \; changes \; at \; a \; constant \; rate$$

Similar results for $v_{avg,y}$ and $v_{avg,z}$

Geometric proof

If $F_{net,x}$ is constant, $p_{fx} = p_{ix} + F_{net,x}\Delta t$ implies that p_x changes at a constant rate. At speeds small compared to the speed of light, $v_x \approx p_x/m$, so a graph of v_x vs. time is a straight line, as in Figure 2.31. Using this graph, we form narrow vertical slices, each of height v_x and narrow width Δt.

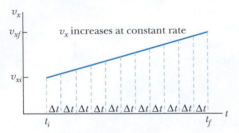

Figure 2.31 Graph of v_x vs. t (constant force), divided into narrow vertical slices each of height v_x and width Δt.

Within each narrow slice v_x changes very little, so the change in position during the brief time Δt is approximately $\Delta x = v_x\Delta t$. Therefore the change in x is approximately equal to the area of the slice of height v_x and width Δt (Figure 2.32).

Figure 2.32 One narrow slice has an area given approximately by $v_x\Delta t$. This is equal to Δx, the displacement of the object.

If we add up the areas of all these slices, we get approximately the area under the line in Figure 2.31, and this is also equal to the total displacement $\Delta x_1 + \Delta x_2 + \Delta x_3 + ... = x_f - x_i$. If we go to the limit of an infinite number of slices, each with infinitesimal width, the sum of slices really *is* the area, and this area we have shown to be equal to the change in position. This kind of sum of an infinite number of infinitesimal pieces is called an "integral" in calculus.

The area under the line is a trapezoid, and from geometry we know that the area of a trapezoid is the average of the two bases times the altitude:

$$area = \frac{(top + bottom)}{2}(altitude)$$

Turn Figure 2.31 on its side, as in Figure 2.32, and you see that the top and bottom have lengths v_{ix} and v_{fx}, while the altitude of the trapezoid is the total time $(t_f - t_i)$. Therefore we have the following result:

$$Trapezoid \; area = x_f - x_i = \frac{(v_{ix} + v_{fx})}{2}(t_f - t_i)$$

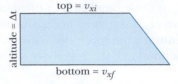

Figure 2.33 The area of the whole trapezoid is equal to the total displacement. The figure in Figure 2.31 has been rotated 90 degrees clockwise.

Dividing by $(t_f - t_i)$, we have this:

$$\frac{x_f - x_i}{t_f - t_i} = \frac{(v_{ix} + v_{fx})}{2}$$

But by definition the x component of average velocity is the change in x divided by the total time, so we have proved that

$$v_{avg,x} = \frac{(v_{ix} + v_{fx})}{2} \; only \; if \; v_x \; changes \; linearly \; with \; time$$

The proof depended critically on the straight-line ("linear") change in velocity, which occurs if $F_{net,x}$ is constant (and $v \ll c$). Otherwise we wouldn't have a trapezoidal area. That's why the result isn't true in general; it's only true in this important but special case.

Algebraic proof using calculus

An algebraic proof using calculus can also be given. We will use the x component of the derivative version of the Momentum Principle (more about this in a later chapter):

$$\Delta\vec{p} = \vec{F}_{net}\Delta t \; implies \; that \; \frac{\Delta\vec{p}}{\Delta t} = \vec{F}_{net}$$

In the limit we have

$$\lim_{\Delta t \to 0} \frac{\Delta\vec{p}}{\Delta t} = \frac{d\vec{p}}{dt} = \vec{F}_{net} \; and \; \frac{dp_x}{dt} = F_{net,x}$$

If $F_{net,x}$ is a constant, the time derivative of p_x is a constant, so we have

$$p_x = F_{net,x}t + p_{xi} \; since \; p_x = p_{ix} \; when \; t = 0$$

You can check this by taking the derivative with respect to time t, which gives the original equation $dp_x/dt = F_{net,x}$. At speeds small compared to the speed of light, $v_x \approx p_x/m$, so we can write

$$v_x = \frac{F_{net,x}}{m}t + v_{ix} \; since \; v_x = v_{ix} \; when \; t = 0$$

But the x component of velocity is the rate at which x is changing:

$$v_x = \frac{dx}{dt} = \frac{F_{net,x}}{m}t + v_{ix}$$

Now the question is, can you think of a function of x that has this time derivative? Since the time derivative of t^2 is $2t$, the following formula for x has the appropriate derivative:

$$x = \frac{1}{2}\frac{F_{net,x}}{m}t^2 + v_{ix}t + x_i \text{ since } x = x_i \text{ when } t = 0$$

You can check this by taking the derivative with respect to t, which gives the equation for v_x, since $d(\frac{1}{2}t^2)/dt = t$ and $d(t)/dt = 1$.

The average velocity which we seek is the change in position divided by the total time:

$$v_{avg,x} = \frac{x - x_i}{t} = \frac{1}{2}\frac{F_{net,x}}{m}t + v_{ix} = \frac{1}{2}(v_{fx} - v_{ix}) + v_{xi}$$

where we have used the equation we previously derived for the velocity:

$$v_{fx} = v_x = \frac{F_{net,x}}{m}t + v_{ix}$$

Simplifying the expression for $v_{avg,x}$ we have the proof:

$$v_{avg,x} = \frac{(v_{ix} + v_{fx})}{2} \text{ only if } v_x \text{ changes at a constant rate.}$$

2.12 *INERTIAL FRAMES

Newton's first law is valid only in an "inertial frame" of reference, one in uniform motion (or at rest) with respect to the pervasive "cosmic microwave background" (see optional discussion at the end of Chapter 1). Since the Momentum Principle is a quantitative version of Newton's first law, we expect the Momentum Principle to be valid in an inertial reference frame, but not in a reference frame that is not in uniform motion. Let's check that this is true.

If you view some objects from a space ship that is moving uniformly with velocity $\vec{v}_s$ with respect to the cosmic microwave background, all of the velocities of those objects have the constant $\vec{v}_s$ subtracted from them, as far as you are concerned. For example, a rock moving at the same velocity as your spacecraft would have $\vec{v}_{rock} = (\vec{v} - \vec{v}_s) = \vec{0}$ in your reference frame: it would appear to be stationary as it coasted along beside your spacecraft.

With a constant spaceship velocity, we have $\Delta\vec{v}_s = \vec{0}$, and the change of momentum of the moving object reduces to the following (for speeds small compared to c):

$$\Delta[m(\vec{v} - \vec{v}_s)] = \Delta(m\vec{v}) - \Delta(m\vec{v}_s) = \Delta(m\vec{v}) \text{ since}$$

$$\Delta(m\vec{v}_s) = \vec{0}$$

Therefore, $\Delta[m(\vec{v} - \vec{v}_s)] = \Delta(m\vec{v}) = \vec{F}_{net}\Delta t$

If the velocity $\vec{v}_s$ of the space ship doesn't change (it represents an inertial frame of reference), the form (and validity) of the Momentum Principle is unaffected by the motion of the space ship.

However, if your space ship increases its speed, or changes direction, $\Delta\vec{v}_s \neq \vec{0}$, an object's motion relative to you changes without any force acting on it. In that case the Momentum Principle is not valid for the object, because you are not in an inertial frame. Although the Earth is not an inertial frame because it rotates, and goes around the Sun, it is close enough to being an inertial frame for many everyday purposes.

2.13 *VELOCITY AND MOMENTUM AT HIGH SPEEDS

If $v \ll c$, $\vec{p} \approx m\vec{v}$ and $\vec{v} \approx \vec{p}/m$. But at high speed it is more complicated to determine the velocity from the (relativistic) momentum. Here is a way to solve for $\vec{v}$ in terms of $\vec{p}$:

$$|\vec{p}| = \frac{1}{\sqrt{1 - (|\vec{v}|/c)^2}}m|\vec{v}|$$

Divide by m and square: $\dfrac{|\vec{p}|^2}{m^2} = \dfrac{|\vec{v}|^2}{1 - (|\vec{v}|/c)^2}$

Multiply by $(1 - (|\vec{v}|/c)^2)$: $\dfrac{|\vec{p}|^2}{m^2} - \left(\dfrac{|\vec{p}|^2}{m^2 c^2}\right)|\vec{v}|^2 = |\vec{v}|^2$

Collect terms: $\left(1 + \dfrac{|\vec{p}|^2}{m^2 c^2}\right)|\vec{v}|^2 = \dfrac{|\vec{p}|^2}{m^2}$

$$|\vec{v}| = \frac{|\vec{p}|/m}{\sqrt{1 + \left(\dfrac{|\vec{p}|}{mc}\right)^2}}$$

But since $\vec{p}$ and $\vec{v}$ are in the same direction, we can write this:

$$\vec{v} = \frac{\vec{p}/m}{\sqrt{1 + \left(\dfrac{|\vec{p}|}{mc}\right)^2}}$$

2.14 *MEASUREMENTS AND UNITS

Using the Momentum Principle requires a consistent way to measure length, time, mass, and force, and a consistent set of units. We state the definitions of the standard Système Internationale (SI) units, and we briefly discuss some subtle issues underlying this choice of units.

Units: meters, seconds, kilograms, coulombs, and newtons

Originally the meter was defined as the distance between two scratches on a platinum bar in a vault in Paris, and a second was 1/86,400th of a "mean solar day." Now however the second is defined in terms of the frequency of light emitted by a cesium atom, and the meter is defined as the distance light travels in 1/299,792,458th of a second, or about 3.3×10^{-9} seconds (3.3 nanoseconds). The speed of light is defined to be exactly 299,792,458 m/s (very close to 3×10^8 m/s). As a result of these modern redefinitions, it is really speed (of light) and time that are the internationally agreed-upon basic units, not length and time.

By international agreement, one kilogram is the mass of a platinum block kept in that same vault in Paris. As a practical matter, other masses are compared to this standard kilogram by using a balance-beam or spring weighing scale (more about this in a moment). The newton, the unit of force, is defined as that force which acting for 1 second imparts to 1 kilogram a velocity change of 1 m/s. We could make a scale for force by calibrating the amount of stretch of a spring in terms of newtons.

The coulomb, the SI unit of electric charge, is defined in terms of electric currents. The charge of a proton is 1.6×10^{-19} coulomb.

Some subtle issues

What we have just said about SI units is sufficient for practical purposes to predict the motion of objects, but here are some questions that might bother you. Is it legitimate to measure the mass that appears in the Momentum Principle by seeing how that mass is affected by gravity on a balance-beam scale? Is it legitimate to use the Momentum Principle to define the units of force, when the concept of force is itself associated with the same law? Is this all circular reasoning, and the Momentum Principle merely a definition with little content? Here is a chain of reasoning that addresses these issues.

Measuring inertial mass

When we use balance-beam or spring weighing scales to measure mass, what we're really measuring is the "gravitational mass," that is, the mass that appears in the law of gravitation and is a measure of how much this object is affected by the gravity of the Earth. In principle, it could be that this "gravitational mass" would be different from the so-called "inertial mass"—the mass that appears in the definition of momentum. It is possible to compare the inertial masses of two objects, and we find experimentally that inertial and gravitational mass seem to be entirely equivalent.

Here is a way to compare two inertial masses directly, without involving gravity. Starting from rest, pull on the first object with a spring stretched by some amount s for an amount of time Δt, and measure the increase of speed Δv_1. Then, starting from rest, pull on the second object with the same spring stretched by the same amount s for the same amount of time Δt, and measure the increase of speed Δv_2. We *define* the ratio of the inertial masses as $m_1/m_2 = \Delta v_2/\Delta v_1$. Since one of these masses could

be the standard kilogram kept in Paris, we now have a way of measuring inertial mass in kilograms. Having defined inertial mass this way, we find experimentally that the Momentum Principle is obeyed by both of these objects in all situations, not just in the one special experiment we used to compare the two masses.

Moreover, we find to extremely high precision that the inertial mass in kilograms measured by this comparison experiment is exactly the same as the gravitational mass in kilograms obtained by comparing with a standard kilogram on a balance-beam scale (or using a calibrated spring scale), and that it doesn't matter what the objects are made of (wood, copper, glass, etc.). This justifies the convenient use of ordinary weighing scales to determine inertial mass.

Is this circular reasoning?

The definitions of force and mass may sound like circular reasoning, and the Momentum Principle may sound like just a kind of definition, with no real content, but there is real power in the Momentum Principle. Forget for a moment the definition of force in newtons and mass in kilograms. The experimental fact remains that any object if subjected to a single force by a spring with constant stretch experiences a change of momentum (and velocity) proportional to the duration of the interaction. Note that it is not a change of *position* proportional to the time (that would be a constant speed), but a change of *velocity*. That's real content. Moreover, we find that the change of velocity is proportional to the amount of stretch of the spring. That too is real content.

Then we find that a different object undergoes a different rate of change of velocity with the same spring stretch, but after we've made one single comparison experiment to determine the mass relative to the standard kilogram, the Momentum Principle works in all situations. That's real content.

Finally come the details of setting standards for measuring force in newtons and mass in kilograms, and we use the Momentum Principle in helping set these standards. But logically this comes after having established the law itself.

2.15 SUMMARY

The Momentum Principle

$$\Delta \vec{p} = \vec{F}_{net}\Delta t \text{ (for a short enough time interval } \Delta t\text{)}$$

Update form: $\vec{p}_f = \vec{p}_i + \vec{F}_{net}\Delta t$

Momentum Principle for multiparticle systems

$$\Delta \vec{p}_{total} = \vec{F}_{net,ext}\Delta t$$

Where $\vec{p}_{total} \equiv \vec{p}_1 + \vec{p}_2 + \vec{p}_3 + ... = M_{total}\vec{v}_{center of mass}$ and

$\vec{F}_{net,ext} \equiv \vec{F}_1 + \vec{F}_2 + \vec{F}_3 + ...$, the sum of all external forces acting on the system

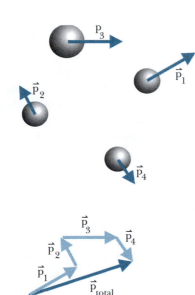

Figure 2.34 The total momentum of the system of four objects is the sum of the individual momenta of each object.

Conservation of momentum

$$\Delta \vec{p}_{system} + \Delta \vec{p}_{surroundings} = \vec{0}$$

To analyze the motion of a real-world system, several steps are required:

1. Momentum Principle

   Choose a system, consisting of a portion of the Universe.

   Make a list of objects in the surroundings that exert significant forces on the chosen system, and make a labeled diagram showing the external forces exerted by the objects in the surroundings.

   Apply the Momentum Principle to the chosen system:
   $$\vec{p}_f = \vec{p}_i + \vec{F}_{net}\Delta t$$

   For each term in the Momentum Principle, substitute any values you know.

2. Apply the position update formula, if necessary:

   $$\vec{r}_f = \vec{r}_i + \vec{v}_{avg}\Delta t$$

3. Check for reasonableness (units, direction, magnitude, etc.).

System is a portion of the Universe acted on by the *surroundings*.

Force is a quantitative measure of interactions; units are newtons.

Impulse is the product of force times time $\vec{F}\Delta t$; momentum change equals net impulse (the impulse due to the net force).

Physical models are tractable approximations/idealizations of the real world.

Special case result for average velocity:

$$v_{avg,x} = \frac{(v_{ix} + v_{fx})}{2} \text{ only if } v_x \text{ changes linearly with time}$$

Similar results for $v_{avg,y}$ and $v_{avg,z}$

2.16 REVIEW QUESTIONS

The Momentum Principle

2.RQ.14 An object is moving in the $+x$ direction. Which, if any, of the following statements do you know must be false?
   A. The net force on the object is in the $+x$ direction.
   B. The net force on the object is in the $-x$ direction.
   C. The net force on the object is zero.

2.RQ.15 You observe three carts moving to the left:
   Cart A moves at nearly constant speed.
   Cart B moves to the left, gradually speeding up.
   Cart C moves to the left, gradually slowing down.
Which cart or carts, if any, experience a net force to the left?

2.RQ.16 The x component of the momentum of an object is observed to increase with time:
   At $t = 0$ s, $p_x = 30$ kg $\cdot$ m/s
   At $t = 1$ s, $p_x = 40$ kg $\cdot$ m/s
   At $t = 2$ s, $p_x = 50$ kg $\cdot$ m/s
   At $t = 3$ s, $p_x = 60$ kg $\cdot$ m/s
   What can you conclude about the x component of the net force acting on the object during this time?
   A. $F_{net,x} = 0$
   B. $F_{net,x}$ is constant.
   C. $F_{net,x}$ is increasing with time.
   D. Not enough information is given.

2.RQ.17 At a certain instant, a particle is moving in the $+x$ direction with momentum $+10$ kg·m/s. During the next 0.1 s, a constant force acts on the particle: $F_x = -6$ N, and $F_y = +3$ N. What is the magnitude of the momentum of the particle at the end of this 0.1 s interval?

2.RQ.18 At $t = 12.0$ seconds an object with mass 2 kg was observed to have a velocity of $\langle 10, 35, -8 \rangle$ m/s. At $t = 12.3$ seconds its velocity was $\langle 20, 30, 4 \rangle$ m/s. What was the average (vector) net force acting on the object?

2.RQ.19 A proton has mass 1.7×10^{-27} kg. What is the magnitude of the impulse required to increase its speed from $0.990c$ to $0.994c$?

2.RQ.20 In order to pull a sled across a level field at constant velocity you have to exert a constant force. Doesn't this violate Newton's first and second laws of motion, which imply that no force is required to maintain a constant velocity? Explain this seeming contradiction.

Conservation of momentum

2.RQ.21 A bullet traveling horizontally at a very high speed embeds itself in a wooden block that is sitting at rest on a very slippery sheet of ice. You want to find the speed of the block just after the bullet embeds itself in the block.

   (a) What should you choose as the system to analyze?
   1. The bullet
   2. The block
   3. The bullet and the block
   (b) Which of the following statements is true?
   1. After the collision, the speed of the block with the bullet stuck in it is the same as the speed of the bullet before the collision.
   2. The momentum of the block with the bullet stuck in it is the same a the momentum of the bullet before the collision.
   3. The momentum of the block with the bullet stuck in it is less than the momentum of the bullet before the collision.

2.RQ.22 One kind of radioactivity is called "alpha decay." For example, the nucleus of a radium-220 atom can spontaneously split into a radon-216 nucleus plus an alpha particle (a helium nucleus, containing two protons and two neutrons).

   Consider a radium-220 nucleus which is initially at rest. It spontaneously decays, and the alpha particle travels off in the +z direction. What can you conclude about the motion of the new radon-216 nucleus? Explain your reasoning.

   A. It is also moving in the +z direction
   B. It remains at rest.
   C. It is moving in the –z direction

2.RQ.23 A bowling ball is initially at rest. A ping pong ball moving in the +z direction hits the bowling ball and bounces off it, traveling back in the –z direction.

   Consider a time interval Dt extending from slightly before to slightly after the collision.

   (a) In this time interval, what is the sign of Δp_z for the system consisting of both balls?
   1. Positive
   2. Negative
   3. Zero —no change in p_z.
   (b) In this time interval, what is the sign of Δp_z for the system consisting of the bowling ball alone?
   1. Positive
   2. Negative
   3. Zero —no change in p_z.

2.17 PROBLEMS

2.P.24 In the space shuttle
A space shuttle is in a circular orbit near the Earth. An astronaut floats in the middle of the shuttle, not touching the walls. On a diagram, draw and label
   (a) the momentum $\vec{p}_1$ of the astronaut at this instant;
   (b) all of the forces (if any) acting on the astronaut at this instant;
   (c) the momentum $\vec{p}_2$ of the astronaut a short time Δt later;
   (d) the momentum change (if any) $\Delta\vec{p}$ in this time interval.
   (e) Why does the astronaut seem to "float" in the shuttle?
It is ironic that we say the astronaut is "weightless" despite the fact that the only force acting on the astronaut is the astronaut's weight (that is, the gravitational force of the Earth on the astronaut).

2.P.25 Crash test
In a crash test, a truck with mass 2200 kg traveling at 25 m/s (about 55 miles per hour) smashes head-on into a concrete wall without rebounding. The front end crumples so much that the truck is 0.8 m shorter than before. What is the approximate magnitude of the force exerted on the truck by the wall? Explain your analysis carefully, and justify your estimates on physical grounds.

2.P.26 Ping-pong ball
A ping-pong ball is acted upon by the Earth, air resistance, and a strong wind. Here are the positions of the ball at several times.
Early time interval:
   At $t = 12.35$ s, the position was $\langle 3.17, 2.54, -9.38 \rangle$ m
   At $t = 12.37$ s, the position was $\langle 3.25, 2.50, -9.40 \rangle$ m
Late time interval:
   At $t = 14.35$ s, the position was $\langle 11.25, -1.50, -11.40 \rangle$ m
   At $t = 14.37$ s, the position was $\langle 11.27, -1.86, -11.42 \rangle$ m
   (a) In the early time interval, from $t = 12.35$ s to $t = 12.37$ s, what was the average momentum of the ball? The mass of the ping-pong ball is 2.7 grams (2.7×10^{-3} kg). Express your result as a vector.
   (b) In the late time interval, from $t = 14.35$ s to $t = 14.37$ s, what was the average momentum of the ball? Express your result as a vector.
   (c) In the time interval from $t = 12.35$ s (the start of the early time interval) to $t = 14.35$ s (the start of the late time interval), what was the average net force acting on the ball? Express your result as a vector.

2.P.27 Proton and HCl molecule
A proton interacts electrically with a neutral HCl molecule located at the origin. At a certain time t, the proton's position is $\langle 1.6 \times 10^{-9}, 0, 0 \rangle$ m and the proton's velocity is $\langle 3200, 800, 0 \rangle$ m/s. The force exerted on the proton by the HCl molecule is $\langle -1.12 \times 10^{-11}, 0, 0 \rangle$ N. At a time $t + (2 \times 10^{-14}$ s), what is the approximate velocity of the proton?

2.P.28 Kick a soccer ball

A 0.4 kg soccer ball is rolling by you at 3.5 m/s. As it goes by, you give it a kick perpendicular to its path. Your foot is in contact with the ball for 0.002 s. The ball eventually rolls at a 20° angle from its original direction. The overhead view in the diagram below is approximately to scale. The arrow represents the force your toe applies briefly to the soccer ball.

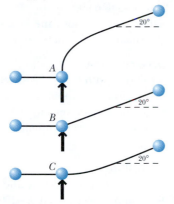

(a) In the diagram, which letter corresponds to the correct overhead view of the ball's path?

(b) Determine the magnitude of the average force you applied to the ball.

2.P.29 Projectile motion

A small dense ball with mass 1.5 kg is thrown with initial velocity $\langle 5, 8, 0 \rangle$ m/s at time $t = 0$ at a location we choose to call the origin ($\langle 0, 0, 0 \rangle$ m). Air resistance is negligible.

(a) When the ball reaches its maximum height, what is its velocity (a vector)? It may help to make a simple diagram.

(b) When the ball reaches its maximum height, what is t? You know how v_y depends on t, and you know the initial and final velocities.

(c) Between the launch at $t = 0$ and the time when the ball reaches its maximum height, what is the average velocity (a vector)? You know how to determine average velocity when velocity changes at a constant rate.

(d) When the ball reaches its maximum height, what is its location (a vector)? You know how average velocity and displacement are related.

(e) At a later time the ball's height y has returned zero, which means that the average value of v_y from $t = 0$ to this time is zero. At this instant, what is the time t?

(f) At the time calculated in part (e), when the ball's height y returns to zero, what is x? (This is called the "range" of the trajectory.)

(g) At the time calculated in part (e), when the ball's height y returns to zero, what is v_y?

(h) What was the angle to the x-axis of the initial velocity?

(i) What was the angle to the x-axis of the velocity at the time calculated in part (e), when the ball's height y returned to zero?

2.P.30 A free throw in basketball

Determine two different possible ways for a player to make a free throw in basketball. In both cases give the initial speed, initial angle, and initial height of the basketball. The rim of the basket is 10 feet (3.0 m) above the floor. It is 14 feet (4.3 m) along the floor from the free-throw line to a point directly below the center of the basket.

2.P.31 A basketball pass

You have probably seen a basketball player throw the ball to a teammate at the other end of the court, 30 m away. Estimate a reasonable initial angle for such a throw, and then determine the corresponding initial speed. For your chosen angle, how long does it take for the basketball to go the length of the court? What is the highest point along the trajectory, relative to the thrower's hand?

2.P.32 The case of the falling flower pot

You are a detective investigating why someone was hit on the head by a falling flowerpot. One piece of evidence is a home video taken in a 4th-floor apartment, which happens to show the flowerpot falling past a tall window. Inspection of individual frames of the video shows that in a span of 6 frames the flowerpot falls a distance that corresponds to 0.85 of the window height seen in the video (note: standard video runs at a rate of 30 frames per second). You visit the apartment and measure the window to be 2.2 m high. What can you conclude? Under what assumptions? Give as much detail as you can.

2.P.33 Tennis ball hits wall

A tennis ball has a mass of 0.057 kg. A professional tennis player hits the ball hard enough to give it a speed of 50 m/s (about 120 miles per hour). The ball hits a wall and bounces back with almost the same speed (50 m/s). As indicated in the diagram, high-speed photography shows that the ball is crushed 2 cm (0.02 m) at the instant when its speed is momentarily zero, before rebounding.

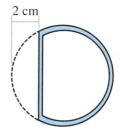

A high-speed tennis ball deforms when it hits a wall.

Making the very rough approximation that the large force that the wall exerts on the ball is approximately constant during contact, determine the approximate magnitude of this force. Hint: Think about the approximate amount of time it takes for the ball to come momentarily to rest. (For comparison note that the gravitational force on the ball is quite small, only about $(0.057 \text{ kg})(9.8 \text{ N/kg}) \approx 0.6 \text{ N}$. A force of 5 N is approximately the same as a force of one pound.)

2.P.34 Mars probe

A small space probe, of mass 240 kg, is launched from a spacecraft near Mars. It travels toward the surface of Mars, where it will land. At a time 20.7 seconds after it is launched, the probe is at the location $\langle 4.30 \times 10^3, 8.70 \times 10^2, 0 \rangle$ m , and at this same time its momentum is $\langle 4.40 \times 10^4, -7.60 \times 10^3, 0 \rangle$ kg · m/s . At this instant, the net force on the probe due to the gravitational pull of Mars plus the air resistance acting on the probe is $\langle -7 \times 10^3, -9.2 \times 10^2, 0 \rangle$ N .

(a) Assuming that the net force on the probe is approximately constant over this time interval, what is the momentum of the probe 20.9 seconds after it is launched?

(b,) What is the location of the probe 20.9 seconds after launch?

2.P.35 Spacecraft navigation

Suppose you are navigating a spacecraft far from other objects. The mass of the spacecraft is 1.5×10^5 kg (about 150 tons). The rocket engines are shut off, and you're coasting along with a constant velocity of $\langle 0, 20, 0 \rangle$ km/s. As you pass the location $\langle 12, 15, 0 \rangle$ km you briefly fire side thruster rockets, so that your spacecraft experiences a net force of $\langle 6 \times 10^4, 0, 0 \rangle$ N for 3.4 s. The ejected gases have a mass that is small compared to the mass of the spacecraft. You then continue coasting with the rocket engines turned off. Where are you an hour later? Also, what approximations or simplifying assumptions did you have to make in your analysis? Think about the choice of system: what are the surroundings that exert external forces on your system?

2.P.36 Electron motion in a CRT

In a cathode ray tube (CRT) used in oscilloscopes and television sets, a beam of electrons is steered to different places on a phosphor screen, which glows at locations hit by electrons. The CRT is evacuated, so there are few gas molecules present for the electrons to run into. Electric forces are used to accelerate electrons of mass m to a speed $v_0 \ll c$, after which they pass between positively and negatively charged metal plates which deflect the electron in the vertical direction (upward in the diagram, or downward if the sign of the charges on the plates is reversed).

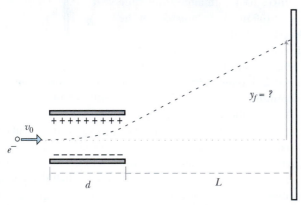

A cathode ray tube.

While an electron is between the plates, it experiences a uniform vertical force F, but when the electron is outside the plates there is negligible force on it. The gravitational force on the electron is negligibly small compared to the electric force in this situation. The length of the metal plates is d, and the phosphor screen is a distance L from the metal plates. Where does the electron hit the screen? (That is, what is y_f?)

2.P.37 The SLAC two-mile accelerator

SLAC, the Stanford Linear Accelerator Center, located at Stanford University in Palo Alto, California, accelerates electrons through a vacuum tube two miles long (it can be seen from an overpass of the Junipero Serra freeway that goes right over the accelerator). Electrons which are initially at rest are subjected to a continuous force of 2×10^{-12} newton along the entire length of two miles (one mile is 1.6 kilometers) and reach speeds very near the speed of light.

(a) Determine how much time is required to increase the electrons' speed from $0.93c$ to $0.99c$. (That is, the quantity $|\vec{v}|/c$ increases from 0.93 to 0.99.)

(b) Approximately how far does the electron go in this time? What is approximate about your result?

2.P.38 Outer space collision

In outer space a small rock with mass 5 kg traveling with velocity $\langle 0, 1800, 0 \rangle$ m/s strikes a stationary large rock head-on and bounces straight back with velocity $\langle 0, -1500, 0 \rangle$ m/s. After the collision, what is the vector momentum of the large rock?

2.P.39 Two rocks collide

Two rocks collide in outer space. Before the collision, one rock had mass 9 kg and velocity $\langle 4100, -2600, 2800 \rangle$ m/s. The other rock had mass 6 kg and velocity $\langle -450, 1800, 3500 \rangle$ m/s. A 2 kg chunk of the first rock breaks off and sticks to the second rock. After the collision the rock whose mass is 7 kg has velocity $\langle 1300, 200, 1800 \rangle$ m/s. After the collision, what is the velocity of the other rock, whose mass is 8 kg?

2.P.40 Two rocks collide

Two rocks collide with each other in outer space, far from all other objects. Rock 1 with mass 5 kg has velocity $\langle 30, 45, -20 \rangle$ m/s before the collision and $\langle -10, 50, -5 \rangle$ m/s after the collision. Rock 2 with mass 8 kg has velocity $\langle -9, 5, 4 \rangle$ m/s before the collision. Calculate the final velocity of rock 2.

2.P.41 Two rocks stick together

In outer space two rocks collide and stick together. Here are the masses and initial velocities of the two rocks:

Rock 1: mass = 15 kg, initial velocity = $\langle 10, -30, 0 \rangle$ m/s
Rock 2: mass = 32 kg, initial velocity = $\langle 15, 12, 0 \rangle$ m/s
What is the velocity of the stuck-together rocks after colliding?

2.P.42 Moving the Earth

Suppose all the people of the Earth go to the North Pole and, on a signal, all jump straight up. Estimate the recoil speed of the Earth. The mass of the Earth is 6×10^{24} kg, and there are about 6 billion people (6×10^9).

2.P.43 Bullet embeds in block

A bullet of mass m traveling horizontally at a very high speed v embeds itself in a block of mass M that is sitting at rest on a nearly frictionless surface. What is the speed of the block after the bullet embeds itself in the block?

2.P.44 Meteor hits a spinning satellite

A satellite which is spinning clockwise has four low-mass solar panels sticking out as shown. A tiny meteor traveling at high speed rips through one of the solar panels and continues in the same direction but at reduced speed. Afterwards, calculate the v_x and v_y components of the center-of-mass velocity of the satellite. Initial data are provided on the diagram.

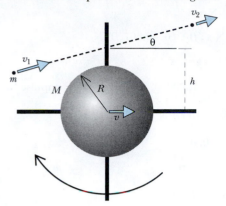

2.P.45 Space junk

A tiny piece of space junk of mass m strikes a glancing blow to a spinning satellite. Before the collision the satellite was moving and rotating as shown in the diagram. After the collision the space junk is traveling in a new direction and moving more slowly. The space junk had negligible rotation both before and after the collision. The velocities of the space junk before and after the collision are shown in the diagram. The satellite has mass M and radius R.

Just after the collision, what are the components of the center-of-mass velocity of the satellite (v_x and v_y)?

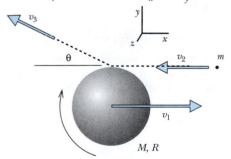

2.P.46 Two balls collide

A ball of mass 0.05 kg moves with a velocity $\langle 17, 0, 0 \rangle$ m/s. It strikes a ball of mass 0.1 kg which is initially at rest. After the collision, the heavier ball moves with a velocity of $\langle 3, 3, 0 \rangle$ m/s.

(a) What is the velocity of the lighter ball after impact?

(b) What is the impulse delivered to the 0.05 kg ball by the heavier ball?

(c) If the time of contact between the balls is 0.03 sec, what is the force exerted by the heavier ball on the lighter ball?

2.P.47 Space station

A space station has the form of a hoop of radius R, with mass M. Initially its center of mass is not moving, but it is spinning. Then a small package of mass m is thrown by a spring-loaded gun toward a nearby spacecraft as shown; the package has a speed v after launch.

Calculate the center-of-mass velocity of the space station (v_x and v_y) after the launch.

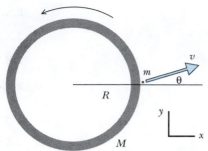

2.P.48 Falling ball

Apply the general results obtained in the full analysis of motion under the influence of a constant force on page 51 to answer the following questions. You hold a small metal ball of mass m a height h above the floor. You let go, and the ball falls to the floor. Choose the origin of the coordinate system to be on the floor where the ball hits, with y up as usual. Just after release, what are y_i and v_{iy}? Just before hitting the floor, what is y_f? How much time Δt does it take for the ball to fall? What is v_{fy} just before hitting the floor? Express all results in terms of m, g, and h. How would your results change if the ball had twice the mass?

2.18 ANSWERS TO EXERCISES

2.X.1 (page 45) $\langle -60, -24, 96 \rangle$ N·s

2.X.2 (page 45)   **(a)** $\langle 0, -2, 0 \rangle$ N·s , **(b)** $\langle 0, -0.667, 0 \rangle$ N·s

2.X.3 (page 49) $\vec{p}_f = \langle 10, 0, 11 \rangle$ kg·m/s

2.X.4 (page 49) $\Delta\vec{p} = \vec{p}_f - \vec{p}_i = \langle -15000, 0, 3000 \rangle$ kg·m/s = impulse by ground

2.X.5 (page 49) $\vec{F}_{net} = \langle -5000, 0, 1000 \rangle$ N

2.X.6 (page 49) $\vec{F}_{net} = \langle -1\times10^4, 0, 0 \rangle$ N

2.X.7 (page 49) $\vec{p}_f = \langle 9.4, 0.3, 0 \rangle$ kg·m/s

2.X.8 (page 54)   (a) $\langle -10, 7.12, -5 \rangle$ m/s

          (b) $\langle 3, 6.036, -8 \rangle$ m

          (c) 8.6 m

          (d) 1.33 s

          (e) 2.66 s

          (f) $\langle -17.5, 0, -18.3 \rangle$ m

2.X.9 (page 62) $\langle 0.04, 0.04, -0.1 \rangle$ kg·m/s $\langle 0.04, 0.04, -0.1 \rangle$ kg·m/s

          $\langle 1, 1, -2.5 \rangle$ m/s

2.X.10 (page 62) $\langle 18, 14, 7 \rangle$ kg · m/s , $\langle 18, 14, 7 \rangle$ kg · m/s

2.X.11 (page 62) $\langle 1\times10^{-21}, -1.6\times10^{-21}, 0 \rangle$ kg · m/s

2.X.12 (page 63)   (a) $-m\vec{v}$, $+m\vec{v}$; (b) $+m\vec{v}$, $-m\vec{v}$; (c) $\vec{0}$, $\vec{0}$

2.X.13 (page 63)   (a) The Earth (b) No. The changes in your momentum and the Earth's momentum are equal and opposite, and add up to zero. As you move down, the Earth moves up. The Earth gets as much magnitude of momentum as you do, but very little velocity because its mass is so huge ($v \approx p/m$.

CHAPTER 3

THE MOMENTUM PRINCIPLE: NON-CONSTANT FORCES

Key concepts:
- Many forces change when object positions change
  - Fundamental forces
    - Electric force
    - Gravitational force
  - Composite forces
    - Spring force
    - Air resistance force
- To predict motion when forces are changing
  - Divide the time interval into many small intervals
  - Apply the momentum principle iteratively (repeatedly)

3.1 CHANGING FORCES

In the previous chapter we applied the Momentum Principle, along with the position-velocity relation, to predict the motion of systems subject to forces that were constant, or approximately constant. Let's see what happens when we try to predict the motion of a system in which force is changing.

Example: Earth and Sun

Let's choose a coordinate system whose origin is at the center of the Sun, and in which the Earth orbits the Sun in the xy plane, as shown in Figure 3.1. We'll pick a coordinate system with its origin at the center of the Sun. At a particular instant, the Earth is at location $\langle 1.5 \times 10^{11}, 0, 0 \rangle$ m, with velocity $\langle 0, 3 \times 10^4, 0 \rangle$ m/s. At this instant, the force exerted on the Earth by the Sun is $\langle -3.6 \times 10^{22}, 0, 0 \rangle$ N (we will see later how to calculate this force). The mass of the Earth is 6×10^{24} kg. What will the location of the Earth be 3 months later?

1. Momentum principle

System: Earth
Surroundings: Sun (ignore all other planets, stars)

$$\Delta t = (3 \text{ mo})\left(30 \frac{\text{days}}{\text{mo}}\right)\left(24 \frac{\text{hr}}{\text{day}}\right)\left(60 \frac{\text{min}}{\text{hr}}\right)\left(60 \frac{\text{sec}}{\text{min}}\right) = 7.8 \times 10^6 \text{s}$$

$$\vec{p}_i = (6 \times 10^{24} \text{ kg})(\langle 0, 3 \times 10^4, 0 \rangle \text{ m/s}) = \langle 0, 1.8 \times 10^{29}, 0 \rangle \text{ kg} \cdot \text{m/s}$$

$$\vec{p}_f = \vec{p}_i + \vec{F}_{\text{net}} \Delta t$$

$$\vec{p}_f = \left(\langle 0, 1.8 \times 10^{29}, 0 \rangle \frac{\text{kg} \cdot \text{m}}{\text{s}}\right) + (\langle -3.6 \times 10^{22}, 0, 0 \rangle \text{ N})(7.8 \times 10^6 \text{s})$$

$$\vec{p}_f = \langle -2.8 \times 10^{29}, 1.8 \times 10^{29}, 0 \rangle \text{ kg} \cdot \text{m/s}$$

2. Update position

$$\vec{v}_f \approx \frac{\vec{p}_f}{m} = \frac{\langle -2.8 \times 10^{29}, 1.8 \times 10^{29}, 0 \rangle \text{ kg} \cdot \text{m/s}}{6 \times 10^{24} \text{ kg}}$$

We know that the distance from the Earth to the Sun is 1.5×10^{11} m. We can figure out the Earth's speed by noting that in 365 days it travels a distance of $2\pi(1.5 \times 10^{11})$ m in its nearly circular orbit, so:

$$v = \frac{2\pi(1.5 \times 10^{11}) \text{ m}}{\left(365 \frac{\text{days}}{\text{yr}}\right)\left(24 \frac{\text{hr}}{\text{day}}\right)\left(60 \frac{\text{min}}{\text{hr}}\right)\left(60 \frac{\text{sec}}{\text{min}}\right)}$$

$$v = 3 \times 10^4 \frac{\text{m}}{\text{s}}$$

$\vec{p} \approx m\vec{v}$ because $v \ll c$

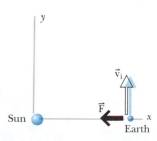

Figure 3.1 Initial location and velocity of the Earth, and force on Earth by Sun. The origin of the coordinate system is at the center of the Sun. Sizes of the Sun and Earth are exaggerated.

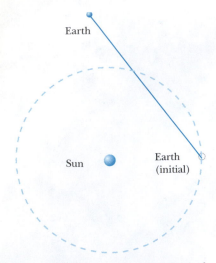

Figure 3.2 The displacement of the Earth predicted by a one-step calculation using the initial value of gravitational force, and the arithmetic average velocity. The dashed line shows the actual path of the Earth around the Sun.

$$\vec{v}_f = \langle -4.6\times10^4, \, p3\times10^4, \, 0\rangle \, \text{m/s}$$

$$v_{\text{avg,x}} \approx \frac{(0 + -4.6\times10^4)}{2} \frac{\text{m}}{\text{s}} = -2.3\times10^4 \text{ m/s}$$

$$v_{\text{avg,y}} \approx \frac{(3\times10^4 + 3\times10^4)}{2} \frac{\text{m}}{\text{s}} = 3\times10^4 \text{ m/s}$$

$$x_f = x_i + v_{\text{avg,x}}\Delta t$$

$$x_f = (1.5\times10^{11} \text{ m}) + (-2.3\times10^4 \text{ m/s})(7.8\times10^6\text{s}) = -3.1\times10^{10} \text{ m}$$

$$y_f = y_i + v_{\text{avg,y}}\Delta t$$

$$y_f = 0 + (3\times10^4 \text{ m/s})(7.8\times10^6\text{s}) = 2.3\times10^{11} \text{ m}$$

$$\vec{r}_f = \langle -3.1\times10^{10}, \, 2.3\times10^{11}, \, 0\rangle \text{ m}$$

3. Check: units ok (meters). Reasonable location? NO! The path we predicted, from initial location to final location, is shown in Figure 3.2. We know the Earth's path around the Sun is nearly circular, so our result can't be correct. We show the Earth getting much farther away from the Sun, which is definitely wrong.

Further discussion

? What went wrong? In the preceding calculation we used the same procedure we used in the previous chapter.

The major error was the tacit assumption that the force on the Earth by the Sun was constant in magnitude and direction over the three month interval. An additional error was introduced by using the arithmetic average velocity as a (poor) approximation to the actual average velocity during this interval—recall that the arithmetic average is equal to the true average velocity only if the force is constant. The direction and magnitude of the gravitational force between two objects changes when the relative positions of the objects change. We need a way to take this into account.

3.2 ITERATIVE PREDICTION OF MOTION

To "iterate" means "to repeat", or "to perform again"; the word "iteration" implies repeating the same process many times. The basic idea we will use in predicting the motion of systems subject to changing forces is to divide up the time interval of interest into many very small time intervals, and iteratively apply the momentum principle over successive intervals. Each time interval must be short enough that we may safely make the approximation that the force on the system does not change significantly during the time interval. We will update momentum and position iteratively like this:

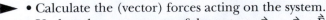

ITERATIVE PREDICTION OF MOTION

- Calculate the (vector) forces acting on the system.
- Update the momentum of the system: $\vec{p}_f = \vec{p}_i + \vec{F}_{\text{net}}\Delta t$.
- Update the position: $\vec{r}_f = \vec{r}_i + \vec{v}_{\text{avg}}\Delta t$.
- Repeat.

This procedure need not be confined to situations in which forces are changing. It can be used even in a constant force situation to predict the entire path along which an object will travel. For example, we could apply this procedure to calculate the trajectory of a ball moving without air resistance.

A graphical representation of three successive steps in the process might look like this (Figure 3.3):

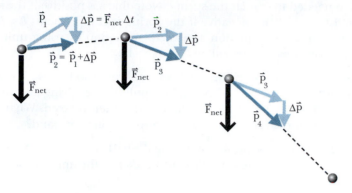

At each location we calculate the net impulse $\vec{F}_{net}\Delta t$, the change in momentum, and the new momentum. We use the new momentum to predict the new position, then iterate. (Here we have used $\vec{v}_f \approx \vec{p}_f / m$ to approximate $\vec{v}_{avg}$.) Even with only three steps the trajectory looks roughly like what is observed in the real world, and you can see the increasing displacement of the ball during each time step, corresponding to increasing speed. If we used smaller time steps, the straight line segments representing displacement would be shorter, and the trajectory would look more and more like the actual curving path of the ball.

A changing force (air resistance)

A more interesting problem would be to predict the path of a ball including the effects of air resistance, because the magnitude of the air resistance force depends on the ball's speed, and the direction of the force depends on the direction of the ball's velocity. (The air resistance force is discussed in more detail in Chapter 6.) The two trajectories shown in Figure 3.4 were computed using this procedure; one including air resistance, the other without air resistance. In these calculations, the total time (5.7 seconds with air resistance, 7.3 seconds without air resistance) was divided into intervals of 0.01 second, so calculating these trajectories required 570 and 730 steps, respectively. Clearly it is advantageous to program a computer to do this many calculations. In the following pages, we will predict the motion of systems subject to changing forces, such as spring forces, gravitational forces, and electric forces.

The spring force law

A "force law" describes mathematically how a force depends on the situation. In this chapter we'll learn about various force laws, including the gravitational force law and the electric force law. For a spring, it is determined experimentally that the magnitude of the force exerted by a spring on an object attached to the spring is given by the following force law:

THE SPRING FORCE LAW

$$\left|\vec{F}_{spring}\right| = k_s |s|$$

$|s|$ is the absolute value of the stretch:

$$s = L - L_0$$

L_0 is the length of the relaxed spring
L is the length of the spring when stretched or compressed
k_s is the "spring stiffness"
The force acts in a direction to restore the spring to its relaxed length.

Figure 3.3 Three steps in the iterative prediction of the path of a ball in the absence of air resistance. In each step the new momentum of the ball is calculated, then this new momentum is used to update the position of the ball. In the calculation represented here, the final velocity was used as an approximation to the average velocity in each step.

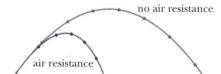

Figure 3.4 Predicted trajectories of a baseball thrown at high speed (about 100 m.p.h.), with and without the effect of air resistance. In the iterative calculations time steps of 0.01 second were taken.

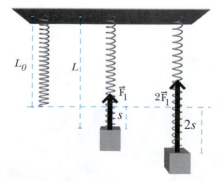

Figure 3.5 The magnitude of the force exerted by a spring is proportional to the absolute value of the stretch of the spring. For an elongated spring, stretch is positive.

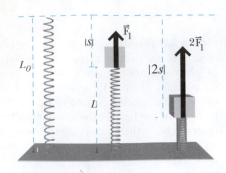

Figure 3.6 The magnitude of the force exerted by a spring is proportional to the absolute value of the stretch of the spring (for a compressed spring, stretch is negative).

Because of the reciprocity of the electric forces between the protons and electrons in the spring and those in your hand, the force you exert on the spring is equal in magnitude and opposite in direction to the force the spring exerts on your hand.

The constant k_s is a positive number, and is a property of the particular spring: the stiffer the spring, the larger the spring stiffness, and the larger the force needed to stretch the spring. Note that s is positive if the spring is stretched ($L > L_0$) and negative if the spring is compressed ($L < L_0$). It is possible to write this equation as a vector equation by using unit vectors. This equation is sometimes called "Hooke's law."

? Suppose a certain spring has been calibrated so that we know that its spring stiffness k_s is 500 N/m. You pull on the spring and observe that it is 0.01 m (1 cm) longer than it was when relaxed. What is the magnitude of the force exerted by the spring on your hand?

The force law gives $\left| \vec{F}_{\text{spring}} \right| = (500 \text{ N/m})(|+0.01 \text{ m}|) = 5 \text{ N}$. Note that the total length of the spring doesn't matter; it's just the amount of stretch or compression that matters.

? Suppose that instead of pulling on the spring, you push on it, so the spring becomes shorter than its relaxed length. If the relaxed length of the spring is 10 cm, and you compress the spring to a length of 9 cm, what is the magnitude of the force exerted by the spring on your hand?

The stretch of the spring in SI units is

$$s = L - L_0 = (0.09 \text{ m} - 0.10 \text{ m}) = -0.01 \text{ m}.$$

The force law gives

$$\left| \vec{F}_{\text{spring}} \right| = (500 \text{ N/m})(|-0.01 \text{ m}|) = 5 \text{ N}.$$

The magnitude of the force is the same as in the previous case. Of course the direction of the force exerted by the spring on your hand is now different, but we would need to write a full vector equation to incorporate this information.

3.X.1 You push on a spring whose stiffness is 11 N/m, compressing it until it is 2.5 cm shorter than its relaxed length. What is the magnitude of the force the spring now exerts on your hand?

3.X.2 A spring is 0.17 m long when it is relaxed. When a force of magnitude 250 N is applied, the spring becomes 0.24 m long. What is the stiffness of this spring?

3.X.3 The spring in the previous exercise is now compressed so that its length is 0.15 m. What magnitude of force is required to do this?

3.X.4 The stiffness of a particular spring is 40 N/m. One end of the spring is attached to a wall. When you pull on the other end of the spring with a steady force of 2 N, the spring elongates to a total length of 18 cm. What was the relaxed length of the spring? (Remember to convert to S.I. units.)

Motion of a block-spring system

If we attach a block to the top of a spring, push down on the block, and then release it, the block will oscillate up and down. (This repetitive motion is described as "periodic.") As the spring stretches and compresses, the force exerted on the block by the spring changes in magnitude and direction. There is also a constant gravitational force on the block. Because the net force on the block is continually changing, we can't use a one-step calculation to predict its motion. We need to apply the Momentum Principle iteratively to predict the location and velocity of the block at any instant.

Example: Block on spring (1D, nonconstant net force)

A spring has a relaxed length of 20 cm (0.2 m) and its spring stiffness is 8 N/m (Figure 3.7). You glue a 60 gram block (0.06 kg) to the top of the spring, and push the block down, compressing the spring so its total length is 10 cm (Figure 3.8). You make sure the block is at rest, then you quickly move your hand away. The block begins to move upward, because the upward force on the block by the spring is greater than the downward force on the block by the Earth. Make a graph of y vs. time for the block during a 0.3 second interval after you release the block.

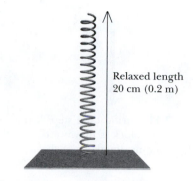

Figure 3.7 A relaxed vertical spring.

To get an approximate answer, let's divide the 0.3 second time interval into three intervals each 0.1 second long. (It would be better to use even shorter intervals, but this would be unduly tedious if done by hand.)

Sign of y component of spring force: When the stretch is negative (spring compressed; $s = L - L_0 < 0$), force is upward (+y). When the stretch is positive (spring stretched; $s = L - L_0 > 0$), force is downward (−y). So we can write the y component of the spring force as:

$$F_{\text{spring},y} = -k_s s$$

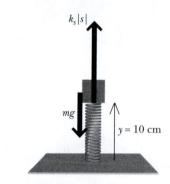

Figure 3.8 You compressed the spring, then released the block; the block heads upward with increasing speed, because the net force is nonzero and upward.

I. First time step:

1. Momentum Principle
System: the block
Surroundings: Earth, spring (neglect air resistance)

$$p_{fy} = p_{yi} + F_{\text{net},y}\Delta t = p_{yi} + (-mg - k_s s)\Delta t$$

Evaluate the net force due to the Earth and the spring:

$$F_{\text{net},y} = (-mg - k_s s)$$
$$= (-(0.06 \text{ kg})(9.8 \text{ N/kg}) - (8 \text{ N/m})(0.1 \text{ m} - 0.2 \text{ m}))$$
$$= 0.212 \text{ N}$$
$$p_{fy} = 0 + (0.212 \text{ N})(0.1 \text{ s})$$
$$p_{fy} = 0.0212 \text{ kg} \cdot \text{m/s}$$

2. Update position

$v_{\text{avg},y} \approx v_{fy}$ (F is not constant, so arithmetic avg. not necessarily better)

$$v_{fy} = \frac{p_{fy}}{m} = \frac{(0.0212 \text{ kg} \cdot \text{m/s})}{0.06 \text{ kg}} = 0.353 \text{ m/s}$$
$$y_f = y_i + v_{fy}\Delta t = 0.1 \text{m} + (0.353 \text{ m/s})(0.1 \text{ s})$$
$$y_f = 0.135 \text{ m}$$

Figure 3.9 During time step 1, the net force on the block is upward, and the block is moving upward.

II. Second time step:

$$p_{fy} = p_{yi} + F_{\text{net},y}\Delta t = p_{yi} + (-mg - k_s s)\Delta t$$
$$F_{\text{net},y} = (-mg - k_s s)$$
$$= (-(0.06 \text{ kg})(9.8 \text{ N/kg}) - (8 \text{ N/m})(0.135 \text{m} - 0.2 \text{ m}))$$
$$= (-0.0707 \text{ N})$$
$$p_{fy} = (0.0212 \text{ kg} \cdot \text{m/s}) + (-0.0707 \text{ N})(0.1 \text{ s})$$
$$p_{fy} = 0.0141 \text{ kg} \cdot \text{m/s}$$

2. Update position

Figure 3.10 During time step 2, the net force on the block is now downward, but the block is still moving upward, though more slowly.

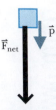

Figure 3.11 During time step 3, the net force on the block is downward, and the block is now moving downward.

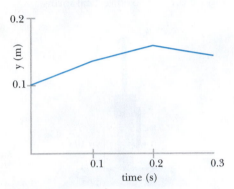

Figure 3.12 A graph of the y component of the block's position vs. time for the three step iterative calculation.

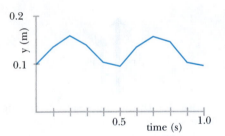

Figure 3.13 A graph of the y component of the block's position vs. time for an iterative calculation carried out for 10 steps of 0.1 second.

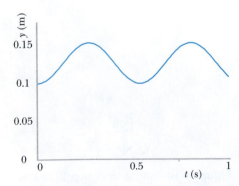

Figure 3.14 A graph of the y component of the block's position vs. time for an iterative calculation using a step size of 0.01 second.

$$v_{fy} = \frac{p_{fy}}{m} = \frac{(0.0141 \text{ kg} \cdot \text{m/s})}{0.06 \text{ kg}} = 0.236 \text{ m/s}$$

$$y_f = (y_i + v_{fy}\Delta t) = 0.135 \text{ m} + (0.236 \text{ m/s})(0.1 \text{ s})$$

$$y_f = 0.159 \text{ m}$$

III. Third time step:

$$p_{fy} = p_{yi} + F_{\text{net}, y}\Delta t = p_{yi} + (-mg - k_s s)\Delta t$$

$$F_{\text{net}, y} = (-mg - k_s s)$$

$$= (-(0.06 \text{ kg})(9.8 \text{ N/kg}) - (8 \text{ N/m})(0.159 \text{ m} - 0.2 \text{ m}))$$

$$= (-0.259 \text{ N})$$

$$p_{fy} = (0.0141 \text{ kg} \cdot \text{m/s}) + (-0.259 \text{ N})(0.1 \text{ s})$$

$$p_{fy} = -0.118 \text{ kg} \cdot \text{m/s}$$

2. Update position

$$v_{fy} = \frac{p_{fy}}{m} = \frac{(-0.118 \text{ kg} \cdot \text{m/s})}{0.06 \text{ kg}} = -0.196 \text{ m/s}$$

$$y_f = (y_i + v_{fy}\Delta t) = 0.159 \text{ m} + (-0.196 \text{ m/s})(0.1 \text{ s})$$

$$y_f = 0.139 \text{ m}$$

The graph of y vs. time is shown in Figure 3.12.

3. Check: units okay. Reasonable? Yes, because we expect the block to oscillate up and down on the spring.

Further discussion

? What approximations were made in this calculation?

We made the approximation that the net force did not change significantly over each small time step. During each time step, we also used the final velocity as an approximation for the average velocity. This is simpler than computing the arithmetic average, and not necessarily worse in a situation where the force is actually changing. (In this case, it turns out that using the arithmetic average would actually have given a less accurate answer.)

Figure 3.13 shows the graph of y vs. *t* produced when the iterative calculation above is carried out for 10 time steps of 0.1 second. Although the graph does show the oscillatory motion of the block, its jagged lines reflect the fact that 0.1 second is too large a step size to produce an accurate result. Using a step size of 0.01 second produces a smoothly oscillating graph of y vs. *t*, as shown in Figure 3.14.

While this iterative scheme is very general, doing it by hand is incredibly tedious. It is not difficult to program a computer to do these calculations repetitively. Computers are now fast enough that it is possible to get high accuracy simply by taking very short time steps, so that during each step the net force and velocity aren't changing much. We'll talk more about computer prediction of motion later in this chapter.

3.3 FUNDAMENTAL FORCES

There are four different kinds of fundamental forces currently known to science, associated with four different kinds of interactions: gravitational, electromagnetic, nuclear (also referred to as the "strong" interaction), and the "weak" interaction.

- The *gravitational* interaction is responsible for an attraction every object exerts on every other object. For example, the Earth exerts a gravitational force on the Moon, and the Moon exerts a gravitational force on the Earth.

- The *electromagnetic* interaction includes electric forces responsible for sparks, static cling, and the behavior of electronic circuits, and magnetic forces responsible for the operation of motors driven by electric current. Protons repel each other electrically, as do electrons, whereas protons and electrons attract each other (Figure 3.15). Electric forces bind protons and electrons to each other in atoms, and are responsible for the chemical bonds between atoms in molecules. The force of a stretched or compressed spring is due to electric forces between the atoms that make up the spring.

- The nuclear or *strong* interaction holds protons and neutrons together in the nucleus of an atom despite the large mutual electric repulsion of the protons (Figure 3.16). (The neutrons are not electrically charged and don't exert electric forces.)

- An example of the *weak* interaction is seen in the instability of a neutron. If a neutron is removed from a nucleus, with an average lifetime of about 15 minutes the neutron decays into a proton, an electron, and a ghostly particle called the antineutrino. This change is brought about by the weak interaction.

We will be mainly concerned with gravitational and electric interactions, but we will occasionally encounter situations where the nuclear or strong interaction plays an important role. We will have little to say about the weak interaction. Nor will we deal with magnetism, the other part of the electromagnetic interaction. The second volume of this textbook deals extensively with both electric and magnetic interactions.

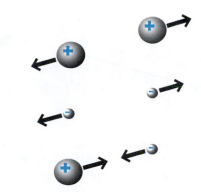

Figure 3.15 The electric force: protons repel each other; electrons repel each other; protons and electrons attract each other.

Figure 3.16 The strong force: the protons in the nucleus of an atom exert repulsive electric forces on each other, but the strong interaction (which involves neutrons as well as protons) holds the nucleus together despite this electric repulsion.

3.4 THE GRAVITATIONAL FORCE LAW

The motion of stars and planets is in important ways simpler than other mechanical phenomena, because there is no friction to worry about. These massive bodies interact through the gravitational force, which is always an attractive force. Studying how to predict the motion of stars and planets is one of the most direct ways to understand in general how the Momentum Principle determines the behavior of objects in the real world. The basic ideas used to predict the motion of stars and planets can be applied later to a wide range of everyday and atomic phenomena.

In the 1600's Isaac Newton deduced that there must be an attractive force associated with a gravitational interaction between any pair of objects. The gravitational force acts along a line connecting the two objects (Figure 3.18), is proportional to the mass of one object and to the mass of the other object, and is inversely proportional to the square of the distance between the *centers* of the two objects (not the gap between their surfaces). The gravitational force exerted on object 2 by object 1 is expressed by this compact vector equation:

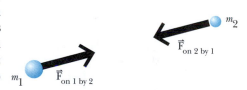

Figure 3.17 As is evident from the gravitational force equation, the force on object 1 by object 2 is equal and opposite to the force on object 2 by object 1 (reciprocity).

THE GRAVITATIONAL FORCE LAW

$$\vec{F}_{grav \text{ on 2 by 1}} = -G\frac{m_1 m_2}{|\vec{r}|^2}\hat{r}$$

$\hat{r} = \vec{r}_2 - \vec{r}_1$ extends from the center of object 1 to the center of object 2

G is a universal constant: $G = 6.7 \times 10^{-11}\ \dfrac{\text{N} \cdot \text{m}^2}{\text{kg}^2}$

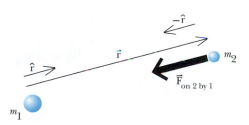

Figure 3.18 The gravitational force exerted on object 2 by object 1.

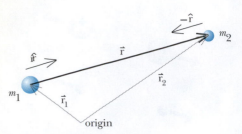

Figure 3.19 $\hat{r} = \vec{r}_2 - \vec{r}_1$, the location of object 2 relative to object 1: "final minus initial." The gravitational force on 2 is in the direction of $-\hat{r}$.

$$\vec{F}_{\text{on 2 by 1}} = \boxed{-}G\,\frac{\boxed{m_1\,m_2}}{|\vec{r}|^2}\,\boxed{\hat{r}}$$

Figure 3.20 The direction of the gravitational force is opposite to the direction of the unit vector $\hat{r}$, which points from object 1 to object 2.

$$\vec{F}_{\text{on 2 by 1}} = -G\,\frac{\boxed{m_1\,m_2}}{|\vec{r}|^2}\,\hat{r}$$

Figure 3.21 The magnitude of the gravitational force depends on the masses of both interacting objects.

$$\vec{F}_{\text{on 2 by 1}} = -G\,\frac{m_1\,m_2}{\boxed{|\vec{r}|^2}}\,\hat{r}$$

Figure 3.22 The gravitational force is an "inverse square" law.

The gravitational force law involves a lot of different symbols and may look pretty intimidating at first. Let's take the law apart and look at the individual pieces to try to make sense of the formula.

The relative position vector

The relative position vector $\hat{r}$ extends from the center of object 1 to the center of object 2, as shown in Figure 3.19. (This vector can also be written as $\hat{r}_{2\text{-}1}$; we will use the notation $\hat{r}$, which is more compact, but requires that you remember that object 1 is the initial location and object 2 is the final location.) The unit vector $\hat{r}$ points in the same direction as $\hat{r}$, but has magnitude 1. The magnitude of $\hat{r}$ is the distance between the centers of the two objects.

The direction of the gravitational force

The direction of the gravitational force on object 2 by object 1 is specified by $-\hat{r}$ (Figure 3.19), which is the unit vector $\hat{r}$, in combination with the minus sign, as indicated in Figure 3.20.

? Why is the minus sign necessary?

The minus sign is necessary because the force on object 2 due to object 1 is in the direction opposite to $\hat{r}$ (Figure 3.19).

Mass and magnitude of the gravitational force

As highlighted in Figure 3.21, the gravitational force is proportional to the product of the two masses, $m_1 m_2$. If you double either of these masses, keeping the other one the same, the force will be twice as big.

? If both masses were doubled, how much larger would the force be?

If you double both of the masses, the force will be four times as big. Since $m_2 m_1 = m_1 m_2$, the magnitude of the force exerted on object 1 by object 2 is exactly the same as the magnitude of the force exerted on object 2 by object 1, but the direction is opposite (see the discussion of reciprocity on page 90).

Distance and magnitude of the gravitational force

The gravitational force is an "inverse square" law. As highlighted in Figure 3.22, the square of the center-to-center distance appears in the denominator. This means that the gravitational force depends very strongly on the distance between the objects. For example, if you double the distance between them, the only thing that changes is the denominator, which gets four times bigger (2 squared is 4), so the force is only 1/4 as big as before.

? If you move the masses 10 times farther apart than they were originally, how does the gravitational force change?

The force goes down by a factor of 100. Evidently when two objects are very far apart, the gravitational forces they exert on each other will be vanishingly small: big denominator, small force.

The gravitational constant G

$$G = 6.7 \times 10^{-11}\,\frac{\text{N} \cdot \text{m}^2}{\text{kg}^2}$$

We say that the gravitational constant G is "universal" because it is the same for any pair of interacting masses, no matter how big or small they are, or where they are located in the universe. Because G is universal, it can be measured for any pair of objects and then used with other pairs of objects. As we describe on page 97, Cavendish was the first person to make such a measurement.

"Factoring" the force into a magnitude and a direction

A useful way to think about the gravitational force law is to factor it into magnitude and direction, like this:

- A vector is a magnitude times a direction: $\vec{F}_{grav} = |\vec{F}_{grav}|\hat{F}_{grav}$

  - Magnitude: $|\vec{F}_{grav}| = G\dfrac{m_2\,m_1}{|\vec{r}|^2}$

  - Direction (unit vector): $\hat{F}_{grav} = -\hat{r}$

It is usually simplest to calculate the magnitude and direction separately, then combine them to get the vector force. That way you can focus on one thing at a time rather than getting confused (or intimidated!) by the full complexity of the vector force law.

3.X.5 Suppose that a star exerts a gravitational force of magnitude 4×10^{25} N on a planet. If the mass of the planet were twice as large, what would the magnitude of the gravitational force on the planet be?

3.X.6 A planet exerts a gravitational force of magnitude 8×10^{20} N on a star. If the distance between the star and planet were three times larger, what would the magnitude of this force be?

3.X.7 Masses M and m attract each other with a gravitational force of magnitude F. Mass m is replaced with a mass $3m$, and it is moved four times farther away. Now what is the magnitude of the force?

3.X.8 A 3 kg ball and a 5 kg ball are 2 m apart, center to center. What is the magnitude of the gravitational force that the 3 kg ball exerts on the 5 kg ball? What is the magnitude of the gravitational force that the 5 kg ball exerts on the 3 kg ball?

3.X.9 Measurements show that the Earth's gravitational force on a mass of 1 kg near the Earth's surface is 9.8 N. The radius of the Earth is 6400 km (6.4×10^6 m). From these data determine the mass of the Earth.

Calculating gravitational force

Calculating the gravitational force on an object due to another object requires several steps.

CALCULATING GRAVITATIONAL FORCE

- Calculate $\vec{r} = \vec{r}_{planet} - \vec{r}_{star}$, the position of the center of object 2 relative to the center of object 1.

- Calculate $|\vec{r}|$, the center-to-center distance between the objects.

- Calculate $Gm_1\,m_2/|\vec{r}|^2$, the magnitude of the force.

- Calculate $-\hat{r} = -\vec{r}/|\vec{r}|$, the direction of the force.

- Multiply the magnitude times the direction to get the vector force.

Example: The force on a planet by a star

Figure 3.23 shows a star of mass 4×10^{30} kg located at this moment at position $\langle 2\times10^{11}, 1\times10^{11}, 1.5\times10^{11}\rangle$ m and a planet of mass 3×10^{24} kg located at position $\langle 3\times10^{11}, 3.5\times10^{11}, -0.5\times10^{11}\rangle$ m. (These are typical values for stars and planets. Notice that the mass of the star is much greater than that of the planet.) In Figure 3.23 we show the x, y, and z components of the positions, to be multiplied by 1×10^{11} m. Make sure you understand how the numbers on the diagram correspond to the positions given as vectors.

(a) Calculate the gravitational force exerted on the planet by the star. (b) Calculate the gravitational force exerted on the star by the planet.

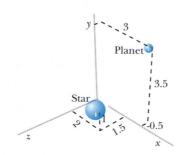

Figure 3.23 A star and a planet interact gravitationally. Distances shown should be multiplied by 1×10^{11} m.

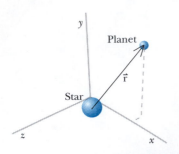

Figure 3.24 The position vector of the planet relative to the star. Initial location: center of star. Final location: center of planet.

This looks like a big force, but it's acting on a planet with a big mass, so it isn't obvious whether this is really a "big" force in terms of what it will do.

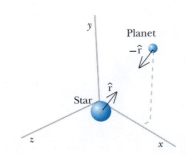

Figure 3.25 Unit vectors.

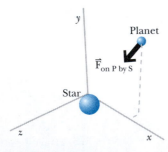

Figure 3.26 Gravitational force on the planet by the star.

? Which object is object 1, and which is object 2?

In this case (a) the planet is object 2 (we want the force on the planet), and the star is object 1.
Relative position vector:

$$\vec{r} = \vec{r}_2 - \vec{r}_1$$

$$\vec{r} = \langle 3\times10^{11}, 3.5\times10^{11}, -0.5\times10^{11} \rangle \text{ m} - \langle 2\times10^{11}, 1\times10^{11}, 1.5\times10^{11} \rangle \text{ m}$$

$$\vec{r} = \langle 1\times10^{11}, 2.5\times10^{11}, -2\times10^{11} \rangle \text{ m}$$

Check: do signs of components make sense? Yes, see Figure 3.24.
Distance:

$$|\vec{r}| = \sqrt{(1\times10^{11})^2 + (2.5\times10^{11})^2 + (-2\times10^{11})^2} \text{ m}$$

$$|\vec{r}| = 3.35\times10^{11} \text{ m}$$

Magnitude of force:

$$|\vec{F}_{\text{on P by S}}| = G\frac{m_1 m_2}{|\vec{r}|^2}$$

$$|\vec{F}_{\text{on P by S}}| = \left(6.7\times10^{-11} \frac{\text{N} \cdot \text{m}^2}{\text{kg}^2}\right)\frac{(3\times10^{24} \text{ kg})(4\times10^{30} \text{ kg})}{(3.35\times10^{11} \text{ m})^2}$$

$$|\vec{F}_{\text{on P by S}}| = 7.16\times10^{21} \text{ N}$$

Direction of force on planet:

$$\hat{F}_{\text{on P by S}} = -\hat{r} = -\frac{\vec{r}}{|\vec{r}|}$$

$$\hat{F}_{\text{on P by S}} = -\frac{\langle 1\times10^{11}, 2.5\times10^{11}, -2\times10^{11} \rangle \text{ m}}{3.35\times10^{11} \text{ m}}$$

$$\hat{F}_{\text{on P by S}} = \langle -0.299, -0.746, 0.597 \rangle$$

Check: units? Okay, unit vector has no units. Direction? Figure 3.25 shows that $\vec{F}_{\text{on P by S}}$ points from planet toward star, which is correct for the direction of the force. Magnitude? $\sqrt{(-0.299)^2 + (-0.746)^2 + (0.597)^2} = 1.001$, okay to 3 significant figures.
Calculate force as a vector:

$$\vec{F}_{\text{on P by S}} = |\vec{F}_{\text{on P by S}}|\hat{F}_{\text{on P by S}}$$

$$\vec{F}_{\text{on P by S}} = (7.16\times10^{21} \text{ N})\langle -0.299, -0.746, 0.597 \rangle$$

$$\vec{F}_{\text{on P by S}} = \langle -2.14\times10^{21}, -5.34\times10^{21}, 4.27\times10^{21} \rangle \text{ N}$$

Check: direction? Figure 3.26 shows force in correct direction. Magnitude? Rough "order of magnitude" check: distance between star and planet is *very* roughly 1×10^{11} m, the mass of the star is *very* roughly 1×10^{30} kg, the mass of the planet is *very* roughly 1×10^{24} kg, and the gravitational constant is *very* roughly 1×10^{-11} N $\cdot$ m²/kg². Therefore we expect the magnitude of the force to be *very* roughly this:

$$\left(1\times10^{-11} \frac{\text{N} \cdot \text{m}^2}{\text{kg}^2}\right)\frac{(1\times10^{24} \text{ kg})(1\times10^{30} \text{ kg})}{(1\times10^{11} \text{ m})^2} = 1\times10^{21} \text{ N}$$

The fact that we get within an order of magnitude of the result 7.16×10^{21} N is evidence that we haven't made any huge mistakes.

(b) The magnitude of the force on the star by the planet will be the same. The only change is in the direction, because $\hat{r}_{\text{S-P}} = -\vec{r}_{\text{P-S}}$

$$\vec{F}_{\text{on P by S}} = \langle 2.14\times10^{21}, 5.34\times10^{21}, -4.27\times10^{21} \rangle \text{ N}$$

Further discussion

Note that the signs of the force components for the force on the star are consistent with Figure 3.27. You might be puzzled that the planet pulls just as hard on the star as the star pulls on the planet. We'll discuss this in more detail later in this chapter.

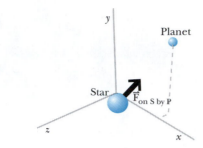

Figure 3.27 The gravitational force exerted on the star by the planet.

3.X.10 At a particular instant, two of the moons of Jupiter, Ganymede and Europa, are aligned as shown in Figure 3.28. Coordinate axes shown in the diagram.

In calculating the gravitational force on Ganymede by Europa:

(a) What is the direction of $\hat{\mathbf{r}}$?

(b) What is the direction of $-\hat{\mathbf{r}}$?

(c) What is the direction of the gravitational force?

In calculating the gravitational force on Europa by Ganymede:

(d) What is the direction of $\hat{\mathbf{r}}$?

(e) What is the direction of $-\hat{\mathbf{r}}$?

(f) What is the direction of the gravitational force?

3.X.11 The mass of the Earth is $6\times10^{24}\,\mathrm{kg}$, and the mass of the Moon is $7\times10^{22}\,\mathrm{kg}$. At a particular instant the Moon is at location $\langle 2.8\times10^8, 0, -2.8\times10^8\rangle$ m, in a coordinate system whose origin is at the center of the Earth.

(a) What is $\vec{\mathbf{r}}_{\mathrm{M\text{-}E}}$ the relative position vector from the Earth to the Moon?

(b) What is $|\vec{\mathbf{r}}_{\mathrm{M\text{-}E}}|$?

(c) What is the unit vector $\hat{\mathbf{r}}_{\mathrm{M\text{-}E}}$?

(d) What is the gravitational force exerted by the Earth on the Moon? Your answer should be a vector.

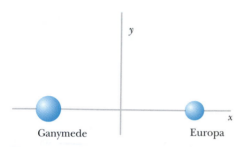

Figure 3.28 Two of Jupiter's moons, aligned along the *x* axis. (Exercise 3.X.10)

3.5 APPROXIMATE GRAVITATIONAL FORCE NEAR THE EARTH'S SURFACE

Earlier we used the expression mg to represent the magnitude of the gravitational force on an object near the Earth's surface. This is an approximation to the actual force, but it is a good one. The magnitude of the gravitational force that the Earth exerts on an object of mass m near the Earth's surface (Figure 3.29) is

$$F_g = G\frac{M_E m}{(R_E + y)^2}$$

where y is the distance of the object above the surface of the Earth, R_E is the radius of the Earth, and M_E is the mass of the Earth.

A spherical object of uniform density can be treated as if all its mass were concentrated at its center (see page 97 for a proof of this). As a result, we can treat the effect of the Earth on the object as though the Earth were a tiny, very dense ball a distance $R_E + y$ away.

The gravitational force exerted by the Earth on an atom at the top of the object is slightly less than the force on an atom at the bottom of the object, because each atom is a slightly different distance from the center of the Earth. How much can this difference really matter in an analysis? Suppose the height of the object is a meter, and the bottom of the object is one meter above the surface. The radius of the Earth is about 6.4×10^6 m. Then

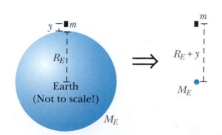

Figure 3.29 Determining the gravitational force by the Earth on an object a height y above the surface. The nearly spherical Earth acts as if all its mass were concentrated at its center.

$$F_{\mathrm{top}} = G\frac{M_E m}{(6400002\text{ m})^2} \quad \text{whereas} \quad F_{\mathrm{bottom}} = G\frac{M_E m}{(6400001\text{ m})^2}$$

For most purposes this difference is not significant. In fact, for all interactions of objects near the surface of the Earth, it makes sense to use the same approximate value, R_E, for the distance from the object to the center of the

Earth. This simplifies calculation of gravitational forces by allowing us to combine all the constants into a single lumped constant, g:

$$F_g \approx \left(G\frac{M_E}{R_E^2} \right) m = gm \text{, so that } g = G\frac{M_E}{R_E^2}$$

? The mass of the Earth is 6×10^{24} kg and the radius of the Earth is 6.4×10^6 m. What is the value of g?

$$g = G\frac{M_E}{R_E^2} = \left(6.7 \times 10^{-11} \ \frac{\text{N} \cdot \text{m}^2}{\text{kg}^2} \right) \frac{(6 \times 10^{24} \text{ kg})}{(6.4 \times 10^6 \text{ m})^2}$$

$$g = 9.8 \ \frac{\text{N}}{\text{kg}}$$

The constant g, called "the magnitude of the gravitational field," has the value $g = +9.8$ newtons/kilogram near the Earth's surface.

Note that g is a *positive* number. We can use this approximate formula for gravitational force in our analysis of any interactions occurring near the surface of the Earth.

The "gravitational field" $\vec{g}$ at a location in space is defined to be the (vector) gravitational force per kg, so a 1 kg mass placed at that location would be $(1 \text{ kg})\vec{g}$, with $\vec{g}$ having units of N/kg. A 2 kg mass would experience a force twice as large at that location.

In general, a mass m will experience a force $m\vec{g}$ of magnitude mg. We will not use the concept of field in this volume, other than using the magnitude of the gravitational field near the Earth's surface ($g = +9.8$ N/kg). The field concept is a major focus of the second volume of this textbook, on electricity and magnetism.

3.X.12 The mass of the Moon is 7×10^{22} kg, and the radius of the Moon is 1.75×10^6 m. (a) What is the value of the constant g at a location on the surface of the Moon? (b) What would be the magnitude of the gravitational force exerted by the Moon on a 70 kg astronaut standing on the Moon? (c) How does this compare to the gravitational force on the same astronaut when standing on the surface of the Earth?

3.X.13 The mass of Mars is 6.4×10^{23} kg and its radius is 3.4×10^6 m. What is the value of the constant g on Mars?

3.X.14 At what height above the surface of the Earth is there a 1% difference between the approximate magnitude of the gravitational field (9.8 N/kg) and the actual magnitude of the gravitational field at that location? That is, at what height y above the Earth's surface is $GM_E/(R_E + y)^2 = 0.99 \, GM_E/R_E^2$?

3.6 PREDICTING MOTION OF GRAVITATIONALLY INTERACTING OBJECTS

The iterative approach we used to predict the motion of a block attached to a spring can be used to predict the motion of objects that interact gravitationally. The only difference is in the force law we use: the gravitational force instead of the spring force. The gravitational force changes direction as the objects move, so at the beginning of every time step we need to recalculate the gravitational force, both magnitude and direction.

At the beginning of this chapter we attempted, unsuccessfully, to predict the motion of the Earth around the Sun. Let's try again, dividing the time interval into several steps.

Example: Earth and Sun

Choose a coordinate system whose origin is at the center of the Sun, and in which the Earth orbits the Sun in the xy plane, as shown in Figure 3.30. The mass of the Sun is 2×10^{30} kg, the mass of the Earth is 6×10^{24} kg. At a particular instant, the Earth is at location $\langle 1.5 \times 10^{11}, 0, 0 \rangle$ m, with velocity $\langle 0, 3 \times 10^4, 0 \rangle$ m/s. What will the location of the Earth be 3 months later?

To get an approximate answer, let's divide the 3 month time interval into three intervals each 1 month long. We'll go through the first two steps of the calculation in detail, displaying results only to 2 significant figures.

1. Momentum principle
System: Earth
Surroundings: Sun (ignore all other planets, stars)

$$\Delta t = (1\,\mathrm{mo})\left(30 \, \frac{\mathrm{days}}{\mathrm{mo}}\right)\left(24 \, \frac{\mathrm{hr}}{\mathrm{day}}\right)\left(60 \, \frac{\mathrm{min}}{\mathrm{hr}}\right)\left(60 \, \frac{\mathrm{sec}}{\mathrm{min}}\right) = 2.6 \times 10^6 \mathrm{s}$$

$$\vec{p}_i = (6 \times 10^{24} \, \mathrm{kg})(\langle 0, 3 \times 10^4, 0 \rangle \, \mathrm{m/s}) = \langle 0, 1.8 \times 10^{29}, 0 \rangle \, \mathrm{kg \cdot m/s}$$

I First time step:

$$\vec{r} = \vec{r}_E - \vec{r}_S = \langle 1.5 \times 10^{11}, 0, 0 \rangle \, \mathrm{m} - \langle 0, 0, 0 \rangle \, \mathrm{m} = \langle 1.5 \times 10^{11}, 0, 0 \rangle \, \mathrm{m}$$

$$|\vec{r}| = 1.5 \times 10^{11} \, \mathrm{m}$$

$$|\vec{F}| = \frac{G m_E m_S}{|\vec{r}|^2} = \frac{\left(6.7 \times 10^{-11} \, \frac{\mathrm{N \cdot m^2}}{\mathrm{kg^2}}\right)(6 \times 10^{24} \, \mathrm{kg})(2 \times 10^{30} \, \mathrm{kg})}{(1.5 \times 10^{11} \, \mathrm{m})^2}$$

$$|\vec{F}| = 3.6 \times 10^{22} \, \mathrm{N}$$

$$\hat{F} = -\hat{r} = -\frac{\vec{r}}{|\vec{r}|} = \frac{-\langle 1.5 \times 10^{11}, 0, 0 \rangle \, \mathrm{m}}{1.5 \times 10^{11} \, \mathrm{m}}$$

$$\hat{F} = \langle -1, 0, 0 \rangle$$

$$\vec{F}_{\text{on E by S}} = |\vec{F}|\hat{F} = (-3.6 \times 10^{22} \, \mathrm{N})\langle -1, 0, 0 \rangle$$

$$\vec{F}_{\text{on E by S}} = \langle 3.6 \times 10^{22}, 0, 0 \rangle \, \mathrm{N}$$

$$\vec{p}_f = \vec{p}_i + \vec{F}_{\text{net}}\Delta t$$

$$\vec{p}_f = \left(\langle 0, 1.8 \times 10^{29}, 0 \rangle \frac{\mathrm{kg \cdot m}}{\mathrm{s}}\right) + \langle -3.6 \times 10^{22}, 0, 0 \rangle \, \mathrm{N}(2.6 \times 10^6 \mathrm{s})$$

$$\vec{p}_f = \langle -2.2 \times 10^{28}, 1.8 \times 10^{29}, 0 \rangle \, \mathrm{kg \cdot m/s}$$

2. Position update

We use the final momentum to find an approximate value for the average velocity of the Earth; because the force is changing this is not necessarily a worse approximation than the arithmetic average would be. Details are not shown (you can work this out yourself). The results of the first iteration are shown in Figure 3.31.

$$\vec{r}_f = \langle 1.1 \times 10^{11}, 7.8 \times 10^{10}, 0 \rangle \, \mathrm{m}$$

II Second time step:

1. Momentum principle
Assume that the Sun's motion was negligible

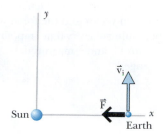

Figure 3.30 Initial location and velocity of the Earth. The origin of the coordinate system is at the center of the Sun. Sizes of the Sun and Earth are exaggerated.

$\vec{p} \approx m\vec{v}$ because $v \ll c$

The initial location and the initial momentum of the objects are called the "initial conditions" of the system.

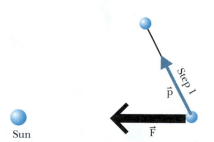

Figure 3.31 Step 1 of 3-step iterative calculation. Gravitational force on Earth at beginning of step and final momentum of Earth are shown.

It is necessary to recalculate the relative position vector and the force vector at the beginning of every time step, since their directions and magnitudes may have changed.

$$\vec{r} = \vec{r}_E - \vec{r}_S = \langle 1.1\times10^{11}, 7.8\times10^{10}, 0 \rangle \text{ m}$$

$$|\vec{r}| = \sqrt{(1.1\times10^{11}\text{ m})^2 + (7.8\times10^{10}\text{ m})^2} = 1.3\times10^{11}\text{ m}$$

$$|\vec{F}| = \frac{Gm_Em_S}{|\vec{r}|^2} = \frac{\left(6.7\times10^{-11}\dfrac{\text{N}\cdot\text{m}^2}{\text{kg}^2}\right)(6\times10^{24}\text{ kg})(2\times10^{30}\text{ kg})}{(1.3\times10^{11}\text{ m})^2}$$

$$|\vec{F}| = 4.4\times10^{22}\text{ N}$$

Now the gravitational force on the Earth has a nonzero y component.

$$\hat{F} = -\hat{r} = -\frac{\vec{r}}{|\vec{r}|} = \frac{-\langle 1.1\times10^{11}, 7.8\times10^{10}, 0 \rangle \text{ m}}{1.3\times10^{11}\text{ m}}$$

$$\hat{F} = \langle -0.82, -0.58, 0 \rangle$$

$$\vec{F}_{\text{on E by S}} = |\vec{F}|\hat{F} = (4.4\times10^{22}\text{ N})\langle -0.82, -0.58, 0 \rangle$$

$$\vec{F}_{\text{on E by S}} = \langle -3.6\times10^{22}, -2.6\times10^{22}, 0 \rangle \text{ N}$$

$$\vec{p}_f = \vec{p}_i + \vec{F}_{\text{net}}\Delta t$$

$$\vec{p}_f = \left(\langle 0, 1.8\times10^{29}, 0 \rangle \frac{\text{kg}\cdot\text{m}}{\text{s}} \right) + (\langle -3.6\times10^{22}, -2.6\times10^{22}, 0 \rangle \text{ N})(2.6\times10^6\text{s})$$

$$\vec{p}_f = \langle -1.9\times10^{29}, 1.1\times10^{29}, 0 \rangle \text{ kg}\cdot\text{m/s}$$

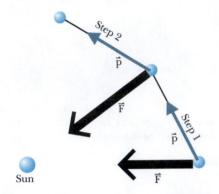

Figure 3.32 Results of first two steps of the iterative calculation. The gravitational force on the Earth is different in the two steps.

2. Position update

We again use the final momentum to find an approximate value for the average velocity of the Earth. Details are not shown (you can work this out yourself). The results from the second time step are shown in Figure 3.32.

$$\vec{r}_f = \langle 2.0\times10^{10}, 7.8\times10^{10}, 0 \rangle \text{ m}$$

III: Third time step:

The results of the third iteration are shown in Figure 3.33

Check: Path reasonable? Yes. Although this was still a rough calculation, it does predict that the Earth will orbit around the Sun. If we had used smaller time steps, our predicted path would have looked more like a circle.

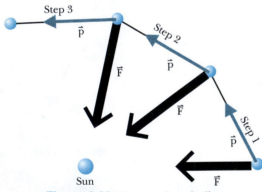

Figure 3.33 The results of all three steps in the iterative calculation of the motion of the Earth under the influence of the Sun.

Telling a computer what to do

In the next chapter we will see that in a few special cases it is possible to derive an analytical (algebraic) solution for the motion of two objects interacting gravitationally. However, in most situations (for example, a situation involving three or more interacting objects), this is actually not possible. Therefore, this iterative approach for predicting motion can be very valuable. Doing iterative calculations with very small time steps is practical only if one uses a computer. This is how a computer program can be structured in order to apply the Momentum Principle repetitively to predict the future:

STRUCTURING ITERATIVE CALCULATIONS ON A COMPUTER

- Define the values of constants such as G to use in the program.
- Specify the masses, initial positions, and initial momenta of the interacting objects.
- Specify an appropriate value for Δt, small enough that the objects don't move very far during one update.
- Create a "loop" structure for repetitive calculations:
  - Calculate the (vector) forces acting on each object.
  - Update the momentum of each object: $\vec{p}_f = \vec{p}_i + \vec{F}_{net}\Delta t$.
  - Update the positions: $\vec{r}_f = \vec{r}_i + \vec{v}_{avg}\Delta t$, where $\vec{v}_{avg} \approx \vec{p}_f / m$.
  - Repeat.

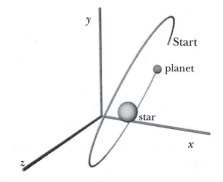

Figure 3.34 Display created by a computer program to calculate the motion of a planet around a star. The star and planet are shown much larger than they really are.

Figure 3.34 shows a display created by such a computer program. The details of the actual program statements depend on what programming language or system you use, and your instructor will provide you with the details of the particular tool you will use.

Why not just use calculus?

You might wonder why we don't simply use calculus to predict the motion of planets and stars. In fact, we actually are using calculus. Step by step, we add up a large number of small increments of the momentum of an object to calculate a large change in its momentum over a long time, and this corresponds to an approximate evaluation of an integral:

$$\Delta\vec{p}_1 + \Delta\vec{p}_2 + \dots \approx \int_i^f d\vec{p} = \int_i^f \vec{F}_{net}\, dt$$

The integral sign is a distorted "S" meaning "sum" of an infinite number of infinitesimal quantities, from the initial time i to the final time f.

A more interesting answer is that the motion of most physical systems actually *cannot* be predicted using calculus in any way other than by this step-by-step, calculus-based approach. In a few special cases calculus does give a general result without going through this procedure. For example, an object subjected to a constant force has a constant rate of change of momentum and velocity, and calculus can be used to obtain a prediction for the position as a function of time, as we saw in the previous chapter in the example of a ball thrown through a vacuum. An analytical solution can be derived for the motion of a mass attached to a very low mass spring, in the absence of air resistance, as we will see in the next chapter. The elliptical orbits of two stars around each other can be predicted mathematically without an iterative approach, although the math is quite challenging.

However, the general motion of three stars around each other has never been successfully analyzed in this way. The basic problem is that it is usually relatively easy to take the derivative of a known function, but it is often impossible to determine in algebraic form the integral of a known function, which is what would be involved in long-term prediction (adding up a large number of small momentum increments due to known forces).

In contrast, a step-by-step procedure of the kind we carried out for the Earth going around the Sun can easily be extended to three or more bodies. It is also possible to include the effects of various kinds of friction and damping in an iterative calculation. This is why we study the step-by-step prediction method in detail, because it is a powerful technique of increasing importance in modern science and engineering, thanks to the availability of powerful computers to do the repetitive work for us.

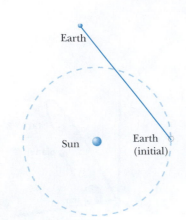

Figure 3.35 A time step of 3 months is too large in this situation.

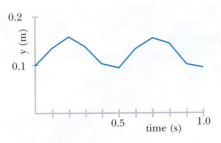

Figure 3.36 A 0.1 second time step gave a rough idea of the motion of the block-spring system.

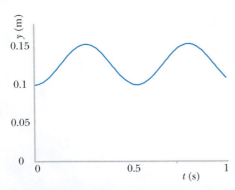

Figure 3.37 A computer calculation using a 0.01 second time step gave a more accurate prediction of the motion of the block-spring system.

3.7 CHOOSING A TIME STEP

In any calculation based on the Momentum Principle, we need to choose a value for the time step Δt short enough that the net force on the object of interest is approximately constant during this time interval. How do we do this? There is not a simple formula for choosing Δt, but there are some useful guidelines.

Distance dependent forces

In many cases the force on the system depends on the distance of the object from some other object. In such a case we need to pick a Δt small enough that this distance won't change very much during this interval. How do we do this?

(a) If the initial velocity of the system is not zero, use $|\vec{v}_i \Delta t|$ to estimate the displacement of the system during one time step. Make sure this displacement is small compared to the relevant distance.

For example, in our initial attempt to predict the motion of the Earth around the Sun on page 73, we used a time step of 3 months, or 7.8×10^6 s. Given an initial speed of 3×10^4 m/s, the Earth's estimated displacement during 3 months would be $(3 \times 10^4 \text{ m/s})(7.8 \times 10^6 \text{s}) = 2.3 \times 10^{11}$ m. This displacement is actually larger than the initial distance between the Earth and the Sun (1.5×10^{11} m), so we should expect serious problems if we use a time step this large, as seen in Figure 3.35.

(b) In a case in which the initial velocity is zero, as in the block and spring example on page 76, we may need to do a quick guess and check estimate, by completing one iteration.

In the block and spring example, the initial stretch of the spring was 10 cm. We guessed that a 0.1 second time step would be reasonable, and after calculating v_{yf} and y_f, we found that the block had moved from 0.1 m above the floor to 0.135 m above the floor: a displacement of 3.5 cm, and a 35% change in the length of the spring. For a rough estimate this was acceptable, and did give us a general idea about the periodic motion of the block and spring, as shown in Figure 3.36. To make more accurate predictions, it was necessary to use a significantly shorter time step, as shown in Figure 3.37. Because the "period" of the system's motion (the time between maximum displacements) was about 0.5 seconds, we could have missed the periodicity of the motion altogether if we had chosen a time step of 0.5 seconds or larger!

Speed dependent forces

Some forces, like air resistance, depend on an object's speed instead of its position. In such a case, we need to make sure that the object's speed does not change very much during the time interval Δt.

Periodic motion

For a system in periodic motion, such as a mass and spring, or an orbiting body, if we know the approximate period, we need to choose Δt to be significantly smaller than the period (the time for the motion to repeat). For an Earth-Sun orbit, we want Δt to be much smaller than a year. For a mass-spring system with a period of 2 seconds, Δt should be much less than 2 seconds.

Time step size in a computer calculation

When doing a calculation by hand, there is a tradeoff between accuracy and time required to do many iterations (the smaller the time step, the more calculations must be done). Since computers are quite fast, it is reasonable to use much smaller time steps in a computer calculation than one would use by hand. However, even a fast computer can take a very long time to do calculations which use an unnecessarily tiny time step—it would not be reasonable to use a time step of 1×10^{-20} s in a prediction of the Earth's motion around the Sun.

A standard method of checking the accuracy of a computer calculation is to decrease the time step and repeat the calculation. If the results do not change significantly, the original time step was adequately small.

3.X.15 A block is hung from a spring, and the spring is stretched 5 cm before the block is released from rest. You need to choose a time step to use in predicting the block's motion. Which of the following would be a reasonable distance for the block to move in the first time step, doing an iterative calculation *by hand?*
(a) 10 cm (b) 5 cm (c) 0.5 cm (d) 0.005 cm

3.X.16 A comet passes near the Sun. When the comet is closest to the Sun, it is 9×10^{10} m from the Sun. You need to choose a time step to use in predicting the comet's motion. Which of the following would be a reasonable distance for the comet to move in one time step, doing an iterative calculation *by hand?*
(a) 1×10^{2} m (b) 1×10^{10} m (c) 1×10^{11} m (d) 1×10^{9} m

3.X.17 Jupiter goes around the Sun in 4333 Earth days. Which of the following would be a reasonable value to try for Δt in a *computer* calculation of the orbit?
(a) 1 day (b) 4333 days (c) 0.01 second (d) 800 days

3.X.18 In a lab experiment you observe that a pendulum swings with a "period" (time for one round trip) of 2 s. In an iterative calculation of the motion, which of the following would NOT be a reasonable choice for Δt?
(a) 1 second (b) 0.1 second (c) 0.05 second (d) 0.01 second

3.8 THE ELECTRIC FORCE LAW: COULOMB'S LAW

There are electric interactions between "charged" particles such as the protons and electrons found in atoms. It is observed that two protons repel each other, as do two electrons, while a proton and an electron attract each other (Figure 3.38; where protons are said to have "positive electric charge" and electrons have "negative electric charge").

The force corresponding to this electric interaction is similar to the gravitational force law, and it is known as "Coulomb's law" to honor the French scientist who established this law in the late 1700's, during the same period when Cavendish measured the gravitational constant G. See Figure 3.39.

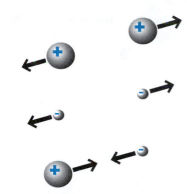

Figure 3.38 Protons repel each other; electrons repel each other; protons and electrons attract each other.

THE ELECTRIC FORCE LAW (COULOMB'S LAW)

$$\vec{F}_{elec\ on\ 2\ by\ 1} = \frac{1}{4\pi\varepsilon_0}\frac{q_1 q_2}{|\vec{r}|^2}\hat{r}$$

$\vec{r} = \vec{r}_2 - \vec{r}_1$ is the position of 2 relative to 1

$\dfrac{1}{4\pi\varepsilon_0}$ is a universal constant: $\dfrac{1}{4\pi\varepsilon_0} = 9\times10^{9}\ \dfrac{\text{N}\cdot\text{m}^2}{\text{C}^2}$

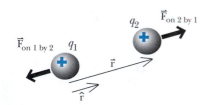

Figure 3.39 The electric force exerted on object 2 by object 1. (The force exerted on object 2 by object 1 has the same magnitude but opposite direction.)

Both the gravitational and electric forces are proportional to the inverse square of the center-to-center distance ("inverse square" laws). The universal electric constant (read as "one over four pi epsilon-zero") is very much larger than the gravitational constant, reflecting the fact that electric interactions are intrinsically much stronger than gravitational interactions. For example, consider a heavy weight that hangs from a thin metal wire. The small number of atoms interacting electrically in the wire have as big an effect on the hanging weight as a very much larger number of atoms in the huge Earth, interacting gravitationally.

The charges q_1 and q_2 must be measured in SI units called "coulombs, and abbreviated "C"." The proton has a charge of $+1.6 \times 10^{-19}$ C and the electron has a charge of -1.6×10^{-19} C.

? The electric force law looks very much like the gravitational force law. What aspects of these force laws are the same?

Both laws involve the relative position vector between two objects, and both involve the unit vector associated with the relative position vector. Like the gravitational force, the electric force is an inverse square force, and is proportional to one over the square of the distance between the objects (Figure 3.40).

$$\vec{F}_{electric} = \frac{1}{4\pi\varepsilon_0} \frac{q_1 \, q_2}{|\vec{r}|^2} \, \hat{r}$$

Figure 3.40 Like the gravitational force, the magnitude of the electric force is inversely proportional to the square of the distance between charged objects, and its direction depends on the unit vector $\hat{r}$.

Both laws also involve the product of a property of each object: mass or charge. In both cases, there is a scalar multiplicative constant: G or $1/(4\pi\varepsilon_0)$. (Figure 3.41).

? Study the three cases in Figure 3.38. Why is no minus sign needed in the force law, unlike the case with the gravitational force law?

$$\vec{F}_{electric} = \left(\frac{1}{4\pi\varepsilon_0}\right) \frac{q_1 \, q_2}{|\vec{r}|^2} \, \hat{r}$$

Figure 3.41 Like the gravitational force, the electric force involves the product of a property of each object (charge), and a scalar constant ($1/(4\pi\varepsilon_0)$).

The difference is that two positively charged particles such as two protons repel each other, whereas masses are always positive but the gravitational force is attractive. Two negatively charged particles such as electrons also attract each other and minus times minus gives plus. Only if the two particles have opposite charges will they attract each other, and then the factor $q_1 q_2$ contributes the necessary minus sign, just as in the gravitational force law.

Interatomic forces

When two objects touch each other they exert forces on each other. At the microscopic level, these contact forces are due to electric interactions between the protons and electrons in one object and the protons and electrons in the other object. In a later chapter we will examine interatomic forces in more detail.

 proton

3.X.19 A proton and an electron are separated by 1×10^{-10} m, the radius of a typical atom. Following a procedure similar to the one you used to calculate the magnitude of a gravitational force, calculate the magnitude of the electric force that the proton exerts on the electron, and the magnitude of the electric force that the electron exerts on the proton.

electron

Figure 3.42 A proton and an electron.

3.X.20 Figure 3.42 shows a proton and an electron. What is the direction of the electric force on the electron by the proton? What is the direction of the electric force on the proton by the electron? How do the magnitudes of these forces compare?

3.X.21 Figure 3.43 shows two electrons. What is the direction of the electric force on electron A due to electron B? What is the direction of the electric force on electron B due to electron A?

electron A

3.9 RECIPROCITY

electron B

Figure 3.43 Two electrons.

An important aspect of the gravitational and electric interactions (including the electric forces of atoms in contact with each other) is that the force that object 1 exerts on object 2 is equal and opposite to the force that object 2 exerts on object 1 (Figure 3.44). That the magnitudes must be equal is clear from the algebraic form of the laws, because $m_1 m_2 = m_2 m_1$ and $q_1 q_2 = q_2 q_1$. The directions of the forces are along the line connecting the centers, and in opposite directions.

This property is called "reciprocity" or "Newton's third law of motion":

RECIPROCITY

$$\vec{F}_{\text{on 1 by 2}} = -\vec{F}_{\text{on 2 by 1}} \quad \text{(gravitational and electric forces)}$$

The force that the Earth exerts on the massive Sun is just as big as the force that the Sun exerts on the Earth, so in the same time interval the momentum changes are equal in magnitude and opposite in direction ("equal and opposite" for short):

$$\Delta \vec{p}_1 = \vec{F}_{\text{on 1 by 2}} \Delta t = -\Delta \vec{p}_2$$

However, the velocity change $\Delta \vec{v} = \Delta \vec{p}/m$ of the Sun is extremely small compared to the velocity change of the Earth, because the mass of the Sun is enormous in comparison with the mass of the Earth. The mass of our Sun, which is a rather ordinary star, is 2×10^{30} kg. This is enormous compared to the mass of the Earth (6×10^{24} kg) or even to the mass of the largest planet in our Solar System, Jupiter (2×10^{27} kg). Nevertheless, very accurate measurements of small velocity changes of distant stars have been used to infer the presence of unseen planets orbiting those stars.

Magnetic forces do not have the property of reciprocity. Two electrically charged particles that are both moving can interact magnetically as well as electrically, and the magnetic forces that these two particles exert on each other need not be equal in magnitude nor opposite in direction. Reciprocity applies to gravitational and electric forces, but not in general to magnetic forces acting between individual moving charges.

Why reciprocity?

The algebraic forms of the gravitational and electric force laws indicate that reciprocity should hold. A diagram may be helpful in explaining *why* the forces behave this way.

Consider two small objects, a 2 gram object made up of two 1 gram balls, and a 3 gram object made up of three 1 gram balls (Figure 3.45). The distance between centers of the two objects is large compared to the size of either object, so the distances between pairs of 1 gram balls are about the same for all pairs.

You can see that in the 2 gram object at the top of Figure 3.45 each ball has three gravitational forces exerted on it by the distant balls in the 3 gram object.

? How many forces act on the 2 gram object?

There is a net force of $2 \times 3 = 6$ times the force associated with one pair of balls.

? Similarly, consider the forces acting on the 3 gram object at the bottom of Figure 3.45. How many forces act on the 3 gram object?

There is again a net force of $3 \times 2 = 6$ times the force associated with one pair of balls.

The effect is that the force exerted by the 3 gram object on the 2 gram object has the same magnitude as the force exerted by the 2 gram object on the 3 gram object. The same reciprocity holds, for the same reasons, for the electric forces between a lithium nucleus containing 3 protons and a helium nucleus containing 2 protons.

3.X.22 A 60 kg person stands on the Earth's surface. (a) What is the approximate magnitude of the gravitational force on the person by the Earth? (b) What is the magnitude of the gravitational force on the Earth by the person?

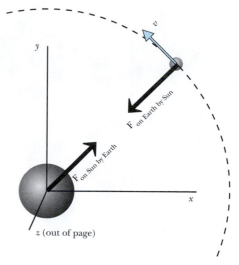

Figure 3.44 The Sun and the Earth exert equal and opposite forces on each other.

2 grams

3 grams

Figure 3.45 Reciprocity: there are 6 forces acting on the 3 gram object and 6 forces acting on the 2 gram object.

3.X.23 A moving electron passes near the nucleus of a gold atom, which contains 79 protons and 118 neutrons. At a particular moment the electron is a distance of 3×10^{-9} m from the gold nucleus. (a) What is the magnitude of the force exerted by the gold nucleus on the electron? (b) What is the magnitude of the force exerted by the electron on the gold nucleus?

3.10 NEWTON AND EINSTEIN

Newton devised a particular explanatory scheme in which the analysis of motion is divided into two distinct parts:

1) Quantify the interaction in terms of a concept called "force." Specific examples are Newton's law of gravitation and Coulomb's law.

2) Quantify the change of motion in terms of the change in a quantity called "momentum." The change in the momentum is equal to the force times Δt.

This scheme, called the "Newtonian synthesis," has turned out to be extraordinarily successful in explaining a huge variety of diverse physical phenomena, from the fall of an apple to the orbiting of the Moon. Yet we have no way of asking whether the Universe "really" works this way. It seems unlikely that the Universe actually uses the human concepts of "force" and "momentum" in the unfolding motion of an apple or the Moon.

We refer to the "Newtonian synthesis" both to identify and honor the particular, highly successful analysis scheme introduced by Newton, but also to remind ourselves that this is not the only possible way to view and analyze the Universe.

Einstein's alternative view

Newton stated his gravitational force law but could give no explanation for it. He was content with showing that it correctly predicted the motion of the planets, and this was a huge advance, the real beginning of modern science.

Einstein made a further huge advance by giving a deeper explanation for gravity, as a part of his general theory of relativity. He realized that the massive Sun bends space and time (!) in such a way as to make the planets move the way they do. The equations in Einstein's general theory of relativity make it possible to calculate the curvature of space and time due to massive objects, and to predict how other objects will move in this altered space and time.

Moreover, Einstein's theory of general relativity accurately predicts some tiny effects that Newton's gravitational law does not, such as the slight bending of light as it passes near the Sun. General relativity also explains some bizarre large-scale phenomena such as black holes and the observed expansion of the space between the galaxies.

Einstein's earlier special theory of relativity established that nothing, not even information, can travel faster than light. Because Newton's gravitational force formula depends only on the distance between objects, not on the time, something's wrong with Newton's formula, since this implies that if an object were suddenly yanked away, its force on another object would vanish instantaneously, thus giving (in principle) a way to send information from one place to another instantaneously. Einstein's theory of general relativity doesn't have this problem.

Since the equations of general relativity are very difficult to work with, and Newton's gravitational law works very well for most purposes, in this textbook we will use Newton's approach to gravity. But you should be aware that for the most precise calculations one must use the theory of general relativity. For example, the highly accurate atomic clocks in the satellites that make up the Global Positioning System (GPS) have to be continually corrected using Einstein's theory of general relativity. Otherwise GPS positions would be wrong by several kilometers after just one day of operation!

3.11 PREDICTING THE FUTURE OF COMPLEX SYSTEMS

We can use the Newtonian method and numerical integration to try to predict the future of a group of objects that interact with each other mainly gravitationally, such as the Solar System consisting of the Sun, planets, moons, asteroids, comets, etc. There are other, non-gravitational interactions present. Radiation pressure from sunlight makes a comet's tail sweep away from the sun. Streams of charged particles (the "solar wind") from the Sun hit the Earth and contribute to the Northern and Southern Lights (auroras). But the main interactions within the Solar System are gravitational.

The Solar System as a group is orbiting around the center of our Milky Way galaxy, and our galaxy is interacting gravitationally with other galaxies in a local cluster of galaxies (Figure 3.46), but changes in these speeds and directions take place very slowly compared to those within the solar system. To a very good approximation we can neglect these interactions while studying the inner workings of the Solar System itself.

We have modeled the motion of planets and stars using the Momentum Principle and the universal law of gravitation. We can use the same approach to model the gravitational interactions of more complex systems involving three, four, or any number of gravitationally interacting objects. All that is necessary is to include the forces associated with all pairs of objects.

In principle we could use these techniques to predict the future of our entire Solar System. Of course our prediction would be only approximate, because we take finite time steps which introduce numerical errors, and we are neglecting various small interactions. We would need to investigate whether the prediction of our model is a reasonable approximation to the real motion of our real Solar System.

Our Solar System contains a huge number of objects. Most of the eight planets have moons, and Jupiter and Saturn have many moons. There are thousands of asteroids and comets, not to mention an uncounted number of tiny specks of dust. This raises a practical limitation: the fastest computers would take a very long time to carry out even one time step for all the pieces of the Solar System.

The three-body problem

As we have seen, in order to carry out a numerical integration we need to know the position and velocity of each object at some time t, from which we can calculate the forces on each object at that time. This enables us to calculate the position and momentum for each object at a slightly later time $t + \Delta t$. We need the net force on an object, which we get from vector addition of all the pairwise forces calculated from Newton's universal law of gravitation.

Suppose that there are three gravitating objects, as shown in Figure 3.47. The force on object m_2, which is acted upon gravitationally by objects m_1 and m_3, is a natural extension of what we did earlier for the analysis of a star and planet. We simply use the superposition principle to include the interaction with m_3 as well as the interaction with m_1:

$$\vec{F}_{\text{net on } m_2} = -G\frac{m_2 m_1}{\left|\vec{r}_{2\text{-}1}\right|^2}\hat{r}_{2\text{-}1} - G\frac{m_2 m_3}{\left|\vec{r}_{2\text{-}3}\right|^2}\hat{r}_{2\text{-}3}$$

We use this force to update the momentum of object m_2. Similar computations apply to the other masses, m_1 and m_3. These calculations are tedious, but let the computer do them! We can carry out a numerical integration for each mass: given the current locations of all the masses we can calculate the vector net force exerted on each mass at a time t, and then we can calculate the new values of momentum and position at a slightly later time $t + \Delta t$.

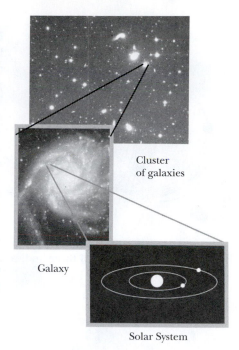

Cluster
of galaxies

Galaxy

Solar System

Figure 3.46 Our Solar System orbits around the center of our galaxy, which interacts with other galaxies.

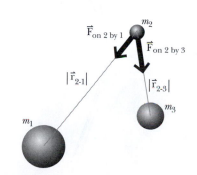

$\vec{F}_{\text{on 2 by 1}}$ m_2

$\vec{F}_{\text{on 2 by 3}}$

$|\vec{r}_{2\text{-}1}|$ $|\vec{r}_{2\text{-}3}|$

m_3

m_1

Figure 3.47 In the "three body problem" each of the three objects interacts with two other objects.

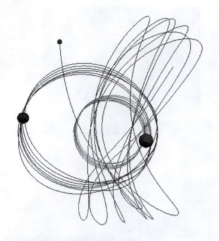

Figure 3.48 One example of three bodies interacting gravitationally.

Figure 3.49 A low mass object orbiting two massive objects.

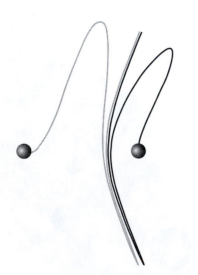

Figure 3.50 Two different orbits (one gray, the other black), starting from rest but from slightly different positions.

While there exist analytical (non-numerical) solutions for two-body gravitational motion, except for some very special cases the "three-body problem" has not been solved analytically. However, in a numerical integration, the computer doesn't mind that there are lots of quantities and a lot of repetitive calculations to be done. You yourself can solve the three-body problem numerically, by adding additional calculations to your two-body computations (see the three-body homework problem 3.P.45 on page 101).

For a single object orbiting a very massive object, or for two objects orbiting each other (with zero total momentum), the possible trajectories are only a circle, ellipse, parabola, hyperbola, or straight line. The motion of a three-body system can be vastly more complex and diverse. Figure 3.48 shows the numerical integration of a complicated three-body trajectory for one particular set of initial conditions. Adding just one more object opens up a vast range of complex behaviors. Imagine how complex the motion of a galaxy can be!

Sensitivity to initial conditions

For a two-body system, slight changes in the initial conditions make only slight changes in the orbit, such as changing a circle into an ellipse, or an ellipse into a slightly different ellipse, but you never get anything other than a simple trajectory.

The situation is very different in systems with three or more interacting objects. Consider a low-mass object orbiting two massive objects which we imagine somehow to be nailed down so that they cannot move (Figure 3.49). This could also represent two positive charges that are fixed in position, with a low-mass negative charge orbiting around them.

In Figure 3.50 are two very different orbits (one shown in black, the other in gray), starting from only slightly different initial conditions. In both of these cases, the low-mass object was released from rest, but at slightly different initial locations. The trajectories are wildly different! (The mass of the object on the left is twice the mass of the object on the right in this computation.)

This sensitivity to initial conditions generally becomes more extreme as you add more objects, and the Solar System contains lots of objects, big and small. Also, we can anticipate that if small errors in specifying the initial conditions can have large effects, so too it is likely that failing to take into account the tiny force exerted by a small asteroid might make a big difference after a long integration time.

The Solar System is actually fairly predictable, because there is one giant mass (the Sun), and the other, much smaller masses are very far apart. This is unlike the example shown here, which deliberately emphasizes the sensitivity observed when there are large masses near each other.

3.12 DETERMINISM

If you know the net force on an object as a function of its position and momentum (or velocity), you can predict the future motion of the object by simple step-by-step calculations based on the Momentum Principle. Newton's demonstration of the power of this approach induced many seventeenth and eighteenth century philosophers and scientists to adopt the view already proposed by Descartes, Boyle, and others, that the Universe is a giant clockwork device, whose future is completely determined by the present positions and momenta of all the macroscopic and microscopic objects in it. Just turn the crank, and predict the future! This point of view is called "determinism," and taken to its extreme it raises the question of whether humans actually have any free will, or whether all of our actions are predetermined. Scientific and technological advances in the twentieth century, however, have led us to see that although the Newtonian approach can be used to predict the long-term future of a simple system or the short-term future of a complicated system, there are both practical and theoretical limitations on predictions in some systems.

Practical limitations

One reason we may be unable to predict the long-term future of even a simple system is a practical limitation in our ability to measure its initial conditions with sufficient accuracy. Over time, even small inaccuracies in initial conditions can lead to large cumulative errors in calculations. This is a practical limitation rather than a theoretical one, because in principle if we could measure initial conditions more accurately, we could perform more accurate calculations.

Another practical limitation is our inability to account for all interactions in our model. Every object in the Universe interacts with every other object. In constructing simplified models, we ignore interactions whose magnitude is extremely small. However, over time, even very small interactions can lead to significant effects. Even the tiny "radiation pressure" exerted by sunlight has been shown to affect noticeably the motion of an asteroid over a long time. With larger and faster computers, we can include more and more interactions in our models, but our models can never completely reflect the complexity of the real world.

A branch of current research that focuses on how the detailed interactions of a large number of atoms or molecules lead to the bulk properties of matter is called "molecular dynamics." For example, one way to test our understanding of the nature of the interactions of water molecules is to create a computational model in which the forces between molecules and the initial state of a large number of molecules—several million—are described, and then to see if running the model produces a virtual fluid that actually behaves like water. However, even with an accurate force law for the (electric) interactions between atoms or molecules, it is not feasible to compute and track the motions of the 10^{25} molecules in a glass of water. Significant efforts are underway to build faster and faster computers, and to develop more efficient computer algorithms, to permit realistic numerical integration of more complex systems.

Chaos

A second kind of limitation occurs in systems that display an extreme sensitivity to initial conditions. We saw something like this in the example of the three-body problem in gravitating systems—small changes in initial conditions produced large changes in behavior. In recent years scientists have discovered systems in which a change in the initial conditions, no matter how small (infinitesimal), can lead to complete loss of predictability: the difference in the two possible future motions of the system diverges exponentially with time. Such systems are called "chaotic." It is thought that over a long time period the weather may be literally unpredictable in this sense.

An interesting popular science book about this new field of research is *Chaos: Making a New Science*, by James Gleick (Penguin 1988). The issue of small changes in initial conditions is a perennial concern of time travellers in science fiction stories. Perhaps the most famous such story is Ray Bradbury's "The Sound of Thunder," in which a time traveller steps on a butterfly during the Jurassic era, and returns to an eerily changed present time.

Breakdown of Newton's laws

There are other situations of high interest where we cannot usefully apply Newton's laws of motion, because these classical laws do not adequately describe the behavior of physical systems. To model systems composed of very small particles such as protons and electrons, quarks (the constituents of protons and neutrons), and photons, it is necessary to use the laws of quantum mechanics. To model in detail the gravitational interactions between massive objects, it is necessary to apply the principles of general relativity. Given the principles of quantum mechanics and general relativity, one might assume that it would be possible to follow a procedure similar to the one we have followed in this chapter, and predict in detail the future of these systems. However, there appear to be more fundamental limitations on what we can know about the future.

Probability and uncertainty

Our understanding of the atomic world of quantum mechanics suggests that there are fundamental limits to our ability to predict the future, because at the atomic and subatomic levels the Universe itself is non-deterministic. We cannot know exactly what will happen at a given time, but only the probability that certain events will occur within a given time frame.

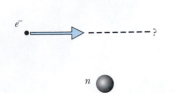

Figure 3.51 The electron may continue to travel in a straight line, but we cannot be certain that it will, because we do not know when the neutron will decay.

A simple example may make this clear. A free neutron (one not bound into a nucleus) is unstable and eventually decays into a positively charged proton, a negatively charged electron, and an electrically neutral anti-neutrino:

$$n \rightarrow p^+ + e^- + \bar{\nu}$$

The average lifetime of the free neutron is about 15 minutes. Some neutrons survive longer than this, and some last a shorter amount of time. All of our experiments and all of our theory are consistent with the notion that *it is not possible to predict when a particular neutron will decay*. There is only a (known) probability that the neutron will decay in the next microsecond, or the next, or the next.

If this is really how the Universe works, then there is an irreducible lack of predictability and determinism of the Universe itself. As far as our own predictions using the Momentum Principle are concerned, consider the following simple scenario. An electron is traveling with constant velocity through nearly empty space, but there is a free neutron in the vicinity (Figure 3.51).

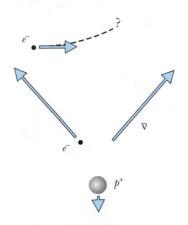

Figure 3.52 One possible path of the electron if the neutron does decay.

We might predict that the electron will move in a straight line, and some fraction of the time we will turn out to be right (the electrically uncharged neutron exerts no electric force on the electron, there is a magnetic interaction but it is quite small, and the gravitational interaction is tiny). But if the neutron happens (probabilistically) to decay just as our electron passes nearby, suddenly our electron is subjected to large electric forces due to the decay proton and electron, and will deviate from a straight path. Moreover, the directions in which the decay products move are also only probabilistic, so we can't even predict whether the proton or the electron will come closest to our electron (Figure 3.52 and Figure 3.53).

This is a simple but dramatic example of how quantum indeterminacy can lead to indeterminacy even in the context of the Momentum Principle.

The Heisenberg uncertainty principle

The Heisenberg uncertainty principle states that there are actual theoretical limits to our knowledge of the state of physical systems. One quantitative formulation states that the position and the momentum of a particle cannot both be simultaneously measured exactly:

$$\Delta x \Delta p_x \geq h$$

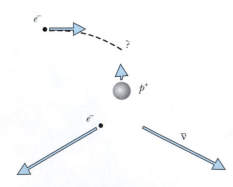

Figure 3.53 Another possible path of the electron if the neutron does decay. The trajectory of the electron depends on the directions of the momenta of the decay products.

This relation says that the product of the uncertainty in position Δx and the uncertainty in momentum Δp_x is equal to a constant h, called Planck's constant. Planck's constant is tiny ($h = 6.6 \times 10^{-34}$ kg·m²/s), so this limitation is not noticeable for macroscopic systems, but for objects small enough to require a quantum mechanical description, the uncertainty principle puts fundamental limits on how accurately we can know the initial conditions, and therefore how well we can predict the future.

3.13 *POINTS AND SPHERES

The gravitational force law applies to objects that are "point-like" (very small compared to the center-to-center distance between the objects). In the second volume of this textbook we will be able to show that any hollow spherical shell with uniform density acts gravitationally on external objects as though all the mass of the shell were concentrated at its center. The density of the Earth is not uniform, because the central iron core has higher density than the outer layers. But by considering the Earth as layers of hollow spherical shells, like an onion, with each shell of nearly uniform density, we get the result that the Earth can be modeled for most purposes as a point mass located at the center of the Earth (for very accurate calculations one must take into account small irregularities in the Earth's density from place to place). Similar statements can be made about other planets and stars. In Figure 3.54, the gravitational force is correctly calculated using the center-to-center distance $|\vec{r}|$.

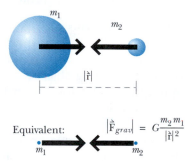

Figure 3.54 A sphere whose density depends only on distance from the center acts like a point particle with all the mass at the center.

This is not an obvious result. After all, in Figure 3.54 some of the atoms are closer and some farther apart than the center-to-center distance $|\vec{r}|$, but the net effect after adding up all the interactions of the individual atoms is as though the two objects had collapsed down to points at their centers. This is a very special property of $1/r^2$ forces, both gravitational and electric, and is not true for forces that have a different dependence on distance.

3.14 *MEASURING THE GRAVITATIONAL CONSTANT *G*

In order to make quantitative predictions and analyses of physical phenomena involving gravitational interactions, it is necessary to know the universal gravitational

constant G. In 1797-1798 Henry Cavendish performed the first experiment to determine a precise value for G (Figure 3.55). In this kind of experiment, a bar with metal spheres at each end is suspended from a thin quartz fiber which constitutes a "torsional" spring. From other measurements, it is known how large a tangential force measured in newtons is required to twist the fiber through a given angle. Large balls are brought near the suspended spheres, and one measures how much the fiber twists due to the gravitational interactions between the large balls and the small spheres.

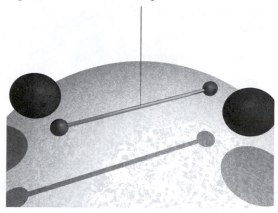

Figure 3.55 The Cavendish experiment.

If the masses are measured in kilograms, the distance in meters, and the force in newtons, the gravitational constant G has been measured in such experiments to be

$$G = 6.7 \times 10^{-11} \ \frac{\text{N} \cdot \text{m}^2}{\text{kg}^2}$$

This extremely small number reflects the fact that gravitational interactions are inherently very weak compared with electromagnetic interactions. The only reason that gravitational interactions are significant in our daily lives is that objects interact with the entire Earth, which has a huge mass. It takes sensitive measurements such as the Cavendish experiment to observe gravitational interactions between two ordinary-sized objects.

3.15 SUMMARY

When forces change as the position of objects change, it is necessary to:

- divide the total time into several smaller time steps
- apply the momentum principle iteratively (repeatedly):

  - Calculate the (vector) forces acting on the system.
  - Update the momentum of the system: $\vec{p}_f = \vec{p}_i + \vec{F}_{net}\Delta t$.
  - Update the position: $\vec{r}_f = \vec{r}_i + \vec{v}_{avg}\Delta t$.
  - Repeat.

Four fundamental interactions:

- gravitational interactions (all objects attract each other gravitationally)
- electromagnetic interactions (electric and magnetic interactions, closely related to each other); interatomic forces are electric in nature
- "strong" interactions (inside the nucleus of an atom)
- "weak" interactions (neutron decay, for example)

The gravitational force law

$$\vec{F}_{grav\;on\;2\;by\;1} = -G\frac{m_1 m_2}{|\vec{r}|^2}\hat{r}$$

$\vec{r} = \vec{r}_2 - \vec{r}_1$ is the position of 2 relative to 1

G is a universal constant: $G = 6.7\times10^{-11}\;\dfrac{\text{N}\cdot\text{m}^2}{\text{kg}^2}$

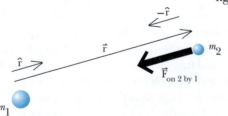

Near the Earth's surface, $|\vec{F}_{grav}| = mg$, where g = +9.8 N/kg

The electric force law (Coulomb's law)

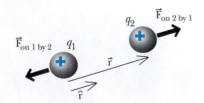

$$\vec{F}_{elec\;on\;2\;by\;1} = \frac{1}{4\pi\varepsilon_0}\frac{q_1 q_2}{|\vec{r}|^2}\hat{r}$$

$\vec{r} = \vec{r}_2 - \vec{r}_1$ is the position of 2 relative to 1

$\dfrac{1}{4\pi\varepsilon_0}$ is a universal constant: $\dfrac{1}{4\pi\varepsilon_0} = 9\times10^9\;\dfrac{\text{N}\cdot\text{m}^2}{\text{C}^2}$

The spring force law

$$\left|\vec{F}_{spring}\right| = k_s|s|$$

$|s|$ is the absolute value of the stretch:

$$s = L - L_0$$

L_0 is the length of the relaxed spring
L is the length of the spring when stretched or compressed
k_s is the "spring stiffness"

The force acts in a direction to restore the spring to its relaxed length.

Reciprocity

$$\vec{F}_{on\;1\;by\;2} = -\vec{F}_{on\;2\;by\;1}\quad\text{(gravitational and electric forces)}$$

This is also called "Newton's third law of motion."

Additional results

Uniform-density spheres act as though all the mass were at the center.

Structuring iterative calculations on a computer

- Define the values of constants such as G to use in the program.
- Specify the masses, initial positions, and initial momenta of the interacting objects.
- Specify an appropriate value for Δt, small enough that the objects don't move very far during one update.
- Create a "loop" structure for repetitive calculations:

  - Calculate the (vector) forces acting on each object.
  - Update the momentum of each object: $\vec{p}_f = \vec{p}_i + \vec{F}_{net}\Delta t$.
  - Update the positions: $\vec{r}_f = \vec{r}_i + \vec{v}_{avg}\Delta t$, where $\vec{v}_{avg} \approx \vec{p}_f/m$.
  - Repeat.

3.16 REVIEW QUESTIONS

Gravitational force

3.RQ.24 At a particular instant the magnitude of the gravitational force exerted by a planet on one of its moons is 3×10^{23} N. If the mass of the moon were three times as large were, what would the magnitude of the force be? If instead the distance between the moon and the planet were three times as large (no change in mass), what would the magnitude of the force be?

3.RQ.25 You hold a tennis ball above your head, then open your hand and release the ball, which begins to fall. At this moment, which statement about the gravitational forces between the Earth and the ball is correct?

A. The force on the ball by the Earth is larger than the force on the Earth by the ball.

B. The force on the Earth by the ball is larger than the force on the ball by the Earth.

C. The forces are equal in magnitude.

D. Not enough information is given.

3.RQ.26 A moon orbits a planet in the *xy* plane, as shown in the diagram. You want to calculate the force on the moon by the planet at each location labeled by a letter (*A,B,C,D*). At each of these locations, what are:
(a) the unit vector $\hat{r}$, and (b) the unit vector $\hat{F}$ in the direction of the force?

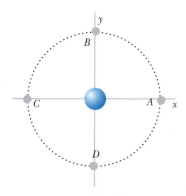

3.RQ.27 Why is the value of the constant *g* different on Earth and on the Moon? Explain in detail.

3.RQ.28 Calculate the approximate gravitational force exerted by the Earth on a human standing on the Earth's surface. Compare with the approximate gravitational force of a human on another human at a distance of 3 meters. What approximations or simplifying assumptions must you make? (See data on the inside back cover.)

Electric force

3.RQ.29 The diagram below shows two positively charged objects and one negatively charged object. What is the direction of the net electric force on the negatively charged object? (If the net force is zero, state this explicitly.)

3.RQ.30 The diagram below shows two negatively charged objects and one positively charged object. What is the direction of the net electric force on the positively charged object? (If the net force is zero, state this explicitly.)

3.RQ.31 An alpha particle contains two protons and two neutrons, and has a net charge of +2*e*. The alpha particle is 1 mm

away from a single proton, which has a charge of +*e*. Which statement about the magnitudes of the electric forces between the particles is correct?

A. The force on the proton by the alpha particle is larger than the force on the alpha particle by the proton.

B. The force on the alpha particle by the proton is larger than the force on the proton by the alpha particle.

C. The forces are equal in magnitude.

D. Not enough information is given.

Reciprocity

3.RQ.32 The windshield of a speeding car hits a hovering insect. Compare the magnitude of the force that the car exerts on the bug to the force that the bug exerts on the car. Which is bigger? Compare the magnitude of the change of momentum of the bug to that of the car. Which is bigger? Compare the magnitude of the change of velocity of the bug to that of the car. Which is bigger? Explain briefly. (Note: the interatomic forces between bug and windshield are electric forces.)

3.17 PROBLEMS

3.P.33 Two books attract each other

Two copies of this textbook are standing right next to each other on a bookshelf. Make a rough estimate of the magnitude of the gravitational force that the books exert on each other. Explicitly list all quantities that you had to estimate, and all simplifications and approximations you had to make to do this calculation. Compare your result to the gravitational force on the book by the Earth.

3.P.34 Compare gravitational and electric forces

Use data from the inside back cover to calculate the gravitational and electric forces two protons exert on each other when they are 1×10^{-10} m apart (about one atomic radius). Which interaction between two protons is stronger, the gravitational attraction or the electric repulsion? If the two protons are at rest, will they begin to move toward each other or away from each other? Note that since both the gravitational and electric force laws depend on the inverse square distance, this comparison holds true at all distances, not just at a distance of 1×10^{-10} m.

3.P.35 Spacecraft and asteroid

At $t = 532.0$ s after midnight a spacecraft of mass 1400 kg is located at position $\langle 3 \times 10^5, 7 \times 10^5, -4 \times 10^5 \rangle$ m, and an asteroid of mass 7×10^{15} kg is located at position $\langle 9 \times 10^5, -3 \times 10^5, -12 \times 10^5 \rangle$ m. There are no other objects nearby.
(a) Calculate the (vector) force acting on the spacecraft.
(b) At $t = 532.0$ s the spacecraft's momentum was $\vec{p}_i$, and at the later time $t = 538.0$ s its momentum was $\vec{p}_f$, Calculate the (vector) change of momentum $\vec{p}_f - \vec{p}_i$.

3.HW.36 The car and the mosquito

A car of mass *M* moving in the *x* direction at high speed *v* strikes a hovering mosquito of mass *m*, and the mosquito is smashed against the windshield. The interaction between the mosquito and the windshield is an electric interaction between the electrons and protons in the mosquito and those in the windshield.

(a) What is the approximate momentum change of the mosquito? Give magnitude and direction. Explain any approximations you make.

(b) At a particular instant during the impact, when the force exerted on the mosquito by the car is F, what is the magnitude of the force exerted on the car by the mosquito?

(c) What is the approximate momentum change of the car? Give magnitude and direction. Explain any approximations you make.

(d) Qualitatively, why is the collision so much more damaging to the mosquito than to the car?

3.HW.37 The bouncing ball

A steel ball of mass m falls from a height h onto a scale calibrated in newtons. The ball rebounds repeatedly to nearly the same height h. The scale is sluggish in its response to the intermittent hits and displays an *average* force F_{avg}, such that $F_{avg} T = F\Delta t$ where $F\Delta t$ is the brief impulse that the ball imparts to the scale on every hit, and T is the time between hits.

Calculate this average force in terms of m, h, and physical constants. Compare your result with what the scale reads if the ball merely rests on the scale. Explain your analysis carefully (but briefly).

3.P.38 Star and planet

A star of mass 7×10^{30} kg is located at $\langle 5 \times 10^{12}, 2 \times 10^{12}, 0 \rangle$ m. A planet of mass 3×10^{24} kg is located at $\langle 3 \times 10^{12}, 4 \times 10^{12}, 0 \rangle$ m and is moving with a velocity of $\langle 0.3 \times 10^4, 1.5 \times 10^4, 0 \rangle$ m/s.

(a) At a time 1×10^6 seconds later, what is the new velocity of the planet?

(b) Where is the planet at this later time?

(c) Explain briefly why the procedures you followed in parts (a) and (b) were able to produce usable results but wouldn't work if the later time had been 1×10^9 seconds instead of 1×10^6 seconds after the initial time. Explain briefly how could you use a computer to get around this difficulty.

3.P.39 Two stars

At $t = 0$ a star of mass 4×10^{30} kg has velocity $\langle 7 \times 10^4, 6 \times 10^4, -8 \times 10^4 \rangle$ m/s and is located at $\langle 2.00 \times 10^{12}, -5.00 \times 10^{12}, 4.00 \times 10^{12} \rangle$ m relative to the center of a cluster of stars. There is only one nearby star that exerts a significant force on the first star. The mass of the second star is 3×10^{30} kg, its velocity is $\langle 2 \times 10^4, -1 \times 10^4, 9 \times 10^4 \rangle$ m/s, and this second star is located at $\langle 2.03 \times 10^{12}, -4.94 \times 10^{12}, 3.95 \times 10^{12} \rangle$ m relative to the center of the cluster of stars.

(a) At $t = 1 \times 10^5$ s, what is the approximate momentum of the first star?

(b) Discuss briefly some ways in which your result for (a) is approximate, not exact.

(c) At $t = 1 \times 10^5$ s, what is the approximate position of the first star?

(d) Discuss briefly some ways in which your result for (b) is approximate, not exact.

3.P.40 Determining the mass of an asteroid

In June 1997 the NEAR spacecraft ("Near Earth Asteroid Rendezvous"; see http://near.jhuapl.edu/), on its way to photograph the asteroid Eros, passed within 1200 km of asteroid Mathilde at a speed of 10 km/s relative to the asteroid. From photos transmitted by the 805 kg spacecraft, Mathilde's size was known to be about 70 km by 50 km by 50 km. It is presumably made of rock. Rocks on Earth have a density of about 3000 kg/m^3 (3 grams/cm^3).

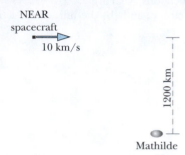

(a) Make a rough diagram to show qualitatively the effect on the spacecraft of this encounter with Mathilde. Explain your reasoning.

(b) Make a very rough estimate of the change in momentum of the spacecraft that would result from encountering Mathilde. Explain how you made your estimate.

(c) Using your result from part (b), make a rough estimate of how far off course the spacecraft would be, one day after the encounter.

(d) From actual observations of the location of the spacecraft one day after encountering Mathilde, scientists concluded that Mathilde is a loose arrangement of rocks, with lots of empty space inside. What was it about the observations that must have led them to this conclusion?

Experimental background: The position was tracked by very accurate measurements of the time that it takes for a radio signal to go from Earth to the spacecraft followed immediately by a radio response from the spacecraft being sent back to Earth. Radio signals, like light, travel at a speed of 3×10^8 m/s, so the time measurements had to be accurate to a few nanoseconds (1 ns = 10^{-9} s).

Computational problems

The following problems require the use of a a computer to model matter, interactions, and motion. Some parts of these problems can be done with almost any tool (spreadsheet, math package, etc.). Other parts are most easily done with a programming language, for which we recommend the free 3D programming language VPython (http://vpython.org). Your instructor will introduce you to an available computational tool and assign problems, or parts of problems, that can be addressed using the chosen tool.

3.P.41 Planetary orbits

In this problem you will study the motion of a planet around a star. To start with a somewhat familiar situation, you will begin by modeling the motion of our Earth around our Sun. Write answers to questions either as comments in your program or on paper, as specified by your instructor.

Planning

(a) The Earth goes around the Sun in a nearly circular orbit, taking one year to go around. Using data on the inside back cover, what initial speed should you give the Earth in a computer model, so that a circular orbit should result? (If your computer model does produce a circular orbit, you have a strong indication that your program is working properly.)

(b) Estimate an appropriate value for Δt to use in your computer model. Remember that t is in seconds, and consider your answer to part (a) in making this estimate. If your Δt is too small, your calculation will require many tedious steps. Explain briefly how you decided on an appropriate step size.

Circular orbit

(c) Starting with speed calculated in (a), and with the initial velocity perpendicular to a line connecting the Sun and the Earth, calculate and display the trajectory of the Earth. Display the whole trail, so you can see whether you have a closed orbit (that is, whether the Earth returns to its starting point each time around). Include the Sun's position $\vec{r}_s = <x_s, y_s, z_s>$ in your computations, even if you set its coordinates to zero. This will make it easier to modify the computation to let the Sun move (Problem 2.2 Binary stars).

Computational accuracy

(d) As a check on your computation, is the orbit a circle as expected? Run the calculation until the accumulated time t is the length of a year: does this take the Earth around the orbit once, as expected? What is the largest step size you can use that still gives a circular orbit and the correct length of a year? What happens if you use a much larger step size?

Noncircular trajectories

(e) Set the initial speed to 1.2 times the Earth's actual speed. Make sure the step size is small enough that the orbit is nearly unaffected if you cut the step size. What kind of orbit do you get? What step size do you need to use to get results you can trust? What happens if you use a much larger step size? Produce at least one other qualitatively different noncircular trajectory for the Earth. What difficulties would humans have in surviving on the Earth if it had a highly noncircular orbit?

Force and momentum

(f) Choose initial conditions that give a noncircular orbit. Continuously display a vector showing the momentum of the Earth (with its tail at the Earth's position), and a different colored vector showing the force on the Earth by the Sun (with its tail at the Earth's position). You must scale the vectors appropriately to fit into the scene. A way to figure out what scale factor to use is to print the numerical values of momentum and force, and compare with the scale of the scene. For example, if the width of the scene is W meters, and the force vector has a typical magnitude F (in newtons), you might scale the force vector by a factor $0.1\,W/F$, which would make the length of the vector be one-tenth the width of the scene. Is the force in the same direction as the momentum? How does the momentum depend on the distance from the star?

3.P.42 Binary stars

About a half of the visible "stars" are actually systems consisting of two stars orbiting each other, called "binary stars." In your computer model of the Earth and Sun (3.P.38), replace the Earth with a star whose mass is half the mass of our Sun, and take into account the gravitational effects that the second star has on the Sun.

(a) Give the second star the speed of the actual Earth, and give the Sun zero initial momentum. What happens? Try a variety of other initial conditions. What kinds of orbits do you find?

(b) One of the things to try is to give the Sun the same magnitude of momentum as the second star, but in the opposite direction. If the initial momentum of the second star is in the $+y$ direction, give the Sun the same magnitude of momentum but in the $-y$ direction, so that the total momentum of the binary star system is initially zero, but the stars are not headed toward each other. What is special about the motion you observe in this case?

3.P.43 Other force laws

Modify your orbit computation to use a different force law, such as a force that is proportional to $1/r$ or $1/r^3$, or a constant force, or a force proportional to r^2 (this represents the force of a spring whose relaxed length is nearly zero). How do orbits with these force laws differ from the circles and ellipses that result from a $1/r^2$ law? If you want to keep the magnitude of the force roughly the same as before, you will need to adjust the force constant G.

3.P.44 The effect of the Moon and Venus on the Earth

In Problem 3.P.41 you analyzed a simple model of the Earth orbiting the Sun, in which there were no other planets or moons. Venus and the Earth have similar size and mass. At its closest approach to the Earth, Venus is about 40 million kilometers away (4×10^{10} m). The Moon's mass is about 7×10^{22} kg, and the distance from Earth to Moon is about 4×10^8 m (400,000 km, center to center).

(a) Calculate the ratio of the gravitational forces on the Earth exerted by Venus and the Sun. Is it a good approximation to ignore the effect of Venus when modeling the motion of the Earth around the Sun?

(b) Calculate the ratio of the gravitational forces on the Earth exerted by the Moon and the Sun. Is it a good approximation to ignore the effect of the Moon when modeling the motion of the Earth around the Sun?

3.P.45 The three-body problem

Carry out a numerical integration of the motion of a three-body gravitational system and plot the trajectory, leaving trails behind the objects. Calculate all of the forces before using these forces to update the momenta and positions of the objects. Otherwise the calculations of gravitational forces would mix positions corresponding to different times.

Try different initial positions and initial momenta. Find at least one set of initial conditions that produces a long-lasting orbit, one set of initial conditions that results in a collision with a massive object, and one set of initial conditions that allows one of the objects to wander off without returning. Report the masses and initial conditions that you used.

3.P.46 The Ranger 7 mission to the Moon

The first U.S. spacecraft to photograph the Moon close up was the unmanned "Ranger 7" photographic mission in 1964. The spacecraft, shown in the illustration at the right (NASA photograph), contained television cameras that transmitted close-up pictures of the Moon back to Earth as the spacecraft approached the Moon. The spacecraft did not have retro-rockets to slow itself down, and it eventually simply crashed onto the Moon's surface, transmitting its last photos immediately before impact.

The image below is the first image of the Moon taken by a U.S. spacecraft, Ranger 7, on July 31, 1964, about 17 minutes before impacting the lunar surface. The large crater at center right is Alphonsus (108 km diameter); above it (and to the right) is Ptolemaeus and below it is Arzachel. The Ranger 7 impact site is off the frame, to the left of the upper left corner.
To find out more about the actual Ranger lunar missions, see
http://nssdc.gsfc.nasa.gov/planetary/lunar/ranger.html

The first Ranger 7 photo of the Moon (NASA photograph).

To send a spacecraft to the Moon, we put it on top of a large rocket containing lots of rocket fuel and fire it upward. At first the huge ship moves quite slowly, but the speed increases rapidly. When the "first-stage" portion of the rocket has exhausted its fuel and is empty, it is discarded and falls back to Earth. By discarding an empty rocket stage we decrease the amount of mass that must be accelerated to even higher speeds. There may be several stages that operate for a while and then are discarded before the spacecraft has risen above

most of Earth's atmosphere (about 50 km, say, above the Earth), and has acquired a high speed. At that point all the fuel available for this mission has been used up, and the spacecraft simply coasts toward the Moon through the vacuum of space.

We will model the Ranger 7 mission. Starting 50 km above the Earth's surface (5×10^4 m), a spacecraft coasts toward the Moon with an initial speed of about 10^4 m/s. Here are data we will need:
mass of spacecraft = 173 kg; mass of Earth $\approx 6 \times 10^{24}$ kg; mass of Moon $\approx 7 \times 10^{22}$ kg; radius of Moon = 1.75×10^6 m; distance from Earth to Moon $\approx 4 \times 10^8$ m (400,000 km, center to center)

We're going to ignore the Sun in a simplified model even though it exerts a sizable gravitational force. We're expecting the Moon mission to take only a few days, during which time the Earth (and Moon) move in a nearly straight line with respect to the Sun, because it takes 365 days to go all the way around the Sun. We take a reference frame fixed to the Earth as representing (approximately) an inertial frame of reference with respect to which we can use the Momentum Principle.

For a simple model, make the Earth and Moon be fixed in space during the mission. Factors that would certainly influence the path of the spacecraft include the motion of the Moon around the Earth, and the motion of the Earth around the Sun. In addition, the Sun and other planets exert gravitational forces on the spacecraft. As a separate project you might like to include some of these additional factors.

(a) Compute the path of the spacecraft, and display it either with a graph or with an animated image. Here, and in remaining parts of the problem, report the step size Δt that gives accurate results (that is, cutting this step size has little effect on the results).

(b) By trying various initial speeds, determine the approximate *minimum* launch speed needed to reach the Moon, to two significant figures (this is the speed that the spacecraft obtains from the multistage rocket, at the time of release above the Earth's atmosphere). What happens if the launch speed is less than this minimum value? (Be sure to check the step size issue.)

(c) Use a launch speed 10% larger than the approximate minimum value found in part (b). How long does it take to go to the Moon, in hours or days? (Be sure to check the step size issue.)

(e) What is the "impact speed" of the spacecraft (its speed just before it hits the Moon's surface)? Make sure that your spacecraft crashes on the surface of the Moon, not at the Moon's center! (Be sure to check the step size issue.)

You may have noticed that you don't actually need to know the mass m of the spacecraft in order to carry out the computation. The gravitational force is proportional to m, and the momentum is also proportional to m, so m cancels. However, nongravitational forces such as electric forces are not proportional to the mass, and there is no cancellation in that case. We kept the mass m in the analysis in order to illustrate a general technique for predicting motion, no matter what kind of force, gravitational or not.

3.P.47 The effect of Venus on the Moon voyage

In the Moon voyage analysis, you used a simplified model in which you neglected among other things the effect of Venus. An important aspect of physical modeling is making estimates of how large the neglected effects might be. Venus and the Earth have similar size and mass. At its closest approach to the Earth, Venus is about 40 million kilometers away (4×10^{10} m). In the real world, Venus would attract the Earth and the Moon as well as the spacecraft, but to get an idea of the size of the effects, imagine that the Earth, the Moon, and Venus are all fixed in position.

Moon

Venus · · Earth

The relative positions of Venus, Earth, and Moon.

If we take Venus into account, make a rough estimate of whether the spacecraft will miss the Moon entirely. How large a sideways deflection of the crash site will there be? Explain your reasoning and approximations.

3.18 ANSWERS TO EXERCISES

3.X.1 (page 76)	**0.275 N**
3.X.2 (page 76)	3571 N/m
3.X.3 (page 76)	71.4 N
3.X.4 (page 76)	0.13 m
3.X.5 (page 81)	8×10^{25} N
3.X.6 (page 81)	8.9×10^{19} N
3.X.7 (page 81)	$(3/16)F = 0.1875F$
3.X.8 (page 81)	2.5×10^{-10} N , 2.5×10^{-10} N
3.X.9 (page 81)	6×10^{24} kg
3.X.10 (page 83)	(a) toward Ganymede, (b) toward Europa, (c) toward Europa, (d) toward Europa, (e) toward Ganymede, (f) toward Ganymede
3.X.11 (page 83)	$\langle -1.3 \times 10^{20}, 0, 1.3 \times 10^{20} \rangle$ N
3.X.12 (page 84)	(a) 1.53 N/kg, (b) 107 N, (c) 0.15 as much (about 1/6)
3.X.13 (page 84)	3.5 N/kg
3.X.14 (page 84)	32242 m = 32.2 km (about 20 miles)
3.X.15 (page 89)	(c) 0.5 m
3.X.16 (page 89)	(b) 1×10^{10} m
3.X.17 (page 89)	(a) 1 day
3.X.18 (page 89)	(a) 1 second
3.X.19 (page 90)	2.3×10^{-8} N
3.X.20 (page 90)	upward, downward, equal
3.X.21 (page 90)	upward, downward
3.X.22 (page 91)	588 N, 588 N
3.X.23 (page 92)	2.4×10^{-7} N , 2.4×10^{-7} N

CHAPTER 4

CONTACT FORCES AND THE MOMENTUM PRINCIPLE

Key ideas:

- A solid object may be modeled as a lattice of balls (atoms) connected by springs (chemical bonds).
- Contact forces between solid objects are due to the compression or stretching of springlike interatomic bonds.
- The magnitude and direction of unknown forces, including contact forces, may be deduced by applying the derivative form of the momentum principle.
- Changing direction of motion is caused by a force component perpendicular to momentum.
- Changing speed is caused by a force component parallel to momentum.

4.1 TARZAN AND THE VINE

Tarzan wants to use a vine to swing across a river. To make sure the vine is strong enough to support him, he tests it by hanging motionless on the vine for several minutes (Figure 4.1). The vine passes this test, so Tarzan grabs the vine and swings out over the river. He is annoyed and perplexed when the vine breaks midway through the swing (Figure 4.2), and he ends up drenched and shivering in the middle of the cold river, to the great amusement of the onlooking apes.

Why did the vine break while Tarzan was swinging on it, but not while he hung motionless from it? To answer this question we will need to understand how objects like vines, wires, and strings exert forces. We will also need to invoke the derivative form of the Momentum Principle, which will be introduced in this chapter.

4.2 A MODEL OF A SOLID: BALLS CONNECTED BY SPRINGS

The great 20th-century American physicist Richard Feynman said:

"If, in some cataclysm, all of scientific knowledge were to be destroyed, and only one sentence passed on to the next generations of creatures, what statement would contain the most information in the fewest words? I believe it is the atomic hypothesis (or the atomic fact, or whatever you wish to call it) that all things are made of atoms—little particles that move around in perpetual motion, attracting each other when they are a little distance apart, but repelling upon being squeezed into one another. In that one sentence, you will see, there is an enormous amount of information about the world, if just a little imagination and thinking are applied." (*The Feynman Lectures on Physics* by R. P. Feynman, R. B. Leighton, & M. Sands, 1965; Palo Alto: Addison-Wesley.) In the quote above, Feynman summarized the basic properties of atoms and of interatomic forces. The main properties of atoms that will be important to us in this chapter are these:

- All matter consists of atoms, whose typical radius is about 10^{-10} meter.
- Atoms attract each other when they are close to each other but not too close.

Figure 4.1 Tarzan hangs motionless from a vine, which does not break.

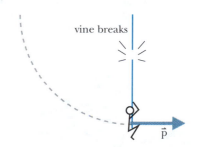

Figure 4.2 The vine breaks midway through Tarzan's swing.

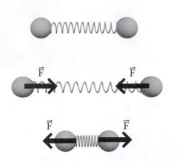

Figure 4.3 Two balls connected by a spring. Top: relaxed spring exerts no forces on the balls. Middle: stretched spring exerts forces to bring balls closer together. Bottom: compressed spring exerts forces to move balls apart.

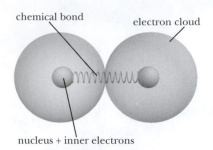

Figure 4.4 The ball-spring model for two atoms connected by a chemical bond (represented by a spring), superimposed on a space-filling model in which the entire electron clouds of the atoms are shown.

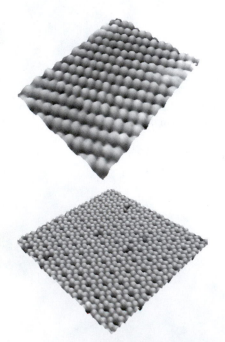

Figure 4.5 STM images of two surfaces of a silicon crystal (courtesy of Randall Feenstra, Carnegie Mellon University).

- Atoms repel each other when they get too close to each other.
- Atoms in solids, liquids, and gasses keep moving even at very low temperatures.

These properties were established during a century of intense study of atoms by physicists and chemists, using a wide variety of experimental techniques.

? What properties of a block of aluminum, observable without special equipment, support the claims that "atoms attract each other when they are not too close to each other" and "atoms repel each other when they get too close to each other"?

Since the block of metal does not spontaneously fall apart or evaporate, the atoms in the block must be attracting each other. It is very difficult to compress the block, so the atoms must resist an attempt to push them closer together. Evidently the atoms in a block of aluminum are normally at just the right distance from each other, not too close and not too far away. This just-right distance is called the "equilibrium" distance between atoms.

A chemical bond is like a spring

Two atoms linked by a chemical bond behave in a manner very similar to two macroscopic balls attached to the ends of a very low mass spring (Figure 4.3). The ball-spring system is a good model for the atomic system. In this model:

- Each ball in the model represents a massive atomic nucleus, surrounded by the inner electrons of the atom. Almost all the mass of an atom is concentrated in the tiny nucleus.
- The spring in the model represents the chemical bond, which is due to the shared outer electrons of both atoms.

The microscopic atomic system behaves much like the macroscopic model system, as long as the stretch or compression is small, as it is in ordinary processes. If the atoms are moved farther apart, they experience forces resisting the separation. If the atoms are pushed closer together, they experience forces resisting the compression. In Figure 4.4 the ball-spring model of two bonded atoms is superimposed on a space-filling representation showing the full electron cloud of each atom.

A ball-spring model for a solid object

A solid object contains many atoms, not just two. We know from x-ray studies and from images produced by scanning tunneling microscopes (STM) and atomic force microscopes (AFM) that many solid objects, such as metals, are crystals composed of regular arrays of atoms, as shown in the STM images of silicon surfaces in Figure 4.5. In these images, the spheres indicate the entire electron cloud associated with each atom, not just the nucleus and inner electrons represented by the balls in our model.

These STM images show only one surface of a solid, but inside the solid, atoms are arranged in 3D patterns, like the balls and springs in the model solid shown in Figure 4.6. Solids in which the atoms are arranged in regular "lattices" like those shown in Figure 4.6 are called crystals; these include metals, quartz, diamond, ice, and table salt (NaCl), but not most organic solids such as plastic or wood.

The lattice shown in Figure 4.6 is the simplest kind of crystal, and is called a "cubic" lattice because the atoms in the crystal are located at the corners of adjacent cubes. More complex lattice arrangements are possible; for example, in one common variant called "body centered cubic" there is an additional atom at the center of each cube. In Figure 4.5 the lower image shows a hexagonal arrangement of atoms. Although most crystals are more complex than the simple cubic lattice, we will use this simple cubic ball-and-spring model because it incorporates all the important features we need.

In a solid at room temperature, the atoms are in motion, continually oscillating around their equilibrium positions. If we heat the solid, these atomic oscillations become more vigorous. One of the most important outcomes of research on the atomic nature of matter was the realization that for many materials, the temperature of an object, as measured by an ordinary thermometer, is just an indicator of the average energy of the atoms; the higher the temperature, the more vigorous the atomic motion. Figure 4.6 is a "snapshot" of the atoms in a model solid at a particular instant.

4.3 TENSION FORCES

By applying the Momentum Principle we can show that an object like a wire, string, or vine can exert a force on an object attached to it. Figure 4.7 shows a heavy iron ball hanging motionless on the end of a wire. The *y* momentum of the ball is not changing, so the *y* component of the net force on the ball must be zero. Since the Earth is pulling down on the ball, the wire must be pulling up. The force exerted by an object like a wire or a string is often called a "tension" force, or sometimes just "the tension in the wire." A tension force always acts along the wire or string.

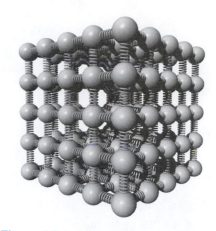

Figure 4.6 A simple model of a solid: tiny balls in constant motion, connected by springs. This figure shows only a small section of a solid object, which has many more atoms than are depicted here.

? If the mass of the ball is 1 kg, how large an upward force does the wire exert on the ball?

System: Ball
Surroundings: Earth, wire
Momentum Principle:

$\Delta p_y = F_{net,y} \Delta t$

Choose a value for Δt: 10 seconds

$$\Delta p_y = (-mg + F_{wire})(10 \text{ s})$$

$$0 = (-mg + F_{wire})(10 \text{ s})$$

$$mg = F_{wire}$$

$$(1 \text{ kg})(9.8 \text{ N/kg}) = 9.8 \text{ N} = F_{wire}$$

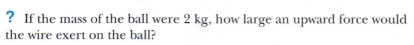

Since the ball remains at rest indefinitely, this is true for any value for Δt (other than zero).

? If the mass of the ball were 2 kg, how large an upward force would the wire exert on the ball?

$$(2 \text{ kg})(9.8 \text{ N/kg}) = 19.6 \text{ N} = F_{wire}$$

This simple result has an interesting implication. Evidently the magnitude of the tension force F_{wire} exerted by the wire depends on the mass of the ball.

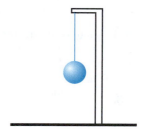

Figure 4.7 A heavy ball hangs motionless at the end of a thin wire.

? How does an inanimate object like a wire "know" how large a force to exert in a given situation?

We have already seen that the force exerted by a spring depends on how much the spring is stretched. A wire can be thought of as a very stiff spring. When a weight is hung on the end of a wire, the wire does stretch, although the stretch is usually very small. The ball-spring model of a solid helps us understand what happens when a wire stretches.

Microscopic view: Tension in a wire

We can model a wire as a chain of balls and springs (atoms connected by springlike chemical bonds). For simplicity, we'll consider a wire that is only one atom thick.

When the wire lies on a table, each of the springs (bonds) is relaxed. When the wire hangs vertically with nothing attached to it, the bonds stretch a tiny amount, just to support the weight of the atoms below them.

Figure 4.8 The interatomic bonds in a wire stretch when a heavy mass is hung from the wire. In this idealized diagram the wire is only one atom thick, and the stretch of each bond is greatly exaggerated.

bond length d

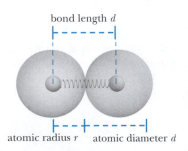

atomic radius r atomic diameter d

Figure 4.9 The length of an interatomic bond is defined as the center-to-center distance between adjacent atoms. This is the same as the diameter of an atom (including the full electron cloud).

However, when a massive object is hung from the wire, the springlike bonds between atoms stretch significantly, because each bond must support the weight of everything below it (Figure 4.8). If we make the approximation that the mass of the atoms in the wire is negligible compared to the mass of the hanging object, then we can say that each bond in the wire is stretched by approximately the same amount, or equivalently, that the tension is the same throughout the wire.

? How does the wire manage to exert a larger upward force when a more massive object is hung from it?

The more massive object stretches the interatomic bonds more than a less massive object. Of course, there is a limit to how much the wire can stretch. A sufficiently heavy weight will break the wire.

4.4 LENGTH OF AN INTERATOMIC BOND

Our goal in the next sections is to determine the stiffness of an interatomic bond, considered as a spring. The first step is to determine the length of an interatomic bond in a particular material. Bond lengths will be slightly different in different materials (for example, aluminum vs. lead), depending on the size of the atoms. We will determine the interatomic bond length in solid copper.

We will define the length of one interatomic bond as the center-to-center distance between two adjacent atoms (Figure 4.9). If we consider the space-filling model of a solid, this distance is equal to twice the radius of an atom, or the diameter of an atom.

The diameter of an atom in a solid is one of the important properties of matter which plays a role in interactions. We can calculate atomic diameters for crystals of particular elements by using the measured density of the material in kilograms per cubic meter, and Avogadro's number (the number of atoms in one mole of the material).

? Does the density of a block of aluminum depend on the dimensions of the block? Does it depend on the mass of the block?

Density does not depend on the size, shape, or mass of an object. Density is a property of the material itself, and the ratio of mass to volume will always be the same for objects made of the same solid material. The density of a wide range of materials has been measured and may easily be looked up in reference materials.

Unit conversion: density

Often densities are given in grams per cubic centimeter rather than in kilograms per cubic meter. For example, the density of liquid water is 1 gram/cm³, the density of aluminum is 2.7 grams/cm³ and the density of lead is 11.4 grams/cm³. We often need to convert these densities to S.I. units, like this:

$$\left(1\frac{\text{gram}}{\text{cm}^3}\right)\left(\frac{(100\ \text{cm})^3}{1\ \text{m}^3}\right)\left(\frac{1\ \text{kg}}{1000\ \text{grams}}\right)$$

$$= 1000\ \text{kg/m}^3$$

Note that since 100 cm = 1 m,

$$(100\ \text{cm})^3 = 1 \times 10^6\ \text{cm}^3 = 1\ \text{m}^3$$

Example: Diameter of a copper atom (length of bond in copper)

One mole of copper (6.02×10^{23} atoms) has a mass of 64 grams (see the periodic table on the inside front cover). The density of copper is 8.94 grams/cm³. What is the approximate diameter, in meters, of a copper atom in solid copper?

$$\left(8.94\frac{\text{g}}{\text{cm}^3}\right)\frac{(1\ \text{kg})}{(1 \times 10^3\ \text{g})}\frac{(1 \times 10^2\ \text{cm})^3}{(1\ \text{m})^3} = 8.94 \times 10^3\ \frac{\text{kg}}{\text{m}^3}\ \text{(S.I. units)}$$

Figure 4.10 shows a cube that contains 5x5x5 = 125 atoms. The number of atoms on one side of the cube is $\sqrt[3]{125} = 5$.

In a cube that is 1 meter on each side, how many copper atoms are there?

$$(8.94 \times 10^3\ \text{kg})\left(\frac{1\ \text{mol}}{0.064\ \text{kg}}\right)\frac{(6.02 \times 10^{23}\ \text{atoms})}{(1\ \text{mol})} = 8.41 \times 10^{28}\ \text{atoms}$$

Along one edge of the cube, which is 1 m long, there are:

$$\sqrt[3]{8.41\times10^{28}} = 4.38\times10^{9} \text{ atoms.}$$

A row of 4.38×10^{9} atoms is 1 m long, so the diameter of one atom is:

$$d = \left(\frac{1 \text{ m}}{4.38\times10^{9} \text{ atoms}}\right) = 2.28\times10^{-10} \text{ m}$$

In a solid block of copper, the length of a bond between two adjacent copper atoms is 2.28×10^{-10} m.

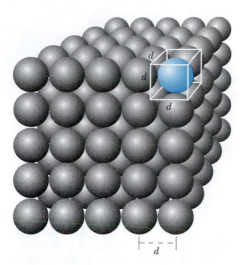

Figure 4.10 A simple cubic arrangement of atoms. The volume of space associated with each atom is a tiny cube d by d by d.

Further discussion

An alternative approach to this calculation involves a "micro" view of density. Since density is independent of the amount of an object, the density of one copper atom ought to be the same as the density of a large block of copper. Figure 4.10 shows that the volume of space taken up by each atom is a tiny cube d on a side, where d is the diameter of an atom in the solid and is also the distance from the center of one atom to the center of a neighboring atom. The micro density is the mass of one atom divided by the volume of the tiny cube associated with each atom. This micro density is the same as the macro density, mass per volume:

$$\text{density} = \frac{\text{mass of } 8.41\times10^{28} \text{ atoms}}{\text{volume of } 8.41\times10^{28} \text{ atoms}} = \frac{\text{mass of 1 atom}}{\text{volume of 1 atom}} = \frac{m_a}{d^3}$$

The mass of one atom can be determined using the mass of one mole and the knowledge that one mole contains 6.02×10^{23} atoms (Avogadro's number):

$$m_a = \frac{\text{mass of one mole}}{6.02\times10^{23} \text{ atoms/mole}}$$

It might be tempting to put these concepts together into a formula of the form "$d = \dots$", but try to avoid the temptation. It is much more memorable and much safer to think through these physical relationships each time you make the calculation, to avoid serious mistakes.

Few solid elements actually have cubic lattices, because a cubic lattice is rather unstable. However, assuming that all crystalline solids have cubic lattices greatly simplifies our model, and gives adequately accurate results.

4.X.1 The density of aluminum is 2.7 grams/cm^3. What is the approximate diameter of an aluminum atom (length of a bond) in solid aluminum?

4.X.2 The density of lead is 11.4 grams/cm^3. What is the approximate diameter of a lead atom (length of a bond) in solid lead?

The diameters of the smallest and largest atoms differ by only about a factor of 8. Most metallic elements have similar radii, on the order of 1.5×10^{-10} m. It is useful to remember that the radius of an "average" atom is on the order of 1×10^{-10} m.

4.5 THE STIFFNESS OF AN INTERATOMIC BOND

Knowing the length of an interatomic bond in solid copper (and the diameter of a copper atom), we can now use experimental data to find the stiffness of the interatomic bond, considered as a spring. You have probably had some experience with ordinary macroscopic springs. A Slinky, for example, is a soft spring; its spring stiffness is small—around 1 N/m. The stiffness of the spring on a Pogo stick is much larger—around 5000 N/m. We can't measure the stiffness of an interatomic bond directly, but we can analyze data from macroscopic (large scale) experiments to determine this quantity.

Figure 4.11 A solid wire considered as many parallel long chains of atoms connected by springs (interatomic bonds). For clarity, the horizontal bonds are not shown; they aren't significant when stretching the wire.

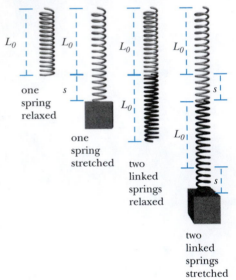

Figure 4.12 Two identical springs linked end-to-end stretch twice as much as one spring when the same force is applied. The combined spring therefore is only half as stiff as the individual springs.

Figure 4.13 Two springs side-by-side supporting a block.

The basic idea is to hang heavy masses on a long wire and measure the stretch of the wire; then, by figuring out how many interatomic "springs" there are in the wire, we can determine the spring constant of a single interatomic bond. To do this, we need to be able to relate the stiffness of an object composed of many springs to the stiffness of each individual spring. We will consider the wire to be composed of many parallel long chains of atoms connected by springs (Figure 4.11).

? (a) How is the stiffness of one of the short springs (bonds) related to the stiffness of an entire chain of springs? (b) How is the stiffness of the entire wire related to the stiffness of one chain of springs?

Two springs linked end-to-end (in series)

Suppose we have a long, low mass spring whose spring stiffness is 100 N/m. If we hang a block whose weight is 100 N (about a 10 kg mass) on the end of the spring, the spring will stretch 1 meter. (This is a big stretch, so the relaxed spring must be several meters long; we are using simple numbers here to make our reasoning easier.)

? We make a longer spring by linking two identical springs end-to-end (in "series"). Each shorter spring has a stiffness of 100 N/m. What will the stiffness of the longer spring be?

When we hang the 100 N block on the end of the longer spring, this longer spring will stretch 2 meters, because each of the individual springs stretches 1 meter (Figure 4.12). We can use the Momentum Principle to get the stiffness of this longer spring. When the block hangs motionless, its momentum is not changing, Choosing a Δt of 10 seconds, we find this:

System: block
Surroundings: spring, Earth
Momentum Principle:

$$\Delta p_y = F_{net,y}\Delta t$$
$$0 = (-mg + k_s s)(10 \text{ s})$$
$$mg = k_s s$$
$$100 \text{ N} = k_s(2 \text{ m})$$
$$k_s = 50 \text{ N/m}$$

A long spring made of two identical springs linked end-to-end is only half as stiff as each of the shorter springs.

4.X.3 If a chain of 20 identical short springs linked end-to-end has a stiffness of 40 N/m, what is the stiffness of one short spring?

Two springs in parallel (side-by-side)

? Suppose we support a 100 N weight with two identical springs side-by-side (Figure 4.13). If the spring stiffness of each spring is 100 N/m, how much will each spring stretch? What is the effective stiffness of the two-spring system?

Informally, we see that each spring supports only 50 N, so each spring will stretch only 0.5 m. You can work out the formal solution yourself using the Momentum Principle, remembering to include two separate upward spring forces.

We can think of the two springs as a single, wider spring. The effective spring stiffness of this "double-width" spring is:

$$k_{s,effective} = \frac{100 \text{ N}}{0.5 \text{ m}} = 200 \text{ N/m}$$

Two springs side-by-side are effectively twice as stiff as a single spring.

4.X.4 9 identical springs are placed side-by-side (in parallel, as in Figure 4.13), and connected to a large massive block. The stiffness of the 9-spring combination is 2700 N/m. What is the stiffness of one of the individual springs?

Cross-sectional area

The cross-sectional area of an object is the area of a flat surface made by slicing through the object (Figure 4.14). A cylindrical object like a round pencil has a circular cross-sectional area (imagine sawing crosswise through a pencil). A cylindrical pencil 10 cm long and 0.5 cm in diameter has a cross-sectional area of

$$A = \pi(0.025 \text{ m})^2 = 1.96 \times 10^{-3} \text{ m}^2$$

Note that the length of the object is irrelevant.

A long object with four flat sides, like a board, has a cross-sectional area that is rectangular, or perhaps square. A wooden board 7 meters long, 5 cm wide, and 3 cm high, has a cross sectional area of

$$A = (0.05 \text{ m})(0.03 \text{ m}) = 1.5 \times 10^{-3} \text{ m}^2$$

Figure 4.14 The cross-sectional area of a cylinder is the area of a circle. The cross sectional area of a rectangular solid is the area of a rectangle.

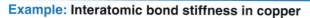

Example: Interatomic bond stiffness in copper

A copper wire is 2 meters long. The wire has a square cross-section (that is, it has four flat sides—it is not round). Each side of the wire is 1 mm in width. Making sure the wire is straight, you hang a 10 kg mass on the end of the wire. Careful measurement shows that the wire is now 1.51 mm longer. From these measurements, determine the stiffness of one interatomic bond in copper.

1. What is the spring stiffness $k_{s,wire}$ of the entire wire, considered as a single macroscopic (large scale), very stiff spring?
System: mass
Surroundings: Earth, wire
Momentum Principle:

$$\Delta p_y = 0 = (-mg + k_{s,wire}s)\Delta t$$

$$k_{s,wire} = \frac{mg}{s} = \frac{(10 \text{ kg})(9.8 \text{ N/kg})}{(1.51 \times 10^{-3} \text{ m})} = 6.49 \times 10^4 \text{ N/m}$$

2. How many side-by-side atomic chains (long springs) are there in this wire? This is the same as the number of atoms on the bottom surface of the copper wire (see Figure 4.15).
Cross-sectional area of wire:

$$A_{wire} = (1 \times 10^{-3} \text{ m})^2 = 1 \times 10^{-6} \text{ m}^2$$

Cross-sectional area of one copper atom (even though an atom itself is spherical, each atom occupies a cubical space in the crystal):

$$A_{1 \text{ atom}} \approx (2.28 \times 10^{-10} \text{ m})^2 = 5.20 \times 10^{-20} \text{ m}^2$$

Number of side-by-side atomic chains in the wire:

$$N_{chains} = \frac{A_{wire}}{A_{1 \text{ atom}}} = 1.92 \times 10^{13}$$

top of wire

edge of wire

bottom of wire

Figure 4.15 A wire modeled as an assembly of side-by-side chains of balls and springs, viewed from the bottom. Colored lines connect the bottom layer of atoms in the wire (only nine ball-spring chains are shown here). The horizontal bonds are not shown; they aren't stretched significantly.

3. How many interatomic bonds are there in one atomic chain running the length of the wire?

$$N_{\text{bonds in 1 chain}} = \frac{L_{\text{wire}}}{d} = \frac{2 \text{ m}}{2.28 \times 10^{-10} \text{ m}} = 8.77 \times 10^9$$

4. What is the stiffness $k_{s,i}$ of a single interatomic spring?

Using our results about springs linked end-to-end and springs arrayed side-by-side:

$$k_{s,i} = \frac{(k_{s,\text{wire}})(N_{\text{bonds in 1 chain}})}{(N_{\text{chains}})}$$

$$k_{s,i} = \frac{(6.49 \times 10^4 \text{ N/m})(8.77 \times 10^9)}{(1.92 \times 10^{13})} = 27 \text{ N/m}$$

An interatomic bond in copper is stiffer than a slinky, but less stiff than a pogo stick. The stiffness of a single interatomic bond is very much smaller than the stiffness of the entire wire (which we found to be around 6×10^4 N/m).

4.6 STRESS, STRAIN, AND YOUNG'S MODULUS

? Suppose we had used a different copper wire, which was 3 meters long, and had a circular cross section, with a diameter of 0.9 mm. Would our result for the interatomic bond stiffness have been different?

The result should not be different, because the interatomic bond stiffness is a property of the material (solid copper). The macroscopic dimensions of the wire will not change the intrinsic properties of copper.

You will not find tables of interatomic spring stiffness for different solid materials in reference libraries. Instead, what is published is a macroscopic quantity called Young's modulus. Like density, and interatomic spring stiffness, Young's modulus is a property of a particular material (for example, copper), and is independent of the shape or size of a particular object made of that material. Young's modulus is a macroscopic measure of the "stretchability" of a solid material. It relates the fractional change in length of an object to the force per square meter of cross sectional area applied to the object.

Strain

The longer a wire is, the more atomic bonds it has along its length, and the more it will stretch when a force is applied to it. We need to take into account not just the measured stretch, but also the original length of the wire. If the wire has a length L, we call the amount of stretch of the wire ΔL (a small increment in the length, what we have called s in a spring), and the fractional stretch $\Delta L/L$ is called the "strain."

DEFINITION OF STRAIN

$$\text{strain} \equiv \frac{\Delta L}{L}$$

Stress

As we saw above, all the chains of atoms help hold up the weight attached to the wire, so we need to take into account not just the tension force F_T, but also the cross-sectional area of the wire A (Figure 4.16). The tension force per unit area is called the "stress."

DEFINITION OF STRESS

$$\text{stress} \equiv \frac{F_T}{A}$$

Young's modulus

As long as the strain isn't too big, the strain is proportional to the stress. At the atomic level, the stress F_T/A can be related to the force that each long chain of atomic bonds must exert, and the strain $\Delta L/L$ can be related to the stretch of the interatomic bond. The ratio of stress to strain is a property of the material and differs for steel, aluminum, etc.

DEFINITION OF YOUNG'S MODULUS

$$Y \equiv \frac{\text{stress}}{\text{strain}} = \frac{\left(\dfrac{F_T}{A}\right)}{\left(\dfrac{\Delta L}{L}\right)}$$

Young's modulus is the ratio of stress to strain for a particular material. Young's modulus is a property of the material, and does not depend on the size or shape of an object. The stiffer the material, the larger is Young's modulus.

Using Young's modulus, a quantitative form of the relation between stress and strain in a wire is written as follows:

$$\frac{F_T}{A} = Y\frac{\Delta L}{L}$$

Note the similarity to the spring force law, $F_T = k_s s \; (= k_s \Delta L)$.

You may have a chance to measure Young's modulus in a laboratory experiment, and from your measurements, to determine the interatomic spring stiffness for a particular material (Problem 4.HW.71).

Limit of applicability of Young's modulus

If you apply too large a stress, the wire "yields" (stretches a great deal) or breaks, and the proportionality of stress and strain is no longer even approximately true. Figure 4.17 shows a graph of the strain $\Delta L/L$ that results as a function of the applied stress F_T/A for a particular aluminum alloy.

You see that for moderate stress the resulting strain is proportional to the applied stress: double the stress, double the strain. But once you reach the yield stress, a further slight increase in the stress leads to a very large increase in the length of the metal. This can be quite dramatic: as you add more weights to the end of a wire the wire gets slightly longer, then suddenly it starts to lengthen a lot, very rapidly, and then the wire breaks. (In materials reference manuals stress is usually plotted against strain, so that the slope of the line is equal to Young's modulus. We plot strain vs. stress to emphasize that stress is the *cause* and strain the *effect*.)

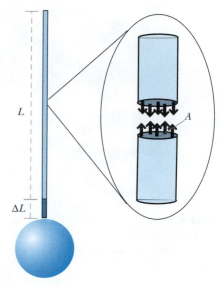

Figure 4.16 The heavy ball stretches the wire an amount ΔL. A is the cross-sectional area—the area of a slice through the wire, perpendicular to its length.

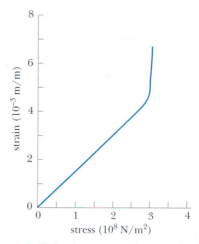

Figure 4.17 Strain vs. stress for a particular aluminum alloy.

4.X.5 In Figure 4.17, what is the approximate value of Young's modulus for this aluminum alloy?

4.X.6 Suppose we hang a heavy ball with a mass of 10 kg (about 22 pounds) from a steel wire 3 m long that is 3 mm in diameter. Steel is very stiff, and Young's modulus for steel is unusually large, 2×10^{11} N/m^2. Calculate the stretch ΔL of the steel wire. This calculation shows why in many cases it is a

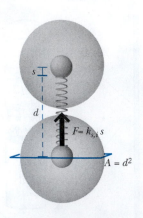

Figure 4.18 A single stretched bond in a solid wire. The effective cross-sectional area of one atom is d^2, and the force acting on that portion is $k_{s,i}s$.

very good approximation to pretend that the wire doesn't stretch at all ("ideal non-extensible wire").

Relating Young's modulus to interatomic spring stiffness

From macroscopic measurements of Young's modulus we can calculate an approximate spring stiffness $k_{s,i}$ for the interatomic bond by again making a "macro-micro" connection.

Consider a single stretched interatomic bond in a wire that is being stretched. Let d represent both the relaxed length of an interatomic bond and the diameter of one atom. The cross-sectional area occupied by one atom is d^2 (Figure 4.18) (recall that even though an atom is approximately spherical, it occupies a cube of space in the crystal lattice).

? What is the magnitude of the stress on one atom in terms of the interatomic force $k_{s,i}s$?

If the stretch of the interatomic bond is s, the atomic stress (force per unit area) is $\dfrac{k_{s,i}s}{d^2}$.

? In terms of the stretch s of the interatomic "spring," what is the strain for one atom (change of height of the atom, divided by its normal height)?

The strain for one atom is s/d, because its height was originally d, and the change in its height is s. Therefore we can express Young's modulus in terms of atomic quantities:

YOUNG'S MODULUS IN TERMS OF ATOMIC QUANTITIES

$$Y = \frac{(k_{s,i}s/d^2)}{(s/d)} = \frac{k_{s,i}}{d}$$

Where $k_{s,i}$ is the stiffness of an interatomic bond in a solid, and d is the length of an interatomic bond (and the diameter of an atom).

If you know Young's modulus for some metal, you can calculate the effective stiffness of the interatomic bond, modeled as a spring: $k_{s,i} = Yd$, where d is the atomic diameter in the metal. This is another example of an important theme, that of relating macroscopic properties to microscopic (atomic-level) properties.

4.7 COMPRESSION (NORMAL) FORCES

A closely related situation is that of a lead brick lying at rest on a metal table (Figure 4.19). In this situation the table is compressed, while in the previous discussion the wire was stretched.

Take the lead brick as the system of interest (Figure 4.20). The Earth pulls down on the brick of mass M, and the magnitude of the gravitational force is Mg. The table pushes up on the brick with a force of magnitude F_{Table}. Since the brick remains at rest, the vertical component of the net force must be zero:

$$0 = +F_{\text{Table}} - Mg$$

We conclude that the force F_{Table} exerted by the table is equal to the weight Mg of the brick, an unsurprising result.

How does the metal table exert a force on the brick? If the lead brick were extremely heavy, we might not be surprised to observe the table sagging under its weight. Even if the table does not sag visibly, the molecules in the top

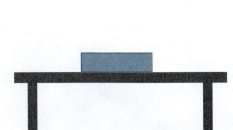

Figure 4.19 A brick lies on a table.

Figure 4.20 Forces acting on the brick.

layers of the table are pushed down by molecules in the bottom layer of the brick. The table actually deforms somewhat—its interatomic "springs" are compressed. (Of course, atoms in the brick are compressed somewhat, too.)

Atoms in the table are squeezed closer than their normal equilibrium positions, compressing the springlike interatomic bonds. Figure 4.21 schematically shows the contact region between the brick and the table.

It would be appropriate to call the upward force of the table F_C for "compression force," but it is common practice to label this kind of force as a "normal" force. The word is used in the mathematical sense, meaning "perpendicular to", because the direction of the force is perpendicular to the surface of the table. You may see "normal" forces denoted F_N, to indicate that they are "normal" forces.

> Note: Neither "normal" force nor "tension" force is a different kind of fundamental force! These are merely names for interatomic electric interactions.

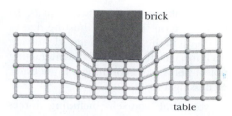

Figure 4.21 A schematic view of a portion of a brick resting on a portion of a table. Interatomic "springs" in the contact region are compressed by the weight of atoms above.

Tarzan and the vine revisited

We now know that interatomic bonds in the vine stretch when Tarzan hangs from the vine. It is not yet clear why the vine breaks when Tarzan swings on it, but not when he hangs motionless. To learn more we need to apply the Momentum Principle.

4.8 CONTACT FORCES AND THE MOMENTUM PRINCIPLE

In the preceding sections we have applied the Momentum Principle to systems in which contact forces (tension or compression forces) were acting on an object. Let's examine one of these situations again, in the light of a more detailed understanding of how tension and compression forces work. A 1 kg ball hangs motionless at the end of a wire. The top of the wire is attached to an iron support. How large an upward force does the wire exert on the ball?

> System: Ball
> Surroundings: Earth, wire

? Why isn't the iron support included in the surroundings? Why doesn't the force diagram show a force due to the support?

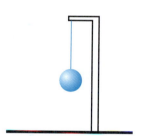

We have chosen the ball to be the system. Only two kinds of forces can act on the ball: forces involving interaction at a distance, like gravitational and electric forces, and contact forces. To exert a contact force, the atoms of one object must be in contact with the atoms of another object.

The support does not touch the ball, so the interatomic bonds in the ball are not stretched or compressed by the support. The support does exert a very tiny gravitational force on the ball (acting over a distance), but this force is too small to have a noticeable effect.

It is of course important for the iron support to be present, but the support does not interact directly with the ball. The support exerts a force on the wire, but the wire is not the system of interest; it does get taken into account indirectly, since the wire would not be under tension if it were not supported by something.

Figure 4.22 A heavy ball hangs motionless at the end of a thin wire.

An interaction involves two objects, so each force on the system must be due to an object in the surroundings. Identifying each force by the name of the interacting object helps avoid counting a force twice, or missing a force.

If you find yourself including forces named "tension", "normal", "gravitational", or "centripetal", you are in danger of double-counting forces, or including non-existent forces. Don't do this.

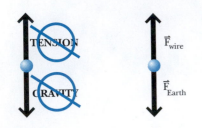

Figure 4.23 To avoid missing or double-counting forces, identify each force with the name of the interacting object in the surroundings.

RULES FOR IDENTIFYING FORCES ON A SYSTEM

Systematically following these rules will help you avoid making errors in an analysis involving contact forces:

1. Choose a system.

2. Include in the surroundings all objects interacting with the system at a distance (gravitationally or electrically). Note that you must list an object; the Earth is an object, but "gravity" is not. Label each distance force with the name of the interacting object.

3. Include in the surroundings all objects that are touching the system. These are the only objects that can exert contact forces (tension or compression) on the system. Label each contact force with the name of the interacting object.

Don't include any other forces on the system—there aren't any!

4.X.7 You hit a baseball with a bat. Choose the baseball as the system. At the instant of contact with the bat, what objects should be included in the surroundings (what objects exert forces on the ball)?

4.X.8 You are playing with a yo-yo. Choose the yo-yo as the system. As the yo-yo moves downward, you pull up on the string. At this moment, what objects should be included in the surroundings (what objects exert forces on the yo-yo?

4.X.9 Tarzan swings on a vine, which is attached to the branch of a tree. Choose Tarzan as the system. What are the objects in the surroundings that exert forces on Tarzan?

4.X.10 You ski down a hill. Choose yourself plus your skis as the system. What objects in the surroundings exert forces on the system?

4.9 DERIVATIVE FORM OF THE MOMENTUM PRINCIPLE

The forms of the Momentum Principle we have used so far

$$\Delta \vec{p} = \vec{F}_{net} \Delta t$$

and

$$\vec{p}_f = \vec{p}_i + \vec{F}_{net} \Delta t$$

are particularly useful when we know the momentum at a particular time and want to predict what the momentum will be at a later time. We use these forms in repetitive computer calculations to predict future motion.

In solving problems involving objects whose momentum was not changing at all, such as a ball hanging motionless on the end of a wire (page 107), we saw that although we had to choose a value for Δt, it didn't matter what value we chose, as long as it was greater than zero. In such problems, we could also have used a different form of the Momentum Principle, which we obtain by dividing by the time interval Δt:

$$\frac{\Delta \vec{p}}{\Delta t} = \vec{F}_{net}$$

Just as we did in finding the instantaneous velocity of an object (see Section 1.7.6), we can find the instantaneous rate of change of momentum by letting the time interval Δt approach zero (an infinitesimal time interval). The ratio of the infinitesimal momentum change (written as $d\vec{p}$) to the infinitesimal time interval (written as dt) is the instantaneous rate of change of the momentum. As the time interval gets very small, $\Delta t \rightarrow 0$, the ratio of the momentum change to the time interval approaches the ratio of the infinitesimal quantities

$$\lim_{\Delta t \to 0} \frac{\Delta \vec{p}}{\Delta t} = \frac{d\vec{p}}{dt}$$

and we obtain the following form:

THE MOMENTUM PRINCIPLE (DERIVATIVE FORM)

$$\frac{d\vec{p}}{dt} = \vec{F}_{net}$$

In words, "the instantaneous time rate of change of the momentum of an object is equal to the net force acting on the object." In calculus terms, since the limit of a ratio of infinitesimal quantities is called a derivative, we can say that "the derivative of the momentum with respect to time is equal to the net force acting on the object."

This form of the Momentum Principle is useful when we know something about the rate of change of the momentum at a particular instant. Knowing the rate of change of momentum, we can use this form of the Momentum Principle to deduce the net force acting on the object, which is numerically equal to the rate of change of momentum. Knowing the net force, we may be able to figure out particular contributions to the net force.

4.X.11 At a certain instant the z component of the momentum of an object is changing at a rate of 4 kg·m/s per second. At that instant, what is the z component of the net force on the object?

4.X.12 If an object is sitting motionless, what is the rate of change of its momentum? What is the net force acting on the object?

4.X.13 If an object is moving with constant momentum $\langle 10, -12, -8 \rangle$ kg·m/s, what is the rate of change of momentum $d\vec{p}/dt$? What is the net force acting on the object?

Acceleration

In previous studies you may have seen a simplified, approximate form of the Momentum Principle, written $ma = F$. Where does this come from, and under what circumstances it is useful?

If we make the following assumptions:
• The mass of the system is constant,
• The speed of the system is much less than the speed of light,

then if we take the approximate derivative of $\vec{p}$ with respect to time, we get:

$$\frac{d\vec{p}}{dt} \approx \frac{d(m\vec{v})}{dt} \quad \text{(assuming } v \ll c\text{)}$$

$$\frac{d(m\vec{v})}{dt} = \left(\frac{dm}{dt}\right)\vec{v} + m\frac{d\vec{v}}{dt} \quad \text{(product rule for derivatives)}$$

$$\frac{d(m\vec{v})}{dt} = 0 + m\frac{d\vec{v}}{dt} \quad \text{(assuming system mass is constant)}$$

The rate of change of velocity with respect to time is called acceleration, and is usually denoted by the symbol $\vec{a}$.

DEFINITION OF ACCELERATION

$$\vec{a} \equiv \frac{d\vec{v}}{dt}$$

Acceleration is the rate of change of velocity with respect to time. It is a vector quantity.

The derivative of a vector

The derivative of a vector is itself a vector. We are already familiar with one such derivative, the instantaneous velocity:

$$\vec{v} = \frac{d\vec{r}}{dt} = \langle \frac{dx}{dt}, \frac{dy}{dt}, \frac{dz}{dt} \rangle$$

Like any other vector, the derivative of a vector has both a magnitude and a direction.

Using this definition we can write the approximate rate of change of momentum as:

$$\frac{d\vec{p}}{dt} \approx m\vec{a}$$

and an approximate form of the Momentum Principle, usually referred to as Newton's second law, as:

$$m\vec{a} \approx \vec{F}_{net} \quad \text{(nonrelativistic form; constant mass)}$$

A highly simplified scalar form of this equation, $ma = F$, deals only with magnitudes, involves a single force, and gives no information about directions.

The definition of acceleration is useful, because it is occasionally important to know the acceleration of an object (the rate at which its velocity is changing). However, we will generally find that the full form of the Momentum Principle, $d\vec{p}/dt = \vec{F}_{net}$, is more appropriate for our analyses.

- $d\vec{p}/dt = \vec{F}_{net}$ is relativistically correct (correct at any speed), and we will sometimes deal with fast-moving particles.
- $d\vec{p}/dt = \vec{F}_{net}$ involves momentum, which is a conserved quantity; we need to work with momentum when we analyze situations such as collisions, as we did in Chapter 2.
- $d\vec{p}/dt = \vec{F}_{net}$ is correct even if the mass of an object is not constant, such as a rocket with exhaust gases ejecting out the back.
- $d\vec{p}/dt = \vec{F}_{net}$ is a vector principle, and contains information about directions. The arrows over the symbols are extremely important; they remind us that there are really three separate component equations, for x, y, and z.
- $d\vec{p}/dt = \vec{F}_{net}$ reminds us that we have to add up all vector forces to give the "net" force, so we write $\vec{F}_{net}$, not just F.

4.X.14 The velocity of a 80 gram ball changes from $\langle 5, 0, -3 \rangle$ m/s to $\langle 5.02, 0, -3.04 \rangle$ m/s in 0.01 s, due to the gravitational attraction of the Earth and to air resistance. What is the acceleration of the ball? What is the rate of change of momentum of the ball? What is the net force acting on the ball?

4.10 MOMENTUM NOT CHANGING (STATICS)

We saw in Chapter 1 that if a system is in uniform motion its momentum is constant, and does not change with time. A special case of uniform motion is a situation in which object is at rest and remains at rest; this situation is called "static equilibrium," or simply "equilibrium."

Net force in equilibrium

? If the system never moves, what is $\dfrac{d\vec{p}}{dt}$?

The momentum of the system isn't changing, so the rate of change of the momentum is zero:

$$\frac{d\vec{p}}{dt} = \vec{0} = \langle 0, 0, 0 \rangle \quad \text{(equilibrium)}$$

? What does this imply about the net force acting on the system?

From the derivative form of the Momentum Principle, $d\vec{p}/dt = \vec{F}_{net}$, we deduce that the net force acting on a system that never moves must be zero:

$$\vec{F}_{net} = \vec{0} = \langle 0, 0, 0 \rangle \quad \text{(equilibrium)}$$

Statics problems

Problems involving equilibrium are called "statics" problems. If you are studying engineering you may take an entire course on this topic. In traditional problems involving equilibrium situations you are usually asked to deduce the unknown magnitude of a contact force by applying the Momentum Principle. We have solved several simple problems of this kind earlier in the chapter (for examples, see page 107, and page 110), although we used the finite time version of the Momentum Principle. Harder statics problems often involve several contact forces, acting in more than one direction, so one must work with x, y, and z components of forces. These problems are not conceptually more difficult; they simply require more algebra.

Forces exerted by strings

Because a string is flexible, the tension force exerted by a string is always in the direction of the string. A string cannot exert a force perpendicular to the string. And, of course, a string cannot push—it can only pull.

Example: **Equilibrium with tension forces**

A 3 kg block hangs from string 1 (Figure 4.24). Then you pull the block to the side by pulling horizontally with a second string (string 2). (a) What is the magnitude F_2 of the force you must apply to string 2 in order for the block to hang motionless with string 1 at an angle of 33° to the vertical, as shown in Figure 4.24? (b) In this situation, what is the tension F_1 in string 1? (c) In which situation are the interatomic bonds in string 1 stretched more: Block hanging straight down, suspended from string 1, or block hanging at an angle, pulled on by string 1 and string 2?

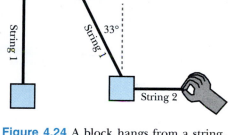

Figure 4.24 A block hangs from a string. Then you pull the block to the side with another string.

System: Block
Surroundings: String 1, string 2, Earth
Momentum Principle:

$$\frac{d\vec{p}}{dt} = \vec{F}_{net}$$

$$\langle 0, 0, 0 \rangle = \vec{F}_1 + \vec{F}_2 + \vec{F}_{Earth}$$

Write out forces as vectors, using direction cosines:

$$\hat{F}_1 = \langle \cos(90° + 33°), \cos 33°, 0 \rangle$$

$$\vec{F}_1 = |\vec{F}_1|\hat{F}_1 = |\vec{F}_1|\langle \cos 123°, \cos 33°, 0 \rangle$$

$$\vec{F}_2 = |\vec{F}_2|\hat{F}_2 = |\vec{F}_2|\langle 1, 0, 0 \rangle$$

$$\vec{F}_{Earth} = \langle 0, -(3 \text{ kg})(9.8 \text{ N/kg}), 0 \rangle = \langle 0, -29.4, 0 \rangle \text{ N}$$

The angle between the $+x$ axis and the string is $(90° + 33°)$.

The procedure for finding a unit vector from a known angle using direction cosines is discussed in Chapter 1.

Separate components:

$$\frac{dp_x}{dt} = 0 = F_{net,x} \text{ so } 0 = |\vec{F}_1|\cos 123° + |\vec{F}_2| + 0$$

$$\frac{dp_y}{dt} = 0 = F_{net,y} \text{ so } 0 = |\vec{F}_1|\cos 33° + 0 - 29.4 \text{ N}$$

$$\frac{dp_z}{dt} = 0 = F_{net,z} \text{ so } 0 = 0 + 0 + 0 \text{ (not very informative)}$$

Recall that the x component of momentum is affected only by the x component of force; same for y and z.

We now have two equations in two unknowns ($|\vec{F}_1|$ and $|\vec{F}_2|$). Solve for the unknown force magnitudes.

(a) $|\vec{F}_1| = \dfrac{29.4 \text{ N}}{\cos 33°} = 35.1 \text{ N}$

Use the y equation to solve for $|\vec{F}_1|$

Use x and y equations to solve for $\left|\vec{\mathbf{F}}_2\right|$

(b) $\left|\vec{\mathbf{F}}_2\right| = \left|\vec{\mathbf{F}}_1\right|(\cos 123°) = (35.1\ \text{N})(\cos 123°) = 19.1\ \text{N}$

(c) When the block hangs straight down, the force exerted by string 1 must be equal in magnitude to the force by the Earth on the block, which is 29.4 N. Evidently the tension in string 1 is greater when the block is pulled to the side than when the block hangs straight down. The interatomic bonds in string 1 must be stretched more when the block is pulled to the side.

Check: Units correct (N) Special cases: consider other angles.

Call the angle between the string and the vertical θ. If $\theta = 0°$ instead of $33°$, the string 1 is vertical, and we expect the tension in this string to be mg. Rewriting the y equation algebraically:

$$\left|\vec{\mathbf{F}}_1\right| = \frac{mg}{\cos \theta}$$

Since $\cos(0°) = 1$, we do indeed find $F_1 = mg$. Also, $F_2 = 0$, as it should be in this case (no pulling).

On the other hand, if $\theta = 90°$, $\cos(90°) = 0$, the strings are horizontal, with $F_1 = mg/0$, which is infinitely large and impossible. This makes sense: if both strings are horizontal, there is no y component to support the block.

Tarzan and the vine: Further insights

From the preceding example it is clear that the tension force exerted by a string (or a wire, or a vine) is not always the same. In different situations the tension force exerted by an object may be different, due to different amounts of stretch in interatomic bonds. You may suspect that the tension in the vine is greater when Tarzan swings on the vine than when he hangs motionless; this will turn out to be true, as we will be able to show later in this chapter.

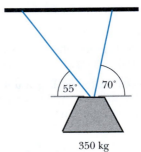

Figure 4.25 A load hangs from two ropes. What are the tensions in the ropes?

Example: Equilibrium with tension forces

A 350 kg load is suspended from two ropes as shown in Figure 4.25. Calculate the tension in each rope (that is, the magnitude of the tension force exerted by each rope).

System: Load
Surroundings: Rope 1, rope 2, Earth
Momentum Principle:

$$\frac{d\vec{\mathbf{p}}}{dt} = \vec{\mathbf{F}}_{\text{net}}$$

$$\langle 0, 0, 0 \rangle = \vec{\mathbf{F}}_1 + \vec{\mathbf{F}}_2 + \vec{\mathbf{F}}_{\text{Earth}}$$

Write out forces as vectors, using direction cosines:

$\hat{\mathbf{F}}_1 = \langle -\cos 55°, \cos(90° - 55°), 0 \rangle$

$\vec{\mathbf{F}}_1 = \left|\vec{\mathbf{F}}_1\right|\hat{\mathbf{F}}_1 = \left|\vec{\mathbf{F}}_1\right|\langle -\cos 55°, \cos 35°, 0 \rangle$

$\hat{\mathbf{F}}_2 = \langle \cos 70°, \cos(90° - 70°), 0 \rangle$

$\vec{\mathbf{F}}_2 = \left|\vec{\mathbf{F}}_2\right|\hat{\mathbf{F}}_2 = \left|\vec{\mathbf{F}}_2\right|\langle \cos 70°, \cos 20°, 0 \rangle$

$\vec{\mathbf{F}}_{\text{Earth}} = \langle 0, -(350\ \text{kg})(9.8\ \text{N/kg}), 0 \rangle = \langle 0, -3430, 0 \rangle\ \text{N}$

Separate components:

$$\frac{dp_x}{dt} = 0 = F_{\text{net},x} \text{ so } 0 = -|\vec{F}_1|\cos 55° + |\vec{F}_2|\cos 70° + 0$$

$$\frac{dp_y}{dt} = 0 = F_{\text{net},y} \text{ so } 0 = |\vec{F}_1|\cos 35° + |\vec{F}_2|\cos 20° - 3430 \text{ N}$$

$$\frac{dp_z}{dt} = 0 = F_{\text{net},z} \text{ so } 0 = 0 + 0 + 0 \text{ (not very informative)}$$

Recall that the x component of momentum is affected only by the x component of force; same for y and z.

We now have two equations in two unknowns ($|\vec{F}_1|$ and $|\vec{F}_2|$). Solve for the unknown force magnitudes.

$$|\vec{F}_1| = \frac{\cos 70°}{\cos 55°}|\vec{F}_2| = 0.596|\vec{F}_2|$$

$$0 = (0.596|\vec{F}_2|)\cos 35° + |\vec{F}_2|\cos 20° - 3430 \text{ N}$$

$$|\vec{F}_2|(0.596\cos 35° + \cos 20°) = 3430 \text{ N}$$

$$|\vec{F}_2| = \frac{3430 \text{ N}}{0.596\cos 35° + \cos 20°} = 2402 \text{ N}$$

$$|\vec{F}_1| = 0.596|\vec{F}_2| = 0.596(2402 \text{ N}) = 1431 \text{ N}$$

Check: Units correct (N). Can plug these results back into the momentum equations to verify that they satisfy those equations.

Uniform motion: Momentum not changing

If a system is at rest and remains at rest, it is clear that its momentum is not changing, and therefore that the net force on the system must be zero. Newton's first law reminds us of another situation in which momentum is not changing: uniform motion. If the momentum of a system is not zero, but remains constant, then the net force on the object must be zero, just as if the object were at rest.

Example: Uniform motion

You place a 4 kg crate on the floor of an elevator which is temporarily stopped. Then the elevator starts moving, briefly speeding up, then moving upward at a constant speed of 2 m/s for many seconds (it is in a very tall building). While the elevator is moving upward at constant speed, what is the magnitude of the force exerted by the elevator floor on the crate?

During the brief time when the elevator was speeding up, the crate's momentum was increasing, and the net force on the crate was not zero.

Now, however, the crate's momentum is constant, and therefore the rate of change of its momentum is zero. In this situation the net force on the crate must now also be zero.

System: Crate
Surroundings: Earth, elevator floor
Momentum Principle:

$$\frac{d\vec{p}}{dt} = \vec{F}_{\text{net}}$$

$$\langle 0, 0, 0 \rangle = \vec{F}_{\text{Floor}} + \vec{F}_{\text{Earth}}$$

$$y: \frac{dp_y}{dt} = 0 = |\vec{F}_{\text{Floor}}| - mg$$

$$|\vec{F}_{\text{Floor}}| = (4 \text{ kg})(9.8 \text{ N/kg}) = 39.2 \text{ N}$$

Check: Units ok (N). Reasonable? Yes, agrees with Newton's first law.

Note that the speed of the crate (2 m/s) does not actually appear anywhere in the solution. The crate and elevator could have been traveling at any speed—even in principle a speed close to the speed of light—and this analysis would still be valid.

Further discussion

While the elevator and the crate move at constant speed, the interatomic springs in the elevator floor will be compressed exactly as much as they were when there was no motion. This is an example of the principle of relativity, that physical laws (in this case, the Momentum Principle) work in the same way in all reference frames that are in uniform motion (that is, constant velocity) with respect to the cosmic microwave background, which is the stage on which phenomena play themselves out.

Preview of the Angular Momentum Principle

In many equilibrium situations, the Momentum Principle is not sufficient for carrying out a full analysis and we also need the Angular Momentum Principle, which we will discuss in a later chapter.

For example, consider two children sitting motionless on a seesaw (Figure 4.26). Choose as the system the two children and the (lightweight) board. For this choice of system, the objects in the surroundings that exert significant (external) forces on the system are the Earth (pulling down) and the support pivot (pushing up). The momentum of the system isn't changing, so the Momentum Principle correctly tells us that the net force must be zero, and the support pivot pushes up with a force equal to the weight of the children (we are neglecting the low-mass board).

However, the Momentum Principle alone doesn't tell us where the children have to sit in order to achieve balance and equilibrium. The force on a child exerts a twist (the technical term is "torque") about the pivot which tends to make the board turn. The torque associated with the forces on the two children have to add up to zero (one tends to twist the board clockwise, the other counterclockwise). Torque is defined as force times lever arm, and a nonzero net *torque* causes changes in a quantity called angular momentum, just as a nonzero net *force* causes changes in the ordinary momentum we have been studying.

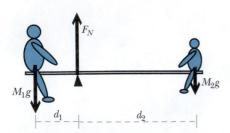

Figure 4.26 Two children sit on a seesaw in equilibrium.

4.11 FINDING THE RATE OF CHANGE OF MOMENTUM

In order to apply the derivative form of the Momentum Principle, it is important to be able to decide if $d\vec{p}/dt$, the instantaneous rate of change of momentum, is zero or nonzero, and if it is nonzero, what its magnitude and direction are. The procedure for finding $d\vec{p}/dt$ is based on the procedure for finding $\Delta\vec{p}$ which you practiced in Chapter 1, and requires that you draw a simple diagram, as shown in Figure 4.27.

1. Draw two arrows:

   one representing $\vec{p}_i$, the momentum of the object a short time before the instant of interest, and

   a second arrow representing $\vec{p}_f$, the momentum of the object a short time after the instant of interest

2. Graphically find $\Delta\vec{p} = \vec{p}_f - \vec{p}_i$ by placing the arrows tail to tail, and drawing the resultant arrow starting at the tip of $\vec{p}_i$ and going to the tip of $\vec{p}_f$

3. This arrow indicates the direction of $\Delta\vec{p}$, which is the same as the direction of $d\vec{p}/dt$, provided the time interval involved is sufficiently small.

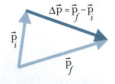

Figure 4.27 Calculation of $\Delta\vec{p}$.

You may wish to review Exercises 1.X.46 - 1.X.48 in Chapter 1, which deal with finding change of momentum.

A system momentarily at rest

It is clear that if a system remains at rest, its momentum is not changing. Likewise, $d\vec{p}/dt = \vec{0}$ for a system whose momentum (magnitude and direction) remains constant. However, situations in which the momentum of an object is only momentarily zero can be confusing if you do not use the procedure for finding $d\vec{p}/dt$ described in the section above. Two examples of such situations are these:

- You hang a block on a spring, stretch the spring downwards, and let go. The block oscillates straight up and down. At the moment the block reaches its lowest point, is the rate of change of its momentum zero or nonzero? If nonzero, what is the direction of $d\vec{p}/dt$?
- Tarzan hangs from a vine, swinging back and forth in a gentle arc. At the moment when he reaches his maximum displacement from the vertical and is momentarily at rest, is the rate of change of Tarzan's momentum zero or nonzero? If nonzero, what is the direction of $d\vec{p}/dt$?

? Before reading further, decide for yourself whether $d\vec{p}/dt$ is zero or nonzero, and if nonzero, what the direction of $d\vec{p}/dt$ is, for each of the situations described in the bulleted paragraphs above. Draw diagrams and use the graphical procedure described above to find $\Delta\vec{p}$ and $d\vec{p}/dt$.

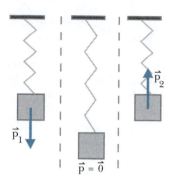

Figure 4.28 The momentum of a block oscillating on a spring, at an instant before it comes to a momentary rest at the bottom of its path (1) and an instant after it starts back upward (2).

Example: **Block oscillating on a spring**

Consider the situation of a block oscillating up and down on a spring, as shown in Figure 4.28. When the block comes to a momentary rest at the bottom of its oscillation, is $d\vec{p}/dt$ zero or nonzero? If nonzero, what is its direction?

Use the procedure for finding $d\vec{p}/dt$ given above:

Immediately before reaching bottom, the block is still moving downward, and the momentum of the block is in the $-y$ direction ($\vec{p}_1$ in Figure 4.28).

Immediately after reaching bottom, the block begins moving upward, and its momentum is in the $+y$ direction ($\vec{p}_2$ in Figure 4.28). $\Delta\vec{p} = \vec{p}_2 - \vec{p}_1$ is therefore in the $+y$ direction (Figure 4.29).

Therefore $d\vec{p}/dt$ is nonzero, and its direction is $+y$.

Figure 4.29 The direction of $d\vec{p}/dt$ is the direction of $\Delta\vec{p} = \vec{p}_2 - \vec{p}_1$.

? At this instant, which is larger in magnitude, the force on the block by the spring or the force on the block by the Earth?

If the net force on the block were zero, the block would remain at rest. Since the only way to change the momentum of a system is to apply a nonzero net force, there must be a nonzero net force on the block at this instant, acting in the direction of $d\vec{p}/dt$. Therefore the upward force on the block by the stretched spring is larger than the downward force on the block by the Earth.

Example: **Tarzan swinging on a vine**

Tarzan hangs from a vine, swinging back and forth. At the moment when he reaches his maximum displacement from the vertical, as indicated in Figure 4.30, Tarzan is momentarily at rest. At this moment, is the rate of change of Tarzan's momentum zero or nonzero? If nonzero, what is the direction of $d\vec{p}/dt$?

Use the procedure for finding $d\vec{p}/dt$.

Both immediately before this moment and immediately after this moment, Tarzan's momentum is nonzero. The direction of his momentum at these instants is shown in Figure 4.31, as is $\Delta\vec{p} = \vec{p}_2 - \vec{p}_1$. Therefore $d\vec{p}/dt$ at the instant Tarzan is motion-

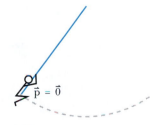

Figure 4.30 At the extreme point of his swing, Tarzan is momentarily at rest.

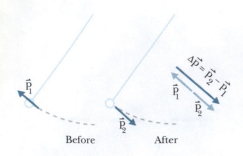

Before   After

Figure 4.31 Tarzan's momentum an instant before and an instant just after he is momentarily at rest, and $\Delta\vec{p}$.

Figure 4.32 The direction of $d\vec{p}/dt$ at the instant Tarzan is motionless.

less is nonzero, and its direction is the direction of $\Delta\vec{p}$, as shown in Figure 4.32.

Note that it was important to consider a very short time interval in order to get the direction of $d\vec{p}/dt$. The direction of $\Delta\vec{p}$ would be different for a longer time interval (and not the same as the direction of $d\vec{p}/dt$).

Since Tarzan's momentum is changing (he does not remain at rest), the net force on Tarzan (which is the vector sum of the tension force by the vine and the force by the Earth) is nonzero. The direction of this net force is the same as the direction of $d\vec{p}/dt$ at this instant.

4.X.15 A block oscillating up and down on a spring comes to the top of its path, where it is momentarily at rest. At the instant it is at rest, is $d\vec{p}/dt$ zero or nonzero? If nonzero, what is its direction?

4.X.16 You throw a ball straight up into the air. At the instant the ball reaches its highest location, is $d\vec{p}/dt$ zero or nonzero? If nonzero, what is its direction?

4.X.17 A basketball travels straight down, hits the floor, and bounces back straight up. At the moment when it is in contact with the floor and its speed is zero, is the rate of change of its momentum zero or nonzero? If nonzero, what is the direction of $d\vec{p}/dt$?

4.X.18 A tennis ball traveling horizontally to the right hits a wall and bounces straight back with nearly the same speed. While the ball is in contact with the wall, is $d\vec{p}/dt$ zero or nonzero? If nonzero, what is its direction?

Momentarily at rest *vs.* uniform motion and equilibrium

From the previous examples and exercises it is clear that a system that is only momentarily at rest is not in the same situation as a system in equilibrium.

? What is the difference in $d\vec{p}/dt$ between a system in equilibrium, a system in uniform motion, and a system that is momentarily at rest?

For a system in equilibrium $d\vec{p}/dt = \vec{0}$

For a system in uniform motion $d\vec{p}/dt = \vec{0}$

For a system that is only momentarily at rest, $d\vec{p}/dt \neq \vec{0}$

? Is the net force on a system that is momentarily at rest zero or nonzero?

Since the only way to change the momentum of a system is to apply a nonzero net force, there must be a nonzero net force acting on such a system.

4.12 PARALLEL AND PERPENDICULAR COMPONENTS

Because momentum is a vector, there are two ways in which momentum can change: the magnitude of momentum can change (corresponding to a change in speed), or the direction of momentum can change. We can see this mathematically by "factoring" momentum into the product of the magnitude of the momentum (a scalar) and a unit vector in the direction of the momentum

$$\vec{p} = |\vec{p}|\hat{p}$$

and then differentiating this expression, by applying the product rule. The resulting expression has two pieces, which correspond to the two ways in which momentum can change:

$$\frac{d\vec{p}}{dt} = \frac{d|\vec{p}|}{dt}\hat{p} + |\vec{p}|\frac{d\hat{p}}{dt}$$

This may look unfamiliar, but we can make sense of it by considering each term separately. To simplify the following discussion we'll assume that the mass of the system of interest is not changing, so a change in magnitude of momentum is due to a change in speed. (This assumption is true for almost all of the systems we will analyze, though it is not true for an object like a rocket, whose mass decreases as exhaust gases leave the rocket.) However, we do not need to make any assumptions about the speed of the system; v may be near the speed of light.

? Which term in the expression above involves a change in magnitude of momentum (change in speed)? Which term involves a change in direction?

$$\frac{d|\vec{p}|}{dt}\hat{p} \quad \text{speed changing}$$

$$|\vec{p}|\frac{d\hat{p}}{dt} \quad \text{direction changing}$$

Rate of change of magnitude of momentum

The rate of change of magnitude (a scalar quantity) should be familiar to you—it is just the instantaneous slope of a plot of the quantity *vs.* time. Physically, if only this term is nonzero it indicates that the speed of the system is changing, but the line along which it is moving is not (Figure 4.33).

The direction of $(d|\vec{p}|/dt)\hat{p}$ is either parallel or antiparallel to $\hat{p}$. If an object is speeding up, the direction of the quantity $(d|\vec{p}|/dt)\hat{p}$ is $\hat{p}$, and if the object is slowing down, the direction of the quantity $(d|\vec{p}|/dt)\hat{p}$ is $-\hat{p}$. This is because the quantity $(d|\vec{p}|/dt)$ is positive if $|\vec{p}|$ is increasing, and negative if $|\vec{p}|$ is decreasing.

Rate of change of direction of momentum

The rate of change of direction of momentum is itself a vector. Physically, if only this term is nonzero, it indicates that the speed of a system is constant, but its direction of motion is changing (Figure 4.34).

For example, if you hold the string of a yo-yo, and swing the yo-yo around and around at constant speed in a horizontal circle, the direction of the yo-yo's momentum is changing, but the magnitude of its momentum remains constant (Figure 4.34).

The direction of $|\vec{p}|(d\hat{p}/dt)$ is perpendicular to the direction of momentum at the given instant. $(d\hat{p}/dt)$ cannot have a component in the direction of $\hat{p}$. If it did, the magnitude of $\hat{p}$ would be changing. This is impossible: the magnitude of a unit vector is always 1. We can also see this graphically by applying the procedure discussed on page 122, as shown in Figure 4.35. Note that $(d\hat{p}/dt)$ does not necessarily have magnitude 1.

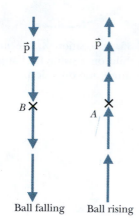

Figure 4.33 Falling ball (left): Magnitude of momentum is increasing, direction of $(d|\vec{p}|/dt)\hat{p}$ is $\hat{p}$. Rising ball (right): Magnitude of momentum is decreasing; direction of $(d|\vec{p}|/dt)\hat{p}$ is $-\hat{p}$.

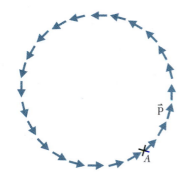

Figure 4.34 Yo-yo swung in a horizontal circle at constant speed (top view): At location A, the direction but not the magnitude of momentum is changing.

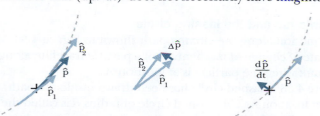

Figure 4.35 Left: Find values of $\hat{p}$ slightly before and slightly after the instant of interest. Center: compute $\Delta\hat{p}$. Right: $d\hat{p}/dt$ is perpendicular to $\hat{p}$.

Figure 4.36 At location *A*, the momentum of a child on a swing is changing in both magnitude and direction.

Figure 4.37 The unit vector $\hat{n}$ (n-hat) is perpendicular to momentum at any instant.

Speed and direction changing

If both terms are nonzero, then both the speed and direction of an object's motion are changing. As a child swings back and forth on a swing (Figure 4.36), both the magnitude and direction of the child's momentum change continuously.

A unit vector in the perpendicular direction

The unit vector in the direction of momentum is simply $\hat{p}$, When considering curving motion, it is useful to define a unit vector in the direction of $(d\hat{p}/dt)$, perpendicular to $\hat{p}$.

UNIT VECTOR N-HAT PERPENDICULAR TO MOMENTUM

$\hat{n}$ (n-hat) is a unit vector in the direction of $\dfrac{d\hat{p}}{dt}$

Since $\dfrac{d\hat{p}}{dt} \perp \hat{p}$, $\hat{n} \perp \hat{p}$ (Figure 4.37)

(Note that $\hat{n}$ is undefined at an instant when $\vec{p} = \vec{0}$, and is also undefined for an object moving in a straight line.)

We could write $\hat{n}$ as $\left(\dfrac{d\hat{p}}{dt}\right)$, or "(d-p-hat-dt)-hat", but this is awkward to say and write.

Rate of change of direction of momentum along a curving path

When Tarzan swings out over the river on a vine, he follows a curving path. We want to understand what happens to the vine to cause it to break (see page 105); this requires applying the derivative form of the Momentum Principle to Tarzan at a particular instant during his motion along this curving path.

When an object moves along a curving path, the direction of its momentum is continuously changing (and the magnitude of its momentum may be changing as well). Using geometry and algebra, we will show that for a particle traveling on a curving path, at any location we have the following:

$$\left|\frac{d\hat{p}}{dt}\right| = \frac{|\vec{v}|}{R} \quad \text{(magnitude of } d\hat{p}/dt\text{)}$$

where $|\vec{v}|$ is the particle's speed and R is the radius of a circle "kissing" the inside of the curve at that location.

We already know that the direction of $d\hat{p}/dt$ is perpendicular to the particle's momentum (i.e. in the direction of $\hat{n}$).

Our derivation will follow these steps:

- Graphically find $\Delta\hat{p}$ over an interval bracketing location *A*, as we did in Figure 4.35.
- Inscribe a "kissing" circle inside the curve.
- Use similar triangles to relate $\Delta\hat{p}$ to the radius of the circle

Change in p-hat and the kissing circle

Consider motion along the curving path shown in Figure 4.38. We want to find the rate of change of momentum of a particle traveling along this path, at the instant when the particle is at location *A*.

In Figure 4.38 a dashed circle has been drawn inside the path, tangent to the path at location A. This dashed circle of radius R is called the "osculatory" or "kissing" circle, because it just "kisses" the curving path at location *A*, fitting into the trajectory as smoothly as possible, with the circle and the trajectory sharing the same tangent ($\hat{p}$ is tangent to the path) and same radius of curvature R at location *A*.

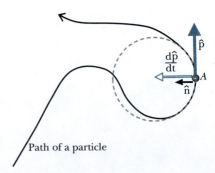

Figure 4.38 For a particle moving along a curving path, $d\hat{p}/dt$ is perpendicular to the particle's momentum, and points to the inside of the curve.

To find $\Delta\hat{p}$, pick a short interval centered on location A, and draw arrows representing the momentum of the particle at the beginning and end of that interval, as shown in Figure 4.39. Then put the initial and final unit vectors tail to tail and draw the vector $\Delta\hat{p} = \hat{p}_f - \hat{p}_i$ (final minus initial).

An important thing to notice is that the *change* in direction of momentum is not in the direction of the momentum but is perpendicular to it. $\Delta\hat{p}$ points to the left, toward the center of the kissing circle. In order to undergo such a change in the momentum, the object would have to experience a net force toward the left.

Similar triangles

We can use geometry to calculate the rate at which the direction of momentum changes. In Figure 4.39 the $\hat{p}$-triangle and the R-triangle are similar to each other, with the same acute angle θ, because the radius vectors in the R-triangle and the momentum vectors in the p-triangle are at right angles to each other. To say it another way, you can rotate one of the triangles through 90 degrees and the small acute angles will clearly be the same. Since $\hat{p}$ is a unit vector, $|\hat{p}_f| = |\hat{p}_i| = 1$.

For short times (small angles), the length of the short side of the R-triangle is approximately equal to the arc length, which is the distance the particle goes in the short time, which is equal to $|\vec{v}|\Delta t$. Since the two triangles are similar, the ratio of the side opposite the acute angle to the side adjacent to the acute angle must be the same for both triangles:

$$\frac{|\Delta\hat{p}|}{|\hat{p}|} = \frac{|\vec{v}|\Delta t}{R}$$

We are interested in the rate of change of $\hat{p}$, so rearrange the formula:

$$\frac{|\Delta\hat{p}|}{\Delta t} = \frac{|\vec{v}|}{R} \text{ since } |\hat{p}| = 1$$

Since the time is increasing, Δt is always positive and therefore the same as $|\Delta t|$, so we can express the magnitude of the rate of change of the momentum like this:

$$\left|\frac{\Delta\hat{p}}{\Delta t}\right| = \frac{|\vec{v}|}{R}$$

In the limit as Δt becomes infinitesimally small,

$$\lim_{\Delta t \to 0}\left|\frac{\Delta\hat{p}}{\Delta t}\right| = \left|\frac{d\hat{p}}{dt}\right| = \frac{|\vec{v}|}{R} \text{ on a curving path}$$

Check units

A simple check on the results of a calculation or derivation is a units check.

? Do the units on both sides of the preceding equation match?

Left hand side: $1/s$; Right hand side: $1/s$

The units match. (If they had not matched, we would know the equation was not correct, and we would have to look for errors in our derivation.)

RATE OF CHANGE OF DIRECTION OF MOMENTUM FOR CURVING MOTION

$$\frac{d\hat{p}}{dt} = \frac{|\vec{v}|}{R}\hat{n}$$

R is the radius of the kissing circle. The unit vector $\hat{n}$ gives the direction of $d\hat{p}/dt$: perpendicular to the particle's momentum, toward the center of the kissing circle (Figure 4.38). (This applies to an object in curving motion, since otherwise $\hat{n}$ is undefined.)

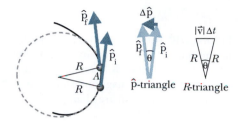

Figure 4.39 Find $\Delta\hat{p} = \hat{p}_f - \hat{p}_i$ graphically, by putting the vectors tail to tail. The two triangles are similar, with the same acute angle θ, because the sides of one triangle were originally perpendicular to the side of the other.

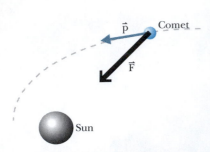

Figure 4.40 The net force on a comet due to the Sun at a particular instant. At this moment the comet is speeding up and turning—both the magnitude and direction of its momentum are changing.

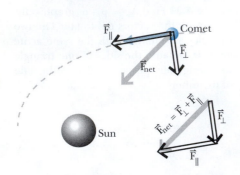

Figure 4.41 The net force on the comet (gray) expressed as the sum of its parallel and perpendicular component vectors. $\vec{F}_\parallel$ changes the speed of the comet, and $\vec{F}_\perp$ changes the direction of its motion.

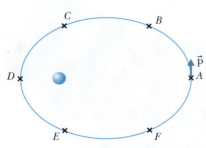

Figure 4.42 The elliptical orbit of a comet around a star. An arrow represents the momentum of the comet when it is at location A.

? Is this result valid for high speeds, when $|\vec{v}| \approx c$?

We did not use the approximate formula for momentum, so our result is valid even if a particle is traveling at high speed.

Parallel and perpendicular components of net force

In the previous discussion we found that the rate of change of momentum could be separated into two components: a component parallel to momentum, which involves a change in magnitude of momentum (speed) but not direction, and a component perpendicular to momentum, which involves a change in direction but not in magnitude.

Similarly, we can resolve the net force acting on a system into two components: one component acting parallel to the instantaneous momentum, which we call $\vec{F}_\parallel$ ("F-parallel"), and another component acting perpendicular to the momentum, which we call $\vec{F}_\perp$ ("F-perpendicular"):

$$\vec{F}_{net} = \vec{F}_\parallel + \vec{F}_\perp$$

An example is shown in Figure 4.40 and Figure 4.41. At a particular instant in a comet's elliptical orbit around the Sun, both the magnitude and direction of its momentum are changing. As one might expect, the change in magnitude of the comet's momentum (increase in speed) is caused by the parallel component of the net force. The change in direction is caused by the perpendicular component of the net force. In general:

$$\frac{d|\vec{p}|}{dt}\hat{p} = \vec{F}_\parallel \quad \text{changing speed due to parallel force component}$$

$$|\vec{p}|\frac{d\hat{p}}{dt} = \vec{F}_\perp \quad \text{changing direction due to perpendicular force component}$$

In summary, the instantaneous (derivative) form of the Momentum Principle can be re-written like this:

THE MOMENTUM PRINCIPLE: PARALLEL AND PERPENDICULAR COMPONENTS

$$\frac{d|\vec{p}|}{dt}\hat{p} = \vec{F}_\parallel \quad \text{and} \quad |\vec{p}|\frac{d\hat{p}}{dt} = \vec{F}_\perp$$

In words: a component of net force parallel to momentum causes a change in magnitude of momentum; a component perpendicular to momentum causes a change in direction.

(This separation is not valid at an instant when $\vec{p} = \vec{0}$, since $\hat{p}$ is undefined at this instant.)

In more informal notation, in terms of magnitudes:

$$\frac{dp}{dt} = F_\parallel \quad \text{and} \quad p\left|\frac{d\hat{p}}{dt}\right| = F_\perp$$

4.X.19 A comet orbits a star in an elliptical orbit, as shown in Figure 4.42. The momentum of the comet at location A is shown in the diagram. At the instant the comet passes each location labeled A, B, C, D, E, and F, answer the following questions:
(a) Draw an arrow representing the gravitational force on the comet by the star. Remember that the magnitude of the gravitational force is proportional to $1/|\vec{r}|^2$.
(b) Is the parallel component of the net force on the comet zero or nonzero?
(c) Is the perpendicular component of the net force on the comet zero or nonzero?

(d) Is the magnitude of the comet's momentum increasing, decreasing, or not changing?

(e) Is the direction of the comet's momentum changing or not changing?

Effect of perpendicular component of force

We have separated the Momentum Principle into two equations, one describing the effect of the component of force perpendicular to momentum (direction change), and the other describing the effect of the component of force parallel to momentum (change in magnitude of momentum, or change in speed). We can apply this to the case of a particle traveling along a curving path:

EFFECT OF PERPENDICULAR COMPONENT OF NET FORCE ON A PARTICLE MOVING ALONG A CURVING PATH

$$|\vec{p}|\frac{d\hat{p}}{dt} = |\vec{p}|\frac{|\vec{v}|}{R}\hat{n} = \vec{F}_{\perp}$$

or in more informal notation, in terms of magnitudes:

$$p\frac{v}{R} = F_{\perp}$$

R is the radius of the kissing circle. The component of net force perpendicular to the particle's momentum changes its direction, but not the magnitude of its momentum (speed).

If the direction but not the magnitude of an object's momentum is changing, this equation tells us that the net force must be perpendicular to the momentum. If the magnitude of the momentum is also changing, there must also be a component of net force parallel to the momentum (in the same direction or the opposite direction), as shown in Figure 4.43.

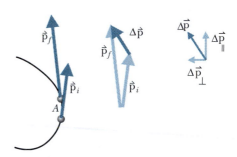

Figure 4.43 If both the magnitude and the direction of the momentum change, then $\Delta\vec{p}$ has a component parallel or antiparallel to the motion as well as a component perpendicular to the motion.

Example: The Moon and the Earth

This example involves curving motion under the influence of a gravitational force, so we can calculate both $d\vec{p}/dt$ and $\vec{F}_{net}$ independently, and compare them.

The Moon, which has a mass of about 7×10^{22} kg, orbits the Earth about once every 28 days, following a path which is nearly circular (Figure 4.44). The distance from the Earth to the Moon is about 4×10^{8} m. The speed of the Moon is nearly constant. (a) Calculate the rate of change of the Moon's momentum. (b) Calculate the gravitational force exerted on the Moon by the Earth, whose mass is about 6×10^{24} kg. Compare these two results.

System: Moon:
Surroundings: Earth
Momentum Principle

$$\frac{d\vec{p}}{dt} = \vec{F}_{net} = \vec{F}_{Earth}$$

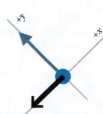

We are free to choose any convenient set of axes, so we'll set the y axis parallel to the Moon's momentum, as shown at right.
Parallel component (along y axis):

$$\frac{d|\vec{p}|}{dt}\hat{p} = \vec{F}_{\parallel} \quad \text{(or, magnitudes: } \frac{dp}{dt} = F_{\parallel}\text{)}$$

$$0 = 0$$

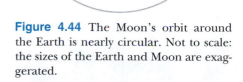

Figure 4.44 The Moon's orbit around the Earth is nearly circular. Not to scale: the sizes of the Earth and Moon are exaggerated.

Since the Moon's speed is not changing, we know that $d|\vec{p}|/dt = 0$.

Since the force exerted on the Moon by the Earth is toward the Earth (in $-x$ direction in Figure 4.44), the parallel component of the net force is also zero in this case.

Perpendicular component (along $-x$ axis):

$$|\vec{p}|\frac{d\hat{p}}{dt} = \vec{F}_\perp$$

(or, since $p\left|\frac{d\hat{p}}{dt}\right| = p\frac{v}{R}$, magnitudes: $p\frac{v}{R} = F_\perp$)

$$v \approx \frac{2\pi(4\times10^8 \text{ m})}{(28 \text{ days})(24 \text{ hr/day})(60 \text{ min/hr})(60 \text{ s/min})} = 1\times10^3 \text{m/s}$$

$p \approx mv = (7\times10^{22} \text{ kg})(1.04\times10^3 \text{m/s}) = 7\times10^{25} \text{ kg}\cdot\text{m/s}$

The kissing circle and the orbital path are the same in this case.

$$p\left|\frac{d\hat{p}}{dt}\right| = p\frac{v}{R} \approx (7\times10^{25} \text{ kg}\cdot\text{m/s})\frac{(1\times10^3\text{m/s})}{(4\times10^8 \text{ m})} = 1.8\times10^{20} \frac{\text{kg}\cdot\text{m}}{\text{s}^2}$$

Direction: $\hat{n}$, toward center of kissing circle.

Right hand side: Magnitude of the gravitational force on the Moon by the Earth:

$$F_\perp = G\frac{Mm}{R^2} \approx \left(6.7\times10^{-11} \frac{\text{N}\cdot\text{m}^2}{\text{kg}^2}\right)\frac{(6\times10^{24} \text{ kg})(7\times10^{22} \text{ kg})}{(4\times10^8 \text{ m})^2}$$

$F_\perp = 1.8\times10^{20}$ N, direction: $\hat{n}$, toward the Earth.

The net force is equal in magnitude to the observed magnitude of the rate of change of the Moon's momentum, as it must be. The directions are also the same: toward the center of the kissing circle.

Further discussion

A similar comparison was first made by Newton, who was the first person to realize that the fall of an apple and the orbit of the Moon were due to the same fundamental cause, the gravitational attraction of the Earth for the apple and for the Moon.

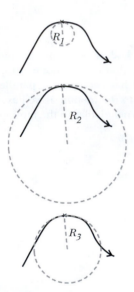

Figure 4.45 Which circle best represents the kissing circle? (Exercise 4.X.20)

Figure 4.46 A person rides on a Ferris wheel at constant speed (Exercise 4.X.23). When the person is at the location shown, what is the direction of the net force acting on the person?

4.X.20 Which of the dashed circles in Figure 4.45 best represents the "kissing circle" tangent to the path of the particle, with the same radius of curvature as that of the path at the location marked by "x"?

4.X.21 Assume that the particle whose path is shown in Figure 4.45 is traveling at constant speed. At the location marked "x", what is the direction of $d\vec{p}/dt$ for the particle?

4.X.22 A child of mass 40 kg sits on a wooden horse on a carousel. The wooden horse is 5 meters from the center of the carousel, which completes one revolution every 90 seconds. What is $(d|\vec{p}|/dt)\hat{p}$ for the child, both magnitude and direction? What is $|\vec{p}|(d\hat{p}/dt)$ for the child? What is the net force acting on the child? What objects in the surroundings exert this force?

4.X.23 A person rides on a Ferris wheel at constant speed (Figure 4.46). When the person is at the location shown, what is $(d|\vec{p}|/dt)\hat{p}$ for the person? What is $|\vec{p}|(d\hat{p}/dt)$ for the person? What must be the direction of the net force acting on the person? What objects in the surroundings exert this force?

4.X.24 The orbit of the Earth around the Sun is approximately circular, and takes one year to complete. The Earth's mass is 6×10^{24} kg, and the distance from the Earth to the Sun is 1.5×10^{11} m. What is $(d|\vec{p}|/dt)\hat{p}$ of the Earth? What is $|\vec{p}|(d\hat{p}/dt)$ of the Earth? What is the magnitude of the gravitational force the Sun (mass 2×10^{30} kg) exerts on the Earth? What is the direction of this force?

4.13 WHY DOES THE VINE BREAK?

We now have all the tools we need to answer the question posed at the beginning of this chapter. In the following problem, we can calculate $d\vec{p}/dt$, but the magnitude of the tension force on Tarzan by the vine is unknown. We can find the unknown $|\vec{F}_{vine}|$ by applying the Momentum Principle.

Example: Tarzan and the broken vine

Tarzan, whose mass is 90 kg, wants to use a vine to swing across a river. To make sure the 8 m long vine is strong enough to support him, he tests it by hanging motionless on the vine for several minutes. The vine passes this test, so Tarzan grabs the vine and swings out over the river. He is annoyed and perplexed when the vine breaks midway through the swing, when his speed is 12 m/s, and he ends up drenched and shivering in the middle of the cold river, to the great amusement of the onlooking apes.

Why did the vine break while Tarzan was swinging on it, but not while he hung motionless from it?

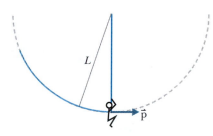

Figure 4.47 Tarzan at the bottom of his swing, just before the vine breaks. The colored line indicates his path up to this moment. The radius of the kissing circle (dotted gray line) is L, the length of the vine.

System: Tarzan
Surroundings: Vine, Earth (neglect air)

Momentum Principle:

$$\frac{d\vec{p}}{dt} = \vec{F}_{net}$$

Parallel component (horizontal):

$$\frac{d|\vec{p}|}{dt}\hat{p} = \vec{F}_{\parallel} \quad \left(\text{or, magnitudes: } \frac{dp}{dt} = F_{\parallel}\right)$$

$$\frac{dp}{dt} = 0 \quad \text{because } F_{\parallel} = 0$$

At this moment we know that the net force has no parallel component. The tension force due to the vine points straight up, along the vine. The gravitational force due to the Earth points straight down.

Perpendicular component (vertical):

Direction of $\hat{n}$ is toward center of kissing circle: straight up.

$$p\left|\frac{d\hat{p}}{dt}\right| = F_{\perp}$$

Left-hand side of equation:

$$p\left|\frac{d\hat{p}}{dt}\right| = p\frac{v}{R} \approx \frac{mv^2}{R}$$

$$p\left|\frac{d\hat{p}}{dt}\right| = \frac{(90 \text{ kg})(12 \text{ m/s})^2}{(8 \text{ m})} = 1620\frac{\text{kg} \cdot \text{m}}{\text{s}^2}$$

$$1620\text{N} = F_{\perp} = F_{vine} - mg$$

$$F_{vine} = 1620\text{N} + (90 \text{ kg})(9.8 \text{ N/kg}) = 2502 \text{ N}$$

1620 N is the net force on Tarzan at this instant, and is due to both the vine and the Earth. The tension force exerted by the vine must be larger than the gravitational force exerted by the Earth.

? When Tarzan hung motionless on the vine, what was the tension in the vine?

$$\frac{dp_y}{dt} = F_{vine} - mg$$

$$0 = F_{vine} - (90 \text{ kg})(9.8 \text{ N/kg})$$

$$F_{vine} = 882 \text{ N}$$

The tension force exerted by the vine was nearly three times as great when Tarzan was midway through his swing as it was when he hung motionless. The interatomic bonds in the vine were stretched three times as much while he swung—evidently too much!

Check: Is the result reasonable? The result shows that the vine has to exert an upward force that is greater than Tarzan's weight. That makes sense. If Tarzan were hanging motionless (equilibrium), the vine would only have to exert an upward force of mg (net force would be zero). But to make Tarzan's momentum turn from horizontal to upward requires that the net force be upward.

Further discussion

Why must the vine exert a greater force when Tarzan's momentum is changing than when he hangs motionless? It is not simply that Tarzan is moving—zero net force would be required if Tarzan's momentum were constant. The "extra" tension in the vine is necessary to change the direction of Tarzan's momentum. To swing upward, increasing his p_y, Tarzan has to pull down harder on the vine, and therefore the vine stretches more, pulling up more on Tarzan.

It may seem odd that at the instant in question, the *magnitude* of Tarzan's momentum is not changing. We know that his speed increased as he swung downward, and his speed would have decreased as he swung upward if the vine had not broken. It is important to keep in mind that it is the instantaneous rate of change of magnitude of momentum (and speed) that is zero. At this instant only the direction of momentum is changing, as the net force is straight up. A nanosecond later, both the momentum and the net force would have changed, and there would be a nonzero parallel force component, so the magnitude of momentum would then be changing as well.

It is important to keep in mind that $\vec{F}_{net}$ and $d\vec{p}/dt$ are not the same thing. In this case the equal sign in the Momentum Principle relates cause ($\vec{F}_{net}$) to effect ($d\vec{p}/dt$). In the example above, we calculated the effect, and from this deduced the cause.

Example: Magnitude and direction of momentum changing

Consider the instant shown in Figure 4.48, before the vine breaks, when the angle between the vine and the vertical is $30°$. At this instant Tarzan's speed is 11.1 m/s, and both the magnitude and direction of his momentum are changing. As before, Tarzan's mass is 90 kg, and the vine is 8 m long. (a) What is the tension in the vine at this moment? (b) What is the rate of change of the magnitude of Tarzan's momentum?

System: Tarzan
Surroundings: Earth, vine (ignore air)

Momentum Principle
(a) Perpendicular (y) component:
Direction of $\hat{n}$: toward center of kissing circle

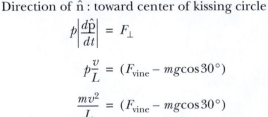

$$p\left|\frac{d\hat{p}}{dt}\right| = F_\perp$$

$$p\frac{v}{L} = (F_{vine} - mg\cos 30°)$$

$$\frac{mv^2}{L} = (F_{vine} - mg\cos 30°)$$

$$\frac{mv^2}{L} + mg\cos 30° = F_{vine}$$

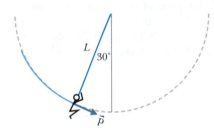

Figure 4.48 Tarzan swinging downward, with a speed of 11.1 m/s.

It is convenient to choose axes aligned along directions parallel and perpendicular to $\hat{p}$, as shown in the force diagram at right, because this simplifies finding $\hat{p}$ and $\hat{n}$.

$$F_{\text{vine}} = \frac{(90 \text{ kg})(11.1 \text{ m/s})^2}{(8 \text{ m})} + (90 \text{ kg})(9.8 \text{ N/kg})\cos 30°$$

$$F_{\text{vine}} = 2150 \text{ N}$$

(b) Parallel (x) component:

$$\frac{dp}{dt} = F_\parallel$$

$$\frac{dp}{dt} = mg\cos(90 - 30)°$$

$$\frac{dp}{dt} = (90 \text{ kg})(9.8 \text{ N/kg})\cos(90 - 30)° = 441 \text{ kg} \cdot \text{m/s}^2$$

The tension in the vine is greater than when Tarzan was hanging motionless, but not as large as it will be at the bottom of the swing.

Further discussion

In this example, both the magnitude and direction of Tarzan's momentum were changing. The net force on Tarzan had nonzero parallel and perpendicular components.

Example: Compression force (curving motion)

A passenger of mass M rides in an airplane flying along a curved path at constant speed v as shown in Figure 4.49, where the radius of the kissing circle at the top of the path is R. At the instant shown, when the airplane is at the top of its arc: How large is the compression force that the seat exerts on the passenger's bottom? Compare this force to the force the seat exerts on the passenger when the airplane moves in a straight line at constant speed. No numbers are given, so your solution should contain symbols only.

System: Passenger
Surroundings: Seat, Earth
Momentum Principle:

$$d\vec{p}/dt = \vec{F}_{\text{net}}$$

Parallel component (along x axis):

$$\frac{dp}{dt} = F_\parallel$$

$$\frac{dp}{dt} = 0 \text{ because } F_\parallel = 0$$

Perpendicular component:

Direction of $\hat{n}$ is toward center of kissing circle: $-y$

$$p\left|\frac{d\hat{p}}{dt}\right| = F_\perp$$

Left-hand side of equation:

$$p\left|\frac{d\hat{p}}{dt}\right| = p\frac{v}{R} = \frac{Mv^2}{R}$$

$$\left(p\left|\frac{d\hat{p}}{dt}\right|\right)_y = -\frac{Mv^2}{R} \quad (y\text{-components})$$

Right-hand side of equation:

$$F_\perp = F_S - Mg$$

Putting it together:

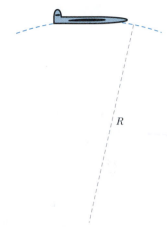

Figure 4.49 An airplane goes over the top of its path with speed v. The kissing circle has radius R at the top of the path.

At this moment we know that the net force has no parallel component. The compression (normal) force due to the seat points straight up. The gravitational force due to the Earth points straight down.

$d\hat{p}/dt$ points toward the center of the kissing circle, so in this situation the y component of $d\hat{p}/dt$ is negative.

$$-\frac{Mv^2}{R} = F_S - Mg$$

$$F_S = Mg - \frac{Mv^2}{R}$$

Check: units okay (newtons)

Physical significance: The minus sign in the result is important. The upward force of the seat F_S on the passenger's bottom is *less* than the passenger's weight Mg. The interatomic bonds in the seat are compressed less than they would be if the airplane's momentum were constant. To the passenger, it feels as if the seat is dropping away from him or her.

? The net force points downward. Does that make sense?

Yes, because the net force should point toward the center of the kissing circle, which is below the airplane.

? But since $F_S = Mg - Mv^2/R$, if the airplane's speed v is big enough, the force exerted by the seat could be zero. Does that make any sense?

Yes, that's the case where the airplane changes direction so quickly that the seat is essentially jerked out from under the passenger, and there is no longer contact between the seat and the passenger. No contact, no force of the seat on the passenger.

The feeling of weight, and of weightlessness

In the preceding example, what does the situation look and feel like to the passenger? From the passenger's point of view it is natural to think that the passenger has been thrown upward, away from the seat. But actually the seat has been yanked out from under the passenger, in which case the passenger is falling toward the Earth (being no longer supported by the seat), but changing velocity less rapidly than the airplane and so the ceiling comes closer to the passenger.

In the extreme case, if the airplane's maneuver is very fast, it may seem as though the passenger is "thrown up against the cabin ceiling." In actual fact, however, it is the cabin ceiling that is yanked down and hits the passenger!

You can see the value of wearing a seat belt, to prevent losing contact with the seat. When the airplane yanks the seat downward, the seat belt yanks the passenger downward. This may be uncomfortable, but it is a lot better than having the cabin ceiling hit you hard in the head.

? How would this analysis change if the airplane goes through the bottom of a curve rather than the top (Figure 4.50)?

Now the center of the kissing circle is above the airplane, so $d\vec{p}/dt$ points upward. That means that the net force must also point upward.

? Which is larger, the force of the seat on the passenger, or the gravitational force of the Earth on the passenger?

The force of the seat is greater than Mg, since the net force is upward. The passenger sinks deeper into the seat. Or to put it another way, the airplane yanks the seat up harder against the passenger's bottom, squeezing the seat against the passenger.

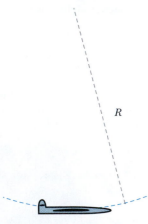

Figure 4.50 An airplane goes through the bottom of its path.

We perceive contact forces

As you sit reading this text, you feel a sensation that you associate with weight. If someone suddenly yanks the chair out from under you, you feel "weightless," which feels funny and odd. It also feels scary, because you know from experience that whenever you lose the support of objects under you (chair, airplane seat, floor, mountain path), something bad is about to happen.

In reality, what we perceive as "weight" is actually not the gravitational force at all but rather the forces of atoms in the chair (or airplane seat or floor or mountain path) on atoms in your skin. You have nerve endings that sense the compression of interatomic bonds in your skin, and you interpret this as evidence for gravity acting on you. But if you were placed in a space-ship that was accelerating, going faster and faster, its floor would squeeze against you in a way that would fool your nerve endings and brain into thinking you were subjected to gravity, even if there are no stars or planets nearby.

If you lose contact with the seat in an airplane that is going rapidly over the top of a curving path, your nerve endings no longer feel any contact forces, and you feel "weightless." Yet this is a moment when the only force acting on you is in fact the Earth's gravitational force. Weightlessness near the Earth paradoxically is associated with being subject only to your weight *Mg*.

Contact forces internal to the body

Nor is it just nerve endings in your skin that give you the illusion of "weight." As you sit here reading, your internal organs press upward on other organs above them, making the net force zero. You feel these internal contact forc-es. If you are suddenly "weightless" a main reason for feeling funny is that the forces one organ exerts on another inside your body are suddenly gone.

To train astronauts, NASA has a cargo plane that is deliberately flown in a curving motion over the top, so that people in the padded cargo bay lose contact with the floor and seem to float freely (in actual fact, they are accel-erating toward the Earth due to the *Mg* forces acting on them, but the plane is also deliberately accelerating toward the Earth rather than flying level, so the people don't touch the walls and appear to be floating). All the nerve endings are crying out for the usual comforting signals and not getting them. This airplane is sometimes called the "vomit comet" in honor of the effects it can have on trainees.

4.14 SPEED OF SOUND IN A SOLID AND INTERATOMIC BOND STIFFNESS

When we have considered objects that exert tension forces, such as vines, strings, and wires, we have made the (very good) approximation that the stretch of the interatomic bonds is nearly the same throughout the length of the object. Likewise, we assume that inside objects exerting compression forces, the interatomic bonds are all compressed nearly the same amount.

The propagation of stretch or compression

The equal stretch or compression of bonds does not, however, occur instan-taneously. When Tarzan first starts pulling to down on the end of a vine, he very slightly lengthens the interatomic bonds between neighboring atoms in the nearby section of the vine (Figure 4.51). As a result of their downward displacement, these atoms stretch the bonds of their neighbors, and very quickly this new interatomic bond-stretching propagates upward, all the way to the other end of the vine, and the whole vine is then in tension.

It is the boundary between the already stretched region of the vine and the newly stretched region that moves upward, along the length of the vine. Individual atoms move extremely short distances, though for clarity in the diagram we have enormously exaggerated the interatomic stretches and the

Einstein came to the theory of general rel-ativity through his realization that the ef-fect of a gravitational force in an inertial reference frame is indistinguishable from the effect of being in an accelerated refer-ence frame.

There are more worked-out example problems involving contact forces in the last section of this chapter (Section 4.18, on page 147).

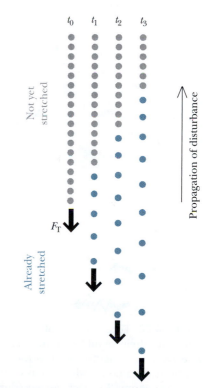

Figure 4.51 At time t_0 Tarzan starts pull-ing downward on the vine with a force F_T. The "disturbance" in the vine propagates upward. At successive times t_1, t_2, and t_3, more and more of the interatomic bonds in the vine are stretched. Atoms in the stretched region are indicated by colored dots. The amount of stretch is greatly exag-gerated here.

distance any individual atom moves. The process is the same for compression.

This process is fast, but it is not instantaneous. The rate of propagation of the boundary between the stretched and unstretched regions is called the "speed of sound" in the material. There are various ways to measure this speed; one of the simplest is to place a microphone at one end of a metal bar, strike the other end of the bar with a hammer, and use a computer interface to measure how long it takes for the disturbance to travel from one end of the bar to the other.

What does this have to do with sound?

What we call "sound" is a disturbance in a material. We are most familiar with sound traveling through air. When you clap your hands, a pulse of sound propagates through the air to your ear (Figure 4.52). This pulse starts as a momentary compression of the air (an increase in the air density) made by your hands. This compression is passed on to neighboring regions of the air, and the air near the hands relapses to its original density.

Eventually your eardrum is hit by the pulse and moves in response to the sudden increase in density (and pressure). The motion of the eardrum is detected by your inner ear and passed on to the brain, which interprets the signal as meaning that you clapped your hands. Note that no individual air molecule moves from near your hands to your ear. Rather it is the disturbance that moves from one place to another.

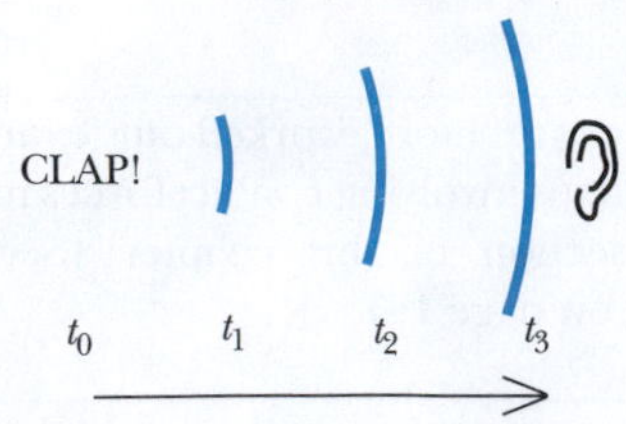

Figure 4.52 Clap your hands—sound travels to your ear as a pulse of air compression.

An iterative model of sound propagation in a metal

A copper rod is modeled as a single horizontal chain of one hundred copper atoms, each with mass $(64)(1.7 \times 10^{-27} \text{ kg})$. The initial distance between any two atoms is the length of the interatomic bond in solid copper $(2.4 \times 10^{-10} \text{ m}$, see page 108), and the stiffness of each interatomic spring is 27 N/m (see page 111). Initially each atom is at rest. The calculation is organized in a way similar to other iterative calculations we have done; the only difference is that we must update the position and momentum of 100 objects after each time step. An outline of the calculation is this:

Displace the leftmost atom by a small amount (a "hit")
Choose a very short time step ($\Delta t \approx 1 \times 10^{-14}$ s)
Set elapsed time = 0
Then, repeatedly:

- From the current positions of the atoms, find the stretch of every spring
- Find the net force on each atom, due to the two springs attached to it
- Calculate the change in momentum of each atom
- Calculate the new momentum of each atom
- Update the position of each atom
- Add Δt to the elapsed time

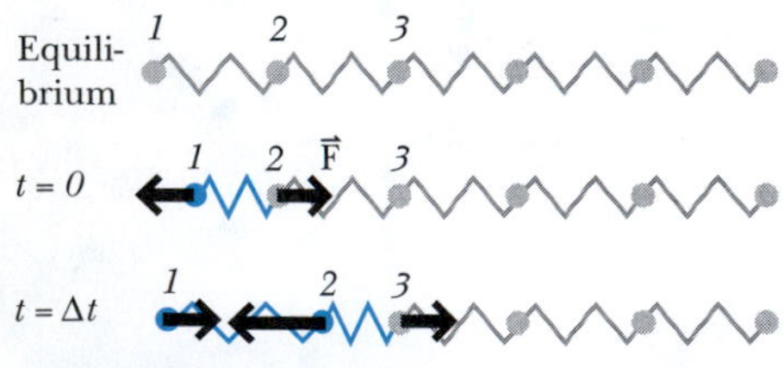

Figure 4.53 Top: a chain of atoms and bonds in their equilibrium position. At time $t=0$ atom 1 is displaced to the right, compressing a bond, which then exerts a force on atom 2 (black arrows denote forces). After applying the Momentum Principle to all atoms, and updating positions of all atoms over a time step Δt, both atoms 1 and 2 are displaced from equilibrium; the second bond is compressed, and the first is now stretched. Iterative updates continue, involving progressively more atoms after each time step.

Stop the calculation when the rightmost atom is displaced.
Divide the distance traveled (100 atoms)$(2.4 \times 10^{-10}$ m/atom) to get the speed at which the disturbance traveled down the chain.

Figure 4.53 shows a visualization of the first step in this process. At first only atoms 1 and 2 are affected; after updating their positions we find that the second bond is now compressed, affecting atom 3, and so on.

Visualizing the propagation of the disturbance can be done in various ways. A particularly useful visualization is to plot the horizontal displacements (y_n) vertically, at the locations of the horizontal equilibrium locations

(x_n). An example is shown in Figure 4.54, where you see that a displacement to the left is plotted below the axis, and a displacement to the right is plotted above the axis. For further details, see problem 4.HW.77 (page 157).

It is clearly desirable to program a computer to do this calculation. When we carry out this calculation for a copper bar, using the values given above, we obtain a value for the speed of sound in copper of about 3530 m/s. Experimentally measured values given in engineering and science reference materials are around 3600 m/s, so our simple model gives a very good result!

An algebraic expression

Although predictions based on iterative calculations can assure us that our understanding of the underlying nature of a phenomenon is correct, it is also useful to have a simple algebraic expression for a central relationship such as that between interatomic bond stiffness and speed of sound. In order to be able to do this, we will need to derive an analytical solution for the oscillations of a single mass and spring. This will give us added insight into the propagation of a disturbance along a chain of masses and springs.

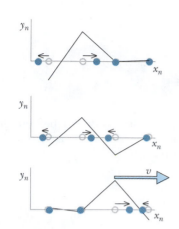

Figure 4.54 Visualizing the propagation by plotting the displacements y_n away from the equilibrium positions x_n.

4.15 ANALYTICAL SOLUTION: SPRING-MASS SYSTEM

In Chapter 3 we applied the finite-time form of the Momentum Principle iteratively to predict the motion of a mass attached to a spring. Now, using the derivative form of the Momentum Principle, we can derive an analytical (algebraic) solution for the motion of an ideal mass-spring system. The reason we can do this is that we know how to calculate the force exerted by the spring on the block, for any location of the block.

Applying the Momentum Principle to a spring-mass system

Consider a block of mass m connected to a spring whose stiffness is k_s (that is, the force exerted by the spring is $k_s s$ when the spring is stretched or compressed an amount s). The other end of the spring is attached to the wall (Figure 4.55). Our goal is to derive an equation of the form:

$$x(t) = \ldots$$

which predicts the position of the block at any time in the future.
We will assume that:

- The block slides with almost no friction on an air table, where it is supported on a cushion of air.
- The mass of the spring is negligible compared to the mass of the block.
- Air resistance is negligible.

We can simplify our calculations by choosing to measure the position of the block with respect to an origin chosen to be at the equilibrium position, the location of the block when the spring is relaxed, so that the stretch s of the spring is equal to the position x of the block. In this case, the x component of the force on the block by the spring is $F_x = -k_s x$, as shown in Figure 4.55.

? What would be the equation for the spring force in terms of the position of the block if the origin were located at the wall, and the relaxed length of the spring is L?

In that case we would write $F_x = -k_s(x - L)$, because what matters in this formula is the stretch (change in length of the spring), which is $(x - L)$. A check on this is the fact that when the block is at location $x = L$, the stretch $(x - L)$ is zero.

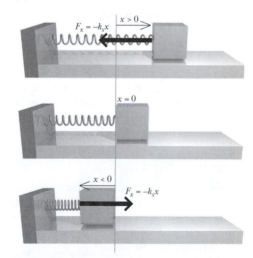

Figure 4.55 A block connected by a spring to a wall slides back and forth with little friction. The x component of the force that the spring exerts on the block is equal to $-k_s x$.

System: Block
Surroundings: Spring, Earth, air table
Momentum principle

$$d\vec{\mathbf{p}}/dt = \vec{\mathbf{F}}_{net}$$

y component:

$$\frac{dp_y}{dt} = F_{AirTable} - mg$$

$$0 = F_{AirTable} - mg$$

$$F_{AirTable} = mg$$

We know that $dp_y/dt = 0$ because the y momentum of the block is zero and remains zero.

x component:

$$\frac{dp_x}{dt} = -k_s x$$

Since $v_x = dx/dt$, dp_x/dt contains the second derivative of x with respect to t.

$$\frac{d(mv_x)}{dt} = \frac{d}{dt}\left(m\frac{dx}{dt}\right) = -k_s x$$

$$m\frac{d^2x}{dt^2} = -k_s x$$

If we ignore the constants and signs for a moment, this equation (which is a "differential equation") says that if we know x as a function of time, taking the second derivative of that function will give us back the original function.

?   What function appears in its own second derivative?

If we differentiate a cosine function twice, we get a cosine again. Further, since the block oscillates back and forth, cosine sounds like a reasonable guess. Let's try this:

$$x = A\cos(\omega t)$$

Guessing a function isn't as peculiar as it may appear. This is actually one of the standard approaches to solving a differential equation.

In this expression, A and ω (lower case Greek omega) are constants whose values we will need to determine. We will look for values that make this functional relation satisfy the Momentum Principle.

We will put this expression for $x(t)$ back into the Momentum Principle equation:

Cosine has values between 1 and –1. Since the block may not move 1 meter, we need to multiply by a constant A to change the maximum displacement to a reasonable value.

$$m\frac{d^2}{dt^2}A\cos(\omega t) = -k_s A\cos(\omega t)$$

$$-Am\omega^2\cos(\omega t) = -k_s A\cos(\omega t)$$

$$m\omega^2 = k_s$$

If the block completed one oscillation in 2π seconds, the constant ω (Greek lowercase omega) would be 1. Otherwise, we need ω to be able to adjust the time taken for a round trip oscillation.

$$\omega = \sqrt{\frac{k_s}{m}}$$

So we have an analytical solution for the position of the mass as a function of time:

AN ANALYTICAL SOLUTION FOR MASS-SPRING

$$x = A\cos(\omega t), \text{ with } \omega = \sqrt{\frac{k_s}{m}}$$

A (the "amplitude") is a constant that depends on the initial conditions. A is equal to the maximum stretch of the spring during an oscillation. The units of ω are radians/s.

Angular frequency, period, and frequency

The constant A is called the "amplitude" of the motion. It is equal to the maximum stretch of the spring. If the spring stretches a maximum of 5 cm, then A will have the value 0.05 m.

The constant ω in $x = A\cos(\omega t)$ is called the "angular frequency" and is measured in radians per second (ω is lower-case Greek omega). When the argument ωt of the cosine increases by 2π radians, the motion repeats (one complete cycle), so if we call the "period" T (the time required for one complete cycle, Figure 4.56), we have $\omega T = 2\pi$.

One other quantity often used to describe oscillating systems is the "frequency" f, which is the number of complete cycles per second: f = (1 cycle) per T seconds. Cycles per second are also called "hertz."

$$\text{Angular frequency } \omega = \frac{2\pi}{T} = \sqrt{\frac{k_s}{m}}, \text{ radians per second}$$

$$\text{Period } T = \frac{2\pi}{\omega}, \text{ seconds}$$

$$\text{Frequency } f = \frac{1}{T} = \frac{\omega}{2\pi}, \text{ cycles per second or "hertz"}$$

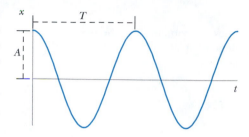

Figure 4.56 Amplitude A and period T, shown on a plot of x vs. t for the mass.

A vertical mass-spring system

It can be shown that this analytical solution is also valid for a vertical mass-spring system, in which the gravitational force on the mass acts parallel or anti parallel to the force on the mass by the spring. The details of this analysis are given on page 150.

Example: **Mass-spring oscillation (analytical solution)**

A 50 gram mass is attached to a low mass horizontal spring whose relaxed length is 25 cm, and whose stiffness is 4 N/m. The block slides on a very slippery, low-friction surface. You stretch the spring so it is 30 cm long, and release the block. (a) What is the amplitude of the oscillations? (b) How long will it take the block to make one full oscillation? (c) What would the amplitude of the oscillations be if the initial length of the spring were 35 cm instead of 25 cm? (d) How long would it take the block to make one full oscillation if the initial length of the spring were 35 cm instead of 25 cm? (e) How many oscillations will the block make in 10 seconds? (f) The initial length of the stretched spring is 35 cm. What will the location of the block be 3.15 seconds after you release it? (Remember to use radians, not degrees.)

System: Mass
Surroundings: Spring, Earth, table

Assumptions for ideal mass-spring system satisfied?
 • Spring mass negligible: yes
 • Friction negligible: yes
 • Air resistance negligible: yes

(a) $A = s_i = 0.3 \text{ m} - 0.25 \text{ m} = 0.05 \text{ m}$

(b) Time for one round trip = T (period)

$$\omega = \sqrt{\frac{k_s}{m}} = \sqrt{\frac{4 \text{ N/m}}{0.05 \text{ kg}}} = 8.94 \text{ s}^{-1}$$

$$T = \frac{2\pi}{\omega} = \frac{2\pi}{8.94 \text{ s}^{-1}} = 0.70 \text{ s}$$

(c) $A = s_i = 0.35 \text{ m} - 0.25 \text{ m} = 0.10 \text{ m}$

Because there is no friction in this idealized system, the amplitude of the motion does not decrease over time.

The expression for the period

$$T = \frac{2\pi}{\omega} = \frac{2\pi}{\sqrt{\dfrac{k_s}{m}}}$$

does not contain A, so the period must not depend at all on the amplitude of the oscillations.

(d) $T = 0.70$ s

(e) $f = \dfrac{1}{T} = \dfrac{1}{0.70 \text{ s}} = 1.43$ cycles/s

   # cycles $= ft = (1.43 \text{ cycles/s})(10 \text{ s}) = 14.3$

(f) $x(t) = A\cos(\omega t)$

   $x(3.15 \text{ s}) = (0.10 \text{ m})\cos((8.94 \text{ s}^{-1})(3.15 \text{ s})) = -0.0994$ m

Check: units ok. Reasonable? Yes, because the origin was put at the end of the relaxed spring, so this means the spring is compressed slightly less than maximum compression (see Figure 4.55).

4.X.25 You have a rubber band whose relaxed length is 8.5 cm. You hang a coffee cup, whose mass is 330 g, from the rubber band, which stretches to a length of 14 cm. Consider the rubber band to be a single spring. (a) What is the stiffness of the rubber band? (b) If you start this mass-spring system oscillating, how many round trip oscillations will the coffee cup make in 5 seconds?

4.X.26 When a particular mass-spring system is started by stretching the spring 4 cm, one complete oscillation takes 2 seconds. If instead the initial stretch had been 6 cm, how long would it take for one complete oscillation?

4.X.27 A particular mass-spring system oscillates with a period of 1 second. If the mass were tripled, what would the period of the system be?

4.X.28 A particular mass-spring system oscillates with a period of 1 second. If the stiffness of the spring were tripled, what would the period of the system be?

Harmonic and anharmonic oscillators

The idealized mass-spring system we have analyzed is called a "harmonic oscillator". The special features of a harmonic oscillator are:
- The position of the mass as a function of time is given by a cosine function
- The period (round-trip time) is independent of the amplitude of the oscillations

The fact that the period of a harmonic oscillator does not depend on the amplitude is surprising. With a larger amplitude, the mass travels farther in one period. However, it also reaches higher speeds, and for an ideal mass-spring system, these two effects exactly cancel.

Not all oscillating systems are harmonic. If you drop a rubber ball and let it bounce, as the bounce amplitude decreases, the period decreases too: the ball bounces faster and faster as the height of the bounces gets lower and lower. The bouncing ball is an example of an "anharmonic oscillator." "Anharmonic" simply means "not harmonic."

Real springs have mass

Real springs have mass, but we have ignored this fact of the real world. It is not easy to include the effect of the mass of the spring, either in a numerical integration or in an analytical treatment. One way to model a real spring with mass is with a chain of point masses connected by massless springs (Figure 4.57).

? Under what circumstances is it likely to be a good approximation to neglect the mass of the spring?

In many cases the spring has much less mass than the objects that are connected to it, in which case the effect of the spring's mass is likely to be small. A contrasting example is an oscillating Slinky, in which the only mass is the mass of the spring.

Figure 4.57 A massive spring may be modeled as a chain of point masses connected by massless springs.

Existence of analytical solutions

For the mass-spring system we were able to obtain an analytical solution—a prediction for all time of the motion of our model spring-mass system not subject to the additional inaccuracies introduced by a finite time step. However, for many systems the differential equation corresponding to the derivative form of the Momentum Principle has a mathematical form that cannot be solved analytically and must be solved numerically. The power of iterative application of the Momentum Principle is that it can be used even in those situations where no analytical solution seems possible.

For example, $x = A\cos(\omega t)$ *cannot* solve the following "nonlinear" differential equation for any value of ω:

$$\frac{dp_x}{dt} = m\frac{d}{dt}\left[\frac{dx}{dt}\right] = -k_s x^3$$

There may or may not exist an analytical solution of this differential equation in terms of well-known mathematical functions, but *any* differential equation can be solved numerically, just as we did for the spring-mass equation. Given x and p at some instant in time, we can calculate the force ($-k_s x^3$) and determine the new values of x and p at a time Δt later, then repeat as often as necessary to reach the time of interest. In fact, you may have explored this possibility in part (e) of Problem 4.HW.73.

It is important to keep clearly in mind that either an analytical or a numerical prediction of the future applies only to our *model* of the actual real-world situation. If our model is a good approximation to the real world, our prediction will be a good approximation to what will actually happen, but even an analytical prediction cannot be exact because of the practical impossibility of knowing exactly the initial conditions and the net force due to all the other objects in the Universe.

In our idealized mass-spring system there is no friction or air resistance. In a real system, both of these are often present. In this case, the idealized model will not give an accurate prediction of the position of the mass as a function of time; the amplitude of the real system will decrease, but the amplitude of the idealized model system will not.

Initial conditions for a mass-spring system

We need to know the initial conditions (position and momentum of the mass) in order to predict the position of the system at some future time. In the previous example, the mass was initially at rest, and the spring was stretched. It is also possible to start an oscillator by placing it at its equilibrium position and giving it a kick—in this case its initial stretch is zero, but its initial momentum is nonzero. In an iterative solution, this is easy to take into account. A discussion at the end of this chapter explains how this is taken into account in the analytical solution.

One might feel that analytical solutions are better than numerical ones, but actually these are not as different as one might think. For example, one of the ways to calculate a table of cosine values is to do a numerical integration of the equation that applies to spring-mass systems! In general, a fruitful way to describe a mathematical function is by specifying the differential equation for which it is a solution, because then a numerical integration of that equation yields numerical values of the function.

4.16 ANALYTICAL EXPRESSION FOR SPEED OF SOUND

Under ordinary conditions, the speed of sound in air is about 340 m/s. If your friend stands 340 meters away and yells, you will hear the sound a second later. The speed of sound in a solid object is much faster than it is in air. In aluminum, the speed of sound is about 5000 m/s. In lead, which is a much softer material than aluminum, the speed of sound is only about 1200 m/s. There are two significant differences between lead and aluminum at the atomic level. First, the mass of a lead atom is about eight times larger than the mass of an aluminum atom. Second, the interatomic bond stiffness in aluminum is greater than the interatomic bond stiffness in lead (see Problem 4.HW.76 on page 157). Since the propagation of a disturbance through a metal rod involves displacements of atoms, it seems reasonable that both of these factors should play a role in determining the speed with which the disturbance propagates. Our goal in the following discussion is to use the ball-spring model of a solid to predict this relationship.

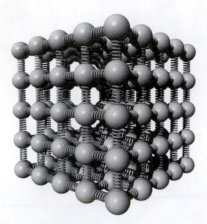

Figure 4.58 A model of a metal: atoms (balls) connected by springs (interatomic bonds).

Qualitative guess

Think of a solid metal rod as a network of atoms (balls) connected by springs (interatomic bonds) as in Figure 4.58.

? How should the speed of sound depend on m_a, the mass of an atom in a material? (Think of hitting one end of a metal rod with a hammer.)

A given applied force produces the same change of momentum for a massive atom like lead or a lighter atom like aluminum. However, the change in speed of the lead atom will be less than the change in speed of the aluminum atom. This suggests that the lower the mass of the atoms, the faster a sound pulse (disturbance) might propagate through a rod.

? How should the speed of sound depend on the stiffness of the interatomic bonds in a material?

The stiffer the bond, the greater the force the "spring" will exert on a neighboring atom in response to the same displacement. Stiffer interatomic bonds should make a sound pulse propagate faster through a rod.

? How should the speed of sound depend on d, the length of interatomic bonds in a material?

The farther apart the atoms are, the farther one atom must move to affect its neighbor. The longer the interatomic bond, the slower a pulse should propagate.

These initial qualitative guesses don't tell us details of the mathematical relationships involved: for example, whether speed of sound is proportional to $1/m_a$, $1/m_a^2$, $1/\sqrt{m_a}$, or some other factor. We need to apply our understanding of mass-spring systems to refine these guesses.

Dimensional analysis

We need to construct a mathematical expression that has the correct units, or "dimensions". Speed has units of m/s, so we need an equation involving interatomic bond stiffness $k_{s,i}$ and atomic mass m, ending up with meters in the numerator and seconds in the denominator.

$$v = \ldots$$

The angular frequency ω involves both $k_{s,i}$ and m_a, and has units of s^{-1}, so ω might appear in the numerator of the expression:

$$v = \omega(?)$$

? Is it reasonable that v should be directly proportional to ω?

Yes. The larger the angular frequency ω is, the faster an atom oscillates back and forth, compressing and stretching intermolecular bonds, which affect neighboring atoms.

? What other quantity should go in the equation?

Since we need to end up with m/s, we need to multiply by something that has units of meters. ω is related to the rate at which one atom hits its neighbor, propagating a change through a distance of one interatomic spacing. Therefore multiplying by the interatomic bond length d might be appropriate.

$$v = \omega d$$

This informal reasoning is an example of "dimensional analysis," in which we identify meaningful quantities in the situation and combine them to guess a formula for some other quantity of interest. While this can be a fruitful procedure, we shouldn't be surprised if we find that the speed of sound is actually given by $v = 2\omega d$, or $v = \omega d/\pi$, for example. Moreover, note that atomic oscillations in the x direction involve two atomic "springs" on each side of an atom, not one, so we may have omitted a factor of 2. Nevertheless, we can guess that $(\sqrt{k_{s,i}/m_a})d$ is proportional to the actual speed of sound, with the correct dependence on $k_{s,i}$ and m_a. As it happens, it can be shown that this *is* the correct formula. The full derivation, which is quite difficult, is not given here; it can be found in advanced textbooks on classical mechanics.

SPEED OF SOUND IN A SOLID

$$v = \omega d = \sqrt{\frac{k_{s,i}}{m_a}}\,d$$

v is the speed of sound (m/s), $k_{s,i}$ is the stiffness of the interatomic bond, m_a is the mass of one atom, d is the length of the interatomic bond.

Example: Speed of sound in copper

We found (page 111) that the stiffness of the interatomic bond in solid copper is about 27 N/m, and that the interatomic bond length in solid copper is about 2.28×10^{-10} m (page 108). There are 64 grams of copper in one mole. Use these data to calculate the speed of sound in solid copper.

$$m_a = \frac{0.064 \text{ kg}}{6.02\times10^{23} \text{ atoms}} = 1.06\times10^{-25} \text{ kg/atom}$$

$$v = \sqrt{\frac{27 \text{ N/m}}{1.06\times10^{-25} \text{ kg}}}(2.28\times10^{-10} \text{ m}) = 3640 \text{ m/s}$$

This agrees well with experimentally measured values of around 3600 m/s.

Speed of sound in different materials

In aluminum, which is very stiff (large k_s) and has a small atomic mass m_a, the speed of sound is very high, about 5000 m/s. If a wire is half a meter long, it takes about

$$(0.5 \text{ m})/(5000 \text{ m/s}) = 10^{-4} \text{ s} \quad \text{(a tenth of a millisecond)}$$

from the time you pull on one end for the other end to become tense. On the other hand, lead is quite soft (small k_s) and has a large atomic mass m_a, and the speed of sound in lead is only about 1200 m/s, much lower than in aluminum. (The speed of sound in solids is much higher than the speed of sound in air, which is about 340 m/s. Because air molecules are usually far from one another, the speed of sound in air is determined by the average speed of the air molecules, not by interatomic forces.)

If possible, it is interesting to measure the speed of sound in a metal bar directly, by hitting one end of a long bar with a short pulse and observing on an oscilloscope when sensors at two locations along the bar detect the sound.

4.X.29 Using your results from Exercise 4.X.1, predict the speed of sound in aluminum and compare with the measured value (about 5000 m/s).

4.X.30 Using your results from Exercise 4.X.2, predict the speed of sound in lead and compare with the measured value (about 1200 m/s).

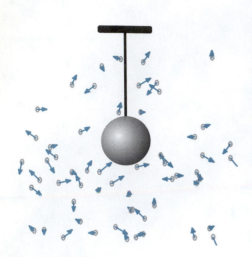

Figure 4.59 Air molecules in motion near a hanging ball.

4.17 CONTACT FORCES DUE TO GASES

Until now we have considered only contact forces due to solids. However, gases can also exert contact forces. In everyday circumstances, the most significant of these is the force that the air exerts on objects—even stationary objects. Consider the surrounding air that is in contact with a ball hanging from a wire, as shown in Figure 4.59. You might think that the air has no effect on the ball, especially since the ball is not moving, so we don't need to worry about air resistance. To be certain, let's consider this issue at the microscopic level.

Buoyancy

Air is a gas, composed of about 80% nitrogen and 20% oxygen, with very small amounts of other gases. The molecules in the air are continuously in motion, moving in random directions. Molecules in the air near the ball will frequently bump into the ball, exerting a small force on it. If there is no wind, the motion of the gas molecules is random, so in a short time period there should be just as many molecules colliding with the right hand side of the ball as with the left hand side, and there should on average be no sideways force on the ball.

However, the situation is not quite the same for collisions with the top and bottom of the ball. The directions of molecular velocities are still random, but there are more air molecules per cubic meter below the ball than there are above it. The density of the Earth's atmosphere decreases as one moves upward away from the Earth's surface, until finally at a sufficient distance (approximately 50 km) there is essentially no air left. You may have observed this variation yourself if you have ever gone from sea level to a location several thousand meters higher; you may feel light-headed if you exercise at high altitude because the density of oxygen is significantly lower.

This variation in air density occurs over kilometers; can it possibly be important over a distance of a few centimeters? Interestingly enough, the surprising answer is yes! There is actually a "buoyant force" upward on the ball, because the number of air molecules hitting the bottom of the ball per second is slightly greater than the number of molecules hitting the top of the ball per second.

A macroscopic view of buoyancy

How can we determine the magnitude of the buoyant force on the ball due to the air? Let's consider a macroscopic viewpoint. The following discussion, based on the Momentum Principle, applies equally well to the buoyant force exerted on an object by any fluid (air, water, other liquids, other gases). The result we will obtain is called the Archimedes principle, for the Greek thinker who first understood it.

Imagine a box filled with air (or water). Consider a ball-shaped region that a ball will eventually occupy, at a moment when the ball is not yet there, so the ball-shaped region is filled with the fluid. Let's choose this ball-shaped mass of fluid as our system, and consider what forces must be acting on it (Figure 4.60). This may seem odd, but it is a perfectly legitimate choice of system.

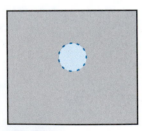

Figure 4.60 A system consisting of a ball-shaped spherical region of fluid, surrounded by more fluid.

> **?** Is the momentum of the indicated sphere of fluid changing? What must the net force on this sphere be?

The momentum of the indicated sphere of fluid is not changing. The Earth exerts a downward gravitational force on it, but the sphere does not sink. Because its momentum isn't changing, the net force on the sphere must be zero, so there must be an upward force that balances the gravitational force.

> **?** What exerts this upward force on the sphere?

The only object in contact with this sphere is the rest of the fluid, so it must be that the rest of the fluid exerts an upward force F_b ("buoyant force") whose magnitude is mg, where m is the mass of the sphere of fluid (Figure 4.61).

Now remove the sphere of fluid and replace it with the ball. It must be that initially the rest of the fluid still exerts the same upward force F_b of magnitude mg (Figure 4.62). This force may not be large enough to counteract the downward gravitational force on the ball; in the case of a ball hanging in air it clearly is not, since if the ball were not supported by a wire it would begin accelerating downward.

The upward force is called a "buoyant" force but is simply an interatomic contact force due to fluid molecules striking atoms in the surface of the ball. The key point is that despite the complexity of the interatomic interactions, the net effect is simply an upward force whose magnitude is the mass of an equivalent volume of fluid, times g. Often it is reasonable to neglect buoyant forces in air (though usually not in water). To see why, let's compare the buoyant force to the gravitational force F_g on a ball hanging in air, where we write V for the volume of the ball:

$$\frac{F_b}{F_g} = \frac{m_{air}g}{M_{ball}g} = \frac{m_{air}}{M_{ball}} = \frac{m_{air}/V}{M_{ball}/V} = \frac{\rho_{air}}{\rho_{ball}}$$

The ratio of the forces is equal to the ratio of ρ_{air} (the density of air) to ρ_{ball} (the density of the ball; ρ is Greek lower-case letter rho). The density of air at STP ("standard temperature and pressure," 0° C and 1 atmosphere) can be calculated if we remember from chemistry that one mole of a gas at STP occupies 22.4 liters. The molecular mass of N_2 is 28 and that of O_2 is 32, which gives an average molecular mass of about 29 for air (which is about 80% nitrogen and 20% oxygen):

$$\rho_{air} \approx \frac{29 \text{ grams/mole}}{22.4 \times 10^3 \text{ cm}^3 / \text{mole}} \approx 1.3 \times 10^{-3} \text{ grams/cm}^3$$

What is the density of the ball? We can estimate it by noting that the density of water is 1 gram/cm^3, and that most solids are more dense than water. Using 1 g/cm^3 as the density of the ball, we find that $F_b/F_g \approx 1.3 \times 10^{-3}/1$.

The buoyant force $m_{air}g$ of air on a solid object is therefore very small compared to the gravitational force mg. In many cases we can safely choose to neglect it. Since the upward force exerted by the air on the ball is small compared to the gravitational force or the tension force exerted by the wire, we can choose to neglect it in an analysis of the forces on a hanging ball. However, it is not always appropriate to neglect the buoyant force; in accurate analytical chemistry it is necessary to correct for buoyancy when weighing a sample of liquid or low-density solid on an analytical balance.

A final note: the buoyant force being equal to $m_{air}g$ is really a time average, because the buoyant force is due to random collisions with air molecules and therefore has an intermittent character. However, the collision rate is so high that the force seems nearly continuous and constant. If, however, the ball is of microscopic size, it may not be possible to ignore the intermittent nature of the collisions. In a microscope it is possible to observe small particles being jostled about by the random collisions with the water molecules. This effect is called "Brownian motion."

4.X.31 Calculate the buoyant force in air on a kilogram of iron (whose density is about 8 grams per cubic centimeter). Compare with the weight *mg* of this much iron.

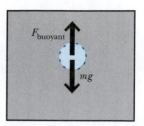

Figure 4.61 Forces on the sphere of fluid. Here m is the mass of the fluid (air or water) in the sphere.

Figure 4.62 A ball hanging from a wire, surrounded by air. The tension force is actually less than the weight of the ball, because the buoyant force F_b due to the air is also in the upward direction.

Figure 4.63 A one-square-meter column of air extending upward through the entire atmosphere.

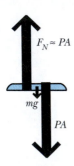

$F_N \approx PA$

mg

PA

Figure 4.64 The weight of the suction cup is very small compared to the downward force by the air or the normal force by the table.

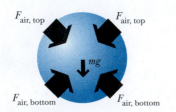

Figure 4.65 A ball hanging in air is subject to large compression forces due to the surrounding air.

Pressure and suction

There is an interesting case where the effect of the air is very significant, and cannot be neglected. Suppose that an object lies on a table and the surfaces of the table and object are so smooth that upon squeezing them together all air is expelled. Such a situation occurs when a suction cup is pressed down onto a smooth surface. Now there are no longer any air molecules striking the bottom surface of the suction cup, while air molecules continue to strike the top surface. The force of the air on the top surface is quite large, about 10^5 newtons on each square meter at sea level! This large force per unit area is called the "pressure" P of the air. The pressure can be thought of as due to the weight of a one-square-meter column of the entire atmosphere, extending upward many kilometers (although with diminishing density), as illustrated in Figure 4.63.

Pressure $P = F/A$ (force per unit area)

If the atmosphere had a constant density ρ like that near the surface (1.3 grams per cubic centimeter), we could calculate the height h of the atmosphere in the following way. The volume of a column of air of height h and cross-sectional area A is Ah, and its mass is $M_{air} = \rho(Ah)$. The pressure, which is about 10^5 newtons per square meter, is the weight $M_{air}g$ divided by the cross-sectional area: $P = M_{air}g/A = \rho gh$.

The height of a constant-density atmosphere, $h = P/(\rho g)$, would be about 8000 meters (about 5 miles), about the height of Mount Everest. In reality air density is not constant but decreases as you go higher, so there is still some low-density air at much higher altitudes than this.

A physics diagram for a suction cup looks like Figure 4.64, where PA represents the force of the air on the area A of the top surface of the suction cup. The normal force exerted upward by the table and the downward force of the air are huge compared to the weight of the suction cup mg.

The normal force $F_N = PA + mg \approx PA$, since the force of the air is so much larger than the weight mg of the suction cup. For example, a rubber suction cup with a diameter of 4 cm has an area of $\pi(0.02 \text{ m})^2 = 1.3 \times 10^{-3} \text{ m}^2$, so the force of the air on the suction cup is

$$PA = (10^5 \text{ N/m}^2)(1.3 \times 10^{-3} \text{ m}^2) = 130 \text{ newtons}$$

whereas its own weight mg might be about 0.01 kg (10 grams) times 9.8 N/kg, or only about 0.1 newton. Note that a force of 130 newtons is the weight of an object whose mass is about 13 kg, or about 29 pounds, which is why it isn't easy to pull even a small suction cup off a flat surface.

Every solid object in air, such as a hanging ball, is subjected to large compression forces by the pressure of the surrounding air (Figure 4.65). The upward buoyant force is the tiny difference between these huge forces. A hollow object must be very strong to support this compression, if air is removed from the inside. There is no problem however if air can flow into and out of the container, because then it is just the solid walls that are compressed, since the air exerts comparable pressure on the inner and outer surfaces.

All of this is even more relevant under water, because the density of water is about 1000 times greater than the density of air, and the compression forces are about 1000 times as strong. A hollow submersible vehicle that dives very deep must have an extremely strong hull. When a scuba diver descends to greater depth where the water pressure is larger, the breathing apparatus automatically increases the pressure of the air supplied to the diver to equal the increased pressure of the outside water at the new depth, to prevent the lungs from being crushed.

4.X.32 It is hard to imagine that there can be enough air between a book and a table so that there is a net upward (buoyant) force on the book despite the large downward force on the top of the book. About how many air molecules are there between a textbook and a table, if there is an average distance of about 0.01 mm between the uneven surfaces of the book and table?

4.18 *MORE CURVING MOTION EXAMPLES

Forward reasoning

A common technique for solving problems is to start by considering what you want to know (or what you have been asked to determine), and to work backward from that to the solution. Physics usually approaches the analysis of physical systems in the opposite way. We begin by organizing our knowledge about the system according to a fundamental principle—the Momentum Principle—and systematically recording our knowledge in equations and diagrams. After we have recorded all the information we know, whatever is missing is what we need to figure out. Figuring out the unknown information may be straightforward, or it may involve several steps.

This way of working problems, sometimes called "forward reasoning," is typical of the way expert physicists solve problems of this kind—starting each new problem from the fundamental principles. This approach works! For the student, however, approaching problems in this way may require a bit of faith—it is not necessarily obvious how the desired answer will emerge from the analysis. If you practice approaching problems in this way, you will develop confidence in the approach and will gain important experience in explaining complex phenomena starting from fundamental principles.

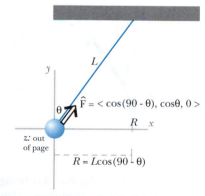

Figure 4.66 A circular pendulum, moving in the *xz* plane.

Example: Circular pendulum (curving motion)

A ball is suspended from a string, and after being given a push, moves along a circular path, as shown in Figure 4.66. You measure the length of the string L and the angle θ, shown in Figure 4.67. You also time the motion, and find that it takes T seconds for the ball to make one complete circular trip. From these measurements, determine the gravitational constant g.

System: Ball
Surroundings: String, Earth
(neglect air resistance)

Figure 4.67 shows a snapshot of a side view of the circular pendulum, at an instant when the string (of length L to the center of the ball) lies in the xy plane (at an angle θ to the vertical), and the ball is at the leftmost point in its circular path of radius $r = L\sin\theta$. At this instant the ball's momentum is in the $+z$ direction (out of the page, toward you).

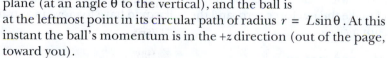

Figure 4.67 The circular pendulum at an instant when the string is in the xy plane, and the ball is moving in the $+z$ direction.

Momentum Principle:

$$d\vec{p}/dt = \vec{F}_{net}$$

$$\hat{F}_T = \langle \cos(90° - \theta), \cos\theta, 0 \rangle$$

$$\vec{F}_T = |\vec{F}_T| \langle \cos(90° - \theta), \cos\theta, 0 \rangle$$

$$\vec{F}_{net} = \langle |\vec{F}_T| \cos(90° - \theta), (|\vec{F}_T| \cos\theta - mg), 0 \rangle$$

Parallel component (+z):

Unit vector in direction of tension force.

We know that there are no forces in the z direction, so we deduce that $dp_z/dt = 0$.

$$\frac{dp_z}{dt} = F_{\text{net},z} = 0$$

Perpendicular components:

We know that $dp_y/dt = 0$ because the ball stays at the same horizontal level; p_y is zero and remains zero.

$$y: \frac{dp_y}{dt} = F_T\cos\theta - mg \text{ but } \frac{dp_y}{dt} = 0$$

$$F_T = \frac{mg}{\cos\theta}$$

$$x: \frac{dp_x}{dt} = F_T\cos(90° - \theta)$$

direction: toward center of circle ($\hat{n} = \langle 1, 0, 0 \rangle$)

$$\frac{|\vec{v}|}{r}|\vec{p}| = F_T\cos(90° - \theta)$$

$$\frac{mv^2}{L\cos(90° - \theta)} = \left(\frac{mg}{\cos\theta}\right)\cos(90° - \theta)$$

Here the fact that $\cos(90° - \theta) = \sin\theta$ is used to simplify the final expression.

$$g = \frac{v^2\cos\theta}{L(\sin\theta)^2}$$

Measuring the speed v, length L and angle θ determines g. v can be found by measuring T, the time to go around once, since $v = (2\pi r)/T$.

Check: units (N/kg). Special cases: If $\theta = 0$ the string hangs vertically and there is no motion, so the tension in the string ought to be mg, which is consistent with the equation for F_T which we found above.

Further discussion

A large circular pendulum was used by Newton to measure g, the magnitude of the gravitational field, with considerable accuracy. You could observe and measure the motion of a circular pendulum yourself, and your analysis of the motion would let you deduce a value for g, which you could compare with the accepted value of 9.8 N/kg.

Note that $\vec{F}_{net}$ and $d\vec{p}/dt$ are not the same thing. In the example above, we knew the x and y components of $d\vec{p}/dt$ (the effect), and from this deduced the x and y components of $\vec{F}_{net}$ (the cause). In the z direction, it was the other way around: we knew that there were no forces in this direction, so we deduced that the z component of momentum was not changing.

Because we systematically identified each force with an object in the surroundings, we did *not* make the mistake of introducing a radially outward "centrifugal force" (Figure 4.68). The only forces on the ball are due to interactions with real objects (string, Earth). If there were such a force, it would make the net force zero, and the ball would have to move in a straight line, not a circle.

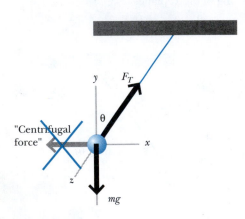

Figure 4.68 There is no "centrifugal force." There is no interaction associated with such a fictitious force. Moreover, if the net force were zero, the ball would have to move in a straight line!

Constrained motion

Note that there could in principle be many possible motions of this system. The ball could swing back and forth in a plane; this is the motion of a "simple pendulum." The ball could go around in a path something like an ellipse, in which case the ball's height would vary, just as with simple pendulum motion. The ball could move so violently that the string goes slack for part of the time, with abrupt changes of momentum every time the string suddenly goes taut. If the string can stretch noticeably, there can be a sizable oscillation superimposed on the swinging

All of these motions are difficult to analyze. In principle, we could use a computer to predict the general motion of the ball hanging from the string if we had a formula for the tension force in the string, as a function of its length. Essentially, we would need the effective spring stiffness for this stiff "spring." However, the string may be so stiff that even a tiny stretch implies a huge tension (a nearly "inextensible" string). The observable length of the string is nearly constant, but the tension force that the string applies to the ball can vary a great deal. This makes it very difficult to do a computer numerical integration, because a tiny error in position makes a huge change in the force.

Example: A turning car (curving motion)

Suppose you're riding as the passenger in a convertible with left hand drive (American or continental European), so you are sitting on the right (Figure 4.69). Assume you have foolishly failed to fasten your seat belt. The driver makes a sudden right turn. Before studying physics you would probably have said that you are "thrown to the left," and in fact you do end up closer to the driver.

? There's something wrong with this "thrown to the left" idea. What object in the surroundings exerted a force to the left to make you move to the left?

There is no such object! To understand what's happening, watch the turning car from a fixed vantage point above the convertible, and observe carefully what happens to the passenger. In the absence of forces to change the passenger's direction of motion, the passenger keeps moving ahead, in a straight line at the same speed as before (Figure 4.70). It is the car that is yanked to the right, out from under the passenger. The driver moves closer to the passenger; it isn't that the passenger moves toward the driver.

The driver may also feel "thrown to the left" against the door. But it is the door that runs into the driver, forcing the driver to the *right*, not to the left.

Sometimes people invent fictitious forces such as the so-called "centrifugal" force to account for the passenger and driver being "thrown to the left." But this just adds a second confusion to the first. The passenger and driver are thrown to the right in a right turn, not to the left, and there are real forces to the right that make them go to the right (for example, the force of the door on the driver's left shoulder).

Moreover, forces are associated with interactions between objects in a chosen system and objects in the surroundings. The "centrifugal" force is just made up and is not associated with any real object.

Example: An amusement park ride (curving motion)

There is an amusement park ride that some people love and others hate in which a bunch of people stand against the wall of a cylindrical room of radius R, and the room starts to rotate at higher and higher speed (Figure 4.71). The surface of the wall is designed to maximize friction between the person and the wall, so it is fuzzy or sticky, not slick. When a certain critical speed is reached, the floor drops away, leaving the people stuck against the wall as they whirl around at constant speed.

Why do the people stick to the wall without falling down? Include a carefully labeled force diagram of a person, and discuss how the person's momentum changes, and why.

Figure 4.69 You are a passenger in a car that makes a sudden right turn.

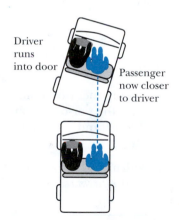

Figure 4.70 The passenger moves straight ahead and is now closer to the driver, who was thrown to the right by the door pushing the driver to the right.

Driver runs into door

Passenger now closer to driver

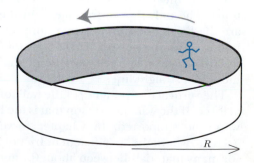

Figure 4.71 An amusement park ride.

System: person
Surroundings: Wall, Earth
(neglect air resistance)

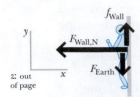

Momentum Principle

$$d\vec{p}/dt = \vec{F}_{net}$$

At this instant the person is stuck to the right wall, with momentum directed into the page ($-z$). The force due to the wall has been separated into two parts: A normal component $\vec{F}_{Wall,N}$, due to the compression of the wall by the person, and a frictional part f_{Wall}, that keeps the person from sliding downward.

Parallel component (z):

No change in speed, so $\dfrac{dp}{dt} = 0$

Perpendicular components:

x: $\dfrac{dp_x}{dt} = -\left|\vec{F}_{Wall,N}\right|$ direction: $\hat{n} = \langle -1, 0, 0 \rangle$ toward center

$$-p\frac{v}{R} = -F_{Wall,N}$$

$$F_{Wall,N} = \frac{mv^2}{R}$$

y: $\dfrac{dp_y}{dt} = 0 = f_{Wall} - mg$ (friction force)

$$f_{Wall} = mg$$

The y component of momentum is not changing, since the person remains at the same horizontal level, and does not start to move up or down.

Check: Units okay. Also, the faster the ride, the bigger the inward force of the wall, which sounds right.

Further discussion: The friction force

We deduce that the wall exerts an unknown force which must have a y component f_{Wall} (because the person isn't falling) and an x component $-F_{Wall,N}$ normal to the wall (because the person's momentum is changing direction). The only significant objects in the surroundings are the Earth and the wall; therefore only these objects interact with the person. There is a momentum change inward; if the net force were zero, the person would move in a straight line.

The vertical component f of the wall force is a frictional force. If the wall has friction that is too low, the person won't be supported. In Chapter 6 we will see that $f \le \mu F_N$, where the "coefficient of friction" μ has a value for many materials between about 0.1 and 1.0. Basically, the more strongly two objects are squeezed together, the harder it is to slide one along the other.

The speed of the ride has to be large enough that $f \le \mu F_N$, so we have

$$mg \le \mu\left(\frac{mv^2}{R}\right)$$

$$\mu \ge \frac{gR}{mv^2}$$

This predicts that the lower the speed v, the greater the coefficient of friction must be to hold the person up on the wall, which sounds right. Conversely, if the ride spins very fast, the wall doesn't have to have a very large coefficient of friction to hold the people against the wall.

4.19 *A VERTICAL SPRING-MASS SYSTEM

To reduce friction, it is easier to experiment with a mass oscillating vertically up and down on a spring (Figure 4.72) rather than sliding on an air table We will show that we expect the same period whether the spring is vertical or horizontal.

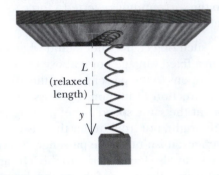

Figure 4.72 A vertical spring-mass system minimizes sliding friction.

We choose a coordinate system with the positive y axis pointing downward (Figure 4.72). We'll measure y downward from the location where the end of the relaxed spring would be (the relaxed length of the spring is L).

Neglecting air resistance and gravitational interactions with objects other than the Earth, the physics diagram of the forces acting on the hanging block is that shown in Figure 4.73.

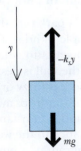

Figure 4.73 Force diagram for the hanging block.

We then have the following:

$$\frac{dp_y}{dt} = -k_s y + mg$$

We choose to analyze only simple vertical motion, with no bouncing side to side, so we know this:

$$p_x = 0 \text{ at all times, so } \frac{dp_x}{dt} = 0$$

We can reduce the p_y equation to a familiar form if we measure a new coordinate, w, from the equilibrium position, when the block is hanging motionless, rather than from the unstretched spring position (Figure 4.74).When the mass hangs motionless from the spring, the spring is stretched an amount $k_s s_0 = mg$, so $s_0 = mg/k_s$. We'll measure w downward from the end of the spring, so

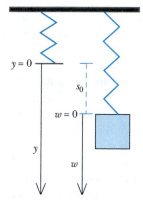

Figure 4.74 Measure w from the place where the block can hang motionless.

$$y = (s_0 + w) = \frac{mg}{k_s} + w$$

and therefore

$$\frac{dp_y}{dt} = -k_s\left(\frac{mg}{k_s} + w\right) + mg = -k_s w$$

This says that if we measure from the end of the spring at its equilibrium location, mg/k_s below the bottom of the unstretched spring, the equation for the motion of the mass is the same as in a horizontal spring-mass system, whose motion we already know. This means that the analytical solution we obtained for the horizontal case is also the solution for the vertical case (substituting y for x in the appropriate places).

The moving mass will oscillate up and down around this equilibrium position, as you can observe if you experiment with a vertical mass-spring system.

4.20 *GENERAL SOLUTION FOR THE MASS-SPRING SYSTEM

It is easy to show that the function $x = A\cos(\omega t + \phi)$ is a general solution for a spring-mass system with arbitrary k_s, m, and initial conditions, where $\omega = \sqrt{k_s/m}$, and where the amplitude A and the "phase shift" ϕ (measured in radians and denoted by the Greek letter phi) are constants determined by the initial conditions, as is explained below. The "phase shift" essentially shifts the starting point of the oscillation. For example, if $\phi = \pi/2$, the cosine function actually represents a negative sine function (Figure 4.75):

$$x = A\cos\left(\omega t + \frac{\pi}{2}\right) = -A\sin(\omega t)$$

Determining the amplitude and phase shift

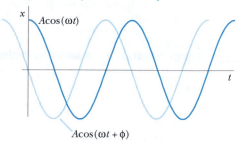

Figure 4.75 Phase shift ϕ.

The amplitude A and phase shift ϕ can be determined from the initial values at time t_0 of position x_0 and velocity v_0:

$$x_0 = A\cos\left(\sqrt{\frac{k_s}{m}}t_0 + \phi\right)$$

$$v_0 = -\sqrt{\frac{k_s}{m}}A\sin\left(\sqrt{\frac{k_s}{m}}t_0 + \phi\right)$$

Using the important trigonometric identity $\sin^2\theta + \cos^2\theta = 1$ (which is the Pythagorean theorem applied to a triangle whose hypotenuse has length 1), we can obtain an equation that can be solved for the amplitude A:

$$x_0^2 + \frac{m}{k_s}v_0^2 = A^2, \text{ so we have } A = \sqrt{x_0^2 + \frac{m}{k_s}v_0^2}$$

For example, if the initial velocity is zero, the amplitude is simply equal to the initial stretch.

The value for A can be plugged back into either the x_0 equation or the v_0 equation to obtain the phase shift ϕ. Alternatively, by dividing the x_0 equation by the v_0 equation we obtain an equation that can be solved for the phase shift ϕ:

$$\frac{v_0}{x_0} = -\sqrt{\frac{k_s}{m}} \tan\left(\sqrt{\frac{k_s}{m}} t_0 + \phi\right)$$

$$\sqrt{\frac{k_s}{m}} t_0 + \phi = \arctan\left(-\sqrt{\frac{m}{k_s}}\frac{v_0}{x_0}\right)$$

$$\phi = \arctan\left(-\sqrt{\frac{m}{k_s}}\frac{v_0}{x_0}\right) - \sqrt{\frac{k_s}{m}} t_0$$

4.X.33 Suppose the system is not oscillating and you strike it with a hammer, applying a large force F for a *very* short time Δt. What is the initial speed v_0? Explain briefly. What will be the amplitude (the maximum displacement away from the equilibrium position)?

4.X.34 In the case you just analyzed (striking with a hammer at time $t_0 = 0$), what is the phase shift ϕ? (Does this make sense?)

Comment: Two constants for a second-order differential equation

Note that whether we specify the starting conditions in terms of x_0 and v_0 or in terms of the amplitude A and the phase shift ϕ, two constants are required. You can see why two constants are needed by looking at the start of the numerical integration. You needed the initial position x_0 in order to be able to calculate the initial force (which lets you step the velocity), and you also needed the initial momentum p_0 or velocity v_0 in order to be able to step the position.

We have been working with what is called a "second-order differential equation." That is, the highest derivative in the equation is a second derivative—the acceleration $dv/dt = d(dx/dt)/dt$. A second-order differential equation normally requires two constants to specify the starting conditions completely.

Linear systems

We saw that doubling the amplitude of a spring-mass system simply multiplies the solution by a factor of two without changing the time dependence of the motion. This is an extremely important property of systems described by "linear" differential equations (equations in which the variables appear only to the first power, not squared, for example). More formally, let x be replaced by the quantity Rx, where R is a constant:

$$m\frac{d}{dt}\left(\frac{d(Rx)}{dt}\right) = -k_s(Rx)$$

The R's cancel, leaving us with the original equation. Therefore starting with a different initial stretch will simply scale up the vertical axis of the graph of x vs. t. Also, since the motion is periodic, a change in the initial velocity corresponds to starting the motion at a different

time. For example, instead of starting with an initial stretch and zero velocity, you could start with zero stretch and some initial velocity, and this would correspond to a point on the original graph where the function crosses the time axis.

In contrast, consider a differential equation in which there is an x^3, and try scaling by R:

$$m\frac{d}{dt}\left(\frac{d(Rx)}{dt}\right) = -b(Rx)^3 = -bR^3 x^3$$

In this case the R's do *not* cancel. The solutions to this "nonlinear" equation do not scale up in the simple way that solutions to a linear equation do.

4.21 SUMMARY

Contact forces:

- A solid object may be modeled as a 3D lattice of balls (atoms) connected by springs (interatomic bonds).
- Based on this model, and given the density and atomic weight of an element, one can find the length of an interatomic bond in a solid metal.
- From macroscopic measurements of stress and strain, one can determine the stiffness of an interatomic bond in a solid material.
- The speed of sound in a solid is the speed at which a disturbance propagates through the material.
- Solid objects exert contact forces because the interatomic bonds in the solid are stretched (a tension force) or compressed (a normal force).

Definitions:

For a long object like a rod, which is stretched or compressed:

$$\text{strain} \equiv \frac{\Delta L}{L}$$

$$\text{stress} \equiv \frac{F_T}{A}$$

L is the original length of the wire, ΔL is the change in length, F_T is the force applied to stretch or compress the object, A is the cross-sectional area of the object.

Young's modulus: $Y \equiv \dfrac{\text{stress}}{\text{strain}} = \dfrac{\left(\dfrac{F_T}{A}\right)}{\left(\dfrac{\Delta L}{L}\right)}$

Macro / micro connections:

Young's modulus in terms of microscopic quantities

$$Y = \frac{k_{s,i}}{d}$$

Y is Young's modulus, $k_{s,i}$ is the stiffness of the interatomic bond, and *d* is the length of the interatomic bond (distance between adjacent nuclei).

$$\text{Speed of sound in a solid: } v = \sqrt{\frac{k_{s,i}}{m_a}}\, d$$

$k_{s,i}$ is the stiffness of the interatomic bond, *d* is the length of the interatomic bond (distance between adjacent nuclei), m_a is the mass of one atom.

Momentum Principle: Derivative form

$$\frac{d\vec{p}}{dt} = \vec{F}_{net}$$

The instantaneous time rate of change of the momentum of an object is equal to the net force acting on the object.

Finding the rate of change of momentum

1. Draw two arrows:

   - the first arrow represents $\vec{p}_i$, the momentum of the object a short time before the instant of interest.
   - the second arrow represents $\vec{p}_f$, the momentum of the object a short time after the instant of interest.

2. Graphically find $\Delta\vec{p} = \vec{p}_f - \vec{p}_i$ by placing the arrows tail to tail, and drawing the resultant arrow starting at the tail of $\vec{p}_i$ and going to the tip of $\vec{p}_f$.

3. This arrow indicates the direction of $\Delta\vec{p}$, which is the same as the direction of $d\vec{p}/dt$, provided the time interval involved is sufficiently small.

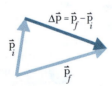

Momentarily at rest vs. uniform motion and equilibrium

For a system in equilibrium $d\vec{p}/dt = \vec{0}$

For a system in uniform motion $d\vec{p}/dt = \vec{0}$

For a system that is momentarily at rest, $d\vec{p}/dt \neq \vec{0}$

The Momentum Principle:
   Parallel and perpendicular Components

$$\left(\frac{d|\vec{p}|}{dt}\right)\hat{p} = \vec{F}_{\parallel} \text{ and } |\vec{p}|\frac{d\hat{p}}{dt} = \vec{F}_{\perp}$$

Valid only for a moving object, because $\hat{p}$ and $d\hat{p}/dt$ are undefined when $\vec{p} = \vec{0}$.

Component of net force parallel to momentum changes magnitude of momentum.

Component of net force perpendicular to momentum changes direction of momentum.

$\hat{n}$ (n-hat) is a unit vector in the direction of $\frac{d\hat{p}}{dt}$

Since $\frac{d\hat{p}}{dt} \perp \hat{p}$, $\hat{n} \perp \hat{p}$

(Note that $\hat{n}$ is undefined at an instant when a particle is not moving along a curve.)

Effect of perpendicular component of net force on a particle moving along a curving path

$$|\vec{p}|\frac{d\hat{p}}{dt} = |\vec{p}|\frac{|\vec{v}|}{R}\hat{n} = \vec{F}_{\perp}$$

R is the radius of the kissing circle. The component of net force perpendicular to the particle's momentum changes its direction, but not the magnitude of its momentum (speed).

Analytical solution for mass-spring oscillations

$$x = A\cos(\omega t) \text{, with } \omega = \sqrt{\frac{k_s}{m}}$$

Assuming: spring's mass is negligible, no friction, no air resistance.

A (the "amplitude") is a constant equal to the maximum stretch of the spring during an oscillation. ω is the "angular frequency", and its units are radians/s. k_s is the spring stiffness, and m is the mass of the object.

The "period" *T* is the time for one complete cycle, in seconds.

$$T = \frac{2\pi}{\omega}$$

The "frequency" *f* is the number of complete cycles per second.

$$f = \frac{1}{T} = \frac{\omega}{2\pi}$$

Contact forces due to gases

Upward buoyancy force = $m_{fluid}g$, where m_{fluid} is mass of equivalent volume of fluid.

Pressure is force per unit area; air pressure at sea level is about 10^5 N/m^2.

4.22 REVIEW QUESTIONS

The nature of solids

4.RQ.35 Approximately what is the radius of a copper atom?

a) 1×10^{-15} m b) 1×10^{-12} m c) 1×10^{-10} m

d) 1×10^{-8} m e) 1×10^{-6} m

4.RQ.36 You hang a 10 kg mass from a copper wire, and the wire stretches by 8 mm. Then you suspend an identical mass from two copper wires, identical to the first. What happens?
a) Each wire stretches 4 mm.
b) Each wire stretches 8 mm.
c) Each wire stretches 16 mm.

4.RQ.37 You hang a 10 kg mass from a copper wire, and the wire stretches by 8 mm. Then you hang the same mass from a second copper wire, whose cross-sectional area is half as large (but whose length is the same). What happens?
a) The second wire stretches 4 mm.
b) The second wire stretches 8 mm.
c) The second wire stretches 16 mm.

4.RQ.38 You hang a 10 kg mass from a copper wire, and the wire stretches by 8 mm. Then you hang the same mass from a second copper wire, which is twice as long, but has the same diameter. What happens?
a) The second wire stretches 4 mm.
b) The second wire stretches 8 mm.
c) The second wire stretches 16 mm.

4.RQ.39 You hang a mass M from a spring, which stretches an amount s_1. Then you cut the spring in half, and hang a mass M from one half. How much does the half-spring stretch?
a) $2s_1$
b) s_1
c) $s_1/2$

4.RQ.40 A spring has stiffness k_s. You cut the spring in half. What is the stiffness of the half-spring?
a) $2k_s$
b) k_s
c) $k_s/2$

4.RQ.41 Lead is much softer than aluminum, and can be more easily deformed or pulled into a wire. What difference between the two materials best explains this?
a) Pb and Al atoms have different sizes
b) Pb and Al atoms have different masses
c) The stiffness of the interatomic bonds is different in Pb and Al

4.RQ.42 Two wires with equal lengths are made of pure copper. The diameter of wire A is twice the diameter of wire B. When 6 kg masses are hung on the wires, wire B stretches more than wire A. You make careful measurements and compute Young's modulus for both wires. What do you find?

a) $Y_A > Y_B$
b) $Y_A = Y_B$
c) $Y_A < Y_B$

4.RQ.43 Two metal rods are made of different elements. The interatomic spring stiffness of element A is three times larger than the interatomic spring stiffness for element B. The mass of an atom of element A is three times greater than the mass of an atom of element b. The atomic diameters are approximately the same for A and B. How does the speed of sound in rod A compare to the speed of sound in rod B?
a) $v_A > v_B$
b) $v_A = v_B$
c) $v_A < v_B$

4.RQ.44 Two rods are both made of pure titanium. The diameter of rod A is twice the diameter of rod B, but the lengths of the rods are equal. You tap on one end of each rod with a hammer and measure how long it takes the disturbance to travel to the other end of the rod. In which rod did it take longer?
a) Rod A
b) Rod B
c) The times were equal

4.RQ.45 Suppose you attempt to pick up a very heavy object. Before you tried to pick it up, the object was sitting still—its momentum was not changing. You pull very hard, but do not succeed in moving the object. Is this a violation of the Momentum Principle? How can you be exerting a large force on the object without causing a change in its momentum? What does change when you apply this force?

Rate of change of momentum

4.RQ.46 The diagram below shows the orbit of the Moon around the Earth. At the instant when the Moon is at each of the locations marked by a number (*1-5*):
- which arrow (a-m) best describes the direction of the Moon's momentum?
- which arrow (a-m) best describes the direction of the rate of change of the Moon's momentum?
- which arrow (a-m) best describes the direction of the net force on the Moon?

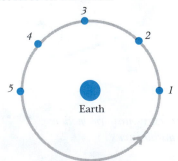

Directions:

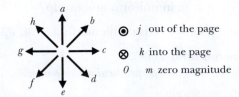

4.RQ.47 A student answering 4.RQ.46 said: "When the Moon is at location 4, there is an inward force due to the Moon and an outward force due to centrifugal force, so the net force on the Moon is zero." Give two or more physics reasons why this is wrong.

4.RQ.48 Tarzan swings from a vine. When he is at the bottom of his swing, as shown in the diagram below:
- Which arrow (a-m) best describes the direction of the rate of change of Tarzan's momentum?
- Which is larger in magnitude, the force by the Earth on Tarzan, the force by the vine (a tension force) on Tarzan, or neither (same magnitude)? Explain how you know this.

Directions:

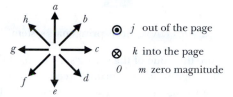

4.RQ.49 Tarzan swings back and forth on a vine. At the microscopic level, why is the tension force on Tarzan by the vine greater than it would be if he were hanging motionless?
a) The interatomic bonds in the vine are stretched less than they would be if the Tarzan's momentum were not changing.
b) The interatomic bonds in the vine are stretched more than they would be if the Tarzan's momentum were not changing.
c) The interatomic bonds are not stretched at all.

4.RQ.50 An airplane flies at constant speed along a curve, as shown in the diagram below. At the top of the arc, what is the direction of $d\vec{p}/dt$ on a passenger riding in the plane? What is the direction of $d\hat{p}/dt$?

4.RQ.51 At the moment shown in the preceding diagram, the magnitude of the upward (normal) force exerted by the seat on a passenger in the airplane is not zero, but it is less than it would be if the plane and passenger were sitting still on the ground. Which of the following statements correctly explains this at a microscopic level?
a) The interatomic bonds in the seat are compressed less than they would be if the passenger's momentum wasn't changing.
b) The interatomic bonds in the seat are compressed more than they would be if the passenger's momentum wasn't changing.
c) The interatomic bonds are not compressed at all.

4.RQ.52 At a particular instant an airplane is flying in the $-x$ direction. At this instant the magnitude of its momentum is changing, but the direction is not changing. Which of the following quantities are not zero? If a quantity is not zero, give a possible direction for that quantity.
a) $d|\vec{p}|/dt$
b) $d\hat{p}/dt$
c) $\vec{F}_\parallel$
d) $\vec{F}_\perp$

4.RQ.53 At a particular instant an airplane is flying in the $-z$ direction. At this instant the direction of its momentum is changing, but the magnitude is not changing. Which of the following quantities are not zero? If a quantity is not zero, give a possible direction for that quantity.
a) $d|\vec{p}|/dt$
b) $d\hat{p}/dt$
c) $\vec{F}_\parallel$
d) $\vec{F}_\perp$

4.RQ.54 You are driving an American car, sitting on the left side of the front seat. You make a sharp right turn. You feel yourself "thrown to the left" and your left side hits the left door. Is there a force that pushes you to the left? What object exerts that force? What really happens? Draw a diagram to illustrate and clarify your analysis.

4.RQ.55 A 30 kg child rides on a playground merry-go-round, 2 m from the center. The merry-go-round makes one complete revolution every 4 seconds. How large is the net force on the child? In what direction does the net force act?

4.RQ.56 The radius of a merry-go-round is 7 meters, and it takes 12 seconds to make a complete revolution.
(a) What is the speed of an atom on the outer rim?
(b) Which statement below (i, ii, or iii) best describes the direction of the momentum of this atom?
(c) Which statement below (i, ii, or iii) best describes the direction of the rate of change of the momentum of this atom?
(i) Inward, toward the center
(ii) Outward, away from the center
(iii) Tangential

4.RQ.57 Figure 4.76 shows the path of a particle. Assuming

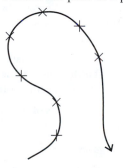

Figure 4.76 Path of a particle traveling at constant speed (4.RQ.57).

the particle's speed does not change as it travels along this path, at each location labeled "x" draw an arrow showing the direction of $d\vec{p}/dt$ for the particle as it passes that location.

Spring-mass systems

For 4.RQ.58 - 4.RQ.68, assume that friction is negligible.

4.RQ.58 In a spring-mass oscillator, when is the magnitude of momentum of the mass largest: When the magnitude of the net force acting on the mass is largest, or when the magnitude of the net force acting on the mass is smallest?

4.RQ.59 For a vertical spring-mass oscillator that is moving up and down, which of the following statements are true? (More than one statement may be true.)

    a) At the lowest point in the oscillation, the momentum is zero.
    b) At the lowest point in the oscillation, the rate of change of the momentum is zero.
    c) At the lowest point in the oscillation, $mg = k_s s$
    d) At the lowest point in the oscillation, $mg > k_s s$
    e) At the lowest point in the oscillation, $mg < k_s s$.

4.RQ.60 Suppose the period of a spring-mass oscillator is 1 s. What will the period be if we double the mass?

4.RQ.61 Suppose the period of a spring-mass oscillator is 1 s. What will be the period if we replace the spring with a spring that is twice as stiff?

4.RQ.62 Suppose the period of a spring-mass oscillator is 1 s. What will be the period if we cut the spring in half and use just one of the pieces?

4.RQ.63 Suppose the period of a spring-mass oscillator is 1 s with an amplitude of 5 cm. What will be the period if we increase the amplitude to 10 cm, so that the total distance traveled in one period is twice as large?

4.RQ.64 Suppose the period of a spring-mass oscillator is 1 s with an amplitude of 5 cm. What will be the period if we take the oscillator to a massive planet where g = 19.6 N/kg?

4.RQ.65 Write the Momentum Principle for a horizontal spring-mass system with spring stiffness k_s and mass m, and write an analytical solution.

4.RQ.66 In terms of k_s and m, what is the angular frequency? The frequency? The period? The amplitude?

4.RQ.67 If you double the amplitude, by what factor does the period change? If you double the mass, by what factor does the period change? If you double the spring stiffness, by what factor does the period change?

4.RQ.68 How should you start the system going at $t = 0$ in order for the motion to be $A\cos(\omega t)$? How should you start the system going at $t = 0$ in order for the motion to be $A\sin(\omega t)$?

4.RQ.69 Describe two examples of oscillating systems that are *not* harmonic oscillators.

Contact forces due to gases

4.RQ.70 Air pressure at the surface of a fresh water lake near sea level is about 10^5 N/m². At approximately what depth below the surface does a diver experience a pressure of 2×10^5 N/m²? How would this be different in sea water, which has higher density than fresh water?

4.23 PROBLEMS

4.HW.71 Experiment—study of strain vs. stress
Measure and graph the strain vs. stress curve for a long metal wire by hanging weights from the end of the wire and carefully measuring the stretch as a function of the hanging weight. If possible, go beyond the "yield" stress and observe the rapid, large increase in the strain. BE CAREFUL! Once you exceed the yield stress, the wire may break and dump the weights on whatever is underneath.

4.HW.72 Experiment—a vertical spring-mass system
You will be given a spring and one or more masses with which to study the motion of a mass hanging from a spring (Figure 4.77).

Figure 4.77 A vertical spring-mass system.

(a) Measure the value of the spring stiffness k_s of the spring in N/m. The magnitude of a spring force is $k_s s$, where s is the stretch of the spring (change from the unstretched length). Explain briefly how you measured k_s. Include in your report the unstretched length of the spring.

(b) Measure the period (the round-trip time) of a mass hanging from the vertical spring. Report the mass that you use, and the amplitude of the oscillation. Amplitude is the maximum displacement, plus or minus, from the equilibrium position (the position where the mass can hang motionless).

(c) With twice the amplitude that you reported in part (b), measure the period again. Since the mass has to move twice as far, one might expect the period to lengthen. Does it?

How to measure the period accurately

There are unavoidable fluctuations in starting and stopping the timing, but you can minimize the error this contributes by timing many complete cycles so that the starting and stopping fluctuations are a small fraction of the total time measured.

It is good practice to count out loud *starting from zero, not one.* To count five cycles, say out loud "Zero, one, two, three, four, five." If on the other hand you say "One, two, three, four, five," you have actually counted only four cycles, not five.

Since the motion is periodic, you can start (say "zero") at any point in the motion. It is best to start and stop when the mass is moving fast past some marker near the equilibrium point, because it is difficult to estimate the exact time when the mass reaches the very top or the very bottom, because it is moving slowly at those turnaround points. Be sure to measure full round-trip cycles, not the half-cycles between returns to the equilibrium point (but going in the opposite direction).

4.HW.73 Numerical integration of a spring-mass system without friction

Carry out a numerical integration of the motion of a horizontal spring-mass system. Use the spring stiffness k_s and mass m that you measured experimentally for a vertical spring-mass system in Problem 4.HW.72. Predict the motion of this spring-mass system sufficiently far into the future to observe several oscillations.

(a) Initial conditions: Make the initial position of the block be such that the stretch of the spring is equal to the amplitude of the oscillations in your experiment, and release the block with zero initial momentum. Display the motion in two ways:

(1) Plot a graph of the position of the block as a function of time. Label the scales on both axes, so your numerical results are clear.

(2) If possible using your computer tool, make an animation of the motion of the block. (You don't need to draw a real spring; it is sufficient to draw a line connecting the block to the wall.)

Very briefly, what are the most important aspects of the predicted motion?

(b) Vary the time step and report the maximum step size that gives reasonable results, in the sense that a smaller step size has little effect.

(c) Read the "period" (the round-trip time) of the oscillation off the graph. How does this compare with the period you measured in your experiment?

(d) Repeat the calculation with double the initial stretch. What happens to the period? Is this what you observed in your experiment?

(e) Temporarily change the force law to be $-k_s x^3$, and observe the period with the amplitude of part (a) and with double this amplitude. What happens to the period when you change the amplitude?

(f) How long would the spring-mass system theoretically continue to oscillate? Would a real block connected to a spring behave that way? What *would* happen?

4.HW.74 Pushing a rod through space

(a) In outer space, a rod is pushed to the right by a constant force F. Describe the pattern of interatomic distances along the rod. Include a specific comparison of the situation at locations A, B, and C. Explain briefly in terms of fundamental principles.

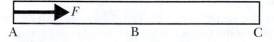

Hint: Consider the motion of an individual atom inside the rod, and various locations along the rod.

(b) After the rod in part (a) reaches a speed v, the object that had been exerting the force on the rod is removed. Describe the subsequent motion of the rod, and the pattern of interatomic distances inside the rod. Include a specific comparison of the situation at locations A, B, and C. Explain briefly.

4.HW.75 Pushing on a block

Two blocks of mass m_1 and m_3, connected by a rod of mass m_2, are sitting on a low-friction surface, and you push to the left on the right block (mass m_1) with a constant force.

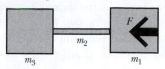

(a) What is the acceleration dv_x/dt of the blocks?

(b) What is the compression force in the rod (mass m_2) near its right end? Near its left end?

(c) How would these results change if you *pull* to the left on the *left* block (mass m_3) with the same force, instead of pushing the right block?

4.HW.76 Interatomic spring stiffness

Young's modulus for aluminum is 6.2×10^{10} N/m^2. The density of aluminum is 2.7 grams/cm^3, and the mass of one mole is 27 grams. If we model the interactions of neighboring aluminum atoms as though they were connected by springs, determine the approximate spring stiffness of such a spring.

Repeat this analysis for lead: Young's modulus for lead is 1.6×10^{10} N/m^2, the density of lead is 11.4 grams/cm^3, and the mass of one mole is 207 grams. Make a note of these results, which we will use for various purposes later on. Note that aluminum is a rather stiff material, whereas lead is quite soft.

4.HW.77 Computer model of the propagation of sounds in a metal

Create a computer model of a long metal bar as many atomic masses connected by interatomic springs, all in a straight line. Give the first atom a sudden push and compute the motions of all the atoms, one time step after another. Calculate all of the forces before using these forces to update the momenta and positions of the objects. Otherwise the calculations of the forces would mix positions corresponding to different times.

After each time step, display the displacements of all the atoms away from their equilibrium positions, and determine the speed of sound (the distance between the first and last atoms, divided by the amount of time it takes for the pulse to reach the last atom). Compare with the prediction that $v = (\sqrt{k_s/m_a})d$.

4.HW.78 The interatomic spring stiffness for titanium

A hanging titanium wire with diameter 2 mm (2×10^{-3} m) is initially 3 m long. When a 5 kg mass is hung from it, the wire stretches an amount 0.4035 mm, and when a 10 kg mass is hung from it, the wire stretches an amount 0.807 mm. A mole of titanium has a mass of 48 grams, and its density is 4.51 grams/cm^3. Find the approximate value of the effective spring stiffness of the interatomic force, and explain your analysis.

4.HW.79 A coiled wire

A certain coiled wire with uneven windings has the property that to stretch it an amount s from its relaxed length requires a force that is given by $F = bs^3$, so its behavior is different from a normal spring. You suspend this device vertically, and its unstretched length is 20 cm.

(a) You hang a mass of 15 grams from the device, and you observe that the length is now 24 cm. What is b, including units? Start your analysis from the Momentum Principle.

(b) You hold the 15 gram mass and throw it downward, releasing it when the length of the spring-like device is 27 cm and the speed of the mass is 4 m/s. One millisecond later (10^{-3} s), what is the approximate change in the stretch of the device, and what is the approximate change in the speed of the mass?

4.HW.80 String and spring

A ball of mass 450 g hangs from a spring whose stiffness is 110 newtons per meter. A string is attached to the ball and you are pulling the string to the right, so that the ball hangs motionless, as shown. In this situation the spring is stretched, and its length is 15 cm. What would the relaxed length of the spring be, if it were detached from the ball and laid on a table?

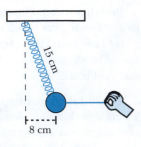

4.HW.81 Helicopter load

A helicopter flies to the right (in the $+x$ direction) at a constant speed of 12 m/s, parallel to the surface of the ocean. A 900 kg package of supplies is suspended below the helicopter by a cable as shown; the package is also traveling to the right in a straight line, at a constant speed of 12 m/s. The pilot is concerned about whether or not the cable, whose breaking strength is listed at 9300 N, is strong enough to support this package under these circumstances.

(a) Choose the package as the system. What is the rate of change of the momentum of the system?

(b) List all the objects that are exerting forces on the package. (Note that "gravity" is not an object.)

(c) Represent the package as a dot. Draw a careful force diagram ("freebody diagram") showing all the forces acting on the package. You should have the same number of arrows as you had objects in your list in part (b), and the tail of each arrow should be on the package. Label each force with the name of the object exerting the force.

(d) What is the magnitude of the tension in the cable supporting the package? Carefully show all steps in your work.

(e) Write the force exerted on the package by the cable as a vector.

(f) What is the magnitude of the force exerted by the air on the package? Carefully show all steps in your work.

(g) Write the force on the package by the air as a vector.

(h) Is the cable in danger of breaking?

4.HW.82 Your weight on a bathroom scale

If you stand on a bathroom scale at the North Pole, the scale shows your "weight" as an amount Mg (actually, it shows the force F_N that the scale exerts on your feet). At the North Pole you are 6357 km from the center of the Earth. If instead you stand on the scale at the equator, the scale reads a different value due to two effects: 1) The Earth bulges out at the equator (due to its rotation), and you are 6378 km from the center of the Earth. 2) You are moving in a circular path due to the rotation of the Earth (one rotation every 24 hours). Taking into account both of these effects, what does the scale read at the equator?

4.HW.83 Two ropes support a load

An 800 kg load is suspended as shown. Calculate the tension in all three ropes (that is, the magnitude of the tension force exerted by each of these ropes).

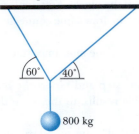

4.HW.84 Looping roller coaster

What is the minimum speed v that a roller coaster car must have in order to make it around an inside loop and just barely lose contact with the track at the top of the loop (see diagram below)? The center of the car moves along a circular arc of radius R. Include a carefully labeled force diagram. State briefly what approximations you make. Design a plausible roller coaster loop, including numerical values for v and R.

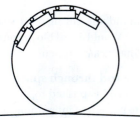

4.HW.85 Experiment—determine g with a circular pendulum

Use a circular pendulum to determine g. You can increase the accuracy of the time it takes to go around once by timing N revolutions and then dividing by N. This minimizes errors contributed by inaccuracies in starting and stopping the clock. It is wise to start counting from zero (0, 1, 2, 3, 4, 5) rather than starting from 1 (1, 2, 3, 4, 5 represents only 4 revolutions, not 5). It also improves accuracy if you start and stop timing at a well-defined event, such as when the mass crosses in front of an easily visible mark.

This was the method used by Newton to get an accurate value of g. Newton was not only a brilliant theorist but also an excellent experimentalist. For a circular pendulum he built a large triangular wooden frame mounted on a vertical shaft, and he pushed this around and around while making sure that the string of the circular pendulum stayed parallel to the slanting side of the triangle.

4.HW.86 Space station

An engineer whose mass is 70 kg holds onto the outer rim of a rotating space station whose radius is 14 m and which takes 30 s to make one complete rotation. What is the magnitude of the force the engineer has to exert in order to hold on? What is the magnitude of the net force acting on the engineer?

4.HW.87 NEAR spacecraft orbits Eros

After the NEAR spacecraft passed Mathilde, on several occasions rocket propellant was expelled to adjust the spacecraft's momentum in order to follow a path that would approach the asteroid Eros, the final destination for the mission. After getting close to Eros, further small adjustments made the momentum just right to give a circular orbit of radius 45 km (45×10^3 m) around the asteroid. So much propellant had been used that the final mass of the spacecraft while in circular orbit around Eros was only 500 kg. The spacecraft took 1.04 days to make one complete circular orbit around Eros. Calculate what the mass of Eros must be.

4.HW.88 At the bottom of a Ferris wheel

A Ferris wheel is a vertical, circular amusement ride. Riders sit on seats that swivel to remain horizontal as the wheel turns. The wheel has a radius R and rotates at a constant rate, going around once in a time T. At the bottom of the ride, what are the magnitude and direction of the force exerted by the seat on a rider of mass m? Include a diagram of the forces on the rider.

4.HW.89 Weightlessness

By "weight" we usually mean the gravitational force exerted on an object by the Earth. However, when you sit in a chair your own perception of your own "weight" is based on the contact force the chair exerts upward on your rear end rather than on the gravitational force. The smaller this contact force is, the less "weight" you perceive, and if the contact force is zero, you feel peculiar and "weightless" (an odd word to describe a situation when the only force acting on you is the gravitational force exerted by the Earth!). Also, in this condition your internal organs no longer press on each other, which presumably contributes to the odd sensation in your stomach.

(a) How fast must a roller coaster car go over the top of a circular arc for you to feel "weightless"? The center of the car moves along a circular arc of radius R (see diagram below). Include a carefully labeled force diagram.

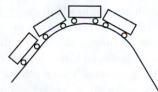

(b) How fast must a roller coaster car go through a circular dip for you to feel three times as "heavy" as usual, due to the upward force of the seat on your bottom being three times as large as usual? The center of the car moves along a circular arc

of radius R (see diagram below). Include a carefully labeled force diagram.

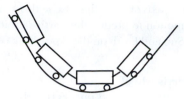

4.HW.90 Swing a ball on a spring

A ball of unknown mass m is attached to a spring. In outer space, far from other objects, you hold the other end of the spring and swing the ball around in a circle of radius 1.5 m at constant speed.

(a) You time the motion and observe that going around 10 times takes 6.88 seconds. What is the speed of the ball?

(b) Is the momentum of the ball changing or not? How can you tell?

(c) If the momentum is changing, what interaction is causing it to change? If the momentum is not changing, why isn't it?

(d) The relaxed length of the spring is 1.2 m, and its stiffness is 1000 N/m. While you are swinging the ball, since the radius of the circle is 1.5 m, the length of the spring is also 1.5 m. What is the magnitude of the force that the spring exerts on the ball?

(e) What is the mass m of the ball?

4.HW.91 Car going over a hill

A sports car (and its occupants) of mass M, is moving over the rounded top of a hill of radius R. At the instant when the car is at the very top of the hill, the car has a speed v. You can safely neglect air resistance.

(a) Taking the sports car as the system of interest, what object(s) exert non-negligible forces on this system?

(b) At the instant when the car is at the very top of the hill, draw a diagram showing the system as a dot, with force vectors whose tails are at the location of the dot. Label the force vectors (that is, give them algebraic names). Try to make the lengths of the force vectors be proportional to the magnitudes of the forces.

(c) Starting from the Momentum Principle, calculate the force exerted by the road on the car.

(d) Under what conditions will the force exerted by the road on the car be zero? Explain.

4.HW.92 Mass on a spring with circular motion

A small block of mass m is attached to a spring with stiffness k_s and relaxed length L. The other end of the spring is fastened to a fixed point on a low-friction table. The block slides on the table in a circular path of radius $R > L$. How long does it take for the block to go around once?

4.HW.93 Relating radius to period for circular orbits

The planets in our Solar System have orbits around the Sun that are nearly circular, and $v \ll c$. Calculate the period T (a "year"—the time required to go around the Sun once) for a planet whose orbit radius is r. This is the relationship discovered by Kepler and explained by Newton. (It can be shown by advanced techniques that this result also applies to elliptical orbits if you replace r by the "semi major axis," which is half the longer, "major" axis of the ellipse.) Use this analytical solution for circular motion to predict the Earth's orbital speed, using the data for Sun and Earth on the inside back cover of the textbook.

4.HW.94 Ferris wheel

A person of mass 70 kg rides on a Ferris wheel whose radius is 4 m. The person's speed is constant at 0.3 m/s. The person's location is shown by a dot in the diagram.

(a) What is the magnitude of the rate of change of the momentum of the person at the instant shown?.

(b) What is the direction of the rate of change of momentum of the person at the instant shown?

(c) What is the magnitude of the *net* force acting on the person at the instant shown? Draw the *net* force vector on the diagram at this instant, with the tail of the vector on the person.

4.HW.95 Circular motion in a magnetic field

When a particle with electric charge q moves with speed v in a plane perpendicular to a magnetic field B, there is a magnetic force at right angles to the motion with magnitude qvB, and the particle moves in a circle of radius r (see diagram below). This formula for the magnetic force is correct even if the speed is comparable to the speed of light. Show that

$$p = \frac{mv}{\sqrt{1 - (|\vec{v}|/c)^2}} = qBr$$

even if v is comparable to c.

This result is used to measure relativistic momentum: if the charge q is known, we can determine the momentum of a particle by observing the radius of a circular trajectory in a known magnetic field.

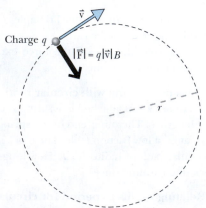

4.HW.96 Dark matter in the Universe

In the 1970's the astronomer Vera Rubin made observations of distant galaxies that she interpreted as indicating that perhaps 90% of the mass in a galaxy is invisible to us ("dark matter").

She measured the speed with which stars orbit the center of a galaxy, as a function of the distance of the stars from the center. The orbital speed was determined by measuring the "Doppler shift" of the light from the stars, an effect which makes light shift toward the red end of the spectrum ("red shift") if the star has a velocity component away from us, and makes light shift toward the blue end of the spectrum if the star has a velocity component toward us. She found that for stars farther out from the center of the galaxy, the orbital speed of the star hardly changes with distance from the center of the galaxy, as is indicated in the diagram below. The visible components of the galaxy (stars, and illuminated clouds of dust) are most dense at the center of the galaxy and thin out rapidly as you move away from the center, so most of the visible mass is near the center.

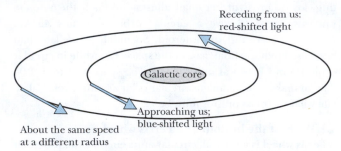

(a) Predict the speed v of a star going around the center of a galaxy in a circular orbit, as a function of the star's distance r from the center of the galaxy, assuming that almost all of the galaxy's mass M is concentrated at the center.

(b) Construct a logical argument as to why Rubin concluded that much of the mass of a galaxy is not visible to us. Reason from principles discussed in this chapter, and your analysis of part (a). Explain your reasoning. You need to address the following issues:

Rubin's observations are not consistent with your prediction in (a).

Most of the *visible* matter is in the center of the galaxy.

Your prediction in (a) assumed that most of the mass is at the center.

This issue has not yet been resolved, and is still a current topic of astrophysics research. Here is a reference to the original work: "Dark Matter in Spiral Galaxies" by Vera C. Rubin, *Scientific American*, June 1983 (96-108). You can find several graphs of the rotation curves for spiral galaxies on page 101 of this article.

4.HW.97 Orbital periods

(a) Many communication satellites are placed in a circular orbit around the Earth at a radius where the period (the time to go around the Earth once) is 24 hours. If the satellite is above some point on the equator, it stays above that point as the Earth rotates, so that as viewed from the rotating Earth the satellite appears to be motionless. That is why you see dish antennas pointing at a "fixed" point in space. Calculate the radius of the orbit of such a "synchronous" satellite. Explain your calculation in detail.

(b) Electromagnetic radiation including light and radio waves travels at a speed of 3×10^8 m/s. If a phone call is routed through a synchronous satellite to someone not very far from you on the ground, what is the minimum delay between saying something and getting a response? Explain. Include in your explanation a diagram of the situation.

(c) Some human-made satellites are placed in "near-Earth" orbit, just high enough to be above almost all of the atmosphere. Calculate how long it takes for such a satellite to go around the Earth once, and explain any approximations you make.

(d) Calculate the orbital speed for a near-Earth orbit, which must be provided by the launch rocket. (Recently large numbers of near-Earth communications satellites have been launched. Their advantages include making the signal delay unnoticeable, but with the disadvantage of having to track the satellites actively and having to use many satellites to ensure that at least one is always visible over a particular region.)

(e) When the first two astronauts landed on the Moon, a third astronaut remained in an orbiter in circular orbit near the Moon's surface. During half of every complete orbit, the orbiter was behind the Moon and out of radio contact with the Earth. On each orbit, how long was the time when radio contact was lost?

4.HW.98 Analytical 3-body orbit

There is no general analytical solution for the motion of a 3-body gravitational system. However, there do exist analytical solutions for very special initial conditions. The diagram below shows three stars, each of mass m, which move in the plane of the page along a circle of radius r. Calculate how long this system takes to make one complete revolution. (In many cases 3-body orbits are not stable: any slight perturbation leads to a break-up of the orbit.)

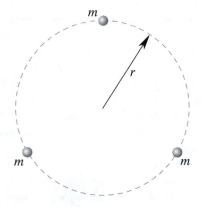

4.HW.99 The center of our galaxy

Remarkable data indicate the presence of a massive black hole at the center of our Milky Way galaxy. The W. M. Keck 10 meter diameter telescopes in Hawaii were used by Andrea Ghez and her colleagues to observe infrared light coming directly through the dust surrounding the central region of our galaxy (visible light is multiply scattered by the dust, blocking a direct view). Stars were observed for several consecutive years to determine their orbits near a motionless center that is completely invisible ("black") in the infrared but whose precise location is known due to its strong output of radio waves, which are observed by radio telescopes. The data were used to show that the

object at the center must have a mass that is huge compared to the mass of our own Sun, whose mass is a "mere" 2×10^{30} kg.

Here are positions from 1995 to 2004 of one of the stars, S0-20, orbiting around the galactic center. The orbit is nearly circular with the radius shown.

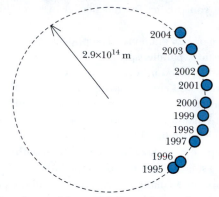

(a) Using the positions and times shown above, what is the approximate speed of this star in m/s? Also express the speed as a fraction of the speed of light.

(b) This is an extraordinarily high speed for a macroscopic object. Is it reasonable to approximate the star's momentum as mv? (Some other stars near the galactic center with highly elliptical orbits move even faster when they are closest to the center.)

(c) Based on these data, calculate the mass of the massive black hole about which this star is orbiting. (This result differs somewhat from that obtained by Ghez and colleagues, who did a careful analysis of the elliptical orbits of several stars.)

(d) How many of our Suns does this represent?
It is thought that all galaxies may have such a black hole at their centers, as a result of long periods of mass accumulation. When many bodies orbit each other, sometimes an object happens in an interaction to acquire enough speed to escape from the group, and the remaining objects are left closer together. Some simulations show that over time, as much as half the mass may be ejected, with agglomeration of the remaining mass. This could be part of the mechanism for forming massive black holes.

For more information, see the home page of Andrea Ghez, http://www.astro.ucla.edu/faculty/ghez.htm. You may see the term "arcseconds," which is an angular measure of how far apart objects appear on the sky, and to "parsecs," which is a distance equal to 3.3 light-years (a light-year is the distance light goes in one year).

4.HW.100 Falling through the Earth

In the approximation that the Earth is a sphere of uniform density, it can be shown that the gravitational force it exerts on a mass m inside the Earth at a distance r from the center is $mg(r/R)$, where R is the radius of the Earth. (Note that at the surface, the force is indeed mg, and at the center it is zero). Suppose that there were a hole drilled along a diameter straight through the Earth, and the air were pumped out of the hole. If an object is released from one end of the hole, how long will it take to reach the other side of the Earth? Include a numerical result.

4.HW.101 Observing spring-mass systems

Using masses and springs similar to those you used, it was found that a 20 gram mass hanging from a spring had an oscillation period of 1.2 seconds.

(a) When two 20 gram masses are hung from this spring, what would you predict for the period in seconds? Explain briefly.

(b) When one 20 gram mass is supported by two of these vertical, parallel springs, what would you predict for the period in seconds? Explain briefly.

(c) Suppose you cut one spring into two equal lengths, and you hang one 20 gram mass from this half spring. What would you predict for the period in seconds? Explain briefly.

(d) Suppose you take a single (full-length) spring and a single 20 gram mass to the Moon and watch the system oscillate vertically there. Will the period you observe on the Moon be longer, shorter, or the same as the period you measured on Earth? (The gravitational field strength on the Moon is about one-sixth that on the Earth.) Explain briefly.

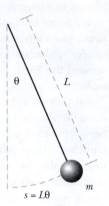

4.HW.102 Vibration of diatomic molecules

In Exercise 4.X.1 and Exercise 4.X.2 we found the effective spring stiffness corresponding to the interatomic force for aluminum and lead. Let's assume for the moment that, *very* roughly, other atoms have similar values.

(a) What is the (very) approximate frequency f for the vibration of H_2, a hydrogen molecule?

(b) What is the (very) approximate frequency f for the vibration of O_2, an oxygen molecule?

(c) What is the approximate vibration frequency f of D_2, a molecule both of whose atoms are deuterium atoms (that is, each nucleus has one proton and one neutron)?

(d) Explain why the *ratio* of the deuterium frequency to the hydrogen frequency is quite accurate, even though you have estimated each of these quantities very approximately, and the effective spring stiffness is normally expected to be significantly different for different atoms. (Hint: what interaction is modeled by the effective "spring"?)

4.HW.103 Speed of sound in nickel

One mole of nickel (6.02×10^{23} atoms) has a mass of 59 grams, and its density is 8.9 grams per cubic centimeter. You have a bar of nickel 2.5 m long, with a square cross section, 2 mm on a side. You hang the rod vertically and attach a 40 kg mass to the bottom, and you observe that the bar becomes 1.2 mm longer.

Next you remove the 40 kg mass, place the rod horizontally, and strike one end with a hammer.

How much time T will elapse before a microphone at the other end of the bar will detect a disturbance? Be sure to show clearly the steps in your analysis.

4.HW.104 Speed of sound in uranium

Uranium-238 (U^{238}) has three more neutrons than uranium-235 (U^{235}). Compared to the speed of sound in a bar of U^{235}, is the speed of sound in a bar of U^{238} higher, lower, or the same? Explain your choice, including justification for assumptions you make.

4.HW.105 The simple pendulum

A "simple" pendulum consists of a small mass of mass m swinging at the end of a low-mass string of length L (in contrast to a pendulum whose mass is distributed, such as a rod swinging from one end).

(a) Show that the momentum of the small mass obeys the following equation, where s is the arc length $L\theta$:

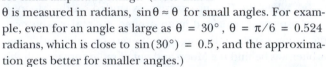

$$\frac{dp}{dt} = -mg\sin\theta = -mg\sin\left(\frac{s}{L}\right)$$

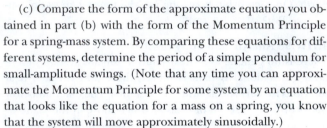

(b) What form does this equation take for small-amplitude swings? (Note that if θ is measured in radians, $\sin\theta \approx \theta$ for small angles. For example, even for an angle as large as $\theta = 30°$, $\theta = \pi/6 = 0.524$ radians, which is close to $\sin(30°) = 0.5$, and the approximation gets better for smaller angles.)

(c) Compare the form of the approximate equation you obtained in part (b) with the form of the Momentum Principle for a spring-mass system. By comparing these equations for different systems, determine the period of a simple pendulum for small-amplitude swings. (Note that any time you can approximate the Momentum Principle for some system by an equation that looks like the equation for a mass on a spring, you know that the system will move approximately sinusoidally.)

(d) Make a simple pendulum and predict its period, then measure the period. Do this for a long pendulum and for a short pendulum. Report your experimental data and results, with comparisons to your theoretical predictions.

(e) For large-amplitude swings, the small-angle approximation is not valid, but you can use numerical integration to calculate the motion. For a mass on a string, the largest possible amplitude is 90°, but if the string is replaced by a lightweight rod, the amplitude can be as large as 180° (standing upside-down, with the mass at the top). It turns out that such a pendulum can be modeled by the same equation as the equation derived in part (a), but the concepts of "torque" and "angular momentum" are required to prove it.

Assume that the equation of part (a) is valid, and use numerical integration to predict the motion of the pendulum. Think of s and p as being like x and p_x so that your program is basically a one-dimensional calculation, with a force $-mg\sin(s/L)$. Plot the position s and momentum p as a function of time for a pendulum whose amplitude is nearly 180°.

Note how different these plots are from sines or cosines, because the system is not well approximated by a "spring-mass" equation. Also note that for this anharmonic oscillator the period increases with increasing amplitude, unlike the situation with a harmonic oscillator.

4.HW.106 Floating objects

(a) A block of wood 20 cm long by 10 cm wide by 6 cm high has a density of 0.7 grams/cm^3 and floats in water (density 1.0

grams/cm^3). How far below the surface of the water is the bottom of the block? Explain your reasoning.

(b) An advertising blimp consists of a gas bag, the supporting structure for the gas bag, and the gondola hung from the bottom, with its cabin, engines, and propellers. The gas bag is about 30 meters long with a diameter of about 10 meters, and it is filled with helium (density about 4 grams per 22.4 liters under these conditions). Estimate the total mass of the blimp, including the helium.

4.24 ANSWERS TO EXERCISES

4.X.1 (page 109) $2.6{\times}10^{-10}$ m

4.X.2 (page 109) $3.1{\times}10^{-10}$ m

4.X.3 (page 110) 800 N/m

4.X.4 (page 111) 300 N/m

4.X.5 (page 113) $Y \approx 6.5{\times}10^{10}$N/m^2

4.X.6 (page 113) $\Delta L \approx 0.2$ mm

4.X.7 (page 116) Bat, Earth, air

4.X.8 (page 116) Earth, string, air

4.X.9 (page 116) Earth, vine, air

4.X.10 (page 116) Earth, snow, air

4.X.11 (page 117) $F_z = 4$ N

4.X.12 (page 117) $\vec{0}$, $\vec{0}$

4.X.13 (page 117) $\vec{0}$, $\vec{0}$

4.X.14 (page 118) $\langle 2, 0, -4 \rangle$ m/s^2

$\langle 0.16, 0, -0.32 \rangle$ (kg · m/s)/s ,

$\langle 0.16, 0, -0.32 \rangle$ N

4.X.15 (page 124) nonzero, downward

4.X.16 (page 124) nonzero, downward

4.X.17 (page 124) nonzero, upward

4.X.18 (page 124) nonzero, in direction of final momentum

4.X.19 (page 128) (a) Forces shown below

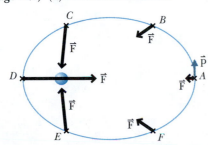

A: (b) zero (c) nonzero (d) not changing (e) changing

B: (b) nonzero (c) nonzero (d) increasing (e) changing

C: (b) nonzero (c) nonzero (d) increasing (e) changing

D: (b) zero (c) nonzero (d) not changing (e) changing

E: (b) nonzero (c) nonzero (d) decreasing (e) changing

F: (b) nonzero (c) nonzero (d) decreasing (e) changing

4.X.20 (page 130) The third (bottom) one (R_3)

4.X.21 (page 130) Downward, toward the center of the kissing circle

4.X.22 (page 130) $(d|\vec{p}|/dt)\hat{p} = \vec{0}$,

$|\vec{p}|(d\hat{p}/dt) = (0.975 \text{ kg} \cdot \text{m/s}^2)\hat{n}$, where $\hat{n}$ points from the child toward the center of the carousel,

$\vec{F}_{net} = (0.975 \text{ N})\hat{n}$ (toward center),

Earth, wooden horse.

4.X.23 (page 130) $(d|\vec{p}|/dt)\hat{p} = \vec{0}$,

$|\vec{p}|(d\hat{p}/dt) = \dfrac{mv}{R}\hat{n}$

where $\hat{n}$ points from the person toward the center of the Ferris wheel,

$\vec{F}_{net}$ is toward the center of the wheel, and is due to the Earth and the seat.

4.X.24 (page 130) $3.57{\times}10^{22}$ (kg · m/s)/s toward the Sun

$3.57{\times}10^{22}$ N toward the Sun

4.X.25 (page 140) (a) 58.8 N/m, (b) 10.6 round trips

4.X.26 (page 140) 2 seconds

4.X.27 (page 140) 1.73 seconds

4.X.28 (page 140) 0.577 seconds

4.X.29 (page 143) 5000 m/s

4.X.30 (page 143) 1200 m/s

4.X.31 (page 145) $F_{buoyant} = 1.6{\times}10^{-3}$ N

$\dfrac{F_{buoyant}}{F_{weight of iron}} = 1.6{\times}10^{-4}$

4.X.32 (page 147) For a book the size of this book, there are approximately 10^{19} air molecules between the book and the table.

4.X.33 (page 152) $v_0 = \dfrac{F}{m}\Delta t$; $A = v_0\sqrt{\dfrac{m}{k}}$

4.X.34 (page 152) 90° ($\pi/2$ radians), corresponding to a sine rather than a cosine (start at x = 0)

CHAPTER 5

THE ENERGY PRINCIPLE

Key concepts

- Energy is a scalar quantity.
- A single particle is the simplest possible system.
  - Particle energy has two parts:
      rest energy (energy an object has even when at rest)
      kinetic energy (energy associated with motion)
- The energy of a system can be changed only by external inputs.
- Work is mechanical energy transfer into or out of a system.
  - Work involves an external force acting through a distance.
  - Work can be positive or negative.
- Energy is conserved: The change in energy of a system plus the change in energy of the surroundings is zero.
- Multiparticle system energy has three parts:
      rest energy
      kinetic energy
      potential energy (the interaction energy of pairs of particles within a multiparticle system)

5.1 THE ENERGY PRINCIPLE

You know a lot about the flow of energy from daily life. You ingest chemical energy in the form of food, and you use up this chemical energy when you are active. You put chemical energy in the form of gasoline into a car, and the car uses up this chemical energy to run the engine. You put electrical energy into a toaster, and the toaster raises the temperature of the bread. You compress a spring, and the mechanical energy stored in the spring can launch a ball.

The Energy Principle is a fundamental principle, just as the Momentum Principle is. The validity of the Energy Principle has been verified through a very wide variety of observations and experiments, involving large and small objects, moving slowly or at speeds near the speed of light, and even undergoing nuclear reactions that change the identity of the objects. It is a summary of the way energy flows in the real world.

THE ENERGY PRINCIPLE

$$\Delta E_{\text{system}} = W_{\text{surr}} + Q$$

Change in energy (E_{system}) of a system is equal to the work done by the surroundings (W_{surr}), and to energy flow between system and surroundings due to a difference in temperature (Q).

ΔE is a *change* in energy, which is the final energy minus the initial energy, $E_{\text{final}} - E_{\text{initial}}$. Q, the flow of energy due to a temperature difference, will be discussed in Chapter 6. The current chapter will consider only situations in which temperature differences do not play a role, and so Q will always be zero, and will be omitted from the Energy Principle equation for the rest of the chapter. Work, the result of a force acting though a displacement, will be discussed in Section 5.3.

The Energy Principle is a Fundamental Principle because:

It applies to every possible system, no matter how large or small (from clusters of galaxies to subatomic particles), or how fast it is moving.

It is true for any kind of interaction (electric, gravitational, strong, weak).

It relates an effect (change in energy of a system) to a cause (an interaction with the surroundings).

As with the Momentum Principle, the left hand side of the Energy Principle equation describes an effect on a system; the right hand side describes the causes, which involve interactions of the system with its surroundings.

Conservation of energy

Energy cannot be created or destroyed, but it can change form. The only way for a system to gain or lose energy is if the surroundings lose or gain the same amount of energy.

CONSERVATION OF ENERGY

$$\Delta E_{\text{system}} + \Delta E_{\text{surroundings}} = 0$$

5.2 THE SIMPLEST SYSTEM: A SINGLE PARTICLE

The simplest possible system consists of a single particle. "A particle" could refer to a proton or an electron, but it could also refer to a baseball or even a planet if during the process of interest there are no significant changes internal to the "particle" such as changes of shape or rotation or vibration or temperature.

In 1905 Einstein discovered that to be consistent with his special theory of relativity and with existing measures of energy, including an already existing definition of mechanical energy transfer to a system, called "work," the total energy of a single-particle system is given by this expression:

ENERGY OF A SINGLE-PARTICLE SYSTEM

$$E_{\text{particle}} \equiv \gamma m c^2$$

$$\text{where } \gamma = \frac{1}{\sqrt{1 - v^2/c^2}}$$

The unit of energy is the joule, abbreviated "J". One joule is equal to one $\text{kg} \cdot (\text{m/s})^2$

The speed of light $c = 3\times10^8 \text{ m/s}$. Later in this chapter we'll see that this definition of particle energy is consistent with the definition of a particle's momentum, $\vec{p} = \gamma m\vec{v}$.

Similarities between particle energy and particle momentum:
- Both contain the same factor γ.
- Both are proportional to the mass m.

Differences between particle energy and particle momentum:
- Particle energy is a scalar.
- Momentum is a vector.

The units of energy are $\text{kg} \cdot (\text{m/s})^2$. This combination of units is given the name "joule" to honor the British physicist James Joule who in the 1840's did some of the earliest experiments to establish the nature of energy. The abbreviation for joule is J.

Rest energy

? If a particle is at rest, what is its energy?

If $v = 0$ then $\gamma = 1$, and we find the famous equation $E = mc^2$. This has the striking interpretation that a particle has energy even when it is sitting still, which was a revolutionary idea when Einstein proposed this. We call mc^2 the "rest energy," the amount of energy a particle has when it is sitting still.

REST ENERGY

$$\text{rest energy} = mc^2$$

The "rest energy" is the energy of a particle at rest.

Even more strikingly, Einstein realized that this means that the mass of an object is its energy content divided by a constant: $m = E/c^2$. For example, a hot object, with more thermal energy, has very slightly more mass than a cold object. In a real sense, mass and energy are the same thing, although for historical reasons we use different units for them: mass in kilograms, energy in joules, with a constant factor of c^2 difference.

Note another difference between energy and momentum. A particle at rest has zero momentum but it has nonzero energy, its rest energy.

Kinetic energy

? If a particle is moving with speed v, is its energy greater than mc^2?

If $v > 0$, $\sqrt{1 - v^2/c^2}$ is less than 1, so $\gamma = 1/\sqrt{1 - v^2/c^2}$ is greater than 1, and the particle energy γmc^2 increases with increasing speed. As a consequence of its motion, the particle has additional, "kinetic" energy K:

KINETIC ENERGY *K* OF A PARTICLE

$$K = \gamma mc^2 - mc^2$$

The kinetic energy K of a particle is the energy a moving particle has in addition to its rest energy.

It is often useful to turn this equation around:

$$E_{\text{particle}} = mc^2 + K$$

If a particle is at rest, its particle energy is just its rest energy mc^2. If the particle moves, it has not only its rest energy mc^2 but also an additional amount of energy which we call the kinetic (motional) energy K (Figure 5.1).

Example: **Energy of a fast moving proton**

A proton in a particle accelerator has a speed of 2.91×10^8 m/s. (a) What is the energy of the proton? (b) What would the energy of the proton be if it were sitting still? (c) What is the kinetic energy of the moving proton?

(a) Moving proton:

$$E = \gamma mc^2$$

$$\gamma = \frac{1}{\sqrt{1 - \left(\dfrac{2.91 \times 10^8}{3 \times 10^8}\right)^2}} = 4.11$$

$$E = (4.11)(1.7 \times 10^{-27} \text{ kg})(3 \times 10^8 \text{ m/s})^2$$

$$E = 6.288 \times 10^{-10} \text{ J}$$

(b) Proton at rest:

$$E = mc^2$$

$$E = (1.7 \times 10^{-27} \text{ kg})(3 \times 10^8 \text{ m/s})^2$$

$$E = 1.53 \times 10^{-10} \text{ J}$$

(c) Kinetic energy of proton:

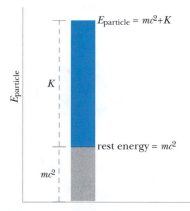

Figure 5.1 Particle energy is equal to rest energy plus kinetic energy *K*. This particle has high speed and its kinetic energy is large compared to its rest energy.

In this example, the kinetic energy of the fast moving proton turns out to be about 3 times as large as the rest energy of the proton—about the ratio shown in Figure 5.1.

$$K = E - mc^2$$

$$K = 6.288 \times 10^{-10} \text{ J} - 1.53 \times 10^{-10} \text{ J} = 4.76 \times 10^{-10} \text{ J}$$

A universal speed limit

Delivering more and more energy to a particle increases its energy more and more, and there is no theoretical limit to how much energy a particle can acquire. As the speed approaches the speed of light $c = 3 \times 10^8$ m/s, the denominator $\sqrt{1 - v^2/c^2}$ becomes very small, and the energy becomes very large.

However, even though you can increase the energy, it becomes very difficult to make large increases in the speed v, because a sizable speed increase would mean a huge energy increase. There is effectively a universal speed limit: a particle can't reach the speed of light c because an infinite amount of energy input would be required to achieve this.

The low-speed approximation for kinetic energy

Next consider the case of a particle moving with a speed v that is small compared to the speed of light ($v \ll c$), the low-speed or "nonrelativistic" approximation. The situation is shown in Figure 5.2, where the kinetic energy is very small compared to the rest energy, so the particle energy is only slightly larger than the rest energy. An approximate formula for the energy at low speeds can be obtained through use of the mathematics of the "binomial expansion," as is shown in Section 5.20 at the end of this chapter:

$$\frac{mc^2}{\sqrt{1 - v^2/c^2}} = mc^2\left[1 + \frac{1}{2}\frac{v^2}{c^2} - \frac{3}{8}\left(\frac{v^2}{c^2}\right)^2 - \frac{5}{16}\left(\frac{v^2}{c^2}\right)^3 + \ldots\right]$$

$$\frac{mc^2}{\sqrt{1 - v^2/c^2}} \approx mc^2 + \frac{1}{2}mv^2 \quad \text{if } v/c \ll 1$$

Since we define kinetic energy through the relation $E_{\text{particle}} = mc^2 + K$, at low speeds the kinetic energy is approximately the following:

LOW-SPEED KINETIC ENERGY K OF A PARTICLE

$$K \approx \frac{1}{2}mv^2 \quad \text{(only if } v \ll c\text{)}$$

Approximate kinetic energy can also be written in terms of approximate magnitude of momentum.

$$K \approx \frac{1}{2}mv^2 \approx \frac{1}{2}m\left(\frac{p}{m}\right)^2 = \frac{p^2}{2m} \quad \text{(if } v \ll c\text{)}$$

Example: Approximate kinetic energy

The solar wind consists of charged particles streaming away from the Sun. The speed of a proton in the solar wind can be as high as 1×10^6 m/s. What is the kinetic energy of such a proton?

? Is it appropriate to use the approximate formula for kinetic energy in this case?

$$\left(\frac{v}{c}\right)^2 = \left(\frac{1 \times 10^6 \text{ m/s}}{3 \times 10^8 \text{ m/s}}\right)^2 = (3.33 \times 10^{-3})^2 = 1.11 \times 10^{-5} \quad \text{so } \gamma \approx 1$$

It is appropriate to use the approximate formula for K.

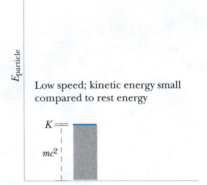

Figure 5.2 At low speeds the kinetic energy K is small compared to the rest energy.

$$K \approx \tfrac{1}{2}mv^2 = \tfrac{1}{2}(1.7\times10^{-27}\ \text{kg})(1\times10^6\ \text{m/s})^2 = 8.5\times10^{-16}\ \text{J}$$

The kinetic energy of this relatively slow moving proton is much less (nearly a million times less) than its rest energy, which was calculated in the previous example.

5.X.1 An electron has mass 9×10^{-31} kg. If the electron's speed is $0.988c$ (that is, $v/c = 0.988$), what is its particle energy? What is its rest energy? What is its kinetic energy?

5.X.2 A commercial airliner flew 20,000 km nonstop from Seattle to Kuala Lumpur at an average speed of 900 km/hr. If its average mass was 140,000 kg what was its average kinetic energy? (We take an average because the mass of an airplane changes as it exhausts burnt fuel.)

5.X.3 It is not very difficult to accelerate an electron to a speed that is 99% of the speed of light, because it has such a very small mass. What is the ratio of the kinetic energy K to the rest energy mc^2 in this case? In the definition of what we mean by kinetic energy K, $E = mc^2 + K$, you must use the full relativistic formula for E, because v/c is not small compared to 1.

5.X.4 A pitcher can throw a baseball at about 100 miles/hour (about 44 m/s). What is the ratio of the kinetic energy to the rest energy mc^2? (Can you use $K \approx \tfrac{1}{2}mv^2$?)

Example: Pull a block

If you pull on a 3 kg block and change its speed from 4 m/s to 5 m/s, what is the minimum change of chemical (food) energy in you? Give magnitude, sign, and units. (Your temperature rises when you do this, involving an increase in thermal energy, which we'll study in the next chapter, so there is more actual chemical cost than the minimum amount you calculate.)

The speed is tiny compared to c, so the change in kinetic energy of the block is

$$\Delta K = \tfrac{1}{2}mv_f^2 - \tfrac{1}{2}mv_i^2 = \tfrac{1}{2}(3\ \text{kg})[(5\ \text{m/s})^2 - (4\ \text{m/s})^2]$$

$$\Delta K = 13.5\ \text{J}$$

Because you increased the kinetic energy of the block by 13.5 J, and your temperature also rose a bit, your store of chemical (food) energy change is at least -13.5 J.

5.X.5 Show that the units of $\dfrac{p^2}{2m}$ and mc^2 are indeed joules. (Note that 1 N is 1 kg $\cdot$ m/s^2.)

5.X.6 An automobile traveling on a highway has an average kinetic energy of 1.1×10^5 J. Its mass is 1.5×10^3 kg; what is its average speed? Convert your answer to miles per hour to see if it makes sense.

5.X.7 If you could use all of the mc^2 rest energy of some fuel to provide the car in the previous exercise with its kinetic energy of 1.1×10^5 J, how much mass of fuel would you need?

Relativistic energy and momentum

Because both relativistic energy and momentum involve the factor

$$\gamma = \frac{1}{\sqrt{1 - (v/c)^2}}$$

it seems likely that they can be related in some way. As you can show in Problem 5.P.85, the relationship turns out to be the following, where p is $|\vec{p}|$:

RELATIVISTIC ENERGY AND MOMENTUM OF A PARTICLE

$$E^2 - (pc)^2 = (mc^2)^2$$

where $E = \gamma mc^2$; $m = $ rest mass

This relation is true in all reference frames, for all particle speeds. Both energy and momentum will change if a phenomenon is viewed in a different reference frame, but this relation will still be true. The quantity $E^2 - (pc)^2$

is therefore called an "invariant" quantity—a quantity that does not vary when the reference frame is changed.

This relation turns out to be $E^2 - (pc)^2 = 0$ for those particles such as photons which always travel at the speed of light. Even though they can never be at rest, it is as though they had zero rest mass. The relation for so-called "massless" particles reduces to $E^2 = (pc)^2$, or $E = pc$. This relation is also approximately true for any particle that is traveling at very high speed, because in that case mc^2 is small compared to E. Neutrinos have extremely small mass and travel at nearly the speed of light.

The relation $E^2 - (pc)^2 = (mc^2)^2$ can be used to find another formula for kinetic energy, valid for both high and low speeds:

The difference of squares $E^2 - (mc^2)^2$ factors into $(E + mc^2)(E - mc^2)$.

Note that $K = E - mc^2$

$$E^2 - (mc^2)^2 = (pc)^2$$
$$(E + mc^2)(E - mc^2) = p^2c^2$$
$$(\gamma mc^2 + mc^2)(K) = p^2c^2$$
$$K = \frac{p^2}{(\gamma + 1)m}$$

At low speeds, $\gamma \approx 1$, and we find as before that $K \approx \dfrac{p^2}{2m}$.

Example: **Proton momentum and energy**

A proton has a mass of 1.7×10^{-27} kg. If its total energy is 3 times its rest energy, what is the magnitude of its momentum?

$$mc^2 = (1.7 \times 10^{-27}\ \text{kg})(3 \times 10^8\ \text{m/s})^2 = 1.5 \times 10^{-10}\ \text{J}$$

$$E_{\text{particle}} = mc^2 + K = 3mc^2 \text{ and } (pc)^2 = (E_{\text{particle}})^2 - (mc^2)^2$$

$$pc = \sqrt{(E_{\text{particle}})^2 - (mc^2)^2} = \sqrt{(3mc^2)^2 - (mc^2)^2} = \sqrt{8(mc^2)^2} = \sqrt{8}(mc^2)$$

$$p = \sqrt{8}(mc) = \sqrt{8}(1.7 \times 10^{-27}\ \text{kg})(3 \times 10^8\ \text{m/s}) = 1.4 \times 10^{-18}\ \text{kg} \cdot \text{m/s}$$

5.3 WORK: MECHANICAL ENERGY TRANSFER

The Momentum Principle relates a change in momentum (effect) to the net impulse applied to a system. Impulse is the product of force and time.

$$\Delta \vec{p} = (\text{net force})\Delta t$$

We might ask what property of a system changes due to forces acting not for a time but through a distance, something like this:

$$\Delta ? = (\text{net force})(\text{distance})$$

The quantity that changes turns out to be the energy of the system, γmc^2 in the case of a single particle. The quantity involving force times distance is called "work".

In Section 5.6 we will provide a proof that this is correct. For now we'll assume this to be true, and after explaining how to calculate work we'll use the Energy Principle for a particle to analyze several kinds of processes.

Magnitude of applied force

Suppose you apply a force with constant magnitude F in a constant direction to a block that was initially at rest on a nearly frictionless surface, pulling it through a displacement Δr (Figure 5.3). The block moves faster and faster, acquiring a certain amount kinetic energy. If you were to pull twice as hard ($2F$) through the same displacement Δr, the block would presumably acquire twice as much kinetic energy (and you would expend twice as much

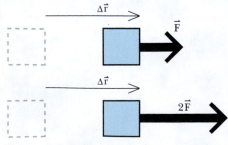

Figure 5.3 Pulling a block over a nearly frictionless surface with a force twice as big (through the same distance) means you expend twice as much chemical energy.

chemical energy). Note that the rest energy mc^2 of the block doesn't change, just the kinetic energy.

Distance through which force acts (displacement)

Similarly, if you lift a heavy box to a height h above the ground, you will have expended a certain amount of chemical energy (Figure 5.4). If you lift the box twice as high (a height $2h$ above the ground), you will presumably have had to expend twice as much energy.

The product of force times displacement

Because the amount of energy expended appears to be proportional both to the magnitude of the force exerted and to the distance through which an object is moved, we'll tentatively define the amount of energy transfer between you and the object as the product of force times displacement, $F\Delta r$. This type of mechanical energy transfer from the surroundings into the system is called "work." Here we use the word "work" as a technical term in the context of physics. This technical meaning is only loosely connected with everyday uses of the word.

The unit of work is newton · meter. From the Momentum Principle we know that a newton is the same as a kg · (m/s)/s, so a newton · meter is the same as a kg · m²/s², which is a joule (same units as mc^2). Work has the same units as energy, as it should if we expect the change in energy of a system to be equal to the work.

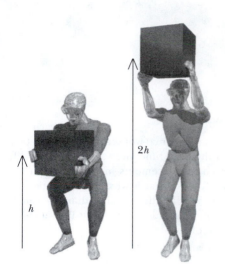

Figure 5.4 Lifting a box twice as high (with the same force) means you expend twice as much chemical energy.

Distance *vs.* time

? Is doubling the displacement the same as doubling the time?

No, it isn't. If you pull a block through twice the displacement, you do twice as much work, and the block acquires twice as much kinetic energy. But because the block is moving faster in the second half of the doubled displacement, the time interval is less than doubled (and the impulse is less than doubled), so the momentum change is not doubled.

Note carefully: work and kinetic energy change are proportional to displacement, while impulse and momentum change are proportional to time interval. In symbols: work $F\Delta r$ is not the same as impulse $F\Delta t$.

Work and displacement in 3D

We need to define work carefully. To see why, suppose you and a co-worker are maneuvering a heavy box on a space station (Figure 5.5). You pull in the x direction with a force F_x through a displacement Δx, so you do work

$$W_{\text{by you}} = F_x \Delta x$$

while at the same time your colleague is pulling in the y direction with a force F_y through a displacement Δy, doing work

$$W_{\text{by co-worker}} = F_y \Delta y$$

The total work done on the box by the two of you is

$$W = F_x \Delta x + F_y \Delta y$$

and this will be equal to the change in the energy of the box.

In general, the work done by a force $\vec{F} = \langle F_x, F_y, F_z \rangle$ acting through a displacement $\Delta \vec{r} = \langle \Delta x, \Delta y, \Delta z \rangle$ is defined as follows:

Your co-worker pulls in the $+y$ direction

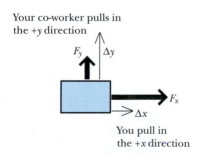

You pull in the $+x$ direction

Figure 5.5 You and a co-worker maneuver a box on a space station.

WORK: MECHANICAL ENERGY TRANSFER

$$W = F_x \Delta x + F_y \Delta y + F_z \Delta z \quad \text{(a scalar quantity)}$$

This is a special kind of product of two vectors, resulting in a scalar result. It is called the "scalar product" of the two vectors $\vec{F}$ and $\Delta\vec{r}$. A compact notation for this kind of product involves a dot between the two vectors. For that reason it is also called the "dot product":

NOTATION: SCALAR PRODUCT OR DOT PRODUCT

$$\vec{F} \bullet \Delta\vec{r} = F_x\Delta x + F_y\Delta y + F_z\Delta z$$

There's no new information in the notation $\vec{F} \bullet \Delta\vec{r}$. It's just shorthand for $F_x\Delta x + F_y\Delta y + F_z\Delta z$. Using this shorthand we can say that the work done by a force acting through a displacement is $\vec{F} \bullet \Delta\vec{r}$.

As we'll discuss in more detail later, work can be positive or negative. If you push or pull in the direction of motion, you do positive work and you increase the energy of the system. If you push or pull in the direction opposite to the motion, you do negative work and you decrease the energy of the system.

You and your co-worker pulled on the box for the same time interval but through different displacements in the directions of the forces. That is why the calculation of work is of a different kind than the calculation of impulse, where Δt is the same for both of you.

Example: **Pull a block**

You pull a block 1.5 m across a table in the $-x$ direction, while exerting a force of $\langle -0.3, 0.25, 0\rangle$ N. How much work do you do on the block?

$$W = \vec{F} \bullet \Delta\vec{r} = (\langle -0.3, 0.25, 0\rangle\ \text{N}) \bullet (\langle -1.5, 0, 0\rangle\ \text{m})$$

$$W = (-0.3\ \text{N})(-1.5\ \text{m}) + (0.25\ \text{N})(0) + 0 \cdot 0 = 0.45\ \text{J}$$

5.X.8 A paper airplane flies from location $\langle 6, 10, -3\rangle$ m to location $\langle -12, 2, -9\rangle$ m. The net force acting on it during this flight, due to the Earth and the air, is nearly constant at $\langle -0.03, -0.04, -0.09\rangle$ N. What is the total work done on the paper airplane by the Earth and the air?

5.X.9 A 2 kg ball rolls off a 30 m high cliff, and lands 25 m from the base of the cliff. Express the displacement and the gravitational force in terms of vectors and calculate the work done by the gravitational force. Note that the gravitational force is $\langle 0, -mg, 0\rangle$, where g is a *positive* number (+9.8 N/kg).

Positive and negative work

The Energy Principle states that the change in energy of a system (in this case, a single particle) is equal to the energy input from work done by the surroundings.

$$\Delta E_{\text{sys}} = W_{\text{surr}}$$

Since the kinetic energy of a particle might increase or decrease, it must be the case that the work done by the surroundings can be positive or negative.

? What is the meaning of negative work?

Suppose you want to slow down a moving object. You push in a direction opposite to the object's motion, and although the object keeps moving in the original direction, it gradually slows down. The object's kinetic energy decreased, so you must have done negative work on the object. Evidently a force acting in a direction opposite to the displacement of a system does negative work.

It is important to determine the correct sign of the work done on a system, because increasing energy is associated with positive work, and decreasing energy is associated with negative work.

Consider four simple situations that bring out the main issues. In Figure 5.6 a box is shown moving to the right or left, acted on by a force to the right or left. Since the forces and displacements are all in the $+x$ or $-x$ direction, in each case $W = F_x\Delta x + F_y\Delta y + F_z\Delta z$ reduces to $W = F_x\Delta x$.

Case 1: The box is speeding up so its energy is increasing. The work should be positive, and we can check the sign: $W = F_x\Delta x = (+)(+) = (+)$.

Case 2: The box is slowing down so its energy is decreasing. The work should be negative, and we can check the sign: $W = F_x\Delta x = (-)(+) = (-)$.

Case 3: The box is speeding up so its energy is increasing. The work should be positive, and we can check the sign: $W = F_x\Delta x = (-)(-) = (+)$.

Case 4: The box is slowing down so its energy is decreasing. The work should be negative, and we can check the sign: $W = F_x\Delta x = (+)(-) = (-)$.

It is important to look at the physical situation first, to see whether the energy is increasing or decreasing, which tells you the sign of the work. Then check that the definition $W = F_x\Delta x + F_y\Delta y + F_z\Delta z$ does in fact give the same sign.

Another way to think about the sign of the work is this: If the force acts in the direction of the motion, it makes the object speed up, and the work is positive (cases 1 and 3). If the force acts opposite to the direction of the motion, it makes the object slow down, and the work is negative (cases 2 and 4). It's not whether the force points left or right, or up or down, but what direction it points *relative to the motion*.

A common mistake

A common mistake that is easy to make is to try to get the sign of the work merely by looking at the direction of the force, without paying attention to the direction of the displacement. For example, in case 3 the force is to the left but the work is positive, because the displacement is also to the left. Always ask yourself whether the force tends to increase or decrease the energy, and that will tell you what the sign of the work should be.

Do the following exercises and check them at the back of the chapter to make sure that you can reliably determine the correct sign of the work.

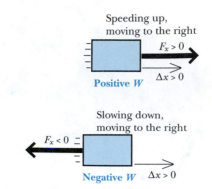
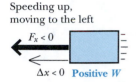
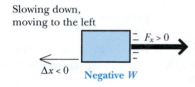

Figure 5.6 Sign of work in four cases: force to the right or left, while moving to the right or the left.

5.X.10 A ball is moving upwards, acted on by a downwards gravitational force. Is the kinetic energy of the ball increasing or decreasing? Is the work done by the gravitational force positive or negative?

5.X.11 A ball is falling downwards, acted on by a downwards gravitational force. Is the kinetic energy of the ball increasing or decreasing? Is the work done by the gravitational force positive or negative?

5.X.12 A car is moving to the left, and Superman pushes on it to the right to slow it down. Is the kinetic energy of the car increasing or decreasing? Is the work Superman does positive or negative?

5.X.13 You throw a ball downwards. Is the kinetic energy of the ball increasing or decreasing? Do you do positive work or negative work?

5.X.14 A boat is coasting toward a dock you're standing on, and as it comes toward you, you push back on it with a force of 300 N. As you do this, you back up a distance of 2 m. The mass of the boat is 500 kg, and its initial speed was 3 m/s. What is its final speed?

Zero work

In Figure 5.7 a puck is sliding with low friction to the right on ice, being speeded up by a hockey stick pushing on it. There are three different objects in the surroundings that are interacting with the puck. The hockey stick pushes horizontally, doing positive work to speed up the puck, the Earth pulls down, and the ice pushes up, supporting the puck.

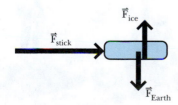

Figure 5.7 A hockey puck slides to the right with low friction on ice, speeded up by a hockey stick pushing on the puck.

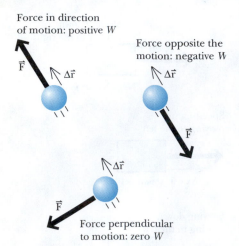

Figure 5.8 Force in direction of motion, opposite the motion, and perpendicular to motion.

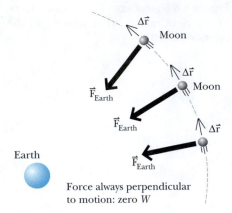

Figure 5.9 The gravitational force of the Earth on the Moon is always perpendicular to the motion, so the work is zero.

Figure 5.10 The floor exerts an upward force through zero distance and does zero work.

? How much work is done by the force exerted downward by the Earth? Does this force speed up or slow down the puck? Is the work done by the Earth positive or negative?

The work done by the Earth's gravitational force is zero. It doesn't make the puck speed up or slow down. We can check this:

$$W = F_x\Delta x + F_y\Delta y + F_z\Delta z = (0)\Delta x + (-mg)(0) + (0)(0) = 0$$

A force perpendicular to the motion does zero work.

? How much work is done by the force exerted upward by the ice? Does this force speed up or slow down the puck? Is the work done by the ice positive or negative?

Again, the work done by the force exerted by the ice is zero. It doesn't make the puck speed up or slow down (remember, we're neglecting any horizontal, frictional component of the force that the ice exerts on the puck).

Sign of work depends on direction of force and direction of motion

These are three important special cases for the sign of work (Figure 5.8):

- Force in direction of motion means work is positive.
- Force opposite the direction of motion means work is negative.
- Force perpendicular to the direction of motion means work is zero.

A particularly striking case of zero work is circular motion at constant speed, such as the Moon going around the Earth under the influence of the Earth's gravitational force, which always acts perpendicular to the motion (Figure 5.9).

? Is the Moon speeding up or slowing down? Therefore what can you say about the work done on the Moon by the Earth's gravitational attraction? Is this consistent with the definition of work?

The Moon's speed and kinetic energy aren't changing (and its rest energy isn't changing), so its energy isn't changing. Therefore the work done should be zero. This makes sense, because the force is always perpendicular to the motion.

? But then what does the Earth's gravitational force do, if it doesn't do any work?

It changes the momentum of the Moon. The momentum is a vector quantity, and change of direction represents a change of the vector momentum. A force is required to change the momentum of the Moon, otherwise it would move in a straight line, not a circle.

Again we see important differences between energy and momentum, and between work and impulse. There is no change in the Earth's energy, since that depends on the *magnitude* of the velocity, but there is a change of the Earth's momentum, since that depends on the *direction* of the velocity.

Another case of zero work

There is another important case of a force doing zero work. Suppose a person crouches down and jumps straight up, and consider the jumper as the system (Figure 5.10). The surroundings are the Earth, the floor, and the air (but we'll neglect the small air resistance force). As the jumper rises, the Earth pulls down, doing negative work. How much work is done by the force of the floor, acting on the jumper's feet? Zero! The force of the floor acts through zero distance and does zero work. The force of the floor contributes to the net force that determines the change in the jumper's momentum, but the floor force does no work. The only significant work done by the surroundings is the negative work done by the force of the Earth (correspondingly, the chemical energy of the jumper decreases).

A similar example is an ice skater who pushes off from a wall. The wall exerts a force on the hands, which changes the skater's momentum, but that force does no work, because at the point of contact there is no displacement. We'll discuss such situations in more detail in Chapter 8.

Work done by parallel and perpendicular forces

Figure 5.11 shows the force exerted by the Sun on a passing comet, with the parallel and perpendicular components of the force. The parallel component of the force does work on the comet, changing the kinetic energy (and changing the magnitude of the momentum). The perpendicular component of the force does zero work but changes the direction of the momentum.

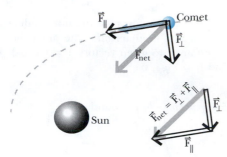

Figure 5.11 The Sun exerts a force on a comet. The parallel component of the force does work, changing the kinetic energy (and the magnitude of the momentum). The perpendicular component does zero work but changes the direction of the momentum.

Work done by a force at an angle to the displacement

Consider the motion of a block sliding to the right, pulled through a small displacement Δx by a string held at an angle θ to the horizontal, as shown in Figure 5.12. We resolve the string force into vertical and horizontal components, as shown in Figure 5.13. We pull at an angle such that the block does not move in the vertical direction.

? In calculating the amount of work done to pull the block, should we use the magnitude of the force F, or just the horizontal component of the force $F_x = F\cos\theta$?

Let's use the definition of work:

$$W = F_x\Delta x + F_y\Delta y + F_z\Delta z = F_x\Delta x + F_y(0) + (0)(0) = F_x\Delta x$$

or alternatively $W = (F\cos\theta)\Delta x$

Only the x component of the force, the component in the direction of the motion, does work. Because only the component of a force parallel to the direction of motion does any work, the work done is given by the product of the component of the force parallel to the displacement times the displacement:

$$\text{work} = F_\parallel\Delta r$$

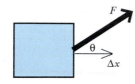

Figure 5.12 Exerting a force of magnitude F at an angle to the motion.

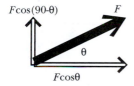

Figure 5.13 x and y components of the force.

The dot product using angles

As we saw above, the component $F_\parallel$ of the force vector $\vec{F}$ in the direction of the displacement $\Delta\vec{r}$ is equal to $F\cos\theta$, where θ is the angle between the two vectors $\vec{F}$ and $\Delta\vec{r}$. The work done by this force is therefore $(F\cos\theta)\Delta r$, or $F\Delta r\cos\theta$. This means that we have another way of evaluating the dot product of the two vectors, as shown in Figure 5.14.

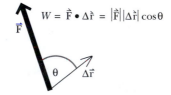

Figure 5.14 θ is the angle between the two vectors.

DOT PRODUCT: FORM USING ANGLE θ

$$W = \vec{F}\bullet\Delta\vec{r} = |\vec{F}||\Delta\vec{r}|\cos\theta \text{ (a scalar quantity)}$$

Remember that the result of the dot product is a *scalar* quantity, a single number, and not a vector.

? Using the definition of the dot product, what is the work done by a force of magnitude F acting through a displacement of magnitude Δr if the angle is 0°? 90°? 180°?

If the force is in the same direction as the displacement (0°), the work is positive: $F\Delta r\cos(0°) = F\Delta r$. If the force is perpendicular to the displacement (90°), the work is zero: $F\Delta r\cos(90°) = 0$. If the force is opposite to the displacement (180°), the work is negative: $F\Delta r\cos(180°) = -F\Delta r$.

The alternative forms of the dot product offer a way to determine the angle between any two given vectors $\vec{a} = \langle a, b, c \rangle$ and $\vec{q} = \langle q, r, s \rangle$.

$\vec{a} \bullet \vec{q} = (aq + br + cs)$ and also
$\vec{a} \bullet \vec{q} = |\vec{a}||\vec{q}| \cos\theta$

So $\cos\theta = \dfrac{(aq + br + cs)}{|\vec{a}||\vec{q}|}$ from which θ can be determined.

Here is a summary of the various equivalent ways to calculate work. Sometimes one of these is more convenient than the others, depending on the situation.

SUMMARY: CALCULATING WORK

$$W = \vec{F} \bullet \Delta\vec{r} = F_x\Delta x + F_y\Delta y + F_z\Delta z$$

$$W = \vec{F} \bullet \Delta\vec{r} = F_{||}\Delta r$$

$$W = \vec{F} \bullet \Delta\vec{r} = |\vec{F}||\Delta\vec{r}| \cos\theta$$

Example: Throw a ball straight up

You throw a 250 gram ball straight up, and it rises 8 meters into the air. Neglecting air resistance, is the work positive or negative? How much work is done by the gravitational force?

   System: Ball
   Surroundings: Earth

Physical reasoning:
The ball slows down on its way up, so its kinetic energy is decreasing. Therefore the work must be negative.

Calculate work:
$F_y = -mg$ and Δy is positive:

$$\text{work} \equiv F_y\Delta y = [-(0.25 \text{ kg})(9.8 \text{ N/kg})](8 \text{ m}) = -19.6 \text{ joule}$$

Negative work is done on the ball, and the ball slows down, losing some kinetic energy.

5.X.15 You pull your little sister across a flat snowy field on a sled. Your sister plus the sled have a mass of 20 kg. The rope is at an angle of 35 degrees to the ground. You pull a distance of 50 m with a force of 30 N. How much work do you do?

5.X.16 A jar of honey with a mass of 0.5 kg is nudged off the kitchen counter and falls one meter to the floor. What force acts on the jar during its fall? How much work is done by this force?

5.X.17 About how much mechanical work is done by the Sun on the Earth during six months? (A separate kind of energy transfer is solar radiation, which warms up the Earth.)

5.X.18 Contrast work and impulse. If the magnitude of the Earth's momentum is P, what is the magnitude of the impulse that the Sun applies to the Earth during six months?

5.4 UPDATE FORM OF THE ENERGY PRINCIPLE

The Energy Principle

$$\Delta E_{\text{sys}} = W_{\text{surr}}$$

Remember that Q is omitted from this equation because in this chapter we consider only situations in which there is not a temperature difference between system and surroundings, so $Q = 0$.

involves a difference in the energy of a system in its final state and the energy in its initial state. We will refer to this as the "difference form" of the Energy Principle. Since $\Delta E_{\text{sys}} = E_{\text{sys},f} - E_{\text{sys},i}$, we can also write the Energy Principle this way:

THE ENERGY PRINCIPLE (UPDATE FORM)

$$E_{\text{sys},f} = E_{\text{sys},i} + W_{\text{surr}} \quad \text{(update form)}$$

Sometimes it's easier to think about the situation using the difference form of the Energy Principle, and sometimes the update form is more appropriate. Use whichever form you're comfortable with, but make sure you understand both forms.

Example: Pushing a box in space

You're outside a spacecraft, pushing on a heavy box whose mass is $m = 2000$ kg. You exert a force $\vec{F} = \langle 300, 500, 0 \rangle$ N while the box moves through a displacement $\Delta \vec{r} = \langle 0.1, -0.3, 0.2 \rangle$ m. Initially the box had a speed $|\vec{v}_i| = 0.7$ m/s. (a) How much work do you do? (b) What is the final kinetic energy of the box? (c) What is the final speed of the box? (d) What is the direction of the final velocity of the box?

System: Box (considered to be a particle)
Surroundings: You

Initial state: when you start pushing the box
Final state: when the box has been displaced the amount $\Delta \vec{r}$

Energy Principle:

$E_f = E_i + W$

$(\cancel{mc^2} + K_f) = (\cancel{mc^2} + K_i) + W$

$K_f = K_i + W$

$K_f = K_i + (F_x \Delta x + F_y \Delta y + F_z \Delta z)$

$K_f = K_i + (300 \text{ N})(0.1 \text{ m}) + (500 \text{ N})(-0.3 \text{ m}) + (0 \text{ N})(0.2 \text{ m})$

$K_f = K_i + (-120 \text{ J})$

$K_f \approx (\frac{1}{2} m v_i^2) + (-120 \text{ J})$

$K_f \approx \frac{1}{2}(2000 \text{ kg})(0.7 \text{ m/s})^2 + (-120 \text{ J})$

$K_f \approx (490 \text{ J}) + (-120 \text{ J}) = 370 \text{ J}$

Final speed:

$\frac{1}{2} m v_f^2 = K_f$

$v_f = \sqrt{\dfrac{2K_f}{m}}$

$v_f = \sqrt{\dfrac{2(370 \text{ J})}{2000 \text{ kg}}} = 0.608 \text{ m/s}$

(a) $W = -120$ J

(b) $K_f \approx 370$ J

(c) $v_f = 0.608$ m/s

(d) We don't have enough information to answer this.

Check: The final speed is smaller than the initial speed, corresponding to the work being negative. We slowed down the box. The units also check.

We use the Energy Principle instead of the Momentum Principle because we know the displacement $\Delta \vec{r}$, not the time interval Δt.

In this case the rest energy of the box mc^2 is the same in the initial and final states, so the mc^2 terms cancel. We know this is so because the identity of the box did not change (for example, it did not turn into solid gold).

You did negative work, pushing against the box to slow it down, so the kinetic energy of the box decreased.

Further discussion

We answered all the questions except for (d), which asked for the direction of the final velocity of the box. We can't answer that question! The Energy

It is interesting that we can find the final *speed* of the box without knowing the exact trajectory of the box, or how long it took the box to move from its initial location to its final location. But because energy is a scalar quantity, we don't learn about the final *direction* of motion.

Principle is a scalar principle, and only the magnitude of the velocity appears in the particle energy, not the direction. In order to figure out the direction of the final velocity we would have to use the Momentum Principle, which is a vector principle. However, note that it was not stated how long the process took, so we can't apply the Momentum Principle directly. We would have to determine the approximate Δt from the known displacement and initial speed. (The actual trajectory is a curve; we've approximated the small displacement by a straight line.)

Simple and powerful problem-solving procedure

Write out all terms in the Energy Principle.

Identify terms that don't change, and therefore cancel.

Substitute in all numerical values you know.

Solve for the unknown quantity.

Example: Throwing a ball

You hold a ball of mass 0.5 kg at rest in your hand, then throw it forward, underhand, so that as it leaves your hand its speed is 20 m/s. How much work did you do on the ball?

> System: the ball
> Surroundings: You (neglect the gravitational effect of the Earth in this short process)
>
> Initial state: At the start of the throw, when the ball's speed is zero
> Final state: When the ball just leaves your hand at 20 m/s
>
> Energy Principle:

No change of identity, so no change in rest energy. mc^2 terms cancel.

$$E_f = E_i + W$$
$$(mc^2 + K_f) = (mc^2 + K_i) + W$$
$$K_f = K_i + W$$
$$\tfrac{1}{2}mv_f^2 = \tfrac{1}{2}mv_i^2 + W$$
$$\tfrac{1}{2}(0.5 \text{ kg})(20 \text{ m/s})^2 = \tfrac{1}{2}m(0)^2 + W$$
$$100 \text{ J} = W$$

The energy of the ball increased; it speeded up from 0 to 20 m/s. You did positive work, pushing the ball to speed it up.

> Check: Work is positive; makes sense because final speed is larger than the initial speed. Units are correct.

Further discussion

We were able to calculate the work you did on the ball even though we didn't know how large a force you applied, nor how large was the displacement of your hand in the throwing motion. This is one aspect of the power of the Energy Principle, that simply by determining the change in energy of a system you know how much energy input there was to that system.

Energy is a conserved quantity: $\Delta E_{\text{system}} + \Delta E_{\text{surroundings}} = 0$. The energy of the system (the ball) increased by 100 J.

?   What was the change in the energy of the surroundings? Can you say what kind of energy it was in the surroundings that changed?

Evidently the change in energy of the surroundings was -100 J. This decrease of energy in the surroundings corresponds to a decrease in the chemical energy stored in your body, coming from food you consumed. In throwing the ball you converted some stored chemical energy into kinetic energy of the ball.

Actually, you need to expend more than 100 J of chemical energy (an energy decrease), because in addition to increasing the kinetic energy of the ball by 100 J, you also raised your body temperature a bit (thermal energy increase). Your muscles aren't 100% efficient; it takes more than 100 J of chemical energy to give the ball 100 J of kinetic energy.

The net energy change of the surroundings is -100 J, which could consist (say) of -180 J of chemical energy change and $+80$ J of thermal energy rise in your body.

5.X.19 In the example above, suppose you wanted the final speed of the ball to be twice as big (40 m/s). How much work would you have to do?

Example: The Energy Principle at high speed

An electron (mass 9×10^{-31} kg) in a particle accelerator is acted on by a constant electric force $\langle 5 \times 10^{-13}, 0, 0 \rangle$ N (this is much greater than the force of gravity, which is only $mg \approx 1 \times 10^{-29}$ N, so we can neglect the effect of the Earth on the electron). The initial velocity of the electron is $\langle 0.995\,c, 0, 0 \rangle$, where c as usual is the speed of light. The electron moves through a displacement of $\langle 5, 0, 0 \rangle$ m. What is its final speed, expressed as a fraction of the speed of light?

System: Electron
Surroundings: Accelerator (which applies the electric force); neglect the gravitational effect of the Earth in this short process.

Initial state: Beginning of given displacement
Final state: End of given displacement

Energy Principle:

$$E_f = E_i + W$$

$$\gamma_f mc^2 = \gamma_i mc^2 + W$$

$$\gamma_f mc^2 = \gamma_i mc^2 + (F_x \Delta x + F_y \Delta y + F_z \Delta z)$$

$$\gamma_f mc^2 = \gamma_i mc^2 + (5 \times 10^{-13}\ \text{N})(5\ \text{m}) + (0\ \text{N})(0\ \text{m}) + (0\ \text{N})(0\ \text{m})$$

$$\gamma_f mc^2 = \gamma_i mc^2 + (2.5 \times 10^{-12}\ \text{J})$$

$$\gamma_f mc^2 = \frac{1}{\sqrt{1 - (0.995\,c/c)^2}}\, mc^2 + (2.5 \times 10^{-12}\ \text{J})$$

> Because the electron's speed is near the speed of light, we can't use an approximate expression for K. We need to use the exact, relativistic equation for particle energy.

$$\gamma_f mc^2 = 10.0\, mc^2 + (2.5 \times 10^{-12}\ \text{J})$$

$$\gamma_f = \frac{10.0\, mc^2 + (2.5 \times 10^{-12}\ \text{J})}{mc^2}$$

> The unknown quantity, final speed, is part of γ_f. It's easiest to find γ_f first, then solve for v_f.

$$\gamma_f = 10.0 + \frac{2.5 \times 10^{-12}\ \text{N}}{(9 \times 10^{-31}\ \text{kg})(3 \times 10^8\ \text{m/s})^2} = 40.9$$

$$\sqrt{1 - v_f^2/c^2} = \frac{1}{40.9} = 2.44 \times 10^{-2}$$

> To solve for v_f it is necessary to invert both sides of the equation.

$$1 - v_f^2/c^2 = (2.44 \times 10^{-2})^2 = 5.98 \times 10^{-4}$$

$$v_f^2/c^2 = 1 - 5.98 \times 10^{-4} = 0.9994$$

$$v_f/c = \sqrt{0.9994} = 0.9997$$

$$v_f = 0.9997\,c$$

Check: The final speed is greater than the initial speed, consistent with positive work. Units look right; in particular, the values for γ were dimensionless, as they should be.

Further discussion

The speed increased only slightly, from $0.995\,c$ to $0.9997\,c$, even though the energy changed a lot, from $10.0\, mc^2$ to $40.9\, mc^2$. The electron's energy increased by about a factor of 4, but the electron's speed increased very little. When a particle is traveling at nearly the speed of light, adding a lot of energy to the particle doesn't change its speed very much.

5.X.20 In the preceding example, at the final speed, $0.9997\,c$, what was the particle energy as a multiple of the rest energy mc^2? (That is, if it was twice

mc^2, write $2mc^2$.) What was the kinetic energy as a multiple of mc^2? Was the kinetic energy large or small compared to the rest energy? At low speeds, is the kinetic energy large or small compared to the rest energy?

5.5 CHANGE OF IDENTITY—CHANGE OF REST ENERGY

As we saw in the preceding examples, if a particle doesn't change its identity, both the initial and the final energies of the system include mc^2 rest-energy terms and these cancel out in the energy equation:

$$mc^2 + K_f = mc^2 + K_i + W$$

$$K_f = K_i + W$$

This is an important and common special case, but it is not the only case. In some processes the rest mass of a particle can change because the identity of the particle changes.

Neutron decay is a good example to illustrate the issues. A free neutron (one not bound into a nucleus) is unstable, with an average lifetime of about 15 minutes. As shown in Figure 5.15, it decays into a proton, electron, and a nearly massless antineutrino, which travels at nearly the speed of light. All three of these particles have kinetic energy. That is, they have energy above and beyond their rest energy; the energy of the antineutrino is nearly all kinetic, because it has almost no rest energy. This is a situation with a change of particle identity, and a change in particle rest energy.

An electron volt (eV) is a unit of energy

The mass of a neutron is 1.6749×10^{-27} kg, so its rest energy mc^2 is about 1.51×10^{-10} J. Because the rest energy of a single particle is such a small number of joules, it is usually expressed in different units: million electron volts, abbreviated "MeV". One electron volt is the amount of energy an electron acquires when moving through an electric potential difference of 1 volt. (Electric potential difference is introduced in Volume 2, Chapter 16; an ordinary flashlight battery has an electric potential difference across it of 1.5 volts).

DEFINITION OF ELECTRON VOLT AND MeV

$$1 \text{ eV} = 1.6 \times 10^{-19} \text{ J}$$

"eV" is the abbreviation for electron volt. $1 \text{ MeV} = 1 \times 10^6$ eV

The rest energy of a neutron is 939.6 MeV, and the rest energy of a proton is 938.3 MeV—slightly less than that of a neutron. An electron is much less massive; its rest energy is 0.511 MeV.

Scientists studying elementary particles often use the unit MeV/c^2 as a unit of mass, because dividing energy by c^2 gives mass. In these units the rest mass of a proton is $938.3 \text{ MeV}/c^2$. The rest mass of a neutron is $939.6 \text{ MeV}/c^2$. The rest mass of the electron is $0.511 \text{ MeV}/c^2$. A neutrino or antineutrino is nearly massless—almost all of its energy is kinetic energy.

Example: **Neutron decay**

In the decay of a free neutron at rest (Figure 5.15), how much kinetic energy (in electron volts) do the decay products (proton, electron, and antineutrino) share? Why can't a proton decay into a neutron?

System: All the particles, both before and after the decay
Surroundings: Nothing

Sidebar (left column):

$$n \to p^+ + e^- + \bar{\nu}$$

$$m_n c^2 = (m_p c^2 + K_p) + (m_e c^2 + K_e) + E_\nu$$

Figure 5.15 A stationary neutron decays into a proton, an electron, and a massless antineutrino. This is an example of a process in which there is a change of identity, and therefore a change in rest energy.

As long as we include all particles in the system, it is okay if some of them change identity during the process.

Initial state: Neutron at rest.

Final state: Proton (p), electron (e), and antineutrino ($\bar{v}$) far away from each other.

Energy Principle:

$$E_f = E_i + W$$

$$(m_p c^2 + K_P) + (m_e c^2 + K_e) + (K_{\bar{v}}) = (m_N c^2 + K_N) + W$$

$$(938.3 \text{ MeV} + K_P) + (0.511 \text{ MeV} + K_e) + (K_{\bar{v}}) = (939.6 \text{ MeV} + 0) + 0$$

$$(K_P + K_e + K_{\bar{v}}) + 938.8 \text{ MeV} = 939.6 \text{ MeV}$$

$$(K_P + K_e + K_{\bar{v}}) = (939.6 - 938.3 - 0.5) \text{ MeV}$$

$$(K_P + K_e + K_{\bar{v}}) = 0.8 \text{ MeV}$$

A proton can't turn into a neutron, because the neutron has more rest energy than a proton, and the kinetic energy of the decay products would turn out to be a negative number, which is impossible. Kinetic energy is always positive.

Check: The kinetic energy of the decay products is positive, which it must be for this reaction to be possible. The units check.

The Earth can be neglected, since the process is so fast that work done by gravitational forces is negligible.

Here the neutron does undergo a change of identity, so there is a change in rest energy, and the mc^2 terms do not cancel.

We found the total kinetic energy of all the particles in the final state. We don't know the separate kinetic energies of the individual particles.

Further discussion

In this example we couldn't cancel the rest energy terms because there was a change of identity. The neutron turned into a lower-mass proton, an electron, and an antineutrino. Because the sum of the rest energies of the these three decay particles is less than the rest energy of the parent neutron, there is energy left over which shows up as kinetic energy shared among the three decay particles. How this kinetic energy is divided up among the three particles varies from one neutron decay to the next, governed by probabilities that can be calculated using the science of quantum mechanics and the properties of the "weak interaction" that is responsible for neutron decay.

In a chemical reaction such as carbon combining with oxygen in the fire of a coal-burning power plant, the resulting kinetic energy (which is responsible for heating water to make steam) is on the order of 1 eV per molecule. Nuclear reactions produce vastly more kinetic energy, in the case of neutron decay almost a million times more (1 MeV/1 eV is one million). This is why some power plants use nuclear rather than chemical reactions.

5.X.21 In the preceding example, what fraction of the original neutron's rest energy was converted into kinetic energy?

Intelligent "plug and chug":

Look back at the last four worked-out examples, and you'll see that the same procedure was used to solve each one. Applying the Energy Principle is basically an intelligent sort of "plug and chug" procedure.

- Specify a system
- Specify surroundings: objects that exert significant forces on the system
- Specify initial state
- Specify final state
- Write out the Energy Principle in detail for this system
- Use given information to evaluate all the terms you can
- Solve for the unknown quantity, which may be inside one of the remaining terms
- Check units, check consistency of signs

? How do you decide whether to use the Momentum Principle or the Energy Principle?

Think about what quantities are given, and what you must find:

Work involves a distance (energy), and impulse involves a time (momentum).

Momentum is a vector, and involves information about directions; energy is a scalar, and has no directional information.

The four examples were quite different, but the same procedure worked in each case. The central idea is to let the terms in the Energy Principle tell you what to calculate, and then go ahead and calculate them.

Often students have the impression that physics consists of "knowing what formula to use" for a particular type of problem, with a different formula for every problem. But look again at the last four examples. There weren't four "formulas" giving the four answers. Rather there was one logical procedure that worked in every case, based on the fundamental Energy Principle that applies to all situations.

In following this procedure, you still need to think, but let your thinking be organized by the principle itself. You need to decide whether to try using the Momentum Principle or the Energy Principle, you need to choose a system and initial and final states intelligently, etc. However, the basic framework never changes.

You should never find yourself saying, "I don't have any idea even where to start on this problem!"

If you do find yourself saying this, take a deep breath, choose a system, identify the objects in the surroundings that exert forces on the system, write down the Energy Principle, choose initial and final states, and flesh out the various terms in the Energy Principle for the particular situation at hand, then solve for the unknown quantities. Physics can be easier than most people think, if you go with the flow. If on the other hand you try to invent a new and different approach for every new problem you encounter, you'll waste huge amounts of time and get very frustrated.

5.6 PROOF OF THE ENERGY PRINCIPLE FOR A PARTICLE

We have asserted that the following is true for a particle:

$$\Delta E = W_{\text{surr}}$$

To prove this, we need to show that $\Delta E = W_{\text{surr}}$ is true given our definitions of particle energy and work. Consider a single point-like particle of mass m. For convenience in the reasoning we are about to do, choose coordinate axes so that the x axis is in the direction of the motion at this instant, so the displacement in the next short time interval is simply Δx (Figure 5.16). There is a net force $\vec{F}$ (magnitude F) acting at an angle θ to the displacement. The physical results we will obtain do not depend on our choice of coordinate axes; we are free to choose axes that simplify our calculations.

The component of the net force that is parallel to the displacement, F_x, affects the speed and hence the kinetic energy of the particle. The perpendicular component F_y merely changes the direction of the momentum.

In the very short displacement Δx, the energy of the particle E_{particle} changes by a small amount, and this is supposed to be equal to the work done on the particle:

$$\Delta E = F_x \Delta x$$

According to the Momentum Principle $F_x = \dfrac{\Delta p_x}{\Delta t}$, so

$$\Delta E = \left(\frac{\Delta p_x}{\Delta t}\right)\Delta x$$

Dividing by Δx, we have this:

$$\frac{\Delta E}{\Delta x} = \frac{\Delta p_x}{\Delta t}$$

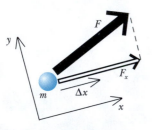

Figure 5.16 A particle of mass m. We choose the x axis to be in the direction of the motion. The net force acting on the particle has a component F_x in the direction of motion.

In the limit as $\Delta t \to 0$, this equation relates the spatial rate of change of energy to the time rate of change of momentum:

$$\frac{dE}{dx} = \frac{dp_x}{dt}$$

This equation reflects the fact that a change in position (Δx or dx) is associated with a change in energy, whereas a change in time (Δt or dt) is associated with a change in momentum.

In words, this says that the change in energy of the particle per unit distance (in the direction of the motion) is equal to the change in the particle's momentum in that direction per unit time. If we can show that this equation is true if $E = \gamma mc^2$, we will have proven that $\Delta E = W$ is correct.

Proof for low speeds

In the special case that the speed of a particle is small compared to the speed of light, the proof is simple. We need to show that if particle energy is given by the approximate equation:

$$E \approx mc^2 + \tfrac{1}{2}mv^2 \quad (v \ll c)$$

then the equation derived above:

$$\frac{dE}{dx} = \frac{dp_x}{dt}$$

is satisfied.

We chose the x axis to be in the direction of the motion, so the particle energy is:

$$E \approx mc^2 + \tfrac{1}{2}mv_x^2 \quad (v \ll c)$$

Left hand side of equation:

$$\frac{dE}{dx} = \frac{d}{dx}\left(\tfrac{1}{2}mv_x^2\right)$$

Since mc^2 is constant, $\dfrac{d}{dx}(mc^2) = 0$

$$\frac{dE}{dx} = \tfrac{1}{2}m\frac{d}{dx}(v_x^2) = mv_x\frac{dv_x}{dx} = m\frac{dx}{dt}\frac{dv_x}{dx} = m\frac{dv_x}{dt}$$

Right hand side of equation:

$$\frac{dp_x}{dt} = \frac{d}{dt}(mv_x) = m\frac{dv_x}{dt} \quad (v \ll c)$$

So:

$$\frac{dE}{dx} = \frac{dp_x}{dt}$$

is indeed true with our definitions of energy and work. We have proved that at least at low speed the Energy Principle for a particle $\Delta E = W$ is correct.

General proof (relativistically correct)

The proof that the relativistically correct expression $E = \gamma mc^2$ satisfies the relation

$$\frac{dE}{dx} = \frac{dp_x}{dt}$$

is given in Section 5.21, at the end of the chapter. It follows the same reasoning as was used to prove the low-speed case, but is algebraically more complex.

5.7 WORK DONE BY A NON-CONSTANT FORCE

If a force changes in magnitude or direction during a process, we can't calculate the work simply by multiplying a constant force times the net displacement. The nonconstant force acts on the object along a path, and we split the path into small increments $\Delta \vec{r}$ (Figure 5.17). This is similar to the way in which we took small time steps Δt when computing the trajectories of objects.

If our increments along the path are sufficiently small, the force is approximately constant in magnitude and direction within one increment. We can write the total work done as a sum along the path:

$$W = \vec{F}_1 \bullet \Delta \vec{r}_1 + \vec{F}_2 \bullet \Delta \vec{r}_2 + \vec{F}_3 \bullet \Delta \vec{r}_3 + \dots \quad \text{(small steps)}$$

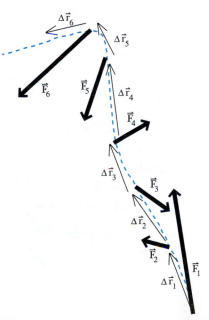

Figure 5.17 A varying force is exerted on an object moving along a curved path.

More compactly, we can use the symbol Σ (Greek capital sigma) to mean "sum":

$$W = \sum \vec{F} \bullet \Delta \vec{r}$$

5.X.22 You push a heavy crate out of a carpeted room and down a hallway with a waxed linoleum floor. While pushing the crate 2.3 m out of the room you exert a force of 30 N; while pushing it 8 m down the hallway you exert a force of 15 N. How much work do you do in all?

5.X.23 An electron traveling through a curving wire in an electric circuit experiences a constant force of 5×10^{-19} N, always in the direction of its motion through the wire. How much work is done on the electron by this force as it travels through 0.5 m of the wire?

5.X.24 One end of a spring whose spring constant is 20 N/m is attached to the wall, and you pull on the other end, stretching it from its equilibrium length of 0.2 m to a length of 0.3 m. Estimate the work done by dividing the stretching process into two stages and using the average force you exert to calculate work done during each stage.

We have refined our definition of work to allow us to calculate the work done by varying forces. This more complete definition of work allows us to apply the Energy Principle in situations where forces vary.

Work as an integral

If we go to the limit where the increments are infinitesimal ($\Delta \vec{r} \rightarrow d\vec{r}$), the sum used for calculating work turns into a sum of an infinite number of infinitesimal contributions. Such a summation is called a "definite integral." It is written as a distorted "S" for "sum," with an indication of the initial position "*i*" and the final position "*f*":

$$W = \int_i^f \vec{F} \bullet d\vec{r}$$

If you have already studied integration in a calculus course, you already know some techniques for evaluating definite integrals such as this; if not, you will learn them soon in calculus. However, in many real-world cases it is not possible to find an analytical form for evaluating an integral, in which case one approximates the infinite sum by a finite sum, with finite increments $\Delta \vec{r}$.

This procedure is called "numerical integration." The iterative calculations you have done with the Momentum Principle, whether by hand or on a computer, have been numerical integration calculations.

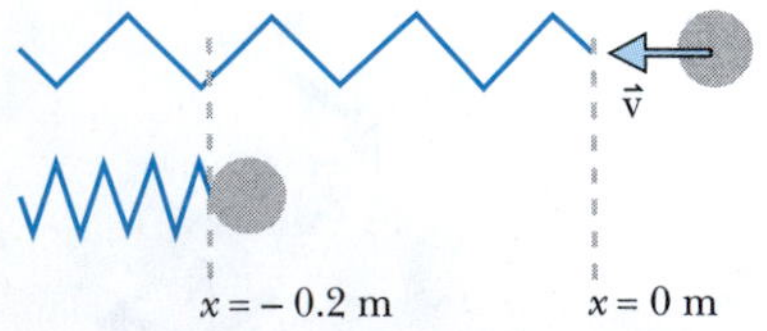

Figure 5.18 A ball runs into a horizontal spring.

Example: **Work done by a spring**

A ball traveling horizontally runs into a horizontal spring, whose stiffness is 100 N/m (Figure 5.18). As the spring is compressed, the ball slows down. If the relaxed length of the spring was 30 cm, and the spring is compressed until its length is 10 cm, how much work does the spring do on the ball?

Initial state: $s = 0$

Final state: $s = (0.10 - 0.30)$ m $= -0.2$ m

Choose origin at end of relaxed spring, so $F_x = -k_s x$

$$W = \int_0^{-0.2} (-100 \text{ N/m}) x \, dx$$

$$W = (-100 \text{ N/m}) \frac{1}{2} x^2 \Big|_{0 \text{ m}}^{-0.2 \text{ m}}$$

$$W = (-100 \text{ N/m}) \left(\frac{1}{2} (-0.2)^2 - 0 \right) = -2 \text{ N} \cdot \text{m}$$

The spring does –2 J of work on the puck.

Check: Sign ok: negative work decreases kinetic energy of puck, Units ok (N · m = J).

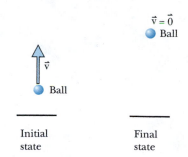

Figure 5.19 Initial and final states for a ball thrown straight upward.

5.8 POTENTIAL ENERGY IN MULTIPARTICLE SYSTEMS

Up to this point, our discussions of energy have been confined to systems consisting of a single particle. However, we saw in the case of momentum that it was often productive to choose a system containing two or more objects. Let's look at the effect of choice of system on the analysis of a simple situation: a ball thrown straight upward. The initial state is just after the ball has left your hand, heading upward, and the final state is the ball at its highest point, momentarily at rest (Figure 5.19).

System: Ball only

If the ball alone is the system, the Earth is part of the surroundings (Figure 5.20). We see that the kinetic energy of the system (ball) decreases, due to negative work done by the Earth on the system (the gravitational force acts in a direction opposite to the displacement of the ball, so the work done is negative).

System: Ball + Earth

If the ball plus the Earth are the system, there is nothing significant in the surroundings (Figure 5.21). As before, we see that the kinetic energy of the system decreases (the change in the Earth's kinetic energy is negligible, but the ball's kinetic energy decreases). However, there is a problem: the surroundings did no work on the system! The change in kinetic energy of the system is not equal to the work done by the surroundings.

Before deciding that the Energy Principle has been violated, we should consider the possibility that we have overlooked a kind of energy that is present in systems containing more than one interacting object. In fact, this is the case.

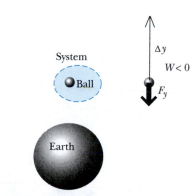

Figure 5.20 If the ball alone is chosen as the system, the Earth is part of the surroundings, and does negative work on the ball, decreasing the kinetic energy of the ball.

Potential energy belongs to pairs of interacting objects

There is a kind of energy associated with pairs of interacting objects inside a system, such as the ball and the Earth. This interaction energy is commonly called "potential energy." A change in potential energy is associated with a change in the separation of interacting objects. As the ball and the Earth get farther apart, the kinetic energy of the system decreases, but the potential energy (interaction energy) increases. The net change in the energy of the system is in fact zero, consistent with the fact that no work was done on the system by the surroundings.

There is observable evidence for the reality of potential energy. If we continue to observe the system, we will see the ball begin to fall, gaining kinetic energy (the Earth also gains some kinetic energy, but only a tiny amount, as we will see later). Since the surroundings are empty, no work is done by the

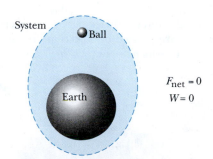

Figure 5.21 If the ball + the Earth are chosen as the system, there are no significant objects in the surroundings. No work is done on the system, but the kinetic energy of the system decreases anyway. The decrease in kinetic energy is accompanied by an increase in the interaction energy (potential energy) of the ball+Earth system.

surroundings. The increasing kinetic energy of the ball must be accompanied by decreasing potential energy of the ball+Earth pair of objects.

In any system containing two or more interacting particles, such as compressed or stretched springs, galaxies of stars interacting gravitationally, or atoms in which the protons and electrons interact electrically, there is energy associated with the interactions between pairs of particles inside the system. This interaction energy is not the same as the rest energies or kinetic energies of the individual particles. Changes in interaction energy are associated with changes in shape of the multiparticle system, such as stretching or compressing a spring, or moving an electron away from a proton.

Interactions internal to a system

When we choose to consider a "system of three particles," as in Figure 5.22, we are explicitly dividing the Universe into two parts: the objects within the system, and everything in the rest of the Universe (the surroundings). Having created this division, it is useful to distinguish between interactions that are internal to the system (that is, interactions between the particles within the system) and interactions with objects external to the system, objects that are in the surroundings.

When initially considering a problem, we are free to define any collection of objects we wish as the "system" of interest. Once we have done this, however, we must stick strictly to this definition. If an object is included in a system, it must be considered this way throughout the entire analysis.

Interaction energy in a multiparticle system

To establish the basic idea of how to calculate interaction energy in general, we will consider a generic multiparticle system.

Consider a system of three particles that exert forces on each other, and which are also acted upon by objects in the surroundings, outside this system, as shown in Figure 5.22. For example, these could be three charged particles that exert electric forces on each other but also have springs connecting them to objects in the surroundings. The forces that are exerted by other particles in the system are called "internal" forces, and are denoted here by lower case "f's". The forces exerted by objects in the surroundings are called "external" forces, and are denoted here by upper case "F's".

The work done by these internal and external forces on each particle changes the particle energy $E_{particle}$ of that particle. Using the symbol W_{int} for work done by internal forces, and W_{surr} for work done by external forces, we can write the Energy Principle for each particle:

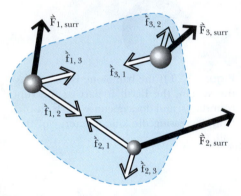

Figure 5.22 A system consisting of three particles. They exert "internal" forces on each other (lower case f's), and additional "external" forces (black arrows, upper case F's) are exerted by objects in the surroundings, which are not shown.

$$\Delta E_1 = (\vec{f}_{1,2} + \vec{f}_{1,3} + \vec{F}_{1,surr}) \bullet \Delta \vec{r}_1 = W_{1,int} + W_{1,surr}$$

$$\Delta E_2 = (\vec{f}_{2,1} + \vec{f}_{2,3} + \vec{F}_{2,surr}) \bullet \Delta \vec{r}_2 = W_{2,int} + W_{2,surr}$$

$$\Delta E_3 = (\vec{f}_{3,1} + \vec{f}_{3,2} + \vec{F}_{3,surr}) \bullet \Delta \vec{r}_3 = W_{3,int} + W_{3,surr}$$

Adding up the changes in the particle energy for each of the three particles (the left sides of the equations) and the work done on each particle (the right sides of the equations), we can write this:

$$\Delta(E_1 + E_2 + E_3) = W_{int} + W_{surr}$$

where W_{int} is the sum of the work done by internal forces on each of the particles, and W_{surr} is the sum of the work done by external forces on each of the particles. Added together, these two terms equal the sum of the changes in the individual particle energies. Note that the work done by each individual external force involves the displacement of that force (equivalently, the displacement of the particle on which that particular force acts).

Separating internal and external energy terms

It would be useful to have all terms relating to energy changes internal to the system on one side of the equation, and terms relating to processes external to the system on the other side. We can do this by making a simple algebraic rearrangement, and moving the work done on the particles by the internal forces, W_{int}, to the other side of the equation:

$$\Delta(E_1 + E_2 + E_3) + (-W_{int}) = W_{surr}$$

We relabel the term $(-W_{int})$ as a change of what is called "potential" energy U, energy associated with the interactions of the particles with each other.

DEFINITION OF CHANGE IN POTENTIAL ENERGY

$$\Delta U \equiv -W_{int}$$

The change in potential energy (ΔU) of a system is equal to the negative of the work done by forces internal to the system.

The left side of the Energy Principle for a multiparticle system now represents all changes in the energy of the system, and the right side represents energy inputs from the surroundings (work).

$$\Delta(E_1 + E_2 + E_3) + \Delta U = W_{surr}$$

The total energy of the three-particle system, E_{system}, is the sum of all the particle energies in the system, plus the potential energy (interaction energy):

$$E_{sys} = (m_1 c^2 + m_2 c^2 + m_3 c^2) + (K_1 + K_2 + K_3) + U$$

Because the "internal work" term involved the forces exerted on each particle by each other particle in the system, the potential energy U is really the sum of the interaction energies of all pairs of particles in the system:

$$U = U_{12} + U_{23} + U_{31}$$

We can now define the energy of a multiparticle system to be the sum of each of the particle energies plus all of the interaction energies:

ENERGY OF A MULTIPARTICLE SYSTEM

$$E_{sys} \equiv (m_1 c^2 + m_2 c^2 + \dots) + (K_1 + K_2 + \dots) + (U_{12} + \dots)$$

The energy of a multiparticle system is the sum of the rest energies and kinetic energies of each particle in the system, plus the sum of the potential energies due to the interactions of all pairs of particles in the system.

Multiparticle systems vs. single-particle systems

The Energy Principle is the same for a multiparticle system as it is for a single-particle system. To the left of the equal sign are quantities that are internal to the system, while to the right of the equal sign are the external effects on the system due to objects in the surroundings. The interpretation is, "if the surroundings do work on a system, the energy of the system changes" (Figure 5.23).

$$\Delta E_{sys} = W_{surr}$$

The difference is in the kinds of energy found inside each kind of system:
- A single-particle system has only rest energy and kinetic energy.
- A multiparticle system has rest energy, kinetic energy, and potential energy (interaction energy).

Figure 5.23 Change of system energy is equal to work. Work can be either positive or negative.

Work done by the surroundings can affect both particle energy and potential energy inside a multiparticle system.

You get to choose what system you want to analyze, but once you have specified what portion of the Universe you're considering to be the system of interest, you must be consistent about what energy flows into or out of the system (work done by the surroundings), versus what energy changes occur inside your system (possible changes in particle energies or potential energy).

It is customary to drop the subscript and write W_{surr} simply as W, with the understanding that the work W is mechanical energy transfer from the surroundings to the system, with a resulting change in the energy of the system:

$$\Delta E_{sys} = W$$

Remember that W can be positive or negative, resulting in either an energy gain or an energy loss in the system of interest.

The experimental basis for the Energy Principle for complex systems

We were able to make a mathematical proof of the Energy Principle for a single particle, starting from the Momentum Principle. For a complex, multiparticle system, the more general Energy Principle ultimately is based on over a hundred years of experiments starting in the early 1800's. Slowly scientists learned how to identify flows of energy in many different kinds of systems, and once the energy was accounted for, they found that conservation of energy was always true, not only in physics but also in chemistry and biology.

Invisible particles: neutrinos

A striking example of this occurred in the early part of the 20th century, when scientists found that apparently energy was not conserved in nuclear reactions called "beta decays." In a beta decay, certain radioactive nuclei spontaneously emit an electron, and at the same time a neutron in the nucleus changes into a proton (similar to the decay of a free neutron outside a nucleus). It was observed that the emitted electrons had a wide range of energies, when only one specific energy was expected, corresponding to the change in mc^2 of the nucleus. The Swiss physicist Wolfgang Pauli made the bold proposal that an unseen particle, later named the "neutrino," was also emitted, with energy shared between it and the electron.

Although at the time some scientists wondered if perhaps energy conservation didn't apply to the new science of nuclear physics, Pauli was so convinced of the universal correctness of energy conservation that he was willing to postulate an unseen particle rather than discard energy conservation. It took many years before experiments were sensitive enough to detect the elusive neutrinos, but eventually Pauli's analysis was confirmed.

Recently neutrinos have been in the scientific news again. For many years it was thought that neutrinos had only kinetic energy, with zero rest energy (and zero rest mass). However, recent careful experiments have verified theoretical predictions that neutrinos should have a very tiny rest mass.

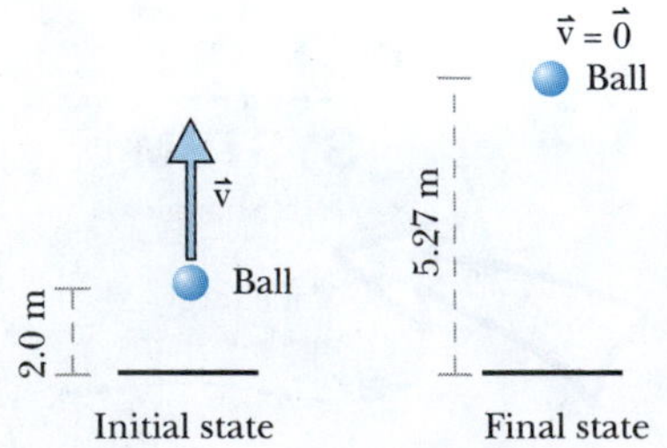

Figure 5.24 A ball is initially moving upward, and eventually comes to a momentary stop.

Example: **Ball and Earth**

Consider again the example of the ball and the Earth discussed on page 185 (Figure 5.24). A ball of mass 100 g is 2 m above the ground, headed upward at a speed of 15 m/s. The ball comes to a momentary stop at a height of 5.27 m above the ground. Consider the ball + Earth as the system. What is the change in potential energy of the system? What is the change in kinetic energy of the system?

System: Ball + Earth
Surroundings: Nothing significant

Energy principle:

$$\Delta(\cancel{m_b c^2 + m_E c^2}) + \Delta K_b + \cancel{\Delta K_E} + \Delta U = \cancel{W}$$

$$0 + \Delta K_b + 0 + \Delta U = 0$$

The force by the Earth is an internal force, so

$$\Delta U = -W_{\text{int}} = -(F_y \Delta y)$$

$$\Delta U = -[-(0.1\,kg)(9.8\text{ N/kg})](5.27 - 2.0)\text{ m}$$

$$\Delta U = +3.2\text{ J}$$

$$\Delta K_b = \left(0 - \frac{1}{2}(0.1\text{ kg})(8\text{ m/s})^2\right) = -3.2\text{ J}$$

There is no change in rest energy in this process. The change in the Earth's kinetic energy is negligible. Work done by the surroundings is 0, because the surroundings are essentially empty.

Further discussion

When the distance between the interacting objects (ball and Earth) increased, the potential energy of the system increased. We will see that this is true in general for attractive interactions such as gravity.

5.X.25 Consider an isolated system, by which we mean a system free of external forces. If the sum of the particle energies in the system increases by 50 joules, what must have been the change ΔU in the system's potential energy?

5.9 PROPERTIES OF POTENTIAL ENERGY

Even without knowing an equation for potential energy related to a particular fundamental interaction, we can deduce the key properties such an equation must have. We will show that:

- Potential energy depends on the separation between pairs of particles, not on their individual positions.
- Potential energy must approach zero as the separation between particles becomes very large.
- If an interaction is attractive, potential energy becomes negative as the distance between particles decreases.
- If an interaction is repulsive, potential energy becomes positive as the distance between particles decreases.

Potential energy depends on separation between a pair of objects

For reciprocal forces such as gravitational and electric forces, the potential energy associated with a pair of interacting particles depends on the separation between the particles, not on their individual positions.

Before proving this in general, we'll look at a concrete example. A big star and a small star are initially at rest (Figure 5.26). They start to move toward each other along the *x* axis due to the gravitational forces they exert on each other, with the same magnitude F (reciprocity). In a short time the big star moves a distance d_1 and the small star moves a distance d_2. The increase in the magnitude of the momentum of each star is the same, because the magnitude of the impulse $F\Delta t$ is the same, so the small star reaches a higher speed ($v \approx p/m$) and moves farther: $d_2 > d_1$, as indicated in Figure 5.26. Therefore more work is done on the small star.

The impulse (and momentum change) have the same magnitude for both stars because both forces act for the same time interval Δt. The work done however is different because the two forces act through different distances. This is an important difference between impulse and work, and therefore between the Momentum Principle and the Energy Principle.

Consider the system consisting of both stars. The internal work W_{int} is the work done by forces internal to the system (not exerted by objects in the surroundings):

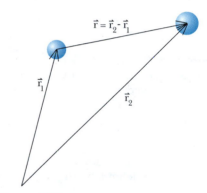

Figure 5.25 The difference in the position vectors of the two stars is the position of star 2 relative to star 1.

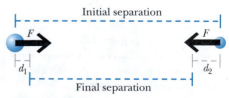

Figure 5.26 Two stars move toward each other. The internal work for the two-star system is F times the change in the separation distance of this pair of objects.

$$W_{\text{int}} = Fd_1 + Fd_2 = F(d_1 + d_2)$$

Change of potential energy is defined as the negative of the internal work:

$$\Delta U = -W_{\text{int}} = -F(d_1 + d_2)$$

Since $\Delta U = -F(d_1 + d_2)$ for the two stars is negative, the potential energy decreased, which would imply that the kinetic energy increased, which is indeed what happened.

The stars got closer to each other; their separation decreased by an amount $(d_1 + d_2)$. We can calculate internal work by multiplying the force one object exerts on another by the change in the separation between the two objects. The individual positions of the objects don't matter. The way the algebra works out, the internal work is calculated as though a single force F acted through a distance equal to the change in the separation of the pair.

Figure 5.27 A system consisting of two stars interacting gravitationally.

Next we'll prove this more generally. Again consider a system of two interacting stars (Figure 5.27). We define the relative position vector $\vec{r}$ to be the difference between the positions of two stars, as shown in Figure 5.28. When the stars move a little closer together, star 1 does work on star 2, and star 2 does work on star 1. Since both stars are inside the system, the change in gravitational potential energy of the system is the negative of the sum of these work terms. Assuming the changes in position are small enough that we can consider the forces to be constant:

$$\Delta U = -W_{\text{int}} = -(\vec{f}_{1,2} \bullet \Delta\vec{r}_1 + \vec{f}_{2,1} \bullet \Delta\vec{r}_2)$$

$$\vec{f}_{1,2} = -\vec{f}_{2,1}, \text{ so } \Delta U = -\vec{f}_{2,1} \bullet (\Delta\vec{r}_2 - \Delta\vec{r}_1)$$

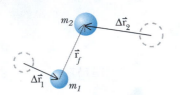

Figure 5.28 Two stars whose relative positions are changing (in this case, the stars move closer together).

The term $(\Delta\vec{r}_2 - \Delta\vec{r}_1)$ involves the change of position of each star (Figure 5.28). We want to see if we can rewrite it in terms of $\Delta\vec{r}$, the change in relative position of the two stars.

$$\Delta\vec{r}_2 - \Delta\vec{r}_1 = (\vec{r}_{2f} - \vec{r}_{2i}) - (\vec{r}_{1f} - \vec{r}_{1i})$$
$$= (\vec{r}_{2f} - \vec{r}_{1f}) - (\vec{r}_{2i} - \vec{r}_{1i})$$
$$= \vec{r}_f - \vec{r}_i$$
$$= \Delta\vec{r}$$

After rearranging terms, we see that according to our definition of $\vec{r}$,

$$\vec{r}_{2f} - \vec{r}_{1f} = \vec{r}_f, \text{ etc.}$$

Using this result we can rewrite the change in potential energy:

$$\Delta U = -\vec{f}_{2,1} \bullet \Delta\vec{r}$$

$\vec{r}$ is the position of star 2 relative to star 1 (Figure 5.27). Therefore we can calculate the amount of change of potential energy ΔU in terms of the *relative displacements* or separation between *pairs of particles*, and any change of potential energy is associated with a change of "shape," with pairs of particles getting closer together or farther apart. The algebra works out in such a way that the force we use in calculating ΔU is just one of the pairs of forces, not both. The result is the same if the interaction is an electric interaction instead of a gravitational interaction, since the reasoning above depended only on reciprocity of forces, and not on any other details of the force.

We write U_{12} for the potential energy of the pair consisting of particles 1 and 2, U_{23} for pair 2 and 3, etc. If a system consists of two particles, there are three energy terms: two particle energies E_1 and E_2, and one potential energy term, U_{12}, since there is only one pair of particles. (Of course the particle energies can be further subdivided into rest energy plus kinetic energy.)

If a system consists of three particles, there are six energy terms: three particle energies E_1, E_2, and E_3, and three potential energy terms, U_{12}, U_{23}, and U_{31}, since there are three pairs of particles.

> For every pair of interacting particles in a system there is there is a potential energy term (U_{12}, U_{23}, etc.). Change of potential energy is associated with change of distance between particles.

? Suppose that two interacting particles move together, with the same velocity (Figure 5.29). In that case, what is $\Delta\vec{r}_{21}$? What is ΔU_{12}?

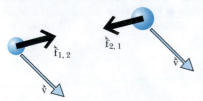

Figure 5.29 The two particles move together, with the same velocity.

In this case the relative positions don't change, $\Delta \vec{r}_{21}$ is zero, and there is no change in the pair-wise potential energy.

? Suppose instead that the two particles rotate around each other (Figure 5.30). What can you say about ΔU_{12}?

This is a bit trickier, because if there is rotation, $\Delta \vec{r}_{21}$ isn't zero, because $\vec{r}_{21}$ is now a rotating vector. But in rotation, $\Delta \vec{r}_{21}$ is perpendicular to $\vec{r}_{21}$, and to the force $\vec{f}_{2,1}$, so the dot product $\vec{f}_{2,1} \bullet \Delta \vec{r}_{21}$ is zero. So again there is no change in the pair-wise potential energy.

? What can you conclude about potential energy for a rigid system?

Evidently U is constant for a rigid system, one whose shape doesn't change. For the potential energy to change, there must be a change of shape or size. Conversely, a change of shape or size is an indicator that U may have changed.

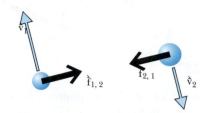

Figure 5.30 The two particles rotate around each other.

5.X.26 If a system contains four particles, how many potential energy pairs U_{12} etc. are there? List them.

$U = 0$ when the particles are very far apart

Consider a system consisting of two identical stars, which attract each other gravitationally. If the stars are at rest, very, very far apart—almost infinitely far apart, so that the force one exerts on the other is nearly zero—then the total energy of the two star system must be simply the sum of their rest energies:

$$E_{\text{sys}} = mc^2 + mc^2$$

This implies that the potential energy associated with the interaction of these two stars must be zero if the stars are "infinitely" far apart.

More generally, the total energy of a multiparticle system is given by the expression $E_{\text{system}} = (E_1 + E_2 + \dots) + U$, where $E_1 = \gamma m_1 c^2$ is the particle energy of particle 1, etc. When the particles are very far apart, we must define U to be zero, so that the total energy of these far-apart particles is simply the sum of the particle energies:

POTENTIAL ENERGY OF A PAIR OF PARTICLES THAT ARE VERY FAR APART

$$U \to 0 \text{ as } r \to \infty$$

Attraction: U becomes negative as r decreases

If the particles start to come closer together, we find that U starts to become negative if the interactions are attractive, while U starts to become positive if the interactions are repulsive. To see this, first consider an isolated system consisting of two stars that are very far apart, initially at rest, so the initial energy of the system is just the sum of the rest energies of the two stars. At first slowly, then more and more rapidly, their kinetic energies increase due to their mutual gravitational attractions (Figure 5.31).

? U was initially zero, so what must be the sign of U when the stars have acquired significant kinetic energies?

If the system is isolated, there are no external forces and hence no external work, so E_{system} does not change. If the relativistic particle energies increase (so the total kinetic energy increases), the potential energy must decrease. Because U was defined to be zero initially, U must become negative. A graph of the energies is shown in Figure 5.32, omitting the very large rest energies of the stars. Since the rest energies don't change, $K+U$ is a constant.

Figure 5.31 Initially two stars are very far apart, at rest, but they attract each other.

$E_{\text{system}} = m_1 c^2 + m_2 c^2 + K + U = \text{constant}$
so $K+U$ is also constant

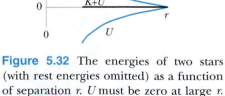

Figure 5.32 The energies of two stars (with rest energies omitted) as a function of separation r. U must be zero at large r. As r decreases, K increases and U decreases. The total energy of the two-star system is constant, and $K+U$ is constant.

Does this negative sign make sense in terms of the definition of potential energy change, $\Delta U = -W_{\text{int}}$? The attractive gravitational forces that the stars exert on each other certainly do a positive amount of work on each star, so the potential energy change must indeed be negative.

Repulsion: *U* becomes positive as *r* decreases

Next consider an isolated system consisting of two protons that are very far apart, and suppose that initially they have high speeds, heading inward (Figure 5.33). Their mutual electric repulsions make the protons slow down (their electric repulsion is enormously larger than their gravitational attraction).

? *U* was initially zero, so what must be the sign of *U* when the protons have slowed down somewhat?

Again, there is no external work done on this isolated multiparticle system, so E_{system} does not change. The particle energies decrease, so the potential energy must increase; *U* becomes positive (Figure 5.34). Also, the repulsive electric forces do negative work on each other, so $\Delta U = -W_{\text{int}}$ is positive. The graph in Figure 5.34 omits the rest energies of the protons. Since the rest energies don't change, $K+U$ is a constant.

More precisely, we will soon show that it is the *change* of U, positive or negative, that indicates whether the interaction is attractive or repulsive. But we wanted to give you a qualitative feel for the issues right away, and in particular to emphasize that there is nothing wrong with a negative value of *U*; this is expected for an attractive force.

Note that the change in potential energy for the two stars or the two protons was associated with a change in shape of the two-particle system. If there is no change in shape or size, there is no change in potential energy.

5.10 GRAVITATIONAL POTENTIAL ENERGY

We have deduced some general properties of potential energy. Now we need to find a specific equation for each type of potential energy, starting from a force law. We will apply this method to find a formula for gravitational potential energy.

Remember that we found that for a small displacement the change in potential energy depends on the relative displacement $\Delta \vec{r}$ of object 2 relative to object 1:

$$\Delta \vec{r} = \Delta \vec{r}_2 - \Delta \vec{r}_1$$

$$\Delta U_{12} = -\vec{f}_{2,1} \bullet \Delta \vec{r}$$

Because the force changes as the relative positions of the objects change, we need to consider a very small displacement $d\vec{r}$, over which the force is nearly constant, and the corresponding small potential energy change dU.

It is the relative displacement that matters, so even if both objects move, we can pretend that object 1 remains stationary, and just calculate the work done on object 2. The force is in a line with $\vec{r}$, so we can evaluate the dot product and write the following, where F_r is the component of the gravitational force on object 2 in the direction of $\vec{r}$, and *r* is the distance from object 1 to object 2 (Figure 5.35):

$$dU = -F_r\, dr$$

Therefore if we already knew the formula for potential energy, we could calculate the associated force like this: $F_r = -dU/dr$. A rate of change of a quantity with respect to position such as this is called a "gradient." For example, if a hill is said to have "a 7% grade," this means that it rises 7 m ver-

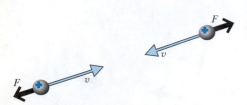

Figure 5.33 Initially two protons are very far apart, heading inward. The mutual repulsions slow the protons down.

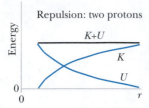

$E_{\text{system}} = m_1 c^2 + m_2 c^2 + K + U = \text{constant}$
so $K+U$ is also constant

Figure 5.34 The energies of two protons as a function of separation (with rest energies omitted). *U* must be zero at large *r*. As *r* decreases, *K* decreases and *U* increases. The total energy of the two-proton system is constant, and $K+U$ is constant.

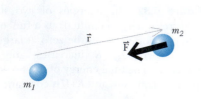

Figure 5.35 The *r* component of the gravitational force, F_r.

tically for every 100 meters horizontally—the tangent of the angle of the hill (the slope) is 0.07 (Figure 5.36).

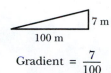

Gradient = $\dfrac{7}{100}$

Figure 5.36 The gradient is the rate of change of *y* with respect to *x*.

FORCE IS NEGATIVE GRADIENT OF POTENTIAL ENERGY

$$F_r = -\frac{dU}{dr}$$

All that remains to be done to find a formula for gravitational energy is to think of an expression whose (negative) derivative is the component of the gravitational force acting on m_2 in the direction of $\hat{r}$. From Figure 5.35 we see that this component of the gravitational force is negative: $F_r = -G(m_1 m_2/r^2)$, where r is the center-to-center distance. The minus sign reflects the fact that the attractive gravitational force on object 2 points back toward object 1.

? Can you think of a function of r whose (negative) derivative with respect to r has this value of F_r? (If necessary, review the appendix on basic calculus at the end of this book.)

The derivative of r^n with respect to r is nr^{n-1}. We have a r^{-2} factor in the gravitational force, so $n = -1$, and the "antiderivative" of r^{-2} (the function whose derivative gives r^{-2}) must be $-r^{-1}$.

? Therefore, what is the formula for gravitational energy?

Being careful about the signs, we find that $U = -Gm_1 m_2/r$

? Check that the negative gradient of this gravitational energy is indeed the force, by differentiating with respect to r.

What about the minus sign on U? Is it correct? Yes, because the potential energy decreases as the particles get closer together. When stars fall toward each other, their kinetic energies increase and the pair-wise potential energies must decrease to more and more negative values. In summary,

GRAVITATIONAL POTENTIAL ENERGY

$$U = -G\frac{m_1 m_2}{r}$$

r is the center-to-center separation of m_1 and m_2

? Could there be an additional constant in this equation?

It is true that the gradient of $U = (-Gm_1 m_2/r + \text{constant})$ would also give us the correct gravitational force, because the derivative of a constant is zero. However, remember that when the objects are very far apart the total energy of the system must be equal just to the sum of the particle energies, and the potential energy must be zero.

? When the particle separation r is very large, what is the value of $-Gm_1 m_2/r$?

When r is very large, $-Gm_1 m_2/r$ is nearly zero. Therefore, the value of the constant in $U = (-Gm_1 m_2/r + \text{constant})$ must be zero, in order for U to be zero at large separations. We have shown that the pair-wise gravitational potential energy must be simply $U = -Gm_1 m_2/r$, with no added constant.

The minus sign
It may seem odd that the expression for gravitational potential energy has a minus sign in it. Figure 5.37 shows that although U is negative, it does in-

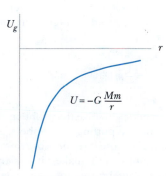

Figure 5.37 Gravitational potential energy is negative. It increases with increasing separation r.

crease with increasing separation r, which means that you have to do work to move two gravitationally attracting objects farther apart.

Here are two little exercises that will help you see why the minus sign is appropriate:

> **5.X.27** If a block falls toward the Earth, the kinetic energy increases and the gravitational potential energy must decrease. In this process, what is the sign of $\Delta(-GMm/r)$ (the change in the potential energy), in which the final separation r_f is less than the initial separation r_i?

> **5.X.28** If a block rises away from the Earth, the kinetic energy decreases and the gravitational potential energy must increase. In this process, what is the sign of $\Delta(-GMm/r)$ (the change in the potential energy), in which the final separation r_f is greater than the initial separation r_i?

Example: A robot spacecraft leaves an asteroid

A robot spacecraft lands on an asteroid, picks up a sample, and blasts off to return to Earth; its total mass is 1500 kg. When it is 200 km (2×10^5 m) from the center of the asteroid, its speed is 5.0 m/s, and the rockets are turned off. At the moment when it has coasted to a distance 500 km (5×10^5 m) from the center of the asteroid, its speed has decreased to 4.1 m/s. You can use these data to determine the mass of the asteroid. Before starting your analysis, draw a diagram of the initial and final states you will consider in your analysis. Label your diagram to show the distances involved. This will help avoid mistakes in your calculation.

(a) Draw and label the diagram.

(b) Calculate the mass of the asteroid. Make an accurate calculation, not a rough, approximate calculation.

(a) Figure 5.38 shows the situation. It is important to keep in mind that r is measured center-to-center in $U = -Gm_1m_2/r$.

System: Asteroid and spacecraft.
Surroundings: Nothing significant

Initial state: Rockets off, 5.0 m/s, 200 km from center of asteroid
Final state: Speed 4.1 m/s, 500 km from center of asteroid

Energy Principle: $E_f = E_i + W$

$$\cancel{Mc^2} + K_{M,f} + \cancel{mc^2} + K_{m,f} + U_f = \cancel{Mc^2} + K_{M,i} + \cancel{mc^2} + K_{m,i} + U_i + \cancel{W}$$

$$\cancel{K_{M,f}} + K_{m,f} + U_f = \cancel{K_{M,i}} + K_{m,i} + U_i$$

$$\cancel{\tfrac{1}{2}Mv_f^2} + \left(-G\frac{M\cancel{m}}{r_f}\right) = \cancel{\tfrac{1}{2}Mv_i^2} + \left(-G\frac{M\cancel{m}}{r_i}\right)$$

$$\tfrac{1}{2}v_f^2 + \left(-G\frac{M}{r_f}\right) = \tfrac{1}{2}v_i^2 + \left(-G\frac{M}{r_i}\right)$$

$$GM\left(\frac{1}{r_i} - \frac{1}{r_f}\right) = \tfrac{1}{2}(v_i^2 - v_f^2)$$

$$M = \frac{\tfrac{1}{2}(v_i^2 - v_f^2)}{G(1/r_i - 1/r_f)}$$

$$M = \frac{\tfrac{1}{2}(5.0^2 - 4.1^2)\ \text{m/s}^2}{(6.7\times10^{-11}\ \text{N}\cdot\text{m}^2/\text{kg}^2)\left(\dfrac{1}{2\times10^5\ \text{m}} - \dfrac{1}{5\times10^5\ \text{m}}\right)} = 2\times10^{16}\ \text{kg}$$

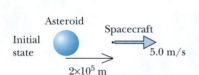

Asteroid
Spacecraft
Initial state
5.0 m/s
2×10^5 m

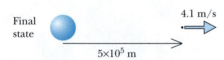

Final state
4.1 m/s
5×10^5 m

Figure 5.38 Initial and final states.

By choosing both the asteroid and the spacecraft to be in the system, we won't have to calculate work, since there are no significant external forces.

No change in rest energies. No work by surroundings (no interacting objects in surroundings).

As discussed below, the kinetic energy of the massive asteroid changes very little: $(K_{M,f} \approx K_{m,i})$.

$v \ll c$ so $K \approx \tfrac{1}{2}mv^2$

It is often better to solve the problem algebraically before putting in numbers, as is done here, because it reduces the number of calculations, and makes it easier to track down algebra errors. Note that m cancelled; we didn't need to know or use the value of 1500 kg for the mass.

Further discussion

We could not have used the Momentum Principle to carry out an accurate analysis without writing an iterative computer program because we didn't know how much time the process took. The velocity was changing at a non-constant rate, so knowing the distance wouldn't give us the time, since we didn't know the average velocity. We were able to use the Energy Principle because we knew something about distances.

5.X.29 What was the advantage of choosing an "inclusive" system (including both the asteroid and the spacecraft)?

5.X.30 Concerning the separation r in $U = -Gm_1m_2/r$: Is this the distance from the spacecraft to the surface of the asteroid, or from the spacecraft to the center of the asteroid?

Multiparticle systems: same time interval, different displacements

In the previous example the spacecraft pulls just as hard on the asteroid as the asteroid pulls on the spacecraft (reciprocity of gravitational forces). Yet we claimed that the kinetic energy of the asteroid hardly changed, compared to the change in the spacecraft's kinetic energy. We can show why by using the Momentum Principle and the Energy Principle together.

First consider the Momentum Principle. The two different forces act on the asteroid and the spacecraft for the same amount of time Δt. Since the magnitude of the two forces is the same, the magnitude of the impulse is the same, and the magnitude of the change of momentum is the same. Since $v = p/M$ at low speeds, the change in speed of the asteroid is extremely small compared to the change in speed of the spacecraft (ratio of m/M), and if the asteroid was initially at rest it will move a very small distance compared to the displacement of the spacecraft.

Next consider the Energy Principle. The work done on the asteroid by the spacecraft is extremely small, because the displacement of the asteroid is extremely small. The two objects exert the same magnitude of force on each other, but through enormously different distances. In the example above, the spacecraft moved 300 km, whereas the asteroid moved roughly (m/M) of that distance, or about $(1500/2{\times}10^{16})(3{\times}10^5 \text{ m}) \approx 2{\times}10^{-8} \text{ m}$, or about 100 atomic diameters!

This illustrates yet another important difference in the role played by forces acting in and on a multiparticle system in the Momentum Principle (all forces act for the same time interval, and the net force is all that matters) and in the Energy Principle (different forces may act through very different distances, and change of shape matters).

Another way of seeing why the asteroid gets very little kinetic energy is to recall that kinetic energy can be expressed as $p^2/(2M)$. The asteroid and spacecraft experience the same change in the magnitude of momentum, but the p^2 term is divided by a huge mass in the case of the asteroid.

Similarly, when a ball falls toward the Earth, the magnitude mg of the force that the Earth exerts on the ball is the same as that which the ball exerts on the Earth, and both experience the same change of momentum in the same time interval. But the massive Earth acquires an extremely small velocity, so the Earth moves an extremely small distance while the ball moves a large distance. Same force mg, very different distances, very different amounts of work, and the Earth gets very little kinetic energy.

Energy diagrams

A graphical representation of energy can be exceptionally helpful in reasoning about a process. In Figure 5.39 on the next page we show an energy graph for the spacecraft leaving the asteroid, the process analyzed in the previous example (the graph continues beyond the separation

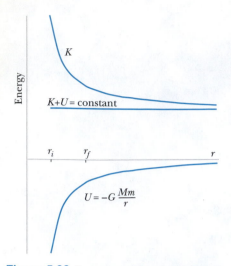

Figure 5.39 Energy vs. the separation distance between the asteroid and the spacecraft. (We omit the rest energies, which aren't changing.)

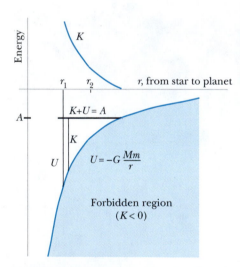

Figure 5.40 Energy vs. the separation distance between a star and a planet. (We omit the rest energies, which aren't changing.)

$r_f = 500$ km). We plot the pair-wise gravitational potential energy U, the spacecraft's kinetic energy K, and the sum $K + U$, as a function of the separation distance r between the center of the asteroid and the spacecraft. We omit the rest energies, which are not changing during the motion

.Important features of the graph are these: The gravitational potential energy U is negative because the interaction is attractive, not repulsive, and the potential energy increases with increasing separation r. As the potential energy increases, the kinetic energy K of the spacecraft decreases (the gravitational force exerted on the spacecraft by the asteroid is slowing it down). Because there is negligible work done on the combined system of asteroid plus spacecraft, $K_f + U_f = K_i + U_i$, so $K + U$ has a constant value at all times, and the graph of $K + U$ is a straight horizontal line on the graph, representing a constant positive value.

As the spacecraft gets farther and farther from the asteroid, you can see from the graph that its kinetic energy K will continue to fall, and the graph of K will approach the $K + U$ line at large separation. When the two objects are very far apart, U is nearly zero, and $K + U$ is just K.

This is an example of an "unbound" system, because the separation between spacecraft and asteroid will increase without bound. An unbound system has $K + U$ greater than zero.

As an example of a "bound" system, consider an isolated system of a star of mass M and a planet of mass m. As a function of the center-to-center star-planet separation r, we again plot a graph of the potential energy $-GMm/r$ (Figure 5.40), and a graph of kinetic energy K. As the planet moves away from the star, its kinetic energy K falls as the pair-wise potential energy U rises. You can see on the graph that K falls to zero, and it cannot go negative, because kinetic energy is always a positive quantity. What happens next? The planet momentarily comes to a stop, then turns around and speeds up toward the star.

We add to the graph a horizontal line whose height above or below the axis represents the total kinetic plus potential energy of the two-particle system. The line on the graph for $K + U$ equal to the value A is horizontal because no external work is done on this isolated system, and there is no change of the rest masses (no change of identity), so $K + U$ is constant during the motion.

? Is A positive or negative for the system shown in Figure 5.40?

The horizontal line representing $K + U = A$ is below the axis representing $U = 0$, so A is negative. Go down from the x axis to the U curve by an amount U, then up by an amount K, to arrive at the horizontal line that represents the fact that $K + U$ is a constant (A). By mentally drawing a vertical line from the potential energy curve up to the horizontal total energy line, you can visualize how the total kinetic energy of the star and planet changes during the motion. Notice that when K falls to zero, $K + U = U$ and the right end of the $K + U$ line touches the U graph.

The shaded area in (Figure 5.40) represents a forbidden region, where the kinetic energy K would be negative, no matter what the value of $K + U$ (more about this in the next section).

5.X.31 The star-planet separation increases from r_1 to r_2 (Figure 5.40). Does the kinetic energy of the star-planet system increase, decrease, or remain constant?

Limits on possible motion

The energy graph shows at a glance the limits on the motion for a given energy. In Figure 5.41, a star and planet with $K + U < 0$ cannot get farther away from each other than r_3, a bound that is set by the potential-energy curve. At a larger separation r_4, the kinetic energy would have to be negative, which is "classically forbidden" (that is, impossible according to the laws of classical, prequantum mechanics). Physically, because kinetic energy is the energy a particle has as a result of being in motion (in addition to its rest energy), kinetic energy cannot be negative.

If the system has the total $K+U$ shown ($A < 0$), the separation of the star and planet can never be larger than r_3, and we say that this is a "bound" system. An energy graph shows at a glance the range of possible positions achievable during the motion, for a given total $K+U$. (We should mention, however, that in the world of atoms where a full analysis requires quantum mechanics, in some cases a system can "tunnel through" a region that is classically forbidden!)

An energy graph can show at a glance whether a system is "bound" or "unbound." An example of a bound system is a planet in circular or elliptical orbit around a star: the planet cannot escape. Another example is an electron bound to an atom.

An example of an unbound system is the Earth plus a spacecraft whose initial speed is great enough that the spacecraft will get away from the Earth and never come back. For another example, if an amount of energy greater than or equal to the ionization energy is supplied to an atom, an electron can become unbound and escape from the atom. An unbound system has total $K + U \geq 0$, which is easy to see on an energy graph because arbitrarily large separations are possible. Only when $K + U \geq 0$ can the system become separated by large distances, because K is never negative, and U goes to zero at large separations.

> *5.X.32* In this energy graph for some system (Figure 5.42), consider the various energy states indicated. Which of these values for the energy $K+U$ (A, B, or C) represent a bound state? Which represent an unbound state (the particle can escape)? Which represents a trapped state with enough energy to be unbound but with a barrier that (classically) prevents escape?

Drawing energy graphs

It is important to be able to draw (create) energy graphs as well as be able to read them. Here is a practical scheme:

- Draw U vs. r for the particular interaction (gravitational, electric).
- At some r where you happen to know K, plot the point (r, K).
- Add that value of K to the value U has at the same separation r.
- Plot $K+U$ at that r, then draw a horizontal line through the point.
- At some other r, find a K which when added to U at that r gives $K+U$.
- Now you have two points on the K graph; sketch the behavior of K vs. r.

Example: Making an energy graph

Let's follow through these steps for the energy graph for the spacecraft leaving the asteroid, which is shown again in Figure 5.43.

- We know the shape of U for gravity; it is $-GMm/r$. If r is small, U is a large negative number. If r is large, U is nearly zero. Plot two such points, then sketch the behavior of U vs. r.
- We know that K is nonzero at large r, because the spacecraft escapes, so at large r, plot some positive K.
- At that distant location, U is nearly zero, so the K you just plotted is nearly the same as $K+U$. Draw a horizontal line representing $K+U$ for all separations.

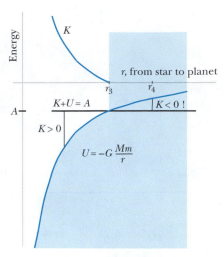

Figure 5.41 The star and planet cannot get farther away from each other than r_3; at r_4 the kinetic energy would be negative. (We omit the rest energies, which aren't changing.)

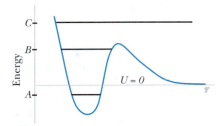

Figure 5.42 An energy graph for some system. Which states are bound? (Exercise 5.X.32)

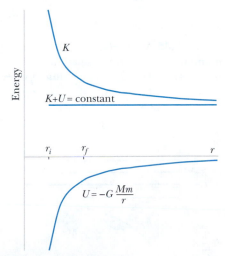

Figure 5.43 Energy vs. the separation distance between the asteroid and the spacecraft. (We omit the rest energies, which aren't changing.)

- At the initial separation r_i, determine graphically a value of K which when added to the value of U at r_i yields the known value of $K+U$. The value of K plus the value of U must be the same as $K+U$.
- Sketch the behavior of K vs. r. Note that its shape is the mirror image of U vs. r.

5.11 APPLYING GRAVITATIONAL POTENTIAL ENERGY

In the next sections we apply the Energy Principle to the analysis of a variety of situations involving gravitational potential energy. We will frequently use energy diagrams as a powerful tool.

Application: Energy graphs and types of orbits

In Figure 5.44 are plotted graphs of potential energy, and potential plus kinetic energy, for three different orbits of a planet and star. The following exercise asks you to interpret the meaning of these graphs.

> **5.X.33** In Figure 5.44, what kind of motion is represented by the situation with $K+U = A$? B? C? Think about the range of r in each situation. For example, C represents a circular orbit (constant r).

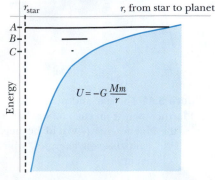

Figure 5.44 What kind of motion is represented by each of these situations? (Exercise 5.X.33).

Application: Escape speed

An analysis based on the Energy Principle will allow us to answer the following question: With what minimum speed v must a spacecraft leave the surface of an airless planet (no air resistance) of mass M and radius R in order that it can coast away without ever coming back (Figure 5.45)? Assume there are no other objects near by.

If you did the Moon voyage problem in Chapter 3 you found that there was a minimum initial speed required to reach the Moon. With an initial speed smaller than this, the spacecraft fell back to Earth without reaching the Moon. Your determination of the minimum initial speed was done by trial and error. However, we can now use energy relationships to calculate such minimum speeds directly.

As is often the case, an energy diagram can be very helpful. Consider Figure 5.46. Motions with $K+U = A$, B, C, or D all start from the surface of an airless planet, with a separation R between the center of the planet and the spacecraft. (Remember that the gravitational force outside a uniform sphere is exactly the same as it would be if all the mass collapsed to the center of the sphere.)

Figure 5.45 A spacecraft is launched with momentum p_i from near an airless planet's surface; it has momentum p_f when it has traveled far away.

? Which of these motions starts with the largest kinetic energy K? Which starts with the least K?

Each of the horizontal lines represents motion with $K+U$ constant. Evidently A is the highest value of $K+U$, and therefore involves the largest initial kinetic energy. D is the lowest value of $K+U$, and starts with the least kinetic energy.

? For which of these motions does the spacecraft "escape" and never come back?

For motions with $K+U = A$, B, or C, no matter how far from the planet the spacecraft gets, there is still some kinetic energy, so the spacecraft will never return. This is another example of an unbound state being associated with a positive value of $K+U$.

In contrast, with motion $K+U = D$ the spacecraft will reach a maximum separation from the planet, at which point its kinetic energy has fallen to zero. The spacecraft will fall back to the planet.

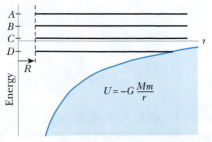

Figure 5.46 What is the minimum K required to escape?

? How would you describe the motion that requires the least possible initial kinetic energy to achieve escape?

Evidently the least costly escape is to have $K = 0$ when the spacecraft has reached a distance very far from the planet. At a large separation, $U = 0$, so $K+U = 0$. This corresponds on the diagram to a horizontal line lying on the axis. Because $K+U$ doesn't change (no external work on the planet-spacecraft system), it must also be true that $K+U = 0$ at the start of the motion:

MINIMAL CONDITION FOR ESCAPE
$$K+U = 0$$

The minimal initial kinetic plus potential energy of the system composed of planet plus spacecraft is this, assuming that the kinetic energy of the planet is negligible:

$$K_i + U_i = \tfrac{1}{2}mv^2_{esc} + \left(-G\frac{Mm}{R}\right) = 0 \ \ (\text{for } v \ll c)$$

This lets us calculate the minimum speed, which is called the "escape speed" v_{esc}, for a planet with mass M and radius R. Though we speak of "escape speed," it is wrong to say that the spacecraft has "escaped from gravity." As long as the spacecraft is a finite distance from the planet, it feels a finite gravitational attraction to the planet.

If you give the spacecraft more than the minimum kinetic energy, it not only escapes but has nonzero kinetic energy when it is far away (for example, see motions A and B in Figure 5.46).

Initial direction does not matter

It is interesting to note that in this analysis, the initial direction of motion doesn't matter. An energy analysis is insensitive to direction. If the initial velocity points directly away from the planet, the path is a straight line. If the initial velocity points in some other direction, the path can be shown to be a parabola if the initial speed is exactly equal to escape speed, and a hyperbola for higher speeds.

If the speed is below escape speed, there is no escape. The various possible orbits (Figure 5.47) can be classified on the basis of their total kinetic plus potential energy, with initial speed v_i:

$$\tfrac{1}{2}mv^2_i + \left(-G\frac{Mm}{R}\right) > 0: \text{escape with } v_\infty > 0\,; \text{straight line or hyperbola}$$

$$\tfrac{1}{2}mv^2_i + \left(-G\frac{Mm}{R}\right) = 0: \text{escape with } v_\infty = 0\,; \text{straight line or parabola}$$

$$\tfrac{1}{2}mv^2_i + \left(-G\frac{Mm}{R}\right) < 0: \text{no escape; straight line, circular, or elliptical}$$

Bound vs. unbound states

This is a specific example of an important general principle. If the kinetic plus potential energy is negative, the system is "bound": the objects cannot become widely separated, because that would require that the kinetic energy become negative, which is impossible (classically). If the kinetic plus potential energy is positive, the system is "unbound" or "free": the objects can become widely separated, with net kinetic energy (and zero potential energy).

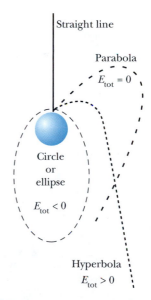

Figure 5.47 Possible orbital trajectories.

BOUND AND UNBOUND STATES

If $K + U < 0$, the system is in a bound state.

If $K + U \geq 0$, the system is unbound (free).

Example: Escape speed

What is escape speed from Earth?

System: Earth + spacecraft
Surroundings: Nothing significant

Initial state: At surface of Earth, with speed
Final state: Spacecraft extremely far away from Earth, nearly at rest

Energy principle:

When we know there is no change in rest energies it is okay to omit these terms from the Energy Principle equation.

Negligible change in kinetic energy of Earth. No work by surroundings.

$$K_{s,f} + \cancel{K_{E,f}} + U_f = K_{s,i} + \cancel{K_{E,i}} + U_i + \cancel{W}$$

$$0 + 0 = \frac{1}{2}mv_i^2 + \left(-\frac{GMm}{r_i}\right)$$

$$v_i = \sqrt{\frac{2GM}{r_i}}$$

$$v_{esc} = \sqrt{\frac{2(6.7\times10^{-11}\ \text{N}\cdot\text{m}^2/\text{kg}^2)(6\times10^{24}\ \text{kg})}{(6.4\times10^6\ \text{m})}} = 1.12\times10^4\ \text{m/s}$$

5.X.34 This escape speed is slightly larger than the launch speed that was found necessary to reach the Moon in the problem in Chapter 3 on the Ranger 7 mission to the Moon. Explain why.

5.X.35 Turn the argument around. If an object falls to Earth starting from rest a great distance away, what is the speed with which it will hit the upper atmosphere? (Actually, a comet or asteroid coming from a long distance away might well have an even larger speed, due to its interaction with the Sun.) Small objects vaporize as they plunge through the atmosphere, but a very large object can penetrate and hit the ground at very high speed. Such a massive impact is thought to have killed off the dinosaurs (see *T. Rex and the Crater of Doom*, Walter Alvarez, Princeton University Press, 1997).

5.X.36 The radius of the Moon is 1750 km, and its mass is 7×10^{22} kg. What is escape speed from the Moon? Why was a small rocket adequate to lift the lunar astronauts back up from the surface of the Moon?

Application: The Moon voyage

We have been applying energy considerations to two-particle systems. A homework problem in Chapter 3 involves predicting the motion of the Ranger 7 spacecraft as it travels from Earth to the Moon (where it crash lands). In our simple model for this voyage there are three "particles": the spacecraft with mass m, the Earth with mass M_{Earth}, and the Moon with mass M_{Moon}.

? How many interaction pairs are there in this three-particle system?

In a three-particle system there are three interaction pairs: U_{12}, U_{13}, and U_{23}. We can write the gravitational potential energy like this:

$$U = \left[-G\frac{M_{\text{Earth}}m}{r_{\text{to Earth}}}\right] + \left[-G\frac{M_{\text{Moon}}m}{r_{\text{to Moon}}}\right] + \left[-G\frac{M_{\text{Earth}}M_{\text{Moon}}}{r_{\text{Earth to Moon}}}\right]$$

In our simple model, the Earth and Moon were fixed in space, so the Earth-Moon term of this expression doesn't change. Also, there is no change of identity of the three particles, so their rest masses do not change. Therefore the energy equation for the system can be written like this, where v is the speed of the spacecraft, the only moving particle in the model system (and $v \ll c$):

$$\Delta(\tfrac{1}{2}mv^2) + \Delta\left[-G\frac{M_{\text{Earth}}m}{r_{\text{to Earth}}}\right] + \Delta\left[-G\frac{M_{\text{Moon}}m}{r_{\text{to Moon}}}\right] = 0$$

The change in the energy of the system is zero because no external work is done on the system (we're ignoring the effects of other objects, including the Sun). We can rearrange the terms and write the energy equation in terms of initial and final energies, which must be equal:

$$\tfrac{1}{2}mv_f^2 + \left[-G\frac{M_{\text{Earth}}m}{r_{\text{to Earth}, f}}\right] + \left[-G\frac{M_{\text{Moon}}m}{r_{\text{to Moon}, f}}\right] = \tfrac{1}{2}mv_i^2 + \left[-G\frac{M_{\text{Earth}}m}{r_{\text{to Earth}, i}}\right] + \left[-G\frac{M_{\text{Moon}}m}{r_{\text{to Moon}, i}}\right]$$

That is, the final (kinetic plus potential) energy of the system is equal to the initial (kinetic plus potential) energy of the system, because no external work is done on the system (and the rest masses don't change; no change of identity).

5.12 GRAVITATIONAL POTENTIAL ENERGY NEAR THE EARTH'S SURFACE

An important special case of gravitational potential energy arises in projectile motion near the surface of the Earth. Suppose you throw a baseball upward (Figure 5.48). When it has reached a height h above you, what is the change in gravitational energy? Let M be the mass of the Earth, m the mass of the ball, and R the radius of the Earth (which is the initial distance from the center of the Earth to the baseball):

$$\Delta U = \Delta\left[-G\frac{Mm}{r}\right] = \left[-G\frac{Mm}{R+h}\right] - \left[-G\frac{Mm}{R}\right] = GMm\left[\frac{(R+h)-R}{R(R+h)}\right]$$

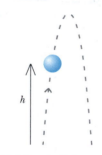

Figure 5.48 A baseball thrown upward has reached a height h.

Because h is very small compared to R, the product $R(R+h)$ is nearly the same as R^2. While we can neglect h in the denominator, we cannot neglect it in the numerator, where it is the largest (indeed the only) term.

In the approximation that $R(R+h)$ is nearly the same as R^2, we have

$$\Delta U \approx GMm\left[\frac{h}{R^2}\right] = \left(G\frac{Mm}{R^2}\right)h$$

Remembering that $GMm/R^2 \approx mg$, and writing $h = \Delta y$, we can rewrite the change in the gravitational potential energy like this:

CHANGE IN GRAVITATIONAL POTENTIAL ENERGY NEAR THE SURFACE OF THE EARTH

$$\Delta U \approx \Delta(mgy)$$

Notice that there is no minus sign in this (approximate) expression for the change in gravitational potential energy. As you go up (farther from the center of the Earth), the change $\Delta(mgy)$ is positive as expected; if you go down, it is negative as expected. In the full formula there is a $(-1/r)$ because moving to larger distance r means an increase (gravitational potential energy *less negative*), as can be seen in Figure 5.49.

In the formula $\Delta U = \Delta(mgy)$, Δy must be small compared to the radius of the Earth (6400 kilometers), because the gravitational force gets smaller as you go farther away from the Earth, inversely proportional to the square of

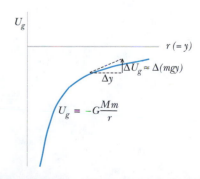

Figure 5.49 A small increase Δy in height y corresponds to a small increase in the (negative) gravitational energy.

the distance from the center of the Earth. But if y doesn't change much, neither does gravity. If you go up 0.1 km (100 meters, the length of a football field), the gravitational force decreases by a tiny factor:

$$\left(\frac{6400 \text{ km}}{6400.1 \text{ km}}\right)^2 \approx 0.99997$$

Example: A falling ball

At a certain instant a ball is falling with a speed of 6 meters/second, and its position is $y = 35$ meters above the Earth's surface. How fast is the ball falling when it has fallen to a position $y = 20$ meters above the Earth's surface, assuming that we can neglect air resistance?

System: Ball + Earth
Surroundings: Nothing significant (neglect air resistance)

Initial state: Speed 6 m/s, 35 m above ground
Final state: 20 m above ground

Energy Principle:

The *change* in potential energy near the surface of the Earth is $\Delta U \approx \Delta(mgy)$, and the potential energy at the surface is $-GMm/R$, a constant. Since this constant doesn't change, we simply omit it and write $U_i = mgy_i$ and $U_f = mgy_f$.

$$K_f + U_f = K_i + U_i + \cancel{W}$$

$$\tfrac{1}{2}mv_f^2 + mgy_f = \tfrac{1}{2}mv_i^2 + mgy_i$$

$$\tfrac{1}{2}mv_f^2 = \tfrac{1}{2}mv_i^2 - (mgy_f - mgy_i)$$

$$v_f^2 = v_i^2 - 2g(y_f - y_i)$$

$$v_f = \sqrt{v_i^2 - 2g(y_f - y_i)}$$

$$v_f = \sqrt{(6 \text{ m/s})^2 - 2(9.8 \text{ N/kg})[(20 \text{ m}) - (35 \text{ m})]} = 18.2 \text{ m/s}$$

Further discussion

There is nothing in our equations to specify whether the ball is initially heading upward or downward. In fact, it doesn't matter. In either case, if we neglect air resistance, the speed will be 18.2 m/s when the ball reaches a height of 20 m above the ground. (If it was initially headed upward, it will again have a speed of 6 m/s when it returns to a height of 35 m.)

5.X.37 Suppose in this situation we measure y from 25 meters above the surface, so that $y_{\text{initial}} = +10$ meters, and $y_{\text{final}} = -5$ meters. Why do we get the same value for the final speed?

5.X.38 Use energy conservation to find the approximate final speed of a basketball dropped from a height of 2 meters (roughly the height of a professional basketball player). Why don't you need to know the mass of the basketball?

5.X.39 A 0.5 kg teddy bear is nudged off a window sill and falls 2 m to the ground. What is its kinetic energy at the instant it hits the ground? What is its speed? What assumptions or approximations did you make in this calculation?

5.X.40 A 1.0 kg flowerpot is nudged off a window sill and falls 2 m to the ground. What is its kinetic energy at the instant it hits the ground? What is its speed? How do the speed and kinetic energy compare to that of the teddy bear in the previous exercise?

5.X.41 Suppose a pitcher can throw a ball straight up at 100 miles per hour (about 45 m/s). Use energy conservation to calculate how high the baseball goes. (See the Answers section for a hint, if necessary.) Explain your work. Actually, a pitcher can't attain this high a speed when throwing straight up, so your result will be an overestimate of what a human can do; air resistance also reduces the achievable height.

Including horizontal motion

Moving vertically up or down involves a change in gravitational energy approximately equal to $\Delta(mgy)$. What about sideways, horizontal movements? There is negligible change in the gravitational energy of the Universe when an object is moved a short distance horizontally (that is, tangent to the Earth's surface). Consider moving a block along a surface with it rolling very easily on low-friction wheels, or a hockey puck sliding easily along the ice. You hardly have to use any energy at all to move the block or puck horizontally, as long as there is little friction.

What if a block is moved horizontally, but at a constant height of 3 meters above the surface? You could support the block with a tall cart, and again there would be little effort involved in moving the block horizontally. Or you could pile up dirt 3 meters deep and place a slick surface on top, and then you could easily move the block horizontally. We conclude that horizontal position doesn't affect the gravitational energy. (Again, we're considering movements over distances small compared with the radius of the Earth, so we can ignore the Earth's curvature.)

Suppose you throw a ball at an angle to the horizontal, and just after it leaves your hand at a height y_i it has a speed v_i (Figure 5.50). At the top of its trajectory, at the instant that it is momentarily traveling horizontally, it has a speed v_f.

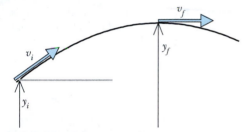

Figure 5.50 Path of a ball thrown at an angle to the horizontal.

5.X.42 What is the height y_f at the top of the trajectory, in terms of the other known quantities, assuming that we can neglect air resistance? Start by picking a system. Choose an initial and a final state, and apply the Energy Principle. If necessary, see the Answers section for a hint.

5.13 ELECTRIC POTENTIAL ENERGY

In this section we will study the potential energy associated with the electric interactions of charged particles, and the potential energy associated with the electric interactions of neutral atoms, which contain charged particles.

Electric potential energy involving charged particles

The electric force law for two charged particles is similar to the gravitational force law:

THE ELECTRIC FORCE

$$F_{\text{electric}} = \frac{1}{4\pi\varepsilon_0}\frac{|q_1 q_2|}{r^2}, \text{ where } \frac{1}{4\pi\varepsilon_0} = 9\times10^9 \frac{\text{N}\cdot\text{m}^2}{\text{coulomb}^2}$$

The constant $1/(4\pi\varepsilon_0)$ is read as "one over four pi epsilon-zero". The quantities q_1 and q_2 represent the amount of charge on the two particles (measured in coulombs), and r is the distance between the two particles. Because this is an inverse square force, like gravitation, the electric potential energy has a formula very similar to that for gravitational potential energy:

ELECTRIC POTENTIAL ENERGY U_{el}

$$U_{\text{el}} = \frac{1}{4\pi\varepsilon_0}\frac{q_1 q_2}{r}$$

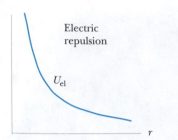

Figure 5.51 Electric potential energy is positive when like charged objects repel each other.

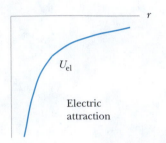

Figure 5.52 Electric potential energy is negative (like gravitational potential energy) when objects attract.

We don't know the initial K of the electron, so assume it is zero (this is a rough approximation).

There is no minus sign in the formula for electric potential energy, because when the electric interaction is attractive, like gravitation, q_1 and q_2 have different signs, and the minus sign shows up automatically. However, if the two charges have the same sign, the interaction is repulsive, and there is a plus sign in the formula for the electric energy (Figure 5.51). If the two charges have opposite signs, the interaction is attractive, and there is a minus sign in the formula for electric energy (Figure 5.52).

A certain amount of energy is required to remove an outer electron from a neutral atom, leaving behind a positive "ion" that is missing one electron and whose charge is $+e$, where $e = 1.6 \times 10^{-19}$ coulomb. (The electron's charge is of course $-e$.) The minimum energy to remove one outer electron is called the "ionization" energy.

Example: Ionization energy

A typical atom has a radius of about 10^{-10} m. Calculate a typical ionization energy in joules. Then convert your result to "electron volts," a unit of energy equal to 1.6×10^{-19} joules. This small energy unit is frequently used in discussions of atomic phenomena.

> System: Two particles, the outer electron with charge $-e$, and the remaining part of the atom, the ion with charge $+e$.
> Surroundings: an object that exerts an external force to remove the electron from the atom.

> Initial state: the electron is bound to the atom, at a distance r_{atom} from the center.
> Final state: The two particles very far from each other, at rest.

> Energy Principle:

$$E_f = E_i + W$$

$$\cancel{K_{ion,f}} + \cancel{K_{e,f}} + \cancel{m_i c^2} + \cancel{m_e c^2} + U_f = \cancel{K_{ion,i}} + \cancel{K_{e,i}} + \cancel{m_i c^2} + \cancel{m_e c^2} + U_i + W$$

$$U_f = U_i + W$$

$$W = U_f - U_i$$

$$W = \frac{1}{4\pi\varepsilon_0}\frac{q_1 q_2}{r_\infty} - \frac{1}{4\pi\varepsilon_0}\frac{q_1 q_2}{r_{atom}}$$

$$W = -\frac{1}{4\pi\varepsilon_0}\frac{(+e)(-e)}{r_{atom}}$$

$$W = \left(9\times10^9 \frac{\text{N}\cdot\text{m}^2}{\text{C}^2}\right)\frac{(1.6\times10^{-19}\,\text{C})^2}{(10^{-10}\,\text{m})}$$

$$W = 2.3\times10^{-18}\,\text{J}$$

$$W = (2.3\times10^{-18}\,\text{J})\left(\frac{1\,\text{eV}}{1.6\times10^{-19}\text{J}}\right) = 14\,\text{eV}$$

> Check. Units correct.

Further discussion

This is a rough estimate because we don't know the electron's initial kinetic energy K_i, but typical ionization energies are indeed a few electron volts.

5.X.43 Two protons are hurled straight at each other, each with a kinetic energy of 0.1 MeV, where 1 mega electron volt is equal to $10^6 \times (1.6\times10^{-19})$

joules. What is the separation of the protons from each other when they momentarily come to a stop?

5.14 THE MASS OF A MULTIPARTICLE SYSTEM

A surprising prediction of Einstein's special theory of relativity (1905) was that mass and energy are actually the same thing. This prediction has been confirmed in many different kinds of experiments. The most dramatic example of mass-energy equivalence is the annihilation reaction that occurs between matter and anti-matter, as in the case of the annihilation of an electron and a positron (a positive electron) into two photons (high-energy quanta of light denoted by lower-case Greek gamma, γ):

$$e^- + e^+ \rightarrow \gamma + \gamma$$

Before the reaction, an electron and a positron can be almost at rest far from each other, a situation we would naturally describe as one where there is mass but seemingly no energy. After the reaction there is light (in the form of two photons), a situation which we would naturally describe as pure energy but seemingly no mass. But from the point of view of special relativity, the mass of the electron-positron system *is* energy, and the energy of the two-photon system *is* mass.

The equivalence of mass and energy is of course visible in Einstein's formula for the energy of a particle:

$$E = \gamma mc^2 = mc^2 + K \ \ (\textit{definition of kinetic energy } K)$$

In this equation m is called the rest mass of the particle. and is a constant. The particle has "rest energy" mc^2 even if it is not moving. If it is moving, it has additional, "kinetic" energy K and additional mass. A valid way of thinking about a fast-moving particle is that it has more mass, because mass is equivalent to energy, so the mass is E/c^2.

What is the mass of a multiparticle system consisting of interacting particles? Consider the simple case where the multiparticle system is overall "at rest." The individual particles are moving around, but the system as a whole is not moving (technically, the total momentum of the system is zero, and its center of mass is at rest). An example would be a box containing air molecules that are moving around rapidly in random directions, but the box isn't going anywhere. Another example is a diatomic molecule such as O_2 that is vibrating but has no overall translational motion.

It would seem natural to suppose that the total mass of such a system is just the sum of the rest masses of the particles: $M = m_1 + m_2 + ...$ However, this is not quite right. Rather, as predicted by the theory of relativity, the energy of the multiparticle system is Mc^2, and therefore its mass is this:

$$M = \frac{E_{\text{system}}}{c^2} = \frac{m_1 c^2 + K_1 + m_2 c^2 + K_2 + ... + U}{c^2}$$

$$M = (m_1 + m_2 + ...) + \left(\frac{K_1 + K_2 + ... + U}{c^2} \right)$$

The mass M of the system may be greater or less than the sum of the individual particle rest masses, because the final term in this equation may be positive or negative.

In most everyday situations the kinetic energies of the individual particles are very small compared to their rest energies (that is, the speeds of the particles are small compared to the speed of light), and the interaction energy U (the sum of all the pair-wise interactions) is also small compared to the rest energies. In that case, the mass of the multiparticle system is very nearly equal to the sum of the individual rest masses. However, this is really only

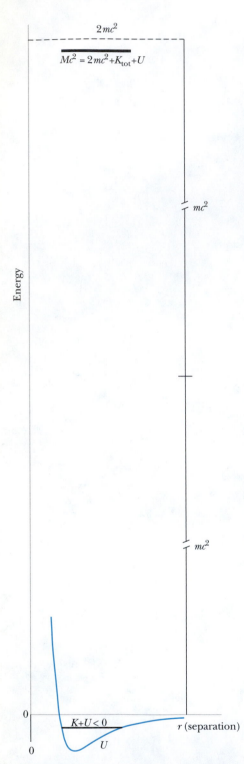

Energy

$2mc^2$

$Mc^2 = 2mc^2 + K_{tot} + U$

mc^2

mc^2

0

$K + U < 0$

U

r (separation)

0

Figure 5.53 A plot of rest energy, kinetic energy, potential energy, and total energy (Mc^2) for O_2. Note that the total energy is less than the sum of the rest energies of the two individual atoms.

an approximation, and it is often not a good approximation in nuclear and particle physics.

The total energy of a multiparticle system

Consider an oxygen molecule (O_2) at rest. Instead of plotting only the kinetic and potential energy for this two-atom system, let's include the rest energy of the atoms, too, as we attempt to show in Figure 5.53. We can't draw the full vertical scale, because the rest energies are enormous compared to the kinetic and potential energies. In order to show the rest energy to the same scale as K and U, we would need a page stretching from the Earth to the Moon!

The total energy of the system (indicated by the thick horizontal line near the top of the graph) is the sum of the rest energies of the two atoms, the potential energy U (which is negative), and K_{tot}, the total kinetic energy of the two atoms. For the bound state drawn on the graph you can see that the rest energy of the two-atom O_2 system is very slightly less than the sum of the individual atomic rest energies, because $K + U < 0$. When the diatomic molecule O_2 forms, about 5 eV of energy are given off (for example, by emitting photons, which carry energy). Correspondingly, it takes an input of about 5 eV of energy to break O_2 apart, if the molecule's energy is near the bottom of the well. This is called the "binding energy."

Two ways of thinking about the energy of the system

Given these aspects of particle energy, there are two rather different ways to think about the energy of a multiparticle system:

- The energy of a multiparticle system consists of the individual particle energies of the particles that make up the system, plus their pair-wise interaction energies.
- A multiparticle system itself has energy, rather like a particle, and if the system is at rest (its center of mass is not moving; its net momentum is zero) its energy E is simply Mc^2, where M is the mass of the system.

The mass of the system

The mass M may be greater or smaller than the sum of the rest masses of the individual particles, depending on whether the total $K+U$ of the particles is positive or negative. The graph in Figure 5.53 indicates that in this case the quantity Mc^2 (where M is the mass of the whole molecule) is less than $2mc^2$. From the graph we can see that:

$$Mc^2 = 2mc^2 + K + U < 2mc^2$$

Mass of a hot object

This way of thinking about mass leads us to conclude that a hot block has a greater mass than a cold block. As we'll discuss in more detail in the next chapter, a hot metal block has more thermal energy than a cold metal block, with more kinetic energy of the atoms and more potential energy in the interactions with neighboring atoms (represented as springs in our simple model of a solid). Although the effect is much too small to measure directly, the equivalence of mass and energy embodied in the special theory of relativity, $M = E/c^2$, leads to the conclusion that a hot block is actually more massive than a cold block! The same impulse $\vec{F}\Delta t$ applied to the two blocks should change the momentum of the cold block very slightly more than the momentum of the hot block.

5.X.44 Calculate the approximate ratio of the binding energy of O_2 (about 5 eV) to the rest energy of O_2. Most oxygen nuclei contain 8 protons and 8 neutrons, and the rest energy of a proton or neutron is about 940 MeV. Do you think you could use a laboratory scale to detect the difference in mass between a mole of molecular oxygen (O_2) and two moles of atomic oxygen?

Application: Nuclear binding

In chemical reactions it is nearly impossible to measure the change of rest mass directly, but in nuclear interactions the changes are easily measurable. Consider the nuclear binding energy of an iron nucleus, which contains 26 protons and 30 neutrons, a total of 56 nucleons. When the protons and neutrons bind to each other through the "strong" (nuclear) force, the resulting iron nucleus has a mass that is about 1% smaller than the total mass of its individual constituents when separated. There is a sizable negative potential energy U associated with the strong force, which makes the multiparticle system's rest energy measurably smaller than the sum of the individual rest energies.

The average binding energy per nucleon is largest in iron. Lighter nuclei (hydrogen through manganese) are less tightly bound and can in principle fuse together to form the more tightly bound iron, with the release of energy. In fact, fusion reactions in stars build up from hydrogen to heavier and heavier nuclei, stopping at iron (heavier elements are made in supernova explosions).

Heavier nuclei (cobalt through uranium and beyond) are also less tightly bound than iron and can in principle fission to form the more tightly bound iron. However, only a few nuclei such as uranium and plutonium do actually spontaneously fission.

Figure 5.54 shows an energy diagram (not to scale) for one of the protons in an iron nucleus, subjected to the forces of all the other protons and neutrons. There is a deep potential well corresponding to the strong or nuclear force, which is essentially the same for protons and neutrons. This potential well has a very steep side corresponding to the very strong nuclear force (force is negative gradient of potential energy). The well is so steep it can be described approximately as a "square well." This deep well extends only out to the radius R of the nucleus, because the nuclear force has a very short range. A proton (but not a neutron) in addition experiences the repulsive electric force of the other protons (the electric U curve outside the nucleus). The total rest mass of the nucleus is about 1% less than the sum of the rest masses of the individual nucleons (protons and neutrons).

Example: Nuclear binding energy

The rest energy of an iron nucleus (26 protons and 30 neutrons) is 52107 MeV. Show that the average binding energy per nucleon is about 1% of the rest energy of a nucleon, which is easily measurable. The rest energy of a proton is 938.3 MeV; of a neutron, 939.6 MeV.

As shown in Figure 5.54, the binding energy is essentially the difference between the rest energy of the multiparticle system (the nucleus) and the sum of the rest energies of its constituent particles.

$$\text{binding energy} = ((26)(938.3)\ \text{MeV} + (30)(939.6)\ \text{MeV}) - 52107\ \text{MeV}$$
$$= 494.8\ \text{MeV}$$

$$\text{energy per nucleon} = \frac{494.8\ \text{MeV}}{26 + 30} = 8.8\ \text{MeV}$$

This is about 1% of the rest energy of one nucleon.

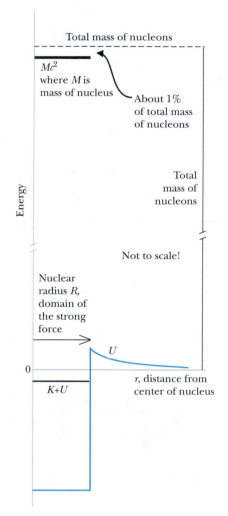

Figure 5.54 An energy diagram for a proton in an iron nucleus, not to scale. The mass of the nucleus is about 1% less than the sum of the masses of the individual nucleons (protons and neutrons).

5.X.45 The deuteron, the nucleus of the deuterium atom ("heavy" hydrogen), consists of a proton and a neutron. It is observed experimentally that a high-energy photon ("gamma ray") with a minimum energy of 2.2 MeV can break up the deuteron into a free proton and a free neutron; this process is called "photo-dissociation." About what fraction of the deuteron rest energy corresponds to its binding energy? The result shows that the deuteron is very lightly bound compared to the iron nucleus.

Application: Nuclear fission

The nuclear fission of uranium leads to an easily measured change in rest masses. The average binding energy per nucleon in uranium is about 7.8 MeV, so the uranium nucleus is less tightly bound than an iron nucleus, whose binding is 8.8 MeV per nucleon. This suggests the possibility of obtaining energy from the fission of a uranium nucleus, corresponding to roughly 0.1% (0.7 MeV per nucleon) of its rest energy (about 940 MeV per nucleon).

In other words, the "mass deficiency" of a medium-mass nucleus is about 1%, with a variation of about a tenth of that. Even though there is only a small difference in average binding energy between iron and uranium, the energy involved is vast compared to the energy of a chemical reaction, which only involves the outer electrons of an atom, not the nucleus.

A uranium-235 nucleus contains 92 protons and 143 neutrons. The 92 protons exert strong electric repulsions on each other, but the large number of neutrons spreads the protons throughout a larger volume and reduces the electric repulsions. The neutrons and the protons also attract neighboring protons with strong nuclear forces. (Low-mass nuclei don't have an overabundance of neutrons; for example, ordinary carbon has 6 protons and 6 neutrons, and iron has only 30 neutrons with its 26 protons.)

Uranium-235 has an unusual property that few other nuclei have. If you hit a U-235 nucleus with a slowly moving neutron, the nucleus may deform into a dumbbell-like form and then split apart because the electric repulsion of the protons can now overcome the short-range attraction of the nucleons for each other (Figure 5.55). (In addition to the two main fission fragments there are typically one or more free neutrons in the final state.)

The fission fragments quickly acquire very high speed, and they leave the original atom's 92 electrons behind. These highly charged nuclei storm through the block of uranium metal, tearing up the material and violently heating the metal. In a nuclear power reactor the reaction is controlled in such a way that the metal merely gets hot enough to boil water. With many uranium-235 nuclei fissioning, the metal can boil water in a nuclear power plant to produce steam and power a turbine to drive an electric generator.

The rest masses of the two fission-fragment nuclei total less than the rest mass of the uranium-235 nucleus. The rest mass difference Δm is only about one-tenth of one percent of the original rest mass, but this rest mass change is easily measurable, and c^2 times this rest mass change is indeed found to be equal to the observed kinetic energy of the fission fragments. This 0.1% change is about ten million times as big an effect as the rest mass changes that occur in chemical reactions, which only involve rearrangements of the outer electrons in the atoms, and do not affect the nucleus.

Nuclear waste

The problem of radioactive waste from nuclear power plants comes from the fact that the fission products are nuclei with a large overabundance of neutrons, and they emit energetic radiations. For example, uranium can fission into two palladium nuclei, each with 46 protons and 71 or 72 neutrons, whereas ordinary stable palladium has only 60 neutrons (see the inside

Figure 5.55 A U-235 nucleus in the process of fissioning.

front cover of this book). A neutron in such a neutron-rich nucleus can change into a proton (Figure 5.56) with the emission of an energetic electron and an antineutrino (the antineutrinos almost never interact with matter, so they aren't dangerous). This process is called "beta decay" (energetic electrons were initially called "beta rays"). After a beta decay the new nucleus, in which a neutron has changed into a proton, may be in an "excited state" and emit a gamma ray (a high-energy photon, see Figure 5.57). Heavy elements such as plutonium also emit alpha particles (the nuclei of helium atoms) and there may be heavy elements mixed in with the rest of the nuclear waste.

When energetic electrons or gamma rays or alpha particles from radioactive nuclei go through matter, energy is absorbed and the matter gets hot. If the radiation goes through living tissue there can be damage to cells, and to the genetic information in the DNA in cells. The range of the radiation is relatively short and it can be stopped by a layer of concrete or other shielding. The big concern is that radioactive wastes might leak out of storage facilities, spread through the ground water, and get into the water supply. Drinking contaminated water would bring radioactive nuclei into the body, close to cells, where damage could occur.

A great deal of engineering and scientific study is under way to try to find a secure storage scheme that would prevent leakage for hundreds or even thousands of years, because some of the radioactive nuclei have very long "half-lives" (the time it takes for half of the nuclei to decay).

$$n \rightarrow p^+ + e^- + \bar{\nu}$$

Figure 5.56 A neutron in a neutron-rich nucleus can change into a proton with emission of an electron and an antineutrino.

$$N^* \rightarrow N + \gamma$$

Figure 5.57 A nucleus in an excited state (called N^*) can drop to a lower energy state and emit a gamma ray (a high-energy photon).

5.15 REFLECTION: WHY ENERGY?

It is appropriate at this point to reflect on why we introduced the topic of energy. Energy methods often make it possible to analyze some aspect of a phenomenon with much less effort than is required by direct application of the Momentum Principle. A striking example is the ability to calculate escape speed, or the impact speed of the Ranger spacecraft on the Moon, without carrying out a lengthy numerical integration.

Limits on the possible
Another powerful capability of energy analyses is that they can easily predict whether some process can occur or not. For example, if there is insufficient energy, the rocket can't escape from the planet, or the atom cannot be ionized, or the biochemical reaction cannot proceed, or a proton can't decay into a neutron.

Less detail
However, energy analyses give less detailed information than is given by a step-by-step numerical integration of the Momentum Principle. In particular, energy analyses tell us nothing about how fast a process will proceed; time does not appear in the statement of conservation of energy. And since the Energy Principle is a scalar principle, it doesn't tell us about change in the direction of motion.

Here is an example that illustrates the strengths and weaknesses of energy analyses. Consider snowboarding down the three hills shown in Figure 5.58, starting from rest. If we can neglect friction, the final speed at the bottom should be the same in all three cases, because the final kinetic energy should be equal to *mgh*, the amount by which the gravitational energy decreases.

? But how will the amount of *time* compare for the ride down the hill?

On the first hill you get going very fast very quickly, and the trip is short. On the second hill you move very slowly for a long time before dropping to the

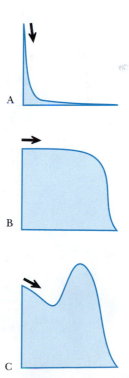

Figure 5.58 The final speed of the snowboarder will be the same at the bottom of each hill.

bottom, and the total trip takes a long time. On the third hill you never get to the bottom, despite the fact that the change in potential energy would be negative, because you would have to pass through a region that is energetically forbidden. So energy methods are powerful, but at the cost of losing some information, including the amount of time required for a process.

5.16 IDENTIFYING INITIAL AND FINAL STATES

Some complex processes involve several identifiable states, not just two. Any two states can be chosen as initial and final states; the rationale for making the choice depends on what unknown quantity you are trying to determine.

The example problem below involves such a complex process. If an energetic alpha particle (a helium nucleus: two protons and two neutrons) collides with a carbon nucleus, a nuclear reaction can occur, forming an oxygen nucleus whose rest energy is higher than an ordinary oxygen nucleus (an "excited state"; this concept is discussed in Chapter 7). After a short time, the excited oxygen nucleus can emit its excess energy as a photon (a particle of light), reducing its rest energy to that of an ordinary oxygen nucleus.

In this process there are four identifiable states:

State 1. Alpha particle and carbon nucleus moving toward each other (Figure 5.59)

State 2. Alpha particle and carbon nucleus just touch (Figure 5.60)

State 3. Excited oxygen nucleus at rest (extra rest energy) (Figure 5.61)

State 4. Oxygen nucleus and photon, heading in opposite directions (Figure 5.62)

Selecting different pairs of initial and final states allows us to answer different questions about the process.

Figure 5.59 State 1. Alpha and C far apart, moving.

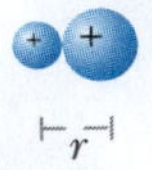

Figure 5.60 State 2. Alpha and C touch.

Figure 5.61 State 3. Excited O* nucleus.

Figure 5.62 State 4. O nucleus and photon.

Example: Reaction between an alpha particle and a carbon nucleus

An alpha particle (the nucleus of a helium atom, containing two protons and two neutrons) moving with magnitude of momentum p_1 is shot toward a carbon-12 nucleus (containing six protons and six neutrons) that is moving with the same magnitude of momentum p_1 toward the alpha particle. The speeds are nonrelativistic.

There is a head-on collision of the alpha particle and the carbon-12 nucleus, and an excited state of an oxygen-16 nucleus is formed at rest. A very short time later the oxygen nucleus drops to its ground state, emitting a photon (called a "gamma ray") whose energy is 10.352 MeV (1 MeV = 10^6 eV). The oxygen nucleus recoils with momentum equal and opposite to the photon's momentum, but negligible kinetic energy.

(a) What is the minimum (nonrelativistic) magnitude of momentum p_1 that the alpha particle must have in order for this sequence of phenomena to occur? Masses of the ground (not excited) states of selected nuclei, where 1 u = 1.6605×10^{-27} kg:

He-4 (2p+2n) = 4.00040868 u
C-12 (6p+6n) = 11.99670852 u
O-16 (8p+8n) = 15.99052636 u

(b) What is the kinetic energy in MeV of an alpha particle with the magnitude of momentum p_1 found in part (a)?

(c) Because the alpha particle and carbon nucleus both have positive electric charges, they repel each other. It could be the case that these particles need to start with extra kinetic energy in order to get close enough to touch and react. Find the minimum initial momentum p_{min} each particle must have to get close enough to touch. Compare this to your answer to part (a) to determine if this reaction can actually occur.

A proton or neutron has a radius of roughly 1×10^{-15} m , and a nucleus is a tightly packed collection of nucleons. Experiments show that the radius of a nucleus containing N nucleons is approximately $(1.3 \times 10^{-15}$ m$) \times N^{1/3}$.

(a) Find p_1:

System: All particles in existence (alpha + C, or excited O, or O + photon).
Surroundings: No external objects.

Initial state: Very far apart, particles moving toward each other (State 1, Figure 5.59).
Final state: O nucleus and photon moving apart (State 4, Figure 5.62).

Energy Principle

$$E_f = E_i + \cancel{W}$$

$$(m_O c^2 + K_O) + K_\gamma + \cancel{U_f} = (m_\alpha c^2 + K_\alpha) + (m_C c^2 + K_C) + \cancel{U_i}$$

$$m_O c^2 + K_\gamma = m_\alpha c^2 + m_C c^2 + \frac{p_1^2}{2 m_\alpha} + \frac{p_1^2}{2 m_C}$$

$$p_1 = \sqrt{\frac{(m_O c^2 + K_\gamma - m_\alpha c^2 - m_C c^2)}{\frac{1}{2 m_\alpha} + \frac{1}{2 m_C}}}$$

In the numerator we need many significant figures to capture small rest mass changes:

$$(m_O - m_\alpha - m_C) = (15.99052636 - 4.00040868 - 11.99670852) \text{ u}$$

$$= -0.00659084 \text{ u}$$

$$(-0.00659084 \text{ u})(1.6605 \times 10^{-27} \text{ kg/u})(3 \times 10^8 \text{ m/s})^2 = -9.8497 \times 10^{-13} \text{ J}$$

$$K_\gamma = (10.352 \times 10^6 \text{ eV})(1.60 \times 10^{-19} \text{ eV/J}) = 16.5632 \times 10^{-13} \text{ J}$$

$$p_1 = \sqrt{\frac{(16.5632 \times 10^{-13} - 9.8497 \times 10^{-13}) \text{ J}}{\frac{1}{2}\left(\frac{1}{4.000} + \frac{1}{11.997}\right)\left(\frac{1}{1.6605 \times 10^{-27} \text{ kg}}\right)}}$$

$$= 8.18 \times 10^{-20} \text{ kg} \cdot \text{m/s}$$

(b) $K_\alpha = \dfrac{p_1^2}{2 m} = \dfrac{(8.18 \times 10^{-20} \text{ kg·m/s})^2}{2(4 \times 1.66 \times 10^{-27} \text{ kg})}\left(\dfrac{10^{-6} \text{ MeV}}{1.6 \times 10^{-19} \text{ J}}\right) = 3.15 \text{ MeV}$

Check. The units for momentum are correct.

(c) Find p_{min} for alpha and carbon to touch.

System: alpha particle and carbon nucleus.
Surroundings: No external objects.

Initial state: Very far apart, particles moving toward each other (State 1, Figure 5.59).
Final state: Particles at rest, touching each other (Figure 5.60). No change of identity yet.

Energy Principle:

$U_f = 0$ because there are no electrical or gravitational interactions (photon has no charge).

$U_i = 0$ because particles are far apart.

Photon (γ) has no rest energy, only kinetic energy.

No change of rest energy yet.

Final state: particles at rest, so $K_f = 0$.

Initial state: $U_i = 0$ because particles far apart.

Approximate K is okay because $v \ll c$.

$$E_f = E_i + W$$

$$(\cancel{m_a c^2} + \cancel{K_{a,f}}) + (\cancel{m_C c^2} + \cancel{K_{C,f}}) + U_f = (\cancel{m_a c^2} + K_{a,i}) + (\cancel{m_C c^2} + K_{C,i}) + \cancel{V_i}$$

$$U_f = K_{a,i} + K_{C,i}$$

$$\frac{1}{4\pi\varepsilon_0}\frac{(2e)(6e)}{r} = \frac{p_{min}^2}{2m_a} + \frac{p_{min}^2}{2m_C}$$

2*e* is charge of alpha particle. 6*e* is charge of carbon nucleus.

$$p_{min} = \sqrt{\frac{\dfrac{1}{4\pi\varepsilon_0}\dfrac{12\,e^2}{r}}{\dfrac{1}{2}\left(\dfrac{1}{m_a} + \dfrac{1}{m_C}\right)}}$$

final separation $r = (1.3\times10^{-15}\text{ m})(4^{1/3} + 12^{1/3}) = 5.04\times10^{-15}$ m

$$p_{min} = \sqrt{\frac{\left(9\times10^9\,\dfrac{\text{N·m}^2}{\text{C}^2}\right)\dfrac{12(1.6\times10^{-19}\text{ C})^2}{(5.04\times10^{-15}\text{ m})}}{\dfrac{1}{2}\left(\dfrac{1}{4} + \dfrac{1}{12}\right)\left(\dfrac{1}{1.66\times10^{-27}\text{ kg}}\right)}} = 7.39\times10^{-20}\text{ kg·m/s}$$

$$K_\alpha = \frac{p_{min}^2}{2m} = \frac{(7.39\times10^{-20}\text{ kg·m/s})^2}{2(4\times1.66\times10^{-27}\text{ kg})}\left(\frac{10^{-6}\text{ MeV}}{1.6\times10^{-19}\text{ J}}\right) = 2.57\text{ MeV}$$

Check. The units for momentum and energy are okay.

Since the momentum required to touch is less than the momentum required to initiate the reaction (answer to part (a)), the reaction will in fact occur.

Further discussion

The potential energy graph for this process is shown in Figure 5.63. The electric potential energy rises (due to a repulsive interaction) as the particles get closer together, up to the point at which they touch. After this, the attractive strong (nuclear) force dominates the interaction, and the electric repulsion between the particles is insignificant.

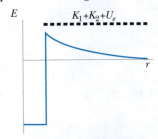

Figure 5.63 The initial kinetic energies are a bit greater than what is needed to overcome the electric potential energy barrier.

This kind of potential energy curve is sometimes described as having a "Coulomb barrier". This means that $K+U$ total must be larger than the maximum value of $U_{electric}$.

We found that the incoming momentum p_{min} required to create the excited oxygen nucleus is greater than the momentum p_1 required just to overcome the electric potential energy barrier and bring the alpha particle and carbon nucleus in contact. It is necessary to bring nuclei into contact for a reaction to proceed, since the nuclear force has a very short range. However, in this reaction just making contact is insufficient. A bit more energy is needed in order to produce the larger-mass oxygen nucleus.

We never did end up using State 3 in our analysis. However, if we needed to find the mass of the excited O* nucleus, we would use this state as either the final or initial state.

5.17 *A PUZZLE

The principle of conservation of energy states that when a system's energy increases, the energy of the surroundings decreases by the same amount. For example, take a block as the system of interest. If you push the block away from you and raise its kinetic energy, your store of chemical energy decreases by the same amount. The flow of energy from you (the "surroundings") to the block takes the form of mechanical energy transfer we call "work." We can demonstrate that there is a puzzle concerning the energy in the surroundings of a star whose kinetic energy increases due to an interaction with another star.

Consider two stars of equal mass initially at rest, far apart. The two stars do work on each other:

$$\Delta E_1 = W_{\text{on }1} \quad \text{star 1 as system}$$

$$\Delta E_2 = W_{\text{on }2} \quad \text{star 2 as system}$$

If we add these two energy equations for the two stars, we have this:

$$\Delta E_1 + \Delta E_2 = W_{\text{on }1} + W_{\text{on }2} \quad \text{sum of two energy equations}$$

Moving $(W_{\text{on }1} + W_{\text{on }2})$ to the left side of the equation and renaming the negative of this quantity ΔU, we have this familiar result:

$$\Delta E_1 + \Delta E_2 + \Delta U = 0 \quad \text{energy equation for two stars}$$

Our interpretation of this equation is that the gravitational potential energy U of the two-star system (calculated from the work on the individual particles) decreases as the particle energies increase. This point of view is valid, yields physically correct results, and is the view that we will take in this introductory textbook.

However, it is not difficult to show that there is a puzzle concerning energy flow between one of the stars and its surroundings. Consider star 1 as the system. Since it is a single object, there is no potential energy U—no pairwise interaction energy. The energy of star 1 increases due to work done on it: $\Delta E_1 = W_{\text{on }1}$. There are very strong reasons to believe in energy conservation, so we expect that when star 1 gains an amount of energy ΔE_1 the surroundings of star 1 should lose that much energy.

? Part of the surroundings is star 2. Does star 2 lose energy?

No! Star 2 doesn't lose energy, it *gains* energy, and of the same amount $\Delta E_2 = \Delta E_1$, due to the symmetry of the situation, with the two stars having equal mass.

What else is there in the surroundings that could experience a change in energy? The only other matter is star 2, but its energy change is inconsistent with the principle of energy conservation. Evidently our model of the world is incomplete. To be able to analyze this situation fully, we need to introduce the abstract idea of a "field." Fields, which are the subject of Volume II of this textbook, are associated with objects having mass (gravitational fields) or charge (electric and magnetic fields), and fields extend throughout all space. It is energy stored in the gravitational field that accounts for the energy in the surroundings of star 1. Fields can also have momentum, as we will see in Volume II.

Despite the fact that we haven't yet defined what a "field" is, we can determine indirectly how much the field energy must change. Since the energy in the surroundings of star 1 changes by an amount $-\Delta E_1$, and the energy of star 2 increases by an amount $+\Delta E_1$, it must be that the energy stored in the gravitational field changes by an amount $-2\Delta E_1$. It is an indication of the power of the principle of conservation of energy that we can calculate the energy change of an entity that we know nothing about! This energy change is precisely the amount of change in the potential energy of the two-star system: $\Delta U = -(\Delta E_1 + \Delta E_2) = -2\Delta E_1$.

Because the potential energy U correctly accounts for the energy changes, we don't have to consider field energy in our calculations. Our analyses in terms of U that focus on particles and their interactions, not their surroundings, are completely correct. However, the difficulty in accounting for the energy in the surroundings of a star suggests that eventually we will need to include fields in our models of the world.

5.18 *GRADIENT OF POTENTIAL ENERGY

Because the (negative) gradient of U is equal to the force, the force is the (negative) slope of a graph of potential energy. This means that you can mentally read the force right off a potential energy graph. Where the curve is steep, the force is large (large gradient), and where the curve is shallow, the force is small. Where the

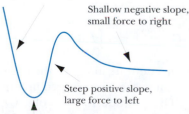

Steep negative slope, large force to right

Shallow negative slope, small force to right

Steep positive slope, large force to left

Zero slope, equilibrium position

slope of the curve is horizontal, the force is zero, and this represents a possible equilibrium position for the system. The component of the force is given by the negative gradient, so a negative slope corresponds to a force to the right, while a positive slope corresponds to a force to the left.

Actually, in determining the x component of the force, we differentiate with respect to x while holding y and z constant. This kind of derivative is called a "partial derivative" and has a special symbol to emphasize that everything else is held constant:

$$F_x = -\frac{\partial U}{\partial x}$$

To get all three components of the force we take partial derivatives of the potential energy with respect to x, y, and z:

FORCE IS NEGATIVE GRADIENT OF POTENTIAL ENERGY

$$F_x = -\frac{\partial U}{\partial x}, \; F_y = -\frac{\partial U}{\partial y}, \; F_z = -\frac{\partial U}{\partial z}$$

$$\vec{F} = \left\langle -\frac{\partial U}{\partial x}, -\frac{\partial U}{\partial y}, -\frac{\partial U}{\partial z} \right\rangle$$

(Another standard mathematical notation for this relation is $\vec{F} = -\vec{\nabla} U$.)

5.X.46 With y pointing upward from the center of the Earth, find the y component of the gravitational force from $-\partial U/\partial y$, where $U = -GMm/y$. Find the x component of the force from $-\partial U/\partial x$.

5.19 *INTEGRALS AND ANTIDERIVATIVES

In determining the formula for gravitational potential energy, we started from these complementary formulas:

$$\Delta U_g = -F_r \Delta r \text{ and } F_r = -\frac{dU_g}{dr}$$

We calculated the change in potential energy from the negative of the internal work:

$$U_{s,f} - U_{s,i} = \Delta\left(-G\frac{m_1 m_2}{r}\right) = -\Sigma F_r \Delta r$$

But if we had let Δr approach zero, the finite sum would have turned into an integral:

$$\Delta\left(-G\frac{m_1 m_2}{r}\right) = -\int_i^f F_r dr = \int_i^f G\frac{m_1 m_2}{r^2} dr$$

You can see that a definite integral (the infinitesimal version of a finite sum) can be evaluated in terms of the antiderivative. The antiderivative of $Gm_1 m_2/r^2$ is $-Gm_1 m_2/r$, and the change in the value of the antiderivative turns out to be equal to the definite integral (the infinite sum of infinitesimal elements).

The reason why this works this way can be traced back to the fact that the force is the derivative of the energy, so the energy must be the antiderivative of the force, yet the energy is also the sum of lots of $F_r \Delta r$'s. Hence the sum is given by the change in the antiderivative.

When the quantity to be integrated (here, the force) is a simple function, it may be easy to find the antiderivative. However, in many practical cases there is no known simple function that is the antiderivative, in which case an integral must be evaluated by numerical integration, as a finite sum.

5.20 *APPROXIMATION FOR KINETIC ENERGY

An approximation for the kinetic energy of a particle at low speeds can be obtained through use of the mathematics of the "binomial expansion":

$$(1+\varepsilon)^n = 1 + n\varepsilon + \frac{n(n-1)}{2}\varepsilon^2 + \frac{n(n-1)(n-2)}{3 \cdot 2}\varepsilon^3 + \dots$$

This formula comes from working through what happens when you multiply the quantity $(1 + \varepsilon)$ times itself n times. The first term is 1 multiplied by itself n times (1^n); the next term is made up of all those products, n of them, in which ε appears only once $[n\varepsilon]$; etc.

Now for the important point that makes the binomial expansion useful in obtaining approximate results. If ε is a number that is very small compared to 1, each term in the expansion is much smaller than the preceding term:

$$n\varepsilon \text{ is much smaller than } 1, \frac{n(n-1)}{2}\varepsilon^2 \text{ is much smaller than } n\varepsilon, \text{ etc.}$$

Therefore we can get a good approximation if we keep just the first few terms and ignore the rest.

Although we have developed the binomial expansion for the case where n is an integer, it can be shown that the formula is valid even for non-integer values of n. In particular,

$$\frac{1}{\sqrt{1-\varepsilon}} = (1-\varepsilon)^{-1/2}$$

$$= \left(1 + (-\tfrac{1}{2})(-\varepsilon) + \frac{(-\tfrac{1}{2})(-\tfrac{3}{2})}{2}\varepsilon^2 + \frac{(-\tfrac{1}{2})(-\tfrac{3}{2})(-\tfrac{5}{2})}{6}\varepsilon^3 + \dots\right)$$

If v/c is small compared to 1, then $(v/c)^2$ is even smaller compared to 1. For example, $1/10$ is small compared to one, but $(1/10)^2 = 1/100$ is very small indeed. Therefore we can write this for low speeds:

$$\frac{mc^2}{\sqrt{1-v^2/c^2}} = mc^2\left[1 + \frac{1}{2}\frac{v^2}{c^2} - \frac{3}{8}\left(\frac{v^2}{c^2}\right)^2 - \frac{5}{16}\left(\frac{v^2}{c^2}\right)^3 + \dots\right]$$

$$\frac{mc^2}{\sqrt{1-v^2/c^2}} \approx mc^2 + \frac{1}{2}mv^2$$

The last term can be rewritten in terms of momentum by inserting $v \approx p/m$ into $K \approx \frac{1}{2}mv^2$:

$$K \approx \frac{1}{2}mv^2 \approx \frac{1}{2}m\left(\frac{p}{m}\right)^2 = \frac{p^2}{2m}$$

5.21 *FINDING THE FORMULA FOR PARTICLE ENERGY

If we choose our axes so that x is in the direction of the motion at this instant we have

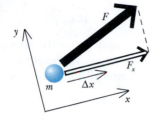

$$\Delta E = F_x \Delta x = \left(\frac{\Delta p_x}{\Delta t}\right)\Delta x$$

$$\text{So } \frac{\Delta E}{\Delta x} = \frac{\Delta p_x}{\Delta t}$$

for the change in energy of a point particle acted on by a net force. In the limit of small displacement and short time interval, the energy E should satisfy the following relationship between two derivatives:

$$\frac{dE}{dx} = \frac{dp_x}{dt}$$

We will show that if we choose the following function for E, the relationship between the two derivatives will be satisfied:

$$E = \frac{mc^2}{\sqrt{1 - v^2/c^2}}$$

The proof proceeds by calculating dE/dx, calculating dp_x/dt, and showing that the two derivatives are equal to each other. First, here is dE/dx:

$$\frac{dE}{dx} = \frac{d}{dx}\left(\frac{mc^2}{\sqrt{1 - v^2/c^2}}\right)$$

$$= \frac{mc^2}{(1 - v^2/c^2)^{3/2}}\left(-\frac{1}{2}\right)\left(-\frac{2v}{c^2}\right)\frac{dv}{dx}$$

$$\frac{dE}{dx} = \frac{mv}{(1 - v^2/c^2)^{3/2}}\frac{dv}{dx}$$

Next we calculate dp_x/dt, keeping in mind that $v_x = v$, because we chose our axes so that x is in the direction of the motion at this instant:

$$\frac{dp_x}{dt} = \frac{d}{dt}\left(\frac{mv}{\sqrt{1 - v^2/c^2}}\right)$$

$$= \left(\frac{m}{\sqrt{1 - v^2/c^2}}\frac{dv}{dt} + \frac{mv}{(1 - v^2/c^2)^{3/2}}\left(-\frac{1}{2}\right)\left(-\frac{2v}{c^2}\right)\frac{dv}{dt}\right)$$

$$\frac{dp_x}{dt} = \frac{m}{(1 - v^2/c^2)^{3/2}}\left[\left(1 - \frac{v^2}{c^2}\right) + \frac{v^2}{c^2}\right]\frac{dv}{dt}$$

$$= \frac{m}{(1 - v^2/c^2)^{3/2}}\frac{dv}{dt}$$

This is almost in the same form as the derivative dE/dx. Now note this:

$$\frac{dv}{dt} = \left(\frac{dv}{dt}\right)\left(\frac{dx}{dx}\right)$$

$$= \left(\frac{dv}{dx}\right)\left(\frac{dx}{dt}\right) = \frac{dv}{dx}v \text{ , because } v = \frac{dx}{dt}$$

Substituting this result for dv/dt into the derivative dp_x/dt, we find that this derivative is indeed equal to the derivative dE/dx if E is defined as

$$E = \frac{mc^2}{\sqrt{1 - v^2/c^2}}$$

Hence the relativistic formula for energy does follow from the relativistic formula for momentum.

It might seem that we could define the particle energy with an additive constant, which would drop out when we took the derivative dE/dx. However, when one considers a variety of particle reactions, it is found that to account for energy in a consistent way requires that the additive constant be zero.

This approach to solving a "differential equation" (an equation containing derivatives), by proposing a solution and showing that it satisfies the equation, is a legitimate technique for solving differential equations, and in fact is used quite frequently.

5.22 SUMMARY

The unit of energy is the joule, abbreviated "J".

Energy of a single particle

The energy of a particle with rest mass m is:

$$E_{\text{particle}} \equiv \gamma mc^2 \text{ where } \gamma = \frac{1}{\sqrt{1 - v^2/c^2}}$$

Particle energy has two parts: rest energy and kinetic energy:

$$E_{\text{particle}} = mc^2 + K$$

The "rest energy" is the energy of a particle at rest.

$$\text{rest energy} = mc^2$$

The kinetic energy K of a particle is the energy a moving particle has in addition to its rest energy.

$$K = \gamma mc^2 - mc^2$$

For a particle whose speed is low, compared to the speed of light, kinetic energy is approximately:

$$K \approx \frac{1}{2}mv^2 = \frac{p^2}{2m} \text{ (if } v \ll c)$$

For a particle with any speed:

$$E^2 - (pc)^2 = (mc^2)^2$$

for a massless particle this reduces to

$$E = pc$$

The Energy Principle

$$\Delta E_{system} = W_{surroundings}$$

The change in energy of a system is equal to the work done by the surroundings.

Work

Work W_F done by a single constant force $\vec{F}$:

$$W_F = \vec{F} \cdot \Delta \vec{r} = F_x \Delta r_x + F_y \Delta r_y + F_z \Delta r_z = F \Delta r \cos\theta$$

Work done by a varying force:

$$W_F = \sum \vec{F} \cdot \Delta \vec{r} \text{ or } W = \int_i^f \vec{F} \cdot d\vec{r}$$

Work can be positive, negative, or zero. The sign of work depends on the relative directions of the force and the displacement of the point of application of the force.

Conservation of energy

$$\Delta E_{system} + \Delta E_{surroundings} = 0$$

Potential energy

Multiparticle system energy has three parts: Rest energy, kinetic energy, and potential energy.

$$E_{sys} \equiv (m_1 c^2 + m_2 c^2 + \dots) + (K_1 + K_2 + \dots) + (U_{12} + \dots)$$

Potential energy U is the energy of interaction of a pair of objects.

$$\Delta U = -W_{internal}$$

Force is negative gradient of potential energy:

$$F_x = -\frac{dU}{dx}$$

Potential energy U depends on the separation r between two objects.

Potential energy U goes to zero as objects get very far apart:

$$U \to 0 \text{ as } r \to \infty$$

For an attractive interaction, U becomes negative as separation decreases.

For a repulsive interaction U becomes positive as separation decreases.

Gravitational potential energy

$$U = -G\frac{m_1 m_2}{r}$$

$$G = 6.7 \times 10^{-11} \text{ N} \cdot \text{m}^2/\text{kg}^2$$

m_1 and m_2 are the masses of the two interacting objects, and r is the center-to-center distance between them.

$$\Delta U \approx \Delta(mgy) \text{ near Earth's surface}$$

Electric potential energy

$$U_{el} = \frac{1}{4\pi\varepsilon_0} \frac{q_1 q_2}{r}$$

$$\frac{1}{4\pi\varepsilon_0} = 9 \times 10^9 \text{ N} \cdot \text{m}^2/\text{coulomb}^2$$

q_1 and q_2 are the charges of the two interacting particles, in coulombs, and r is the center-to-center distance between them.

Nuclear binding:

Heavy nuclei can fission, light nuclei can fuse, with energy released in both cases.

Electron volt: $1 \text{ eV} = 1.6 \times 10^{-19} \text{ J}$

5.23 REVIEW QUESTIONS

Work

5.RQ.47 You push a crate 3 m across the floor with a 40 N force whose direction is 30° below the horizontal. How much work do you do?

5.RQ.48 Give brief explanations for your answers to each of the following questions:
(a) You hold a 1 kg book in your hand for 1 minute. How much work do you do on the book?
(b) In a circular pendulum, how much work is done by the string on the mass in one revolution?
(c) For a mass oscillating horizontally on a spring, how much work is done by the spring on the mass in one complete cycle? In a half cycle?

Kinetic energy

5.RQ.49 Approximately what is the kinetic energy of a baseball pitcher's fast ball? Of a basketball passed from one player to another?

5.RQ.50 What is the speed of an electron whose total energy is equal to the total energy of a proton that is at rest? What is the kinetic energy of this electron?

Work versus impulse, K versus $\vec{p}$

5.RQ.51 You pull a block of mass m across a frictionless table with a constant force. You also pull with an equal constant force a block of larger mass M. The blocks are initially at rest. If you pull the blocks through the same distance, which block has the greater kinetic energy, and which block has the greater momentum? If instead you pull the blocks for the same amount of time, which block has the greater kinetic energy, and which block has the greater momentum?

Energy conservation

5.RQ.52 You throw a ball straight up, and it reaches a height of 20 meters above your hand before falling back down. What was the speed of the ball just after it left your hand?

Gravitational energy

5.RQ.53 A 1 kg block rests on the Earth's surface. How much energy is required to move the block very far from the Earth, ending up at rest again?

Graphs of potential energy

5.RQ.54 The gravitational force inside the Earth is approximately $(GMm/R^3)r = gr/R$, where r is the distance from the center and R is the radius of the Earth; see Problem 5.P.77 (page 219) for additional discussion. Draw the potential energy as a function of r, from $r = 0$ to very large r $(\gg R)$. Note that the potential energy curve must be continuous, with no breaks.

Mass and energy

5.RQ.55 In positron-emission tomography (PET) used in medical research and diagnosis, compounds containing unstable nuclei that emit positrons are introduced into the brain, destined for a site of interest in the brain. When a positron is emitted, it goes only a short distance before coming nearly to rest. It forms a bound state with an electron, called "positronium," which is rather similar to a hydrogen atom. The binding energy of positronium is very small compared to the rest energy of an electron. After a short time the positron and electron annihilate. In the annihilation, the positron and the electron disappear, and all of their rest energy goes into two photons (particles of light) which have zero mass; all their energy is kinetic energy. These high energy photons, called "gamma rays," are emitted at nearly 180° to each other. What energy of gamma ray (in MeV, million electron volts) should each of the detectors be made sensitive to? (The mass of an electron or positron is 9×10^{-31} kg.)

5.RQ.56 One often hears the statement, "Nuclear energy production is fundamentally different from chemical energy production (such as burning of coal), because the nuclear case involves a change of mass." Critique this statement. Discuss the similarities and differences of the two kinds of energy production.

5.24 PROBLEMS

5.P.57 Work in maneuvering a boat
Jack and Jill are maneuvering a 3000 kg boat near a dock. Initially the boat's position is $\langle 2, 0, 3 \rangle$ m and its speed is 1.3 m/s. As the boat moves to position $\langle 4, 0, 2 \rangle$ m , Jack exerts a force $\langle -400, 0, 200 \rangle$ N and Jill exerts a force $\langle 150, 0, 300 \rangle$ N .

(a) How much work does Jack do?
(b) How much work does Jill do?
(c) Without doing any calculations, say what is the angle between the (vector) force that Jill exerts and the (vector) velocity of the boat. Explain briefly how you know this.
(d) Assuming that we can neglect the work done by the water on the boat, what is the final speed of the boat?

5.P.58 Slowing down an electron
An electron traveling at a speed $0.99c$ encounters a region where there is a constant electric force directed opposite to its momentum. After traveling 3 m in this region, the electron's speed was observed to decrease to $0.93c$. What was the magnitude of the electric force acting on the electron?

5.P.59 Pulling and pushing
(a) You bring a boat toward the dock by pulling on a rope with a force of 130 newtons through a distance of 7 meters. How much work do you do?
(b) Next you slow the boat down by pushing against it with a force of 40 newtons, opposite to the boat's movement of 5 meters. How much work do you do?
(c) What is the total amount of work that you do?

5.P.60 Work in several parts
You push a box out of a carpeted room and along a hallway with a waxed linoleum floor. While pushing the crate 2 m out of the room you exert a force of 34 N; while pushing it 6 m along the hallway you exert a force of 13 N. To slow it down you exert a force of 40 N through a distance of 2 m, opposite to the motion. How much work do you do in all?

5.P.61 Moving a crate on a space station
A crate with a mass of 100 kg glides through a space station with a speed of 3.5 m/s. An astronaut speeds it up by pushing on it from behind with a force of 180 N, continually pushing with this force through a distance of 6 m. The astronaut moves around to the front of the crate and slows the crate down by pushing backwards with a force of 170 N, backing up through a distance of 4 m. After these two maneuvers, what is the speed of the crate?

5.P.62 Work done by a spring
A mass of 0.12 kg hangs from a vertical spring in the lab room. You grab hold of the mass and throw it vertically downward. The speed of the mass just after leaving your hand is 3.40 m/s.
(a) While the mass moves downward a distance of 0.07 m, how much work was done on the mass by the Earth?
(b) The speed of the mass has decreased to 2.85 m/s. How much work was done on the mass by the spring?

5.P.63 Launching from Mars
The radius of Mars (from the center to just above the atmosphere) is 3400 km (3.4×10^6 m), and its mass is 6×10^{23} kg . An object is launched straight up from just above the atmosphere of Mars.
(a) What initial speed is needed so that when the object is far from Mars its final speed is 2000 m/s?
(b) What initial speed is needed so that when the object is far from Mars its final speed is 0 m/s? (This is called the "escape speed.")

5.P.64 Launch from an asteroid
The escape speed from an asteroid whose radius is about 10 km is only 10 m/s. If you throw a rock away from the asteroid at a speed of 20 m/s, what will be its final speed?

5.P.65 Launch with known escape speed

The escape speed from a very small asteroid is only 24 m/s. If you throw a rock away from the asteroid at a speed of 35 m/s, what will be its final speed?

5.P.66 Work done in Moon shot

If you have already done the problem in Chapter 3 on the Ranger 7 mission to the Moon, you can use the program you wrote for that problem, and go directly to part (b). Otherwise:

(a) Write a program to model the journey of a spacecraft coasting from the Earth to the Moon. Start the spacecraft at a height of 50 km above the Earth's surface with a speed of 1.3×10^4 m/s. This is approximately the speed it would have after all the rockets have fired, and is high enough to be above most of the Earth's atmosphere. Include the spacecraft's interactions with both the Earth and the Moon. Use a *dt* of 5 seconds. Make sure you stop the program when the spacecraft reaches the Moon's surface (not its center!). The data you need may be found on the inside back cover of this book; the mass of the spacecraft was 173 kg. If you are interested in more of the details of this trip, see the problem in Chapter 3 on the Ranger 7 mission to the Moon.

(b) Add a calculation of the work done by the gravitational forces of the Earth and the Moon to your analysis of sending a spacecraft to the Moon. You need to approximate the work by adding up the amount of work done by gravitational forces along each step of the path:

$$W = \sum \vec{F} \bullet \Delta \vec{r} = \vec{F}_1 \bullet \Delta \vec{r}_1 + \vec{F}_2 \bullet \Delta \vec{r}_2 + \vec{F}_3 \bullet \Delta \vec{r}_3 + \dots$$

Compare the numerical value of the work with the change in the kinetic energy (final kinetic energy just before crashing on the Moon, minus initial kinetic energy when released above the Earth's atmosphere). Try shorter time steps to make sure that the errors introduced by a finite time step are not significant. Report your results.

5.P.67 Energy in the Moon shot

Energy conservation is a powerful check on the accuracy of a numerical integration. Modify the program for the Chapter 3 problem on the Ranger 7 mission to the Moon to plot graphs of kinetic energy, of gravitational potential energy, and of the sum of the kinetic energy and the gravitational potential energy, *vs.* position. Does the kinetic plus potential energy remain constant? What if you vary the step size (which varies the accuracy of the numerical integration)? Vary the launch speed, and explain the effect that this has on your graphs.

5.P.68 Calculating the impact speed

Use energy conservation to calculate analytically (that is, without doing a numerical integration) the final speed of the spacecraft just before it hits the Moon. Include the gravitational effect of the Moon. Use a launch speed of 1.3×10^4 m/s. Modify your program for the Chapter 3 problem on the Ranger 7 mission to the Moon to print out the speed of the spacecraft when it hits the surface of the Moon, and compare this value to your analytical result. What questions that could be addressed in the numerical integration are you *not* able to answer by doing this energy calculation?

5.P.69 Coasting toward Mars

A spacecraft is coasting toward Mars. The mass of Mars is 6.4×10^{23} kg and its radius is 3400 km (3.4×10^6 m). When the spacecraft is 7000 km (7×10^6 m) from the center of Mars, the spacecraft's speed is 3000 m/s. Later, when the spacecraft is 4000 km (4×10^6 m) from the center of Mars, what is its speed? Assume that the effects of Mars's two tiny moons, the other planets, and the Sun are negligible. Precision is required to land on Mars, so make an accurate calculation, not a rough, approximate calculation.

5.P.70 Boosting a satellite orbit

Calculate the speed of a satellite in an orbit near the Earth (just above the atmosphere). If the mass of the satellite is 200 kg, what is the minimum energy required to move the satellite very far away from the Earth?

5.P.71 Comet orbit

A comet is in an elliptical orbit around the Sun. Its closest approach to the Sun is a distance of 4×10^{10} m (inside the orbit of Mercury), at which point its speed is 8.17×10^4 m/s. Its farthest distance from the Sun is far beyond the orbit of Pluto. What is its speed when it is 6×10^{12} m from the Sun? (This is the approximate distance of Pluto from the Sun.)

5.P.72 An electron in an accelerator

An electron is traveling at a speed of $0.95c$ in an electron accelerator. An electric force of 1.6×10^{-13} N is applied in the direction of motion while the electron travels a distance of 2 m. What is the new speed of the electron?

5.P.73 Throwing from an asteroid

You stand on a spherical asteroid of uniform density whose mass is 2×10^{16} kg and whose radius is 10 km (10^4 m). These are typical values for small asteroids, although some asteroids have been found to have much lower average density and are thought to be loose agglomerations of shattered rocks; see the problem "Determining the mass of an asteroid" in Chapter 3. The asteroid is rotating a bit faster than once per day, so that objects on the surface have a speed of 1 m/s, including you and a rock you are holding.

(a) How fast (relative to you) do you have to throw the rock so that it never comes back to the asteroid and ends up traveling at a speed of 3 m/s when it is very far away? Explain briefly, including how you will take advantage of the asteroid's rotation.

(b) Sketch graphs of the kinetic energy of the rock, the gravitational potential energy of the rock plus asteroid, and their sum, as a function of separation (distance from center of asteroid to rock). Label the graphs clearly.

5.P.74 Collision with Jupiter

This problem is closely related to the spectacular impact of the comet Shoemaker-Levy with Jupiter in July 1994 (http://www.jpl.nasa.gov/sl9/sl9.html). A rock far outside our Solar System is initially moving very slowly relative to the Sun, in the plane of Jupiter's orbit around the Sun. The rock falls toward the Sun, but on its way to the Sun it collides with Jupiter.

Calculate the rock's speed just before colliding with Jupiter. Explain your calculation and any approximations that you make.

$$M_{Sun} = 2\times10^{30} \text{ kg}, \; M_{Jupiter} = 2\times10^{27} \text{ kg}$$
Distance, Sun to Jupiter = 8×10^{11} m
Radius of Jupiter = 1.4×10^{8} m

5.P.75 Particle-antiparticle annihilation

(a) A particle with mass M and charge $+e$ and its antiparticle (same mass M, charge $-e$) are initially at rest, far from each other. They attract each other and move toward each other. On axes like those, graph and label the various energies involved in this process, as a function of the distance r between the two particles. Be sure to include the rest energy of the particles.

Plot various energies.

(b) When the particle and antiparticle collide, they annihilate and produce a different particle with rest mass m (much smaller than M) and charge $+e$ and its antiparticle (same rest mass m, charge $-e$). When these two particles have moved far away from each other, how fast are they going? Is this speed large or small compared to c?

(c) Now take the specific case of a proton and antiproton colliding to form a positive and negative pion. Each pion has a rest mass of 2.5×10^{-28} kg. When the pions have moved far away, how fast are they going?

(d) How far apart must the two pions be (in meters) for their electric potential energy to be negligible compared to their kinetic energy? Be explicit and quantitative about your criterion and your result.

5.P.76 Four protons

Four protons, each with mass M and charge $+e$, are initially held at the corners of a square that is d on a side. They are then released from rest. What is the speed of each proton when the protons are very far apart?

5.P.77 Gravitational energy inside the Earth

In the rough approximation that the density of the Earth is uniform throughout its interior, the gravitational field strength (force per unit mass) inside the Earth at a distance r from the center is gr/R, where R is the radius of the Earth. (In actual fact, the outer layers of rock have lower density than the inner core of molten iron.) Using the uniform-density approximation, calculate the amount of energy required to move a mass m from the center of the Earth to the surface. Compare with the amount of energy required to move the mass from the surface of the Earth to a great distance away.

5.P.78 Alpha decay

Many heavy nuclei undergo spontaneous "alpha decay," in which the original nucleus emits an alpha particle (a helium nucleus containing two protons and two neutrons), leaving behind a "daughter" nucleus that has two fewer protons and two fewer neutrons than the original nucleus. Consider a Radium-220 nucleus that is at rest before it decays to Radon-216 by alpha decay.

The mass of the Radium-220 nucleus is 219.96274 u (unified atomic mass units) where 1 u = 1.6603×10^{-27} kg (approximately the mass of one nucleon). The mass of a Radon-216 nucleus is 215.95308 u, and the mass of an alpha particle is 4.00151 u. Radium has 88 protons, Radon 86, and an alpha particle 2.

(a) Make a diagram of the final state of the Radon-216 nucleus and the alpha particle when they are far apart, showing the momenta of each particle to the same relative scale. Explain why you drew the lengths of the momentum vectors the way you did.

(b) Calculate the final kinetic energy of the alpha particle. For the moment, assume that its speed is small compared to the speed of light.

(c) Calculate the final kinetic energy of the Radon-216 nucleus.

(d) Show that the nonrelativistic approximation was reasonable.

5.P.79 The SLAC two-mile accelerator

SLAC, the Stanford Linear Accelerator Center, located at Stanford University in Palo Alto, California, accelerates electrons through a vacuum tube two miles long (it can be seen from an overpass of the Junipero Serra freeway that goes right over the accelerator). Electrons which are initially at rest are subjected to a continuous force of 2×10^{-12} N along the entire length of two miles (one mile is 1.6 kilometers) and reach speeds very near the speed of light. A similar analysis in a previous chapter required numerical integration, but with the new techniques of this chapter you can analyze the motion analytically.

(a) Calculate the final energy, momentum, and speed of the electron.

(b) Calculate the approximate time required to go the two-mile distance.

5.P.80 Determining precise deuteron mass

A proton (1.6726×10^{-27} kg) and a neutron (1.6749×10^{-27} kg) at rest combine to form a deuteron, the nucleus of deuterium or "heavy hydrogen." In this process, a gamma ray (high-energy photon) is emitted, and its energy is measured to be 2.2 MeV (2.2×10^{6} eV).

(a) Keeping all five significant figures, what is the mass of the deuteron? Assume that you can neglect the small kinetic energy of the recoiling deuteron.

(b) Momentum must be conserved, so the deuteron must recoil with momentum equal and opposite to the momentum of the gamma ray. Calculate approximately the kinetic energy of the recoiling deuteron and show that it is indeed small compared to the energy of the gamma ray.

5.P.81 Nuclear fission

Uranium-235 fissions when it absorbs a slow-moving neutron. The two fission fragments can be almost any two nuclei whose charges Q_1 and Q_2 add up to 92e (where e is the charge on a proton, $e = 1.6 \times 10^{-19}$ coulomb), and whose nucleons add up to 236 protons and neutrons (U-236; U-235 plus a neutron). One of the possible fission modes involves nearly equal fragments, palladium nuclei with $Q_1 = Q_2 = 46e$. The rest masses of the two palladium nuclei add up to less than the rest mass of the original nucleus. (In addition to the two main fission fragments there are typically one or more free neutrons in the final state; in your analysis make the simplifying assumption that there are no free neutrons, just two palladium nuclei.) The rest mass of the U-236 nucleus is 235.996 u (unified atomic mass units), and the rest mass of each Pd-118 nuclei is 117.894 u, where 1 u = 1.7×10^{-27} kg (approximately the mass of one nucleon).

(a) Calculate the final speed v, when the palladium nuclei have moved far apart (due to their mutual electric repulsion). Is this speed small enough that $p^2/(2m)$ is an adequate approximation for the kinetic energy of one of the palladium nuclei? (It is all right to go ahead and make the nonrelativistic assumption first, but you then must check that the calculated v is indeed small compared to c.)

(b) Using energy considerations, calculate the distance between centers of the palladium nuclei just after fission, when they are starting from rest.

(c) A proton or neutron has a radius of roughly 1×10^{-15} m , and a nucleus is a tightly packed collection of nucleons. Experiments show that the radius of a nucleus containing N nucleons is approximately $(1.3 \times 10^{-15}$ m$) \times N^{1/3}$. What is the approximate radius of a palladium nucleus? Draw a sketch of the two palladium nuclei in part (b), and label the distances you calculated in parts (b) and (c).

5.P.82 Nuclear fusion

One of the thermonuclear or fusion reactions that takes place inside a star such as our Sun is the production of helium-3 (^{3}He, with two protons and one neutron) and a gamma ray (high-energy photon) in a collision between a proton (^{1}H) and a deuteron (^{2}H, the nucleus of "heavy" hydrogen, consisting of a proton and a neutron):

$$^1H + {}^2H \rightarrow {}^3He + \gamma$$

The rest mass of the proton is 1.0073 u (unified atomic mass unit, 1.7×10^{-27} kg), the rest mass of the deuteron is 2.0136 u, the rest mass of the helium-3 nucleus is 3.0155 u, and the gamma ray is massless.

(a) The strong interaction has a very short range and is essentially a contact interaction. For this fusion reaction to take place, the proton and deuteron have to come close enough together to touch. The approximate radius of a proton or neutron is about 1×10^{-15} m . What is the approximate initial total kinetic energy of the proton and deuteron required for the fusion reaction to proceed, in joules and electron volts (1 eV = 1.6×10^{-19} joule)?

(b) Given the initial conditions found in part (a), what is the kinetic energy of the ^{3}He plus the energy of the gamma ray, in joules and in electron-volts?

(c) The net energy released is the kinetic energy of the ^{3}He plus the energy of the gamma ray found in part (b), minus the

energy input that you calculated in part (a). What is the net energy release, in joules and in electron volts? Note that you do get back the energy investment made in part (a).

(d) Which of the following potential energy curves is a reasonable representation of the interaction in this fusion reaction? Why?

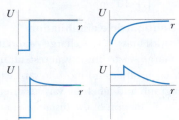

Potential energy curves.

As we will study later, the average kinetic energy of a gas molecule is $\frac{3}{2}kT$, where k is the "Boltzmann constant," 1.4×10^{-23} joule/Kelvin (J/K), and T is the absolute or Kelvin temperature, measured from absolute zero (so that the freezing point of water is 273 K). The approximate temperature required for the fusion reaction to proceed is very high. This high temperature, required because of the electric repulsion barrier to the reaction, is the main reason why it has been so difficult to make progress toward thermonuclear power generation. Sufficiently high temperatures are found in the interior of the Sun, where fusion reactions take place.

5.P.83 Alpha decay

A nucleus whose mass is 3.917268×10^{-25} kg undergoes spontaneous "alpha" decay. The original nucleus disappears and there appear two new particles: a He-4 nucleus of mass 6.640678×10^{-27} kg (an "alpha particle" consisting of two protons and two neutrons) and a new nucleus of mass 3.850768×10^{-25} kg (note that the new nucleus has less mass than the original nucleus, and it has two fewer protons and two fewer neutrons). When the alpha particle has moved far away from the new nucleus (so the electric potential energy is negligible), what is the combined kinetic energy of the alpha particle and new nucleus?

5.P.84 Potential energy and a pendulum

A pendulum consists of a very light but stiff rod of length L hanging from a nearly frictionless axle, with a mass m at the end of the rod.

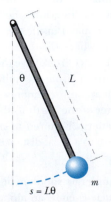

(a) Calculate the gravitational potential energy as a function of the angle θ, measured from the vertical.

(b) Sketch the potential energy as a function of the angle θ, for angles from −210° to +210°.

(c) Let $s = L\theta$ = the arc length away from the bottom of the arc. Calculate the tangential component of the force on the mass by taking the (negative) gradient of the energy with respect to s. Does your result make sense?

(d) Suppose you hit the stationary hanging mass so that it has an initial speed v_i. What is the minimum initial speed in order that the pendulum go over the top (θ = 180°)? On your sketch of the potential energy (part b), draw and label energy levels for the case where the initial speed is less than, equal to, or greater than this critical initial speed.

5.P.85 Relativistic energy and momentum
Show the validity of the relation $E_{particle}^2 - (pc)^2 = (mc^2)^2$ in the case where $m \neq 0$, by making these substitutions:

$$E_{particle} = \frac{mc^2}{\sqrt{1 - v^2/c^2}} \text{ and } p = \frac{mv}{\sqrt{1 - v^2/c^2}}$$

5.25 ANSWERS TO EXERCISES

5.X.1 (page 169) 5.24×10^{-13} J; 8.10×10^{-14} J; 4.43×10^{-13} J

5.X.2 (page 169) 4.4×10^9 J

5.X.3 (page 169) about 6

5.X.4 (page 169) about 1×10^{-14}

5.X.9 (page 172) $\langle 25, -30, 0 \rangle$ m, $\langle 0, -19.6, 0 \rangle$ N, 588 J

5.X.10 (page 173) decreasing; negative

5.X.11 (page 173) increasing; positive

5.X.12 (page 173) decreasing; negative

5.X.13 (page 173) increasing; positive

5.X.14 (page 173) 2.57 m/s

5.X.15 (page 176) 1230 J

5.X.16 (page 176) 4.9 N downward, 4.9 J
Positive work makes the jar speed up.

5.X.17 (page 176) zero

5.X.18 (page 176) $2P$

5.X.19 (page 178) 400 J, four times as much work

5.X.20 (page 179) $40.9\,mc^2$; $39.9\,mc^2$; very large; very small

5.X.21 (page 181) **0.00085, approximately 0.1%**

5.X.22 (page 184) 189 J

5.X.23 (page 184) 2.5×10^{-19} J

5.X.24 (page 184) 0.1 J

5.X.5 (page 169) $kg \cdot m^2/s^2 = (kg \cdot m/s^2) \cdot m = N \cdot m = J$

5.X.6 (page 169) 12 m/s, 27 m.p.h.

5.X.7 (page 169) 1.2×10^{-12} kg (an extremely small amount of mass!)

5.X.25 (page 189) $\Delta U = -50$ joules

5.X.26 (page 191) 6 potential energy pairs:
$$U_{12} + U_{13} + U_{14} + U_{23} + U_{24} + U_{34}$$

5.X.27 (page 194) − change, because more negative

5.X.28 (page 194) + change, because less negative

5.X.29 (page 195) There is no W term, because there are no significant external forces acting on this system.

5.X.30 (page 195) center-to-center

5.X.31 (page 196) K has decreased

5.X.32 (page 197) A is bound;
C is unbound;
B is trapped. In a quantum system in state B there is some probability of "tunneling" through the barrier and getting out of the trap!

5.X.33 (page 198) A: go out in a straight line and stop, then fall back

B: elliptical (motion with varying separation)

C: nearly circular (nearly constant separation)

5.X.34 (page 200) v to reach Moon < v_{esc}, because didn't go to infinity

5.X.35 (page 200) 1.1×10^4 m/s

5.X.36 (page 200) 2.4×10^3 m/s, much smaller than v_{esc} from Earth

5.X.37 (page 202) same ΔU

5.X.38 (page 202) 6.3 m/s; the mass cancels from both sides of the energy equation

5.X.39 (page 202) 9.8 J, 6.3 m/s, neglect air friction

5.X.40 (page 202) 19.6 J, 6.3 m/s; same speed, twice the kinetic energy

5.X.41 (page 202) Hint: At top of trajectory the ball's vertical speed is momentarily zero. Answer below.

5.X.42 (page 203) Hint: $E_f = E_i$. Answer below.

5.X.43 (page 204) 7.2×10^{-15} m

Note: the radius of the proton is about 1×10^{-15} m.

5.X.44 (page 206) About 1.5×10^{-10}; accuracy of laboratory scale is far below what would be needed to detect such a tiny difference

5.X.45 (page 208) About 0.1%

5.X.46 (page 214) $-GMm/y^2$; 0

Additional answers

5.X.41 (page 202) 101 m

5.X.42 (page 203) $y_f = y_i + (v_i^2 - v_f^2)/2g$

CHAPTER 6

ENERGY IN MACROSCOPIC SYSTEMS

Key concepts

- Macroscopic systems are composed of many interacting particles.
- The potential energy associated with stretching or compressing interatomic bonds is similar to the potential energy of macroscopic springs.
- Thermal energy is the kinetic and potential energy of the atoms and interatomic bonds within an object.
    - Temperature is related to thermal energy.
    - A temperature difference between system and surroundings leads to energy transfer (Q).
    - Specific heat capacity is the amount of energy required to raise the temperature of a given amount of a substance by 1 kelvin.
    - Energy can be converted to thermal energy through dissipative mechanisms such as friction and air resistance.
- The potential energy difference between two states does not depend on how a system gets from the initial to the final state (path independence).

6.1 POTENTIAL ENERGY OF MACROSCOPIC SPRINGS

In Chapter 4 we used the behavior of idealized, macroscopic springs as a model for the dynamic behavior of the springlike interatomic bonds in solid objects. This was a productive model, because it allowed us to understand the nature of contact forces (tension and compression forces) and because using this model we were able to predict the speed of sound in different materials.

This model continues to be useful in understanding the flow of energy in macroscopic systems (objects composed of many atoms, such as wires, tables, vines, and humans). To apply it we need to find an equation for the potential energy of a mass-spring system.

An ideal spring

A real spring has a nonzero mass, and can be damaged in various ways: it can be deformed or broken, the metal can fatigue, and so on. There is a limit to how much a real spring can be stretched (it will eventually break) or compressed (the coils will eventually touch each other, turning the spring into a hollow rod). An idealized spring has none of these problems. There are no limits on the stretch s in the ideal spring force law:

$$F_s = -k_s s$$

Nonetheless, an ideal spring proves to be a useful model for both real macroscopic springs and for springlike interatomic bonds.

To determine the potential energy associated with the spring force, remember that the (negative) gradient of potential energy is equal to the associated force. So to find the spring potential energy U_s we need a solution to this equation:

$$-\frac{dU_s}{ds} = -k_s s$$

? Can you think of a function of s whose derivative with respect to s is $+k_s s$?

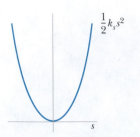

Figure 6.1 Potential energy as a function of stretch *s* for an ideal spring.

The derivative with respect to *s* of $\frac{1}{2}k_s s^2$ is $+k_s s$, and the negative gradient of $\frac{1}{2}k_s s^2$ is the force $-k_s s$.

IDEAL SPRING POTENTIAL ENERGY

$$U_s = \frac{1}{2}k_s s^2$$

s is stretch, measured from equilibrium point

The function $\frac{1}{2}k_s s^2$ is an upturned parabola (Figure 6.10). Note that the stretch *s* appears squared in the formula for spring energy. This reflects the fact that either a lengthening (positive stretch) or a compression (negative stretch) of the spring involves increased potential energy.

Peculiar features of the ideal spring potential energy
This ideal spring potential energy curve has several peculiar features:
* As the absolute value of stretch becomes large, *U* becomes infinite, instead of approaching zero. (This curve is sometimes called an "infinite potential energy well.")
* For a system of a mass and ideal spring, *K+U* is always greater than zero, so apparently there can't be any bound states of the system.

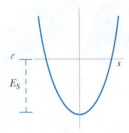

Figure 6.2 Subtracting a constant E_S shifts the ideal spring potential energy curve downward.

So this ideal potential energy function can't provide an adequate description of energy for any real system, whether it is a macroscopic spring you can hold in your hand, or an interatomic bond.

One possible modification to the equation would be to subtract a constant E_S. Such a function would still satisfy the gradient equation above, but it would allow for bound states of the system, as in (Figure 6.2).

$$U_s = \frac{1}{2}k_s s^2 - E_S$$

For the ideal spring, the constant is conventionally set to zero, but the negative constant is needed when analyzing real systems.

Real springs
We are primarily interested in helical (spiral) springs. A modest force applied to a helical spring gives a substantial change in the length of the spring, even though the total length of the coiled wire hardly changes. A large stretch of the spring with a small force implies a much smaller spring stiffness than the wire would have when straightened out.

If you stretch a helical spring far enough, it straightens out into a long straight wire, and it suddenly becomes very difficult to lengthen the wire any further. If you continue to stretch the straight wire, eventually it yields and then breaks. If you compress a helical spring so far that its coils run into each other, it suddenly gets extremely hard to compress the spring any further. Therefore, the linear relation between stretch and force, $F_s = -k_s s$, is valid only over a limited region of stretch, although this linear range is much wider than for the microscopic "spring."

In many situations the entire phenomenon takes place within the valid range of the linear approximation. Or one can speak of an "infinite" well as shown as the dashed curve in Figure 6.3, which is an idealization of the real situation. Within the linear range where $F_s = -k_s s$ is the negative gradient of $\frac{1}{2}k_s s^2$, we approximate the real curve by a parabola. The actual bottom of the parabola is at a negative value of the potential energy. Work would be required to stretch the wire out straight. However, in our calculations with macroscopic springs we're only interested in *changes* in the potential energy, so we shift the origin to the bottom of the well and measure U_s from there.

In Figure 6.3 (which is still an idealization of the real situation) there are very steep sides of the potential well, corresponding to the large forces re-

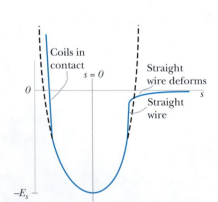

Figure 6.3 The dashed curve represents the spring potential energy

$$U_s = \frac{1}{2}k_s s^2 - E_S$$

within the linear range where $F_s = -k_s s$. The colored solid curve represents a real macroscopic spring.

quired to compress a spring whose coils are in contact, and the large forces required to stretch the spring after it has straightened out into a straight wire. Since force is the negative gradient of the potential energy, a steep slope corresponds to a large force.

Energy in an oscillating spring-mass system

In an oscillating spring-mass system such as the one discussed in Chapter 4, the energy of the system is continually changing from kinetic energy to potential energy, and back again.

Consider a block on a low friction surface, connected by a helical spring to a wall. If you compress the spring and then let go, the block acquires kinetic energy (Figure 6.4), with a loss of spring potential energy. If you stretch the spring and then let go, the block also acquires kinetic energy, again with a loss of spring potential energy (Figure 6.5).

> **?** At the moment that the spring is released, what is the kinetic energy of the spring-mass system? What is the total energy of the system?

Since the block is released from rest, its kinetic energy is zero. The energy of the system is equal to the energy stored in the spring, $\frac{1}{2}k_s s^2$ (omitting the constant rest energy and constant $-E_S$).

> **?** At what point in the oscillation is the kinetic energy of the system highest? At that moment, what is the potential energy of the system?

When the contracting spring reaches its equilibrium length, the energy stored in the spring is a minimum, $\frac{1}{2}k_s s^2 = 0$. The maximum possible amount of spring potential energy has now been converted to kinetic energy, so this is the instant when the kinetic energy of the system is highest.

> **?** At what point in the oscillation will the kinetic energy of the system be lowest?

At either endpoint of the oscillation—spring fully compressed or spring fully extended—the instantaneous speed of the mass is zero. The kinetic energy of the system is lowest here (zero in fact), and all the energy has been momentarily converted back into spring potential energy. Since the absolute value of the stretch is highest at these locations, the potential energy $\frac{1}{2}k_s s^2$ is also highest.

Does the wall do work on the system?

For the system of spring+block, the surroundings include the Earth, the low friction table, and the wall. Both the table and the wall touch the system and therefore exert contact forces on it.

The table and Earth exert forces perpendicular to the block's motion, so they do zero work on the system.

> **?** The force by the wall on the spring is parallel to the block's motion. Does it do nonzero work on the system?

This force does zero work, because the point at which the force acts (the left end of the spring) undergoes zero displacement. So

$$F_{wall,x}\Delta x = 0 \text{ because } \Delta x = 0 \text{ for the left end of the spring.}$$

This force does have an effect on the momentum of the system, however, because the time interval over which it acts is not zero. The momentum of the mass-spring system is changed by the force on the spring by the wall. (Think about what would happen if the wall were not there.) This is another example of the difference between the Momentum Principle (all forces act for the same time interval) and the Energy Principle (different forces may act over different distances).

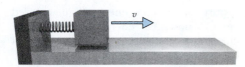

Figure 6.4 As the compressed spring expands the kinetic energy of the block increases, and spring potential energy decreases.

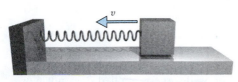

Figure 6.5 As the stretched spring contracts the kinetic energy of the block increases, and spring potential energy decreases.

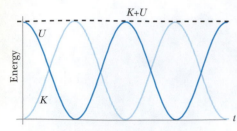

Figure 6.6 Kinetic energy (light colored curve), potential energy (dark colored curve), and the sum of kinetic + potential energy (dashed black line) for an oscillating spring-mass system.

From the discussion in the preceding section, we know that no work is done by the surroundings, because the point where the wall force is applied does not move. We also know that when the stretch is maximum, the block is momentarily at rest.

We could use any known state as the initial state in part (b), because the energy of the system is constant.

Flow of energy in the mass-spring system

Since the surroundings do no work on the system, the total energy of the system must be constant (in the absence of friction), and energy flows from potential to kinetic and back again, as shown in (Figure 6.6).

Example: Energy in a spring-mass system

A mass of 0.2 kg is attached to a horizontal spring whose stiffness is 12 N/m. Friction is negligible. At $t = 0$ the spring has a stretch of 3 cm and the mass has a speed of 0.5 m/s. (a) What is the amplitude (maximum stretch) of the oscillation? (b) What is the maximum speed of the block?

System: mass and spring
Surroundings: Earth, table, wall (neglect friction, air)

Initial state: $s = 3$ cm, $v = 0.5$ m/s
Final state: $v = 0$ (maximum stretch)

(a) Find maximum stretch:

Energy principle:
$$E_f = E_i + W$$
$$\cancel{K_f} + U_f = K_i + U_i + \cancel{W}$$
$$0 + \tfrac{1}{2}k_s s_f^2 = \tfrac{1}{2}mv_i^2 + \tfrac{1}{2}k_s s_i^2$$
$$\tfrac{1}{2}k_s s_f^2 = (\tfrac{1}{2})(0.2 \text{ kg})(0.5 \text{ m/s})^2 + (\tfrac{1}{2})(12 \text{ N/m})(0.03 \text{ m})^2$$
$$\tfrac{1}{2}k_s s_f^2 = 0.03\text{J}$$
$$s_f = \sqrt{2(0.03\text{J})/(12 \text{ N/m})} = 0.07 \text{ m}$$
$$s_f = 0.07 \text{ m}$$

(b) Find maximum speed:

Initial state: Same as above
Final state: Maximum speed (zero stretch)
$$K_f + 0 = 0.03\text{J}$$
$$\tfrac{1}{2}mv_i^2 = 0.03\text{J}$$
$$v_{\text{max}} = \sqrt{2(0.03\text{J})/(0.2 \text{ kg})} = 0.55 \text{ m/s}$$

Further discussion

Assuming no energy dissipation by friction or air resistance, the energy of the system never changes. If can we find $K+U$ for one state, we can always use this as the initial state in any calculation.

6.X.1 About how many joules of energy can you store in a spring like the one you used in Chapter 4, starting from a relaxed spring? Assume ($k_s \approx 0.6$ N/m) and a stretch of about 20 cm.

6.X.2 A horizontal spring with stiffness 0.5 N/m has a relaxed length of 15 cm. A mass of 20 grams is attached and you stretch the spring to a total length of 25 cm. The mass is then released from rest. What is the speed of the mass at the moment when the spring returns to its relaxed length of 15 cm?

Non-helical macroscopic springs

We saw in Chapter 4 that a bar of metal can be treated as a spring, since stretching or compressing the bar stretches or compresses the interatomic bonds (Figure 6.7). Double the force produces double the change in length, as long as we're in the range where the spring approximation to the interatomic force is adequate. This is usually described for a bar in terms of Young's modulus Y, which we studied in Chapter 4, where the tension force F_T (per unit area) is proportional to the stretch ΔL (per unit length):

$$\frac{F_T}{A} = Y\frac{\Delta L}{L}$$

Since ΔL, like s, represents a change in length, and F_T is a spring-like force (tension force), this equation can be rewritten to parallel the spring force law:

$$F_T = \left(\frac{YA}{L}\right)|\Delta L| \quad \text{is like} \quad F = (k_s)|s|$$

The effective "spring stiffness" of the metal bar is YA/L.

If you stretch the bar too much, two things can make the bar stop behaving in a spring-like manner. You might exceed the interatomic stretch for which $k_s s$ is a good approximation to the magnitude of the interatomic force. Also, the regular array of atoms may be disrupted by dislocations of the crystal structure, leading to large-scale slippage of crystal planes. Suddenly the bar "yields" and grows very much longer with little applied force.

A straight object is not usually used as a longitudinal spring but may be used in a bending mode. For example, when a diving board bends under a load, atomic bonds in the upper part of the board are stretched, and atomic bonds in the lower part are compressed (Figure 6.8). The net effect is a spring-like behavior of the diving board, with the amount of bend proportional to the applied force.

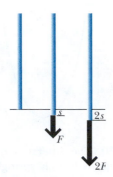

Figure 6.7 A bar of metal can be treated as a spring. For a limited range, double the force gives double the stretch.

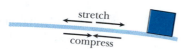

Figure 6.8 The bending of a diving board stretches and compresses interatomic bonds.

6.2 POTENTIAL ENERGY OF A PAIR OF NEUTRAL ATOMS

In the preceding chapter we obtained the formula for electric potential energy for pairs of charged particles. In Chapter 4 we studied the spring-like forces between electrically *neutral* atoms (a neutral object has equal amounts of positive and negative charge, so it has zero net charge). These interatomic forces are the superposition of electric attractions and repulsions among the positively charged protons and negatively charged electrons of which the atoms are made.

When two electrically neutral atoms are far from each other (Figure 6.9), they exert almost no force on each other because the attractions between unlike charges (electrons in atom one with protons in atom two, and vice versa) are nearly equal to the repulsions between like charges. But when the atoms come quite near each other, the electron clouds distort in such a way that the two atoms attract each other. Molecules and solid objects made of two or more atoms are bound together by these electric forces.

However, if you try to push two atoms even closer together, you eventually encounter a rapidly increasing repulsion. The attraction at a small distance and the repulsion when very close are due ultimately to the superposition of the electric forces of the various protons and electrons in the atoms, with the probable locations of these particles governed by the laws of quantum mechanics.

Figure 6.9 Two neutral atoms interact very little when far apart, attract each other at intermediate distances, and repel each other at very short distances.

Potential energy for spring-like interatomic bonds

Since an interatomic bond behaves like a spring as long as it is not stretched or compressed excessively, we can guess that the equation for interatomic

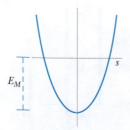

Figure 6.10 The interatomic potential energy is approximately springlike if the bond is not stretched or compressed too much.

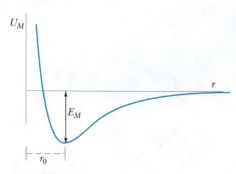

Figure 6.11 The Morse potential energy function U_M, representing the potential energy of two interacting neutral atoms.

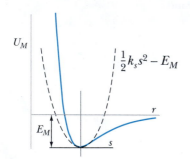

Figure 6.12 The bottom of the interatomic potential energy curve may be approximated by a parabola.

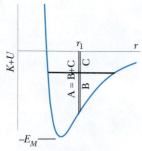

Figure 6.13 A horizontal line represents constant kinetic plus potential energy.

potential energy has the same form as the potential energy for an ideal spring (Figure 6.10):

APPROXIMATE INTERATOMIC POTENTIAL ENERGY

$$U = \tfrac{1}{2}k_s s^2 - E_M$$

What about large separations?

Of course, the potential energy of a real interatomic bond can't become infinitely large at large separation—it needs to approach zero. A more realistic potential energy function for the interatomic bond is called the Morse potential energy, and is shown in Figure 6.11. The Morse potential energy is described by this:

$$U_M = E_M[1 - e^{-\alpha(r - r_{eq})}]^2 - E_M$$

The Morse approximation is designed to model the basic behavior of the interatomic force, and is in approximate agreement with measurements of the interatomic force. The parameter E_M gives the depth of the potential energy where it is a minimum, and the parameter r_{eq} (on the order of 1×10^{-10} m) is the distance at which the potential energy curve is a minimum. The parameter α is adjusted to make the curve have the right width. When the interatomic separation r is very large, $U_M = 0$, as it should.

The depth of the potential well, E_M, is typically a few electron volts, where 1 eV is equal to 1.6×10^{-19} joule, a tiny energy unit often used in describing the world of atoms. For example, the potential energy well is about 5 eV deep for oxygen, O_2. (As you saw in the previous chapter, a typical ionization energy is also in the range of electron volts). However, for the noble gases such as helium and argon the well is so shallow that these atoms don't form stable diatomic molecules at room temperature, because collisions easily provide enough energy to break the molecules apart.

? What is the physical significance of the minimum of the curve?

At the minimum of the potential energy curve, the slope is zero, so the force between the two atoms is zero. This is the position of stable equilibrium.

? What is the direction of the force on an atom at a distance smaller than the equilibrium distance?

The left side of the curve has a large negative slope, so the force will be large and to the right (in the $+r$ direction), away from the other atom. This makes sense; when the atoms are too close together they push each other apart.

Likewise, if the atom moves to the right (farther away) there should be an attractive restoring force to the left. The slope of this region of the curve is positive, so the sign of the force is negative, meaning its direction is to the left, as expected.

As indicated by the dashed curve in Figure 6.12, the bottom of the interatomic potential energy curve can be approximated by the spring-like potential energy function:

$$U = \tfrac{1}{2}k_s s^2 - E_M$$

Because we can make this approximate fit, we know that for ordinary small oscillations around the equilibrium point the two atoms will act as though they are connected by a spring with a $-k_s s$ force. This is the rationale for modeling a solid as a collection of balls connected by springs.

Bound and unbound states with interatomic potential energy

In Figure 6.13, a horizontal line represents motion in which the sum of the kinetic energies and the potential energy of the two-atom system does not change. A horizontal line corresponds to a possible state of the system when it is isolated from external forces.

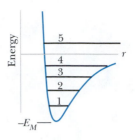

> **6.X.3** At separation r_1 in Figure 6.13, what is the physical significance of the quantity A? Of the quantity B? Of the quantity C?
>
> **6.X.4** In Figure 6.14, which of the states are bound states? Which are unbound states?

6.3 PATH INDEPENDENCE OF POTENTIAL ENERGY

In situations describable by a potential energy, the associated interaction depends solely on position, not on speed or direction. For example, spring energy depends on the stretch s, and gravitational and electric energy depend on the separation r. It is easy to prove that the following must be true:

PATH INDEPENDENCE OF POTENTIAL ENERGY

Change in potential energy doesn't depend on the path taken.

In a round trip, the potential energy doesn't change.

Here is the proof: In Figure 6.15, change in potential energy U of the system when one of its particles moves along some path from location A to location B depends solely on the initial and final values of r or s, so it doesn't matter what path the particle follows to get from A to B. Similarly, in Figure 6.16 if one of the particles in a multiparticle system starts at point A, moves around and comes back to the starting point, the change in potential energy U of the multiparticle system must be zero, since the initial and final potential energy depends on pair-wise distances r or s, and these return to their original values in a round trip.

There is a connection between the fact that change of potential energy is independent of path, and that a round trip gives zero. In Figure 6.17, suppose that along path 1 from A to B the change in potential energy is +5 J.

? What will the change in potential energy be along the path 1 in the opposite direction, from B to A?

Going backwards from B to A along this same path the change would be –5 J. Since the change is independent of the path, you could take *any* path to return from B to A, such as path 2, with a change of –5 J. Any round trip from A to B and back to A will yield a change of zero.

A simple example is projectile motion when air resistance can be neglected (Figure 6.18). You throw a ball up in the air with speed v, and when it returns to your hand it has regained the same speed v that you gave it (the gravitational potential energy change of the ball+Earth system is zero, so the change in the kinetic energy is zero). Another example is the motion of a mass on a spring in the absence of friction: every time the mass passes a particular position, it has the same speed (same s implies same v).

In both of these examples, although the speed doesn't change, the velocity may change. The ball was headed up but is now headed down; a mass on a spring may be going to the left and then going to the right with the same speed. The key point is this:

Potential energy depends only on position,
not on the direction of motion.

> **6.X.5** During one complete oscillation of a mass on a spring (one period), what is the change in potential energy of the mass + spring system, in the absence of friction?

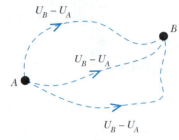

Figure 6.14 Which are bound states? Which are unbound states?

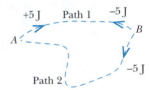

Figure 6.15 Potential energy difference is independent of the path between the initial and final locations.

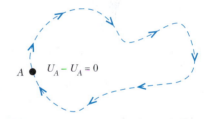

Figure 6.16 A particle moves from A along a path and returns to A.

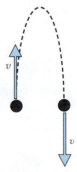

Figure 6.17 Forward and reversed paths, and a round trip.

Figure 6.18 A ball goes up and down with negligible air resistance.

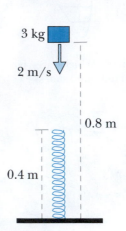

Figure 6.19 A block rebounds from a vertical spring.

Both Earth and spring are part of the system, so there are two potential energy terms, one for (block+spring) and one for (block+Earth). Since the mass of the spring is negligible, there is not a (spring+Earth) potential energy term.

We could equally well have used the final state from part (a) as the initial state in part (b).

Example: Example: A rebounding block

A metal block of mass 3 kg is moving downward with speed 2 m/s when the bottom of the block is 0.8 m above the floor (Figure 6.19). When the bottom of the block is 0.4 m above the floor, it strikes the top of a relaxed vertical spring 0.4 m in length. The stiffness of the spring is 2000 N/m.

(a) The block continues downward. When the bottom of the block is 0.3 m above the floor, what is its speed?

(b) The block eventually heads back upwards, loses contact with the spring, and continues upward. What is the maximum height reached by the bottom of the block above the floor?

(c) What approximations did you make?

(a) Find speed of block 0.3 m above floor.

System: Earth, block, spring
Surroundings: Nothing significant

Initial state: Block 0.8 m above floor, moving downward, $v = 2$ m/s, spring relaxed
Final state: Block 0.3 m above floor, spring compressed

Energy principle:
$$E_f = E_i + W$$
$$K_f + U_{s,f} + U_{g,f} = K_i + \cancel{U_{s,i}} + U_{g,i} + \cancel{W}$$
$$\tfrac{1}{2}mv_f^2 + mgy_f + \tfrac{1}{2}k_s s_f^2 = mgy_i + \tfrac{1}{2}mv_i^2$$
$$v_f^2 = v_i^2 + 2g(y_i - y_f) - \frac{k_s}{m}s_f^2$$
$$v_f = \sqrt{(2 \text{ m/s})^2 + 2(9.8 \text{ N/kg})(0.5 \text{ m}) - \frac{(2000 \text{ N/m})}{(3 \text{ kg})}(0.1 \text{ m})^2}$$
$$v_f = 2.7 \text{ m/s}$$

(b) Find maximum height

Initial state: Same as in part (a)
Final state: Block at highest point, $v = 0$, spring relaxed

Energy principle:
$$E_f = E_i + W$$
$$\cancel{K_f} + \cancel{U_{s,f}} + U_{g,f} = K_i + \cancel{U_{s,i}} + U_{g,i} + \cancel{W}$$
$$mgy_f = mgy_i + \tfrac{1}{2}mv_i^2$$
$$y_f = y_i + \frac{v_i^2}{2g} = (0.8 \text{ m}) + \frac{(2 \text{ m/s})^2}{2(9.8 \text{ N/kg})} = 1.0 \text{ m}$$

(c) Approximations: Negligible air resistance and dissipation in the spring. Approximate expression $U_g = mgy$ near Earth's surface. The gravitational and spring forces on the Earth and on the block are equal and opposite, so the momentum changes are equal and opposite. But since $K = p^2/(2m)$, for the same momentum p the kinetic energy of the Earth is extremely small.

Another way to look at this is to say that while the forces on the Earth and the block have the same magnitudes, the Earth hardly moves, so the force does extremely little work on the Earth and gives it little kinetic energy.

6.4 THERMAL ENERGY

In our idealized models of spring-mass systems, the sum of kinetic plus spring potential energy is constant. In real spring-mass systems, however, the oscillations die away ("damp down") with time, which means that the kinetic plus spring potential energy must be decreasing. The decrease in the amplitude of the oscillations is mainly due to air resistance or to friction with a solid, and some energy is transferred to the surroundings, to the air or to a solid along which the mass slides. On the microscopic level, if we could look inside the solid, we would see increased atomic motion—the atoms in the solid have gained energy. A macroscopic indication of this increased atomic motion is an increase in the temperature of the object.

We can measure "macroscopic" spring potential energy and kinetic energy with simple measurements that don't require a microscope. But when the temperature of a solid object increases, the increased energy inside the solid is not visible to the naked eye, and we face the problem of how to evaluate the increase in the "thermal" energy.

Microscopic kinetic and spring potential energy

The increased "microscopic" energy in a solid is in two forms. First, the atoms on the average are moving around faster, with increased kinetic energy $\frac{1}{2}mv^2$. Second, there is increased spring potential energy $\frac{1}{2}k_s s^2$ in the interatomic bonds ("springs" in our model of solids), where k_s is the stiffness of an interatomic spring (Figure 6.20).

In principle the microscopic, thermal energy in a solid could be evaluated by simultaneously measuring at some instant the momenta of *all* of the atoms, and the stretches or compressions of *all* of the springs (that is, changes in atomic positions away from their equilibrium positions, together with knowledge of the spring stiffness). Since there are about 1×10^{23} atoms (and even more "springs") in an ordinary-sized object, it is in practice impossible to evaluate the energy this way.

Perhaps we could make microscopic measurements of just one atom (and its attached "springs"), assume that this is the average energy of each atom, and then multiply by the number of atoms there are in the object to get the total energy. That doesn't work either, because at some instant we may find an atom that is momentarily sitting motionless at its equilibrium position (zero energy), and an instant later it is displaced away from its equilibrium position and moving rapidly. Energy keeps getting passed back and forth among the many atoms and springs, and measuring the energy of just one atom and its attached springs at one instant doesn't give us the correct average energy we would need to determine the total energy of the object.

Temperature

In the 1800's it was realized that for many systems the temperature is essentially a measure of the average energy of the atoms in the system:

> Temperature and the average energy of atoms are related

Given the impossibility of evaluating the energy of each individual atom, we can simply use a thermometer to measure thermal energy. A familiar example is the mercury thermometer (silver colored) or the alcohol thermometer (tinted red for visibility). In these thermometers, a very thin column of liquid expands when heated. By tradition, marks on the thermometer are placed to represent "degrees of temperature," but we could just as well place marks to indicate the average energy of the molecules in the thermometer, in joules (Figure 6.21). One kelvin (or one Celsius degree of temperature) is equivalent to an average molecular energy of about 10^{-23} joules.

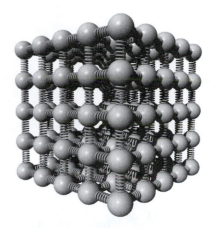

Figure 6.20 In our model solid we must account for the kinetic energy of every ball and the spring energy of every spring.

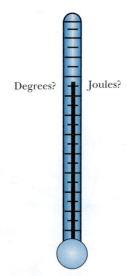

Degrees?   Joules?

Figure 6.21 A thermometer might be calibrated in joules instead of in degrees.

This is not the whole story. In a later chapter we will make more precise the relationship between temperature and energy, and we will see that temperature is more directly related to a quantity called "entropy" than it is to energy. Nevertheless, for most ordinary systems at room temperature it is true that temperature is a good measure of the average energy of the atoms.

The usefulness of a thermometer lies in the important observation that when two objects are in good contact with each other, the average kinetic energies of the molecules in the two objects slowly come to be equal. Where the two objects touch each other, molecules collide with each other, and the faster molecules on average lose kinetic energy to the slower molecules in a collision. Once the average kinetic energies of molecules in both objects have come to be equal, additional collisions on average make no further change.

To put it in terms of temperature (which is related to the average molecular kinetic energy), the hotter object gets cooler, and the cooler object gets hotter. Everyday examples abound: a cold drink warms up in a hot room, and warm water placed in a refrigerator cools down.

To measure the microscopic energy of an object, we place a thermometer in contact with the object. Then we wait for the temperatures (average molecular kinetic energies) of the object and thermometer to equilibrate, and we read the thermometer. It is of practical importance that the thermometer have a relatively small mass compared to the object of interest, so that attaching the thermometer doesn't add or subtract much energy to or from the object. Another advantage of a small thermometer is that it reaches thermal equilibrium quicker than a large thermometer does.

How was the calibration between one kelvin (or one Celsius degree) and energy established? In one of a series of classic experiment performed by Joule in the 1840's, a paddle wheel in water was turned by a falling weight. The work done by the Earth's gravitational force on the falling weight (mgh), was accompanied by an increase in the thermal energy of the water (presumably also mgh, assuming that energy is conserved), with a measurable rise in the water temperature measured in kelvins (after the paddle wheel had stopped turning). It was found for water that the energy required to raise the temperature of one gram of water by one kelvin (1 K) is 4.2 joules. The "heat capacity" of an object is the amount of energy required to raise its temperature one kelvin. The "specific heat capacity" is the heat capacity on a per-gram or per-mole or per-atom basis, and is a property of the material. The specific heat capacity of water is 4.2 joules per gram per kelvin (Figure 6.22). Other materials have different specific heat capacities. We will study an atomic theory of specific heat capacity in a later chapter.

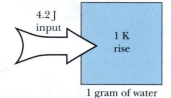

Figure 6.22 Energy input of 4.2 joules into a gram of water raises the temperature by 1 K. We say that the specific heat capacity of water is 4.2 J/gram/K.

Volume and temperature

An ordinary thermometer depends on the effect that higher temperature makes the material expand. Why does this happen?

For large oscillations of atoms whose interaction is described by a potential energy such as that shown in Figure 6.23, the average stretch increases. Because the actual potential energy curve is not really symmetric around the equilibrium point, with increasing energy the average interatomic bond length gets slightly longer. This is why an object typically expands at higher temperature. Higher temperature implies larger amplitude of oscillations and higher energy, and there is a slight shift in the center of the oscillations at higher energy.

In the interior of the solid the net potential curve for an atom is symmetric, not asymmetric, because the potential curve is associated with forces exerted by atoms to the left and to the right. But since the bond lengthening can start at the surface and propagate into the interior, a solid ultimately expands at higher temperatures.

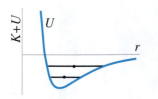

Figure 6.23 For larger oscillations the average stretch increases.

Other kinds of thermometers

There are many other kinds of thermometers. All materials change in some way when they get hotter or colder, and some of these effects are the basis for useful temperature indicators. A new kind of thermometer that is increasingly used is the liquid crystal thermometer. In a liquid crystal there is some molecular order over long distances in the liquid, unlike a true liquid in which there is hardly any long-range order. In one form of liquid crystal thermometer, helical molecules remain roughly parallel to each other, and the distance between coils in the helix changes with temperature. Because the distance between coils is comparable to the wavelength of visible light, this distance affects the way light is reflected from the liquid crystal. The effect is that the material changes color with changing temperature.

Calculations using specific heat capacity

Joule found that it takes 4.2 J of energy input to a gram of water to raise its temperature 1 K. If such experiments are done on other materials, the temperature rise is found to be different. For example, the specific heat capacity of ethanol is found to be 2.4 J/gram/K, and the specific heat capacity of copper is only 0.4 J/gram/K. One of the many unusual and useful properties of water is that it has a very large specific heat capacity on a per-gram basis, which means that it is difficult to change its temperature. Here is the meaning of specific heat capacity on a per-gram basis, where $\Delta E_{thermal}$ is the rise in the energy of the system, in the form of increased atomic kinetic and potential energy:

SPECIFIC HEAT CAPACITY C ON A PER-GRAM BASIS

$$\Delta E_{thermal} = mC\Delta T, \text{ or } C = \frac{\Delta E_{thermal}}{m\Delta T} \quad (m \text{ in grams})$$

In principle, it should be possible to calculate the specific heat capacity of a material if something is known about its atomic structure, since a temperature rise is a measure of increased energy at the atomic level. Comparisons of calculations with experimental values of the specific heat capacity data are a good test of our understanding of atomic models. In later chapters you will learn how to calculate the specific heat capacity of a gas, and of a solid, based on the ball-and-spring model of a solid.

If the specific heat capacity of a material is known, the amount of energy transfer into an object can be determined by measuring the temperature rise.

Example: Energy input raises the temperature

You stir 12 kg of water vigorously, doing 36000 joules of work. If the water is well insulated (so that all of your energy input goes into increasing the energy of the water), what temperature rise would you expect?

$$\left(\frac{36000 \text{ J}}{12000 \text{ grams}}\right)\left(\frac{1}{4.2 \text{ J/gram/K}}\right) = 0.7 \text{ K}$$

Including units helps avoid making calculational mistakes. Note that a lot of work is required to make a small temperature change.

6.X.6 Niagara Falls is about 50 meters high. What is the temperature rise in kelvins of the water from just before to just after it hits the rocks at the bottom of the falls, assuming negligible air resistance during the fall and that the water doesn't rebound but just splats onto the rock? It is useful (but not necessary) to consider a 1-gram drop of water.

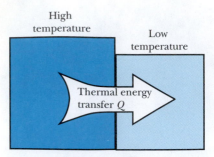

Figure 6.24 A hot block in contact with a cold block.

Energy transfer due to a temperature difference

Work is mechanical energy transfer into or out of a system involving forces acting through macroscopic displacements. When a hot object is placed in contact with a cold object, energy is transferred from the hot object to the cold one, but there are no macroscopic forces or displacements, so we don't refer to this kind of energy transfer as work.

Instead, we speak of "energy transfer due to a temperature difference" Q. At the microscopic level there is actual work; when a hot block is placed in contact with a cold block (Figure 6.24), at the interface the atoms in the two blocks collide with each other, and do work on each other. The atoms in the hot block have greater average kinetic energy than the atoms in the cold block, so in an individual collision it is likely that a fast-moving atom in the hot block loses energy to a slow-moving atom in the cold block.

It can happen that the atom in the hot block happens to be moving slowly and gains energy when it is hit by an atom in the cold object that happens to be moving fast, but this is less likely. On average there is energy flow ("microscopic work") from the hot block to the cold block. These energy changes will diffuse throughout the blocks. If left to themselves, the two blocks will eventually come to the same temperature, intermediate between their two initial temperatures.

> ### Q = ENERGY TRANSFER DUE TO A TEMPERATURE DIFFERENCE
>
> The symbol Q represents an amount of energy that flows from the surroundings into a system, due to a temperature difference between the system and the surroundings. One can also call this process "microscopic work."

If we choose just the initially cold block as our system of interest, this system gains energy (at the expense of the surroundings, the initially hot block). At the atomic level, this energy increase is the result of work done on the atoms at the surface of the cold block. Usually we are unable to observe these atomic-level interactions directly. We can infer the amount of energy transfer into the system by observing the temperature rise of that system, if we know the specific heat capacity of the material.

It is common practice to denote energy transfer due to a temperature difference (microscopic work) by the letter Q and work (macroscopic, mechanical energy transfer) by the letter W. The energy equation for an "open" (not isolated) system is then

$$\Delta E_{system} = Q + W$$

That is, energy transfer due to a temperature difference and mechanical energy transfers into or out of the system produce a change in the energy *of* the system. The energy of the system includes particle energy and potential energy, at the microscopic or macroscopic level.

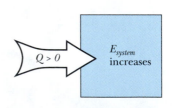

Figure 6.25 If energy flows from the surroundings into the system, Q is positive.

Sign of Q

Like W, Q can be negative. If there is energy transfer out of a system, we give Q a negative sign, because the change in the energy of the system is a decrease. See Figure 6.25 and Figure 6.26. Typically such transfer would be due to the system having a higher temperature than the surroundings. Similarly, if the system does work on its surroundings rather than the other way round, the sign of W is negative. (Some older textbooks on thermal physics reverse the sign of W, counting energy outputs from the system as positive.)

We believe that for the Universe as a whole Q and W are zero, and the energy of the Universe does not change. But even a small system may be ef-

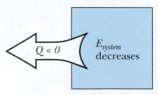

Figure 6.26 If energy flows from the system into the surroundings, Q is negative.

fectively closed, with its energy unchanging, if it is thermally well insulated (to prevent the flow of energy due to a temperature difference) and if there are no other energy transfers from the surroundings.

Terminology issues

The technical meaning of the word "heat" in science is not supposed to be the same as its everyday meaning, as is also the case with the word "work." In science the word "heat" is supposed to be reserved to refer to energy transfer due to a temperature difference (microscopic work) across a system boundary (Q). Rather than saying "there is heat in the object," one says there is energy in the object. The amount of energy in the object might increase due to "heat" (microscopic work). However, even professional scientists sometimes slip in their usage of the word "heat"!

To avoid confusion, we will avoid the use of the word "heat" as a noun. Instead, we will speak of "energy transfer due to a temperature difference" between system and surroundings (Q), and change of "thermal energy" inside the system $\Delta E_{\text{thermal}}$ (in the form of increased atomic kinetic energy and potential energy).

> *6.X.7* Suppose you warm up 500 grams of water (half a liter, or about a pint) on a stove while doing 5×10^4 J of work on the water with an electric beater. The temperature of the water is observed to rise from 20° C to 80° C. What was the change in the thermal energy of the water? Taking the water as the system, how much transfer of energy Q due to a temperature difference was there across the system boundary? What was the energy change of the surroundings?

Other kinds of input

Work and energy transfer due to a temperature difference are very common kinds of energy transfers into or out of an open system. Mass transfer is another mechanism for changing the energy of a system; a simple example is putting gasoline in a car.

Also, electromagnetic radiation, including visible light, infrared light, gamma rays, x-rays, and ultraviolet light, can carry energy into or out of an open system. Sunlight absorbed by the Earth raises the temperature (and the thermal energy) of the daylight side of the Earth. We will typically account for this kind of energy transfer in terms of photons (packets of energy) crossing the system boundary, which is a kind of change of identity for the system.

Example: **Specific heat capacity**

A 300 gram block of aluminum at temperature 500 K is placed on a 650 gram block of iron at temperature 350 K in an insulated enclosure. At these temperatures the specific heat capacity of aluminum is approximately 1.0 J/K/gram, and the specific heat capacity of iron is approximately 0.42 J/K/gram. Within a few minutes the two metal blocks reach the same common temperature T. Calculate T.

> System: the two blocks
> Surroundings: due to the insulation, no objects exchange energy with the chosen system
>
> Initial state: different temperatures, not in contact
> Final state: blocks have come to thermal equilibrium, same T.
>
> Energy principle: $\Delta E_{Al} + \Delta E_{Fe} = 0$

(The total energy of the two blocks does not change, because there is no energy transfer from or to the surroundings.)

$$(300 \text{ gram})(1.0 \text{ J/K/gram})(T - 500) +$$

$$(650 \text{ gram})(0.42 \text{ J/K/gram})(T - 350) = 0$$

Solving for the final temperature, we find $T = 429 \text{ K}$.

In words, what happens in this process is that the aluminum temperature falls from 500 K to 429 K, and the iron temperature rises from 350 K to 429 K. The thermal energy decrease in the aluminum is equal to the thermal energy increase in the iron.

6.5 REFLECTION: FORMS OF ENERGY

There are three fundamental kinds of energy in a multiparticle system:
- rest energy
- kinetic energy
- potential energy

All other forms of energy are examples of these basic forms. For example, chemicals in a system can react with each other and raise the temperature (and thermal energy) of the system, so we speak of "chemical energy," of which food is an example, with the molecule ATP providing energy storage in the body. The "chemical energy" is the kinetic energy and electric potential energy of molecules. Change of shape (configuration) is associated with change of potential energy, and there may also be changes of the kinetic energy of the molecules.

Nuclear energy is similar. Nucleons can be more or less tightly bound to particular nuclei, associated with different amounts of nuclear potential energy (change of shape) and different amounts of kinetic energy of the nucleons. Also, a change of identity in a nuclear reaction can be associated with a (huge) kinetic energy change associated with a change of rest mass.

6.6 POWER: ENERGY PER UNIT TIME

In technical usage, the word "power" is defined to mean "energy per unit time." From the point of view of energy usage, it makes no difference whether you take a minute or a month to climb a stairs, yet from a practical point of view you certainly do notice the *rate* of energy usage, called power. The units of power are joules per second, or watts (honoring James Watt, the developer of the first efficient steam engine).

POWER

Energy per unit time (J/s or watts)

There is a useful formula for the instantaneous power associated with the work done by a force:

INSTANTANEOUS POWER

$$\text{power} = \frac{dW}{dt} = \frac{\vec{F} \bullet d\vec{r}}{dt} = \vec{F} \bullet \frac{d\vec{r}}{dt} = \vec{F} \bullet \vec{v}$$

Example: Light bulb power

How much energy is required to run a 100-watt light bulb for an hour?

$$(100J/s)(1 \text{ hr})\left(60\frac{\text{min}}{\text{hr}}\right)\left(60\frac{\text{s}}{\text{min}}\right) = 3.6 \times 10^5 \text{ J}$$

6.X.8 In the Niagara Falls hydroelectric generating plant, the energy of falling water is converted into electricity. The height of the falls is about 50 meters. Assuming that the energy conversion is highly efficient, approximately how much energy is obtained from one kilogram of falling water? Therefore, approximately how many kilograms of water must go through the generators every second to produce a megawatt of power (10^6 watts)?

6.X.9 A vehicle with a mass of 1000 kg has an engine whose maximum power output is 50 kilowatts (about 67 horsepower; one horsepower is 746 watts). At a speed of 20 m/s (about 45 miles/hour), what is the maximum acceleration possible?

6.7 OPEN AND CLOSED SYSTEMS

We have already had some experience with the difference between a closed system and an open system. Because of the importance of this distinction, in this section we will go more deeply into the issues. Consider this diagram showing transfers of money into and out of a bank during a particular time period, and the corresponding change in accounts inside the bank (Figure 6.27).

? We have shown a situation where more money is transferred into the bank than is transferred out. During these transactions, does the total amount of money inside the bank remain unchanged?

We say that the bank is an "open system"—a portion of the world open to transfers in and out, and therefore subject to changes in its internal amount of money. During times when the bank does not permit transfers, the bank is temporarily a "closed system" and its total internal amount of money is unchanged, although there may be changes in the form of money inside the bank such as transfers between checking accounts and savings accounts.

Now consider Figure 6.28 which shows energy transfers during a particular time period in the winter, into and out of a house that is heated by gas, and the corresponding change in the energy inside the house. This is a situation where more energy comes into the house than goes out, and as a result the temperature inside the house rises.

? During these energy transactions, the total amount of energy in the whole Universe is unchanged, but is the amount of energy inside the house unchanged?

The amount of energy inside the house increased by 2000 joules. We say that the house is an "open system"—a portion of the Universe open to energy transfers, and therefore subject to changes in its internal amount of energy. If the house could be perfectly insulated against heat leakage (and solar radiation), and the gas (and electricity) turned off, the house would be a "closed system" with respect to energy and its total internal amount of energy would be unchanged, although there may be changes in the form of energy inside the house such as the family dog converting chemical energy into kinetic energy when chasing its tail.

These considerations lead to the following scheme for keeping track of energy. Choose some portion of the Universe and mentally surround it by a dashed line marking the boundary (Figure 6.29). Then the energy transfer into the system across the system boundary minus the energy transfer

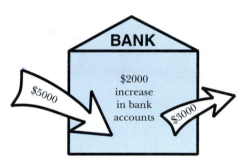

Figure 6.27 A bank is an open system.

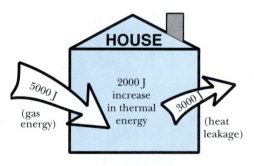

Figure 6.28 A house is an open system.

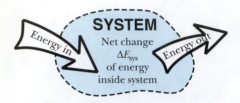

Figure 6.29 The change of energy of the system is equal to energy transferred into the system across the system boundary minus energy transferred out of the system across the system boundary.

out of the system across the system boundary is equal to the change of the energy inside the system.

The change in energy *inside* the system, ΔE_{sys}, can be either a positive or a negative amount of energy. For example, take an electric car as the system of interest:

Charge the battery: ΔE_{sys} is positive (more energy stored in the battery).

Do work on the car by pushing it: ΔE_{sys} is positive (more kinetic energy).

Run the headlights (radiate electromagnetic energy): ΔE_{sys} is negative (less energy stored in the battery).

The car is an open system whose total amount of energy changes due to energy transfers into or out of the system: the energy of an *open* system is *not* constant.

In contrast, the total energy of the Universe remains constant, because there are compensating energy changes in the surroundings of the car: the electric company lost a store of chemical energy by running its generators to charge the car, you used up some chemical energy to push the car, and the light from the headlights is absorbed by the surroundings and leads to a rise in temperature of the surroundings. The measurements that we are able to make confirm the premise that the Universe is a closed system whose total amount of energy never changes, although the form of the energy may change.

For any closed system, *inflow* = *outflow* = 0; $\Delta E_{sys} = 0$.
The energy of a closed system does not change.

The Universe as a whole is the most important example of a closed system, but it is often the case that a portion of the Universe can be considered to be a closed system, at least approximately. For example, put hot water and ice cubes into a very well-insulated container. During the short time that the ice melts and the water gets somewhat cooler, we can neglect the small amount of energy leakage through the walls of the insulated container. There is an increase in the energy of the ice cubes (which changed from solid to liquid), and a decrease in the energy of the hot water, but negligible net change in the energy inside the container, which is approximately a closed system whose total energy is (approximately) unchanged.

6.8 THE CHOICE OF SYSTEM AFFECTS ENERGY ACCOUNTING

What you choose to include in your chosen system affects the form of the energy equation for that system. For example, when we analyze a system consisting of an airless Earth plus a ball moving upward, the energy of the system does not change, because there are no energy transfers into or out of our chosen system (Figure 6.30).

System of Earth + ball

System: Earth + ball

Surroundings: Nothing significant

Initial state: Ball moving upward

Final state: Ball has risen a distance h, and is momentarily at rest

Energy Principle:

$$E_f = E_i + W$$

$$K_f + U_f = K_i + \cancel{U_i} + \cancel{W}$$

$$0 + mgh = \frac{1}{2}mv_f^2$$

Since only the difference in U matters, we can set $U_i = 0$.

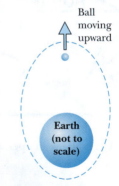

Ball moving upward

Earth (not to scale)

Figure 6.30 System: Earth + ball.

System: Ball

On the other hand, when we analyze a system consisting of just the ball (Figure 6.31), the Earth (which is now outside our chosen system) does work on the ball, and the energy of the ball increases:

System: Ball
Surroundings: Earth
Energy Principle:

$$E_f = E_i + W$$

$$K_f = K_i + W$$

$$0 = \frac{1}{2}mv_f^2 - mgh$$

There are no potential energy terms for a single particle system. The Earth does negative work because force is opposite to displacement.

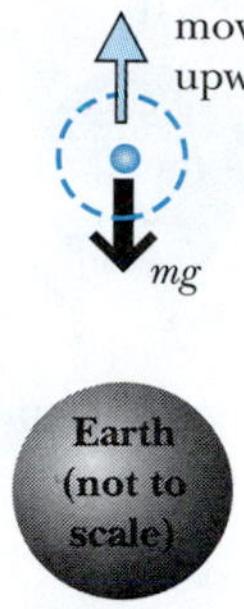

Energy terms move back and forth across the equal sign in the energy equation, depending on our choice of the system of interest. It is extremely important to be clear about the choice of system, because this affects the form of the energy equation, and helps avoid serious mistakes.

Figure 6.31 System: ball only.

Warning: avoid double counting!

? A student said, "The Earth exerts a downward force mg as the ball rises a distance h and does an amount of work $-mgh$, and there is also an increase in gravitational potential energy $+mgh$, so the kinetic energy decreases by an amount $-2mgh$." What is wrong with the student's statement?

The problem here is double counting due to a lack of clarity about the choice of system. If the Earth is part of the system, the force mg is internal to the system and does no external work on the system; rather, there is work done by internal forces on both the Earth and the ball, and the negative of this work by internal forces is the change in potential energy. On the other hand, if the ball alone is the system, there is no shape change of the ball and no change of gravitational potential energy, but there is an external force mg which does an amount of work $-mgh$.

In the next section we will illustrate the issue of choice of system by analyzing the same problem three times, using three different choices of system. We will find that the form of the Energy Principle will be different in each case, but we obtain the same physical results each time.

The woman, the barbell, and the Earth

To help avoid double-counting mistakes, it is useful practice to write the energy equation for various choices of system in a slightly more complicated situation, that of a woman applying a constant force F to lift a barbell of mass m from rest through a distance h, at which point the barbell is not only higher above the Earth but has also acquired some speed v. We're considering a time when the barbell is still headed upward, before the woman has brought the barbell to a stop above her.

For convenience in the following discussion, let E_w represent the following energy terms:

E_w = chemical energy of woman +
        kinetic energy of woman (moving arm) +
        gravitational energy of woman and Earth +
        thermal energy of woman

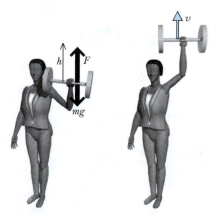

Figure 6.32 A woman lifts a barbell a distance h, exerting a force F. In the final state the barbell is moving upward with speed v.

System: Woman + barbell + Earth

System: Woman + barbell + Earth (Figure 6.33)
Surroundings: Nothing significant

Initial state: Barbell at rest
Final state: Barbell has moved upward a distance h, and has speed v

Figure 6.33 System: woman + barbell + Earth.

Energy Principle:

$$E_f = E_i + W$$

$$K_f + U_f + E_{W,f} = \cancel{K_i} + \cancel{U_i} + E_{W,i} + \cancel{W}$$

$$\frac{1}{2}mv_f^2 + mgh + E_{W,f} = E_{W,i}$$

If we choose the system to consist of the woman, the barbell, and the Earth, there are no energy transfers into or out of our system, so energy does not change, and the energy equation looks like this as the woman lifts the barbell a distance h, starting from rest (we neglect the tiny kinetic energy of the recoiling Earth). This allows us to calculate the change in the woman's energy from our final equation:

$$\frac{1}{2}mv^2 + mgh + \Delta E_w = 0$$

The energy term ΔE_w represents changes in energy associated with the woman, mainly a decrease in her stored chemical energy, because she must perform chemical reactions with food (or with chemicals stored in body tissues after eating food) in order to be able to move her arm. This term also includes an increase in the kinetic energy of her arm, an increase in gravitational potential energy of woman and Earth associated with her arm being higher, and an increase in her thermal energy (because her temperature rises a bit when she exerts herself).

? The energy term ΔE_w includes one decrease and some increases. What is the net sign of ΔE_w, positive or negative?

The net effect must be a decrease in her energy, because the barbell and the Earth are farther apart, and the barbell goes faster, both of which represent energy increases. The woman has to burn more energy than the rest of the system gets, since she has to supply energy to the rest of the system and also supply her own energy increases (increased kinetic energy and gravitational potential energy associated with her arm, and increased thermal energy).

System: Barbell

System: Barbell (Figure 6.34)
Surroundings: Woman and Earth

Initial state: Barbell at rest
Final state: Barbell has moved upward a distance h, and has speed v

Energy Principle:

$$E_f = E_i + W$$

$$K_f = \cancel{K_i} + W_{Woman} + W_{Earth}$$

$$\frac{1}{2}mv_f^2 = Fh - mgh$$

Figure 6.34 System: barbell only.

If we choose the system to consist just of the barbell, the woman does (positive) work $+Fh$ on the system, and the Earth does (negative) work $-mgh$ on the system (because mg is down and the displacement h is up). Compare this equation to the equation we obtained when considering the system of woman+barbell+Earth. We see that energy terms have moved to the other side of the equation as a result of our different choice of system.

? Which is bigger, F or mg? Why?

Since the barbell's upward momentum increases, F must be larger than mg, since the net force in the upward direction is $F-mg$.

? Compare the equations for system 1 (woman + barbell + Earth) and system 2 (barbell only). In terms of F, calculate ΔE_w.

$$\tfrac{1}{2}mv^2 + mgh + \Delta E_w = 0 \quad \text{and} \quad \tfrac{1}{2}mv_f^2 = Fh - mgh$$

By comparing the two equations, we see that the change in the energy associated with the woman must be $-Fh$. The sign seems right: we had concluded earlier that her energy change must be negative.

System: Barbell + Earth

System: Barbell+Earth
Surroundings: Woman

Initial state: Barbell at rest
Final state: Barbell has moved upward a distance h, and has speed v

Energy Principle:
$$E_f = E_i + W$$
$$K_f + U_f = \cancel{K_i} + \cancel{U_i} + W_{Woman}$$

$$\tfrac{1}{2}mv_f^2 + mgh = Fh$$

If we choose the system to consist of the barbell plus the Earth (Figure 6.35), this system changes shape as the woman pushes the two pieces of the system apart. It is almost as though she is stretching an invisible spring connecting the two objects.

? Does the woman do any work *on the Earth*? Why or why not?

The woman's feet push down on the Earth but do negligible work, because there is essentially no displacement there (the Earth hardly moves). Work done by a force is calculated by taking into account the displacement of the point where the force is applied. In this case, the point of application hardly moves, so the force does hardly any work.

Further discussion

Looking back over the equations obtained for different choices of system, you can see how energy terms that represent transfers across the system boundary for one choice of system become changes of energy *inside* the system for a different choice of system. Comparing equations for different choices of system can be useful in determining an unknown quantity (such as the energy change in the woman, $-Fh$, in the example you just worked through). Also, analyzing a process for more than one choice of system is a good check on your calculations and your understanding.

> *6.X.10* Consider a harmonic oscillator (mass on a spring without friction). Taking the mass alone to be the system, how much work is done on the system as the spring contracts from its maximum stretch A to its relaxed length? What is the change in kinetic energy of the system during this motion? For what choice of system does energy remain constant during this motion?

6.9 ENERGY DISSIPATION

The total energy of the Universe does not change, but useful energy is often "dissipated" into forms less useful to us. Push a chair across the floor, and some of the work that you do goes into raising the temperature of the floor and the chair, rather than into increasing the macroscopic kinetic energy of the chair. Throw a ball up into the air, and some of the initial energy is

Warning! It is tempting, but wrong, to include a term for the change in the gravitational energy of the barbell and to write the energy equation for the barbell alone like this:

$$\left(\tfrac{1}{2}mv^2\right) + (mgh) = (+Fh) + (-mgh)$$

The error here is in assigning some gravitational energy to the barbell alone. Gravitational potential energy is associated with a change in the shape of a multi object system; in particular, to changes in the separations of pairs of interacting objects. The barbell alone is not a pair of objects, and does not change shape.

Figure 6.35 System: barbell+Earth.

dissipated into increased microscopic energy of the air. Sliding friction, air resistance, viscous friction—all of these phenomena are examples of energy dissipation discussed in the rest of the chapter.

6.10 AIR RESISTANCE

Figure 6.36 A sequence of video frames of a falling metal ball, which travels farther in each frame, reflecting the increase in its speed.

We begin our study of energy dissipation by considering air resistance. Air resistance isn't a major factor when you drop a metal ball a short distance. A video sequence of a falling metal ball shows that the ball moves faster and faster as it falls, an effect that is hard to observe by eye alone. The time between adjacent frames in Figure 6.36 is 1/15 second, and the increasing distances between heights of the ball in adjacent frames show that the speed is getting faster and faster. (Also note the increasing blur due to faster motion of the ball.) The gravitational attraction of the Earth acting on the ball makes the momentum of the ball continually increase. A curve is drawn along the tops of the ball images. The major visible marks on the vertical meter sticks are 10 cm apart. In 7/15 second the ball falls about 1 meter.

Terminal speed

In contrast, it is observed that although a sky diver initially speeds up due to the gravitational force acting downward, the sky diver's speed does *not* keep getting bigger and bigger. Rather the sky diver eventually reaches a "terminal speed" and falls thereafter at constant speed despite the gravitational force. This terminal speed for falling humans is so high (about 60 m/s, or about 135 miles per hour) that hitting the ground without a parachute is normally fatal, although there have been cases of people hitting deep snow at terminal speed and surviving.

Similarly, if you drop a bowl-shaped paper coffee filter, a video sequence of the falling filter shows that the filter's speed does increase at first, but instead of continuing to gain speed it quickly reaches a constant terminal speed despite the gravitational force acting on it:

Figure 6.37 A sequence of video frames of a falling coffee filter, which speeds up briefly, then falls at constant speed.

In the video sequence shown above the time between adjacent frames is again 1/15 second. In 16/15 second the filter falls about 1 meter (the dense ball took only 7/15 second to fall that far). A curve is drawn through the centers of the coffee filter images. The nearly straight line in the later part of the motion indicates motion at constant speed.

The restraining effect of the air is called "air resistance" or "drag." You yourself have probably observed that it is harder to move something quickly through a fluid such as water or air than to move it slowly. This suggests that air resistance acting on a falling object might depend on the speed of the object.

? Consider a falling coffee filter at two different times: 1) while the filter's speed is still increasing, and 2) after terminal speed has been reached. How do you know the air resistance is larger at the second instant than at the first instant?

While the filter's speed is increasing there must be a nonzero net force, which means that the air resistance is smaller than the gravitational force (Figure 6.38). But later the net force is zero, because the speed is no longer increasing, which means that the force of the air has grown to be as large as the gravitational force (which is nearly constant at mg and in fact increases very slightly as the coffee filter gets closer to the Earth). Since the change in the density of the air is small over the distance the filter falls, and the change in the gravitational force is very small, we can conclude that the change in the air resistance is due to a speed dependence of that force.

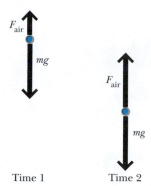

Dependence on cross-sectional area

We observe that the effect of air resistance increases as the cross sectional area of an object increases—a sky diver falls much more slowly with an open parachute than with a closed parachute. The low-density paper coffee filter has a large cross-sectional area, and a small gravitational force acts on it, so air resistance plays a major role. In contrast, a high-density metal ball has a small cross-sectional area, but a large gravitational force acts on it, and air resistance may be negligible in comparison with gravity.

Figure 6.38 Forces acting on a falling coffee filter at Time 1 when the filter's speed is increasing and at Time 2 when terminal speed has been reached.

Applications

Drop a ball. At the instant when the ball has nearly reached the ground, the gravitational potential energy of the Universe has decreased, with a corresponding increase in the kinetic energy of a portion of the Universe (the ball). In the presence of air, kinetic plus gravitational potential energy is only approximately constant. As the ball falls, it has to push aside the air underneath it, and the air molecules acquire additional kinetic energy K_{air}.

> *6.X.11* Therefore, in air will the *actual* final speed of the ball be greater or less than you would predict from the constancy of kinetic plus gravitational energy? Why?

> *6.X.12* A coffee filter of mass 1.4 grams dropped from a height of 2 m reaches the ground with a speed of 0.8 m/s. How much kinetic energy K_{air} did the air molecules gain from the falling coffee filter?

Mechanism of air resistance

We would like to understand the details of the interaction of the air with a falling object. In particular, we'd like to see if we can find a reason that the air resistance would depend on the speed of the object.

Here is a possible microscopic model to explain air resistance. When you hold a coffee filter stationary, air molecules hit it nearly equally from above and below, as seen in Figure 6.39. (Actually there are slightly more collisions per second on the lower surface than on the upper surface, leading to a small upward buoyant force which is very small compared to air resistance.) But when the filter is moving downward due to the gravitational attraction of the Earth, the bottom side of the filter is running into the air molecules, while the top side is moving away from the air molecules.

On average, the bottom side will have a significantly increased number of collisions per second with air molecules, and with greater impact. On the top side there will be a reduced number of collisions per second, and with less impact. There is a net upward push on the filter. This air resistance force increases with increasing downward speed of the filter, because higher speed increases the rate and impact of collisions on the bottom, and decreases the rate and impact of collisions on the top.

Eventually, when the downward speed of the coffee filter has become big enough, the net upward push by the air molecules becomes as large as the downward gravitational pull of the Earth. From then on the filter falls at a constant speed (terminal speed).

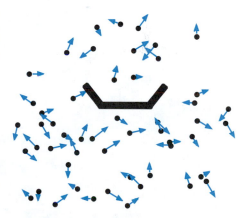

Figure 6.39 Air molecules collide with a falling coffee filter.

While this model captures the qualitative aspects of the situation, a simple quantitative calculation along these lines (using techniques discussed in a later chapter on gases) predicts a terminal speed that is *a thousand times smaller* than is observed! This is a spectacular failure of a simple model. The problem is that the falling coffee filter establishes macroscopic "wind"-like motions in the air, and the random motions of the air molecules at any particular location are relative to the bulk wind motion at that location. The techniques of "fluid dynamics" are capable of analyzing such phenomena, but these techniques are mathematically quite difficult and well beyond the scope of this introductory textbook. (Our later study of gases will address simpler situations, where there is no large-scale wind motion.)

Dependence of air resistance on speed

We established that the force of air resistance is dependent on the speed of a falling object.

? Did we establish that the force F_{air} of the air is proportional to the speed v of the falling object?

All we know is that the air resistance increases with increasing speed, but it need not be simply proportional to v. For all we know the force could be proportional to $\sqrt{v}$ or v^2 or v^3. There is a simple experiment you can do to determine the form of the speed dependence, at least for low speeds. This is the topic of 6.P.39 Experiment—falling coffee filters .

Other aspects of air resistance

? If you could double the cross-sectional area without changing the mass or the shape of the filter (by making the filter out of thinner paper), what change would you expect in the magnitude of the air resistance force at a particular speed? If you double the density of the air through which the filter falls, what change would you expect?

Measurements of a variety of ordinary-sized objects moving through air or other fluids show that the friction force is in fact proportional to the cross-sectional area A (hitting more molecules per second.) Air resistance is also proportional to the density ρ of the air (again, hitting more molecules per second), so there is less air resistance at higher altitudes. For ordinary-sized blunt objects such as a coffee filter moving at ordinary speeds, the force is found to be approximately proportional to the square of the speed, which is probably what you found in your own experiment. Combining these effects, the air resistance is approximately proportional to $\rho A v^2$.

? There is another important effect. From your own experience, would you expect larger air resistance (at a particular speed) for a pointed object or a blunt object?

As you might expect, the air resistance force depends not only on density, cross-sectional area, and speed, but also on the shape of the object. There is a larger air resistance force on a blunt object such as a coffee filter than on a ball or a streamlined object such as an arrow. This effect, and the dependence on v^2 , are not easily explainable in quantitative detail in terms of a molecular model without performing a large molecular dynamics calculation.

The shape effect is usually taken into account by measuring the air resistance force (for example, by measuring the terminal speed) and determining an experimental "drag coefficient" C such that the following formula expresses the experimental results for the friction force:

APPROXIMATE AIR RESISTANCE FORCE (EMPIRICAL)

$F_{air} \approx \frac{1}{2} C\rho A v^2$, where $0.3 \le C \le 1.0$ (blunt objects, ordinary speeds)

The direction is opposite to the velocity: $\vec{F}_{air} \approx -\frac{1}{2} C\rho A v^2 \hat{v}$

Blunter objects have a higher value of C.

It is important to keep in mind that this is *not* a fundamental force law like the law of gravitation. Air resistance is the average result of a huge number of momentary contact electric interactions of air molecules hitting atoms on the surfaces of the falling object, and the average net air resistance is very difficult to calculate from fundamental principles at the molecular level.

Motion through Earth's atmosphere

How important is the effect of air resistance in the everyday world? In Chapter 2 we analyzed the motion of a ball thrown through the air. However, we neglected air resistance, which can have a sizable effect. In Problem 6.P.40 you are asked to include the air resistance force given above. You find that a baseball thrown at high speed by a professional baseball pitcher goes only about half as far in air as it would go in a vacuum.

A spinning ball

Even when we include air resistance in our calculations, we still have ignored a force that can have a major effect on the trajectory. If a ball has spin, there is an effect of fluid flow around the ball that raises the air pressure on the side where the rotational motion is in the same direction as the ball's velocity, and lowers the air pressure on the other side, where the rotational motion is in the opposite direction to the velocity. If the force points to the side, the ball curves; this is an important effect in baseball pitching. If the force points upward (due to "backspin," as in Figure 6.40), it lifts the ball and extends the range. If the force points downward (due to "topspin") it decreases the range. The calculations involved in modeling this effect are quite complex, and involve "fluid dynamics" (the study of fluid flow).

Viscous friction

Very small particles such as dust or droplets of fog falling slowly through air, or small objects moving through a thick liquid such as honey, experience a friction force that is proportional to v rather than v^2. This is called "viscous" friction. The approximate dependence on v or v^2 for air resistance is roughly valid only for specific situations. The formulas for friction do not have the wide applicability of such major physical laws as the law of gravitation. However, even approximate formulas for friction make it possible to make much better predictions than we could make if we ignore friction completely.

6.11 MOTION ACROSS A SURFACE: SLIDING FRICTION

Sliding friction between two solid objects is a familiar interaction that involves energy dissipation. We'll consider it in some detail. Connect a spring to a block and stretch the spring an amount s to the right without moving the block (Figure 6.41). We need to identify interactions and draw forces on the block on a physics diagram.

Non-contact forces

? What objects exert non-contact forces on the block?

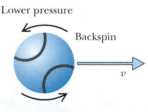

Figure 6.40 A spinning ball experiences additional forces.

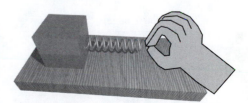

Figure 6.41 Stretch the spring without moving the block.

Figure 6.42 Physics diagram (incomplete) showing all non-contact forces.

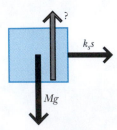

Figure 6.43 Physics diagram (incomplete) showing some contact forces as well as non-contact forces.

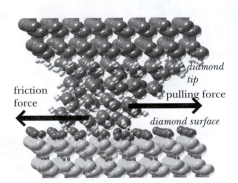

Figure 6.44 A computer simulation of sliding friction between a diamond tip and diamond surface. Image courtesy of Judith A. Harrison, U.S. Naval Academy.

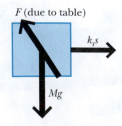

Figure 6.45 A complete physics diagram for the block.

The Earth exerts a force on the block straight down, in the $-y$ direction (Figure 6.42). Your hand doesn't touch the block, so if it exerts a force on the block it must exert a non-contact force. How could it do this? The hand can exert a very small gravitational force on the block, but this force is so small that we can neglect it. The hand and block are not electrically charged. There is no mechanism for a significant interaction between the hand and the block, so we can't list the hand as an object that interacts with the block.

Contact forces

? What objects exert contact forces on the block?

The spring touches the block, and exerts a force to the right (in the $+x$ direction) of magnitude $k_s s$. The table touches the block. What is the direction of the force exerted on the block by the table? Clearly the table exerts a force on the block upward, in the $+y$ direction (Figure 6.43). In Chapter 4 we examined the interactions between a brick and a table, and found that the table exerted a force upward on the brick as the atoms in the table were compressed by the brick. A similar interaction will happen between the block and the table, and the upward force by the table on the block can balance the downward gravitational force exerted on the block by the Earth.

Write the Momentum Principle and motion information

? Is the momentum of the block changing? Is your answer consistent with the system diagram in Figure 6.43?

We are left with a problem. Although the force exerted on the block by the table is upward, we can understand why the y component of the block's momentum is not changing, because there is a downward gravitational force. However, there appears to be a component of the net force on the block to the right, yet the block is at rest—the block's momentum is not changing. This is not consistent with the Momentum Principle. We must have failed to identify a force.

What could be exerting a force to the left on the block? The only object that can possibly be exerting a force to the left to balance the spring force is the table. How does it do that? Atoms of the block "run into" atoms in the table, and the resulting interactions stretch and compress interatomic bonds in the objects. Figure 6.44 provides an atomic-level picture of such an interaction.

This image is a single frame from a molecular dynamics computer simulation of sliding friction between two objects, performed by Judith A. Harrison and coworkers at the United States Naval Academy in Annapolis, Maryland. The computations are similar to the numerical integrations you have done, but involving a large number of objects (atoms). The forces between the atoms are modeled by spring-like forces.

In this image, a diamond tip is dragged to the right across a diamond surface. The large spheres represent carbon atoms, and the small ones are hydrogen atoms on the carbon surfaces. Note that the projecting tip has "caught" on the lower surface, causing interatomic bonds in the two objects to deform. The stretching and compression of interatomic bonds produces the horizontal component of the force exerted on the moving object by the lower object. For more information about Harrison's work, see

http://www.chemistry.usna.edu/jah/ResearchStuff/Friction_page.html

Figure 6.45 shows a complete physics diagram for the block. The force exerted by the table acts at an angle. It is common practice to replace the force vector exerted by the table with two equivalent vectors, a compression force F_N normal to the surface and a friction force f parallel to the surface, as shown in Figure 6.46. This turns out to be very useful for many reasons, but it is important to remember that an atom in the table exerts one interatomic force on a neighboring atom in the block, not two. It is for our own convenience that we split this force into two equivalent forces.

We can construct the component equations for the Momentum Principle by reading the forces off of our physics diagram. We choose axes with x to the right and y up (z is outward). We know that the block remains at rest even though we are pulling on it, so we can write motion information in addition to the Momentum Principle:

$$\frac{d\vec{p}}{dt} = \langle (k_s s - f), (F_N - Mg), 0 \rangle$$

$$p_x = 0 \text{ at all times, so } dp_x/dt = 0$$

$$p_y = 0 \text{ at all times, so } dp_y/dt = 0$$

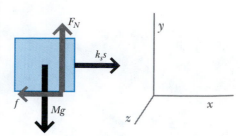

Figure 6.46 Complete physics diagram for the block; the force on the block by the table has been resolved into components normal to and parallel to the surface.

Solve equations

? What is the magnitude of the compression force F_N? What is the magnitude of the friction force f?

From $(F_N - Mg) = 0$ we deduce that the compression force is just Mg; the table supports the block. From $(k_s s - f) = 0$ we deduce that the friction force must be equal to the spring force. If you reduce the stretch of the spring, the friction force f gets smaller, too, because the sideways interatomic forces are smaller. As long as the block remains at rest, the friction force has a range of values equal to the range of spring forces applied to the block. However, this is not the case when the block is in motion.

Note that this is another example of a situation where we don't have a force law for one of the forces, the force exerted by the table (which we split into two components). We are able to deduce that $F_N = Mg$ and $f = k_s s$ thanks to the motion information that supplements our incomplete force information.

Block in motion with sliding friction

Now slowly stretch the spring more and more, and eventually the block moves to the right, which means that during the instant when it began moving, its momentum changed. By the Momentum Principle, that implies that the spring force $k_s s$ to the right became greater than the friction force f to the left. Evidently there must be a maximum possible friction force f_{max} for this object on this table (it is different for different surfaces).

If we apply a larger spring force than this maximum friction force, we break temporary interatomic bonds between the block and the table, and we can speed up the block. Measurements on sliding objects show that the friction force remains approximately equal to f_{max}, independent of the speed of sliding. This implies that the friction force is related to the strength of the interatomic bonds, which are continually made and continually broken as the block slides along.

In contrast, remember that air resistance depends on the speed of the moving object. The mechanism for air resistance is very different from that of sliding friction and leads to a dependence on the speed.

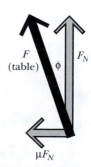

Figure 6.47 Sliding friction approximately independent of the speed.

$$\tan \phi \approx \frac{\mu F_N}{F_N} = \mu$$

Coefficient of friction

It is approximately true for many objects on many surfaces that the maximum friction force f_{max} is proportional to the amount of compression F_N, with a proportionality constant μ that is a property of the two surfaces that are in contact (Figure 6.47):

$$f_{max} \approx \mu F_N$$

Typical values for the "coefficient of friction" μ for wooden blocks sliding on wooden tables are in the range of 0.25 to 0.5 (μ is Greek lowercase mu).

Qualitatively, the bigger the compression force, the deeper the atoms of one object penetrate into the other object, and the harder it is to drag one object across the other.

These approximate properties of sliding friction do not have the status of fundamental physical laws such as the law of gravitation. For example, the details of motion with friction depend critically on how clean the surface is. Very flat, very clean metal or glass plates may bind to each other so strongly that it is almost impossible to slide them.

When two very clean and flat metal surfaces are brought in contact, the interface region looks much like the interior of the metals, and the two metal objects may behave pretty much like a single block. Also, when surfaces are very flat, it is possible that air will be squeezed out from between the objects, in which case air pressure on the outer surfaces raises the compression force F_N to a very large value, and the maximum friction force μF_N is then very large. This is an example of a situation where F_N is much greater than the weight Mg of the block. What matters in the formula $f_{max} = \mu F_N$ is how much the two surfaces are squeezed together, not the weight of the block.

Dragging with\ constant speed

You can experience some of the vagaries of sliding friction by trying to drag a block at a constant speed. You will need to keep the stretch of the spring constant, so $k_s s = f_{max}$, and there should be no acceleration—no change in the constant velocity. You may find that it is very difficult to maintain a constant stretch of the spring (and a constant velocity of the block), in part because the block skips and loses some contact with the table. It can be easier to maintain constant velocity if the object is quite massive and therefore less subject to sudden large changes in velocity.

You may observe that you have to pull harder to break the block loose from the table and start it moving than you do to keep it moving at a constant speed. For some surfaces the maximum friction force for a block at rest is larger than the maximum friction force for a moving block. We will look at sliding friction in more detail in a later chapter, but a full understanding of this complex phenomenon is the subject of ongoing contemporary research.

An intriguing recent experimental result is that with extremely clean surfaces ("atomically" clean), it matters whether the interatomic spacings of the two objects are commensurate or not; that is, whether the grid of atoms of one surface fit into the grid of atoms of the other surface. If they do fit, like nested egg cartons, it takes a sizable force to pull the block up out of the atomic grid, and a smaller force to keep it going. But if there isn't a good fit, the slightest force will move the block.

Dragging with continually increasing speed

To drag with continually increasing speed, it is necessary for the pulling force to be greater than the friction force (Figure 6.48). Try dragging a block across a table by stretching the spring enough that $k_s s$ is greater than f_{max}, and try to maintain a constant stretch s in order to obtain a constant nonzero dv_x/dt. You may find that it is quite difficult to maintain a constant acceleration for very long. A practical difficulty is that as the speed increases, you need to move your hand faster and faster to maintain a constant large stretch in the spring, and you also may run out of room on the table. In Figure 6.48 we have assumed, based on experiments, that during sliding the frictional force remains approximately equal to f_{max}.

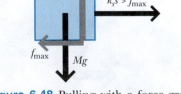

Figure 6.48 Pulling with a force greater than the maximum friction force.

> *6.X.13* If the mass of the block is M, what is the acceleration $d\vec{v}/dt$ of the block? (Apply the Momentum Principle along both the x and y axes.)

Energy dissipation

If you pull the block with a constant force F at constant speed through a distance d, you do an amount of work Fd, but the block's speed doesn't increase. The making and breaking of interatomic bonds at the sliding interface increases the oscillation of atoms around their equilibrium positions. This increased energy is then shared with many other atoms in the block and table, with the result that the average energy of atoms in both objects increases. This is the same thing as saying that the temperature rises in the block and in the table. You can observe such a rise in temperature by rubbing your hands together.

The energy is dissipated throughout the block and table, and cannot easily be recovered. There is a kind of irreversibility in a process where friction plays a role, because the dissipated or dispersed energy can't be brought back to the interface region and used to move the block. Note that no energy is actually lost. Rather the energy is dissipated among a large number of atoms and can't be re-ordered to make all the atoms move in the same direction.

A paradox involving friction

If we are careful in our energy bookkeeping for a system involving motion with friction, we encounter a paradox that cannot easily be resolved. Consider the situation shown in Figure 6.49.

You exert a constant force F, and pull a block a distance d across a table. Because there is friction between the block and the table, the block travels at constant speed. Since the momentum of the block does not change, we conclude correctly that the net force on the block is zero, and that therefore the magnitude of the friction force must also be F.

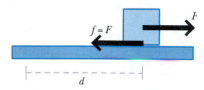

Figure 6.49 Forces in the x direction on a block pulled with constant speed across a table. (Forces in the y direction are not shown).

In terms of energy inputs, you have done an amount of work Fd on the block. The friction force, acting in the opposite direction, has apparently done an amount of work $-Fd$. The net work done on the block therefore would seem to be zero. This is consistent with the fact that the kinetic energy of the block did not seem to change.

It is not, however, consistent with the fact that both the block and the table get hotter as they rub together! If the net work done on the block was zero, where did that increased thermal energy come from?

This is not a "trick question," but a genuine puzzle. We will return to this issue in Chapter 8, when we will deal with it in enough detail to resolve the paradox.

6.12 FRICTION IN A MODEL OF A SPRING-MASS SYSTEM

One of the most obvious things about the motion of a real macroscopic spring-mass system is that the amplitude gets smaller and smaller with time. Our model of a spring-mass system that we developed in Chapter 4 is too simple: we should include the effects of friction. See 6.P.41 Numerical integration of spring-mass system with friction .

In the case of "viscous" friction, where the friction force is $\vec{f} = -c\vec{v}$, it is possible with elaborate math (or by guessing the solution and plugging it into the Momentum Principle for the system) to find an analytical solution, in which a cosine function is multiplied by an exponential factor:

$$Ae^{-\frac{c}{2m}t}\cos(\omega t)$$

The amplitude dies away exponentially with time. The angular frequency is

$$\omega = \sqrt{\frac{k_s}{m} - \left(\frac{c}{2m}\right)^2}$$

which is approximately $\sqrt{k_s/m}$ for low friction.

There is no such simple analytical solution if the friction force is independent of the speed (sliding friction) or proportional to v^2 (air resistance for large blunt objects).

6.13 FRICTION AND IRREVERSIBILITY

When there is friction, processes are not reversible, in contrast with situations describable in terms of potential energy. If energy dissipation is negligible, a bouncing ball continually returns to nearly its original position and momentum, and the process would run just as well backwards as forwards. A movie of the bouncing ball could be played backwards and it would look the same as when the movie is played forwards. But in the real world, energy is dissipated in the collision with the floor, and due to air resistance, so that the bouncing ball does *not* return to its initial position and momentum, and the process is not reversible. A movie of the process played backward would look quite odd, because the ball would bounce higher and higher.

Energy dissipation

In Section 6.3 we saw that potential energy is independent of path. In the absence of air resistance, a ball thrown up into the air will return to your hand with exactly the same kinetic energy it had when it left your hand. The situation is entirely different in the presence of friction. You throw the ball up in the air, and as it goes up it loses some energy to the air (air resistance), and on the way down it loses some more energy to the air. When it returns to your hand the net change in the kinetic energy is negative despite the fact that the net change in the gravitational energy is zero (same position). The round trip difference in the kinetic energy is *not* zero.

Similarly, in the presence of friction, when a mass attached to a spring returns to the same s, it has a smaller v. Again, the change in the kinetic energy of the mass during the round trip is not zero, even though the spring potential energy has not changed.

A particularly compelling case involving friction is that of pulling a block at constant speed to the right across a table with friction and then to the left back to its starting point. On the way to the right the block and table get hot, and on the way back to the left the block and table get even hotter (Figure 6.50 and Figure 6.51). In order to have no change in the round trip, the block and table would have had to cool down as you pulled the block to the left; this would be a very surprising effect indeed.

Note that the friction force exerted by the table points to the left on the way out, and then to the right on the way back. The direction of the friction force depends on the direction of the motion, not merely on position. An interaction that we can associate with potential energy would not behave this way. The force would have the same direction on the way out and on the way back ($F_x = -dU/dx$); one energy change would be positive and the other negative, making a net change of zero. Friction is a good example of an interaction for which we cannot define a potential energy, because the friction force depends on the direction of motion.

Forces which depend only on position, and for which it is possible to define an associated potential energy (with the force equal to the negative gradient of the potential energy), are called "conservative forces." Friction is not a conservative force.

A key issue is that we define potential energy in terms of the interactions of point particles, or particle-like objects whose internal state doesn't change. But a block being dragged across a table is a complicated multiparticle system. Individual atoms interact through electric forces associated with the interatomic potential energy, and at the atomic level these interac-

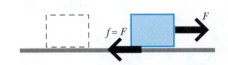

Figure 6.50 Pull to the right; the block and table get hot.

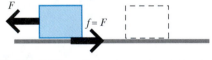

Figure 6.51 Pull back to the left; the block and table get even hotter.

tions are reversible. But energy spreads throughout the block and floor, and macroscopically this looks to us like energy dissipation.

Irreversibility has far-reaching consequences for the statistical or probabilistic behavior of macroscopic objects consisting of large numbers of atoms. We will consider issues of reversibility and irreversibility in more detail in a later chapter.

> *6.X.14* You move a block slowly to the right a distance of 0.2 m on a table. You apply a constant force of 20 N. How much work do you do? You then move slowly back to the starting point; how much work do you do in this second move? What change has there been in the thermal energy of the block and table?
>
> *6.X.15* A block is attached to a spring whose stiffness is 40 N/m. You move the block slowly from $s = 0$ to $s = 0.2$ m, with negligible friction present. How much work do you do? You then move slowly back to the starting point; how much work do you do in this second move? What change has there been in the thermal energy of the block?

6.14 *RESONANCE

Imagine pushing somebody in a swing. If you time your pushes to match the natural frequency of the swing, you can build up an increasingly large amplitude of the swing. But larger amplitude means higher speed and a higher rate of energy dissipation.

Right frequency
Big response

? You keep pushing with constant (small) amplitude, and the swing eventually reaches a "steady state" with constant (large) amplitude. What determines the size of this steady-state amplitude?

The steady state is established when the energy dissipated through air resistance on each cycle has grown to be equal to the energy input you make on each cycle. The large steady-state oscillation needs only little pushes at the right times from you to make up for energy dissipation, and the amplitude of the oscillation can be much larger than the amplitude of the pushes.

On the other hand, if you deliberately push at the wrong times, not matching the natural frequency of the swing, the amplitude of the swing won't continually build up. To get a large response from an oscillating system, inputs should be made at the natural, free-oscilla-

tion frequency of that system. This important feature of driven oscillating systems is called "resonance."

Wrong frequency
Small response

*Analytical treatment of resonance

Problem 6.P.60 is a numerical integration of a sinusoidally driven spring-mass oscillator, as shown below.

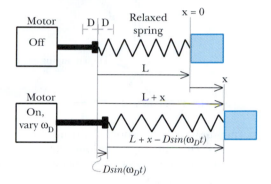

This problem lets you study the buildup of the steady state as well as the steady state itself. For the steady-state portion of the motion, it is possible to obtain an analytical solution, if the energy dissipation is due to viscous friction. Here is the Momentum Principle for the driven oscillator in the diagram, with viscous friction force $-cv$:

$$\frac{dp}{dt} = -k_s[x - D\sin(\omega_D t)] - cv$$

where x is the position of the mass; D is the amplitude of the sinusoidal motion and ω_D is the variable angular frequency of the driven end of the spring. For more details on how this equation was obtained, see the discussion in Problem 6.P.60.

In the steady state x must be a sinusoidal function, with amplitude A and phase shift ϕ depending strongly on the driving angular frequency ω_D:

$$x = A\sin(\omega_D t + \phi)$$

By taking derivatives of this expression to obtain $v = dx/dt$ and dv/dt as a function of time t, and plugging these derivatives into the Momentum Principle, it can be shown that this expression for x as a function of time is a solution in the steady state if the amplitude A and phase shift ϕ have values determined by the following expressions:

$$A = \frac{\omega_F^2}{\sqrt{(\omega_F^2 - \omega_D^2)^2 + \left(\frac{c}{m}\omega_D\right)^2}} D$$

$$\cos\phi = \frac{(\omega_F^2 - \omega_D^2)}{\sqrt{(\omega_F^2 - \omega_D^2)^2 + \left(\frac{c}{m}\omega_D\right)^2}}$$

The detailed proof is rather complicated, so we omit it. Instead, in the exercises below we give you the opportunity to see that these expressions for amplitude and phase shift are in agreement with your experimental observations and the computer modeling of Problem 6.P.60. The diagram below is a graph of the amplitude A as a function of the driving angular frequency ω_D, for two different values of the viscous friction parameter c.

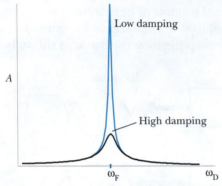

Amplitude as a function of driving frequency. The damping in the lower curve is five times greater than in the upper curve.

6.X.16 Using the formula for the amplitude A, show that if the viscous friction is small, the amplitude is large when ω_D is approximately equal to ω_F.

6.X.17 Show that with small viscous friction, the amplitude A drops to $1/\sqrt{2}$ of the peak amplitude when the driving angular frequency differs from resonance by this amount:

$$|\omega_F - \omega_D| \approx \frac{c}{2m}\omega_F$$

(Hint: note that near resonance $\omega_D \approx \omega_F$, so $\omega_F + \omega_D \approx 2\omega_F$.)

6.X.18 Given the results of the previous exercise, how does the width of the resonance peak depend on the amount of friction? What would the resonance curve look like if there were very little friction?

6.X.19 Using the formula involving the phase shift ϕ, show that the phase shift is approximately 0° for very low driving frequency ω_D, approximately 180° for very high driving frequency ω_D, and 90° at resonance, consistent with your experiment.

Resonance in other systems

We have discussed resonance in the context of a particular system—a mass on a spring driven by a motor. We study resonance because it is an important phenomenon in a great variety of situations.

When you choose a radio or television station you adjust the parameters of an electrically oscillating circuit so

that the circuit has a narrow resonance at the chosen station's main frequency, and other stations with significantly different frequencies have little effect on the circuit.

Magnetic resonance imaging (MRI) is based on the phenomenon of nuclear magnetic resonance (NMR) in which nuclei acting like toy tops precess in a strong magnetic field. The precessing nuclei are significantly affected by radio waves only at the precession frequency. Because the precession frequency depends slightly on what kind of atoms are nearby, resonance occurs for different radio frequencies, which makes it possible to identify different kinds of tissue and produce remarkably detailed images.

6.15 SUMMARY

Potential energy of an ideal spring-mass system:

$$U_s = \frac{1}{2}k_s s^2$$

Approximate interatomic potential energy:

$$U = \frac{1}{2}k_s s^2 - E_M$$

A better approximation to interatomic potential energy:

$$U_M = E_M[1 - e^{-\alpha(r - r_{eq})}]^2 - E_M$$

(Morse potential energy)

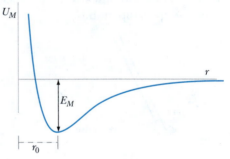

Thermal energy: Average energy of atoms in an object, proportional to temperature of the object

Power = $\dfrac{\text{energy}}{\text{time}}$ (J/s = watts); also calculable as $\vec{F} \bullet \vec{v}$

Energy transfer due to a temperature difference, Q

Specific heat capacity = energy per gram to raise temperature 1K

$\Delta E_{\text{thermal}} = mC\Delta T$; $C = \dfrac{\Delta E_{\text{thermal}}}{m\Delta T}$ (C in J/K per gram, m in grams)

Open system: energy flows into and/or out of system.

Closed system: no energy flows into or out of system.

The choice of system (open or closed) affects energy accounting.

Dissipation of energy

Terminal speed

Air resistance: collisions between air molecules and moving object

Sliding friction force due to forming and breaking of bonds between two solid objects

Path independence of potential energy
Impossibility of describing friction in terms of potential energy

Resonance: sensitivity of driven oscillator to driving frequency

Air resistance

Air resistance is approximately proportional to speed squared, and in a direction opposite to the velocity:

$$\vec{F}_{air} \approx -\frac{1}{2}C\rho A v^2 \hat{v}$$

Viscous friction

For small particles ("viscous" friction), fluid friction is approximately proportional to v.

Sliding friction

Sliding friction force $f \leq \mu F_N$; the coefficient of friction μ depends on the surfaces. F_N is the normal force compressing the two surfaces together. Sliding friction force does not depend on speed (approximate).

*Driven spring-mass system

Analytical solution for driven spring-mass system: $x = A\sin(\omega_D t + \phi)$, where $\omega_F = \sqrt{k_s/m}$; viscous friction cv, driven at ω_D:

$$A = \frac{\omega_F^2}{\sqrt{(\omega_F^2 - \omega_D^2)^2 + \left(\frac{c}{m}\omega_D\right)^2}}D$$

$$\cos\phi = \frac{(\omega_F^2 - \omega_D^2)}{\sqrt{(\omega_F^2 - \omega_D^2)^2 + \left(\frac{c}{m}\omega_D\right)^2}}$$

6.16 REVIEW QUESTIONS

Definitions

6.RQ.20 An oil company included in its advertising the following phrase: "Energy—not just a force, it's power!" In technical usage, what are the differences among the terms energy, force, and power?

Energy conservation

6.RQ.21 50 grams of water whose temperature is 25° Celsius are added to a thin glass containing 800 grams of water at 20° Celsius (about room temperature). What is the final temperature of the water? What simplifying assumptions did you have to make in order to determine your approximate result?

Spring potential energy

6.RQ.22 A spring has a relaxed length of 6 cm and a stiffness of 100 N/m. How much work must you do to change its length from 5 cm to 9 cm?

Power

6.RQ.23 Electricity is billed in kilowatt-hours. Is this energy or power? How much is one kilowatt-hour in standard physics units? (The typical cost of one kilowatt-hour is 5 to 10 cents.)

Open and closed systems

6.RQ.24 State which of the following are open systems with respect to energy, and which are closed: a car; a person; an insulated picnic chest; the Universe; the Earth. Explain why.

6.RQ.25 A horse whose mass is M gallops at constant speed v up a long hill whose vertical height is h, taking an amount of time t to reach the top. The horse's hooves do not slip on the rocky ground, so the work done by the force of the ground on the hooves is zero (no displacement of the force). When the horse started running, its temperature rose quickly to a point at which from then on, energy transfer from the horse to the air due to the temperature difference keeps the horse's temperature constant.

(a) First consider the horse as the system of interest: What objects exert forces on the horse? How much work is done on the horse by each of these forces? What non-work energy transfers are there? What is the change in the energy of the horse? In what forms?

(b) Next consider the Universe as the system of interest: What objects exert forces on this system? How much work is done on the system by each of these forces? What non-work energy transfers are there? What is the change in the energy of the system? In what forms?

Specific heat capacity

6.RQ.26 Substance A has a large specific heat capacity (on a per gram basis), while substance B has a smaller specific heat capacity. If the same amount of energy is put into a 100 gram block of each substance, and if both blocks were initially at the same temperature, which one will now have the higher temperature?

Energy dissipation

6.RQ.27 When a falling object reaches terminal speed, its kinetic energy reaches a constant value. However, the gravitational energy of the system consisting of object plus Earth continues to decrease. Does this violate the principle of conservation of energy? Explain why or why not.

Air resistance

6.RQ.28 Describe a situation in which it would be appropriate to neglect the effects of air resistance.

6.RQ.29 Describe a situation in which neglecting the effects of air resistance would lead to significantly wrong predictions.

6.RQ.30 List all the approximations that are made in order to find a simple analytical solution for the motion of a projectile near the Earth's surface.

6.RQ.31 You are standing at the top of a 50 m cliff. You throw a rock as hard as you can, in the horizontal direction. Approximately where does it hit on the flat plain below? Is your answer too large or too small?

6.RQ.32 Write the computational statements required at the heart of a numerical integration of projectile motion that includes air resistance.

6.RQ.33 If you let a mass at the end of a string start swinging, at first the maximum swing decreases rather quickly, but once the swing has become small it takes a long time for further significant decrease to occur. Try it! Explain this simple observation.

Sliding friction

6.RQ.34 You drag a block with constant speed v across a table with friction. Explain in detail what you have to do in order to change to a constant speed of $2v$ on the same surface. (That is, the puzzle is to explain how it is possible to drag a block with sliding friction at different constant speeds.)

Damped oscillators

6.RQ.35 Sketch a graph of position vs. time for a spring-mass oscillator subject to friction (not over-damped).

Potential energy

6.RQ.36 A particle moves inside a circular glass tube under the influence of a tangential force of constant magnitude F. Explain why we cannot associate a potential energy with this force (which is therefore a "nonconservative" force). How is this situation different from the case of a block on the end of a string, which is swung in a circle?

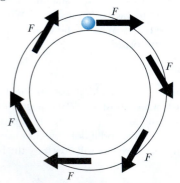

A particle moves in a circular glass tube.

6.RQ.37 Calculate the change in electric energy along two different paths in moving charge q away from charge Q from A to B along a radial path, then to C along a circle centered on Q, then to D along a radial path. Also calculate the change in energy in going directly from A to D along a circle centered on Q. Specifically, what are $U_B - U_A$, $U_C - U_B$, $U_D - U_C$, and there sum? What is $U_D - U_A$? Also, calculate the round-trip difference

in the electric energy when moving charge q along the path from A to B to C to D to A.

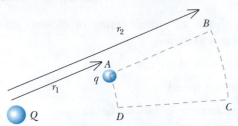

Change in electric energy along two different paths.

6.17 PROBLEMS

6.P.38 Energy in a spring-mass system
Energy calculations provide a powerful check on the accuracy of a numerical integration. Modify your numerical integration of the motion of a mass on a spring that you did in Chapter 4. In addition to plotting graphs of position versus time, also plot graphs of kinetic energy, of spring energy, and of the sum of the kinetic energy and the spring energy, versus time during the motion. It is up to you to choose some value for the potential energy U_0 at the bottom of the well. Label the energy graphs with numerical values and units. Make sure that the energy values are reasonable for your system.

(a) Does the sum of kinetic and potential energy remain constant as a function of time? What if you vary the step size (which varies the accuracy of the numerical integration)?

(b) Increase the energy by increasing the initial speed. How does the amplitude A change? How does the angular frequency ω change?

6.P.39 Experiment—falling coffee filters
A paper coffee filter in the shape of a truncated cone falls in a stable way and quickly reaches terminal speed because it has a large area (big air resistance) but low mass (small gravitational force). Use a stopwatch to time the drop of coffee filters from as high a starting position as you can conveniently manage. If a stairwell is available, drop the coffee filters there to time a longer fall. This experiment is easier to do with a partner.

(a) By stacking coffee filters you can change the mass of a falling object without changing the shape. With this scheme you can explore how the terminal speed depends on the mass. Time the fall for different numbers of stacked filters, taking some care that the shape of the bottom filter is always the same (the filters tend to flatten out when removed from a stack). Start the measurement after the filters have fallen somewhat, to allow them to reach terminal speed. Average the results of several repeated measurements of time and height. Can you think of a simple experiment you can do to verify that the coffee filters do in fact reach terminal speed before you start timing?

(b) We want to know quantitatively how air resistance depends on speed. Plot your data for the air resistance force *vs.* the terminal speed. (The air resistance force is equal to the gravitational force when terminal speed has been reached, so it is proportional to the number of filters in a stack). How does the air resistance force depend on v?

There is an important constraint on the graph of speed dependence: Should the curve of the air resistance force vs. terminal speed go through the origin (force and terminal speed both very small), or not?

Analysis hint: If you suspect that the force is proportional to v^3, you might plot force vs. v^3 and see whether a straight line fits the data.

6.P.40 Projectile motion with air resistance

Use a computer to carry out the following step-by-step quantitative calculations for a baseball. Remember that the direction of the air resistance force on an object is opposite to the direction of the object's velocity.

A baseball has a mass of 155 grams and a diameter of 7 cm. The drag coefficient C for a baseball is about 0.35, and the density of air at sea level is about 1.3×10^{-3} grams/cm^3, or 1.3 kg/m^3.

(a) Including the effect of air resistance, compute and plot the trajectory of a baseball thrown or hit at an initial speed of 100 miles per hour, at an angle of 45° to the horizontal. How far does the ball go? Is this a reasonable distance?

(Continued on next page.)

(A baseball field is about 400 feet from home plate to the fence in center field. An outfielder cannot throw a baseball in the air from the fence to home plate.)

(b) Plot a graph of $(K + U_g)$ vs. time.

(c) Temporarily neglect air resistance (set $C = 0$ in your computations) and determine the range of the baseball. Check your computations by using the analytical solution for motion of a projectile without air resistance. Compare your result with the range found in part (a), where the effect of air resistance was taken into account. Is the effect of air resistance significant? (The effect is surprisingly large—about a factor of 2!)

(d) Neglecting air resistance, plot a graph of $(K + U_g)$ vs. time.

(e) In Denver, a mile above sea level, the air is about 83% as dense as the air at sea level. Including the effect of air resistance, use your computer model to predict the trajectory and range of a baseball thrown in Denver. How does the range compare with the range at sea level?

6.P.41 Numerical integration of spring-mass system with friction

Modify your numerical integration of the spring-mass system to include the effects of friction. Such a system is called a "damped oscillator."

Simulate the effect of all three kinds of friction: sliding friction (independent of v), viscous friction (proportional to v), and air resistance for large blunt objects (proportional to v^2). Choose friction to be small enough that you get at least two full cycles of oscillation, but large enough that you see an effect. If the friction is so large that the graph never crosses the axis, the system is said to be "over-damped." For each kind of friction, display the following:

(1) Display an animation of the motion of the system
(2) Plot a graph of position vs. time

(3) Plot graphs of K, U, and $(K+U)$ vs. time

On the energy graphs, when is the energy loss rate large and small, and why?

6.P.42 Experiment—observing resonance

You can observe the main effects of resonance with very simple experiments. Hold a spring vertically with a mass suspended at the other end, and observe the frequency of "free" oscillations with your hand kept still. Then stop the oscillations, and move your hand *extremely* slowly up and down in a kind of slow sinusoidal motion. You will see that the mass moves up and down with the same very low frequency.

(a) How does the amplitude (plus or minus displacement from the center location) of the mass compare with the amplitude of your hand? (Notice that the phase shift of the oscillation is 0°; the mass moves up when your hand moves up.)

(b) Next move your hand up and down at a significantly higher frequency than the free-oscillation frequency. How does the amplitude of the mass compare to the amplitude of your hand? (Notice that the phase shift of the oscillation is 180°; the mass moves down when your hand moves up.)

(c) Finally, move your hand up and down at the free-oscillation frequency. How does the amplitude of the mass compare with the amplitude of your hand? (It is hard to observe, but the phase shift of the oscillation is 90°; the mass is at the midpoint of its travel when your hand is at its maximum height.)

(d) Change the system in some way so as to increase the air resistance significantly. For example, attach a piece of paper to increase drag. At the free-oscillation frequency, how does this affect the size of the response?

A strong dependence of the amplitude and phase shift of the system to the driving frequency is called resonance.

6.P.43 Energy in a spring-mass system

A horizontal spring-mass system has low friction, spring stiffness 200 N/m, and mass 0.4 kg. The system is released with an initial compression of the spring of 10 cm and an initial speed of the mass of 3 m/s.

(a) What is the maximum stretch during the motion?
(b) What is the maximum speed during the motion?
(c) Now suppose that there is energy dissipation of 0.01 J per cycle of the spring-mass system. What is the average power input in watts required to maintain a steady oscillation?

6.P.44 Vertical spring-mass system

Write an equation for the total energy of a system consisting of a mass suspended vertically from a spring, and include the Earth in the system. Place the origin for gravitational energy at the equilibrium position of the mass, and show that the changes in energy of a vertical spring-mass system are the same as the changes in energy of a horizontal spring-mass system.

6.P.45 Hot water

400 grams of boiling water (temperature 100° C, specific heat capacity 4.2 J/gram/K) are poured into an aluminum pan whose mass is 600 grams and initial temperature 20° C (the specific heat capacity of aluminum is 0.9 J/gram/K). After a short time, what is the temperature of the water? Explain. What simplifying assumptions did you have to make?

6.P.46 Place a mass on a vertical spring

A spring with stiffness k_s and relaxed length L stands vertically on a table. You hold a mass M just barely touching the top of the spring.

(a) You *very slowly* let the mass down onto the spring a certain distance, and when you let go, the mass doesn't move. How much did the spring compress? How much work did *you* do?

(b) Now you again hold the mass just barely touching the top of the spring, and then let go. What is the maximum compression of the spring? State what approximations and simplifying assumptions you made.

(c) Next you push the mass down on the spring so that the spring is compressed an amount s, then let go, and the mass starts moving upward and goes quite high. When the mass is a height of $2L$ above the table, what is its speed?

6.P.47 Bungee jumping

Design a "bungee jump" apparatus. A bungee jumper falls from a high platform with two elastic cords tied to the ankles. The jumper falls freely for a while, with the cords slack. Then the jumper falls an additional distance with the cords increasingly tense. Assume that you have cords that are 10 m long, and that the cords stretch in the jump an additional 20 m for an average jumper.

(a) Make a series of simple diagrams, like a comic strip, showing the platform, the jumper, and the two cords at various times in the fall and the rebound. On each diagram, draw and label vectors representing the forces acting on the jumper, and the jumper's velocity. Make the relative lengths of the vectors reflect their relative magnitudes.

(b) At what instant is there the greatest tension in the cords? How do you know?

(c) What is the jumper's speed at this instant?

(d) Is the jumper's momentum changing at this instant or not? (That is, is dp/dt nonzero or zero?) Explain briefly.

(e) Focus on this instant, and use the principles of this chapter to determine the spring stiffness k_s for each cord. Explain your analysis. Give numerical values for your design.

(Continued on next page.)

(f) What is the maximum tension that each cord must support without breaking? Give numerical values for your design.

(g) What is the maximum acceleration (in "g's") that the jumper experiences? What is the direction of this maximum acceleration?

(h) State clearly what approximations and estimates you have made in your design.

6.P.48 Experiment—your power capability

(a) If you follow a diet of 2000 food calories per day (2000 kilocalories), what is your average power consumption in watts? (A food or "large" calorie is a unit of energy equal to 4.2×10^3 J; a regular or "small" calorie is equal to 4.2 J.) Compare with the power consumption in a table lamp.

(b) You can produce much higher power for short periods. Make appropriate measurements as you run up some stairs, and report your measurements. Use these measurements to estimate your power output (this is in addition to your basal metabolism—the power needed when resting). Compare with a horsepower (which is about 750 watts) or a toaster (which is about 1000 watts).

(c) How many days of a diet of 2000 large calories are equivalent to the gravitational energy difference for you between sea level and the top of Mount Everest, 8848 m above sea level? (However, the body is not anywhere near 100% efficient in converting chemical energy into change in altitude. Also note that this is in addition to your basal metabolism.)

6.P.49 Human energy consumption

(a) A typical candy bar provides 280 calories (one "food" or "large" calorie is equal to 4.2×10^3 joules). How many candy bars would you have to eat to replace the chemical energy you expend doing 100 sit-ups? Explain your work, including any approximations or assumptions you make. (In a sit-up, you go from lying on your back to sitting up.)

(b) Humans have about 60 milliliters (60 cm^3) of blood per kilogram of body mass, and blood makes a complete circuit in about 20 seconds, to keep tissues supplied with oxygen. Make a crude estimate of the *additional* power output of your heart (in watts) when you are standing compared with when you are lying down. Note that we're not asking you to estimate the power output of your heart when you are lying down, just the change in power when you are standing up. You will have to estimate the values of some of the relevant parameters. Because we're only looking for a crude estimate, try to construct a model that is as simple as possible. Describe what approximations and estimates you made.

6.P.50 Drop coffee filters from a tall building

You drop a single coffee filter from a very tall building, and it takes 52 seconds to reach the ground. Next you drop a stack of 3 of these coffee filters. About how long will they take to hit the ground? Explain briefly, including any approximations or simplifying assumptions you had to make.

6.P.51 Graph of energy with dissipation

Here is a portion of a graph of energy terms vs. time for a mass on a spring, subject to air resistance. Identify and label the three curves as to what kind of energy each represents. Explain briefly how you determined which curve represented which kind of energy.

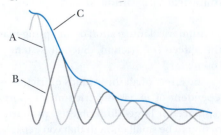

Which curve represents which kind of energy?

6.P.52 Heating a house

During three hours one winter afternoon, when the outside temperature was 0° C (32° F), a house heated by electricity was kept at 20° C (68° F) with the expenditure of 45 kwh (kilowatt·hours) of electric energy. What was the average energy leakage in joules per second through the walls of the house to the environment (the outside air and ground)?

The rate at which energy is transferred between two systems is often proportional to their temperature difference. Assuming this to hold in this case, if the house temperature had been kept at 25° C (77° F), how many kwh of electricity would have been consumed?

6.P.53 Launch a package from an asteroid

A package of mass m sits at the equator of an airless asteroid of mass M, radius R, and spin angular speed ω. We want to launch the package in such a way that it will never come back, and when it is very far from the asteroid it will be traveling with speed v. We have a powerful spring whose stiffness is k_s. How much must we compress the spring?

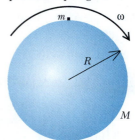

Launch a package from an asteroid.

6.P.54 A mass hangs from a spring

A mass of 0.3 kg hangs motionless from a vertical spring whose length is 0.8 m and whose unstretched length is 0.65 m. Next the mass is pulled down so the spring has a length of 0.9 m and given an initial speed upwards of 1.2 m/s. What is the maximum length of the spring during the following motion? What approximations or simplifying assumptions did you make?

6.P.55 A sky diver

Terminal speed for a falling sky diver has been measured to be about 60 m/s. Using your computer model, determine how far (in meters) and how long (in seconds) a sky diver falls before reaching terminal speed. (You will need to think about the meaning of "terminal speed" and how to test for it.)

Plot graphs of speed *vs.* time and position *vs.* time.

6.P.56 Experiment—coefficient of friction

Design and carry out an experiment to determine the coefficient of sliding friction for a block on a flat plate.

6.P.57 Experiment—speed dependence of sliding friction

For this problem you will need measurements of the position vs. time of a block sliding on a table, starting with some initial velocity, slowing down, and coming to rest. If you do not have an appropriate laboratory setup for making these measurements, your instructor will provide you with such data. Analyze these data to see how well they support the assertion that the force of sliding friction is essentially independent of the speed

of sliding. (Recall that a reason was given for this assertion—the friction force is determined by the strength of temporary interatomic bonds that must be broken.)

6.P.58 Block dragged at an angle

A block of mass M is dragged along a table by a force applied at an angle θ to the horizontal by a spring of stiffness k_s, stretched a constant amount s (Figure). The coefficient of friction between block and table is μ. At $t = 0$, the speed of the block was v_0. What is its speed v at time t?

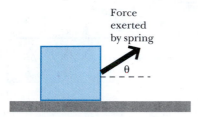

A block is dragged by a force at an angle to the horizontal.

6.P.59 Two neutral atoms

Here is a potential energy curve for the interaction of two neutral atoms. The two-atom system is in a vibrational state indicated by the heavy solid horizontal line.

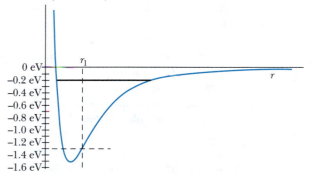

Potential energy diagram.

(a) At $r = r_1$, what are the approximate values of the kinetic energy K, the potential energy U, and the quantity $K+U$?

(b) What minimum energy must be supplied to cause these two atoms to separate?

(c) In some cases, when r is large, the interatomic potential energy can be expressed approximately as $U = -a/r^6$. For large r, what is the algebraic form of the magnitude of the force the two atoms exert on each other in this case?

6.P.60 Computer modeling of resonance

You can study resonance in a driven oscillator in detail by modifying your computation for the spring-mass system with friction. Let one end of the spring be moved back and forth sinusoidally by a motor, with a motion given by $D\sin(\omega_D t)$ (Figure). Here D is the amplitude of the motor motion and ω_D is the angular frequency of the motor, which can be varied. (The free-oscillation angular frequency $\omega_F = \sqrt{k_s/m}$ of the spring-mass system has a fixed value, determined by the spring stiffness k_s and the mass m.)

We need an expression for the stretch s of the spring in order to be able to calculate the force $-k_s s$ of the spring on the mass. We have to do a bit of geometry to figure out the length of the spring when the mass is displaced a distance x from the equilibrium position, and the motor has moved the other end of the spring by an amount $D\sin(\omega_D t)$.

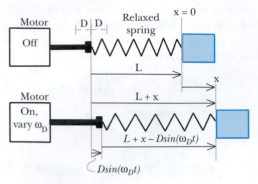

A driven spring-mass system.

On the diagram, the spring gets longer when the mass moves to the right ($+x$), but the spring gets shorter when the motor moves to the right ($-D\sin(\omega_D t)$). The new length of the spring is $L + x - D\sin(\omega_D t)$, and the net stretch of the spring is this quantity minus the unstretched length L, yielding $(s = x - D\sin(\omega_D t))$. A check that this is the correct expression for the net stretch of the spring is that if the motor moves to the right the same distance as the mass moves to the right, the spring will have zero stretch. Replace x in your computer computation with the quantity $[x - D\sin(\omega_D t)]$.

(a) Use viscous damping (friction proportional to v). The damping should be small. That is, without the motor driving the system ($D = 0$), the mass should oscillate for many cycles. Set ω_D to $0.9\omega_F$ (that is, 0.9 times the free-oscillation angular frequency, $\omega_F = \sqrt{k_s/m}$), and let x_0 be 0. Graph position x vs. time for enough cycles to show that there is a transient buildup to a "steady state." In the steady state the energy dissipation per cycle has grown to exactly equal the energy input per cycle.

Vary ω_D in the range from 0 to $2.0\omega_F$, with closely spaced values in the neighborhood of $1.0\omega_F$. Have the computer plot the position of the motor end as well as the position of the mass as a function of time. For each of these driving frequencies, record for later use the steady-state amplitude and the phase shift (the difference in angle between the sinusoids for the motor and for the mass). If a harmonic oscillator is lightly damped it has a large response only for driving frequencies near its own free-oscillation frequency.

(b) Is the steady-state angular frequency equal to ω_F or ω_D for these various values of ω_D? (Note that during the transient buildup to the steady state the frequency is not well-defined, because the motion of the mass isn't a simple sinusoid.)

(c) Sketch graphs of the steady-state amplitude and phase shift vs. ω_D. Note that when $\omega_D = \omega_F$ the amplitude of the mass can be much larger than D, just as you observed with your hand-driven spring-mass system. Also note the interesting variation of the phase shift as you go from low to high driving frequencies—does this agree with the phase shifts you observed with your hand-driven spring-mass system?

(d) Repeat part (c) with viscous friction twice as large. What happens to the resonance curve (the graph of steady-state amplitude vs. angular frequency ω_D)?

6.18 ANSWERS TO EXERCISES

6.X.1 (page 226)   If $k_s = 0.6$ N/m and $s = 0.2$ m, $U_s = 0.012$ J

6.X.2 (page 226)   0.5 m/s

6.X.5 (page 229)   No change; same stretch

6.X.6 (page 233)   0.12 K (that is, 0.12 kelvin)

6.X.7 (page 235) 1.26×10^5 J; 7.6×10^4 J; -1.26×10^5 J

6.X.8 (page 237)   About 500 J; about 2000 kg per second

6.X.9 (page 237)   2.5 m/s^2

6.X.10 (page 241) $\frac{1}{2}kA^2$; $\frac{1}{2}kA^2$; spring + mass

6.X.11 (page 243)   less; some of the initial energy is now in the air

6.X.12 (page 243) $[2.7\times10^{-2} - 4.5\times10^{-4}]$ J $\approx 2.7\times10^{-2}$ J

6.X.13 (page 248) $\dfrac{dv_x}{dt} = \dfrac{(k_s s - f_{max})}{M}$ and $\dfrac{dv_y}{dt} = 0$.

6.X.14 (page 251)   +4 J; +4 J; 8 J

6.X.15 (page 251)   +0.8 J; –0.8 J; 0 J

6.X.16 (page 252)   Denominator gets very small when $\omega_D = \omega_F$.

6.X.17 (page 252)   When $\omega_D = \omega_F$, denominator $= \dfrac{c}{m}\omega_D$.

Denominator $= \sqrt{2}\dfrac{c}{m}\omega_D$ when

$$\left|(\omega_F^2 - \omega_D^2)\right| = \frac{c}{m}\omega_D.$$

But since near resonance $\omega_D \approx \omega_F$, we have

$$(\omega_F^2 - \omega_D^2) = (\omega_F + \omega_D)(\omega_F - \omega_D) \approx 2\omega_D(\omega_F - \omega_D)$$

Hence when $\left|\omega_F - \omega_D\right| \approx \dfrac{c}{2m}\omega_D \approx \dfrac{c}{2m}\omega_F$,

amplitude down by a factor of $1/\sqrt{2}$.

6.X.18 (page 252)   Larger friction, wider resonance peak. If friction very small, resonance peak is a very narrow spike.

6.X.19 (page 252)   Low frequency,

$$\cos\phi \approx \frac{(\omega_F^2)}{\sqrt{(\omega_F^2)^2 + \left(\dfrac{c}{m}\omega_D\right)^2}} \approx 1 \text{ so } \phi \approx 0°.$$

High frequency,

$$\cos\phi \approx \frac{(-\omega_D^2)}{\sqrt{(-\omega_D^2)^2 + \left(\dfrac{c}{m}\omega_D\right)^2}} \approx -1,$$

so $\phi \approx 180°$.

Resonance frequency, $\cos\phi = 0$, so $\phi \approx 90°$.

CHAPTER 7

ENERGY QUANTIZATION

Key concepts

- The energy of microscopic systems is quantized; that is, the energy of the system can have only certain specific energies ("energy levels"); the system is never observed to have energies between these levels.
- Photons are particles of light with different energies corresponding to different portions of the electromagnetic spectrum.
- An atom in an excited state can drop to a lower energy level, and emit a photon whose energy is the difference of the two energy levels.
- An atom in a low-energy state (normally the ground state) can absorb a photon and jump to an excited state, if the photon has just the right energy (the difference in the two energy levels).
- Energetic electrons can excite atoms to higher energy states, leaving the electron with less kinetic energy than it had initially.
- Energy quantization is found in atomic electronic states, vibration and rotation of diatomic molecules, and states of nuclei.

7.1 ENERGY QUANTIZATION

In the early 20th century, as scientists learned more about the detailed behavior of atoms they were excited to discover that small systems such as atoms can have only certain energies, called energy "levels," and cannot have energies in between these discrete levels. The energy changes associated with interactions therefore come in finite packets, or "quanta."

Discrete energy levels were not noticed for a long time because the energy differences between these levels are quite small. A typical separation between energy levels in an atom is about 10^{-19} joules. The energies of macroscopic objects are on the order of a joule; an energy increase of 10^{-19} joules is completely undetectable even by the most sensitive measurements of motion or temperature. However, it is possible to detect these tiny differences by observing the interaction of light and matter, as we will see.

The discovery of discrete energy levels is analogous to the discovery of the atomic nature of matter. We now know that matter is not continuous but lumpy, being made up of atoms. For most of human history this lumpiness was not noticed because the atoms are so tiny. Similarly, the noncontinuous, lumpy nature of energy was not noticed because the quantized energies are so very close together that energy seemed continuous.

Electronic energy levels

An example of discrete energy levels is found in the "electronic" energy associated with the motion of electrons around the nucleus of an atom. You probably know from your study of chemistry that electronic energy levels are discrete—an atom can have only certain energies. The electronic energy levels in atomic hydrogen are shown in Figure 7.1 (atomic hydrogen is a single H atom, composed of one proton and one electron, not the usual diatomic molecule H_2).

The diagram shows the familiar electric potential energy of the proton-electron system. The horizontal lines represent the possible bound energy states for the hydrogen atom; bound states with other values of $(K+U)$ are not observed. Unbound states however $(K+U \geq 0)$ are not quantized.

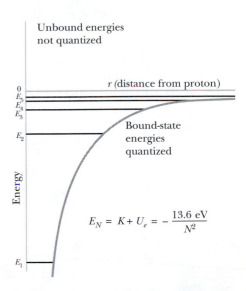

Figure 7.1 Electronic energy levels for a hydrogen atom. The potential energy curve describes the potential energy of the electron-proton system as a function of distance.

The electric potential energy of an electron and a proton bound together in a hydrogen atom is negative for all separations:

$$U_e = \frac{1}{4\pi\varepsilon_0} \frac{q_1 q_2}{r} = \frac{1}{4\pi\varepsilon_0} \frac{(+e)(-e)}{r} = -\frac{1}{4\pi\varepsilon_0} \frac{e^2}{r}$$

A quantum mechanical calculation of the energy levels of a system involves solving a differential equation called the Schrödinger equation, a form of the Energy Principle consistent with the nature of microscopic systems. This equation includes both a potential energy term and a kinetic energy term. Solving this equation for the hydrogen atom predicts that the discrete energy levels for bound states of hydrogen (ignoring the rest energies of the proton and the electron) will be these:

ENERGY LEVELS FOR THE HYDROGEN ATOM

$$E_N = K + U_e = -\frac{13.6 \text{ eV}}{N^2}, N = 1, 2, 3, \text{ etc.}$$

An electron volt (eV) is an energy unit equal to $1.6 \times 10^{-19} \text{ J}$.

This arrangement of hydrogen energy levels has been verified by observations of the interaction of light and hydrogen, as we will see.

These energy levels denote different bound states of the electron+proton system. Recall that a bound state has a negative value of $K+U$, since a potential energy of zero is associated with a very large separation between the electron and the proton. We have seen that bound gravitational orbits are also associated with negative values of $K+U$. As you can see from the energy diagram, in higher energy states the average location of the electron is farther from the proton. Note that there is a lowest energy state; a hydrogen atom cannot have $K+U$ less than -13.6 eV.

Before the development of the fully quantum mechanical Schrödinger equation, Niels Bohr made a simple model for the hydrogen atom which also yielded the energy formula given above. We will study the Bohr model in Chapter 10.

Electron volts (eV)

In our analysis of quantum systems we will frequently be considering very small changes in energy, on the order of 10^{-19} joules. A common unit of energy used in descriptions of quantum phenomena is the "electron volt," abbreviated "eV" and introduced in an earlier chapter. In terms of joules:

$$1 \text{ eV} = 1.6 \times 10^{-19} \text{ J}$$

Here are some typical energies measured in electron volts:

Typical chemical reaction: about 1 eV.
Typical nuclear reaction: about 1 MeV (mega-electron-volt, or 10^6 eV).
Rest energy of the electron: about 0.5 MeV.
Rest energy of the proton: about 1 GeV (giga-electron-volt, or 10^9 eV).

Example: Raising the energy of a hydrogen atom

How much energy in electron volts is required to raise the energy of a hydrogen atom from its lowest level ($N = 1$) to the next lowest level ($N = 2$)?

Calculate the required increase in energy:

$$\Delta E = E_2 - E_1 = \left(-\frac{13.6 \text{ eV}}{2^2}\right) - \left(-\frac{13.6 \text{ eV}}{1^2}\right) = 10.2 \text{ eV}$$

Note that you cannot add less than 10.2 eV to a hydrogen atom in its lowest energy level; any smaller amount fails to be absorbed.

7.X.1 How much energy in electron volts is required to ionize a hydrogen atom (that is, remove the electron from the proton), if initially the atom is in its lowest energy level?

7.2 THE QUANTUM MODEL OF INTERACTION OF LIGHT AND MATTER

What is the evidence for energy quantization? A key piece of evidence is that atomic systems emit and absorb light only with quantized packets of energy, implying a quantized loss or gain of energy in the system. The study of the interaction of light and matter is called spectroscopy.

A particle model of light

Early in the 20th century it was discovered that light can be considered to be lumpy, being made up of packets of energy called photons. As with atoms and energy, the noncontinuous, lumpy nature of light was not noticed for a long time because the amount of energy carried by one photon in an ordinary beam of light is extremely small compared to the total energy of the beam. It has now become routine to detect individual photons in a very weak beam of light, and to measure the energy of an individual photon.

You may already be familiar with the classical model of light, in which light is described as a travelling wave. The classical wave model of light makes it possible to account for important aspects of the behavior of light, including interference and diffraction phenomena. To account for other properties of light, including the observation that packets of light corresponding to certain fixed energies can be absorbed or emitted by quantum systems, we need a quantum model of light that describes light as particles (photons). In this volume we will emphasize the quantum model of light, and we will be concerned with the energy of photons emitted or absorbed as quantum systems go from one energy state to another. The wave model of light is treated in Volume II of this textbook.

The electromagnetic spectrum

Visible light is an example of electromagnetic radiation, a phenomenon associated with the production and detection of radiative electric and magnetic fields to be studied in Chapter 23 of Volume II of this textbook. The instruments needed to detect and measure electromagnetic radiation depend on the photon energies involved. The spread of observable energies is called the electromagnetic spectrum (Figure 7.2).

That part of the electromagnetic spectrum for which your eye is a good detector is called visible light (Figure 7.3). Its energy ranges from around 1.8 eV per photon (red light) to 3.1 eV per photon (violet light). Photon energies greater than 3 eV include ultraviolet, x-rays, and gamma rays. Electromagnetic radiation with photon energies less than 1.8 eV includes infrared, radio waves, and microwaves. Sometimes the term "light" is used to refer just to visible electromagnetic radiation, that part of the spectrum detectable by human eyes.

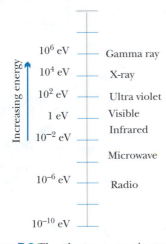

Figure 7.2 The electromagnetic spectrum.

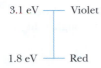

Figure 7.3 The visible portion of the electromagnetic spectrum.

7.X.2 The photon energy for green light lies between that for red and violet light. What is the approximate energy of the photons in green light?

7.X.3 The intensity of sunlight is about 1400 watts (J/s) per square meter. That is, when sunlight hits perpendicular to a square meter of area, about 1400 watts of energy can be absorbed. Using the average photon energy found in the previous exercise, about how many photons per second strike an area of one square meter? (This is why the lumpiness of light was not noticed for so long.)

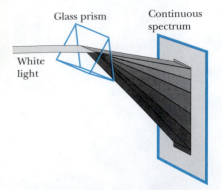

Figure 7.4 A prism separates a narrow beam of white light into a continuous rainbow of colors.

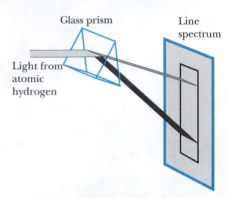

Figure 7.5 The light emitted by excited atomic hydrogen contains only a few separate colors. In a spectrum, light of a particular color and energy is called a "spectral line" (due to the line-like shape of the light coming through a slit as seen here).

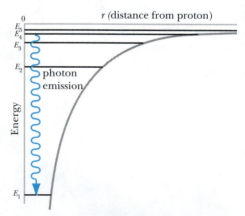

Figure 7.6 An atom can drop to a lower energy level by emitting a photon whose energy corresponds to the difference between the energies of the levels. By convention a squiggly line with an arrowhead indicates a photon.

Viewing a spectrum of visible light

One can separate a beam of visible light into its component energies by passing a narrow beam of light through a prism. You have probably seen the continuous rainbow spectrum produced when ordinary white light goes through a prism, as in Figure 7.4. Your eye reacts differently to photons of different energies, which is why you see a colorful spectrum.

Emission spectra

In contrast to white light, when a narrow beam of light emitted by electrically excited atomic hydrogen is passed through a prism (Figure 7.5) it does not look like a continuous spectrum but rather appears as a few widely spaced bright lines of particular colors. Such discontinuous spectra are strong evidence for energy quantization in the systems that are emitting the light.

When a hydrogen atom drops from a higher energy level E_4 to a lower energy level E_1 it emits a photon with the associated energy $E_4 - E_1$ (Figure 7.6). A collection of hydrogen atoms can emit photons whose energies correspond to the differences in energy between any two electronic energy levels of the hydrogen atom. The light emitted by excited atomic hydrogen is called the "hydrogen emission spectrum," and the bright lines of light of particular energies in the spectrum are called "spectral lines." The observed energies of light emitted by excited atomic hydrogen are consistent with the energy level differences predicted by quantum mechanics. Other atoms have different quantized energy levels, and a line spectrum can be used to identify the emitting atoms by the pattern of emitted photon energies.

If an isolated hydrogen atom is initially in a high energy level, over time it will spontaneously emit photons and drop to lower and lower states, eventually ending up in the lowest energy level, which is called the "ground state." Once an atom has dropped to the ground state, it cannot lose any more energy, nor emit photons, because there is no lower energy level than the ground state. This is quite different from the classical (prequantum) picture, in which a system can have an arbitrarily low energy (arbitrarily strong binding energy).

Example: Emission from an atom

Different atoms have different quantized energy levels. Consider a particular atom (not hydrogen) whose first excited state is 6.0 eV above the ground state, and the second excited state is 8.5 eV above the ground state. What is the energy of a photon emitted when such an atom drops from the second excited state to the first excited state?

System: everything (atom and photon)
Surroundings: nothing significant

Initial state: Atom in 2nd excited state | Final state: Atom in 1st excited state + photon

Energy Principle:

$$E_f = E_i + W$$

$$(E_0 + 6.0 \text{ eV}) + E_{photon} = (E_0 + 8.5 \text{ eV})$$

$$E_{photon} = 2.5 \text{ eV}$$

$E_0 + 8.5$ eV ——— 2nd excited state
$E_0 + 6.0$ eV ——— 1st excited state

E_0 ——— Ground state

The Energy Principle is valid even for atomic systems with quantized energy. Writing down the Energy Principle explicitly helps to avoid the common mistake of confusing the energy levels of the atom with the energy of the photon. Note that in the example above that the photon energy (2.5 eV) is the *difference* between the energies of the second and third energy levels.

Example: Emission from hydrogen atoms

Consider a collection of hydrogen atoms. Initially each individual atom is in one of the lowest three energy states ($N = 1, 2, 3$). What will be the energies of photons emitted by this collection of hydrogen atoms?

There are three possibilities: transitions from $N = 3$ to $N = 2$, from $N = 2$ to $N = 1$, or from $N = 3$ to $N = 1$. The photon energies for the three transitions are these:

$N = 3$ to $N = 2$: $\left(-\dfrac{13.6\ \text{eV}}{3^2}\right) - \left(-\dfrac{13.6\ \text{eV}}{2^2}\right) = 1.9\ \text{eV}$

$N = 2$ to $N = 1$: $\left(-\dfrac{13.6\ \text{eV}}{2^2}\right) - \left(-\dfrac{13.6\ \text{eV}}{1^2}\right) = 10.2\ \text{eV}$

$N = 3$ to $N = 1$: $\left(-\dfrac{13.6\ \text{eV}}{3^2}\right) - \left(-\dfrac{13.6\ \text{eV}}{1^2}\right) = 12.1\ \text{eV}$

7.X.4 How many different photon energies would emerge from a collection of hydrogen atoms which occupy the lowest four energy states ($N = 1, 2, 3, 4$)? (You need not calculate the energies of these transitions.)

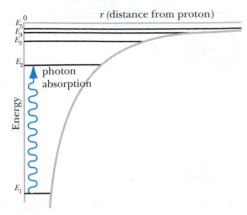

Figure 7.7 An atom can gain energy by absorbing a photon whose energy corresponds to the difference between two energy levels.

Absorption spectra

An atom can drop from a higher energy level to a lower energy level and emit a photon. An atom can also absorb a photon of an appropriate energy and jump from a lower energy level to a higher energy level (Figure 7.7). If an atom is in its ground state, only photon absorption is possible; photon emission cannot occur. Energy absorption can occur if the photon energy corresponds exactly to the difference between two energy levels of the atom. If we shine white light (containing a mixture of all photon energies in the visible spectrum) through a collection of atoms (such as a container of hydrogen gas), only photons whose energy is the difference of atomic energy levels will be strongly absorbed, while other photons can pass right through the material. In this case, the spectral lines are dark lines (missing energies) on a light background, instead of the bright lines on a dark background observed in an emission spectrum. This is called an "absorption spectrum" (Figure 7.8).

Photon absorption may be followed almost immediately by emission from the excited state to a lower state, so at any instant there are few atoms in excited states which could absorb photons. The result is that in dark-line absorption spectra only absorption from the ground state is observed, not from higher-level states.

The atoms that have jumped to higher energy levels through photon absorption will eventually drop to lower energy levels with the emission of photons. However, these photons are emitted in all directions, not just in the direction of the original beam of light, so the light that has passed through the material will have little intensity at photon energies corresponding to the strongly absorbed energies.

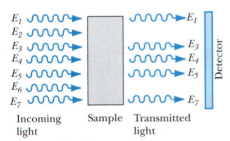

Figure 7.8 Measuring an absorption spectrum. A sample of a material is placed in a transparent container. Some energies of incoming light are absorbed. The outgoing light is depleted of the absorption energies (in this case E_2 and E_6).

The energy absorbed by the sample may be quickly emitted as photons with energies E_2 and E_6. However, these emitted photons go in all directions in 3D, so almost none of them reach the detector.

Example: Why is hydrogen gas transparent?

At room temperature in a gas of atomic hydrogen almost all of the atoms are in the ground state. Why might you expect the gas to be completely transparent to visible light?

As we saw in the example on page 260, the energy required to go from the ground state of atomic hydrogen to the first excited state is 10.2 eV, but visible light consists of photons whose energies lie between 1.8 eV and 3.1 eV. So visible light cannot be absorbed by ground-state hydrogen atoms. (In a collision between a photon of visible light and an atom, the atom can acquire some kinetic energy, but no internal energy change is possible.)

7.X.5 Suppose you had a collection of hypothetical quantum objects whose individual energy levels were −4.0 eV, −2.3 eV, and −1.6 eV. If nearly all of the individual objects were in the ground state, what would be the energies of dark spectral lines in an absorption spectrum if visible white light (1.8 to 3.1 eV) passes through the material?

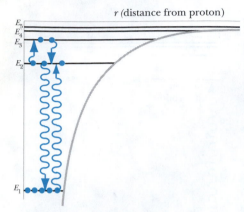

Figure 7.9 A dot represents the energy state of one hydrogen atom in a group of atoms. Energy is exchanged with the surroundings through photon emission and absorption.

The effects of temperature

When we observe atoms we are usually observing a collection of atoms, not an isolated single atom. Each atom in the collection (the "sample") interacts with other atoms and perhaps with a container. If an atom absorbs radiation coming from elsewhere, or absorbs energy from being jostled by a neighboring atom, it may jump to a higher energy level, and eventually emit photons and drop again to a lower level (Figure 7.9). In thermal equilibrium with its surroundings, an atom's energy continually goes up and down, but with an average energy determined by the temperature of the collection of atoms.

In Figure 7.9 each dot represents one hydrogen atom in a container of atomic hydrogen. If three dots are on a line, this means three atoms are in this particular state. Each hydrogen atom has the same allowed energy levels, and at any particular instant some of the atoms will be in the ground state, some in the first excited state, etc.

For a particular temperature, at any instant there is a certain probability that one particular atom will be in state 1, 2, 3, etc. At very low temperature almost all atoms are in the ground state (Figure 7.10). At high temperature the various atoms are in various states, from the ground state up to a high energy level, though the largest fraction will still be in the ground state. As we will see in Chapter 11 on "statistical mechanics," the population of a level whose energy is E above the ground state is proportional to the "Boltzmann factor" $\exp(-E/kT)$, where k is the Boltzmann constant (1.38×10^{-23} J/K) and T is the absolute temperature in kelvins (K), where 0° C is 273.15 K.

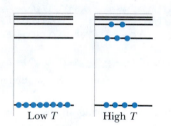

Figure 7.10 At a low temperature (system on the left) all atoms are in low energy states. At a high temperature (system on the right) some atoms are in higher energy states, though the largest fraction will still be in the ground state.

? Can you observe an emission spectrum if the material is very cold? Can you observe an absorption spectrum if the material is very cold?

Very cold matter will normally be in the ground state and will not emit, but an absorption spectrum will show gaps (dark lines) corresponding to photons being absorbed and lifting an atom above its ground state. (In order to observe an absorption spectrum you would of course need to use a beam of photons whose energies are sufficiently high to raise the system energy at least to the first excited state above the ground state.)

Room temperature is "very cold" for transitions involving visible light

In a sample at room temperature (about 300 K), almost all atoms are in their atomic ground state. For example, if the first excited state is 1 eV above the ground state, the Boltzmann factor is

$$\exp(-E/kT) = \exp\left(-\frac{1.6 \times 10^{-19} \text{ J}}{(1.4 \times 10^{-23} \text{ J/K})(300 \text{ K})}\right) = \exp(-38) = 3 \times 10^{-17}$$

This extremely tiny number represents the fact that at room temperature hardly any atoms will be excited thermally to 1 eV above the ground state, and all the atoms in an ordinary room-temperature absorption measurement can be considered to be in the ground state.

On the other hand, the surface temperature of the Sun is about 5000 K, and with $E = 1$ eV the Boltzmann factor is $\exp(-E/kT) = \exp(-2.3) = 0.1$, so a sizable fraction of atoms can be in an excited state 1 eV above the ground state, due to thermal excitation.

Electron excitation

In order for an atom to emit a photon, it first has to absorb enough energy to raise the atom to an excited state, above the ground state. As we have just seen, it requires very high temperatures for collisions with other atoms to provide energy at the level of electron volts.

A common nonthermal way to excite atoms in a gas is to run energetic electrons through the gas. Because electrons are (negatively) charged electrically, they interact strongly with atoms, and an electron that passes near an atom may excite the atom from the ground state to an excited state ΔE_{atom} higher, with a corresponding loss of kinetic energy of the electron. Unlike photon absorption, which almost always occurs only if the photon energy exactly matches the difference between energy levels, an electron can give up some kinetic energy and have some kinetic energy left over. The process is illustrated in Figure 7.11.

? For example, suppose an electron with kinetic energy of 11 eV collides with a hydrogen atom which is in the ground state. What can happen?

The first excited state of hydrogen is 10.2 eV above the ground state. This 11 eV electron has enough kinetic energy to excite the hydrogen atom to the first excited state, giving up 10.2 eV of energy. The electron moves away with 0.8 eV of kinetic energy remaining. The hydrogen atom will quickly drop back down to the ground state, emitting a 10.2 eV photon. If a beam containing a large number of energetic electrons passes through a gas of atomic hydrogen, there will be continual electron excitation and continual photon emission. But as soon as you turn off the electron beam, the hydrogen will stop emitting photons.

? What happens if you send a beam of electrons with kinetic energy of 7 eV through the gas of atomic hydrogen?

7 eV isn't enough energy to go from the ground state to the first excited state in atomic hydrogen, so the electrons can pass right through without losing any kinetic energy except for ordinary collisions that can give the atoms some kinetic energy but leave them still in the ground state.

? What happens if you send a beam of electrons with kinetic energy of 25 eV through a gas of atomic hydrogen?

Because the ground state of hydrogen has $K + U = -13.6$ eV, only 13.6 eV is needed to raise a hydrogen atom to any of its excited states, or even to break the atom apart into a proton and an electron. Therefore a beam containing lots of 25 eV electrons passing through the atomic hydrogen gas will keep many excited states populated, and you'll observe many different photon energies being emitted as atoms drop from excited states to lower excited states or to the ground state.

The Franck-Hertz experiment

The German physicists James Franck and Gustav Hertz performed an experiment in 1914 that provided early dramatic evidence for quantized energies in atoms. Electrons with varying amounts of initial kinetic energy were sent through a gas of mercury vapor (Hg), and the number of electrons per sec-

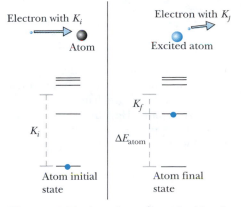

Figure 7.11 An electron with kinetic energy K_i hits an atom in its ground state. Part of the electron's kinetic energy is absorbed by the atom, which goes to an excited state, leaving the electron with decreased kinetic energy K_f.

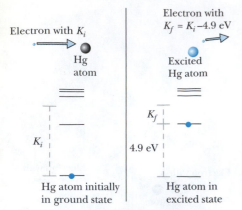

Figure 7.12 The Franck-Hertz experiment, in which electrons with initial kinetic energy of more than 4.9 eV excited mercury atoms (Hg) from the ground state to the first excited state. The electrons lost 4.9 eV of kinetic energy.

Note that every element has a different set of quantized energies. For example, the first excited state in mercury is 4.9 eV above the ground state, whereas the first excited state in hydrogen is 10.2 eV above the ground state. These differences make it possible to identify elements from emission or absorption spectra, or measurements of the Franck-Hertz type.

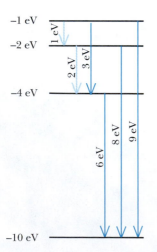

Figure 7.13 A possible energy-level scheme that fits the emission data. Lighter arrows indicate photons not detectable by this detector.

ond making it through the gas was measured (the "electron current"). When the initial kinetic energy was less than 4.9 eV, there was plenty of current. But when the initial kinetic energy reached 4.9 eV the current dropped sharply. The Franck-Hertz experiment was correctly interpreted to mean that the first excited state of a mercury atom is 4.9 eV above the ground state, and an electron with kinetic energy of 4.9 eV can excite the atom, leaving the electron with almost no kinetic energy, so the number of electrons that made it through the gas became very small (Figure 7.12).

? What could the final kinetic energy of an electron be if its initial kinetic energy is 7.0 eV?

An electron can lose 4.9 eV to a mercury atom, leaving the electron with 2.1 eV of remaining kinetic energy.

? What happens if the initial kinetic energy of the electron is 9.8 eV?

In this case the electron could excite one mercury atom and still have 4.9 eV of kinetic energy available to excite a second mercury atom. In the experiment it was observed that the electron current again dropped sharply when the initial kinetic energy of the electrons was raised to 9.8 eV.

7.X.6 The first excited state of a mercury atom is 4.9 eV above the ground state. A moving electron collides with a mercury atom, and excites the mercury atom to its first excited state. Immediately after the collision the kinetic energy of the electron is 0.3 eV. What was the kinetic energy of the electron just before the collision?

7.3 SUMMARY OF THREE IMPORTANT PROCESSES

- Electron excitation: A beam containing many electrons with high kinetic energy can maintain a gas of atoms in many excited states. The electron energy need not match the energy difference between the excited state and the ground state.

- Photon emission: Atoms in excited states emit photons as the atoms drop from higher energy states to lower energy states.

- Photon absorption: Atoms in cold material, in the absence of an electron beam, can absorb photons and go from the ground state to an excited state, but only if the photon's energy exactly matches the energy difference between the excited state and the ground state. Absorption leads to a "dark-line" spectrum.

Example: Observed photon emission

You observe photon emissions from a collection of quantum objects, each of which is known to have just four quantized energy levels. The collection is kept at a high temperature, and you detect emitted photons using a detector sensitive to photons in the energy range from 2.5 eV to 30 eV. With this detector, you observe photons emitted with energies of 3 eV, 6 eV, 8 eV, and 9 eV, but no other energies.

(a) It is known that the ground state energy of each of these objects is −10 eV. Propose two possible arrangement of energy levels that are consistent with the experimental observations. Explain in detail, using diagrams.

(b) You obtain a second detector that is sensitive to photon energies in the energy range from 0.1 eV to 2.5 eV. What additional photon energies do you observe to be emitted? Explain briefly.

(c) You cool the collection of objects to a very low temperature and send a beam of photons with a wide range of energies through the material. Using both detectors to determine the absorption spectrum, what are the photon energies of the dark lines? How can this information be used to choose between your two proposed energy level schemes? Explain briefly.

Fundamental principle: the Energy Principle.

System: everything—some unknown object and a photon
Surroundings: nothing significant

Initial state: object in excited state (no photon)
Final state: object in lower energy state, plus photon

Energy Principle: $E_{obj,f} + E_{photon} = E_{obj,i}$

So the energy of any emitted photon must be equal to the difference in energy between an excited state and a lower energy state.

(a) Figure 7.13 shows a possible energy-level scheme that fits the emission data. We found this scheme essentially by trial and error. We knew the energy differences between levels, so we tried various arrangements. The arrows are labeled with the energies of photons emitted in transitions between the two states indicated. The 1 eV and 2 eV emissions would not be detected by a detector that is sensitive to the photon energy range 2.5 eV to 30 eV.

Figure 7.14 shows another possible scheme that fits the emission data. This is somewhat less likely, because it is more common to find the energy level spacing decreasing at higher energies, as shown in Figure 7.13.

(b) With this detector we observe the 1 eV and 2 eV photon emissions.

(c) At very low temperature almost all of the objects are in the ground state. If the energy levels are as shown in Figure 7.13, the dark lines will be at 6 eV, 8 eV, and 9 eV, corresponding to absorption in the ground state. If the energy levels are as shown in Figure 7.14, the dark lines will be at 1 eV, 3 eV, and 9 eV. Therefore the low temperature absorption spectrum allows us to distinguish between the two possible energy level schemes.

7.X.7 The energy levels of a particular quantum object are −8 eV, −3 eV, and −2 eV. If a collection of these objects is at a high enough temperature that some objects are in each excited state, what are the energies of the photons that will be emitted?

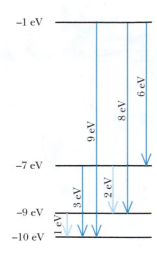

Figure 7.14 Another possible energy-level scheme that fits the emission data.

Note that since this was an energy analysis, we didn't need to know what kind of object this was, or whether the energy levels were electronic, vibrational, or rotational energy levels.

7.4 QUANTIZED VIBRATIONAL ENERGY LEVELS

For concreteness we have described energy quantization for the electronic states of atomic hydrogen. Many other kinds of microscopic systems have quantized energy. We will next discuss vibrational energy quantization in atomic oscillators.

Vibrational energy in a classical oscillator

We have modeled a solid as a network of classical harmonic oscillators, balls and springs (Figure 7.15). Although this classical model explains many aspects of the behavior of solids, the predictions it makes regarding the thermal properties of solids do not correspond to experimental measurements at low temperatures. The classical model also fails to explain some aspects of the interaction of light with solid matter. To explain these phenomena we need to make our model of a solid more sophisticated, by incorporating the quantum nature of these atomic oscillators into our model.

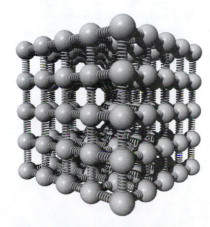

Figure 7.15 Model of a solid as a network of balls and springs.

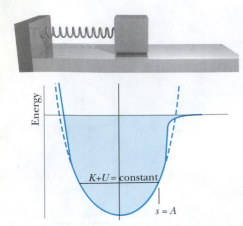

Figure 7.16 A classical harmonic oscillator (spring-mass system) can vibrate with any amplitude, and hence can have any energy. $K+U$ of the system could have any value in the colored region.

The vibrational energy of a classical (non-quantum) spring-mass system can have *any* value (the colored region in Figure 7.16). Equivalently, the amplitude A (maximum stretch) can have *any* value. We can express the energy as $E = \frac{1}{2}k_s A^2$, because at the maximum stretch A the speed is zero, so the kinetic energy is zero, and the energy $K+U$ is just the potential energy at that instant. Classically, the amplitude A can have any value whatsoever, and the energy is continuous, not quantized.

Vibrational energy in a quantum mechanical oscillator

Historically, the first hint that energy might be quantized came from an attempt by the German physicist Max Planck in 1900 to explain some puzzling features of the spectrum of light emitted through a small hole in a hot furnace (so-called "blackbody radiation"). Planck found that he could correctly predict the properties of blackbody radiation if he assumed that transfers of energy in the form of electromagnetic radiation were quantized—that is, an energy transfer could take place only in fixed amounts, rather than having a continuous range of possible values.

Later, with the full development of quantum mechanics, it was recognized that one could model an atom in the solid wall of the furnace as a tiny spring-mass system whose energy levels are discrete (Figure 7.17). Classically, the energy of a mass on a spring is related to the amplitude of its oscillation ($E = \frac{1}{2}k_s A^2$), so saying that a spring-mass oscillator can only have certain discrete amounts of energy is equivalent to saying that only certain amplitudes are allowed. To see this, note that we have

$$K + U = \frac{1}{2}\frac{p^2}{m} + \frac{1}{2}k_s s^2 = \frac{1}{2}k_s A^2$$

since when the stretch s is equal to the amplitude A (the maximum stretch), the momentum is zero.

By solving the quantum mechanical Schrödinger equation for a system involving an atom and a chemical bond, one can show that the spacing ΔE between the allowed, quantized energies has the following value:

$$\Delta E = \hbar\sqrt{\frac{k_s}{m}}, \text{ or } \Delta E = \hbar\omega_0, \text{ where } \omega_0 = \sqrt{\frac{k_s}{m}}$$

In this case, the quantity $k_s = k_{s,i}$, the interatomic spring stiffness, which we estimated in Chapter 4, $m = m_a$, the mass of one atom. $\omega_0 = \sqrt{k_s/m}$ is the angular frequency of the oscillator (in radians per second), as defined in Chapter 4. The quantity $\hbar$, pronounced "h-bar," is equal to h, "Planck's constant," divided by 2π. Planck's constant h has an astoundingly small value:

$$h = 6.6\times10^{-34} \text{ joule} \cdot \text{second}$$

The quantized energies for an atomic harmonic oscillator are these:

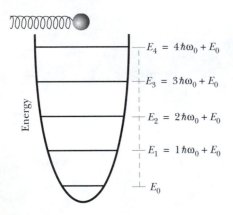

Figure 7.17 A quantum harmonic oscillator can have only certain energies, indicated by the horizontal lines on the graph. This idealized potential energy well represents the situation near the bottom of a real well (which is at a negative energy).

QUANTIZED VIBRATIONAL ENERGY LEVELS

$E_N = N\hbar\omega_0 + E_0$, $N = 0, 1, 2$, etc., where $\omega_0 = \sqrt{k_s/m}$

In other words, the energy can have only the following values:
$E = E_0, \ 1\hbar\omega_0 + E_0, \ 2\hbar\omega_0 + E_0, \ 3\hbar\omega_0 + E_0$, etc.

For an atom and a springlike interatomic bond: $k_s = k_{s,i}$ (interatomic spring stiffness) and $m = m_a$ (mass of an atom)

$$\hbar = \frac{h}{2\pi} = 1.05\times10^{-34} \text{ joule} \cdot \text{second} \quad \text{("h-bar")}$$

It is an unusual and special feature of the quantized harmonic oscillator that the spacing between adjacent energy levels is the same for all levels ("uniform spacing"). This is quite different from the uneven spacing of energy levels in the hydrogen atom. These energy levels are predicted by quantum mechanics, and experiments confirm this prediction.

It is important to understand that in both the quantum and the classical picture, the angular frequency $\omega_0 = \sqrt{k_s/m}$ of a particular oscillator is fixed and is the same whether the energy of the oscillator is large or small. A useful way to think about the situation is to say that the frequency is determined by the parameters k_s and m, while the energy is determined by the initial conditions (or equivalently by the amplitude A). In the quantum world, it is as though the amplitude can have only certain specific values.

7.X.8 Show that $(\hbar\sqrt{k_s/m})$ has units of joules.

7.X.9 Suppose a collection of quantum harmonic oscillators occupies the lowest 4 energy levels, and the spacing between levels is 0.4 eV. What is the complete emission spectrum for this system? That is, what photon energies will appear in the emissions? Include all energies, whether or not they fall in the visible region of the electromagnetic spectrum.

Spring stiffness k_s

The energy quantum depends on the stiffness k_s of the "spring." A stiffer spring (larger value of k_s, Figure 7.18) means a steeper, narrower potential curve and larger spacing between the allowed energies. A smaller value of k_s (Figure 7.19) means a shallower, wider potential curve and smaller spacing between the allowed energies.

7.X.10 You may have measured the properties of a simple spring-mass system in the lab. Suppose you found $k_s = 0.7\ \text{N/m}$ and $m = 0.02\ \text{kg}$, and you observed an oscillation with an amplitude of 0.2 m. What is the approximate value of N, the "quantum number" for this oscillator? (That is, how many levels above the ground state is this oscillator?)

7.X.11 Summarize the differences and similarities between different energy levels in a quantum oscillator. Specifically, for the first two levels in Figure 7.18, compare the angular frequency $\sqrt{k_s/m}$, the amplitude A, and the kinetic energy K at the same value of s. (In a full quantum-mechanical analysis the concepts of angular frequency and amplitude require reinterpretation. Nevertheless there remain elements of the classical picture. For example, larger amplitude corresponds to a higher probability of observing a large stretch.)

7.X.12 What is the energy of the photon emitted when a harmonic oscillator with stiffness k_s and mass m loses 3 of its quanta of energy?

The ground state of a quantum oscillator

The lowest possible quantized energy E_0 is not zero relative to the bottom of the potential curve, but is $\frac{1}{2}(\hbar\sqrt{k_s/m})$ above the bottom of the well. This is closely related to the "Heisenberg uncertainty principle." When applied to the harmonic oscillator, this principle implies that if you knew that the mass on the spring were exactly at the center of the potential, its speed would be completely undetermined, whereas if you knew that the speed of the mass were exactly zero, the position of the mass could be literally anywhere! The minimum energy $\frac{1}{2}(\hbar\sqrt{k_s/m})$ above the bottom of the well can be viewed as a kind of a compromise, with neither the speed nor the position fixed at zero. The minimum energy is called the ground state of the system.

In our work with the quantized oscillator we will measure energy from the ground state rather than from the bottom of the potential well, so the offset from the bottom of the well won't come into our calculations.

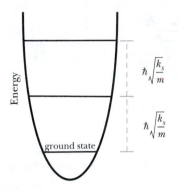

Figure 7.18 A quantum oscillator with a large k_s has widely spaced energy levels.

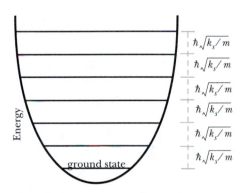

Figure 7.19 A quantum oscillator with a small k_s has closely spaced energy levels.

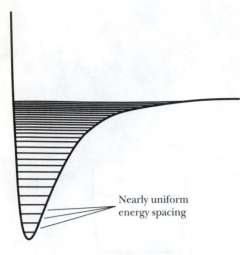

Figure 7.20 Energy levels for the interatomic potential energy, which describes the interaction of two neighboring atoms. This diagram is not to scale; the number of vibrational energy levels in the uniform region may be very large.

Figure 7.21 The vibrational energy of a diatomic molecule is quantized.

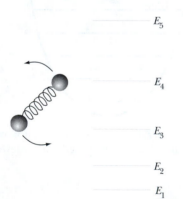

Figure 7.22 Rotational kinetic energy levels of a diatomic molecule.

Energy levels for the interatomic potential energy

If we model the atoms in a solid as simple harmonic oscillators, whose potential energy curve is a parabola, then the energy levels of these systems are uniformly spaced. However, if we use a more realistic interatomic potential energy such as is shown in Figure 7.20, the energy levels of the system follow a more complex pattern. The lowest energy levels are nearly evenly spaced, corresponding to the fact that for small amplitude the bottom of the potential energy curve can be fit rather well by a curve of the form $\frac{1}{2}k_s s^2$. However, for large energies the energy spacing is quite small. And of course there are no bound states for energies larger than the binding energy. The spacing of the vibrational energy levels of a diatomic molecule (Figure 7.21) typically corresponds to photons in the infrared region of the spectrum (see Problem 7.P.25).

Energy spacing in various systems

Most quantum objects have energy level spacings that decrease as you go to higher levels. The harmonic oscillator potential energy is unusual in leading to uniform energy level spacing. Even less common is the situation where the energy level spacing actually increases with higher energy. One such exception is the so-called "square well" with very steep sides and a flat bottom. The energy levels in a square well actually get farther and farther apart for higher energies.

Rotational energy levels

A molecule in the gas phase has not only electronic energy and vibrational energy, but rotational energy as well. Like electronic and vibrational energy levels, rotational energy levels are also discrete—only certain rotational energies are permitted. As we will see in Chapter 10, the angular momentum of a molecule is quantized, and this quantization gives rise to discrete rotational energy levels. In this case all the energy of rotation is kinetic energy—there is no potential energy involved. Rotational energy levels for a diatomic molecule such as N_2 are shown in Figure 7.22. Note that the spacing between levels actually increases with increasing energy, which is the opposite of the trend seen in many atomic systems. Rotational energy levels are typically close together; transitions between these levels usually involve photons in the microwave region of the electromagnetic spectrum. Microwave ovens excite rotational energies of water molecules in food. These excited molecules then give up their energy to the food, making the food hotter.

7.5 CASE STUDY: THE DIATOMIC MOLECULE

Analyzing a diatomic molecule such as N_2 provides a good review of three important types of quantized energy levels: electronic, vibrational, and rotational.

Only certain configurations of the electron clouds in a diatomic molecule are possible, corresponding to discrete electronic energy levels. The spacings of the low-lying molecular electronic energy levels are typically in the range of 1 eV or more, like the spacings of the low-lying energy levels in the hydrogen atom.

For each major configuration of the electron clouds, many different vibrational energy states are possible, with energy spacings of the order of 10^{-2} eV (see Problem 7.P.25).

For each major configuration of the electron clouds, and each vibrational energy state, many different rotational kinetic energy states are possible, with energy spacings of the order of 10^{-4} eV, as we will see in Chapter 10.

These three different families of energy levels, with their very different energy level spacings, give rise to energy "bands" and a distinctive "band" spectrum of emitted photons. See Figure 7.23, which is not to scale due to the large differences in level spacing for the three kinds of quantized energy levels. An emitted photon can have the energy of an electronic transition plus a smaller-energy vibrational transition plus an even smaller-energy rotational transition. The various possible energy differences give rise to "bands" in the observed emission spectrum.

7.6 NUCLEAR ENERGY LEVELS

Just as atoms have discrete states corresponding to allowed configurations of the electron cloud, so nuclei have discrete states corresponding to allowed configurations of the nucleons (protons and neutrons). An excited nucleus can drop to a lower state with the emission of a photon. The energy spacing of nuclear levels is very large, of the order of 10^6 eV (1 MeV), so the emitted photons have energies in the range of millions of electron volts. Such energetic photons are called "gamma rays."

> *7.X.13* A gamma ray with energy greater than or equal to 2.2 MeV can dissociate a deuteron (the nucleus of deuterium, heavy hydrogen) into its constituents, a proton and a neutron. What is the nuclear binding energy of the deuteron?

7.7 HADRONIC ENERGY LEVELS

Atoms have energy levels associated with the configurations of electrons in the atom, and nuclei have energy levels associated with the configurations of nucleons in the nucleus. Hadrons are particles made of quarks. In a simple model we represent baryons (particles like protons) as a combination of three quarks, and mesons (particles like pions) as a combination of a quark and an antiquark. Different configurations of quarks correspond to different energy levels of the multiquark system. These different configurations are the different hadrons we observe, and the corresponding energy levels are the observed rest energies of these particles. The energy spacing is of the order of hundreds of MeV, or about 10^8 eV. (Leptons, which include electrons, muons, and neutrinos, are not made of quarks.)

An example of a hadron is the Δ^+ particle, which has a rest energy of 1232 MeV. It can be considered an excited state of the three-quark system whose ground state is the proton, whose rest energy is 938 MeV. The Δ^+ particle can decay into a proton with the emission of a very high energy photon. However, this "electromagnetic" decay occurs in only about 0.5% of the decays. In most cases the Δ^+ decays into a proton plus a pion, a strong-interaction decay which has a much higher rate, corresponding to the much stronger interaction.

An excited electronic state of hydrogen is still called hydrogen, but for historical reasons we give excited quark states distinctive names, such as Δ^+.

> *7.X.14* What is the approximate energy of the photon emitted when the Δ^+ decays electromagnetically?

7.8 COMPARISON OF ENERGY LEVEL SPACINGS

Figure 7.24 is a summary of typical energy scales for various quantum objects. It is important to distinguish among hadronic, nuclear, electronic, vibrational, rotational, and other kinds of quantized energy situations. We can generalize the main issues to any kind of generic quantum object. The key features of all these different kinds of objects are these:

• discrete energy levels

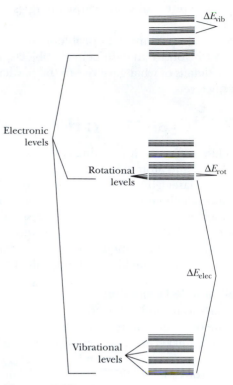

Figure 7.23 The energy bands of a diatomic molecule (not to scale). This energy level structure gives rise to a band spectrum of emitted photons.

Type of state	Typical energy level spacing
hadronic	10^8 eV
nuclear	10^6 eV
electronic (atoms, molecules)	1 eV
vibrational (molecules)	10^{-2} eV
rotational (molecules)	10^{-4} eV

Figure 7.24 Typical energy level spacings in various quantum objects.

• discrete emissions whose energies equal differences in energy levels

Some of the homework problems at the end of the chapter ask you to analyze generic objects, independent of the details of what gave rise to the particular discrete energy levels.

7.9 *CASE STUDY: HOW A LASER WORKS

A laser offers an interesting case study of the interaction of light and atomic matter. We will give a basic introduction to lasers, but we can only scratch the surface of a vast topic which continues to be an active field of research. Laser action depends on a process called "stimulated emission" and on creating an "inverted population" of quantum states, which can lead to a chain reaction and produce a special kind of light, called "coherent" light.

Stimulated emission

If an atom can be placed into an excited state, that is, a state of higher energy than the ground state, normally it drops quickly to a lower energy state and emits a photon in the process. For example, in a neon sign high-energy electrons collide with ground-state neon atoms and raise them to an excited state. An excited neon atom quickly drops back to the ground state, emitting a photon whose energy is the difference in energy between the two quantized energy levels, and you see the characteristic orange color corresponding to these photons.

We call this emission process "spontaneous emission." Spontaneous emission happens at an unpredictable time. All that quantum mechanics can predict is the *probability* that the excited atom will drop to the ground state in the next small time interval. It is truly not possible to predict the exact time of the emission, just as it is impossible to predict exactly when a free neutron will decay into a proton, electron, and antineutrino.

A collection of excited atoms such as those in a neon sign emit spontaneously, at random times. In the full quantum-mechanical description of light, light has both the properties of a particle (the photon picture) and the properties of a wave; see Section 7.10. Emissions at random times have random phases in a wave description: sine, cosine, and all phases between a sine and a cosine. We say that such light is "incoherent."

"Stimulated emission" of light is a different process which produces coherent light. Consider an atom in an energy level E above the ground state. A photon of this same energy E can interact with the atom in a remarkable way, causing the atom to drop immediately to the ground state with the emission of a second, "clone" photon. Where there had been one photon of energy E, there are now two photons of exactly the same energy E, and in a wave description they have exactly the same phase (Figure 7.25). That is, the two waves are both sines, or both cosines, or both with exactly the same phase somewhere between a sine and a cosine. Such light is called coherent light, and it has valuable properties for such applications as making holograms.

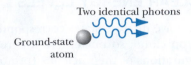

Figure 7.25 Stimulated emission produces two identical photons from one.

Chain reaction

The phenomenon of stimulated emission can be exploited to create a chain reaction. Start with one photon of the right energy E to match the energy difference between the ground state and an excited state of the atoms in the system. If that photon causes stimulated emission from an excited atom, there are now two identical photons, both capable of triggering stimulated emission of other excited atoms, which yields four identical photons, then 8, 16, 32, 64, 128, 256, 512, 1024, etc.

However, the chain reaction won't proceed if all of the photons escape from the system, so some kind of containment mechanism is required. In a common type of gas laser the gas is in a long tube with mirrors at both ends:

Figure 7.26 A gas laser with mirrors at the ends. Photons leak through one of the mirrors.

Photons that happen to be going in the direction of the tube are continually reflected back and forth through the tube, which increases the probability that they will interact with excited atoms and cause stimulated emission. Since the stimulated-emission photons are clones, they too are headed in the direction of the tube and they too are likely to interact. One of the mirrors is deliberately made to be an imperfect reflector, so a fraction of the photons leaks out and provides a coherent beam of light.

Unfortunately, the stimulated-emission photons also have just the right energy E to be absorbed by an atom in the ground state, in which case they can no longer contribute to the chain reaction. For this reason we need an "inverted population," with few atoms in the ground state to absorb the photons and stop the chain reaction. We discuss this issue next.

Inverted population

As we have discussed earlier in this chapter, in a normal collection of atoms more atoms are in the ground state than in any other state, and the number of atoms decreases in each successively higher state. In fact, for a collection of atoms in thermal equilibrium the fraction of atoms in a particular energy state decreases exponentially with the energy above the ground state, as we will see in Chapter 11.

By clever means it is possible to invert this population scheme, so that there are more atoms in an excited state than in the ground state (this is not a state of thermal equilibrium). An inverted population is crucial to laser action, so that stimulated emission with cloning of photons will dominate over simple absorption of photons by atoms in the ground state.

There are many different schemes for creating an inverted population. For example, the scheme used in the common helium-neon laser is to "pump" atoms to a high energy level:

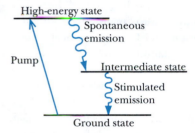

Figure 7.27 A three-state scheme for sustaining an inverted population.

There are diverse energy input schemes, including electric discharges and powerful flashes of ordinary light. Atoms in this high energy level rapidly drop to an intermediate energy level, higher than the ground state. This intermediate state happens to have a rather long lifetime for spontaneous emission, so that it is possible to sustain an inverted population, with more atoms in the intermediate state than in the ground state. Stimulated emission drives these excited atoms down to the ground state, from which they are again pumped up to the high-energy state. Spontaneous emission from the high-energy state to the intermediate state maintains the inverted population.

7.10 *WAVELENGTH OF LIGHT

In this chapter we have emphasized the particle model of light, in terms of photons and their energy. For completeness we will mention a connection between the particle model and the wave model of light. Light consisting of photons whose energy is E can also be treated as a wave, and the energy E and wavelength λ are related as follows:

ENERGY AND WAVELENGTH OF LIGHT

$$E_{\text{photon}} = \frac{hc}{\lambda_{\text{light}}}$$

where h is Planck's constant.

A model for light that incorporates both the features of a wave and a localized particle is a "wave packet," a wave whose magnitude is nonzero only in a small region of space (Figure 7.28) We will return to this topic in Chapter 24 in Volume II of this textbook.

Figure 7.28 A wave packet is a travelling wave of finite spatial extent.

7.11 SUMMARY

Fundamental Physical Principles

At the microscopic level the energy of bound states is discrete, not continuous. (Unbound states have a continuous range of energies.)

New concepts

Quantized systems can have only certain energies. They can emit or absorb photons when they change energy levels.

The temperature is a measure of the average energy of a collection of objects, so high temperature is associated with significant probability of finding particles in any of many states, from the ground state up to a high energy level. Very low temperature is associated with little probability of the particles being in any state other than the ground state.

Discrete hydrogen atom energy levels

$E_N = -(13.6 \text{ eV})/N^2$, where N is a nonzero positive integer (1, 2, 3...)

Discrete harmonic oscillator energy levels

$$E_N = N\hbar\omega_0 + E_0, \quad N = 0, \ 1, \ 2, \ etc., \ where$$

$$\omega_0 = \sqrt{k_s/m}$$

Planck's constant $h = 6.6 \times 10^{-34}$ joule $\cdot$ second

$$\hbar = \frac{h}{2\pi} = 1.05 \times 10^{-34} \text{ joule} \cdot \text{second}$$

$1 \text{ eV} = 1.6 \times 10^{-19} \text{ J}$

Photon energy and wavelength: $E_{\text{photon}} = \frac{hc}{\lambda_{\text{light}}}$

7.12 REVIEW QUESTIONS

7.RQ.15 If you double the amplitude, what happens to the frequency in a classical (non-quantum) harmonic oscillator? In a quantum harmonic oscillator?

7.RQ.16 If you try to increase the amplitude of a quantum harmonic oscillator by adding an amount of energy

$$\frac{1}{2}\hbar\sqrt{\frac{k_s}{m}}$$

the amplitude doesn't increase. Why not?

7.RQ.17 When starlight passes through a cold cloud of hydrogen gas, some hydrogen atoms absorb energy, then reradiate it in all directions. As a result, a spectrum of the star shows dark absorption lines at the energies for which less energy from the star reaches us. How does the spectrum of dark absorption lines for very cold hydrogen differ from the spectrum of bright emission lines from very hot hydrogen?

7.13 PROBLEMS

7.P.18 Determining energy levels
Suppose we have reason to suspect that a certain quantum object has only three quantum states. When we excite such an object we observe that it emits electromagnetic radiation of three different energies: 2.48 eV (green), 1.91 eV (orange), and 0.57 eV (infrared).
(a) Propose two possible energy-level schemes for this system.
(b) Explain how to use an absorption measurement to distinguish between the two proposed schemes.

7.P.19 Predicting a spectrum
Suppose a hypothetical object has just four quantum states, with the energies shown on this potential energy diagram.

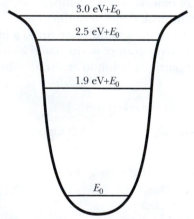

- 3.0 eV+E_0
- 2.5 eV+E_0
- 1.9 eV+E_0
- E_0

(a) Suppose that the temperature is high enough that in a material containing many such objects, at any instant some objects are found in all of these states. What are all the energies of photons that could be strongly emitted by the material? (In actual quantum objects there are often "selection rules" that forbid certain emissions even though there is enough energy; assume that there are no such restrictions here.)
(b) If the temperature is very low and electromagnetic radiation with a wide range of energies is passed through the mate-

rial, what will be the energies of photons corresponding to missing ("dark") lines in the spectrum? (Assume that the detector is sensitive to a wide range of photon energies, not just energies in the visible region.)

7.P.20 Emission and absorption
A certain material is kept at very low temperature. It is observed that when photons with energies between 0.2 and 0.9 eV strike the material, only photons of 0.4 eV and 0.7 eV are absorbed. Next the material is warmed up so that it starts to emit photons. When it has been warmed up enough that 0.7 eV photons begin to be emitted, what other photon energies are also observed to be emitted by the material? Explain briefly.

7.P.21 Emission from hot metal
A hot bar of iron glows a dull red. Using our simple model of a solid, answer the following questions, explaining in detail the processes involved. The mass of one mole of iron is 56 grams.
(a) What is the energy of the lowest-energy spectral emission line? Give a numerical value.
(b) What is the approximate energy of the highest-energy spectral emission line? Give a numerical value.
(c) What is the quantum number of the highest-energy occupied state?
(d) Predict the energies of two other lines in the emission spectrum of the glowing iron bar.
(Note: our simple model is too simple—the actual spectrum is more complicated. However, this simple analysis gets at some important aspects of the phenomenon.)

7.P.22 Potential energy in a molecule
(a) The diagram below is a graph of potential energy vs. interatomic distance for a particular molecule. What is the direction of the associated force at location A? At location B? At location C? Rank the magnitude of the force at locations A, B, and C (that is, which is greatest, which is smallest, are any of these equal to each other). For the energy level shown on the graph, draw a line whose height is the kinetic energy when the system is at location D.

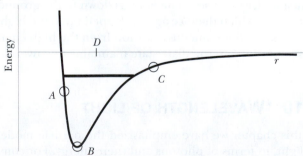

(b) The diagram below shows all of the quantized energies (bound states) for one of these molecules. The energy for each state is given on the graph, in electron volts (1 eV = 1.6×10^{-19} J). How much energy is required to break a molecule apart, if it is initially in the ground state? (Note that the final state must be an unbound state; the unbound states are not quantized.)

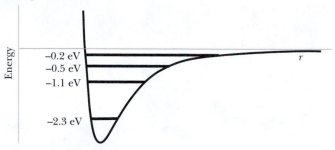

(c) If the temperature is high enough, in a collection of these molecules there will be at all times some molecules in each of these states, and light will be emitted. What are the energies in electron volts of the emitted light?

(d) The "inertial" mass of the molecule is the mass that appears in Newton's second law, and which determines how much acceleration will result from applying a given force. Compare the inertial mass of a molecule in the top energy state and the inertial mass of a molecule in the ground state. If there is a difference, briefly explain why and calculate the difference. If there isn't a difference, briefly explain why not.

7.P.23 A microscopic oscillator

Consider a microscopic spring-mass system whose spring stiffness is 50 N/m, and the mass is 4×10^{-26} kg.

(a) What is the smallest amount of vibrational energy that can be added to this system?

(b) What is the difference in mass (if any) of the microscopic oscillator between being in the ground state and being in the first excited state?

(c) In a collection of these microscopic oscillators, the temperature is high enough that the ground state and the first three excited states are occupied. What are possible energies of photons emitted by these oscillators?

7.P.24 An experiment with quantum objects

Some material consisting of a collection of microscopic objects is kept at a high temperature. A photon detector capable of detecting photon energies from infrared through ultraviolet observes photons emitted with energies of 0.3 eV, 0.5 eV, 0.8 eV, 2.0 eV, 2.5 eV, and 2.8 eV. These are the only photon energies observed.

(a) Draw and label a possible energy-level diagram for one of the microscopic objects, which has 4 bound states. On the diagram, indicate the transitions corresponding to the emitted photons. Explain briefly.

(b) Would a spring-mass model be a good model for these microscopic objects? Why or why not?

(c) The material is now cooled down to a very low temperature, and the photon detector stops detecting photon emissions. Next a beam of light with a continuous range of energies from infrared through ultraviolet shines on the material, and the photon detector observes the beam of light after it passes through the material. What photon energies in this beam of light are observed to be significantly reduced in intensity ("dark absorption lines")? Explain briefly.

7.P.25 Quantized energy for a diatomic molecule

(a) A nitrogen molecule, N_2, can be considered to be a quantized harmonic oscillator, with quantized vibrational energy levels that are evenly spaced. Make a rough estimate of this uniform energy spacing in electron volts (where $1 \text{ eV} = 1.6 \times 10^{-19}$ J). You will need to make some rough estimates of atomic properties based on prior work. For comparison with the spacing of these vibrational energy states, note that the spacing between quantized energy levels for "electronic" states such as in atomic hydrogen is of the order of several electron volts.

(b) List several photon energies that would be emitted if a number of these vibrational energy levels were occupied due to electron excitation. To what region of the spectrum (x-ray, visible, microwave, etc.) do these photon belong? (See Figure 7.2 on page 261.)

7.P.26 The spectrum of atomic hydrogen

The eye is sensitive to photons with energies in the range from about 1.8 eV, corresponding to red light, to about 3.1 eV, corresponding to violet light. White light is a mixture of all the energies in the visible region. If you shine white light through a slit onto a glass prism, you can produce a rainbow spectrum on a screen, because the prism bends different colors of light by different amounts.

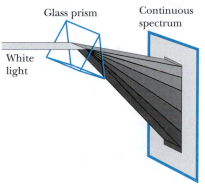

If you replace the source of white light with an electric-discharge lamp containing excited atomic hydrogen, you will see only a few lines in the spectrum, rather than a continuous rainbow.

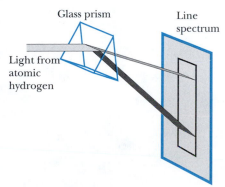

Predict how many lines will be seen in the visible spectrum of atomic hydrogen. Specify the atomic transitions that are responsible for these lines, given that the energies of the quantized states in atomic hydrogen are given by $E_N = -(13.6 \text{ eV})/N^2$, where N is a nonzero positive integer (1, 2, 3...).

7.14 ANSWERS TO EXERCISES

7.X.1 (page 261) 13.6 eV

7.X.2 (page 261) About 2.5 eV

7.X.3 (page 261) About 3.5×10^{21} photons per square meter per second.

7.X.4 (page 263) Six different photon energies, corresponding to transitions from 4 to 3, 4 to 2, 4 to 1, 3 to 2, 3 to 1, and 2 to 1.

7.X.5 (page 264) Just one dark line at 2.4 eV

7.X.6 (page 266) 5.2 eV

7.X.7 (page 267) 6 eV, 5 eV, 1 eV

7.X.9 (page 269) 0.4 eV (three transitions), 0.8 eV (two transitions), and 1.2 eV (one transition)

7.X.10 (page 269) $N \approx 10^{31}$. A system in a very high quantum state behaves like a classical system.

7.X.11 (page 269) Upper level has same ω, larger A, larger K at same s.

7.X.12 (page 269) $3\hbar \sqrt{\dfrac{k_s}{m}}$

7.X.13 (page 271) 2.2 MeV

7.X.14 (page 271) 294 MeV

CHAPTER 8

MULTIPARTICLE SYSTEMS

Key Concepts

- The kinetic energy of a non-rigid multiparticle system has two parts:
  - Translational kinetic energy (due to motion of center of mass)
  - Kinetic energy relative to the center of mass (rotation + vibration)
- Treating a non-rigid multiparticle system as if it were a point particle lets us calculate changes in translational kinetic energy.
- Analyzing the real, non-rigid system lets us calculate changes in the other energy terms (rotational, vibrational, potential).

8.1 THE MOTION OF A MULTIPARTICLE SYSTEM

There are some major complications in applying the Momentum Principle to systems consisting of many particles. Suppose you pull on a thin wire that is attached to a metal block, and you accelerate the block across a low-friction surface (Figure 8.1).

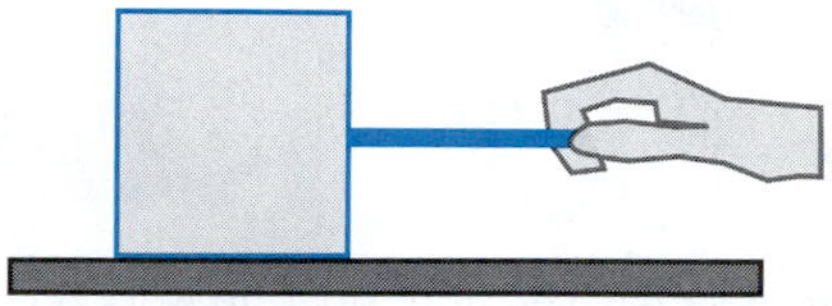

? Actually, you are not touching the block and are not exerting a force on the atoms in the block. How do these atoms know that they should accelerate?

When you first start pulling on your end of the wire, you very slightly lengthen the interatomic bonds between neighboring atoms in the nearby section of the wire (Figure 8.2). As a result of your pulling, these atoms stretch the bonds of their neighbors, and very quickly this new interatomic bond-stretching propagates all the way to the other end of the wire, and the whole wire is then in tension.

This process is very fast, but it is not instantaneous. These changes in the average lengths of the interatomic bonds propagate at the speed of sound in the material. (See Chapter 4.)

Eventually the interatomic bonds at the left end of the wire will be stretched. The left-most atoms, which are attached to the metal block, will pull on atoms in the block. Now the block itself begins to stretch. As with the wire, initially, only those atoms of the block that are in direct contact with the wire are pulled to the right, but quickly (at the speed of sound in the material that the block is made of) these effects propagate throughout the block. Eventually it must be the case that all of the interatomic bonds in the block are stretched, but not equally.

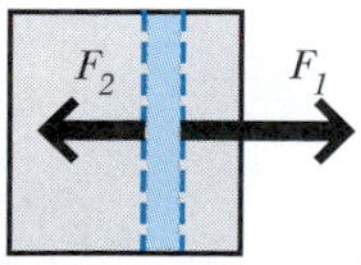

Consider a slice of the block, indicated by dashed lines in Figure 8.3. Since this slice is accelerating to the right, the interatomic forces acting on the right face of this slice (F_1) must be larger than those on the left face (F_2). This requires that the interatomic bonds on the right be stretched more than those on the left. While the block accelerates to the right, the pattern of strain in the block must look something like Figure 8.4 (strain is $\Delta L/L$, or fractional stretch). Actually, even this description is oversimplified, because there will be maximum strain near the point where the wire is attached, and the lines of strain will be somewhat curved.

For a metal block, the strain is normally too small to see with the unaided eye. Nevertheless, the situation is pretty complicated! And there is a similar

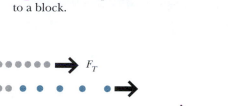

Figure 8.1 You pull on a wire attached to a block.

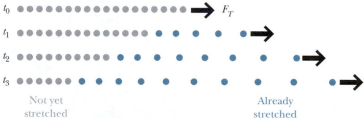

Not yet stretched · Already stretched

Figure 8.2 Strain in the wire increases with time.

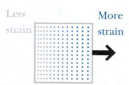

Figure 8.3 Forces acting on a slice of the block. $F_2 < F_1$

Less strain · More strain

Figure 8.4 Pattern of strain in a block accelerating to the right.

gradient of strain in the wire. There is also random thermal motion of the atoms within the block. Given all this complexity, what hope is there of treating the block as though it were just one big, rigid particle?

> *8.X.1* If you let go and allow the block to coast along at nearly constant speed, what must happen to the strain throughout the block? What is the mechanism for that happening at the atomic level?

Energy in multiparticle systems

Just as there are great complexities in dealing with forces in multiparticle systems, so also there are puzzling aspects of work and energy. For example, when you jump upwards the floor pushes up on your feet, yet this force does no work, because there is no displacement of the point of contact (Figure 8.5). However, it is clear that your kinetic energy has increased. Evidently there are some subtle issues here that we need to address.

Figure 8.5 The floor does no work on the jumper, yet *K* increases.

8.2 THE MOMENTUM PRINCIPLE FOR MULTIPARTICLE SYSTEMS

Earlier we derived an equation that predicts the overall motion of a complex multiparticle system, in terms of its total momentum $\vec{P}_{tot}$:

THE MOMENTUM PRINCIPLE FOR MULTIPARTICLE SYSTEMS

$$\frac{d\vec{P}_{tot}}{dt} = \vec{F}_{net,surr}$$

The basis for this result is that internal forces between pairs of particles cancel for gravitational and electric forces (reciprocity; Newton's third law). This result is what makes it possible for many purposes to treat a complex system as though it were a simple point particle. A concept that simplifies analyses of this kind is the "center of mass," which we discuss next.

Center of mass

For a system of particles moving slowly compared to the speed of light, with total mass *M*, we will show that $\vec{P}_{tot} \approx M\vec{v}_{cm}$, where $\vec{v}_{cm}$ is the velocity of the "center of mass" of the system. Since $d\vec{P}_{tot}/dt = \vec{F}_{net,surr}$, we can treat a complicated system as though it were a single particle of mass *M*, located at the center of mass of the system.

The center of mass can be thought of as a weighted average of the locations of all the particles in the system. In many cases it is simply the geometrical center of a system. For example, the center of mass of a disk or sphere or cube is the geometrical center of the object, if the object has uniform density. Let $\vec{r}_1$ be the location (relative to some arbitrary origin) of mass m_1, $\vec{r}_2$ be the location of mass m_2, etc., as shown in Figure 8.6 The center of mass is located at position $\vec{r}_{cm}$, defined like this:

CENTER OF MASS.

$$\vec{r}_{cm} \equiv \frac{m_1\vec{r}_1 + m_2\vec{r}_2 + m_3\vec{r}_3 + ...}{m_1 + m_2 + m_3 + ...}$$

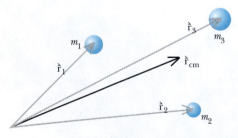

Figure 8.6 Location of the center of mass of the three-particle system.

Let's see whether this makes sense as a definition for the location of the center of mass. We'll consider several simple concrete situations and check that this definition yields results that are consistent with what we would intuitively think of as the center of mass.

? Consider the case where there are three masses very close to each other. Does the definition

$$\vec{r}_{cm} \equiv \frac{m_1\vec{r}_1 + m_2\vec{r}_2 + m_3\vec{r}_3 + \dots}{m_1 + m_2 + m_3 + \dots}$$

make sense in this case?

In this case, $\vec{r}_1 = \vec{r}_2 = \vec{r}_3$, and the definition of the location of the center of mass leads to the result that $\vec{r}_{cm} = \vec{r}_1$, which makes sense. If all three masses are located at the origin, then according to the definition the center of mass is also at the origin, which makes sense.

? Consider the case where there are two particles, with equal masses, located at $\langle -5, 0, 0 \rangle$ and $\langle +5, 0, 0 \rangle$. Does the definition make sense in that case?

In this case we find that $\vec{r}_{cm} = 0$, which makes sense; the center of mass is equidistant between the two equal masses.

? What if there is a very small mass located at $\langle -5, 0, 0 \rangle$ and a very large mass located at $\langle +5, 0, 0 \rangle$. Does the definition make sense in that case?

In this case the formula yields $\vec{r}_{cm} \approx \langle +5, 0, 0 \rangle$, corresponding to the fact that almost all of the total mass is concentrated at that location.

If M is the total mass of the system, we can write these two equivalent equations:

$$\vec{r}_{cm} = \frac{m_1\vec{r}_1 + m_2\vec{r}_2 + m_3\vec{r}_3 + \dots}{M}$$

$$M\vec{r}_{cm} = m_1\vec{r}_1 + m_2\vec{r}_2 + m_3\vec{r}_3 + \dots$$

Center of mass of several large objects

Often we need to know the location of the center of mass of a system consisting of several large objects, each of whose center of mass is known. In Figure 8.7, $\vec{R}_1$ is the location (relative to the origin) of the center of mass of the object whose total mass is M_1, etc.

For each of the large objects we have an equation for its center of mass of the form $M_1\vec{R}_1 = \vec{r}_{11} + m_{12}\vec{r}_{12} + \dots$, $M_2\vec{R}_2 = m_{21}\vec{r}_{21} + m_{22}\vec{r}_{22} + \dots$, etc.

Group terms in the calculation of the location of the center of mass of the total system:

$$\vec{r}_{cm} \equiv \frac{(m_{11}\vec{r}_{11} + m_{12}\vec{r}_{12}) + \dots + (m_{21}\vec{r}_{21} + m_{22}\vec{r}_{22}) + \dots}{(m_{11} + m_{12} + \dots) + (m_{21} + m_{22} + \dots) + \dots}$$

$$\vec{r}_{cm} = \frac{M_1\vec{R}_1 + M_2\vec{R}_2 + M_3\vec{R}_3 + \dots}{M_1 + M_2 + M_3 + \dots}$$

This looks just like the equation for finding the center of mass of a bunch of atoms, except that the $\vec{R}$ vectors refer to the centers of mass of large objects.

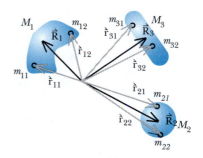

Figure 8.7 Finding the center of mass of a system consisting of several large objects.

8.X.2 Determine the location of the center of mass of a "T" whose thin vertical and horizontal members have the same length L and the same mass M. Use the formal definition to find the x and y coordinates, and check your result by doing the calculation with respect to two different origins, one at the bottom of the vertical member and one at the left end of the horizontal member.

8.X.3 Relative to an origin at the center of the Earth, where is the center of mass of the Earth-Moon system? The mass of the Earth is 6×10^{24} kg, the mass of the Moon is 7×10^{22} kg, and the distance from the center of the Earth to the center of the Moon is 4×10^{8} m. The radius of the Earth is 6400 km. One can show that the Earth and Moon orbit each other around their center of mass (for example, see Problem 8.P.22).

Total momentum and center of mass motion

Now that we have introduced the concept of center of mass, we can make a useful connection between the total momentum and the motion of the center of mass. The velocity of the center of mass of a system is obtained by taking the time derivative of the definition of the position of the center of mass, where $\vec{v}_1 = d\vec{r}_1/dt$, etc. Take the important case where all the individual masses are constants. In that case, since

$$M\vec{r}_{cm} = m_1\vec{r}_1 + m_2\vec{r}_2 + m_3\vec{r}_3 + \dots$$

we have this:

$$M\vec{v}_{cm} = M\frac{d\vec{r}_{cm}}{dt} = m_1\vec{v}_1 + m_2\vec{v}_2 + m_3\vec{v}_3 + \dots \text{ (constant masses)}$$

But $(m_1\vec{v}_1 + m_2\vec{v}_2 + m_3\vec{v}_3 + \dots)$ is the total (nonrelativistic) momentum of the system, so we have shown that

$$\vec{P}_{tot} \approx M\vec{v}_{cm} \text{ (if } v \ll c, \text{ and constant mass)}$$

In summary, we restate the important results:

THE MOMENTUM PRINCIPLE FOR MULTIPARTICLE SYSTEMS

$$\frac{d\vec{P}_{tot}}{dt} = \vec{F}_{net,surr} \text{ (in general)}$$

$$\vec{P}_{tot} \approx M\vec{v}_{cm} \text{ (if } v \ll c, \text{ and constant mass)}$$

For example, if you apply an external force F to one side of a block (through the tension in a wire, for example), the center of mass of the block moves as though that force were applied at the center of mass, producing an acceleration of the center of mass $dv_{cm}/dt = F/M$, even though the detailed motions of the many atoms in the block aren't simple (different amounts of strain in different parts of the block, thermal agitation of the atoms, etc.).

? A high diver jumps outward, then tucks into a tight ball and spins rapidly, then straightens out before hitting the water. What is the path of the diver's center of mass?

Although the diver's overall motion is very complex, the center of mass of the diver must move like that of a simple projectile! We do not yet have enough tools to analyze the diver's complicated rotation and stretching relative to the center of mass, yet we can easily analyze the motion of the center of mass.

Application: An ice skater pushes off from a wall

All of this may sound quite plausible, but it can lead to some rather odd consequences. Consider a woman on ice skates who pushes off from a wall with a nearly constant force F_N (normal to the wall) and moves backwards with increasing speed. Her center of mass is marked on each frame of the following sequence:

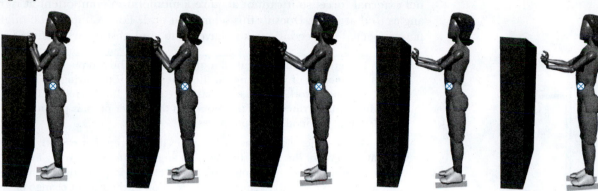

What would we predict for the motion of the skater's center of mass? If M is her total mass, and if we can assume that the frictional force of the ice on the skates is negligibly small, we predict this from the force diagram (Figure 8.8):

$$\frac{d\mathrm{P}_{tot}}{dt} = F_N$$

What's odd about this simple and intuitively appealing result is that during the time that the wall exerts a force F_N on the skater's hand, her hand does not move! (Note that the work done by the force exerted by the wall is zero; more about this later.) The place where the force F_N is applied is very far away from the mathematical point (the center of mass) whose rate of change of speed we predict. It is only because the electric interactions between neighboring atoms in the skater's body obey the principle of reciprocity (Newton's third law) that these internal forces cancel and lead to a simple prediction for the motion of the center of mass:

$$\frac{d\vec{P}}{dt} = \vec{F}_{net,surr}$$

Note that as the skater straightens her arms, the location in her body of the center of mass shifts toward her hands a little; it is not a point fixed at some place in her body. The motion that we predict is the motion of this mathematical center of mass point.

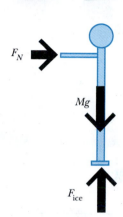

Figure 8.8 Force diagram for the skater.

Application: Pull on two hockey pucks

We'll look at another odd consequence of this multiparticle version of the Momentum Principle. Tie a string to the center of a hockey puck, and wrap another string around the outside of a second hockey puck, as shown in Figure 8.9. Then two people pull on the two strings so that both strings have the same tension F_T. As shown in the figure, in a time Δt hand 2 pulling the bottom puck has traveled farther, because the string has unrolled from the puck a bit.

The pucks have the same mass M. Assume that friction with the ice rink is negligible. It would be reasonable to suppose that it would make a difference where you attach the string to the puck.

? What will be the velocity (direction and magnitude) of the center of the first puck after a short time Δt? Of the center of the second puck? How do you justify your predictions?

Both pucks are subjected to exactly the same external force, so their centers of mass must move in exactly the same way. After a short time Δt, the momentum of the center of mass must be $F_T \Delta t$, in the direction of the string, for both pucks. Despite the fact that one puck is pulled from the side, the

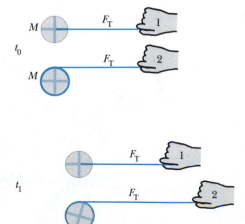

Figure 8.9 One puck is pulled by a string attached to its center. The other puck is pulled by a string wrapped around its edge, which unrolls as the puck is pulled.

change in momentum of its center of mass must be in the direction of the net external force, so it cannot acquire a momentum component at right angles to that force. Doesn't this seem a bit odd? For example, one might have expected the second puck to have some kind of sideways motion.

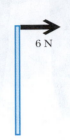

Figure 8.10 Pull one end of a meter stick that is lying on ice.

8.X.4 Is there any difference at all in the motion of the two pucks? Think about the motion of your hands at the other ends of the strings. Which hand does more work? Which puck has more energy? In what forms? Does the Momentum Principle for the motion of the center of mass predict all the details of the motion of both pucks?

8.X.5 A meter stick whose mass is 300 grams lies on ice (Figure 8.10). You pull at one end of the meter stick, at right angles to the stick, with a force of 6 newtons. The ensuing motion of the meter stick is quite complicated, but what are the initial magnitude and direction of the rate of change of the momentum of the stick, $d\mathbf{P}_{tot}/dt$, when you first apply the force? What is the magnitude of the initial acceleration of the center of the stick?

8.3 ENERGY IN MULTIPARTICLE SYSTEMS

We have seen that we can predict the motion of a multiparticle object by treating it as if all its mass were concentrated at the center of mass, acted upon by a force that is the (vector) sum of all the external forces acting on the object. In a number of important situations we find that the energy of a multiparticle system can also be analyzed very simply.

Gravitational energy of a multiparticle system

Our first example is the gravitational energy of a block of mass M and the Earth, near the Earth's surface. The gravitational energy associated with the i-th atom in the block is $m_i g y_i$ (Figure 8.11).

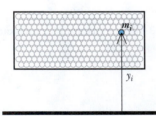

Figure 8.11 The i-th atom in the block is a distance y_i from the Earth's surface.

We calculate the total gravitational energy by adding up the gravitational energy associated with the interaction of each of the N atoms with the Earth:

$$U_g = m_1 g y_1 + m_2 g y_2 + m_3 g y_3 + ... = g(m_1 y_1 + m_2 y_2 + m_3 y_3 + ...)$$

But the last expression in parentheses appears in the calculation of the y component of the location of the center of mass:

$$M y_{cm} = m_1 y_1 + m_2 y_2 + m_3 y_3 + ...$$

Therefore for a block made up of many atoms, near the Earth's surface, we have this:

GRAVITATIONAL ENERGY U_g

$$U_g = M g y_{cm} \text{ near the Earth's surface}$$

In other words, the gravitational energy of a block of material and the Earth, near the Earth's surface, can be evaluated as though the block were a tiny particle of mass M located at the center of mass of the block.

This simple formula is not valid if the object is so large that the strength of the gravitational field is significantly different at different locations in the object, but in that case mg would not be a good approximation for the gravitational force, nor mgy a good approximation for the gravitational energy associated with each atom.

? In a calculation of gravitational energy in the voyage of a spacecraft to the Moon is it a valid approximation to treat the spacecraft as if all its mass were concentrated at one point?

This is a good approximation, because the spacecraft is very small compared to the distances to the Earth and Moon, so the gravitational force on each atom is nearly the same.

Kinetic energy of a multiparticle system

We can show that the total kinetic energy K_{tot} of a multiparticle system (that is, the sum of the kinetic energies of all the particles in the system) can usefully be split into two parts: the "translational" kinetic energy $K_{trans} = P_{tot}^2/(2M)$ of a fictitious particle of mass M located at the center of mass and moving with total momentum P_{tot} (which is equal to Mv_{cm}), plus the kinetic energy K_{rel} of the various atoms relative to the center of mass.

MULTIPARTICLE KINETIC ENERGY

$$K_{tot} = K_{trans} + K_{rel}, \text{ where } K_{trans} = \frac{P_{tot}^2}{2M} = \tfrac{1}{2}Mv_{cm}^2$$

For example, the kinetic energy of a rotating, vibrating diatomic molecule such as oxygen (O_2) is the sum of the kinetic energy that the molecule would have if it were not rotating or vibrating (this simple motion is called "translational" motion), plus additional kinetic energy of rotation around the center of mass and kinetic energy of vibration relative to the center of mass (Figure 8.12).

This split of the total kinetic energy into different parts sounds reasonable. A formal derivation of this result is given in Section 8.8 at the end of this chapter.

Rotation and translation

As an example, suppose you spin a bicycle wheel on its axis, and hold the axle stationary, as in Figure 8.13. The spinning wheel has some kinetic energy K_{rel}, which in this case we will call K_{rot}, because the atoms are moving in rotational motion relative to the center of mass. Almost all of the mass M of the wheel is in the rim, so if the rim is traveling at a (tangential) speed v_{rel}, the kinetic energy of the wheel is approximately $K_{rot} \approx \tfrac{1}{2}Mv_{rel}^2$. (This isn't exact, because there is some mass in the spokes, and the atoms in the spokes are moving at speeds smaller than v.)

With the wheel still spinning, now move the axle with some speed v_{cm}, without touching the wheel to the ground, as in Figure 8.14. When the center of mass moves, we say that there is *translational* motion. At the top of the wheel the speed of an atom is $v_{cm} + v_{rel}$, because it shares the overall translational motion of the wheel but is also moving forward relative to the axle. At the bottom of the wheel the speed of an atom is $v_{cm} - v_{rel}$, because it shares the overall translational motion of the wheel but is also moving backward relative to the axle. The motions of atoms at other locations around the rim are somewhat complicated. Nevertheless, when we add up all the kinetic energies of all the atoms, we find that $K_{tot} = \tfrac{1}{2}Mv_{cm}^2 + K_{rot}$. The wheel has kinetic energy corresponding to its overall motion (this is kinetic energy that it would have even if it weren't rotating) plus kinetic energy corresponding to its rotation (this is kinetic energy that it would have even if the center of mass weren't moving). It is appropriate to call the rotational kinetic energy a kind of "internal energy" of the system.

If you drop the spinning bicycle wheel, the Energy Principle can be expressed like this:

$$\left(\tfrac{1}{2}Mv_{cm}^2 + K_{rot} + Mgy_{cm}\right)_f = \left(\tfrac{1}{2}Mv_{cm}^2 + K_{rot} + Mgy_{cm}\right)_i$$

If you simply drop the wheel, nothing will make the wheel spin faster or slower around the axle, so K_{rot} doesn't change, and the energy equation for

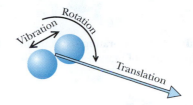

Figure 8.12 A diatomic molecule such as O_2 can vibrate and rotate while also translating.

Figure 8.13 A bicycle wheel spinning on its axle, which is at rest. Atoms in the rim have a speed v_{rel} relative to the center of mass.

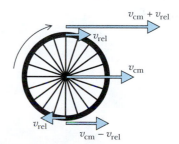

Figure 8.14 The velocity of an atom in the rim is the vector sum of the center-of-mass velocity and the velocity relative to the center of mass.

this very complicated motion, involving both translation and rotation, reduces to the Energy Principle for a simple point particle:

$$\left(\tfrac{1}{2}Mv_{cm}^2 + Mgy_{cm}\right)_f = \left(\tfrac{1}{2}Mv_{cm}^2 + Mgy_{cm}\right)_i$$

Vibration and translation

Another kind of energy that is internal to a system is vibrational energy, both elastic and kinetic. Consider an oxygen molecule (O_2) that has no translational motion (that is, the center of mass is stationary), but is vibrating (Figure 8.15), with elastic energy and kinetic energy continually interchanging, but with the sum of the two energies remaining constant, like the mass and spring you modeled in Chapter 4.

If in addition the oxygen molecule is translating (Figure 8.16), the total energy of the molecule is the sum of the translational kinetic energy, the vibrational kinetic energy (in terms of velocities of the two atoms relative to the center of mass), and the elastic energy of the "spring" holding them together, and the rest energies of the constituent atoms:

$$E_{tot} = \tfrac{1}{2}Mv_{cm}^2 + K_{vib} + \tfrac{1}{2}k_s s^2 + 2mc^2$$

Note that the elastic energy $\tfrac{1}{2}k_s s^2$ is clearly unaffected by the translational motion of the center of mass, because it depends only on the stretch of the distance between the two atoms.

A related example is the motion of a hot block of metal. In addition to its translational kinetic energy $\tfrac{1}{2}Mv_{cm}^2$ there is vibrational kinetic and elastic energy of the atoms around their equilibrium positions, which we call thermal energy (or internal energy E_{int}).

Rotation, vibration, and translation

The most general motion of an oxygen molecule involves rotation, vibration, and translation. It is easy to show that the internal (non-translational) kinetic energy separates cleanly into rotational and vibrational contributions. Consider one atom with momentum p relative to the center of mass, and resolve its momentum into "tangential" and "radial" components (Figure 8.17). By "radial" component we mean a component in the direction of the line from the center of mass to the particle. By "tangential" component we mean a component perpendicular to that line. If the particle is in circular motion around the center of mass, the radial component is zero, so tangential momentum is associated with rotation. Radial momentum is associated with vibration.

Find the kinetic energy is in terms of these two momentum components:

$$\frac{(p^2)}{2m} = \frac{(p^2{}_{tangential} + p^2{}_{radial})}{2m} = K_{rot} + K_{vib}$$

Therefore the kinetic energy relative to the center of mass splits into two terms, rotational energy and vibrational energy:

ROTATIONAL + VIBRATIONAL ENERGY

$$K_{rel} = K_{rot} + K_{vib}$$

The total energy of a translating, rotating, vibrating oxygen molecule can be written like this:

$$E_{tot} = K_{trans} + K_{rot} + K_{vib} + \tfrac{1}{2}k_s s^2 + 2mc^2, \text{ where } K_{trans} = \frac{P_{tot}^2}{2M} = \tfrac{1}{2}Mv_{cm}^2$$

(If there are changes in height near the Earth's surface, there is also a term of the form Mgy_{cm} for the energy of the system of Earth plus molecule.)

Figure 8.15 A vibrating oxygen molecule whose center of mass is at rest.

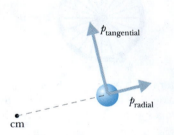

Figure 8.16 A translating, vibrating oxygen molecule.

Figure 8.17 Radial and tangential components of momentum of a single atom, relative to the center of mass of an object.

If the object can fly apart into separate pieces, we might call the kinetic energy relative to the center of mass $K_{explosion}$! For example, a fireworks rocket has translational kinetic that doesn't change from just before to just after its explosion, and the pieces move at high speeds relative to the center of mass of the rocket. These large kinetic energies relative to the center of mass come from the chemical energy used up in the explosion.

8.4 THE "POINT PARTICLE SYSTEM"

In analyzing the motion and energy of deformable multiparticle systems (systems whose shape can change), we encounter situations in which it would be extremely convenient to be able to calculate only the change in the translational kinetic energy K_{trans} of the system. Perhaps this might be all we are interested in, or perhaps we might wish to subtract this energy change from the change of the total energy of the system to find out how much energy has gone into vibration or rotation. Fortunately, it is quite simple to calculate ΔK_{trans} by itself.

We have established that the center of mass of a multiparticle system moves exactly like a simple "point" particle whose mass is the mass of the entire system, under the influence of the net external force applied to the entire system. Pretend you could crush the real system down to a very tiny ball and apply to this "point particle" forces with the same magnitudes and directions to those that acted on the real system, but applied directly to the point particle rather than at their original locations. What would the motion of this point particle system look like? It would look exactly like the motion of the center of mass of the real system (Figure 8.18).

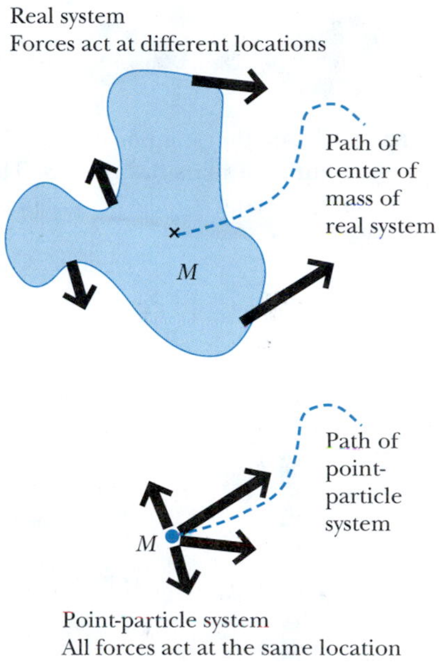

These two different systems have the same total mass M, and both are acted on by the same net force, so the two paths are exactly the same. The difference is that the real system may rotate and stretch and vibrate due to the effects of the forces acting at different locations on the extended object. In contrast, the point particle system has no rotational motion, no vibrational motion, no internal energy of any kind. All of the forces act at the location of the point particle, and these forces don't stretch or rotate it. The only energy the point particle system can have is translational kinetic energy, and this is exactly the same as the translational kinetic energy of the real multiparticle system:

$$K_{trans} = \frac{P_{tot}^2}{2M} = \frac{1}{2}M_{tot}v_{cm}^2 \quad \text{for the real system or the point particle system}$$

An analysis of the point particle system gives the translational kinetic energy of the real system. This is why the point particle system is a useful concept. The force and energy equations for the fictitious point particle are these:

POINT PARTICLE SYSTEM (MASS *M* AT CM LOCATION)

$$\frac{d\vec{P}_{tot}}{dt} = \vec{F}_{net,surr}$$

$$\Delta K_{trans} = \Delta(\tfrac{1}{2}Mv_{cm}^2) = \int_i^f \vec{F}_{net,surr} \bullet d\vec{r}_{cm}$$

The integral is the work done on the point particle system by the net force, which acts at the location of the point itself. In Chapter 5 we showed that starting from the Momentum Principle for a single point particle we could derive the result that the change in the kinetic energy of the particle is equal to the work done on the particle by the net force. Here, we consider the net force to be applied to the fictitious point particle located at the center-of-

Real system
Forces act at different locations

Path of
center of
mass of
real system

M

Path of
point-
particle
system

M

Point-particle system
All forces act at the same location

Figure 8.18 The path of the point particle system is exactly the same as the path of the center of mass of the real system.

mass point, doing work to increase the kinetic energy of the point particle. A more formal derivation of the energy equation for the point particle system may be found at the end of this chapter.

These equations tell us about the motion of the center of mass of a multiparticle system, but they tell us nothing about the vibration or rotation or thermal changes of the system. For a full treatment we need to combine analyses of the point particle system with analyses of the real system.

Application: Jumping up

As an example of how to exploit the point particle system, we'll analyze jumping (Figure 8.19). You crouch down and then jump straight upward. We want to calculate how fast your center of mass v_{cm} is moving at the instant that your feet leave the floor. Let h be the distance through which your center of mass rises in this process. Note that there is a change of shape of the system.

Let F_N be the electric contact force that the atoms in the floor exert on the bottom of your foot, and let M be your total mass. We don't know exactly how the floor force F_N varies with time, but to get an idea of what happens let's make the crude approximation that it is constant as long as you are in contact with the floor (of course it falls suddenly to zero as your foot leaves the floor). Consider the different point particle system shown in Figure 8.20, which does not change shape.

Figure 8.19 As the jumper leaves the floor, the center of mass has risen h. The system changes shape.

Jumper: Point particle system

? What is the net force on you? What is the energy equation for the point particle system?

The net force in the upward direction is $(F_N - Mg)$ (Figure 8.20), and the energy equation for the point particle system is

$$\Delta K_{\text{trans}} = \tfrac{1}{2} M v_{cm}^2 - 0 = \int_i^f \vec{F}_{\text{net,surt}} \bullet d\vec{r}_{cm} = (F_N - Mg)h$$

As we will see, this is not the same as the energy equation for the real system consisting of you, the jumper! The reason F_N shows up in the point particle energy equation is that it contributes to the net force and impulse, which affects the momentum (which is related to the motion of the center of mass). The work done on the point particle system involves the net force multiplied by the displacement of the center-of-mass point.

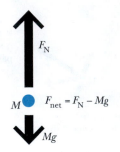

Figure 8.20 Force diagram for your body, considered as a point particle. This system does not change shape.

Jumper: Real system

? What are some of the energy changes that occur during the jump that are not represented in this point particle energy equation?

There is change of kinetic energy of your moving arms and legs relative to the center of mass, decrease of chemical energy inside you, and increase of thermal energy inside you (your temperature rises somewhat).

? In the real system, how much work is done by the floor force F_N?

The floor force F_N is applied to the bottom of your foot, and the contact point does not move during the entire process leading to lift-off, so there is no work. The definition of the work done by a force involves (the parallel component of) the displacement of the point of contact, at the place where the force is applied to the object. No displacement, no work done on you.

The energy equation for the real system (you, the jumper) looks something like the following, ignoring Q, transfer of energy due to a temperature difference between you and the surrounding air:

$$\Delta E_{sys} = W = \sum_i \int_i^f \vec{F}_i \cdot d\vec{r}_i \quad (W \text{ is sum of work done by all external forces})$$

$$\Delta K_{trans} + \Delta K_{other} + \Delta E_{thermal} + \Delta E_{chemical} = -Mgh$$

The floor force F_N does not appear in the energy equation for the real system, because this force does no work (no displacement of the point of contact; $d\vec{r}_{floor} = 0$). (The floor force does however appear in the energy equation for the point particle system.) As before, the gravitational force does negative work on you.

ΔK_{other} includes the rotation of your upper and lower legs and the swinging of your arms. $\Delta E_{thermal}$ corresponds to the higher temperature that your body runs due to the exertion (which ultimately will lead to transfer of energy Q out of your body into the surroundings, due to the higher temperature compared to the cooler air). $\Delta E_{chemical}$ represents the burning of chemical energy stored in your body, in order to support this activity. None of these terms appear in the energy equation for the point particle system, because the point particle equation deals just with the motion of the mathematical center of mass point, which doesn't rotate or stretch or get hot. A crucial difference is that you change shape (legs unbend, etc.), but the point particle system doesn't.

The point particle equation gives you $\Delta K_{trans} = (F_N - Mg)h$. Substitute this into the energy equation for the real system and solve for the other terms, $\Delta K_{other} + \Delta E_{thermal} + \Delta E_{chemical}$.

Comparison of the point particle and real energy equations

The important thing to notice is that the energy equation for the real system and for the point particle system are quite different. Most energy terms do not appear at all in the point particle equation, and the work that appears in the energy equation for the real system is not the same as the work that appears in the energy equation for the point particle system.

? Which of the energy changes in the energy equation shown above for the real system are positive? Which are negative?

All of these energy changes of your body are positive except for the change in chemical energy, which decreases to pay for all the other increases.

Application: Stretching a spring

Let's check to make certain this kind of analysis gives us a sensible answer in a simple situation. You pull on one end of a spring (Figure 8.21) with a force F_L to the left, making a short displacement Δr_L to the left, and you pull on the other end with an equal but opposite force F_R and equal but opposite short displacement Δr_L. The system (the spring) changes shape. Our analysis should yield the expected result: the spring should not gain translational kinetic energy!

Spring: Real system

Consider the spring itself as the system.

? Does your right hand do any work on the spring? Of what sign? Does your left hand do any work on the spring? Of what sign? Is there any change in the energy of the spring?

Your left hand exerts a force through a short displacement Δr_L and does some (positive) work $W_L = F_L \Delta r_L \cos(0°) = +F_L \Delta r_L$.

Your right hand also exerts a force through a short displacement Δr_R and does some (positive) work $W_R = F_R \Delta r_R \cos(0°) = +F_R \Delta r_R$.

The total work increases the energy of the spring, in the form of spring (elastic) energy corresponding to increased stretch of the spring. Note care-

Figure 8.21 You pull on both ends of a spring with equal and opposite forces. The system (the spring) changes shape.

fully that both hands do positive work. The energy equation for the real system undergoing this process (the spring) is this:

$$\Delta[\tfrac{1}{2}k_s s^2] = W_L + W_R$$

Spring: Point particle system

Next consider the "point particle system" consisting of a fictitious particle located at the center of mass of the spring (Figure 8.22). This point particle system does not change shape.

? What is the net force $\vec{F}_{\text{net,surr}}$? How far does the center of mass move? How much work $\sum \vec{F}_{\text{net,surr}} \cdot \Delta \vec{r}_{\text{cm}}$ is done? What is the change in $K_{\text{trans}} = \tfrac{1}{2}Mv_{\text{cm}}^2$?

The net force is zero ($F_R - F_L = 0$). Reassuringly, the center of mass does not move. The work done on the fictitious point particle is zero, so there is no change in the speed of the center of mass, so no change in the translational kinetic energy $K_{\text{trans}} = \tfrac{1}{2}Mv_{\text{cm}}^2$. The energy equation reduces to 0 = 0, which is certainly correct but provides no detailed information.

Comparison of the point particle and real energy equations

The key point here is that in the real system, work is calculated by adding up the amount of work done by *each* force: the work done by your right hand *plus* the work done by your left hand. The work done by each force is the force times the actual displacement of the point of application of that force, $F_L \Delta r_L$ and $F_R \Delta r_R$, both of which are positive.

But in the point particle system, work is calculated by first adding up all the forces to get the *net* force before calculating the work, and the net force is multiplied by the displacement of the center of mass. In the present example the net force is zero, and the displacement of the center of mass is zero, so the work done on the point particle system is zero.

Note also that the actual work done by an individual force involves the displacement of the point where the force is applied. In contrast, work done on the point particle system involves the displacement of the center of mass point, a point which may be very far away from any of the actual locations where the forces are applied, with a motion that is quite different from the motion of the point of application of an individual force.

Another way of looking at the situation is that the real multiparticle system (the spring) changes shape, and there is potential energy associated with that change of shape. The point particle system does not change shape, and there is no change in potential energy.

The energy equation for the real system and the energy equation for the point particle system are both valid and give correct results for any process, but they provide different kinds of information:

Real system: Info on change of total energy of system
Point Particle system: Info on change of translational kinetic energy

The differences are what make it useful to analyze certain kinds of phenomena both ways. Note that the two energy equations are different for this spring example because the system is deformable and changes shape, which leads to very different calculations of the work in the two systems.

8.X.6 If a system such as a large block does not change shape (or rotate faster and faster due to tangential forces, and there are no changes of identity or temperature), the energy equations for the real system and for the point particle system are identical. Explain why.

8.X.7 A runner whose mass is 50 kg accelerates from a stop to a speed of 10 m/s in 3 seconds. (A good sprinter can run 100 meters in about 10

Figure 8.22 Force diagram for the "point particle system," which does not change shape.

seconds, with an average speed of 10 m/s.) What is the average horizontal component of the force that the ground exerts on the runner's shoes?

8.X.8 How much displacement is there of the force that acts on the sole of the runner's shoes, assuming that there is no slipping? Therefore, how much work is done on the real system (the runner) by the force you calculated in the previous exercise? How much work is done on the point particle system by this force?

8.X.9 The kinetic energy of the runner in the previous exercises increases—what kind of energy decreases? By how much?

8.X.10 A very similar situation is the acceleration of a car on dry pavement, if there is no slipping. The axle moves at speed v, and the outside of the tire moves at speed v relative to the axle. The instantaneous velocity of the bottom of the tire is zero. How much work is done by the force exerted on the tire by the road? What is the source of the energy that increases the car's translational kinetic energy?

8.5 ANALYZING POINT PARTICLE AND REAL SYSTEMS

Here is a procedure to use in analyzing the point particle and real systems.

- Imagine crushing the real system and its forces down to a point at the center of mass of the real system, including moving the tails of the force arrows to that point. Draw diagrams of the initial and final states, representing the point particle system as a dot, and showing all the forces that act on the real system.

- The tails of all the forces must be on the point particle itself.

- Write the Energy Principle for the point particle, paying attention to the fact that each force acts through the distance that the point particle moves, *not* the distance the tail of the force moves in the real system. The Energy Principle will include the term ΔK_{trans}.

- Draw diagrams of the initial and final states of the real system.

- Draw arrows representing the forces acting on the real system. It is very important to place the tail of each arrow at the location where that force acts on an object.

- Look carefully at how far the tails of the force arrows move, from the initial state to the final state. Use this information to calculate the work done by each force. Add up these work terms to find the work done on the real system.

- Write the Energy Principle for the real system, which will include the term ΔK_{trans}.

- Replace ΔK_{trans} in the equation for the real system with the ΔK_{trans} you found from the equation for the point particle.

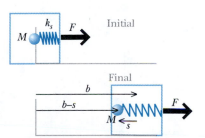

Figure 8.23 A constant force acts on a thin box containing a spring and a ball of clay. The box moves a distance b, but the ball (center of mass of the system) moves only a distance b-s. Tails of arrows representing forces are drawn at the point at which they act.

Example: A box containing a spring

A thin box in outer space contains a large ball of clay of mass M, connected to an initially relaxed spring of stiffness k_s (Figure 8.23). The mass of the box and spring are negligible compared to M. The apparatus is initially at rest. Then a force of constant magnitude F is applied to the box. When the box has moved a distance b, the clay makes contact with the left side of the box and sticks there, with the spring stretched an amount s. See the diagram for distances. (a) Immediately after the clay sticks to the box, how fast is the box moving? (b) What is the increase in thermal energy of the clay?

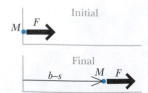

Figure 8.24 The point particle system for the apparatus. Tails of arrows representing forces are attached to the point particle, which moves the same distance as the center of mass of the real system.

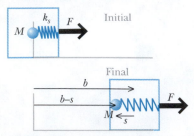

Figure 8.25 A constant force acts on a thin box containing a spring and a ball of clay. The box moves a distance b, but the ball (center of mass of the system) moves only a distance $b-s$. Tails of arrows representing forces are drawn at the point at which they act.

Assume that the process takes place so quickly that there isn't time for any significant energy transfer Q due to a temperature difference between the (real) system and surroundings

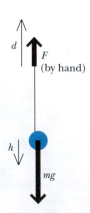

Figure 8.26 You pull up with a force of magnitude F through a distance d on the string of a yo-yo while the yo-yo moves downward a distance h.

(a) Point particle system (Figure 8.24)
It is important to draw this diagram of the "crushed" system. Place the tails of the forces on the point particle in the diagram.

System: point particle of mass M
Surroundings: external object exerting force F

Initial state: point particle at rest
Final state: point particle moving with (unknown) speed v

Energy Principle (point particle only has K_{trans}):
$$\Delta K_{trans} = F\Delta x_{cm}$$

Read Δx_{cm} off the diagram (Figure 8.24); it is $(b-s)$.

$$(\tfrac{1}{2}Mv^2 - 0) = F(b-s)$$

$$v = \sqrt{\frac{2F(b-s)}{M}}$$

(b) Real system (Figure 8.25, repeated from previous page)

System: mass, box, and spring
Surroundings: external object exerting force F

Initial state: system at rest
Final state: box moving with (known) speed v (same as clay)

Energy Principle: $\Delta E_{sys} = W_{surr}$
$$\Delta(K_{trans} + U_{spring} + E_{thermal}) = Fb$$

Substitute ΔK_{trans} obtained from point particle analysis:

$$F(b-s) + \tfrac{1}{2}k_s s^2 + \Delta E_{thermal} = Fb$$

$$\Delta E_{thermal} = Fs - \tfrac{1}{2}k_s s^2$$

In the energy equation for the point particle system there is nothing about the spring or thermal energy, because the point particle system doesn't stretch or get hot. In Figure 8.24 the tail of the force vector moves a distance $(b-s)$, moving with the point particle, so the work done on the point particle system is $F(b-s)$.

In the real system, the tail of the force vector moves a longer distance b, so the work done on the real system is Fb. More work is done, and the extra work goes into spring potential energy and thermal energy.

Good labeled diagrams of the two choices of system are extremely important, in order to determine the distances through which forces act in the two cases. Review the procedure on the previous page and see how these steps were implemented in the Example given above.

Example: A yo-yo

You're playing with a yo-yo of mass m on a low-mass string (Figure 8.26). You pull up on the string with a force of magnitude F, and your hand moves up a distance d. During this time the mass falls a distance h (and some of the string reels off the yo-yo's axle). (a) What is the change in the translational kinetic energy of the yo-yo? (b) What is the change in the rotational kinetic energy of the yo-yo, which spins faster?

(a) Point particle system (Figure 8.27)
It is important to draw this diagram of the "crushed" system. Place the tails of the forces on the point particle in the diagram.

System: point particle of mass m
Surroundings: Earth and hand

Initial state: point particle with initial translational kinetic energy
Final state: point particle with final translational kinetic energy

Energy Principle (point particle only has K_{trans}):
$$\Delta K_{trans} = (mg - F)\left|\Delta y_{cm}\right|$$
Read Δy_{cm} off the diagram (Figure 8.27); it is $-h$.
$$\Delta K_{trans} = (mg - F)h$$

(b) Real system (Figure 8.26)

System: mass and string
Surroundings: Earth and hand

Initial state: initial rotational and translational kinetic energy
Final state: final rotational and translational kinetic energy

Energy Principle: $\Delta E_{sys} = W_{surr}$
$$\Delta K_{trans} + \Delta K_{rot} = F_1 \Delta y_1 + F_2 \Delta y_2 = Fd + (-mg)(-h)$$

Substitute ΔK_{trans} obtained from point particle analysis:

$$(mg - F)h + \Delta K_{rot} = Fd + mgh$$

$$\Delta K_{rot} = F(d + h)$$

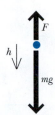

Figure 8.27 The point particle system for the yo-yo. Tails of arrows representing forces are attached to the point particle, which moves the same distance as the center of mass of the real system.

The key issue is that the distance through which the force F moves is different in the real system and the point particle system. In the real system, your hand applies an upward force of magnitude F through an upward displacement d, doing an amount of work $+Fd$. But in the other "crushed" system, where all the mass is located at the center of mass of the real system and the tails of all the force vectors are moved to that point, the tail of the upward-pointing force vector for F is applied to the point particle as the point particle drops a distance h. Therefore the work done by the force F on the point particle is $-Fh$. (The work done by the Earth's force is $+mgh$ for both systems, because this force is applied to the center of mass of the real system, which is also the center of mass of the crushed system.) The diagrams labeled with distances are critical to calculating the work correctly for the two different systems.

It is important to understand that the point particle or "crushed" system is a different system, with forces acting on it that are equal in magnitude and direction to the real forces acting on the real system, but applied directly to the point particle, so the displacements through which these forces act are different from the displacements of the real forces acting on the real system.

In this yo-yo example the difference between the two systems is quite striking, because the work done by the force F on the real system is positive, but the work done by the force F on the point particle system is negative.

Note that the change of the rotational kinetic energy, $\Delta K_{rot} = F(d + h)$, is the same as the work that would be done to increase the purely rotational kinetic energy of a yo-yo whose axle was fixed in space. The distance $(d + h)$ is the length of the extra string that reels off the axle.

8.6 *MODELING FRICTION IN DETAIL

We can use what we've learned about complex systems to model the nature of sliding friction, which is a complicated phenomenon. In Chapter 6 we described a seeming paradox involving friction, that when you drag a block at constant speed across a table it seems that no work is done on the block, yet its temperature increases, indicating an increase in its thermal energy. With the new tools developed in this chapter we are in a position to analyze this phenomenon in more detail and resolve the paradox.

Sliding block: Point particle system

A key issue that we will address is the question of how much work is done on the block by the friction force f exerted by the table. One of the new tools is the energy equation for the point particle system (Figure 8.28 and Figure 8.29). You apply a force F to the right and the table exerts a force f to the left. (Pretend for the moment that you don't know the magnitude of the friction force f.) The change in the kinetic energy of the fictitious center-of-mass point particle is given by the product of the net force $(F-f)$ times the displacement d of the center-of-mass point:

$$\Delta K_{trans} = (F-f)d$$

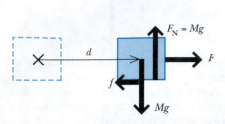

Figure 8.28 The center of mass of the block moves a distance d under the influence of a net force F–f.

Since the speed of the block is constant, $\Delta K_{trans} = 0$, so $F-f = 0$, which shows that the friction force is equal and opposite to the force you apply: $f = F$. The thermal energy does not appear in this energy equation for the point particle system, because this equation deals merely with the motion of the (mathematical) center of mass, not with the many kinds of energy in the real system.

We could also have obtained the result $f = F$ directly from the Momentum Principle: $dP_{tot}/dt = F-f = 0$. The Momentum Principle is closely related to the energy equation for the point particle system, because both involve the net force acting on a system.

Sliding block: Real system

The energy equation for the real system of the block can be written as follows, where Fd is the work done by you, and W_{fric} is the work done by the table:

$$\Delta K_{trans} + \Delta E_{th} = Fd + W_{fric}$$

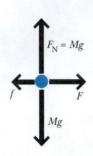

Figure 8.29 The block considered as a point particle.

where ΔE_{th} is the rise in the thermal energy of the block. We're assuming that the process takes a short enough time that there is negligible transfer of energy Q due to a temperature difference between the system of the block and the surroundings (the table), because energy transfer due to a temperature difference between table and block is a relatively slow process. (To remove such energy transfer as a possible complication, we could consider a block sliding not on a table but on an identical block. Then the symmetry of the situation is such that there cannot be any net transfer of energy due to a temperature difference into or out of the upper block. This tactic—reducing the complexity of a model—is a useful approach to complex problems.)

Since the speed of the center of mass v_{cm} does not change, $\Delta K_{trans} = 0$, and the energy equation reduces to

$$\Delta E_{th} = Fd + W_{fric}$$

If we conclude that the friction force does an amount of work $-Fd$, then we would have to conclude that the thermal energy of the block doesn't change, which is absurd. The block definitely gets hotter, indicating an increase in its thermal energy. We need to find a way around the entirely plausible but apparently incorrect conclusion that the friction force does an amount of work $-Fd$.

? Considering the sign of the change of thermal energy, must the magnitude of the work done by the friction force be greater or less than Fd?

Since the thermal energy change ΔE_{th} is surely positive (the block gets hotter, not colder), the magnitude of the work done by the friction force must be less than Fd. Since the friction force is definitely equal to F, the friction force must act through some effective distance d_{eff} *that must be less than d!*

Evidently the energy equation for the real system has the following form, with d_{eff} less than d:

$$\Delta E_{th} = Fd - Fd_{eff}$$

8.7 *A PHYSICAL MODEL FOR DRY FRICTION

How can the effective distance through which the friction force does work be less than the distance through which the block moves? We can understand this by looking at a microscopic picture of what happens on the surfaces in contact.

When a metal block slides on a metal surface, the block is supported by as few as three protruding "teeth," called "asperities" in the literature on friction (Figure 8.30). The very high load per unit area on these teeth makes the material partially melt and flow, and high local temperatures produced during sliding lead to adhesion (welding) in the contact regions. The frictional force divided by the tiny contact area corresponds to the large "shear" (sideways) stress required to break these welds. This shearing of contact welds is the dominant friction mechanism for a dry metal sliding on the same metal.

Because the tooth tips can become stronger than the bulk metal due to a process called "work-hardening" (which introduces dislocations in the otherwise regular geometrical arrangements of atoms), shearing often occurs in the weaker regions of the teeth, away from the tip. This is a major effect when the two objects are made of the same material, and chunks of metal can break off and embed in the other surface. Nevertheless, this wear will be ignored in the further discussion. It is in any case a symmetrical effect for identical blocks. If the metal surfaces have oxide coatings, this can reduce the shear stress required to break the temporary weld (which reduces the friction force) and can prevent the breaking off of chunks of metal, if the oxide contact area is the weakest section.

This is the model of dry friction developed in a classic treatise on friction, *The Friction and Lubrication of Solids*, Part 1 and Part II, F. P. Bowden and D. Tabor (Oxford University Press, 1950 and 1964). The physics and chemistry of friction continues to be an active field of research, because the effects of friction can be quite complex, and there is high practical interest in controlling friction. A recent textbook is *Friction, Wear, Lubrication*, K. C Ludema, (CRC Press, 1996).

Having briefly reviewed a basic model of dry friction, we proceed to use this model to calculate the work done by frictional forces exerted at the contact points. The key issue is that the surface is deformable, which leads to differences in the energy equation for the real system compared with the energy equation for the point particle system.

Actual work done by friction forces

We'll consider a microscopic picture of the contact region between two identical blocks. Figure 8.31 shows in a schematic way two teeth that have temporarily adhered to each other. The vertical scale has been greatly exaggerated for clarity—machined surfaces have much gentler slopes

As the top block is dragged to the right, the teeth continue to stick together for a while. Both teeth must deform as a result; the top tooth is stretched backward, and the bottom tooth is stretched forward. In Figure 8.32 we see that when the top block has moved a distance d to the right, the point where

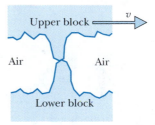

Figure 8.30 A block slides on a small number of protruding teeth (vertical scale exaggerated).

Figure 8.31 Two projecting teeth of identical blocks, which have temporarily adhered to each other. (Exaggerated vertical scale).

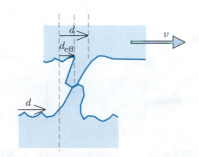

Figure 8.32 The top block has moved a distance d to the right, but the point of contact has moved only d_{eff}.

the teeth touch each other has moved a distance of only d_{eff}, because the bottom block hasn't moved. For two identical blocks made of the same material, on average the two teeth will bend the same amounts, and d_{eff} will be equal to $d/2$.

> *This is the key issue:* The point of contact where the friction force is applied moves a shorter distance (d_{eff}) than the block itself moves (d). This means that the work done by the friction force is less than one might expect.

Eventually the weld breaks, and the top tooth snaps forward, so all the atoms in the tooth now catch up with the rest of the top block, but during this part of their journey, there is no external force acting on them (Figure 8.33).

The time average of the contact forces is indeed F (as determined by the fact that the net force must be zero, $F - F$), but the effective displacement d_{eff} at the point of contact of the frictional force is less than the displacement of the center of mass of the upper block. For two identical blocks we expect to find that $d_{\text{eff}} = d/2$.

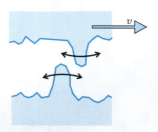

Figure 8.33 After the temporary weld breaks, the top tooth catches up to the rest of the block. Now no force acts on the teeth. Both teeth vibrate.

Internal energy of the blocks

Once the weld has broken, the teeth can vibrate, as we saw in Figure 8.33. The vibration of the tooth belonging to the top block and thermal conduction upward from the hot tip into the main body of the block contribute to the increase in the thermal energy of the upper block. Similarly, vibration and thermal conduction in the tooth belonging to the bottom block end up as increase in the thermal energy of the bottom block.

A more physical picture

Having two long teeth in contact is rather unphysical. A better picture is the one shown in Figure 8.34 (again with a greatly exaggerated vertical scale), where we show one of the top block's longest teeth in contact with the average surface level of the bottom block, and one of the bottom block's longest teeth similarly in contact with the average surface level of the top block. The two teeth shown here are to be taken as representative of time and space averages of the sliding friction. The frictional force F exerted on the top block is divided on the average into two forces each of magnitude $F/2$ at the ends of the two sets of long teeth.

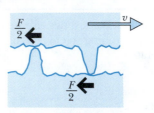

Figure 8.34 A more representative picture of contact between two surfaces, with projecting teeth in contact with flat areas.

In Figure 8.35 we see that when the top block moves a distance d to the right the upper contact also moves a distance d, whereas the lower contact does not move at all. The frictional work is therefore $(-F/2)(0) + (-F/2)(d)$, which is $-Fd/2 = -Fd_{\text{eff}}$, so d_{eff} is equal to $d/2$.

Summary of dry friction

The fundamental reason why the friction force can act through a distance less than d is that the block is deformable. All atoms in the top block eventually move the same distance d, but because of the stick/slip contact between the blocks the friction force F acts only for a portion of the displacement ($d/2$ for identical blocks). The point of contact for the friction force does not move in the same way as the center of mass. Note again that the energy equation for the point particle system is not the same as the energy equation for the real system if the system changes shape.

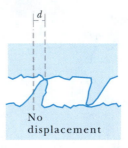

Figure 8.35 During this time interval the upper contact moves a distance d, while the lower contact does not move at all.

A model-independent calculation of the effective distance

In the special case of a block sliding on an identical block we can calculate d_{eff}, independent of the particular model of the surfaces. Consider a system consisting of both blocks together (Figure 8.36). The bottom block is held stationary by applying a force to the left (to prevent it being dragged to the right). This constraining force acts through no distance (the bottom block stands still), so this force does no work. The only work done on the two-

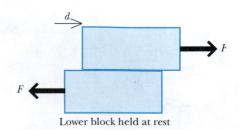

Figure 8.36 The lower block is held at rest, while the upper block slides a distance d. Forces in the vertical direction are not shown.

block system is done by you, of magnitude *Fd*. So the energy equation for the real two-block system is this:

$$\Delta E_{th,1} + \Delta E_{th,2} = Fd$$

Since the two blocks are identical, half of this increased thermal energy shows up in the upper block ($Fd/2$), and half in the lower block ($Fd/2$). We can plug this into our earlier energy equation for the upper block:

$$\Delta E_{th,1} = \frac{Fd}{2} = Fd - Fd_{eff}$$

? Calculate d_{eff}.

We find that the effective distance through which the friction force acts is half the distance *d* that the block moves: $d_{eff} = d/2$. This is in agreement with the model of dry friction with bending teeth that we examined earlier, but our result is a general one that applies to any kind of model of friction surfaces for two symmetrical blocks.

Lubricated friction

It is interesting to see that the model-independent result $d_{eff} = d/2$ is also consistent with the case of lubricated friction. We separate the two blocks with a film of viscous lubricating oil, so that the two blocks do not make direct contact, as shown in Figure 8.37.

It is a property of simple fluid flow that fluid layers immediately adjacent to the blocks are constrained to share the motion of the blocks. Also, for a common type of flow called "laminar" flow, the displacement profile in the oil is linear. In particular, at the midplane the fluid moves half as far as the top block moves. As a result, as the top block is pulled a distance *d* to the right, the top layer of the oil is dragged along and moves a distance *d* to the right, the bottom layer of oil doesn't move, and the layer in the midplane (halfway between the blocks) moves a distance *d/2*.

If we take as the symmetrical systems of interest the top block with the upper half of the oil, and the bottom block with the lower half of the oil, we see that the shear force between the two systems (at the midplane in the oil) acts through a distance which is again half the displacement of the top block: $d_{eff} = d/2$. (Of course the magnitude of the friction force is much reduced by the lubrication, and the applied force *F* must be much smaller if the velocity is to be constant.)

This discussion of friction is based on an article "Work and heat transfer in the presence of sliding friction," by B. Sherwood and W. Bernard, *American Journal of Physics* volume **52**, number 11, Nov. 1984, pages 1001-1007, which in turn draws on an earlier article "Real work and pseudowork," by B. Sherwood, *American Journal of Physics* volume **51**, number 7, July 1983, pages 597-602.

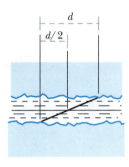

Figure 8.37 Two objects separated by a film of oil, which is modeled as a stack of fluid layers that can move relative to each other.

8.8 *DERIVATION: KINETIC ENERGY OF A MULTIPARTICLE SYSTEM

In this section we derive the important result that $K_{tot} = K_{trans} + K_{rel}$, where $K_{trans} = \frac{1}{2}Mv_{cm}^2$ and K_{rel} is the kinetic energy relative to the center of mass. This result sounds entirely plausible, but the formal proof is rather difficult.

As in the case of calculating the gravitational energy of a multiparticle object, the derivation of the kinetic energy of a multiparticle system hinges on the definition of the center of mass point of a collection of atoms. The (vector) location of the *i*-th atom of the object can be expressed as the sum of two vectors, one from the origin to the center of mass ($\vec{r}_{cm}$), plus another from the center of mass to the *i*-th atom ($\vec{r}_i$).

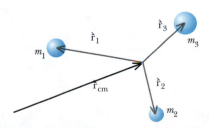

Figure 8.38 Center of mass of a multiparticle system.

The kinetic energy $K_i = \frac{1}{2}mv_i^2$ of the i-th atom is this, where we write $\vec{v}_i = d\vec{r}_i/dt$, the velocity of the i-th particle relative to the center of mass:

$$K_i = \frac{1}{2}m_i \left| \frac{d}{dt}(\vec{r}_{cm} + \vec{r}_i) \right|^2$$

$$= \frac{1}{2}m_i \left| \vec{v}_{cm} + \vec{v}_i \right|^2$$

This can be expanded, using the fact that the magnitude squared of a vector can be written as a vector dot product, $|\vec{C}|^2 = \vec{C} \bullet \vec{C} = C^2\cos(0) = C^2$:

$$K_i = \frac{1}{2}m_i[\vec{v}_{cm} + \vec{v}_i] \bullet [\vec{v}_{cm} + \vec{v}_i]$$

$$= \frac{1}{2}m_i v_{cm}^2 + 2\frac{1}{2}m_i\vec{v}_{cm} \bullet \vec{v}_i + \frac{1}{2}m_i v_i^2$$

Now that we have the kinetic energy of the i-th atom in terms of its position $\vec{r}_i$ relative to the center of mass, we need to add up the total kinetic energy of all the atoms. The total kinetic energy K_{tot} can be written in the following compact way, where Σ (Greek capital sigma) means "sum," and the sum goes from $i = 1$ through $i = N$:

$$K_{tot} = \sum_{i=1}^{N} \frac{1}{2}m_i v_{cm}^2 + \sum_{i=1}^{N} 2(\frac{1}{2}m_i)\vec{v}_{cm} \bullet \vec{v}_i + \sum_{i=1}^{N} \frac{1}{2}m_i v_i^2$$

$$\text{First term} \qquad \text{Second term} \qquad \text{Third term}$$

First term

The first term in this summation turns out to be the kinetic energy of a single particle of mass M, moving at the speed of the center of mass:

$$\sum_{i=1}^{N} \frac{1}{2}m_i v_{cm}^2 = \frac{1}{2}v_{cm}^2 \left(\sum_{i=1}^{N} m_i \right) = \frac{1}{2}v_{cm}^2(m_1 + m_2 + m_3 + ...)$$

$$= \frac{1}{2}Mv_{cm}^2 = \frac{P_{tot}^2}{2M}$$

Second term

The second term can be shown to be zero:

$$\sum_{i=1}^{N} 2(\frac{1}{2}m_i)\vec{v}_{cm} \bullet \vec{v}_i = \vec{v}_{cm} \bullet \sum_{i=1}^{N} m_i\vec{v}_i = \vec{v}_{cm} \bullet \frac{d}{dt}\left(\sum_{i=1}^{N} m_i\vec{r}_i \right) = 0$$

This result follows from the way that we calculate the location of the center of mass:

$$\vec{r}_{cm} = \frac{m_1\vec{r}_1 + m_2\vec{r}_2 + m_3\vec{r}_3 + ...}{m_1 + m_2 + m_3 + ...} = \frac{\sum_{i=1}^{N} m_i\vec{r}_i}{M}$$

We are measuring the location $\vec{r}_i$ of the i-th atom relative to the center of mass, and since the distance from the center of mass to the center of mass is of course zero, we have

$$\vec{r}_{cm} = 0 = \frac{\sum_{i=1}^{N} m_i\vec{r}_i}{M} \quad \text{and therefore} \quad \sum_{i=1}^{N} m_i\vec{r}_i = 0.$$

Third term

The third term by definition is the kinetic energy of the atoms relative to the (possibly moving) center of mass:

$$\sum_{i=1}^{N} \tfrac{1}{2} m_i v_i^2 = \tfrac{1}{2} m_1 v_1^2 + \tfrac{1}{2} m_2 v_2^2 + \tfrac{1}{2} m_3 v_3^2 + \ldots = K_{rel}$$

Putting the three pieces together, we find that the total kinetic energy splits into two parts: a term associated with the overall motion of the center of mass, plus the kinetic energy relative to the center of mass:

$$K_{tot} = K_{trans} + K_{rel}, \text{ where } K_{trans} = \frac{P_{tot}^2}{2M} = \tfrac{1}{2} M v_{cm}^2$$

8.9 *DERIVATION: THE POINT PARTICLE ENERGY EQUATION

In this chapter we showed that the energy equation for the point particle version of the system follows from the fact that the motion of the center of mass is just like that of a point particle with the total mass of the real system and subjected to the *net* force acting on the real system. Here we give a more formal derivation of this important result.

Start from the x component of the Momentum Principle for a multiparticle system whose center of mass is moving at nonrelativistic speed:

$$\frac{dP_{tot,x}}{dt} = M_{tot} \frac{dv_{cm,x}}{dt} = F_{net,surr,x}$$

Integrate through the x displacement of the center of mass (dropping the "surroundings" subscript):

$$M_{tot} \int_i^f \frac{dv_{cm,x}}{dt} dx_{cm} = \int_i^f F_{net,x} dx_{cm}$$

Switch dv and dx:

$$M_{tot} \int_i^f \frac{dx_{cm}}{dt} dv_{cm,x} = \int_i^f F_{net,x} dx_{cm}$$

But dx_{cm}/dt is the x component of the center of mass velocity:

$$M_{tot} \int_i^f v_{cm,x} dv_{cm,x} = \int_i^f F_{net,x} dx_{cm}$$

The integral on the left can be carried out:

$$M_{tot} \left[\tfrac{1}{2} v_{cm,x}^2 \right]_i^f = \int_i^f F_{net,x} dx_{cm}$$

$$\Delta \left[\tfrac{1}{2} M_{tot} v_{cm,x}^2 \right] = \int_i^f F_{net,x} dx_{cm}$$

We could repeat exactly the same argument for the y and z motions:

$$\Delta \left[\tfrac{1}{2} M_{tot} v_{cm,y}^2 \right] = \int_i^f F_{net,y} dy_{cm}$$

$$\Delta \left[\tfrac{1}{2} M_{tot} v_{cm,z}^2 \right] = \int_i^f F_{net,z} dz_{cm}$$

Note that

$$\tfrac{1}{2} M_{tot} v_{cm,x}^2 + \tfrac{1}{2} M_{tot} v_{cm,y}^2 + \tfrac{1}{2} M_{tot} v_{cm,z}^2 = \tfrac{1}{2} M_{tot} v_{cm}^2.$$

Moreover,

$$F_{net,x} dx_{cm} + F_{net,y} dy_{cm} + F_{net,z} dz_{cm} = \vec{F}_{net} \bullet d\vec{r}_{cm}$$

Adding the three equations together (the sum of the left sides is equal to the sum of the right sides), we have the following equation for the translational kinetic energy of the system:

$$\Delta \left[\tfrac{1}{2} M_{tot} v_{cm}^2 \right] = \int_i^f \vec{F}_{net} \bullet d\vec{r}_{cm}$$

In words, the change in the translational kinetic energy of a system is equal to the integral of the *net* force acting through the displacement of the center of mass point.

The derivation that we have just carried out shows that although this equation looks like an energy equation, it is actually closely related to the Momentum Principle from which it was derived. The common element is the *net* force.

In contrast, the actual energy equation for the real system involves the work done by each individual force through the displacement of the point of application of that force. If the system deforms or rotates, these displacements of the individual forces need not be the same as the displacement of the center of mass.

8.10 SUMMARY

Center of Mass

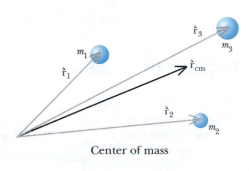

Center of mass

$$\vec{r}_{cm} \equiv \frac{m_1 \vec{r}_1 + m_2 \vec{r}_2 + m_3 \vec{r}_3 + \ldots}{m_1 + m_2 + m_3 + \ldots}$$

Problem Solving Techniques

Combined use of the energy equations for the real system and for the fictitious "point particle system" in analyzing complex phenomena involving multiparticle systems.

Point Particle system: Info on change of translational kinetic energy

Real system: Info on change of total energy

The Momentum Principle for multiparticle systems, derived from Newton's second and third laws of motion, was extended by defining the center of mass:

The Momentum Principle for multiparticle systems:

$$\frac{d\vec{P}_{tot}}{dt} = \vec{F}_{net,surr}$$

$$\vec{P}_{tot} \approx M\vec{v}_{cm} \text{ (if } v \ll c, \text{ and constant mass)}$$

Gravitational energy of multiparticle systems plus Earth, near the Earth's surface:

$$U_g = Mgy_{cm}$$

Kinetic energy of multiparticle systems:

$$K_{tot} = K_{trans} + K_{rel}, \text{ where } K_{trans} = \frac{P_{tot}^2}{2M} = \frac{1}{2}Mv_{cm}^2$$

Kinetic energy relative to center of mass can be split into two terms:

$$K_{rel} = K_{rot} + K_{vib}$$

Point particle system

$$\Delta K_{trans} = \Delta(\tfrac{1}{2}Mv_{cm}^2) = \int_i^f \vec{F}_{net,ext} \cdot d\vec{r}_{cm}$$

Sliding friction can deform the contact surfaces, with the result that the frictional force may act through a distance that is different from the distance through which the center-of-mass point moves. As with any deformable system, the energy equations for the real system and for the point particle system differ (though both are correct).

8.11 REVIEW QUESTIONS

The motion of a multiparticle system

8.RQ.11 Discuss qualitatively the motion of the atoms in a block of steel that falls onto another steel block. Why and how do large-scale vibrations damp out?

Center of mass

8.RQ.12 Can you give an example of a system that has no atoms located at its center of mass?

The Momentum Principle for multiparticle systems

8.RQ.13 Two people with different masses but equal speeds slide toward each other with little friction on ice with their arms extended straight out to the side (so each has the shape of a "t"). Her right hand meets his right hand, they hold hands and spin 90°, then release their holds and slide away. Make a rough sketch of the path of the center of mass of the system consisting of the two people, and explain briefly. (It helps to mark equal time intervals along the paths of the two people, and of their center of mass.)

8.RQ.14 The Momentum Principle for multiparticle systems would seem to say that the center of mass of a system moves just as though the system were a point particle. This led Chris to ask, "Wouldn't that mean that a piece of paper ought to fall in the same way as a small metal ball, if they have the same mass?" Explain carefully to Chris the resolution of this puzzle.

Gravitational energy of a multiparticle system

8.RQ.15 Consider the voyage to the Moon that you studied in Chapter 3. Would it make any difference, even a very tiny difference, whether the spacecraft is long or short, if the mass is the same? Explain briefly.

Kinetic energy of a multiparticle system

8.RQ.16 Outline how you would calculate the kinetic energy of the Earth as it moves through space.

The point particle system

8.RQ.17 Under what conditions does the energy equation for the point particle system differ from the energy equation for the real system? Give two examples of such a situation. Give one example of a situation where the two equations look exactly alike.

8.12 PROBLEMS

8.P.18 Experiment: Jumping straight up
(a) Crouch down and jump straight up, as high as you can. Estimate the location of your center of mass, and measure its height at three stages in this process: in the crouch, at lift-off, and at the top of the jump. Report your measurements. You may need to have a friend help you make the measurements.
(b) Analyze this process as fully as possible, using all the theoretical tools now available to you, especially the concepts in this chapter. Include a calculation of the force of the floor on your feet, the change in your chemical energy, and the time of contact from the beginning of the jump to lift-off. Be sure to explain clearly what approximations and simplifying assumptions you made in modeling the process.

8.P.19 Ice skater pushes away from a wall
On page 280 we discussed a woman ice skater who pushes away from a wall.
(a) Estimate the speed she can achieve just after pushing away from the wall. Then estimate the average acceleration during this process. How many "g's" is this? (That is, what fraction or multiple of 9.8 m/s² is your estimate?) Be sure to explain clearly what approximations and simplifying assumptions you made in modeling the process.
(b) For this process, choose the woman as the system of interest and discuss the energy transfers, and the changes in the

various forms of energy. Estimate the amount of each of these, including the correct signs.

8.P.20 Translation and rotation

A hoop of mass M and radius R rolls without slipping down a hill, as shown in the diagram below. The lack of slipping means that when the center of mass of the hoop has speed v, the tangential speed of the hoop relative to the center of mass is also equal to v, since in that case the instantaneous speed is zero for the part of the hoop that is in contact with the ground ($v - v = 0$).

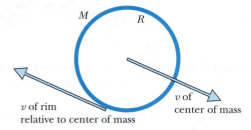

(a) The initial speed of the hoop is v_i, and the hill has a height h. What is the speed v_f at the bottom of the hill?
(b) Replace the hoop with a bicycle wheel whose rim has mass M and whose hub has mass m, as shown in the diagram below. The spokes have negligible mass. What would the bicycle wheel's speed be at the bottom of the hill?

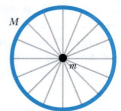

8.P.21 Pulling a disk from the side

A string is wrapped around a disk of mass M and radius R. Starting from rest, you pull the string with a constant force F along a nearly frictionless surface. At the instant when the center of the disk has moved a distance d, a length L of string has unwound off the disk.

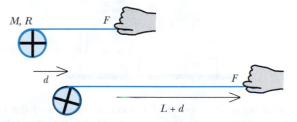

(a) At this instant, what is the speed of the center of mass of the disk?
(b) At this instant, how much rotational kinetic energy does the disk have relative to its center of mass?

8.P.22 Moving in a canoe

A man whose mass is 80 kg and a woman whose mass is 50 kg sit at opposite ends of a canoe 5 m long, whose mass is 30 kg.
(a) Relative to the man, where is the center of mass of the system consisting of man, woman, and canoe? (Hint: Choose a specific coordinate system with a specific origin.)

(b) Suppose the man moves quickly to the center of the canoe and sits down there. How far does the canoe move in the water? Explain your work and your assumptions.

8.P.23 Two disks colliding

Two disks are initially at rest, each of mass M, connected by a string between their centers, as shown below. The disks slide on low-friction ice as the center of the string is pulled by a string with a constant force F through a distance d. The disks collide and stick together, having moved a distance b horizontally.

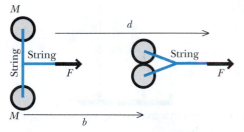

(a) What is the final speed of the stuck-together disks?
(b) When the disks collide and stick together, their temperature rises. Calculate the increase in thermal energy of the disks, assuming that the process is so fast that there is insufficient time for there to be much transfer of energy to the ice due to a temperature difference. (Also, ignore the small amount of energy radiated away as sound produced in the collisions between the disks.)

8.P.24 An uncoiling chain

A chain (mass M) of metal links is coiled up in a tight ball on a low-friction table. You pull on a link at one end of the chain with a constant force F. Eventually the chain straightens out to its full length L, and you keep pulling until you have pulled your end of the chain a total distance d.

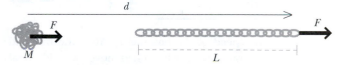

(a) What is the speed of the chain at this instant?
(b) In straightening out, the links of the chain bang against each other, and their temperature rises. Calculate the increase in thermal energy of the chain, assuming that the process is so fast that there is insufficient time for the chain there to be much transfer of energy to the table due to a temperature difference. (Also, ignore the small amount of energy radiated away as sound produced in the collisions among the links.)

8.P.25 Falling person

You hang by your hands from a tree limb that is a height L above the ground, with your center of mass a height h above the ground, and your feet a height d above the ground, as shown in the diagram below. You then let yourself fall. You absorb the shock by bending your knees, ending up momentarily at rest in a crouched position with your center of mass a height b above the ground. Your mass is M. You will need to draw labeled physics diagrams for the various stages in the process.

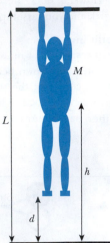

(a) What is the net internal energy change ΔE_{int} in your body (chemical plus thermal)?
(b) What is your speed v at the instant when your feet first touch the ground?
(c) What is the approximate average force F exerted by the ground on your feet during the time when your knees are bending?
(d) How much work is done by this force F?

8.P.26 Pulling a box containing an apparatus

A box and its contents have a total mass M. A string passes through a hole in the box, and you pull on the string with a constant force F (this is in outer space—there are no other forces acting).

(a) Initially the speed of the box was v_i. After the box had moved a long distance w, your hand had moved an additional distance d (a total distance of $w + d$), because additional string of length d came out of the box. What is now the speed v_f of the box?
(b) If we could have looked inside the box, we would have seen that the string was wound around a hub that turns on an axle with negligible friction, as shown below. Three masses, each of mass m, are attached to the hub at a distance r from the axle. Initially the angular speed of this apparatus was ω_i. In terms of the given quantities, what is the final angular speed ω_f of the apparatus?

8.P.27 Moving blocks connected by a spring

Two identical 0.1 kg blocks (labeled 1 and 2) were at rest on a nearly frictionless surface, connected by an unstretched spring whose stiffness is 100 N/m. Then a constant force of 5 N to the right was applied to block 2, and at a later time the blocks are in the new positions shown in the lower diagram.

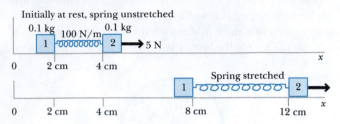

(a) At this later time, what is $K+U$ of the two-block system?
(b) What is the translational kinetic energy of the two-block system?
(c) What is the speed of the center of mass of the two-block system?
(d) What is the vibrational kinetic energy of the two-block system? Note that the spring is now stretched.

8.P.28 Spring and two masses

You exert an upward force of $2Mg$ to hold an object at rest that consists of two masses, each of mass M, connected by a low-mass spring whose stiffness is k_s, as shown below. The initial stretch of the spring is s_i (equal to Mg/k_s). Then you suddenly start applying a larger force, of constant magnitude $F > 2Mg$. The diagram shows the situation some time later, when the blocks have moved upward, and the spring stretch has increased to s_f.

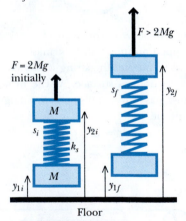

(a) What is now the speed of the center of mass of the two blocks? As part of your explanation, on a diagram show all of the forces that are acting. Also, be clear and explicit about what system you are analyzing.
(b) What is the vibrational kinetic energy of the two blocks (their kinetic energy relative to the center of mass)? Be clear and explicit about what system you are analyzing.

8.P.29 Friction

It is sometimes claimed that friction forces always slow an object down, but this is not true. If you place a box of mass M on a moving horizontal conveyor belt, the friction force of the belt acting on the bottom of the box speeds up the box. At first there is some slipping, until the speed of the box catches up to the speed v of the belt. The coefficient of friction between box and belt is μ.

(a) What is the distance d (relative to the floor) that the box moves before reaching the final speed v? Use energy arguments, and explain your reasoning carefully.

(b) How much time does it take for the box to reach its final speed?

(c) The belt and box of course get hot. Is the effective distance through which the friction force acts greater or less than d? Give as quantitative an argument as possible. You can assume that the process is quick enough that you can neglect transfer of energy Q due to a temperature difference between the belt and the box. Do not attempt to use the *results* of the friction analysis in this chapter; rather, apply the *methods* of that analysis to this different situation.

(d) Explain the result of part (c) qualitatively from a microscopic point of view, including physics diagrams.

8.P.30 A binary star system

(a) About half of the visible "stars" are actually binary star systems, two stars that orbit each other with no other objects nearby. Describe the motion of the center of mass of a binary star system. Briefly explain your reasoning.

(b) For a particular binary star system, telescopic observations repeated over many years show that one of the stars (whose unknown mass we'll call M_1) has a circular orbit with radius $R_1 = 6 \times 10^{11}$ m, while the other star (whose unknown mass we'll call M_2) has a circular orbit of radius $R_2 = 9 \times 10^{11}$ m about the same point. Make a sketch of the orbits, and show the positions of the two stars on these orbits at some instant. Label the two stars as to which is which, and label their orbital radii. Indicate on your sketch the location of the center of mass of the system, and explain how you know its location, using the concepts and results of this chapter.

(c) This double star system is observed to complete one revolution in 40 years. What are the masses of the two stars? (For comparison, the distance from Sun to Earth is about 1.5×10^{11} m, and the mass of the Sun is about 2×10^{30} kg.) This method is often used to determine the masses of stars. The mass of a star largely determines many of the other properties of a star, which is why astrophysicists need a method for measuring the mass.

8.P.31 Problem 7.14 Kinetic energy of a two-particle system

By calculating numerical quantities for a multiparticle system, one can get a concrete sense of the meaning of the relationships $\vec{p}_{tot} = M_{tot}\vec{v}_{cm}$ and $K_{tot} = K_{trans} + K_{rel}$. Consider an object consisting of two balls connected by a spring, whose stiffness is 400 N/m. The object has been thrown through the air and is rotating and vibrating as it moves. At a particular instant the spring is stretched 0.3 m and the two balls at the ends of the spring have the following masses and velocities:

1: 5 kg, $\langle 8, 14, 0 \rangle$ m/s

2: 3 kg, $\langle -5, 9, 0 \rangle$ m/s

(a) For this system, calculate $\vec{p}_{tot}$.

(b) Calculate $\vec{v}_{cm}$.

(c) Calculate K_{tot}.

(d) Calculate K_{trans}.

(e) Calculate K_{rel}.

(f) Here is a way to check your result for K_{rel}. The velocity of a particle relative to the center of mass is calculated by subtracting $\vec{v}_{cm}$ from the particle's velocity. To take a simple example, if you're riding in a car that's moving with $v_{cm,x} = 20$ m/s, and you throw a ball with $v_{rel,x} = 35$ m/s, relative to the car, a bystander on the ground sees the ball moving with $v_x = 55$ m/s. So $\vec{v} = \vec{v}_{cm} + \vec{v}_{rel}$, and therefore we have $\vec{v}_{rel} = \vec{v} - \vec{v}_{cm}$. Calculate $\vec{v}_{rel} = \vec{v} - \vec{v}_{cm}$ for each mass and calculate the corresponding K_{rel}. Compare with the result you obtained in part (e).

8.13 ANSWERS TO EXERCISES

8.X.1 (page 278): No acceleration, no force, no strain. Once you are no longer exerting a force on the right-most atoms in the block, atoms just to the left catch up and the interatomic force relaxes. This relaxation propagates to the left through the block, until finally there is no strain, and the block coasts along at a constant speed.

8.X.2 (page 279) In the vertical member, $3L/4$ up from its bottom.

8.X.3 (page 279): 4600 km (so inside the Earth)

8.X.4 (page 282): The puck with the string wound around it is rotating and so has additional kinetic energy. You have to pull farther, by the amount of string that unwinds, so you do more work, which corresponds to this puck having more kinetic energy. The Momentum Principle for the motion of the center of mass tells us nothing about the rotational aspects of the motion.

8.X.5 (page 282): 6 N to the right; 20 m/s^2

8.X.6 (page 288): All of the applied forces act through the same displacement, which is also the displacement of the center of mass. So it makes no difference whether you calculate the work done by each individual force, and then add them up, or first add up all the forces, and then calculate the work. Also, the only energy change in such a system is the overall kinetic energy.

8.X.7 (page 288): Average force = 167 newtons.

8.X.8 (page 289) Work done on real system = 0 joules. Work done on point particle system = 2500 joules.

8.X.9 (page 289) Change in chemical energy of runner = −2500 joules

8.X.10 (page 289) The instantaneous speed of the bottom of the tire is zero. The force by the road on the tire does no work. The increased translational kinetic energy of the car comes from a decrease in the chemical energy of the gasoline.

CHAPTER 9

COLLISIONS

Key concepts

- Collisions are brief interactions involving large internal forces.
- Momentum is *conserved:* the change of momentum of the system plus the change of momentum of the surroundings is zero (see Chapter 2).
  - The momentum of an isolated system does not change.
- To analyze collisions in detail, we must use the Energy Principle in combination with the Momentum Principle.
- Collisions in which there is no change in internal energy are called "elastic." Other collisions are "inelastic."
- Changing to a reference frame moving with the center of mass of the system simplifies the analysis of a collision.

DEFINITION OF COLLISION

An interaction that takes place in a very short time, with little interaction before or after interaction, is called a "collision".

The objects need not come into actual contact during the collision, if they interact via long range forces such as the gravitational force or the electric force.

Neutral molecules in a gas experience collisions—the interatomic electric force is effectively short range, so there is little interaction until the molecules come nearly into contact. Similarly, collisions between macroscopic objects are quite common: before and after a ball hits a wall it has little interaction with the wall, and the contact between ball and wall lasts a very short time.

In Chapter 2 we were able to analyze some aspects of collisions in terms of the conservation of momentum. The total momentum of a system doesn't change in a collision if external forces are zero (or the momentum change due to external forces may be negligible if the external forces are negligible compared to the internal forces). In this chapter we will revisit collisions, using the Energy Principle to obtain additional information not obtainable from the Momentum Principle alone.

We will see what we can learn about interactions by observing the probability distribution of deflection angles. We will investigate "elastic" collisions (no change of total kinetic energy) and "inelastic" collisions (some energy is changed from kinetic to other forms of energy). Our study of collisions will lead naturally into studying how processes look from different reference frames.

9.1 INTERNAL INTERACTIONS IN COLLISIONS

In a collision, interactions between objects in the system ("internal" interactions) are much larger than interactions between the system and its surroundings. In a collision, the effect of the net impulse delivered to the system by the surroundings is very small compared to the effects the objects within the system have on each other. Therefore, for a system consisting of two colliding objects, the total momentum of the system is approximately constant during the collision:

$$\vec{p}_f = \vec{p}_i + \vec{F}_{net}\Delta t$$

$$\vec{p}_f \approx \vec{p}_i$$

The effect of forces due to the surroundings is negligible during a collision.

Similarly, energy inputs from the surroundings are negligible during a collision, because the work done by external forces is very small, and because the collision happens fast enough that energy flow due to temperature differences between system and surroundings is negligible:

$$E_f = E_i + \cancel{W} + \cancel{Q}$$
$$E_f \approx E_i$$

Comparing internal and external forces

In a collision one of the objects receives a very short-duration impulse $\vec{F}\Delta t$ due to the other object. Its momentum changes by an amount $\Delta \vec{p} = \vec{F}\Delta t$. The following example shows that in a collision the forces between the colliding objects can be very large compared to external forces.

Example: Ball hits floor

A ball whose mass is 0.1 kg falls and hits the floor with a speed of 6 m/s, then rebounds upward with a speed of 5 m/s. (a) What was the magnitude and direction of the impulse delivered to the ball by the floor? (Remember that momentum and impulse are vectors.) (b) If the ball was in contact with the floor for 1 ms (1×10^{-3} s), compressing a distance of about 3 mm (average speed about 3 m/s), what was the average magnitude of the force exerted on the ball by the floor? (c) Compare with the force on the ball by the Earth.

During a collision, energy inputs from the surroundings are negligible.

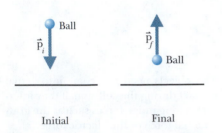

Ball $\vec{p}_i$

$\vec{p}_f$ Ball

Initial   Final

Figure 9.1 A collision between a ball and the floor.

System: Ball
Surroundings: Floor

Momentum Principle:

$$\Delta \vec{p} = \vec{F}_{net}\Delta t$$

Left hand side:

$$\vec{p}_i = (0.1 \text{ kg})(\langle 0, -6, 0\rangle \text{ m/s})$$
$$\vec{p}_f = (0.1 \text{ kg})(\langle 0, 5, 0\rangle \text{ m/s})$$
$$\vec{p}_f - \vec{p}_i = \langle 0, +1.1, 0\rangle \text{ kg·m/s}$$

(a) Impulse:

$$\langle 0, +1.1, 0\rangle \text{ kg·m/s} = (\vec{F}_{avg}\Delta t)$$

(b) Average force:

$$|\vec{F}_{avg}| = \frac{1.1 \text{ kg·m/s}}{1\times10^{-3} \text{ s}} = 1100 \text{ N}$$

(c) The magnitude of the force exerted on the ball by the Earth is
$$mg = (0.1\text{kg})(9.8 \text{ N/kg}) \approx 1 \text{ N}$$

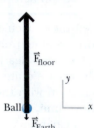

$\vec{F}_{floor}$

Ball

$\vec{F}_{Earth}$

Although the actual force the floor exerts on the ball is not constant during the collision, we can express the impulse in terms of the average force exerted by the floor.

Further discussion

This illustrates the class of interactions we're talking about. We can't neglect the gravitational *mg* force before or after the collision, when it's the only force acting on the ball, but during the brief collision the force exerted by the floor is so large (1100 times as large!) that other forces are negligible.

Effect of external forces

As long as Δt is very short, and the net external force is of ordinary magnitude, the external impulse will be small and the change it causes in the total momentum of the two-object system will usually be negligible. An example is a collision of two baseballs in midair (Figure 9.2).

? Look at the curving trajectories before the collision. Does the total momentum of the two baseballs change during the long period before the collision? What is the cause?

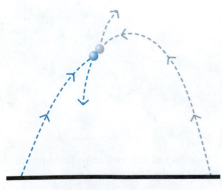

Figure 9.2 Midair collision between two baseballs.

The total momentum of the two-ball system changes, because gravitational and air-resistance forces act on the system. The collisional contact lasts a very short time, and involves very large contact forces, as is evident from the abrupt changes of directions of both balls.

? How does the external impulse $\vec{F}_{net,ext}\Delta t$ (gravitational force, air resistance) acting on just one of the baseballs compare with the impulse due to the other baseball, during contact? Is it a good approximation to say that from just before contact to just after contact, the total momentum of the two balls is (nearly) constant?

The external impulse involves an ordinary force for a very short time, while the impulse due to the other ball involves a huge force acting for that same short time. Therefore the momentum change of one of the balls is very nearly the opposite of the momentum change of the other ball. During the short contact time, the total momentum of the two-ball system is approximately constant.

These considerations justify an approach to the analysis of collisions that proves fruitful. Up until the collision, and after the collision, we do need to take into account external forces, which affect the total momentum. But during the brief time interval of the collision itself, we can use the approximation that the external impulse is small, and the change in total momentum of the interacting objects is negligible.

"Collisions" without contact

Consider the electric interaction of a proton and electron, or the gravitational interaction of two asteroids, as shown in Figure 9.3. They started out far from each other. There is no contact—no touching—yet we call such an interaction a "collision." A collision is any process in which there is little interaction before and after a short time interval, and large interactions during that short time interval. Since electric and gravitational forces depend on $1/r^2$, the forces are small when the objects are far apart.

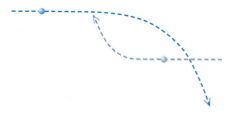

Figure 9.3 Electric or gravitational interactions without contact; these too are considered to be "collisions."

9.2 ELASTIC AND INELASTIC COLLISIONS

Internal energy

Often we want to focus on the changes in translational kinetic energy that occur during a collision between two objects. We will use the term "internal energy" to refer to the other kinds of energy objects may have, for example electronic energy, vibrational energy, rotational energy, or the randomized thermal energy of a solid object whose atoms and chemical bonds are vibrating.

Elastic collisions

We call a collision "elastic" if there is no change in internal energy of the interacting objects: no thermal energy rise (associated with a temperature increase), no springs newly compressed, no new rotations or vibrations, etc.

ELASTIC COLLISION

In an "elastic" collision, the internal energy of the objects in the system does not change: $\Delta E_{int} = 0$.

When atomic systems with quantized energies collide, the collision will be perfectly elastic if there is insufficient energy available to raise the systems to excited quantum states (Figure 9.4). For example, as discussed in Chapter 7, the energy of the first excited electronic state of a mercury atom is 4.9 eV above the ground state. If an electron whose kinetic energy is only 3 eV collides with a ground state mercury atom, the collision will be elastic,

2nd excited state

1st excited state

Energy too small to be absorbed

Ground state

Figure 9.4 Energy levels in an atomic system. If the energy input is too small to reach an excited state, the energy cannot be absorbed.

because there is not enough energy available to change the internal energy of the mercury atom.

With macroscopic systems there are no perfectly elastic collisions, because there is always some dissipation, but many macroscopic collisions are very nearly elastic. For example, collisions between billiard balls or steel balls can be nearly elastic.

Carts used to demonstrate collisions in physics lectures often contain magnets to repel each other without actually making contact, and such collisions are nearly elastic. Collisions between carts that interact through soft springs may also be nearly elastic.

Inelastic collisions

We call a collision "inelastic" if it isn't elastic; that is, there is some change in the internal energy of the colliding objects. They get hot, or deform, or rotate, or vibrate, etc.

INELASTIC COLLISION

In an "inelastic" collision, the internal energy of the objects in the system changes: $\Delta E_{\text{int}} \neq 0$

As stated above, collisions of macroscopic objects are always at least a bit inelastic, but in many common situations the collision may be nearly elastic. When atomic systems collide, one or more of the colliding objects may undergo a change of quantized energy level, and we call that an inelastic collision. For example, if an electron with kinetic energy 5 eV collides with a mercury atom in its ground state, the mercury atom may be excited to a higher electronic state 4.9 eV above the ground state, leaving only 0.1 eV of kinetic energy in the system.

Maximally inelastic collisions

The most extreme inelastic collision is the "maximally inelastic" collision in which there is maximum dissipation. That doesn't mean that the objects stop dead, because momentum must be conserved. It can be shown that a collision in which the objects stick together is a maximally inelastic collision: the only remaining kinetic energy is present only because the total momentum can't change.

The simplest example of a maximally inelastic collision is that of two objects that have equal and opposite momenta, so that before the collision the total momentum of the combined system is zero. The momentum afterwards must also be zero, and if they are stuck together that means that they must be at rest.

MAXIMALLY INELASTIC COLLISION

In a "maximally inelastic" collision, there is maximum dissipation, and the objects stick together.

Identifying inelastic collisions

When deciding whether or not it is reasonable to treat a collision as approximately elastic, one can look for indications of inelasticity, such as:
- objects stuck together after the collision
- an object is deformed after the collision
- objects are hotter after the collision
- there is more vibration or rotation after the collision
- an object is in an excited state after the collision

9.3 A HEAD-ON COLLISION OF EQUAL MASSES

We will start by considering simple 1-D (head-on) collisions in order to see clearly how to apply both the Energy Principle and the Momentum Principle. Consider a head-on collision between two carts rolling or sliding on a track with low friction (Figure 9.5). Similar situations are billiard balls or hockey pucks or vehicles hitting each other head-on. Cart 1 with mass m moves to the right with x-momentum p_{1xi}, and it runs into cart 2 which has the same mass m and is initially sitting still. Taking the two carts as the system, and neglecting the small frictional force exerted by the track, all that the Momentum Principle can tell us is that after the collision the total final x-momentum $p_{1xf} + p_{2xf}$ must equal the initial total x-momentum p_{1xi}. By itself, the Momentum Principle doesn't tell us how this momentum will be divided between the two carts. We need to consider conservation of energy as well as conservation of momentum in order to predict what will happen.

Before the collision, there was rest energy of both carts and kinetic energy of cart 1, K_{1i}. After the collision, there is rest energy of both carts and kinetic energy of both carts, $K_{1f} + K_{2f}$, and possibly something else:

? What other type of energy might be present after the collision?

There could be additional internal energy in the carts as a result of the collision. The temperature of the carts might be a bit higher now, with some of the initial kinetic energy having been dissipated into thermal energy. How much energy dissipation occurs depends on the details of the contact between the two interacting carts.

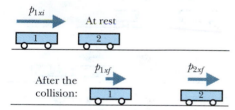

Figure 9.5 Cart 1 runs into stationary cart 2. The carts have equal masses. Friction and air resistance are negligible.

Example: Elastic collision of two identical carts

Cart 1 collides with stationary cart 2, which is identical (Figure 9.5). Suppose the collision is (nearly) elastic, as it will be if the carts repel each other magnetically, or interact through soft springs. In this case there is no change of internal energy. What are the final momenta of the two carts?

> System: both carts
> Surroundings: Earth, track, air (neglect friction and air resistance)
>
> Initial situation: Just before collision
> Final situation: Just after collision
>
> Momentum Principle:
>
> $$\vec{p}_f = \vec{p}_i + \vec{F}_{net}\Delta t$$
>
> $$p_{1xf} + p_{2xf} = p_{1xi} + 0$$

Negligible effect of surroundings.
Only x components change.

> Energy Principle:
>
> $$E_f = E_i + W + Q$$
>
> $$K_{1f} + K_{2f} + E_{int1f} + E_{int2f} = K_{1i} + K_{2i} + E_{int1i} + E_{int2i}$$
>
> $$K_{1f} + K_{2f} = K_{1i} + K_{2i}$$

Negligible effect of surroundings.
Negligible change in internal energy of the carts.

> Combine momentum and energy equations:
>
> $$\frac{p_{1xf}^2}{2m} + \frac{p_{2xf}^2}{2m} = \frac{(p_{1xf} + p_{2xf})^2}{2m}$$
>
> $$p_{1xf}^2 + p_{2xf}^2 = p_{1xf}^2 + 2p_{1xf}p_{2xf} + p_{2xf}^2$$
>
> $$2p_{1xf}p_{2xf} = 0$$

Recall that $K = \frac{1}{2}mv^2 = \frac{1}{2}m\left(\frac{p}{m}\right)^2 = \frac{1}{2}\frac{p^2}{m}$
Expand the quadratic.
Cancel similar terms.

There are two possible solutions to this equation. The term $p_{1xf}p_{2xf}$ can be zero if $p_{1xf} = 0$ or if $p_{2xf} = 0$.

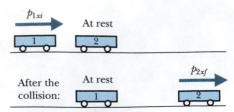

Figure 9.6 If the collision is elastic, cart 1 stops and cart 2 has all the momentum cart 1 used to have.

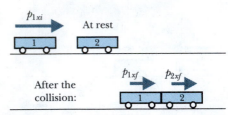

Figure 9.7 Cart 1 sticks to cart 2, so they have the same final momentum.

External forces are negligible during the collision, so the total momentum of the system is constant.

Since we know the speed of the carts we can calculate their kinetic energy.

We know the initial and final kinetic energies of the system, so we can use the Energy Principle to find the change in internal energy.

The Momentum Principle is still valid even though the collision is inelastic. Fundamental principles apply in all situations.

If $p_{1xf} = 0$, the physical situation is that cart 1 came to a complete stop. In that case we see from the momentum equation $p_{1xf} + p_{2xf} = p_{2xf} = p_{1xi}$ that cart 2 now has the same momentum that cart 1 used to have. There has been a complete transfer of momentum from cart 1 to cart 2. There has also been a complete transfer of kinetic energy from cart 1 to cart 2. See Figure 9.6.

If $p_{2xf} = 0$, what is the physical situation? In that case $p_{1xf} + p_{2xf} = p_{1xf} = p_{1xi}$, and cart 1 just keeps going, missing cart 2! Of course that won't happen if the carts are on the same narrow track, but the algebra doesn't know that.

It is not possible for both final momenta to be zero, since the total final momentum of the system must equal the nonzero total initial momentum of the system.

Example: Maximally inelastic collision of two identical carts

Consider the opposite extreme—a maximally inelastic collision of the two identical carts, one initially at rest. That means the carts stick together (perhaps they have sticky material on their ends), and each has the same final momentum $p_{1xf} = p_{2xf}$ (Figure 9.7). (a) Find the final momentum, final speed, and final kinetic energy of the carts. (b) What is the change in internal energy of the two carts?

Since the y and z components of momentum don't change, we can work with only x components.

System: both carts
Surroundings: Earth, track, air (neglect friction and air resistance)

Momentum Principle (x components):

$$p_{1xf} + p_{2xf} = p_{1xi}$$
$$2p_{1xf} = p_{1xi}$$
$$p_{1xf} = \frac{1}{2}p_{1xi}$$

The final speed of the carts is half the initial speed: $v_f = \frac{1}{2}v_i$

Final kinetic energy:

$$(K_{1f} + K_{2f}) = 2\left(\frac{1}{2}mv_f^2\right)$$

$$(K_{1f} + K_{2f}) = 2\left(\frac{1}{2}m\left(\frac{1}{2}v_i\right)^2\right) = \frac{1}{4}mv_i^2$$

$$(K_{1f} + K_{2f}) = \frac{K_{1i}}{2}$$

Energy Principle:

$$K_{1f} + K_{2f} + E_{int,f} = K_{1i} + E_{int,i}$$
$$E_{int,f} - E_{int,i} = K_{1i} - (K_{1f} + K_{2f})$$
$$\Delta E_{int} = \frac{1}{2}mv_i^2 - \frac{1}{4}mv_i^2$$
$$\Delta E_{int} = \frac{1}{4}mv_i^2 = \frac{K_{1i}}{2}$$

The final kinetic energy of the system is only half of the original kinetic energy, which means that the other half of the original kinetic energy has been dissipated into increased internal energy ΔE_{int} of the two carts.

Further discussion

In both cases (elastic collision and inelastic collision) we needed to apply both the Momentum Principle and the Energy Principle to find all the unknown quantities. One equation alone was not enough.

We have shown the two extremes of energy conservation in the head-on collision of the moving cart with the stationary cart:

- If the collision is elastic, cart 1 stops and cart 2 moves with the speed cart 1 used to have.
- If the collision is maximally inelastic, the carts stick together and move with half the original speed. Half of the original kinetic energy is dissipated into increased internal energy.

Between these two extremes we have an inelastic but not maximally inelastic collision, in which some amount of the original kinetic energy (less than half) is dissipated into increased internal energy.

> *9.X.1* A 6 kg mass traveling at speed 10 m/s strikes a stationary 6 kg mass head-on, and the two masses stick together. (a) What was the initial total kinetic energy? (b) What is the final speed? (c) What is the final total kinetic energy? (d) What was the increase in internal energy of the two masses?

9.4 HEAD-ON COLLISIONS BETWEEN UNEQUAL MASSES

Suppose a ping-pong ball hits a stationary bowling ball head-on, nearly elastically. Imagine this takes place in outer space or floating in an orbiting spacecraft, so there is no friction—no significant external forces act on the combined system during the collision.

? What do you expect? How will the ping-pong ball move after hitting the bowling ball? How will the bowling ball move after the collision?

You'll see the ping-pong ball bounce straight back with very little change of speed (Figure 9.8), and we'll show this is consistent with conservation of momentum and conservation of energy. (It is possible to show algebraically that this must be true, but we won't go through the algebra here.) Less obvious is that the bowling ball does move, although very slowly.

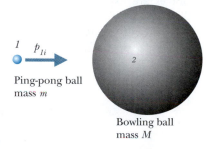
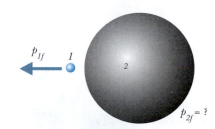

Figure 9.8 A ping-pong ball hits a bowling ball head-on and bounces off elastically.

Example: A ping-pong ball hits a stationary bowling ball head-on

What are the (a) momentum, (b) speed, and (c) kinetic energy of the bowling ball after the collision? Assume little change in the speed of the ping-pong ball, and assume the collision is elastic. The momenta of both objects have only x components; to reduce clutter we will omit the "x" in all subscripts. The initial x-momentum of the ping-pong ball is p_{1i}. The ping-pong ball (object 1) has mass m and the bowling ball (object 2) has mass M. Find the final momentum, final speed, and the final kinetic energy of each object.

System: Ping-pong ball and bowling ball
Surroundings: Nothing that exerts significant forces

Momentum Principle:

$$\vec{P}_{system,f} = \vec{P}_{system,i}$$

All of these equations involve x components of momentum. We have omitted the "x" in subscripts to make equations more readable.

It may be surprising that the bowling ball ends up with about twice the momentum of the ping-pong ball. One way to understand this is that the final x-momentum of the ping-pong ball is approximately $-p_{1i}$, so the change in the ping-pong ball's momentum is approximately

$$(-p_{1i}) - (p_{1i}) = -2p_{1i}$$

The ping-pong ball's speed hardly changed, but its momentum changed a great deal. Because momentum is a vector, a change of direction is just as much a change as a change of magnitude. This big change is of course due to the interatomic electric contact forces exerted on the ping-pong ball by the bowling ball. By reciprocity, the same magnitude of interatomic contact forces are exerted by the ping-pong ball on the bowling ball, which undergoes a momentum change of $2p_{1i}$.

Assume that the speed of the ping-pong ball does not change significantly in the collision, so $p_{1f} \approx -p_{1i}$.

$$-p_{1i} + p_{2f} = p_{1i}$$
$$p_{2f} = 2p_{1i}$$

(a) The final momentum of the bowling ball is twice the initial momentum of the ping pong ball.

(b) Final speed of bowling ball:

$$v_{2f} \approx \frac{p_{2f}}{M} = \frac{2p_{1i}}{M} = \frac{2mv_{1i}}{M} = 2\left(\frac{m}{M}\right)v_{1i}$$

This is a very small speed since $m \ll M$.

(c) Kinetic energies:

$$K_2 = \frac{(2p_{1i})^2}{2M} \quad \text{and} \quad K_1 = \frac{p_{1i}^2}{2m}$$

Because the mass of the bowling ball (about 5 kg) is much larger than the mass of the ping pong ball (about 0.001 kg), the kinetic energy of the bowling ball is much smaller than the kinetic energy of the ping pong ball.

Further discussion

Although the kinetic energy of the slowly moving bowling ball is very small, it is not zero, which means that the ping-pong ball does lose a little speed in the collision. That's why our assumption that the ping-pong ball's speed hardly changes is an approximation.

We made the plausible assumption that the ping-pong ball bounced back with nearly its original speed, and showed that this is consistent with momentum conservation if the bowling ball gets twice the momentum of the ping-pong ball, and also consistent with energy conservation, because the bowling ball acquires very little kinetic energy. With a lot of algebra, it is also possible to solve the momentum and energy equations exactly for the unknown final momentum of the ping-pong ball, and when one makes the approximation that the mass of the bowling ball is very much larger than the mass of the ping-pong ball, the results are those given above. (Problem 9.P.19 (page 329) involves analyzing a head-on collision without making this approximation.)

When a ping-pong ball hits a stationary bowling ball head-on, elastically:
• The ping-pong ball bounces back with almost the same speed.
• The ping-pong ball has a large change of momentum $-2p_{1i}$.
• The bowling ball has a large change of momentum $2p_{1i}$ (Figure 9.9).
• The bowling ball's final speed is very small, because it has large mass but comparable momentum.
• The bowling ball's final kinetic energy is very small, because it has large mass but comparable momentum.

Similar situations include a ball bouncing off a wall, or falling and bouncing off the ground, in which case the entire Earth plays the role of the bowling ball. The Earth gets twice the momentum of the ball but an extremely small speed and extremely small kinetic energy. If a bowling ball sits on a table and is hit by a ping-pong ball, and the table exerts a large enough frictional force to prevent the bowling bowl from moving, the bowling ball is essentially part of the Earth.

Although we've looked at the very special case of a head-on collision, it is generally true that when low-mass projectiles collide elastically with station-

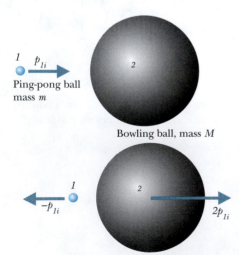

Figure 9.9 The bowling ball gets twice the momentum of the ping-pong ball (but its speed and kinetic energy are very small).

ary large masses, the low-mass projectile bounces off with little change of speed and kinetic energy but may undergo a large change of momentum. The large-mass target can pick up lots of momentum, but little speed or kinetic energy.

9.X.2 We can use our results for head-on elastic collisions to analyze the recoil of the Earth when a ball bounces off a wall embedded in the Earth. Suppose a professional baseball pitcher hurls a baseball ($m = 155$ grams) with a speed of 100 miles per hour ($v_1 = 44$ m/s) at a wall, and the ball bounces back with little loss of kinetic energy. What is the approximate recoil speed of the Earth ($M = 6{\times}10^{24}$ kg)?

9.X.3 In the previous Exercise, calculate the approximate recoil kinetic energy of the Earth and compare to the kinetic energy of the baseball. The Earth gets lots of momentum (twice the momentum of the baseball) but very little kinetic energy.

Frame of reference

A frame of reference is a perspective from which a system is observed. A frame of reference may be stationary or it may move at constant velocity. It is sometimes simpler to analyze phenomena from a moving reference frame.

As a simple example, imagine that you are standing on a street corner and a bus passes you. In your stationary frame of reference, a passenger sitting on the bus is moving with the same velocity as the bus. However, the bus driver may observe the passenger from a moving reference frame "anchored" to the bus. In the moving bus reference frame, the driver sees the passenger sitting still.

Evidently, the velocity $\vec{v}'$ ("v-prime") of an object viewed from a moving reference frame is found by subtracting $\vec{v}_{\text{frame}}$, the velocity of the moving frame, from $\vec{v}$, the velocity of the object relative to a stationary, or "laboratory" reference frame:

$$\vec{v}' = \vec{v} - \vec{v}_{\text{frame}}$$

This is a very powerful technique which is discussed in more depth later in this chapter. A change of reference frame often makes a problem much simpler to analyze, or reduces the problem to one you've already analyzed.

Bowling ball hits ping-pong ball: Change of reference frame

In the previous example we analyzed a collision in which a moving ping pong ball hit a stationary bowling ball head-on. What happens when a bowling ball of large mass M with initial x-velocity V hits elastically head-on a stationary ping-pong ball of small mass m (Figure 9.10)? What is the final speed of the ping-pong ball?

System: Ping-pong ball and bowling ball

Surroundings: nothing that exerts significant forces

We can perform a change of reference frame that reduces this problem to the previous one. Suppose this experiment is being done in a spacecraft, away from the walls. The room is dark, and only the glowing bowling ball and the ping pong ball are visible; the walls are not visible.

Jack and Jill are two crew members in the spacecraft. Jack hangs onto a wall, remaining stationary, and watches the bowling ball go by, while Jill launches herself away from the wall with x-velocity V and glides along beside the bowling ball. In Jack's stationary reference frame, the ping pong ball is stationary, and the bowling ball is moving, as described above.

What is the speed of the bowling ball as measured by Jill? For Jill, the bowling ball isn't moving. It is always the same distance from her. In Jill's reference frame, the bowling ball is at rest.

If the velocity of a reference frame is not constant, the frame is "non-inertial", and the laws of physics as we have stated them do not apply. We'll stick to inertial frames in this textbook.

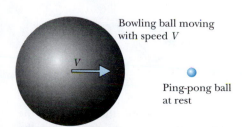

Figure 9.10 In a stationary reference frame, Jack sees a bowling ball with speed V run into a stationary ping-pong ball.

Bowling ball moving with speed V

V

Ping-pong ball at rest

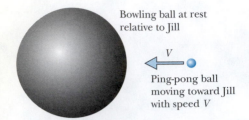

Figure 9.11 Moving with the bowling ball, at speed V, Jill sees the bowling ball at rest, and the ping-pong ball moving toward her with speed V.

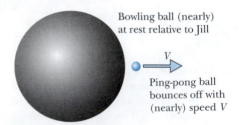

Figure 9.12 In Jill's moving frame of reference, she sees the ping-pong ball bounce off the bowling ball with nearly its original speed V.

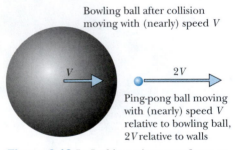

Figure 9.13 In Jack's stationary reference frame, the ping-pong ball has a speed $2V$ after the collision.

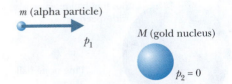

Figure 9.14 Before the collision, the alpha particle (mass m) approaches the gold nucleus (mass M).

What does Jill see the ping-pong ball doing? In Jill's reference frame, where the bowling ball is at rest, she sees the ping-pong ball moving with x-velocity $-V$ (Figure 9.11).

According to the principle of relativity, in any reference frame in uniform motion one can apply physics principles and make correct predictions, even though the details of the calculations look different from the details in some other reference frame in uniform motion. If the walls are dark, and the ping-pong ball and bowling ball glow in ultraviolet light, Jill truly cannot tell whether she is moving and the ping-pong ball is stationary relative to the walls, or whether she is stationary and the ping-pong ball is moving relative to the walls. This is at the heart of the principle of relativity.

By changing reference frames, we've reduced this situation to one very similar to the previous example. In Jill's reference frame, the ping-pong ball moves in the $-x$ direction with speed V and hits a stationary bowling ball elastically, head-on. It bounces off in the $+x$ direction with nearly the same speed V, in Jill's reference frame (Figure 9.12).

After the collision, how fast is the ping-pong ball moving in Jack's stationary reference frame? Jill and the bowling ball are moving with speed V in the $+x$ direction relative to Jack, and the ping-pong ball is moving with speed V in the $+x$ direction relative to Jill and the bowling ball, so it is moving with speed $2V$ relative to Jack and the walls (Figure 9.13). By changing reference frame we've analyzed the problem without doing any significant amount of calculation.

When a bowling ball hits a stationary ping-pong ball head-on, elastically:

- The ping-pong ball moves with twice the speed of the bowling ball.
- We won't go through the details, but you can appreciate that the change in speed of the bowling ball is insignificant. It just plows straight ahead, having lost very little kinetic energy to the ping-pong ball. The small amount of momentum lost by the bowling ball was gained by the ping-pong ball ($2mV$).

9.X.4 A uranium atom traveling at speed 4×10^4 m/s collides elastically with a stationary hydrogen molecule, head-on. What is the approximate final speed of the hydrogen molecule?

9.5 SCATTERING: COLLISIONS IN 2-D AND 3-D

Scattering experiments are often used study the structure and behavior of atoms, nuclei, and other small particles. In a scattering experiment, a beam of particles collides with other particles. In atomic or nuclear collisions, we can't observe in detail the curving trajectories inside the tiny interaction region. We only observe the trajectories before and after the collision, when the particles are far apart and their mutual interaction is very weak, so they are traveling in nearly straight lines. An example of scattering is the collision of an alpha particle (helium nucleus) with the nucleus of a gold atom.

Momentum and energy equations

An alpha particle collides with a gold nucleus, and is deflected through some final angle θ. The gold nucleus recoils at some other final angle ϕ. The final angles are measured relative to the initial direction of the alpha particle (Figure 9.14 and Figure 9.15). We say that the alpha particle "scattered" through an angle θ.

We take the "before" time early enough and the "after" time late enough that the two particles are sufficiently far away from each other that we can neglect their electric potential energy. As for conservation of momentum, we must write two equations, for the x and y components of momentum conservation.

? Try to write the two momentum-conservation equations, and the energy-conservation equation. All the speeds are small compared to the speed of light. Choose x in the direction of the incoming alpha particle, and y at right angles to it. Consider ϕ to be a positive angle. Write equations in terms of the magnitudes of the momenta, p_1, p_3, and p_4; the gold nucleus is initially at rest.

We use direction cosines to get the x and y components of momenta:

$$p_1 = p_3 \cos\theta + p_4 \cos\phi \quad \text{x-momentum conservation}$$

$$0 = p_3 \cos(90° - \theta) - p_4 \cos(90° - \phi) \quad \text{y-momentum conservation}$$

$$\frac{p_1^2}{2m} = \frac{p_3^2}{2m} + \frac{p_4^2}{2M} \quad \text{energy conservation (scalar)}$$

Make absolutely sure that you understand every detail of these three equations! In particular, pay attention to the fact that since momentum is a vector, the momentum equations are written in terms of x and y components of the momentum.

? It is fruitful to reflect on which of the quantities in these three equations are known and which are unknown. We know the masses of the alpha particle and the gold nucleus, and we know the initial kinetic energy of the alpha particle (so we know its speed and momentum). Which quantities in these three equations are unknown?

You should have identified the two final momenta and the two final angles as unknown quantities (p_3, p_4, θ, and ϕ, all of which represent positive numbers). But if there are four unknowns and only three equations, how can we solve for the unknowns? The simple answer is, we can't!

Nevertheless, there are many situations where these three equations can be exploited. For example, if we measure the final direction of motion of the alpha particle, the angle θ, we can solve for the other three unknown quantities and thereby predict the final momenta p_3 and p_4 of the two particles and the direction of motion ϕ of the gold nucleus, even without measuring all of these quantities.

It is worth mentioning that in experimental particle physics the momentum is often measured in a very direct way. Charged particles curve in a magnetic field due to magnetic forces that act on moving charges, and the radius of curvature is a direct measure of the momentum of the particle, if you know how much charge it has, which you do know if you know what kind of particle it is. Usually the momentum of a particle can be measured more accurately than the velocity, though once you have measured the momentum you can deduce the velocity if you know what kind of particle it is, since in that case you know its mass.

Figure 9.15 After the collision, when the particles are far apart and moving with nearly uniform velocities. Final angles are measured relative to the initial direction of the alpha particle.

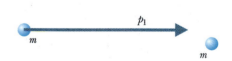

Figure 9.16 Before a collision of two identical particles, one initially at rest.

Example: Elastic collision: Identical particles, one initially at rest

An interesting special case of an elastic collision is one involving particles of equal mass, such as an elastic collision between two alpha particles, or two gold nuclei, or two billiard balls (whose collision is nearly elastic), in which one of the two objects (the "target") is initially at rest (Figure 9.16 and Figure 9.17). In this special case, show that the angle A must be 90° if the speeds are small compared to the speed of light.

Because the vector momentum of the system is constant, we can write $\vec{p}_1 = \vec{p}_3 + \vec{p}_4$, then calculate p_1^2 by expanding the dot product:

$$p_1^2 = \vec{p}_1 \bullet \vec{p}_1 = (\vec{p}_3 + \vec{p}_4) \bullet (\vec{p}_3 + \vec{p}_4) = p_3^2 + p_4^2 + 2p_3 p_4 \cos A$$

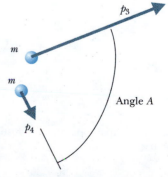

Figure 9.17 After the collision of two identical particles, one initially at rest.

Divide by $2m$:

$$\frac{p_1^2}{2m} = \frac{p_3^2}{2m} + \frac{p_4^2}{2m} + \frac{2p_3p_4\cos A}{2m}$$

Since the kinetic energy K at low speed can be written as $p^2/2m$,

$$K_1 = K_3 + K_4 + \frac{2p_3p_4\cos A}{2m}$$

But energy conservation for this elastic collision says that $K_1 = K_3 + K_4$, so we have

$$\cos A = 0 \text{ or } A = 90°$$

In this special situation, an elastic collision at low speeds between equal particles, one of which is initially at rest, the angle between the final velocities must be 90° as long as neither p_3 nor p_4 is zero, in which case the equation reduces to $0\cos A = 0$, and $\cos A$ is indeterminate. If one of the final momenta is indeed zero, the other particle must have all the momentum of the initial particle. This can happen either because the incoming particle missed the target entirely and kept going in the original direction, or because the incoming particle hit the target dead center, in which case the only consistent solution of the equations is for the incoming particle to come to rest and the target particle to move forward with the initial momentum, as we saw in the head-on collision of two carts earlier in the chapter.

Impact parameter

The distance between centers perpendicular to the incoming velocity is called the "impact parameter" and is often denoted by b. A head-on collision has an impact parameter of zero. Here and on the following page are possible elastic collisions between two billiard balls, for various impact parameters:

If the impact parameter, b, is too large, there is no interaction at all between two billiard balls.

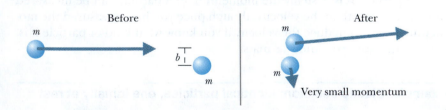

A grazing impact results in small-angle scattering.

Smaller impact parameters lead to larger scattering unless $b = 0$:

A medium impact parameter results in symmetric scattering.

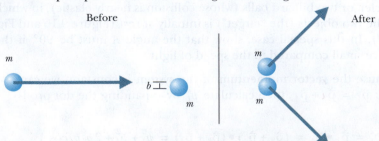

In a head-on collision the impact parameter $b = 0$. In this case all the momentum is transferred to the target.

Before

After

m

m m

m At rest

The smaller the impact parameter, the more severe is the collision, and the larger the deflection angle of the incoming particle (larger "scattering"), except for a head-on collision, where if the masses are equal the incoming ball stops dead and the target ball gets the entire momentum.

9.6 DISCOVERING THE NUCLEUS INSIDE ATOMS

A famous experiment involving collisions led to the discovery of the nucleus inside the atom. It was carried out in 1911 by a group in England led by the New Zealander Ernest Rutherford. In the experiment high-speed alpha particles (now known to be helium nuclei, consisting of two protons and two neutrons) were shot at a thin gold foil (whose nuclei consist of 79 protons and 118 neutrons).

When Rutherford and his coworkers decided to perform a scattering experiment in order to probe the internal structure of matter, they were familiar with the relationships of momenta in collisions that we have just studied, and they reasoned that by shooting microscopic particles at a thin layer of metal, they might be able to learn something about the microscopic structure of the metal by looking at patterns of scattering.

In the Rutherford experiment, the velocity of the alpha particles was not high enough to allow them to make actual contact with the nucleus. It is nonetheless appropriate to think of the process as a collision, because there are sizable electric forces acting between the alpha particle and the gold nucleus for a very short time. We call any process of this kind a collision, even if there is no direct contact but only gravitational, electric, or magnetic interactions at a distance. Figure 9.18 shows what the trajectories look like. We say that the alpha particle projectile is "scattered" (deflected) by the interaction with the initially stationary gold nucleus target, and such experiments are called "scattering" experiments.

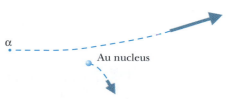

α Au nucleus

Figure 9.18 A collision that is not head-on.

Elastic collisions in the Rutherford experiment

The conditions of this experiment were such that there was no change in the internal energy of either the alpha particle or the gold nucleus (elastic collision). There do exist "excited" states of nuclei, in which there is a change in the configuration of the nucleons corresponding to a higher energy than the ground state, the normal configuration of the nucleus. The internal energy of the nucleus is quantized, and there are no states with energy intermediate between these quantized energies (Figure 9.19). In the Rutherford experiment, the interaction did not raise the gold nucleus nor the alpha particle to an excited state; both nuclei remained in their ground states. This happy accident made it much easier for Rutherford to analyze the experiment.

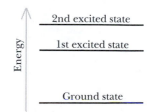

2nd excited state

1st excited state

Energy

Ground state

Figure 9.19 Energy levels in a nucleus. The potential energy curve is not shown.

Electric potential energy in the Rutherford experiment

What about the electric potential energy associated with the electric forces? Long before the collision the two nuclei were far from each other, and long after the collision they are again far from each other. Like gravitational energy, electric energy is proportional to $1/r$, where r is the distance between the two charged objects. Therefore from long before the collision to long

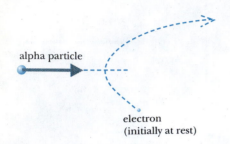

Figure 9.20 The alpha particle sweeps away the low-mass electrons and plows straight on ("bowling ball hits ping-pong ball").

alpha particle

electron
(initially at rest)

after the collision the change in electric energy would be very nearly zero ($\Delta U_{el} \approx 0$) if the alpha particle and gold nucleus are completely isolated from other particles. However, what about all the electrons surrounding a gold nucleus?

As the alpha particle plows through the gold foil it knocks low-mass electrons away (Figure 9.20), much as a moving bowling ball knocks ping-pong balls away (the alpha particle does of course lose a little energy in each such interaction). Therefore we can start our analysis when the alpha particle is deep inside the atom, with the electrons of the gold atom mostly far outside the region of interest. An atomic diameter is about 10^{-10} m, whereas the diameter of a gold nucleus is much smaller, less than 10^{-14} m, so we might start our analysis of the collision when the particles are about 10^{-12} m apart, a distance that is very small compared to an atom and very large compared to a nucleus. At this distance the electric potential energy is

$$U_{el} = \frac{1}{4\pi\varepsilon_0} \frac{(2e)(79e)}{r} = \left(9\times10^9 \frac{\text{N}\cdot\text{m}^2}{\text{C}^2}\right) \frac{(2)(79)(1.6\times10^{-19}\text{ C})^2}{(10^{-12}\text{ m})}$$

$$U_{el} = (3.6\times10^{-14}\text{ J}) \frac{(1\text{ eV})}{(1.6\times10^{-19}\text{ J})} = 0.2\times10^6\text{ eV} = 0.2\text{ MeV}$$

The radioactive source used by the Rutherford group provided alpha particles whose kinetic energy was 10 MeV, so if we take as the initial state a separation of 10^{-12} m the electric energy is only 2% of the kinetic energy. And if we also take as the final state a separation of 10^{-12} m we have $\Delta U_{el} = 0$. Evidently the only significant energy changes in this experiment are changes in the kinetic energies of the alpha particle and the gold nucleus.

The gold nuclei in the gold foil are about 2×10^{-10} m apart, the approximate diameter of a typical atom in a metal, so when an alpha particle is 10^{-12} m from a gold nucleus, it is comparatively very far from any other gold nucleus. This explains why Rutherford could analyze the experiment as though isolated alpha particles scattered off isolated gold nuclei; electrons and other gold nuclei were far away.

The plum pudding model of an atom

For about a decade, the prevailing model of an atom had been what was called a "plum pudding" model proposed by J. J. Thomson. Electrons were discovered in 1897 by Thomson, using a cathode ray tube, a device similar to a television picture tube. Thomson was able to show that the "cathode rays" that he and others had observed in electric gas discharges were in fact particles that were negatively charged, by the already existing convention for labeling electrical charges. This was the first conclusive demonstration of the existence of what we now call "electrons," one of the major constituents of all atoms.

Since these "electrons" were extracted from ordinary matter, which is normally not electrically charged, Thomson's discovery implied that matter also contained something having a positive charge. Because the electrons apparently had little mass, it seemed likely that most of the mass in ordinary matter was charged positively. Thomson proposed what was called a "plum pudding" model of atoms: a positively charged substance of uniform, rather low density, with tiny electrons distributed like raisins throughout it. Rutherford's group expected to gather evidence supporting this model from their collision experiments. Rutherford was extremely surprised by the results of these experiments, which led to a radical revision of the model.

The Rutherford experiment

A new source of fast-moving particles was provided through the discoveries of natural radioactivity by the French scientists Henri Becquerel and Marie and Pierre Curie. These newly discovered particles included high-speed, massive, positively charged "alpha particles," which we now know to be the nuclei of helium atoms, consisting of two protons and two neutrons bound together. Alpha particles are emitted in the radioactive decay of heavy elements. Alpha particles were convenient microscopic projectiles, and Rutherford's group decided to shoot alpha particles at a very thin piece of gold foil and observe where they reappeared.

? Based on the plum pudding model, what did Rutherford and his colleagues expect to see in their collision experiments?

The researchers expected the alpha particles to be deflected only very slightly by interactions with the low-mass electrons and with the low-density "pudding" of positively charged matter in the gold atoms. They could detect a deflected ("scattered") alpha particle by a pulse of light that was emitted when the particle struck a fluorescent material (zinc sulfide). They moved their detector around at various angles to the incoming beam of alpha particles and measured what fraction of the alpha particles were scattered through various angles (Figure 9.21).

Rutherford and his coworkers were astounded to observe that an alpha particle, which is much more massive than an electron, sometimes bounced straight backwards, as though it had encountered an even more massive yet very tiny positive particle inside the gold foil, rather than being slightly deflected in its passage through the "pudding" of positive charge. In other words, the experimenters thought they were throwing bowling balls at ping-pong balls, and they were amazed to see what they thought were bowling balls come bouncing back.

"It was quite the most incredible event that has ever happened to me in my life," Rutherford said. "It was almost as incredible as if you fired a 15-inch [artillery] shell at a piece of tissue paper and it came back and hit you."

He continued, "On consideration, I realized that this scattering backward must be the result of a single collision, and when I made the calculations I saw that it was impossible to get anything of that order of magnitude unless you took a system in which the greater part of the mass of the atom was concentrated in a minute nucleus. It was then that I had the idea of an atom with a minute massive center, carrying a charge." (See Figure 9.22.)

Computer modeling of the Rutherford experiment

We can make a computational model of the Rutherford experiment; see Problem 9.P.16. Remember that Coulomb's law for electric forces is very similar to Newton's law of gravitation. The electric force $\vec{F}_{el}$ between two small charged particles carrying amounts of charge q_1 and q_2, and located a distance r apart is

$$\vec{F}_{el} = \frac{1}{4\pi\varepsilon_0}\frac{q_1 q_2}{r^2}\hat{r}$$

The charges q_1 and q_2 are measured in units of coulombs, and the proportionality factor is measured to be $1/(4\pi\varepsilon_0) = 9\times10^9\,\mathrm{N\cdot m^2/coulomb^2}$. The electric charge of the proton is $e = +1.6\times10^{-19}$ coulomb, and the electric charge of the electron is $-e = -1.6\times10^{-19}$ coulomb. In the Rutherford experiment the charge on the alpha particle is $2e$ (2 protons and 2 neutrons), and the charge on the gold nucleus is $79e$ (79 protons and 118 neutrons).

9.7 *DISTRIBUTION OF SCATTERING ANGLES

In general, we have three equations describing an elastic collision—two momentum equations (in the x and y directions) and one energy equation, but there are four unknown quantities (the outgoing speeds and directions for

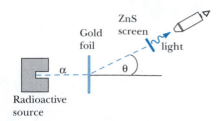

Figure 9.21 Schematic diagram of the Rutherford experiment. A radioactive source emits alpha particles, which are directed at a thin piece of gold foil. Particles scattered by gold nuclei in the gold foil hit a screen coated with ZnS, which fluoresces, giving off light, when hit by an energetic particle. The light was observed by the experimenter.

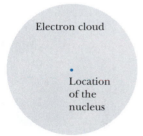

Figure 9.22 This diagram of an atom is not to scale; the radius of the nucleus is around 1×10^{-15} m, which is only $1/100000$ the radius of the electron cloud (about 1×10^{-10} m), so the nucleus would not be visible at all on this scale.

each particle). The concept of impact parameter provides an additional quantity, and you have seen on page 314 how the impact parameter determines the scattering angle θ of the incoming particle, from which you can predict the other final values (p_3, p_4, and φ).

In a macroscopic situation, such as shooting billiard balls, you can choose an impact parameter simply by carefully aiming at a point a distance b from the center of the target. Unfortunately, in experiments where you fire alpha particles or other subatomic projectiles at nuclei, you can't aim at a specific impact parameter because the target is so incredibly small. All you can do is fire lots of projectiles at the subatomic target and expect a probabilistic distribution of impact parameters

When we fire a projectile at this microscopic target, the probability of hitting one particular ring (whose radius corresponds to the impact parameter) is proportional to the area of that ring (Figure 9.23). If you calculate the range of scattering angles that corresponds to hitting somewhere within a particular ring, you can predict the angular distribution of scattering angles that will be observed.

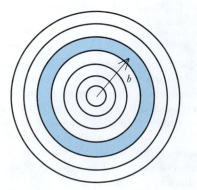

Figure 9.23 The probability of hitting a particular ring is proportional to the area of the ring.

*Cross section

If a target particle such as a gold nucleus has a radius R, we say that its "geometrical cross section" is πR^2. A high-speed alpha particle that hits within this cross-sectional area will interact strongly with the gold nucleus. Quantum-mechanical effects can make the effective target size bigger or smaller than the geometrical cross section, and the effective target size is called the "cross section" for interactions of this kind. Cross sections are measured experimentally by observing how often projectiles are deflected by an interaction, instead of passing straight through the material without interacting. Cross sections have units of square meters.

A "differential" cross section for scattering within a particular range of angles is the area in square meters of a ring bounded by impact parameters corresponding to the given angular range.

*Conservation laws *vs.* details of the interaction

We used general principles to determine the final momenta of two colliding particles (conservation of momentum, conservation of energy). We didn't have to say what kind of force was involved. But doesn't it make a difference whether the interaction is electric or nuclear? Yes, the type of interaction determines the *distribution* of scattering angles. For example, in what fraction of the collisions is the scattering angle between 0 and 10 degrees? Between 10 and 20 degrees? Between 20 and 30 degrees? These fractions are different for different forces.

To put it another way, if the scattering angle is 20 degrees, the momenta are determined. The question is, how often is the scattering angle in fact near 20 degrees? Measurement of the distribution of scattering angles in collisions at modern particle accelerators is one of the ways used to study the nature of the strong and weak interactions.

Rutherford used Coulomb's law of electric force to make a detailed prediction of the distribution of scattering angles, assuming that the positive nucleus was point-like and very massive. That is, he predicted what fraction of the alpha particles would be scattered by less than 10 degrees, by an angle between 10 and 20 degrees, by an angle between 20 and 30 degrees, etc. The observed distribution of scattering angles agreed with the predicted distribution (Figure 9.24 shows a portion of this distribution).

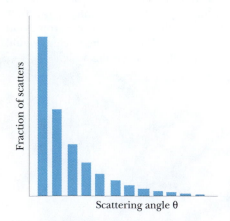

Figure 9.24 For an electric interaction, the fraction of scatters at different scattering angles falls rapidly with angle.

The calculation that Rutherford did involves a lot of complicated geometry, algebra, and probability (with a random distribution of impact parameters), so we won't go through it. But the main features of Figure 9.24 are easily understood. Because the gold nucleus is very small, most of the time the alpha particle will miss the nucleus by a large distance. With a large impact parameter the electric force is very small, and the deflection (scattering angle) is very small. For that reason small scattering angles are common. Large scattering angles occur only for nearly head-on collisions (small impact parameter), and these events are rare.

Since Rutherford's prediction assumed a point-like nucleus, the observations were consistent with the nucleus being very small. Also, the good fit between prediction and observation strengthened the belief that the interaction was purely electric in nature (the strong interaction played no role because it is a very short-range interaction, and the alpha particle was not going fast enough to overcome the electric forces and contact the gold nucleus).

If the alpha particle has high enough energy, it can overcome the electric repulsion and make contact with the gold nucleus. In that case the strong interaction comes into play, and the angular distribution changes. The radius of the gold nucleus can be determined by finding out how much energy an alpha particle has to have in order to obtain an angular distribution that deviates from what is expected due just to the electric interaction.

? For a particular scattering angle, can we use momentum and energy conservation to determine the unknown angle and momenta, even if the alpha particle makes contact with the gold nucleus?

Yes. These conservation principles are completely general and apply to all types of interactions. But they cannot by themselves predict the angular distribution—that is, how often a particular range of scattering angles will occur. To do that you have to have a mathematical description for the specific interaction, such as the $1/r^2$ force law for electric interactions.

9.8 RELATIVISTIC MOMENTUM AND ENERGY

Rutherford's original experiments were carried out with alpha particles whose kinetic energy was about 10 MeV (10 million electron-volts), with a corresponding speed that was small compared to the speed of light. In order to probe the nucleus more deeply, physicists later built special machines, called "particle accelerators," which use electric forces to accelerate charged particles to very much higher energies, with speeds that may be as much as 99.9999% of the speed of light. For such high speeds we cannot use the low-speed approximate formulas for momentum and energy.

Physicists analyzing collisions produced by high-energy accelerators routinely do calculations based on the relativistic energy and momentum formulas. Consider a collision in which a particle of mass m_1 and magnitude of momentum p_1 strikes a stationary particle of mass m_2 (Figure 9.25). There is a change of identity, with the production of two new particles of mass m_3 and m_4, with momenta p_3 and p_4, at angles θ and ϕ to the horizontal (Figure 9.26).

We will write down relativistically correct equations for momentum conservation and energy conservation, in terms of momentum p and energy E, knowing that we can determine E from p, since $E^2 - (pc)^2 = (mc^2)^2$:

$p_1 = p_3 \cos\theta + p_4 \cos\phi$ x-momentum conservation

$0 = p_3 \cos(90° - \theta) - p_4 \cos(90° - \phi)$ y-momentum conservation

$E_1 + m_2 c^2 = E_3 + E_4$ energy conservation (scalar)

The energy equation is equivalent to

$$\sqrt{(p_1 c)^2 + (m_1 c^2)^2} + m_2 c^2 = \sqrt{(p_3 c)^2 + (m_3 c^2)^2} + \sqrt{(p_4 c)^2 + (m_4 c^2)^2}$$

Figure 9.25 Before a high-energy collision.

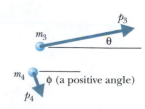

Figure 9.26 After the collision.

? Suppose the initial momentum p_1 of the projectile is known, and all four masses are known (because we were able to identify the particles). How many unknown quantities are there, considering that we can write the energy equation in terms of p?

There are four unknown quantities, the magnitudes of the momenta and the directions of the two outgoing particles (p_3, p_4, θ, and ϕ), but only three equations. A typical situation in a high-energy particle experiment is that one of these unknown quantities is measured, and momentum and energy conservation are used to calculate the other three quantities.

As in the Rutherford experiment, measurements of the outgoing particle momenta alone do not tell us anything about the detailed nature of the interaction involved in the collision, because they are determined by general momentum and energy conservation laws. The identity changes of the particles tell us a lot, and the distribution of scattering angles also tells us a lot.

Photons and neutrinos

If a particle has zero mass (the photon and to a very good approximation the neutrino), the relationship $E^2 - (pc)^2 = (mc^2)^2$ reduces simply to $E = pc$. For the photon we can't write its energy as $mc^2/\sqrt{1 - v^2/c^2}$, because both the numerator and denominator are zero, and $0/0$ is indeterminate. Nevertheless, $E = pc$ is valid: a photon of energy E carries momentum $p = E/c$.

Identifying a particle

A variation on the relativistic collision analysis can be used to identify one of the outgoing particles. Suppose we measure p_3 and θ, and we know m_3 but not m_4. We can use our three equations in the three unknowns (m_4, p_4, and ϕ) to solve for the unknown mass m_4. This is a very powerful technique when we are unable to observe the particle directly. For example, neutrinos don't have electromagnetic or strong interactions and therefore typically fail to register in ordinary detectors. But if we measure the other outgoing particle we can determine the properties of the unseen particle.

The discovery of the neutrino

A specific example of such analyses was the original motivation for proposing the existence of the neutrino. The neutron (either free or when bound into some nuclei) was observed to decay with the creation of a proton and an electron, both charged particles that are easy to detect. But the amount of energy and momentum carried by the proton and electron varied from one decay to the next, which made no sense:

$$n \rightarrow p^+ + e^-,\text{ violating momentum and energy conservation?}$$

It was proposed that there was an unseen third particle, called the neutrino (or antineutrino in neutron decay), which carried energy and momentum that would salvage the conservation laws:

$$n \rightarrow p^+ + e^- + \bar{\nu},\text{ conserving momentum and energy}$$

It was possible to solve for the properties of the unseen particle by the techniques outlined above, and it was found that the neutrino must have nearly zero mass and must travel at nearly the speed of light. (Recent experiments show that the neutrino has a very small nonzero mass and must travel at slightly less than the speed of light.)

For many years the neutrino was just a theoretical construct which balanced the accounts, but eventually the neutrino was detected directly. The neutrino is very hard to detect because it does not have electromagnetic or strong interactions, only weak interactions, so the most probable thing for it to do is to pass right through any detector you place in its way. In fact, neutrinos emitted in nuclear reactions in the Sun mostly pass right through the entire Earth! However, with extremely intense neutrino beams produced in nuclear reactors or accelerators, or with gigantic underground detectors used to identify neutrinos coming from the Sun, it is possible to have a measurable rate of neutrino reactions.

Example: Photodissociation of the deuteron

A photon of energy E is absorbed by a stationary deuteron. As a result, the deuteron breaks up into a proton and a neutron. The photon energy is sufficiently low that the proton and neutron have speeds small compared to the speed of light, but there is a change of rest mass. The proton is observed to move in a direction at an angle θ to the direction of the incoming photon.

Write equations that could be solved to determine the unknown magnitudes of the proton momentum p_p and neutron momentum p_n, and the positive angle α of the neutron to the direction of the incoming photon. Let the precise masses of the particles be M_d, M_p, and M_n for the deuteron, proton, and neutron. Explain clearly on what principles your equations are based. Provide a physics diagram of the reaction and label the angles. Do not attempt to solve the equations.

> System: all of the particles (Figure 9.27)
> Surroundings: Nothing significant
>
> Momentum conservation (Figure 9.27): $\vec{p}_{\text{photon}} = \vec{p}_p + \vec{p}_n$
>
> A photon with energy E has momentum $p = E/c$:
>
> $$\left\langle \frac{E}{c}, 0, 0 \right\rangle = \langle p_p\cos\theta + p_n\cos\alpha,\ p_p\cos(90° - \theta) - p_n\cos(90° - \alpha),\ 0 \rangle$$
>
> This is equivalent to two component equations:
>
> $$\frac{E}{c} = p_p\cos\theta + p_n\cos\alpha$$
>
> $$0 = p_p\cos(90° - \theta) - p_n\cos(90° - \alpha)$$
>
> Energy conservation (small proton and neutron speeds):
>
> $$E + M_d c^2 = \left(M_p c^2 + \frac{p_p^2}{2M_p} \right) + \left(M_n c^2 + \frac{p_n^2}{2M_n} \right)$$
>
> We have three equations (two momentum component equations and one energy equation) in three unknowns: p_p, p_n, and α. So in principle we can solve for the unknown quantities, though the algebra is messy.

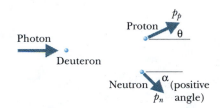

Figure 9.27 Physics diagram for photodissociation of deuteron.

We could also have written the energy equation relativistically, using the relationship $E^2 - (pc)^2 = (mc^2)^2$ which gives $E = \sqrt{(pc)^2 + (mc^2)^2}$:

$$E + M_d c^2 = \sqrt{(p_p c)^2 + (M_p c^2)^2} + \sqrt{(p_n c)^2 + (M_n c^2)^2}$$

9.9 INELASTIC COLLISIONS AND QUANTIZED ENERGY

The Rutherford experiment involved elastic collisions, because the internal energies of the gold nucleus and the alpha particle didn't change during the interaction. When there is an internal energy change, or the production of additional particles or the emission of light, we call the collision "inelastic."

2nd excited state

1st excited state

ΔE

Ground state

Figure 9.28 Energy levels for a gold nucleus. A sufficiently energetic alpha particle may have enough energy to raise the nucleus above its ground state.

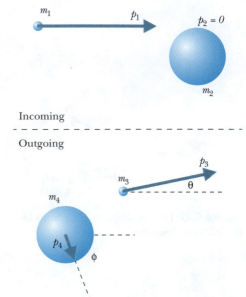

Incoming

Outgoing

Figure 9.29 You measure the incoming and outgoing momenta of the alpha particle and deduce something about the energy levels in the gold nucleus.

If we shoot alpha particles at a gold nucleus with higher energy than the energy of alpha particles in the Rutherford experiment, we may be able to excite the gold nucleus to a higher quantum energy level, an amount ΔE above the ground state of the gold nucleus (Figure 9.28). Such a state corresponds to a different configuration of the nucleons inside the gold nucleus.

Suppose that you know the incoming momentum p_1 of the alpha particle, and you are able to measure its outgoing momentum p_3 and scattering angle θ (Figure 9.29).

? What is the form of the energy-conservation and momentum-conservation equations in this case? Don't assume that the speeds are small compared to the speed of light. Note that the rest energy of the recoiling gold nucleus is $M_4c^2 + \Delta E$ because it is in an excited state.

Here are the relevant equations:

$$p_1 = p_3\cos\theta + p_4\cos\phi,$$
where p_4 and ϕ refer to the gold nucleus, which is not observed in the experiment.

$$0 = p_3\cos(90° - \theta) - p_4\cos(90° - \phi)$$

$$E_1 + Mc^2 = E_3 + E_4, \text{ where } E_4 = \sqrt{(p_4c)^2 + (M_4c^2 + \Delta E)^2}$$

? How many unknown quantities are there in the energy and momentum equations? Is there enough information to be able to determine the internal energy difference ΔE? (This is one of the methods used to study the energy levels of nuclei. This energy level determination can be verified by measuring the energy of a gamma-ray photon emitted when the nucleus drops back to the ground state.)

There are three unknowns (p_4, ϕ, and ΔE), and there are three equations (energy, x and y momentum), so we could determine ΔE.

Change of identity

Instead of merely exciting the target particle to a higher internal energy state, suppose that there is a change of identity in which the target particle turns into some other particle. An example is the following particle reaction, which was used in the 1950's to discover the Δ^+ particle of mass m_Δ, which is a particle more massive than a proton:

$$\pi^- + p^+ \to \pi^- + \Delta^+$$

A beam of negative pions (π^-) of known mass m_π and known kinetic energy, produced by a particle accelerator, is directed at a container of hydrogen gas, and the targets for the collisions are the protons of mass m_p that are the nuclei of the hydrogen atoms. Sometimes an incoming pion interacts through the strong interaction with a proton to produce a Δ^+ particle. The momentum of the scattered pion can be measured by measuring the radius of curvature of its curving trajectory in a magnetic field.

? What is the form of the energy-conservation and momentum-conservation equations in this case? (Don't assume that the speeds are small compared to the speed of light.)

Here are the relativistically correct equations:

$$p_1 = p_3\cos\theta + p_4\cos\phi$$

$$0 = p_3\cos(90° - \theta) - p_4\cos(90° - \phi)$$

$$E_1 + m_pc^2 = E_\pi + E_\Delta, \text{ where } E^2 - (pc)^2 = (mc^2)^2$$

? How many unknown quantities are there in the energy and momentum equations? Is there enough information to be able to determine the mass of the Δ^+ particle?

There are three unknowns (p_4, ϕ, and E_Δ) and three equations, so it is possible to determine the mass of the Δ^+ particle.

What we learn from inelastic collisions

The analysis for finding the mass change in the preceding exercises is essentially the same as the analysis for determining the internal energy change of an excited gold nucleus. In fact, we can say that the Δ^+ particle is an excited state of three quarks, with the proton representing the ground state of three quarks (Figure 9.30). The excited state (the Δ^+ particle) can decay to the ground state (the proton) with the emission of a neutral pion (π^0), which is a quark-antiquark pair. The extra kinetic energy is equal to $(m_\Delta c^2 - m_p c^2)$. Scattering experiments have been an important tool for discovering new particles (or new excited states of quark systems).

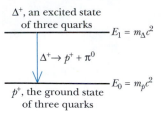

Figure 9.30 The Δ^+ particle can be considered to be an excited state of three quarks, with the proton representing the ground state. A decay to the ground state can occur with the emission of a neutral pion (π^0).

A sticking collision

To illustrate the wide applicability of the Momentum Principle, we'll consider an inelastic collision in which two objects stick together when they make contact. A macroscopic example is the collision of a truck and a car in an icy intersection (Figure 9.31). In the figure we have shown velocity vectors rather than momentum vectors; a typical truck would have much more momentum than the car, given the truck's much greater mass.

Let M be the mass of the truck, m the mass of the car, and θ the angle of their mutual final velocity, as shown in the diagram. Also let v_1 be the initial speed of the truck, v_2 the initial speed of the car, and v_3 the final speed when they're stuck together and sliding on the ice.

? How does the fact that the intersection is icy affect our ability to apply conservation of momentum to the truck-car collision?

Friction with the road is small if the intersection is icy, so we can presumably ignore road friction and air resistance during the short time of the actual collision. Therefore the total momentum of the truck+car system should be (approximately) unchanged. Momentum conservation is really three equations, one each for the x, y, and z directions. In the z direction (vertically upward) the ice simply supports the weights of the vehicles, so the net force is zero, and the zero z component of momentum remains zero throughout the collision.

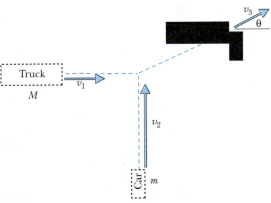

Figure 9.31 A sticking collision between a car and a truck on an icy road.

? In terms of the named quantities, write two equations for the x and y components of the momentum conservation equation, equating momentum components before the collision to momentum components after the vehicles stick together.

Here are the x and y components of the momentum conservation:

$$Mv_1 = (M+m)v_3\cos\theta$$
$$mv_2 = (M+m)v_3\cos(90° - \theta), \text{ where } \cos(90° - \theta) = \sin\theta$$

Assume that we know the initial speeds of the truck and car (v_1 and v_2). Then we have two equations in two unknowns (v_3 and θ), and we can solve for the final speed and angle. First, divide the y equation by the x equation:

$$\frac{(M+m)v_3\sin\theta}{(M+m)v_3\cos\theta} = \frac{mv_2}{Mv_1} = \tan\theta, \text{ so } \theta = \arctan\left(\frac{mv_2}{Mv_1}\right)$$

Next square the equations and add them:

$$(Mv_1)^2 + (mv_2)^2 = [(M+m)v_3]^2[(\sin\theta)^2 + (\cos\theta)^2]$$

$$\sqrt{(Mv_1)^2 + (mv_2)^2} = (M+m)v_3$$

$$v_3 = \frac{\sqrt{M^2 v_1^2 + m^2 v_2^2}}{M+m}$$

We used the trig identity $\sin^2\theta + \cos^2\theta = 1$, which corresponds to the Pythagorean theorem applied to a triangle whose hypotenuse has length 1.

Energy in a sticking collision

A sticking collision is unusual in that momentum conservation alone gives us enough equations (two) to be able to solve for the unknown final speed and unknown direction. If the car and truck bounce off each other instead of sticking, there are four unknowns instead of two, because there are two final velocities, each with a speed and a direction to be calculated. In that case momentum conservation alone doesn't provide enough equations to be able to analyze the situation fully.

There is an important issue associated with energy in a sticking collision such as this one. Using our result for the final speed, we can write the kinetic energy after the collision:

$$\frac{1}{2}(M+m)v_3^2 = \frac{1}{2}(M+m)\left[\frac{(Mv_1)^2 + (mv_2)^2}{(M+m)^2}\right]$$

After minor rearrangements we get the following:

$$\frac{1}{2}(M+m)v_3^2 = \left(\frac{M}{M+m}\right)\left(\frac{1}{2}Mv_1^2\right) + \left(\frac{m}{M+m}\right)\left(\frac{1}{2}mv_2^2\right)$$

Since $M/(M+m)$ is less than 1, and $m/(M+m)$ is also less than 1, the final kinetic energy is less than the initial kinetic energy.

? Where is the "missing" kinetic energy?

Part of the energy was radiated away as sound waves in the air, produced when the vehicles collided with a bang. Some of the energy went into raising the temperature of the vehicles, so that their thermal energy is now higher than before (larger random motions of the atoms).

Some of the energy went into deforming the metal bodies of the truck and car. A bent fender has a higher internal energy, associated with a change in the configuration of the atoms. Sometimes you can get such energy back (as when you compress a spring, changing the configuration of the atoms in the spring, and later let it push back, returning to its original configuration), and we speak of "elastic" energy. But sometimes you can't get the energy back, because the configuration change gets "locked in."

A sticking collision is an example of an "inelastic" collision, referring to the effective "loss" of kinetic energy into forms that are inaccessible.

9.X.5 Two cars collide, one headed north and one headed west. They stick together and leave skid marks on the pavement, which show that car 1 was deflected 30° (so car 2 was deflected 60°. What can you conclude about the cars before the collision?

9.10 COLLISIONS IN OTHER REFERENCE FRAMES

We can show that a truck-car collision process looks a lot simpler from the vantage point of a helicopter flying over the scene with the final velocity of the truck+car combined system. This will lead to new insights. Suppose the accident takes place at night, and from the helicopter all you can see are lights mounted on top of the two vehicles. Figure 9.32 shows a multiple exposure of the scene that you would see from a helicopter that is hovering over the intersection.

Suppose instead that your helicopter flies over the intersection with the same velocity that the stuck-together truck and car share after their collision, as shown in Figure 9.33. The view thus obtained is in an important sense "simpler," as shown in Figure 9.34. From the moving observation platform you see the truck and car run into each other and then stand still (since you and they now have the same velocity). In particular, in this (moving) reference frame it is obvious that there is a loss of kinetic energy into other forms of energy, because after the collision the kinetic energy is clearly zero. In contrast, we had to go through a fair amount of algebra in the non-moving reference frame in order to establish the loss of kinetic energy.

The reason that we imagine making these observations at night is so that you won't be distracted by the moving background. If you actually view this collision from a moving helicopter in daylight, there is a strong tendency to compensate mentally for your motion and to put yourself (mentally) into the reference frame of the road, in which case you could miss the important fact that *relative to you*, the car and truck are stationary after the collision.

You may have had an experience related to this situation. You're sitting in a bus or train, waiting to leave. You think you've started forward, but then you realize that you were fooled by a neighboring bus or train moving backwards. For a moment you were mentally in the reference frame of the moving vehicle.

Total momentum is zero in the center-of-mass frame

Viewing a collision from a moving reference frame can provide insights into all kinds of collisions, not just sticking collisions. The key point that leads to a useful generalization is that the total momentum of the truck+car system after their collision is zero when the vehicles are motionless relative to your helicopter.

? In this moving reference frame, if the total momentum is zero after the collision, what was the total momentum before the collision? Why?

The helicopter is moving at constant velocity (constant speed and direction), and we know that the Momentum Principle is valid in this uniformly moving reference frame. So the principle of momentum conservation still holds, and the initial total momentum of the car+truck system must be zero if the final total momentum is zero. That's what makes the process look so simple: the total momentum is zero before and after the collision in the moving reference frame.

In viewing the scene from the moving helicopter, we have subtracted from each of the velocities an unchanging, constant vector velocity, the velocity $\vec{v}_h$ of the helicopter, and this subtraction makes no change in the Momentum Principle for the car (or truck):

$$\frac{d[\,m(\vec{v}-\vec{v}_h)\,]}{dt} = \frac{d[\,m\vec{v}\,]}{dt} = \vec{F}_{\text{net}}$$

The net force is the same in both reference frames because the contact electric interaction forces depend on interatomic distances, which are the same in both reference frames.

Evidently we gain insight into a collision by viewing the collision in a moving reference frame where the total momentum is zero. How do we determine the appropriate velocity for such a reference frame? Let this velocity be $\vec{v}_{\text{fr}}$, which is to be subtracted from all the original velocities to obtain the

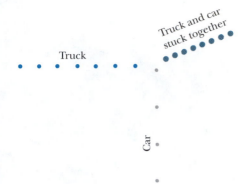

Figure 9.32 View of collision from hovering (stationary) helicopter.

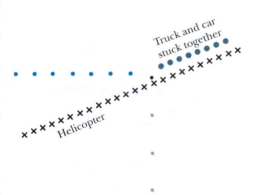

Figure 9.33 Path of moving helicopter that produces the simple view seen in Figure 9.34.

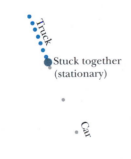

Figure 9.34 View of collision from helicopter moving with velocity of stuck-together wreck.

velocities as viewed in the moving reference frame. We want the total momentum to add up to zero in the moving frame:

$$\vec{P}_{\text{tot,new frame}} = m_1(\vec{v}_1 - \vec{v}_{\text{fr}}) + m_2(\vec{v}_2 - \vec{v}_{\text{fr}}) = 0$$

$$(m_1 + m_2)\vec{v}_{\text{fr}} = m_1\vec{v}_1 + m_2\vec{v}_2$$

$$\vec{v}_{\text{fr}} = \frac{m_1\vec{v}_1 + m_2\vec{v}_2}{m_1 + m_2}$$

We have solved for a particular velocity for the moving reference frame that makes the total momentum in that frame be zero. But look at the form of this equation.

? Does this equation look familiar?

Evidently $\vec{v}_{\text{fr}} = \vec{v}_{\text{cm}}$, the velocity of the center of mass of the system, since

$$\vec{r}_{\text{cm}} = \frac{m_1\vec{r}_1 + m_2\vec{r}_2}{m_1 + m_2}$$

We have obtained an important result. In a reference frame moving with the center of mass of a system, the total momentum of the system is zero. If no external forces act, the total momentum is unchanged, and the total momentum is zero before and after a collision.

We can also express the frame velocity directly in terms of the total momentum:

$$\vec{v}_{\text{cm}} = \frac{m_1\vec{v}_1 + m_2\vec{v}_2}{m_1 + m_2} = \frac{\vec{P}_{\text{tot}}}{M_{\text{tot}}}$$

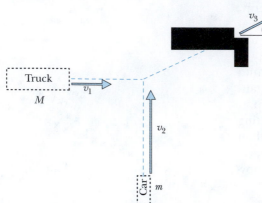

Figure 9.35 Collision of car and truck.

This provides another way to calculate the frame velocity and shows why this is often called the "center-of-momentum" velocity.

Let's reanalyze the car-truck collision from this new point of view. Again let M be the mass of the truck, m the mass of the car, and θ the angle of their mutual final velocity, as shown in Figure 9.35. Also let v_1 be the initial speed of the truck, v_2 the initial speed of the car, and v_3 the final speed when they're stuck together and sliding on the ice.

? What are the x and y components of $\vec{v}_{\text{cm}}$ in terms of v_1 and v_2?

$$\vec{v}_{\text{cm}} = \langle \frac{Mv_1}{M + m}, \frac{mv_2}{M + m}, 0 \rangle$$

? In the moving center-of-momentum reference frame, what are the x and y components of the car's velocity before the collision? Remember that you find the car's velocity (or its components) in the center-of-momentum frame by subtracting the center-of-momentum velocity (or its components) from that of the car. (That is, the view from the helicopter is obtained by subtracting the helicopter's velocity from the velocities of the car and truck.)

$$\langle -\frac{Mv_1}{M + m}, v_2 - \frac{mv_2}{M + m}, 0 \rangle = \langle -\frac{Mv_1}{M + m}, \frac{Mv_2}{M + m}, 0 \rangle$$

? In the moving center-of-momentum reference frame, what are the x and y components of the truck's velocity before the collision? Show that this velocity is in a direction opposite to the velocity of the car, and that the total momentum is zero.

$$\langle v_1 - \frac{Mv_1}{M + m}, -\frac{mv_2}{M + m}, 0 \rangle = \langle \frac{mv_1}{M + m}, -\frac{mv_2}{M + m}, 0 \rangle$$

The *x* and *y* components of velocity are in the opposite direction to those of the car (multiply the car's velocity by $-m/M$ and you have the truck's velocity). Also, the total momentum is zero in the center-of-momentum frame:

$$\left\langle \left[m\left(-\frac{Mv_1}{M+m}\right) + M\left(\frac{mv_1}{M+m}\right) \right], \left[m\left(\frac{Mv_2}{M+m}\right) + M\left(-\frac{mv_2}{M+m}\right) \right], 0 \right\rangle = 0$$

? In the moving center-of-momentum reference frame, what is the speed of the car after the collision? Of the truck? Why?

Both are zero; in the center-of-mass frame the total momentum is zero, and with the car and truck stuck together the only way for this to be true is for them to be at rest. Because of this, the kinetic energy loss to other kinds of energy is equal to the total kinetic energy in the center-of-momentum frame before the collision.

? Now, after the collision, transform back to the (non-moving) reference frame of the intersection, by adding back in the $\vec{v}_{cm}$ that had been subtracted. What are the speed and direction of the car after the collision? Of the truck? Compare with the results obtained in our earlier analysis of this collision.

$$\vec{v}_3 = \left\langle \frac{Mv_1}{M+m}, \frac{mv_2}{M+m}, 0 \right\rangle, \text{ which yields}$$

$$v_3 = \frac{\sqrt{M^2 v_1^2 + m^2 v_2^2}}{M+m}; \ \theta = \arctan\left(\frac{mv_2}{Mv_1}\right)$$

These are the same results we obtained in our earlier analysis.

Energy in the two reference frames

As far as energy is concerned, note that the kinetic energy of the car+truck system is as usual

$$K_{tot} = K_{trans} + K_{rel}, \text{ where } K_{trans} = \frac{P_{tot}^2}{2M_{tot}}$$

In the original reference frame, $K_{trans} > 0$ but is the same before and after the collision, since P_{tot} doesn't change. In the center-of-momentum reference frame, $K_{trans} = 0$ both before and after the collision, because $P_{tot} = 0$ in the center-of-momentum reference frame.

In either reference frame, the kinetic energy relative to the center of mass, K_{rel}, decreases to zero in the collision, because the car and truck after the collision are traveling with the center of mass. It is ΔK_{rel} that represents the loss of kinetic energy in this inelastic collision, and this quantity is the same in both reference frames.

Also note that distances between atoms are the same in both reference frames, so potential energy associated with bending of metal parts is the same in both frames.

9.X.6 A 1000 kg car moving east at 30 m/s runs head-on into a 3000 kg truck moving west at 20 m/s. The vehicles stick together. Use the concept of the center-of-momentum frame to determine how much kinetic energy is lost.

9.11 SUMMARY

New concepts

Elastic and inelastic collisions
Scattering, and scattering distributions
Impact parameter (0 if head-on)
The center-of-mass reference frame

$$\vec{v}_{cm} = \frac{m_1 \vec{v}_1 + m_2 \vec{v}_2}{m_1 + m_2} = \frac{\vec{P}_{tot}}{M_{tot}}$$

Simple results for head-on elastic collisions between ping-pong balls and bowling balls

Evidence for the nuclear model of an atom

Problem solving techniques

Conservation of momentum and conservation of energy applied to collisions

Simplification of calculations in the center-of-mass reference frame

9.12 REVIEW QUESTIONS

Momentum and impulse

9.RQ.7 A force $\langle 30, 0, 0 \rangle$ N acts for 0.2 seconds on an object of mass 1.2 kg whose initial velocity was $\langle 0, 20, 0 \rangle$ m/s. What is the new velocity?

Conservation of momentum

9.RQ.8 Under what conditions is the momentum of a system constant? Can the *x* component of momentum be constant even if the *y* component is changing? In what circumstances? Give an example of such behavior.

Collisions

9.RQ.9 What happens to the velocities of the two objects when a high-mass object hits a low-mass object head-on? When a low-mass object hits a high-mass object head-on?

9.RQ.10 What properties of the alpha particle and the gold nucleus in the original Rutherford experiment were responsible for the collisions being elastic collisions?

9.RQ.11 Give an example of what we can learn about matter through the use of momentum and energy conservation applied to scattering experiments. Explain what it is that we cannot learn this way, for which we need to measure the distribution of scattering angles.

Sticking collisions

9.RQ.12 In an elastic collision involving known masses and initial momenta, how many unknown quantities are there after the collision? How many equations are there? In a sticking collision involving known masses and initial momenta, how many unknown quantities are there after the collision? Explain how you can determine the amount of kinetic energy change.

9.RQ.13 In order to close a door, you throw an object at the door. Which would be more effective in closing the door, a 50 gram tennis ball or a 50 gram lump of sticky clay? Explain clearly what physics principles you used to draw your conclusion.

9.RQ.14 A bullet of mass *m* traveling horizontally at a very high speed *v* embeds itself in a block of mass *M* that is sitting at rest on a nearly frictionless surface.

(a) What is the speed of the block after the bullet embeds itself in the block?

(b) Calculate the total translational kinetic energy before and after the collision. Compare the two results and explain why there is a difference.

Changing reference frame

9.RQ.15 What is it about analyzing collisions in the center-of-mass frame that simplifies the calculations?

9.13 PROBLEMS

9.P.16 The Rutherford experiment

Use numerical integration techniques to explore possible trajectories of an alpha particle when it collides with a gold nucleus at different impact parameters. Your program should display the paths of the alpha particle and of the gold nucleus, and also plot the *x* and *y* components of momentum of both particles as a function of time. We will consider motion only in two dimensions (a slice through a 3-D beam of alpha particles).

An alpha particle is a helium nucleus (two protons and two neutrons, charge +2e); in the Rutherford experiment its initial kinetic energy was about 10 MeV (remember that 1 eV = 1.6×10^{-19} J). The gold nucleus is inside a piece of gold foil and is initially stationary; its charge is +79e and it contains 197 nucleons. Initially we will ignore the interactions of the alpha particle with the cloud of electrons surrounding the nucleus, which are in fact small.

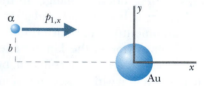

(a) In this model it is important to pick reasonable initial conditions and display scales—otherwise you may not see anything meaningful. Before beginning your program, calculate or estimate reasonable values for the following and report your results:

• *Closest approach.* For a head-on collision, calculate approximately the distance of closest approach, given that the alpha particle and the gold nucleus repel each other. Assume that the gold atom's electrons are scattered away by the incoming alpha particle.

• *Initial position of the alpha particle.* Given the distance of closest approach, what would be a reasonable initial *x* for the alpha particle? What would be a good range to explore for the impact parameter *b* (initial *y* coordinate of the alpha particle)? Use these estimates to decide on a scale for the coordinates used to display the trajectories

• *Value of Δt.* Trial and error alone will not be very helpful in estimating Δt. You can calculate the initial speed of the alpha particle, and you know the approximate size scale for the collision. Choose a value of Δt such that the alpha particle doesn't move very far in one step.

• *Time scale.* To plot p_x and p_y vs. t during the interaction, you must select a scale for the time axis that allows you to observe the entire interaction. About how long should the time axis be?

• *Momentum scale.* What should the maximum value of the p_x axis be? You should scale the p_y axis to the same value, to make comparisons easy.

• *Exit criterion.* You will need to exit from your computational loop when the particles have moved far apart. How will you test for this?

(b) Write a program that displays:

• The paths of the alpha particle and the gold nucleus.

• The value of the impact parameter b, and the value of the corresponding scattering angle (angle between the initial path of the alpha particle and its final path). The angle can be calculated by taking the arctangent of p_y/p_x, using the final values.

• A plot of p_x for each particle and total p_x, and a similar plot for p_y. (Consider what these plots should look like if the program is working properly.)

Report what value of Δt gives adequate accuracy. What is your criterion?

Report values of the impact parameter b that produce the following:

• A very small scattering angle
• An angle around 90 degrees
• A large deflection angle (back-scattering)

(c) What other physical quantity could you plot that would provide a check on the accuracy of your calculation? Carry out this check and report what value of Δt provides good accuracy according to this check.

(d) To see what a different model of matter would have predicted, replace the gold nucleus in your model by a proton. Describe briefly what you observe. (Note that if the protons were uniformly spread out throughout the volume of the atom, they would be far enough apart that on average an alpha particle would interact with only one of them.)

9.P.17 Deviations from Rutherford scattering

(a) A gold nucleus contains 197 nucleons packed tightly against each other. A single nucleon (proton or neutron) has a radius of about 10^{-15} m. Calculate the approximate radius of the gold nucleus.

(b) Rutherford correctly predicted the angular distribution for 10-MeV (kinetic energy) alpha particles colliding with gold nuclei. What kinetic energy must alpha particles have in order to make contact with a gold nucleus? (In this case the angular distribution will deviate from that predicted by Rutherford, which was based solely on electric interactions.)

9.P.18 The Rutherford experiment in the center-of-mass frame

Redo the analysis of the Rutherford experiment, this time using the concept of the center-of-momentum reference frame. Let m = the mass of the alpha particle and M = the mass of the gold nucleus. Consider the specific case of the alpha particle rebounding straight back. The incoming alpha particle has a momentum p_1, the outgoing alpha particle has a momentum p_3, and the gold nucleus picks up a momentum p_4.

(a) Determine the velocity of the center of momentum of the system.

(b) Transform the initial momenta to that frame (by subtracting the center-of-momentum velocity from the original velocities).

(c) Show that if the momenta in the center-of-momentum frame simply turn around (180°), with no change in their magnitudes, both momentum and energy conservation are satisfied, whereas no other possibility satisfies both conservation principles. (Try drawing some other momentum diagrams.)

(d) After the collision, transform back to the original reference frame (by adding the center-of-momentum velocity to the velocities of the particles in the center-of-mass frame). Although using the center-of-momentum frame may be conceptually more difficult, the algebra for solving for the final speeds is much simpler.

9.P.19 Head-on elastic collision

A projectile of mass m_1 moving with speed v_1 in the $+x$ direction strikes a stationary target of mass m_2 head-on. The collision is elastic. Use the Momentum Principle and the Energy Principle to determine the final velocities of the projectile and target, making no approximations concerning the masses. After obtaining your results, see what your equations would predict if $m_1 \gg m_2$, or if $m_2 \gg m_1$. Verify that these predictions are in agreement with the analysis in this chapter of the ping-pong ball hitting the bowling ball, and of the bowling ball hitting the ping-pong ball.

9.P.20 Gamma-ray emission

A ^{57}Fe nucleus is at rest and in its first excited state, 14.4 keV above the ground state (14.4×10^3 eV, where 1 eV = 1.6×10^{-19} J). The nucleus then decays to the ground state with the emission of a gamma ray (a high-energy photon).

(a) What is the recoil speed of the nucleus?

(b) Calculate the slight difference in eV between the gamma-ray energy and the 14.4 keV difference between the initial and final nuclear states.

(c) The "Mössbauer effect" is the name given to a related phenomenon discovered by Rudolf Mössbauer in 1957, for which he received the 1961 Nobel prize for physics. If the ^{57}Fe nucleus is in a solid block of iron, occasionally when the nucleus emits a gamma ray the entire solid recoils as one object. This can happen due to the fact that neighboring atoms and nuclei are connected by the electric interatomic force. In this case, repeat the calculation of part (b) and compare with your previous result. Explain briefly.

9.P.21 Collisions and springs

(a) A spring has an unstretched length of 0.32 m. A block with mass 0.2 kg is hung at rest from the spring, and the spring becomes 0.4 m long. Next the spring is stretched to a length of 0.43 m and the block is released from rest. Air resistance is negligible. How long does it take for the block to return to where it was released?

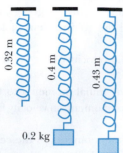

(b) Next the block is again positioned at rest, hanging from the spring (0.4 m long). A bullet of mass 0.003 kg traveling at a speed of 200 m/s straight upward buries itself in the block, which then reaches a maximum height h above its original position. What is the speed of the block immediately after the bullet hits?

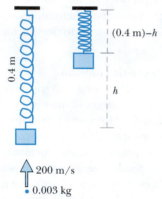

(c) Now write an equation that could be used to determine how high the block goes after being hit by the bullet (a height h), but you need not actually solve for h.

9.P.22 Solar sailing

It has been proposed to propel spacecraft through the Solar System with a large sail that is struck by photons from the Sun.

(a) Which would be more effective, a black sail that absorbs photons or a shiny sail that reflects photons back toward the Sun? Explain briefly.

(b) Suppose N photons hit a shiny sail per second, perpendicular to the sail. Each photon has energy E. What is the force on the sail? Explain briefly.

9.P.23 Slowing down neutrons

In a nuclear fission reactor, each fission of a uranium nucleus is accompanied by the emission of one or more high-speed neutrons which travel through the surrounding material. If one of these neutrons is captured in another uranium nucleus, it can trigger fission, which produces more fast neutrons, which could make possible a chain reaction. However, fast neutrons have low probability of capture and usually scatter off uranium nuclei without triggering fission. In order to

sustain a chain reaction, the fast neutrons must be slowed down in some material, called a "moderator". For reasons having to do with the details of nuclear physics, slow neutrons have a high probability of being captured by uranium nuclei.

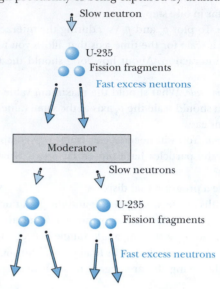

A slow neutron induces fission of U-235, with the emission of additional (fast) neutrons. The moderator is some material that slows down the fast neutrons, enabling a chain reaction.

In the following analyses, remember that neutrons have almost no interaction with electrons. Neutrons do however interact strongly with nuclei, either by scattering or by being captured and made part of the nucleus. Therefore you should think about neutrons interacting with nuclei (through the strong force), not with entire atoms.

(a) Based on what you now know about collisions, explain why fast neutrons moving through a block of uranium experience little change in speed.

(b) Explain why carbon should be a much better moderator of fast neutrons than uranium.

(c) Should water be a better or worse moderator of fast neutrons than carbon? Explain briefly.

Background: The first fission reactor was constructed in 1941 in a squash court under the stands of Stagg Field at the University of Chicago by a team led by the physicist Enrico Fermi. The moderator consisted of blocks of graphite, a form of carbon. The graphite had to be exceptionally pure because certain kinds of impurities have nuclei that capture neutrons with high probability, removing them from contributing to the chain reaction. Many reactors use ordinary "light" water as a moderator, though sometimes a proton captures a neutron and forms a stable deuterium nucleus, in which case the neutron is lost to contributing to the chain reaction. Heavy water, D_2O, in which the hydrogen atoms are replaced by deuterium atoms, actually works better as a moderator than light water, because the probability of a deuterium nucleus capturing a neutron to form tritium is quite small.

9.P.24 Fusion reaction revisited

Here is a modified version of a problem from Chapter 5, before we had discussed collisions in detail. One of the thermonuclear or fusion reactions that takes place inside a star such as our Sun is the production of helium-3 (^{3}He, with two protons and one neutron) and a gamma ray (high-energy photon) in a collision between a proton (^{1}H) and a deuteron (^{2}H, the nucleus of "heavy" hydrogen, consisting of a proton and a neutron):

$$^1\text{H} + {}^2\text{H} \rightarrow {}^3\text{He} + \gamma$$

The rest mass of the proton is 1.0073 u (unified atomic mass unit, 1.66×10^{-27} kg), the rest mass of the deuteron is 2.0136 u, the rest mass of the helium-3 nucleus is 3.015 u, and the gamma ray is a massless photon.

The strong interaction has a very short range and is essentially a contact interaction. For this fusion reaction to take place, the proton and deuteron have to come close enough together to touch. The approximate radius of a proton or neutron is about 10^{-15} m.

(a) In the center-of-momentum frame (that is, the frame in which the total momentum is zero), draw a diagram showing the momentum vectors before the reaction and after the reaction. Clearly label the momentum vectors.

(b) In the center-of-momentum frame, what minimum kinetic energy must the proton have, and what minimum kinetic energy must the deuteron have, in order for the reaction to take place? Express your results in eV. (In the lab frame where the deuteron was at rest, the proton must have more kinetic energy, because the deuteron moves away due to electric repulsion and the proton has to catch up to it.)

(c) In this case, still in the center-of-momentum frame, what will be the kinetic energy of the helium-3 nucleus and the energy of the gamma ray in eV? Hint: You will find that the speed of the helium-3 nucleus is very small compared to the speed of light. You may find it useful to use the relationship $E^2 - (pc)^2 = (mc^2)^2$. Remember that the fundamental definition of the center-of-momentum frame is that it is the frame in which the total momentum is zero. This is the definition that must be used when dealing with zero-mass particles such as photons.

9.P.25 Particle decay

There is an unstable particle called the "sigma-minus" (Σ^-), which can decay into a neutron and a negative pion (π^-): $\Sigma^- \rightarrow n + \pi^-$. The mass of the Σ^- is 1196 MeV/c^2, the mass of the neutron is 939 MeV/c^2, and the mass of the π^- is 140 MeV/c^2. Write equations that could be used to calculate the momentum and energy of the neutron and the pion. You do not need to solve the equations, which would involve some messy algebra. But be clear in showing that you have enough equations that you could in principle solve for the unknown quantities in your equations.

It is advantageous to write the equations not in terms of v but rather in terms of E and p; remember that $E^2 - (pc)^2 = (mc^2)^2$.

9.P.26 Pion production of a particle

A beam of high energy π^- (negative pions) is shot at a flask of liquid hydrogen, and sometimes a pion interacts through the strong interaction with a proton in the hydrogen, in the reaction $\pi^- + p^+ \rightarrow \pi^- + X^+$, where X^+ is a positively-charged particle of unknown mass.

The incoming pion momentum is 3 GeV/c (1 GeV = 1000 MeV = 10^9 electron-volts). The pion is scattered through 40°, and its momentum is measured to be 1510 MeV/c (this is done by observing the radius of curvature of its circular trajectory in a magnetic field). A pion has a rest energy of 140 Mev, and a proton has a rest energy of 938 Mev.

What is the rest mass of the unknown X^+ particle, in MeV/c^2? Explain your work carefully.

It is advantageous to write the equations not in terms of v but rather in terms of E and p; remember that $E^2 - (pc)^2 = (mc^2)^2$.

9.14 ANSWERS TO EXERCISES

9.X.1 (page 309)   (a) 300 J; (b) 5 m/s; (c) 150 J; (d) 150 J

9.X.2 (page 311) 2.3×10^{-24} m/s (!) So it is an excellent approximation to say that the Earth doesn't recoil.

9.X.3 (page 311) $K_{\text{Earth}} = 1.6 \times 10^{-23}$ J; $K_{\text{baseball}} = 150$ J

9.X.4 (page 312) 8×10^4 m/s

9.X.5 (page 324)   Car 1 must have had more momentum than car 2

9.X.6 (page 327)   We outline the entire solution.

First find $v_{\text{cm}, x}$:

$$v_{\text{cm}, x} = \frac{(1000 \text{ kg})(30 \text{ m/s}) + (3000 \text{ kg})(-20 \text{ m/s})}{(4000 \text{ kg})} = -7.5 \text{ m/s}$$

Transform to center-of-momentum frame:

$$v_{\text{car}, x} = (30 \text{ m/s}) - (-7.5 \text{ m/s}) = 37.5 \text{ m/s}$$

$$v_{\text{truck}, x} = (-20 \text{ m/s}) - (-7.5 \text{ m/s}) = -12.5 \text{ m/s}$$

Check to make sure that the momenta are equal and opposite in this frame:

$$p_{\text{car}, x} = (1000 \text{ kg})(37.5 \text{ m/s}) = 37500 \text{ kg·m/s}$$

$$p_{\text{truck}, x} = (3000 \text{ kg})(-12.5 \text{ m/s}) = -37500 \text{ kg·m/s}$$

In this frame the kinetic energy before the collision is this;

$$K = \frac{(37500 \text{ kg·m/s})^2}{2(1000 \text{ kg})} + \frac{(37500 \text{ kg·m/s})^2}{2(3000 \text{ kg})} = 9.4 \times 10^5 \text{ J}$$

After the collision, the velocities are zero in this frame, so kinetic energy in this frame goes to zero, and there is an increase in the internal energy of the mangled car and truck of amount 9.4×10^5 J. There must be the same internal energy increase in the original reference frame, so the kinetic energy lost in the original reference frame was 9.4×10^5 J.

CHAPTER 10

ANGULAR MOMENTUM

Key concepts

- Angular momentum is a measure of rotational motion.
- Angular momentum has two parts:
  - Rotational (around the center of mass)
  - Translational (relative to a specific location)
    Even an object moving in a straight line can have translational angular momentum.
- The change in angular momentum of a system is equal to the net angular impulse applied, which involves:
  - The total torque (twist) exerted by all objects not in the system
  - The time interval over which the interaction occurs
- Angular momentum and torque are vectors involving cross products.
- Angular momentum is conserved: The change in angular momentum of a system plus the change in of angular momentum of the surroundings is zero.
- Angular momentum in microscopic systems is quantized.

10.1 ANGULAR MOMENTUM

A playground ride consists of a disk of mass M and radius R mounted on a low-friction axle (Figure 10.1). A child of mass m runs at speed v on a line tangential to the disk and jumps onto the outer edge of the disk. The disk spins, but at what rate? How many revolutions per minute, or radians per second?

Figure 10.1 A child jumps onto a playground ride, and the disk spins. What is the rate of spin?

The Momentum Principle can be used to determine the impulse exerted by the axle on the disk, which prevents the disk from translating sideways, but it tells us nothing about the spin rate. The Energy Principle can be used to determine how much thermal energy is produced in the inelastic sticking collision of the child and the disk, but it tells us nothing about the spin rate. We need a new principle: the Angular Momentum Principle.

In Figure 10.2 we see a top view of the child heading toward the disk along several different paths, with different effects. Evidently the child's effectiveness in spinning the disk depends on the impact parameter $r_\perp$, which measures how far off-center is the child's initial path. If the child heads straight for the center of the disk, the disk won't rotate. If the child heads for the edge the disk will spin faster than if the child heads toward a smaller radius of the disk. Also, if the child has more mass m and/or more speed v, we expect the spin rate to be greater.

Putting these influences together, the spin rate of the disk should be proportional to $r_\perp mv$, and we call this the magnitude of the "angular momentum" of the child relative to the center of the disk. The term "angular momentum" is used to reflect the fact that it is often associated with something rotating through an angle (in this case, the disk), and it is proportional to the familiar momentum mv. When necessary to emphasize the difference between angular momentum and ordinary momentum, the latter is called "linear momentum."

It is important to see that even though the child is running in a straight line in Figure 10.1, the child does have angular momentum, because the disk is caused to rotate when hit off-center by the child.

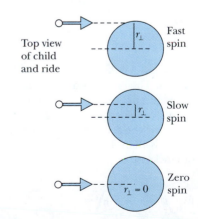

Figure 10.2 How much the disk spins depends on the impact parameter $r_\perp$.

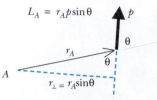

$$L_A = r_A p \sin\theta$$

Figure 10.3 Magnitude of the angular momentum of a particle relative to location *A*.

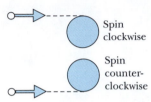

Figure 10.4 The disk will rotate clockwise or counterclockwise, depending on the child's path.

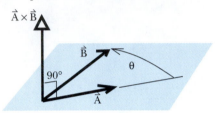

Figure 10.5 The cross product produces a vector that is perpendicular to the two original vectors.

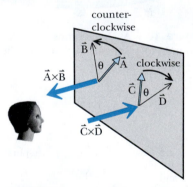

Figure 10.6 The cross product is out of or into the plane defined by the two vectors, depending on whether rotating the first vector toward the second vector (through an angle less than 180°) is counterclockwise or clockwise.

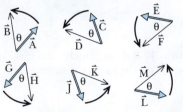

Figure 10.7 All these cross products have the same magnitude and direction (out of the page, toward you.)

We define L_A, the magnitude of the angular momentum of a particle relative to a location *A* like this, as shown in Figure 10.3:

MAGNITUDE OF ANGULAR MOMENTUM OF A PARTICLE RELATIVE TO LOCATION *A*

$$L_A = r_\perp p = r_A p \sin\theta$$

Here $\vec{r}_A$ is a position vector of the particle measured from location *A* to the particle, $\vec{p}$ is the ordinary "linear" momentum of the particle, and θ is the angle between the position vector and the momentum vector. $r_\perp$ is the perpendicular distance from location *A* to the line of motion.

We will find that the sum of the angular momentum of the child plus that of the disk is constant, which is what makes angular momentum useful.

The vector cross product

The magnitude of angular momentum isn't everything: the disk may rotate clockwise or counterclockwise, depending on the child's path (Figure 10.4). How do we take this into account in a definition of the child's angular momentum? The angular momentum of a particle is defined as a vector, using the "cross product," to be explained in a moment:

ANGULAR MOMENTUM OF A PARTICLE RELATIVE TO LOCATION *A*

$$\vec{L}_A = \vec{r}_A \times \vec{p}$$

The cross product $\vec{A} \times \vec{B}$ is a vector in a direction perpendicular to the plane defined by $\vec{A}$ and $\vec{B}$. In Figure 10.5, both $\vec{A}$ and $\vec{B}$ lie in the plane, and the vector $\vec{A} \times \vec{B}$ is perpendicular to the plane. There are two ways of calculating the cross product of two vectors $\vec{A}$ and $\vec{B}$. The first is an algebraic vector calculation; the second involves finding the magnitude and direction separately, using geometry. Both are useful.

CROSS PRODUCT $\vec{A} \times \vec{B}$

Vector: $\vec{A} \times \vec{B} = \langle (A_y B_z - A_z B_y), (A_z B_x - A_x B_z), (A_x B_y - A_y B_x) \rangle$

Magnitude of vector: $|\vec{A} \times \vec{B}| = |\vec{A}||\vec{B}| \sin\theta$

Direction of vector: see below

In the particular use of the cross product to describe angular momentum, angular momentum is a vector $\vec{L}_A = \vec{r}_A \times \vec{p}$ that is perpendicular to $\vec{r}_A$ and perpendicular to $\vec{p}$. The angular momentum vector lies along the axis about which there would be rotation if the particle struck a disk whose axle is at location *A*, and which way it points along this rotation axis is related to whether the disk would rotate counterclockwise or clockwise. The magnitude of the angular momentum vector is proportional to how much spin the particle would produce in the disk.

To determine the direction and magnitude of a vector cross product $\vec{A} \times \vec{B}$ geometrically, do the following:

- Place the tails of the vectors $\vec{A}$ and $\vec{B}$ together. These vectors define a plane. Look at this plane (Figure 10.6).
- Imagine rotating $\vec{A}$ toward $\vec{B}$, through the smaller of the two possible angles (never more than 180°). If the rotation is counter-clockwise, $\vec{A} \times \vec{B}$ is out of the plane (toward the person in Figure 10.6). If the rotation is clockwise the result is into the plane ($\vec{C} \times \vec{D}$ in Figure 10.6 is into the plane, pointing away from the person)

- Find the magnitude of $\vec{A} \times \vec{B}$ by $|\vec{A} \times \vec{B}| = |\vec{A}||\vec{B}| \sin \theta$, where the angle θ is the angle between $\vec{A}$ and $\vec{B}$. All the cross products in Figure 10.7 have the same magnitude (and the same direction).

The right-hand rule

There is another way to determine the direction of the cross product, using a "right-hand rule":

- Point fingers of right hand in direction of first vector $\vec{A}$ (Figure 10.8)
- Rotate wrist, if necessary, to make it possible to
  - Bend fingers of right hand toward second vector $\vec{B}$ through an angle θ less than 180° (Figure 10.9)
  - Stick out thumb, which points in direction of cross product $\vec{A} \times \vec{B}$

In the case of angular momentum $\vec{L}_A = \vec{r}_A \times \vec{p}$, the fingers of your right hand curl in the direction that a disk would rotate if struck by the particle. It may seem odd to have the angular momentum vector perpendicular to the motion, but the point is that it lies along the (possible) axis of rotation.

Wrist rotation

The rotation of the wrist is an important part of the right-hand rule. Consider the situation in Figure 10.10. With your right hand in this position, you can't bend your fingers backwards from the first vector $\vec{A}$ toward the second vector $\vec{B}$ — it is a physical impossibility. You need to rotate your wrist into a position from which it is possible to bend the fingers, as shown in Figure 10.11. Pay attention to the size of the angle through which you bend your fingers. This is the angle whose sine is part of the definition of the magnitude of the cross product. This angle should never be more than 180°. If it is, you have made a mistake in orienting your hand and are in danger of piercing your palm with your nails! Probably you need to rotate your wrist.

Try this right-hand rule on the examples in Figure 10.6. When looking at $\vec{A} \times \vec{B}$ in the plane in Figure 10.6 you will find that your fingers bend counterclockwise and your thumb points out of the plane. When looking at $\vec{C} \times \vec{D}$ in the plane in Figure 10.6 you will find that you have to rotate your wrist before you can bend your fingers, and your thumb points into the plane. So this right-hand rule and the counterclockwise/clockwise rule used in Figure 10.6 are equivalent. In Figure 10.7 you'll find that in all cases the cross product is out of the plane.

Two-dimensional projections

Because it is more difficult to sketch a situation in three dimensions, whenever possible we will work with two-dimensional projections onto the x-y plane. If $\vec{A}$ and $\vec{B}$ lie in the x-y plane, the cross product vector $\vec{A} \times \vec{B}$ points in the $+z$ direction (out of the page, $\odot$) or in the $-z$ direction (into the page, $\otimes$).

Calculating the magnitude of the cross product

It is often easiest to calculate the magnitude $|\vec{r}_A \times \vec{p}| = r_A p \sin \theta$ of the angular momentum by associating the sine of the angle with r_A or with p. In Figure 10.12 on the next page, note that the perpendicular component $r_\perp$ of the "lever arm" is $r_A \sin \theta$, so the magnitude of the cross product is $r_\perp p$. Alternatively, the perpendicular component $p_\perp$ of the momentum is $p \sin \theta$, so the magnitude of the cross product can also be expressed as $r p_\perp$. Study Figure 10.12 to learn two efficient methods for calculating the magnitude of the angular momentum of a particle and try these techniques in the following exercises.

Do the following exercises before going on, because you need to be able to use the concept of cross product quickly and easily in later discussions.

Figure 10.8 Open right hand, fingers point in direction of $\vec{A}$.

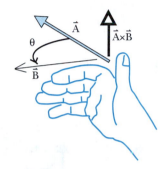

Figure 10.9 Bend fingers less than 180° toward lining up with $\vec{B}$. Thumb points in direction of $\vec{A} \times \vec{B}$.

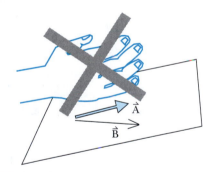

Figure 10.10 You can't bend your fingers backward. You must rotate the wrist into a position that lets you bend the fingers.

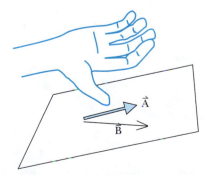

Figure 10.11 After rotating the wrist, it is possible to bend the fingers.

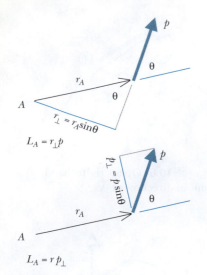

Figure 10.12 Two different ways to calculate the magnitude of the cross product.

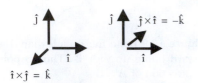

Figure 10.13 Cross products of unit vectors.

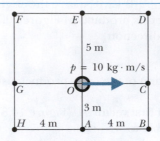

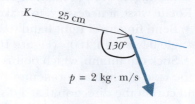

10.X.1 Determine both the direction and magnitude of the angular momentum of the particle at location *O* relative to each point: *A, B, C, D, E, F, G,* and *H.*

10.X.2 What are the magnitude and direction of the angular momentum about location *K* below?

The cross product and unit vectors

It is also possible to evaluate the cross product in terms of unit vectors along the three axes (Figure 10.13). Recall that $\hat{\imath} = \langle 1, 0, 0 \rangle$, etc. First, note that $\hat{\imath} \times \hat{\imath} = 0$, $\hat{\jmath} \times \hat{\jmath} = 0$, and $\hat{k} \times \hat{k} = 0$, since when we cross a vector with itself the angle between the two vectors is zero, and $\sin 0° = 0$.

Second, $\hat{\imath} \times \hat{\jmath} = \hat{k}$, since the angle is 90° and the right-hand rule gives a result in the +z direction (out of the page; Figure 10.13). On the other hand, $\hat{\jmath} \times \hat{\imath} = -\hat{k}$, because the right-hand rule gives a result in the −z direction (into the page). Similarly, $\hat{\jmath} \times \hat{k} = \hat{\imath}$, $\hat{k} \times \hat{\jmath} = -\hat{\imath}$, $\hat{k} \times \hat{\imath} = \hat{\jmath}$, and $\hat{\imath} \times \hat{k} = -\hat{\jmath}$. Putting this all together, we obtain the following general result:

$$\vec{A} \times \vec{B} = (A_y B_z - A_z B_y)\hat{\imath} + (A_z B_x - A_x B_z)\hat{\jmath} + (A_x B_y - A_y B_x)\hat{k} \text{ or}$$

$$\vec{A} \times \vec{B} = \langle (A_y B_z - A_z B_y), (A_z B_x - A_x B_z), (A_x B_y - A_y B_x) \rangle$$

This approach to calculating a cross product is particularly useful in computer calculations. Note the cyclic nature of the subscripts: *xyz, yzx, zxy.*

10.X.3 Given that $\vec{r}_A = \langle x, y, z \rangle$ and $\vec{p} = \langle p_x, p_y, p_z \rangle$, show that angular momentum can be calculated in the following way:

$$\vec{L}_A = \langle (y p_z - z p_y), (z p_x - x p_z), (x p_y - y p_x) \rangle$$

10.2 ANGULAR MOMENTUM IN MULTIPARTICLE SYSTEMS

Later we will show that when the child in Figure 10.1 jumps on the disk-shaped playground ride, the initial angular momentum of the child is equal to the sum of the final angular momenta of the child and disk. In order to calculate the angular momentum of a multiparticle system such as the disk, we need additional tools.

Recall from Chapter 8 that the kinetic energy of a multiparticle system can be split into overall translational kinetic energy plus rotational and vibrational kinetic energy relative to the center of mass:

$$K_{\text{tot}} = K_{\text{trans}} + K_{\text{rel}} = K_{\text{trans}} + K_{\text{rot}} + K_{\text{vib}}$$

In an analogous fashion, angular momentum can be split into a term associated with the motion of the center of mass (which we will call translational angular momentum, sometimes called "orbital" angular momentum) plus a term associated with rotation relative to the center of mass (which we will call rotational angular momentum, sometimes called "spin" angular momentum):

$$\vec{L}_A = \vec{L}_{\text{trans},A} + \vec{L}_{\text{rot}}$$

Splitting the angular momentum into these two terms is very useful in analyzing multiparticle systems such as the Earth, which orbits the Sun but also rotates on its axis. You may also know that electrons orbiting the nucleus of an atom have both translational and rotational angular momentum.

We will calculate the angular momentum relative to location A of a 3-mass system (Figure 10.14). We specify the positions of each mass relative to location A by going from A to the center of mass (vector $\vec{r}_{\text{cm}}$) and then to the individual masses. (The location A need not be in the same plane as the masses, nor are the momentum vectors necessarily in the plane of the masses.)

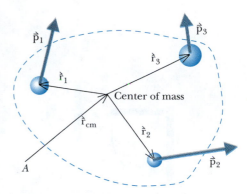

Figure 10.14 Position vectors relative to the center of mass.

The position of mass m_1 relative to location A is $\vec{r}_{\text{cm}} + \vec{r}_1$, the position of mass m_2 is $\vec{r}_{\text{cm}} + \vec{r}_2$, and the position of mass m_3 is $\vec{r}_{\text{cm}} + \vec{r}_3$ (Figure 10.15). Using these vectors we calculate the total angular momentum relative to location A:

$$\vec{L}_A = (\vec{r}_{\text{cm}} + \vec{r}_1) \times \vec{p}_1 + (\vec{r}_{\text{cm}} + \vec{r}_2) \times \vec{p}_2 + (\vec{r}_{\text{cm}} + \vec{r}_3) \times \vec{p}_3$$

$$\vec{L}_A = [\vec{r}_{\text{cm}} \times (\vec{p}_1 + \vec{p}_2 + \vec{p}_3)] + [\vec{r}_1 \times \vec{p}_1 + \vec{r}_2 \times \vec{p}_2 + \vec{r}_3 \times \vec{p}_3]$$

Since $(\vec{p}_1 + \vec{p}_2 + \vec{p}_3)$ is the total momentum $\vec{P}_{\text{tot}}$, we have the following:

$$\vec{L}_A = [\underbrace{\vec{r}_{\text{cm}} \times \vec{P}_{\text{tot}}]}_{\text{"translational"}} + [\underbrace{\vec{r}_1 \times \vec{p}_1 + \vec{r}_2 \times \vec{p}_2 + \vec{r}_3 \times \vec{p}_3]}_{\text{"rotational"}}$$

The first term in the total angular momentum,

$$\vec{r}_{\text{cm}} \times \vec{P}_{\text{tot}}$$

is the angular momentum the system would have if all the mass were concentrated at the center of mass. We call this term the "translational" angular momentum $\vec{L}_{\text{trans},A}$. This term is also called the "orbital" angular momentum.

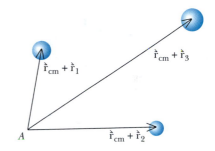

Figure 10.15 Position vectors relative to location A.

The second term in the total angular momentum,

$$\vec{r}_1 \times \vec{p}_1 + \vec{r}_2 \times \vec{p}_2 + \vec{r}_3 \times \vec{p}_3$$

is the angular momentum of the system relative to the center of mass of the system. We call this term the "rotational" angular momentum $\vec{L}_{\text{rot}}$. An example of rotational angular momentum is the angular momentum due to the rotation of the Earth on its axis. This term is also called the "spin" angular momentum. In Figure 10.16 we show both the translational and rotational angular momentum vectors for the Earth relative to the Sun.

TRANSLATIONAL PLUS ROTATIONAL ANGULAR MOMENTUM

$$\vec{L}_A = \vec{L}_{\text{trans},A} + \vec{L}_{\text{rot}}$$

$$\text{Translational: } \vec{L}_{\text{trans},A} = \vec{r}_{\text{cm},A} \times \vec{P}_{\text{tot}}$$

$$\text{Rotational: } \vec{L}_{\text{rot}} = \vec{r}_1 \times \vec{p}_1 + \vec{r}_2 \times \vec{p}_2 + \dots$$

The translational angular momentum differs for different choices of the location A, so we need the "A" subscript on the translational angular momentum and the total angular momentum to remind us of this dependence. The rotational angular momentum doesn't need a subscript because it is calculated relative to the center of mass, and this calculation is unaffected by our choice of the point A, and by the motion of the center of mass, which may even be accelerating.

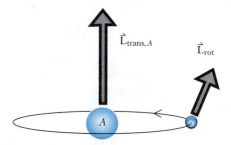

Figure 10.16 Translational angular momentum of the Earth relative to the Sun, with rotational angular momentum relative to the Earth's center of mass. At this instant the linear momentum of the Earth is into the page. (The Earth's axis is tilted $23°$, so the two angular momenta are not parallel.)

For an ideal point particle, or the "collapsed" point-particle version of a multiparticle system, the translational angular momentum is the only kind of angular momentum there is, just as translational kinetic energy is the only kind of kinetic energy a point particle can have.

10.X.4 A rod rotates in the vertical plane around a horizontal axle. A wheel is free to rotate on the rod, as shown in Figure 10.17. A vertical stripe is painted on the wheel. As the rod rotates clockwise, the vertical stripe on the wheel remains vertical. Is the translational angular momentum of the wheel zero or nonzero? If nonzero, what is its direction? Is the rotational angular momentum of the wheel zero or nonzero? If nonzero, what is its direction?

10.X.5 Consider a system similar to that in the previous exercise, but with the wheel welded to the rod (not free to turn). As the rod rotates clockwise, does the stripe on the wheel remain vertical? Is the translational angular momentum of the wheel zero or nonzero? If nonzero, what is its direction? Is the rotational angular momentum of the wheel zero or nonzero? If nonzero, what is its direction?

10.X.6 Pinocchio rides a horse on a merry-go-round turning counterclockwise as viewed from above, with his long nose always pointing forward, in the direction of his velocity. Is Pinocchio's translational angular momentum zero or nonzero? If nonzero, what is its direction? Is his rotational angular momentum zero or nonzero? If nonzero, what is its direction?

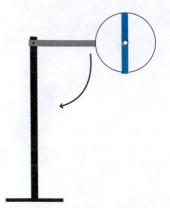

Figure 10.17 A rod rotating in the vertical plane with a wheel attached.

Rotational angular momentum: moment of inertia

An important special case of rotational motion is one where a system is rotating on an axis, with all the atoms in the system sharing the same "angular speed" in radians per second but with different linear speeds in meters per second, depending on their distances from the axis (Figure 10.18). This is the situation for a rigid object.

Angular speed, normally denoted by ω (lowercase Greek omega), is a measure of how fast something is rotating. If an object makes one complete turn of 360 degrees (2π radians) in a time T, we say that its angular speed is $\omega = 2\pi/T$ radians per second. The time T is called the period.

The speed v of an atom rotating around an axis at a distance r from the axis can be expressed in terms of the angular speed:

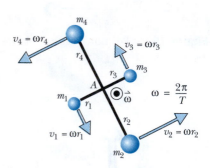

Figure 10.18 A case of rigid rotation about an axis with angular speed ω. Note the different speeds, with $v_i = \omega r_i$.

$$v = \frac{2\pi r}{T} = \left(\frac{2\pi}{T}\right)r = \omega r$$

We measure angular speed ω in the "natural" units of radians per second rather than degrees per second, because the fundamental geometrical relationship between angle and arc length (arc length = $r\theta$) is valid only if the angle θ is measured in radians.

An advantage of using angular speed can be seen in calculating the rotational angular momentum of a rotating rigid object, all of whose parts share a common angular speed ω, but have different linear speeds. We define an angular velocity vector:

ANGULAR VELOCITY VECTOR $\vec{\omega}$

magnitude: radians/second

direction: see Figure 10.19

Figure 10.19 Finding the direction of the angular velocity vector.

Start with your fingers headed radially outward, bend them in the direction of the rotation, and your thumb points in the direction of the angular velocity, along the axis of rotation (angular velocity is also shown pointing out of the page in Figure 10.18).

The magnitude of the angular momentum of each piece is this, where $r_\perp$ is the perpendicular distance of a piece from the axis:

$$|\vec{r} \times \vec{p}| = r_\perp mv = r_\perp m(\omega r_\perp) = mr_\perp^2 \omega$$

So the magnitude of the rotational angular momentum is this:

$$\left|\vec{L}_{\text{rot}}\right| = [\, m_1 r_{\perp 1}^2 + m_2 r_{\perp 2}^2 + m_3 r_{\perp 3}^2 + m_4 r_{\perp 4}^2 \,]\omega$$

The quantity in brackets is called the "moment of inertia" and is usually denoted by the letter I:

MOMENT OF INERTIA

$$I = m_1 r_{\perp 1}^2 + m_2 r_{\perp 2}^2 + m_3 r_{\perp 3}^2 + m_4 r_{\perp 4}^2 + \dots$$

Since the rotational angular momentum points in the direction of $\vec{\omega}$, we have the following for the angular momentum relative to the center of mass, even if the center of mass is moving in some complicated way:

ROTATIONAL ANGULAR MOMENTUM

$$\vec{L}_{\text{rot}} = I\vec{\omega}$$

This way of evaluating rotational angular momentum is often easier to use than the more basic expression in terms of cross products $\vec{r} \times \vec{p}$, especially because the moments of inertia for common objects can be looked up in reference books. For example, the moment of inertia about an axis passing through the center of a uniform solid disk of radius R is $\frac{1}{2}MR^2$ (see Problem 10.P.42 (page 365)), and the moment of inertia about an axis passing through the center of a uniform solid sphere of radius R is $\frac{2}{5}MR^2$.

> *10.X.7* What is the moment of inertia of a nitrogen molecule around its center of mass if the mass of each atom is M and the distance between nuclei is d? (Note that the electrons hardly contribute to the moment of inertia because their mass is so small.)
>
> *10.X.8* A bicycle wheel has almost all its mass M located in the outer rim at radius R. If it rotates on a stationary axle with angular speed ω, what is the angular momentum of the wheel? (Hint: It's helpful to think of dividing the wheel into the atoms it is made of.)

If the masses don't lie in a plane

In Figure 10.18 all the masses lie in the same plane. What if they don't? We'll show that for simple rotating systems the moment of inertia is still calculated by adding up terms like $mr_\perp^2 \omega$, where the perpendicular distances are measured from the axis of rotation.

In Figure 10.20 two equal masses are mounted to a rotating shaft, and we want to determine the angular momentum relative to the center of the device, marked C. We show the directions of the angular momentum $\vec{L}_C = \vec{r}_C \times \vec{p}$ of each mass. You can see that the individual contributions have components that are not along the shaft. But these components cancel for pairs of masses above and below each other. The effect is that the total angular momentum vector lies along the shaft, and can be calculated by adding up $mr_\perp^2 \omega$ for each mass, where $r_\perp$ is the perpendicular distance from the axis.

Here is the proof: the magnitude of the angular momentum of mass 1 about location C is $\left|\vec{L}_1\right| = r_1 mv_1 = rm\omega r_\perp$. The y component of this angular momentum is $\left|\vec{L}_1\right|\cos(90° - \theta) = rm\omega r_\perp \sin\theta = m\omega r_\perp^2$ since $r\sin\theta = r_\perp$. Similarly, the y component of the angular momentum of mass

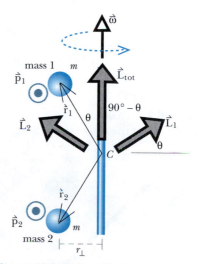

Figure 10.20 The off-axis components of the angular momenta of the two masses cancel. Their total angular momentum is along the axis of rotation.

2 about location C is also $m\omega r_\perp^2$. Therefore the total angular momentum of the two masses is in the y direction and is given by

$$L_{C,y} = mr_\perp^2\omega + mr_\perp^2\omega = (mr_\perp^2 + mr_\perp^2)\omega = I\omega \text{ where}$$

$$I = mr_\perp^2 + mr_\perp^2$$

For example, the moment of inertia of a uniform thin disk rotating on its axle turns out to be $I_{\text{disk}} = \frac{1}{2}MR^2$, where M is the mass of the disk and R is its radius. Because only the distances of atoms from the axle matter, the moment of inertia of a long cylinder has exactly the same form, $I_{\text{disk}} = \frac{1}{2}MR^2$, and it doesn't matter how long the cylinder is (or to put it another way, how thick the disk is). See Figure 10.21.

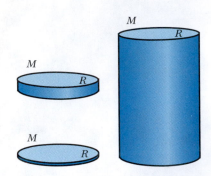

Figure 10.21 These disks and cylinders all have moment of inertia $\frac{1}{2}MR^2$ about their axes.

In this introductory treatment of angular momentum we will deal with simple, symmetrical rotating systems for which the moment of inertia of can be calculated simply as $I = m_1 r_{\perp 1}^2 + m_2 r_{\perp 2}^2 + m_3 r_{\perp 3}^2 + m_4 r_{\perp 4}^2 + \dots$. For a more complicated situation, see the optional Section 10.11 on page 363.

Rotational kinetic energy

The kinetic energy of the system shown in Figure 10.22, which is a rotating rigid object, is $\frac{1}{2}m_1 v_1^2 + \frac{1}{2}m_2 v_2^2 + \frac{1}{2}m_3 v_3^2 + \dots$. Since the speed of each mass is $v = \omega r_\perp$, we can write the rotational kinetic energy like this:

$$K_{\text{rot}} = \frac{1}{2}[m_1(r_{\perp 1}\omega)^2 + m_2(r_{\perp 2}\omega)^2 + m_3(r_{\perp 3}\omega)^2 + \dots]$$

$$= \frac{1}{2}[m_1 r_{\perp 1}^2 + m_2 r_{\perp 2}^2 + m_3 r_{\perp 3}^2 + \dots]\omega^2$$

We again see the moment of inertia showing up in the bracket, so:

$$K_{\text{rot}} = \frac{1}{2}I\omega^2 = \frac{1}{2}\frac{(I\omega)^2}{I} = \frac{L_{\text{rot}}^2}{2I}$$

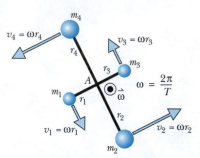

Figure 10.22 A case of rigid rotation about an axis with angular speed ω. Note the different speeds, with $v_i = \omega r_i$.

ROTATIONAL KINETIC ENERGY

$$K_{\text{rot}} = \frac{1}{2}I\omega^2 = \frac{L_{\text{rot}}^2}{2I}$$

10.X.9 A uniform-density disk whose mass is 10 kg and radius is 0.4 m makes one complete rotation every 0.2 s. What is its kinetic energy? (The moment of inertia of a uniform-density disk about its center is $\frac{1}{2}MR^2$.)

10.X.10 A comet orbits the Sun (Figure 10.23). When it is at location 1 it is a distance d_1 from the Sun, and has momentum $\vec{p}_1$. Location A is at the center of the Sun.
(a) When the comet is at location 1, what is the direction of $\vec{L}_A$?
(b) When the comet is at location 1, what is the magnitude of $\vec{L}_A$?
When the comet is at location 2, it is a distance d_2 from the Sun, and has momentum $\vec{p}_2$.
(c) When the comet is at location 2, what is the direction of $\vec{L}_A$?
(d) When the comet is at location 2, what is the magnitude of $\vec{L}_A$?

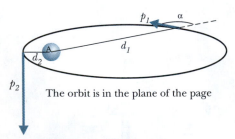

Figure 10.23 A comet orbits the Sun. Location A is at the center of the Sun.

10.X.11 A barbell spins around a pivot at its center at A (Figure 10.24). The barbell consists of two small balls, each with mass m, at the ends of a very low mass rod of length d. The barbell spins clockwise with angular speed ω_0 radians/s.
(a) Calculate $\vec{L}_{\text{trans},1,A}$ (both direction and magnitude).
(b) Calculate $\vec{L}_{\text{tot},A}$ (both direction and magnitude).
(c) Calculate I for the barbell.
(d) What is the direction of $\vec{\omega}_0$?
(e) Calculate $\vec{L}_{\text{rot}}$ (both direction and magnitude).
(f) How does $\vec{L}_{\text{rot}}$ compare to $\vec{L}_{\text{tot},A}$?
(g) Calculate K_{rot}.

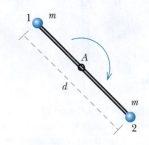

Figure 10.24 A rotating barbell.

10.X.12 The barbell in the previous exercise is mounted on the end of a low mass rigid rod of length b (Figure 10.25). The apparatus is started in such a way that although the rod rotates clockwise with angular speed ω_1 rad/s, the barbell maintains its vertical orientation.
(a) Calculate $\vec{L}_{rot}$ (both direction and magnitude).
(b) Calculate $\vec{L}_{trans,B}$ (both direction and magnitude).
(c) Calculate $\vec{L}_{tot,B}$ (both direction and magnitude).

10.X.13 The apparatus in the previous exercise is restarted in such a way that it again rotates clockwise with angular speed ω_1 rad/s, but in addition, the barbell rotates clockwise about its center, with an angular speed ω_2 rad/s (Figure 10.26).
(a) Calculate $\vec{L}_{rot}$ (both direction and magnitude).
(b) Calculate $\vec{L}_{trans,B}$ (both direction and magnitude).
(c) Calculate $\vec{L}_{tot,B}$ (both direction and magnitude).

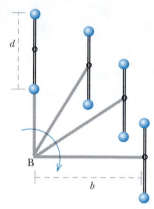

Figure 10.25 A barbell pivoted on a low-mass rotating rod. The barbell does not rotate.

10.3 THE ANGULAR MOMENTUM PRINCIPLE

So far we have concentrated on defining and calculating angular momentum. Next we will show that an off-center force that applies a twist to a system can change the angular momentum of the system.

For a single particle, the Momentum Principle says that the time derivative of the momentum is the net force: $d\vec{p}/dt = \vec{F}_{net}$. Let's see what the time derivative of angular momentum turns out to be. By applying the product rule for derivatives, we have

$$\frac{d(\vec{r}_A \times \vec{p})}{dt} = \frac{d\vec{r}_A}{dt} \times \vec{p} + \vec{r}_A \times \frac{d\vec{p}}{dt}$$

Since $\dfrac{d\vec{r}_A}{dt} = \vec{v}$ and $\vec{p} = \dfrac{m\vec{v}}{\sqrt{1-(v^2/c^2)}}$, we have

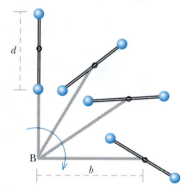

$$\frac{d\vec{r}_A}{dt} \times \vec{p} = \frac{\vec{v} \times m\vec{v}}{\sqrt{1-(v^2/c^2)}} = 0$$

since $|\vec{v} \times \vec{v}| = v^2 \sin(0°) = 0$. Therefore we have proved that the following is true for a single particle:

$$\frac{d(\vec{r}_A \times \vec{p})}{dt} = \vec{r}_A \times \vec{F}_{net}$$

We have previously defined the quantity $\vec{L}_A = \vec{r}_A \times \vec{p}$ to be the angular momentum of a particle relative to location A. We call the quantity $\vec{\tau}_A = \vec{r}_A \times \vec{F}_{net}$ the "torque" exerted about location A (Figure 10.27). Torque is usually symbolized by τ, Greek lowercase letter tau. Torque is a word of Latin origin and simply means "twist."

Figure 10.26 A barbell pivoted on a low-mass rotating rod. The barbell rotates.

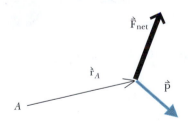

Figure 10.27 Torque relative to location A is defined as $\vec{\tau}_A = \vec{r}_A \times \vec{F}_{net}$.

DEFINITION OF TORQUE

$$\vec{\tau}_A \equiv \vec{r}_A \times \vec{F}_{net}$$

Torque is defined about a particular location, A. $\vec{r}_A$ is a vector from location A to the point of application of the force (Figure 10.27).

Applying a torque, or twist, to a system changes the angular momentum of the system.

THE ANGULAR MOMENTUM PRINCIPLE FOR A PARTICLE

$$\frac{d\vec{L}_A}{dt} = \vec{\tau}_A \quad \text{(derivative form)}$$

The rate of change of the angular momentum of a particle relative to location A is equal to the net torque applied to the particle about location A.

$$\Delta\vec{L}_A = \vec{\tau}_A \Delta t \quad \text{(finite time form)}$$

The change in angular momentum of a particle relative to location A is equal to the net angular impulse $\vec{\tau}_A \Delta t$ applied.

The Angular Momentum Principle is similar to the Momentum Principle $d\vec{p}/dt = \vec{F}_{net}$ or $\Delta\vec{p} = \vec{F}_{net}\Delta t$. The term $\vec{\tau}_A\Delta t$ is the "angular impulse." We will apply this new principle after practicing how to calculate torques.

Calculating torque

Torque and angular momentum are both calculated by using the cross product. In this section we offer some additional ways to think about cross products, in the context of practice in calculating torques.

? In Figure 10.28, which way of pushing on a wrench is more effective in twisting the nut?

You can see that a force parallel to the handle is completely ineffective in twisting the nut. Evidently it is only the perpendicular component of a force that is effective in twisting an object.

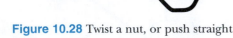

Figure 10.28 Twist a nut, or push straight at it?

? In Figure 10.29, where should you push on the wrench—at the end of the wrench or near the nut?

The farther away from the nut we push, the more effective we are in twisting the nut. The advantage of a long wrench is to provide more leverage in twisting.

Suppose we apply a force at an angle θ to the radius (Figure 10.30). Evidently the component of the force that is effective in twisting is $F\sin\theta$, the perpendicular component. Moreover, the bigger the lever arm (r_A, the distance from the nut at location A), the more effective we'll be. To capture both effects, we define the magnitude of "torque" (which means "twist") exerted by a force $\vec{F}$ relative to a location A as follows:

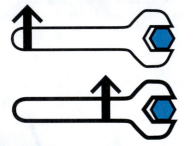

Figure 10.29 Push at the end, or near the nut?

MAGNITUDE OF TORQUE RELATIVE TO LOCATION *A*

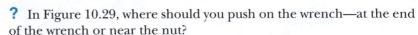

$$\tau = r_A F \sin\theta$$

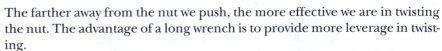

For a purely perpendicular force, $\theta = 90°$, $\sin\theta = 1$, and the torque is $r_A F$. For a force that is parallel to the lever arm, $\theta = 0°$, $\sin\theta = 0$, and the torque is zero. See Figure 10.28 for these extreme examples.

Direction of torque

The direction of the torque is given by the cross product and associated right-hand rule.

? In Figure 10.30, is the torque vector into the page or out of the page? Use the right-hand rule.

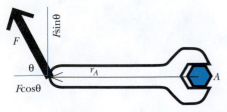

Figure 10.30 Push on the wrench at an angle.

With your right hand at the location of the nut, point your fingers toward the point of application of the force (at the end of the wrench). Next curl your fingers toward the direction of the force vector. The only way you can do this results in your extended right thumb pointing into the page. If you try to keep your extended right thumb pointing out of the page, you will find that you can't curl your fingers toward the direction of the force.

Most nuts and bolts have what are called "right-handed threads." In Figure 10.30 a right-handed nut will advance in the direction of the torque vector, into the page. If you reverse the direction of the force, so the torque vector points out of the page, a right-handed nut will advance out of the page. (Exceptions include left-handed threads on gas cylinders containing explosive gases; the unexpected behavior alerts users to danger.)

The fact that a physical right-handed nut advances in the direction of the torque vector is another justification for drawing torque cross product vectors along an axis of rotation, perpendicular to the plane of the motion.

10.X.14 If $r_A = 3$ m, $F = 4$ N, and $\theta = 30°$, what is the magnitude of the torque about location A, including units?

10.X.15 If the force in Figure 10.30 were perpendicular to $\hat{r}_A$, but gave the same torque as in the preceding exercise, what would its magnitude be?

10.X.16 At $t = 15$ s, a particle has angular momentum $<3, 5, -2>$ kg·m^2/s relative to location A. A constant torque $<10, -12, 20>$ N·m relative to location A acts on the particle. At $t = 15.1$ s, what is the angular momentum of the particle?

Example: Falling object

Let's compare the Momentum Principle and the Angular Momentum Principle in a simple situation. Consider a mass m falling near the Earth (Figure 10.31). We choose a location A off to the side, on the ground.

Neglecting air resistance, the Momentum Principle gives $dp_y/dt = -mg$, yielding $dv_y/dt = -g$ (nonrelativistic). What does the Angular Momentum Principle yield?

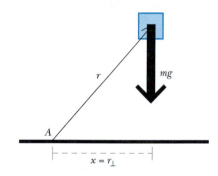

> System: mass m
> Surroundings: Earth, air (neglect air resistance)
>
> Angular Momentum Principle (z component):
>
> $$\frac{dL_{z, A}}{dt} = \tau_{z, A}$$
>
> $$\frac{d}{dt}(xmv_y) = -xmg$$
>
> $$\frac{dv_y}{dt} = -g$$

Figure 10.31 Analyze a falling mass using force and momentum, and using torque and angular momentum.

On the diagram you can see that $x = r_\perp$, so the magnitude of the torque on the mass relative to point A is $r_\perp mg = xmg$. The right-hand rule shows that the direction of the torque is into the page ($\otimes$), so $\tau_{z, A} = -xmg$.

The z component of the angular momentum of the mass relative to point A is xmv_y, which is positive (out of the page) if v_y is positive (ball headed up) and negative if v_y is negative (ball headed down). Both the Momentum Principle and the Angular Momentum Principle for the falling mass give the result $dv_y/dt = -g$. This isn't a big surprise, since we derived the new equation starting from the Momentum Principle.

10.4 CONSERVATION OF ANGULAR MOMENTUM

Since we just found that the Angular Momentum Principle didn't tell us anything new about a falling object, what's the point of all this? An important point is that the angular momentum is constant under certain circumstances. If the net torque around a location A is zero, the angular momentum about that location doesn't change. There may be forces acting, causing changes in the linear momentum, but if these forces don't exert any torques, the rate of change of angular momentum is zero, in which case angular momentum is constant.

Angular momentum is a conserved quantity:

> ### CONSERVATION OF ANGULAR MOMENTUM
>
> $$\Delta \vec{L}_{A,\,\text{system}} + \Delta \vec{L}_{A,\,\text{surroundings}} = 0$$

If a system gains angular momentum, the surroundings lose that amount. An important special case that merits separate attention is the case of no torque acting on the system, in which case the angular momentum does not change. Comet orbits provide a nice example. Most comets have very long elliptical orbits around the Sun (Figure 10.32). We can treat the comet as a point particle, because it is tiny compared to the distances involved.

10.X.17 Relative to the center of the Sun, explain why the torque exerted by the Sun's gravitational force on the comet is zero at every point along the orbit. What does that say about the angular momentum of the comet relative to the center of the Sun?

10.X.18 In terms of the comet's mass m and speed v_1 when nearest the Sun (distance r_1 from the Sun), what is the comet's angular momentum relative to the center of the Sun (direction and magnitude)? What is the comet's angular momentum relative to the center of the Sun (direction and magnitude) when it is farthest from the Sun (distance r_2 from the Sun) and traveling at speed v_2? What can you conclude about the relationship between v_1 and v_2?

Comments about comets

Some comets go far beyond Pluto and spend most of their time there, because they're traveling very slowly, as you have just shown. We see them when they come near the Sun, but only for a few months, because they're now traveling fast.

The result $r_1 v_1 = r_2 v_2$ at the closest and farthest points in the orbit is correct no matter what kind of "central" force is involved. For any force that acts along a line connecting two objects, there is no torque about a point at the center of one of the objects, so angular momentum is constant about that location. In the particular case of the gravitational force, we also have an energy equation relating v_1 and v_2:

$$\tfrac{1}{2}mv_1^2 - GMm/r_1 = \tfrac{1}{2}mv_2^2 - GMm/r_2$$

Historical note: Kepler and elliptical orbits

Based on careful, accurate naked-eye measurements made by Tycho Brahe before the invention of the telescope, Johannes Kepler in 1609 announced his discovery that the planets follow elliptical orbits around the Sun.

Kepler also stated that he had found that "a radius vector joining any planet to the Sun sweeps out equal areas in equal lengths of time." This is equivalent to conservation of angular momentum, as can be seen with the aid of Figure 10.33. The area swept out in a time Δt is the area of a triangle whose base is $v\Delta t$ and whose altitude is $r\sin\theta$. This area is $\tfrac{1}{2}(rv\sin\theta)\Delta t$, which is proportional to the angular momentum $rmv\sin\theta$.

In 1618 Kepler announced his discovery that the square of the time it takes a planet to go around the Sun is proportional to the cube of the mean

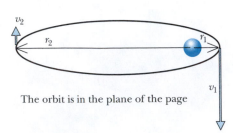

The orbit is in the plane of the page

Figure 10.32 A long elliptical orbit of a comet around the Sun.

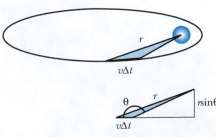

Figure 10.33 "Equal area in equal time."

distance from the Sun (a result you could derive in Chapter 4 for the simpler case of circular orbits).

Later in the 1600's Newton explained all three of these discoveries as derivable from his second law of motion plus his universal law of gravitation. Kepler's insights provided important tests for Newton's theories.

10.5 MULTIPARTICLE SYSTEMS

The Angular Momentum Principle has already given us new insight into the motion of a single particle acted on by a torque due to a single force. The equation becomes a really powerful tool when we extend it to apply to multiparticle systems acted on by multiple torques. For example, we will be able to understand the counterintuitive behavior of spinning tops and gyroscopes.

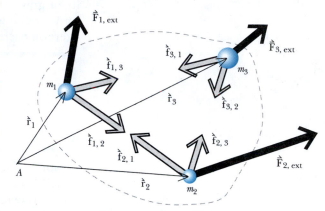

We will now derive a multiparticle version of the Angular Momentum Principle, following closely the derivation in Chapter 2 of a multiparticle version of the Momentum Principle.

To be as concrete as possible, we'll consider a system consisting of just three particles. It is easy to see how this generalizes to larger systems, including a block consisting of an astronomically huge number of atoms.

Figure 10.34 The Angular Momentum Principle in a multiparticle system.

In Figure 10.34 we show all of the forces acting on each particle, where the lower case $\vec{f}$'s are internal forces, and the upper case $\vec{F}$'s are external forces, such as the gravitational attraction of the Earth. We write the Angular Momentum Principle applied to each of the three particles. We measure angular momenta and torques relative to location A, but to avoid clutter we don't write the subscript A in any of these equations until the end. ($\vec{L}_1$ is the angular momentum of m_1 relative to A.)

$$\frac{d\vec{L}_1}{dt} = \vec{r}_1 \times (\vec{F}_{ext,1} + \vec{f}_{1,2} + \vec{f}_{1,3})$$

$$\frac{d\vec{L}_2}{dt} = \vec{r}_2 \times (\vec{F}_{ext,2} + \vec{f}_{2,1} + \vec{f}_{2,3})$$

$$\frac{d\vec{L}_3}{dt} = \vec{r}_3 \times (\vec{F}_{ext,3} + \vec{f}_{3,1} + \vec{f}_{3,2})$$

Nothing new so far. But now we add up these three equations. That is, we create a new equation by adding up all the terms on the left sides of the three equations, and adding up all the terms on the right sides, and setting them equal to each other. In doing so, we take into account that many of these terms cancel. By the reciprocity of electric and gravitational forces (Newton's third law), we have

$$\vec{f}_{1,2} = -\vec{f}_{2,1}$$

$$\vec{f}_{1,3} = -\vec{f}_{3,1}$$

$$\vec{f}_{2,3} = -\vec{f}_{3,2}$$

Therefore the torques of the internal forces cancel. For example, consider this piece of the sum:

$$\vec{r}_1 \times \vec{f}_{1,2} + \vec{r}_2 \times \vec{f}_{2,1} = \vec{r}_1 \times \vec{f}_{1,2} - \vec{r}_2 \times \vec{f}_{1,2} = (\vec{r}_1 - \vec{r}_2) \times \vec{f}_{1,2}$$

The vector $\vec{r}_1 - \vec{r}_2$ points from m_2 toward m_1, so the angle between it and $\hat{r}_{1,2}$ is either $0°$ or $180°$, and $\sin\theta$ is zero. Consequently all of the torques associated with the internal forces cancel, and all that remains is this:

$$\frac{d(\vec{L}_1 + \vec{L}_2 + \vec{L}_3)}{dt} = \vec{r}_1 \times \vec{F}_{ext,1} + \vec{r}_2 \times \vec{F}_{ext,2} + \vec{r}_3 \times \vec{F}_{ext,3}$$

The right side of this equation represents the net torque due to external forces, $\vec{\tau}_{net,\,ext,\,A}$. The great importance of this equation is that the reciprocity of gravitational and electric forces has allowed us to get rid of all the torques due to internal forces, the torques that the particles in the system exert on each other. We rewrite our result like this:

THE ANGULAR MOMENTUM PRINCIPLE
FOR A MULTIPARTICLE SYSTEM

$$\frac{d\vec{L}_{tot,\,A}}{dt} = \vec{\tau}_{net,\,ext,\,A} \quad \text{or} \quad \Delta\vec{L}_{tot,\,A} = \vec{\tau}_{net,\,ext,\,A}\Delta t$$

In words, the rate of change of the "total angular momentum" of a system relative to a location A, $\vec{L}_{tot,A} = \vec{L}_{1,A} + \vec{L}_{2,A} + \vec{L}_{3,A} + ...$, is equal to the net torque due to external forces exerted on that system, relative to location A. (Or, the change in the angular momentum is equal to the angular impulse.) The internal forces and torques do not appear in this multiparticle version of the Angular Momentum Principle.

Often it pays to choose location A to be at the place where the center of mass happens to be at that instant. In that case, the Angular Momentum Principle simplifies:

$$\frac{d\vec{L}_{cm}}{dt} = \frac{d}{dt}[(\vec{r}_{cm,cm} \times \vec{P}_{tot}) + \vec{L}_{rot}] = \frac{d\vec{L}_{rot}}{dt}, \text{ since } \vec{r}_{cm,cm} = 0$$

That is, the position of the center of mass, relative to the center of mass itself, is of course a zero vector. Therefore we have the important and useful special case for the motion relative to the center of mass

---

THE ANGULAR MOMENTUM PRINCIPLE
RELATIVE TO THE CENTER OF MASS

$$\frac{d\vec{L}_{rot}}{dt} = \vec{\tau}_{net,cm} \quad \text{or} \quad \Delta\vec{L}_{rot} = \vec{\tau}_{net,cm}\Delta t$$

The child and the playground ride

We began the chapter with a child running and jumping onto a playground ride. The axle about which the disk turns may exert a force on the disk, but this force has a very small lever arm measured from the center of the axle. If we consider the combined multiparticle system of child plus disk, we expect that the angular momentum to be (nearly) constant. The initial angular momentum of the child should be equal to the sum of the final angular momentum of the child (on the now rotating disk) and the angular momentum of the disk.

10.X.19 Consider a rotating star far from other objects. Its rate of spin stays constant, and its axis of rotation keeps pointing in the same direction. Why?

10.X.20 Because the Earth is nearly perfectly spherical, gravitational forces act on it effectively through its center. Explain why the Earth's axis points at the North Star all year long. Also explain why the Earth's rotation speed stays the same throughout the year (one rotation per 24 hours). In your analysis, does it matter that the Earth is going around the Sun?

In actual fact, the Earth is not perfectly spherical. It bulges out a bit at the equator, and tides tend to pile up water at one side of the ocean. As a result, there are small torques exerted on the Earth by other bodies, mainly the Sun and the Moon. Over many thousands of years there are changes in what portion of sky the Earth's axis points toward (change of direction of rotational angular momentum), and changes in the length of a day (change of magnitude of rotational angular momentum). See page 362.

Summary: Angular momentum and torque

At last we have all the tools we need to investigate a variety of situations involving angular momentum. Let's summarize these new tools and concepts:

- A force exerts a torque relative to a chosen location A: $\vec{\tau}_A = \vec{r}_A \times \vec{F}$.
- Torque causes changes in angular momentum; the rate of change of the angular momentum is equal to the net torque:

$$\frac{d\vec{L}_A}{dt} = \vec{\tau}_{\text{net,ext,A}}$$

- If the net torque relative to A is zero, the angular momentum relative to A does not change. (And if we know the angular momentum isn't changing about some location, as in equilibrium, we know that the net torque must be zero about that location.)
- The angular momentum can be divided into two pieces: translational angular momentum of the system as a whole (as though all the mass were concentrated at the center-of-mass point), plus rotational angular momentum relative to the center of mass: $\vec{L}_A = \vec{L}_{\text{trans},A} + \vec{L}_{\text{rot}}$.
- The rate of change of the rotational angular momentum is equal to the torque relative to the center of mass:

$$\frac{d\vec{L}_{\text{rot}}}{dt} = \vec{\tau}_{\text{net,ext,cm}}$$

- Rotational angular momentum is given by moment of inertia times angular velocity: $\vec{L}_{\text{rot}} = I\vec{\omega}$, where $I = m_1 r_{\perp 1}^2 + m_2 r_{\perp 2}^2 + m_3 r_{\perp 3}^2 + \dots$
- Rotational kinetic energy about the center of mass can be calculated as

$$K_{\text{rot}} = \tfrac{1}{2} I \omega^2 = \frac{1}{2}\frac{(I\omega)^2}{I} = \frac{L_{\text{rot}}^2}{2I}$$

10.6 THREE FUNDAMENTAL PRINCIPLES OF MECHANICS

We now have three fundamental principles that relate a change in some property of a system to the interactions of the system with external objects. In the special case in which there are no external interactions, these principles indicate that a property of the system doesn't change (momentum or energy or angular momentum).

Momentum	Angular momentum	Energy
$\dfrac{d\vec{P}}{dt} = \vec{F}_{\text{net,ext}}$	$\dfrac{d\vec{L}_A}{dt} = \vec{\tau}_{\text{net,ext,A}}$	$\Delta E = W + Q$
If external forces: momentum changes.	If external torques: angular momentum changes.	If energy inputs: energy of system changes.

Momentum	Angular momentum	Energy
If no external forces: the momentum of a system is constant.	If no external torques: the angular momentum of a system is constant.	If no energy inputs: the energy of a system is constant.
Location of object does not matter.	Location of object relative to point *A* is important.	Location of object does not matter.

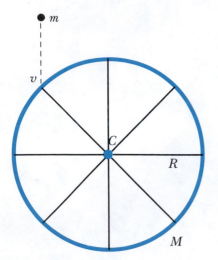

Figure 10.35 A lump of clay falls onto a wheel mounted on a frictionless axle.

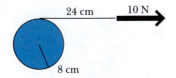

Figure 10.36 Pull a disk on a frictionless surface, unwinding the string.

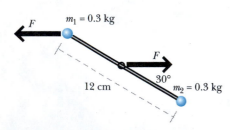

Figure 10.37 Forces applied to a barbell.

10.X.21 A stationary bicycle wheel of radius *R* is mounted in the vertical plane on a horizontal frictionless axle (Figure 10.35). The wheel has mass *M*, all concentrated in the rim (the spokes have negligible mass). A lump of clay with mass *m* falls and sticks to the outer edge of the wheel at the location shown. Just before the impact the clay has a speed *v*.

(a) Just before the impact, what is the angular momentum of the combined system of wheel plus clay about the center *C*?

(b) Just after the impact, what is the angular momentum of the combined system of wheel plus clay about the center *C*?

(c) Just after the impact, what are the magnitude and direction of the angular velocity of the wheel?

(d) Qualitatively, what happens to the linear momentum of the combined system? Why?

10.X.22 A disk of radius 8 cm is pulled along a frictionless surface with a force of 10 N by a string wrapped around the edge (Figure 10.36). 24 cm of string has unwound off the disk.

What are the magnitude and direction of the torque exerted about the center of the disk at this instant?

10.X.23 In Figure 10.36, the uniform solid disk has mass 0.4 kg (moment of inertia $I = \frac{1}{2}MR^2$). At the instant shown, the angular velocity is 20 radians/s into the page.

(a) At this instant, what are the magnitude and direction of the angular momentum about the center of the disk?

(b) At a time 0.2 s later, what are the magnitude and direction of the angular momentum about the center of the disk?

(c) At this later time, what are the magnitude and direction of the angular velocity?

10.X.24 A barbell is mounted on a nearly frictionless axle through its center (Figure 10.37). At this instant, there are two forces of equal magnitude applied to the system as shown, with the directions indicated, and at this instant the angular velocity is 60 radians/s, counterclockwise. In the next 0.001 s, the angular momentum relative to the center increases by an amount 2.5×10^{-4} kg · m²/s .

What is the magnitude of each force?

10.7 SYSTEMS WITH ZERO TORQUE

In a situation in which there is zero external torque, the angular momentum of a system does not change, even if the moment of inertia of the system does change, due to a change in the shape or configuration of the system. Athletes like figure skaters, divers, and dancers can use this fact to change the rotation of their bodies.

You may have seen an ice skater spin vertically on the tip of one skate, with her arms and one leg outstretched, then pull her leg in and bring her arms to a vertical position above her head. She then spins much faster.

What's going on? There is some frictional force of the ice on the tip of the skate, but this force is applied so close to the axis of rotation that the torque is small.

> **?** During the short time when the skater quickly changes her configuration, what can you say about the rotational angular momentum? Why?

Small torque implies small change in angular momentum per unit time, so in a short time the angular momentum of the skater will hardly change.

> **?** But if the rotational angular momentum hardly changes, how can the skater spin faster?

When she changes her configuration, the moment of inertia decreases, because some parts of her body are now closer to the axis of rotation. For the rotational angular momentum not to change, the angular velocity must increase to compensate for the decreased moment of inertia: $I_1\omega_1 = I_2\omega_2$, and therefore $\omega_2 = I_1\omega_1/I_2$.

Now you may quite legitimately be puzzled about how this actually works! *Why* does moving her arms and leg closer to the spin axis increase her angular speed? At one level of discussion, you can close your eyes (and maybe hold your nose) and say, "Well, we did all that general analysis of the effect of torques on the angular momentum of multiparticle systems, so if that's what we get when we apply these general principles, I guess that's that." But at another level it would be very nice to get a better sense of the detailed mechanisms involved in this odd phenomenon.

One approach is to analyze the changes in energy involved in this skating maneuver. The angular momentum L_{rot} of the skater doesn't change, so as her moment of inertia decreases, the kinetic energy of the skater $L_{\text{rot}}^2/(2I)$ actually increases.

> **?** Where does this increased energy come from?

Evidently the skater has to expend chemical energy in order to increase her kinetic energy. In fact, at high spin rates it takes a noticeable effort to pull her arms and leg toward the spin axis. This is even more dramatic when holding heavy weights.

A popular physics demonstration is to sit on a rotating stool holding a dumbbell in each hand. Start spinning slowly with the arms held out, then pull your hands in toward your chest. It requires considerable effort to pull the dumbbells in. After all, the dumbbells would tend to move in straight lines, and you must exert a radial force just to keep turning them in a circle (though in that case you do no mechanical work, because the motion of the dumbbells has no radial component in the direction of the force you apply). To move the dumbbells into a smaller radius requires applying an even larger force which does work on the dumbbells, because there is now a radial component of the motion, in the direction of the radial force you exert.

A high dive

You may have seen a skilled diver leap off a high board, tuck himself into a tight ball (holding onto his ankles), and rotate quite fast for a few turns. Then he straightens out and enters the water like a knife, hardly ruffling the surface (Figure 10.38).

Figure 10.38 While in the air, a diver can change his moment of inertia by curling into a ball or straightening out.

? What can you say about the diver's rotational angular momentum while he is in the air? Why?

We can probably neglect air resistance, because the diver is very far from reaching terminal speed. Then the only significant force acting is the Earth's gravitational force, which effectively acts through the center of mass. Therefore there is negligible torque about the center of mass, and the rotational angular momentum does not change.

? Early in the dive the diver is spinning rapidly. If rotational angular momentum is constant during the dive, where did that rapid spin come from?

When the diver jumps off the board, he must thrust with his feet in such a way that the force of the board on his feet has a sizable lever arm about his center of mass, so that by the time his feet lose contact with the board, the diver already has acquired a sizable rotational angular momentum.

? When the diver pulls out of the tucked position, why does he stop spinning rapidly?

There is a large increase in the moment of inertia, because many atoms are now much farther from the center of mass than they were when the diver was in the tucked position. Larger moment of inertia implies smaller angular speed, since the rotational angular momentum $I\omega$ does not change.

? Can his body go straight from then on?

His body can't really go completely straight, because he still has rotational angular momentum. But his angular speed may be so small that you hardly notice it, especially in comparison with the very rapid spin in the preceding tucked position. Moreover, his body could approximately follow the curving path of his center of mass, with the body rotating just enough to stay tangent to the trajectory. This enhances the illusion of straight motion. The most important aspect for good form is to arrange that the body rotate into the vertical position at the time of entering the water, so as to make little splash.

See Problem 10.P.63, which asks you to analyze a diver's motion.

Example: A satellite with solar panels

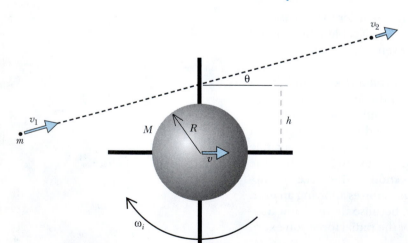

A satellite has four low-mass solar panels sticking out as shown. The satellite can be considered to be approximately a uniform solid sphere. Originally it is traveling to the right with speed v and rotating clockwise with angular speed ω_i.

A tiny meteor traveling at high speed v_1 rips through one of the solar panels and continues in the same direction but at reduced speed v_2. Afterwards, calculate the v_x and v_y components of the center-of-mass velocity of the satellite, and its angular velocity ω_f (magnitude and direction). Additional data are provided on the diagram.

What fundamental principles are useful starting points? The Momentum Principle and the Angular Momentum Principle. The Energy Principle would not be a useful starting point, because we don't know how much thermal energy increase is associated with the meteor ripping through the panel. Take the satellite and meteor together as the system of interest, in which case the momentum and the angular momentum will remain constant, there being no external forces or torques.

System: both the satellite and the meteor
Surroundings: nothing during the collision

Initial state: just before collision
Final state: just after collision

Momentum Principle:

x components:

$$p_{xf} = p_{xi}$$

$$Mv_x + mv_2\cos\theta = Mv + mv_1\cos\theta$$

$$v_x = v + \frac{m}{M}(v_1 - v_2)\cos\theta$$

y components:

$$p_{yf} = p_{yi}$$

$$Mv_y + mv_2\cos(90° - \theta) = m_1 v_1\cos(90° - \theta)$$

$$v_y = \frac{m}{M}(v_1 - v_2)\sin\theta$$

Angular Momentum Principle, about the location in space where the center of the satellite was initially

Component into page $(-z)$:

$$I\omega_i + hmv_1\sin(90° - \theta) = I\omega_f + hmv_2\sin(90° - \theta)$$

$$I = \tfrac{2}{5}MR^2$$

$$\omega_f = \omega_i + \frac{hm}{(\tfrac{2}{5}MR^2)}(v_1 - v_2)\cos\theta \text{, clockwise, into the page}$$

Choosing times just before and just after the collision makes it particularly easy to see how to calculate the translational angular momentum of the meteor, since it is directly above the center of the satellite at those instants.

10.8 SYSTEMS WITH NONZERO TORQUES

In the remainder of this chapter we will analyze systems where there are external torques applied to the system, which can change the angular momentum of the system.

Example: A meter stick on the ice

Consider a meter stick whose mass is 300 grams and which lies on ice (in Figure 10.39 we're looking down on the meter stick). You pull at one end of the meter stick, at right angles to the stick, with a force of 6 newtons. Assume that friction with the ice is negligible. What is the rate of change of the center-of-mass speed v_{cm}? What is the rate of change of the angular speed ω?

System: stick
Surroundings: your hand (pulling); ice (negligible effect)

Momentum Principle:

$$d\vec{P}/dt = d(m\vec{v}_{cm})/dt = \vec{F}_{net}$$

$$dv_{cm}/dt = (6\text{ N})/(0.3\text{ kg}) = 20\text{ m/s}^2$$

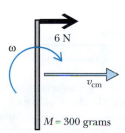

Figure 10.39 Pull one end of a meter stick that is lying on ice (negligible friction).

The moment of inertia I around the center of mass of a uniform rod of mass M and length L can be shown to be $ML^2/12$ (or looked up in a handbook); $L = 1$ m here.

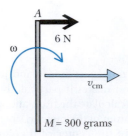

Figure 10.40 Pull one end of a meter stick that is lying on ice (negligible friction).

There is zero torque about location A, because there is zero distance from A to where the force is applied.

Angular Momentum Principle about center of mass:

$$d\vec{L}_{rot}/dt = \vec{\tau}_{net,ext,cm}$$

Component into page ($-z$ direction):

$$I d\omega/dt = (0.5 \text{ m})(6 \text{ N})\sin 90° = 3 \text{ N} \cdot \text{m}$$

$$d\omega/dt = (3 \text{ N} \cdot \text{m})/[(0.3 \text{ kg} \cdot \text{m}^2)/12] = 120 \text{ radians/s}^2$$

In vector terms, $d\vec{\omega}/dt$ points into the page, corresponding to the fact that the angular velocity points into the page and is increasing.

Alternative analysis—taking torques around the end of the stick

It is interesting to re-analyze the motion of the meter stick by calculating torque and angular momentum about the end of the stick where the force is applied (location A in Figure 10.41), rather than about the center of mass. To be cautious and correct, we should say that we are taking torques about a location fixed in the ice next to the place where the end of the stick is momentarily located. This is a fixed location, not tied to the moving stick.

System: stick
Surroundings: your hand (pulling); ice (negligible effect)

Momentum Principle unchanged, which gives this:

$$dv_{cm}/dt = (6 \text{ N})/(0.3 \text{ kg}) = 20 \text{ m/s}^2$$

Angular Momentum Principle about location A:

$$d\vec{L}_A/dt = \vec{\tau}_{net,A}$$

Component into page ($-z$ direction):

$$\frac{d}{dt}\left[\frac{ML^2}{12}\omega - \left(\frac{L}{2}\right)Mv_{cm}\right] = 0$$

$$\frac{d\omega}{dt} = \frac{6}{L}\frac{dv_{cm}}{dt} = \frac{6}{(1 \text{ m})}(20 \text{ m/s}^2) = 120 \text{ radians/s}^2$$

This agrees with our analysis where we took torques about the center of mass of the stick, but the details of the calculation are rather different. It is a good check on an analysis involving the Angular Momentum Principle to do the problem for two different choices of the location about which to calculate the net torque.

A puck with string wound around it

Wrap a string around the outside of a hockey puck. Then pull on the string with a constant tension F_T (Figure 10.41). The puck has mass M and radius R. Assume that friction with the ice rink is negligible. Evidently $dv_{cm}/dt = F_T/M$.

? The moment of inertia of a uniform solid puck of mass M and radius R can be shown to be $MR^2/2$ (see Problem 10.P.42; or this result can be looked up in a handbook). What is the initial rate of change $d\omega/dt$ of the angular speed ω?

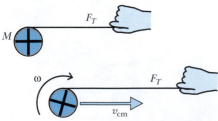

Figure 10.41 Wrap a string around a hockey puck, then pull along ice with negligible friction, with a constant tension F_T.

The torque about the center of mass is RF_T, into the page (down into the ice). The rotational angular momentum is $(MR^2/2)\omega$, into the page (down into the ice). Therefore $(MR^2/2)d\omega/dt = RF_T$, and we have $d\omega/dt = (2F_T)/(MR)$.

10.X.25 Redo the analysis, calculating torque and angular momentum relative to a fixed location in the ice anywhere underneath the string (similar to the analysis of the meter stick around one end). Show that the two analyses of the puck are consistent with each other.

Equilibrium

If the net torque on a system is zero, its angular momentum is constant. Conversely, if we know that the angular momentum is not changing, we can conclude that the net torque must be zero. For example, if a system is in equilibrium, not only must the net force on the system be zero; the net torque must also be zero. This allows us to make conclusions about the individual torques whose vector sum is zero.

> *10.X.26* As a simple application, consider a seesaw that is not rotating around its frictionless support, so that the angular momentum is not changing and is in fact zero (Figure 10.42). Write a force equation, and a torque equation around the support, assuming that the board itself has negligible mass compared to the people sitting on it.
>
> *10.X.27* Write another torque equation around a location fixed in space where the person on the left is sitting, and show that it is in fact equivalent to the torque equation around the axle, when you take the force equation into consideration.
>
> It is often convenient to choose your fixed location so that some of the forces create no torque around that location and therefore don't appear in the Angular Momentum Principle.

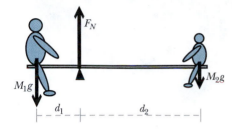

Figure 10.42 A seesaw in equilibrium.

10.9 ANGULAR MOMENTUM QUANTIZATION

Many elementary particles have rotational angular momentum (or, if you wish to be very cautious, these particles behave as though they have rotational angular momentum, which comes down to essentially the same thing). Electrons bound in an atom may have translational angular momentum relative to the nucleus as well as their own rotational angular momentum. Atoms as a whole may have angular momentum, as do many nuclei. The total angular momentum is as expected the sum of the translational angular momenta plus the rotational angular momenta.

The surprising thing about angular momentum at the atomic and sub-atomic level is that it is quantized. The basic quantum of angular momentum is this:

The angular momentum quantum $= \hbar = \dfrac{h}{2\pi} = 1.05 \times 10^{-34}$ joule $\cdot$ second

where h is "Planck's constant," 6.63×10^{-34} joule-second. Whenever you measure a vector component of the angular momentum (that is, along the x or y or z axis), you get either a half-integer or integer multiple of this quantum of angular momentum.

> *10.X.28* Show that $\hbar$ and angular momentum have the same units.

The Bohr model of the hydrogen atom

As discussed in the previous chapter, in 1911 Rutherford and his group discovered the nucleus in atoms. Stimulated by this discovery, in 1913 the Danish physicist Niels Bohr made a bold conjecture that a hydrogen atom could be modeled as an electron going around a proton in circular orbits, but only in those orbits whose translational angular momentum is an integer multiple of $\hbar$. Consequently, these orbits would have only certain radii, and only certain values of energy. The differences in energies between these quantized orbits match the observed energies of photons emitted by atomic hydrogen (Figure 10.43).

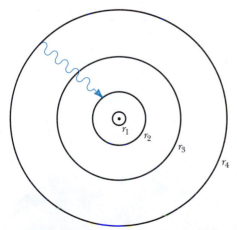

Figure 10.43 In the Bohr model of the hydrogen atom, electrons can be in only those circular orbits for which angular momentum is a multiple of $\hbar$. In a transition between allowed orbits, a photon is emitted.

Bohr's basic hypothesis—that the angular momentum of electrons in a atom is quantized—has proven to be an important insight, and has been retained as a fundamental tenet of the quantum mechanical model of an atom. As this model has been refined, a probabilistic view of the motion of electrons has been adopted. In more sophisticated models electrons do not have precise trajectories, but only a "probability density"—a probability of being found at a particular location. These more complex quantum mechanical models explain a much wider range of atomic and molecular phenomena than does Bohr's original model.

The simple Bohr model does, however, predict correctly the allowed electronic energy levels for atomic hydrogen (as determined from the emission and absorption spectra of hydrogen atoms). We will work with this model to see how quantization of angular momentum leads to the prediction of electronic energy levels in a hydrogen atom.

Allowed radii of electron orbits

Because the proton has much more mass than an electron (about 2000 times as much), we'll make the approximation that the proton is at rest, with the electron in a circular orbit around the proton.

? If the electron momentum is p, and the radius of the circular orbit is r, what is the translational angular momentum?

The translational angular momentum of the electron relative to the location of the proton is $L_{\text{trans}, C} = rp\sin(90°) = rp$ (Figure 10.44). Bohr proposed that the only possible states of the hydrogen atom are those where the electron is in a circular orbit whose translational angular momentum is an integer multiple of $\hbar$:

BOHR: ANGULAR MOMENTUM IS QUANTIZED

$$\left|\vec{L}_{\text{trans}, C}\right| = rp = N\hbar, \text{ where } N \text{ is an integer } (1, 2, 3,...)$$

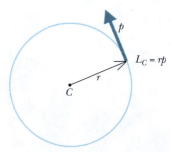

Figure 10.44 The angular momentum of the electron in a circular Bohr orbit relative to the proton. L_C is out of the page.

The electric force and a circular orbit

? What is the magnitude of the electric force that the proton exerts on the electron?

$$F_{\text{el}} = \frac{1}{4\pi\varepsilon_0}\frac{e^2}{r^2}$$

where $+e$ is the electric charge on the proton ($-e$ for the electron).

? Use the Momentum Principle to relate this force to the circular motion of the electron.

In a circular orbit at constant speed we know that the rate of change of the momentum is $(v/r)p$, so we have in the approximation that $v \ll c$, where m is the mass of the electron,

$$\left(\frac{v}{r}\right)p = \frac{mv^2}{r} = F = \frac{1}{4\pi\varepsilon_0}\frac{e^2}{r^2}$$

Solving for r

We are looking for the allowed values of the orbit radius r, so we look for ways to express v in terms of r. Bohr's angular momentum condition gives us v in terms of r:

$$\left|\vec{L}_{\text{trans}, C}\right| = rp = N\hbar \text{ leads to } v = \frac{p}{m} = \frac{N\hbar}{mr}$$

Putting this result for r into the result obtained from the Momentum Principle, $mv^2/r = F$, we have

$$\frac{m}{r}\left(\frac{N\hbar}{r}\right)^2 = \frac{1}{4\pi\varepsilon_0}\frac{e^2}{r^2}$$

Solving the latter relation for r, we obtain the following result (Figure 10.45):

ALLOWED BOHR RADII FOR ELECTRON ORBITS

$$r = N^2\frac{\hbar^2}{\left(\dfrac{1}{4\pi\varepsilon_0}\right)e^2 m}$$

Let's evaluate this result numerically, so that later we can compare with measurements of the spectrum of light emitted by excited atomic hydrogen. Using

$$\hbar = 1.05\times10^{-34}\ \text{J·s},$$

$$1/4\pi\varepsilon_0 = 8.99\times10^9\ \text{N·m}^2/\text{coulomb}^2,$$

$$e = 1.60\times10^{-19}\ \text{coulomb},$$

and the mass of the electron $m = 9.11\times10^{-31}$, we find this:

$$r = N^2(0.53\times10^{-10}\ \text{m})$$

This is a striking result. The simple Bohr model predicts that the smallest permissible electron radius ($n = 1$) is 0.53×10^{-10} m, and atoms are in fact observed to have radii of approximately this size.

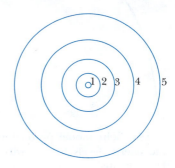

Figure 10.45 Bohr radii for $N=1$ through 5. The nucleus is not shown.

Energy for a circular orbit

We cannot observe directly the radius of the orbit of an electron in an atom. What we can observe are photons emitted by excited atoms, which tell us the differences in energies between various energy levels of the atom. Using the Bohr model we can predict the energies of these energy levels, and see if the differences between these levels match the energies of photons observed in an atomic spectrum.

Given a set of possible values for r, we can calculate the possible values for energy of the electron, starting from this observation:

The result $mv^2/r = F$ leads to $mv^2 = \dfrac{1}{4\pi\varepsilon_0}\dfrac{e^2}{r}$.

? Now write a formula for the kinetic energy plus electric potential energy of the hydrogen atom, using the Bohr model.

Kinetic energy:

$$K = \tfrac{1}{2}mv^2 = \tfrac{1}{2}\left(\frac{1}{4\pi\varepsilon_0}\frac{e^2}{r}\right) \quad \text{(from previous equation)}$$

Electric potential energy:

$$U_{\text{el}} = -\frac{1}{4\pi\varepsilon_0}\frac{e^2}{r}$$

So the energy is this (omitting the rest energies):

$$E = K + U_{\text{el}} = -\tfrac{1}{2}\left(\frac{1}{4\pi\varepsilon_0}\frac{e^2}{r}\right)$$

According to the Bohr model only certain radii will actually occur. Inserting our previous expression for r, we get:

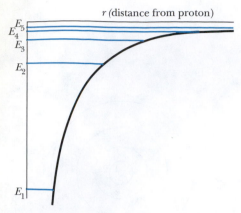

r (distance from proton)

Figure 10.46 Bohr model prediction of electronic energy levels for a hydrogen atom. The potential energy curve for the electron–proton system is also shown.

BOHR MODEL ENERGY LEVELS

$$E = -\frac{\left(\dfrac{1}{4\pi\varepsilon_0}\right)^2 e^4 m}{2N^2\hbar^2}$$

Evaluating this expression we get

$$E = -\frac{2.17\times10^{-18}\,\mathrm{J}}{N^2}$$

Converting to electron-volts (1 eV = 1.6×10^{-19} J) we have (Figure 10.46)

$$E = -\frac{13.6\text{ eV}}{N^2},\ N = 1, 2, 3,...$$

This prediction for the quantized energies of atomic hydrogen agrees well with the observed electronic spectrum of hydrogen. We quoted this result for the quantized energy levels in Chapter 7. The energies are negative, corresponding to bound states.

Photons emitted by atomic hydrogen

Bohr proposed that electromagnetic radiation would be emitted when there was a sudden change from a higher energy level to a lower one. We saw in Chapter 7 that the Bohr energy formula does correctly predict the energies of photons emitted by excited atomic hydrogen.

? In the Bohr model, does photon emission correspond to the radius of the circular orbit getting larger or smaller? Does the quantum number N increase or decrease?

The kinetic plus potential energy is negative, corresponding to a bound state, so a higher energy is one that is less negative and therefore corresponds to a larger radius (since the energy is proportional to $-1/r$). Also, higher energy corresponds to a larger value of N, which is also associated with a larger radius. Therefore photon emission is associated with a decrease in r and a decrease in N (Figure 10.47). Photon absorption involves an increase in r and N.

Limitations of the Bohr model

The Bohr model was an important predecessor to current quantum mechanical models of the atom. It predicts the main aspects of the hydrogen energy levels correctly (neglecting various small effects). The idea that photon emission is associated with a drop from a higher energy level to a lower energy level, with the photon energy equal to the difference in the hydrogen energy levels, is fundamental to the extended quantum-mechanical treatment.

The notion of deterministic circular orbits has been abandoned and replaced with a picture of a probabilistic electron "cloud," but even in more sophisticated models the average radius of this electron cloud is larger for higher energies, just as in the Bohr model. From your study of chemistry you are probably familiar with the shapes of the "probability density" functions or "orbitals" describing the electron's location in the atom in different energy states, some of which are spherical, others dumbbell shaped.

The idea that angular momentum is quantized has far-reaching consequences and is a major element in modern quantum mechanics. However, the actual quantization rules in a hydrogen atom are more complex than those that are assumed in the Bohr model. For example, the translational angular momentum in the ground state ($N = 1$) is actually zero, not $\hbar$, and

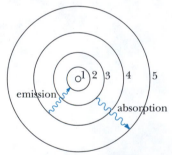

Figure 10.47 A photon is emitted in a transition from a higher to a lower energy level. A photon is absorbed in a transition from a lower to a higher energy level.

for the next higher state ($N = 2$), the z component of translational angular momentum can be either 0 or $\hbar$.

Particle spin

An important example of angular momentum quantization is the observation that many elementary particles have rotational angular momentum, and it is quantized. We know this in part because of observations of the interaction of these particles with other particles having nonzero angular momentum; in these interactions total angular momentum is constant.

For example, the electron, muon, and neutrino are said to have spin 1/2, because measurement of a component of their rotational angular momentum always yields $\pm(1/2)\hbar$. Every quark has spin 1/2, and particles built out of three quarks, such as protons and neutrons, necessarily have half-integral spin. The spin of the proton and neutron is 1/2, as though two of the quarks have their spins opposed ($\Uparrow\Downarrow\Uparrow$), with no translational angular momentum of the quarks. Some short-lived three-quark particles have spin 3/2. One way to get this is for all the quarks to have their spins aligned ($\Uparrow\Uparrow\Uparrow$), or there can be some translational angular momentum of the quarks. While rotational angular momentum can be half-integral, translational angular momentum is always integral (0, 1, 2, etc.).

Particles made out of one quark and one antiquark, called mesons, have integral spin. For example, the pion spin is zero (quark and antiquark spins opposed, $\Uparrow\Downarrow$), and the spin of the "rho" meson is 1 (quark and antiquark spins aligned, $\Uparrow\Uparrow$).

Presumably the angular momentum of a macroscopic object such as a baseball is also quantized, but the angular momentum of a baseball is huge compared to Planck's constant, and you don't notice the quantization, which is on an exceedingly fine scale (10^{-34} J·s!). Only at the atomic level are the effects of quantization really evident.

Consequences of angular momentum quantization

The quantization of angular momentum has far-reaching consequences. Angular momentum quantization plays a major role in determining the structure of atoms, and the nature of the chemical periodic table. The two lowest-energy-state electrons in an atom always have zero translational angular momentum and zero rotational angular momentum (because the two rotational angular momenta are always oppositely aligned and add up exactly to zero).

There is a very deep connection between angular momentum and the statistical behavior of particles. Particles such as the electron and the proton that have half-integral spin are called "fermions" and exhibit the peculiar behavior that two fermions cannot be in the exact same quantum state. This is why only two electrons can be in the lowest energy state of an atom, with their spins opposed. Their spin directions are forbidden to be the same, because then their energy and angular momentum would be exactly the same, which is forbidden for fermions. This prohibition is called the "Pauli exclusion principle."

On the other hand, particles with integral spin, called "bosons," are not subject to the Pauli exclusion principle, and there is no limit to the number of bosons that can be in the same energy state. In fact, there is a special state of matter called a "Bose-Einstein condensation" in which very large numbers of particles end up in exactly the same quantum state. This state of matter was predicted long before it was actually created and observed in 1995.

In a diatomic gas molecule such as oxygen (O_2), the kinetic energy associated with rotation of the molecule is of course $L_{rot}^2/(2I)$. But the rotational angular momentum L_{rot} is quantized, so the rotational energies of oxygen are quantized. The phenomenon of angular momentum quantization affects the specific heat of diatomic gases at low temperatures, as will be discussed in a later chapter. (You can of course refer to the angular momentum of a oxygen molecule as translational angular momentum of the two atoms instead of rotational angular momentum of the molecule.)

It is an important prediction of quantum mechanics that the x, y, or z component of the angular momentum (L_x, L_y, or L_z) can only be an integer or half-integer multiple of $\hbar$, whereas the square magnitude of the angular momentum has the quantized values $l(l+1)\hbar^2$, where l has integer or half-integer values, depending on whether a component has integer or half-integer values:

QUANTIZED VALUES OF L^2

$$L^2 = l(l+1)\hbar^2, \text{ where } l \text{ is integer or half-integer}$$

All nuclei that have an even number of protons and an even number of neutrons ("even-even" nuclei) have a total angular momentum of zero. Examples include carbon-12 (6 protons and 6 neutrons) and oxygen-16 (8 protons and 8 neutrons). The spins of protons and neutrons in even-even nuclei are paired up in the lowest-energy state of a nucleus with each other in such a way as to produce zero net angular momentum. This would be an exceedingly unlikely arrangement if angular momentum weren't quantized.

10.10 *GYROSCOPES

A gyroscope is a fascinating device, and its unusual properties have been exploited to stabilize ships and spacecraft. Its behavior provides a good model and analogy for some important aspects of the quantum mechanical behavior of atoms and nuclei. Magnetic resonance imaging (MRI) is based on the gyroscope-like behavior of spinning nuclei.

A gyroscope has a spinning disk mounted on an axle. If you place one end of the axle on a vertical support, you may observe a very complex motion (Figure 10.48). The gyroscope rises and falls ("nutation") as it revolves around the support ("precession").

One fruitful approach in modeling a complex physical system is to try to analyze the simplest motion that the system is capable of. For example, we can observe a pure precession of a gyroscope, with no nutation. It is possible with care to start the gyroscope with special initial conditions so that it merely precesses, without bobbing up and down. We'll try to analyze this special kind of gyroscope motion. In fact, we'll start with a particularly simple form of precession, with the rotational axis horizontal (Figure 10.49).

We define some variables to make it easier to describe and discuss the situation. The gyroscope disk rotates on its axis with some spin angular speed that we'll call ω, and the disk has a moment of inertia I. The rotational angular momentum $\vec{L}_{rot}$ of the disk always points horizontally but is continually changing direction as the gyroscope precesses (if we neglect friction, the angular speed ω is constant). The gyroscope revolves around the support with angular speed Ω (Greek uppercase omega), constant in magnitude and direction (if we neglect friction). Note that $v_{cm} = \Omega r$ and $\Omega = (v_{cm}/r)$, where r is the distance from the support to the center of mass.

In Chapter 4 we derived the formula for the time rate of change of the unit momentum vector $\hat{p}$, which had the magnitude (v/r), which is the same as angular speed ω. In general, the magnitude of the time rate of change of any rotating unit vector $\hat{X}$ is $\left| d\hat{X}/dt \right| = \omega$, and the magnitude of the rate of change of a rotating vector $\vec{X} = \left| \vec{X} \right|\hat{X}$ whose magnitude isn't

Figure 10.48 A gyroscope can exhibit both "nutation" and "precession."

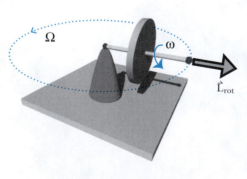

Figure 10.49 A gyroscope which is precessing horizontally, with no nutation.

changing is $\omega\left|\vec{X}\right|$. Since the angular speed of the gyroscope's precession is Ω, the magnitude of the rate of change of the (linear) momentum of the gyroscope is ΩP, and the magnitude of the rate of change of the angular momentum is ΩL.

? What forces act on the gyroscope? What relationships are there among these forces?

The support pushes up with a force we'll call F_N, and the Earth pulls down with a force Mg through the center of mass (Figure 10.50). Since the center of mass stays at the same height all the time in the case of simple precession, it must be that the vertical component of the net force must be zero. Therefore $F_N = Mg$.

This isn't quite the whole story, though. Note that the center of mass of the gyroscope is moving in a circle. That requires that there be a radially inward force to make the momentum vector rotate (with angular speed Ω). The only object that can exert this force is the support. There must be a small horizontal frictional force f such that $\left|d\vec{P}/dt\right| = \Omega P = f$, where $P = Mv_{cm} = M(\Omega R)$; R is the radius of the circle traveled by the center of mass. This force, $f = MR\Omega^2$, is small if the precession rate Ω is small.

It is easy to observe with a toy gyroscope that as the spin of the disk slows down due to friction (smaller ω), the precession actually speeds up (larger Ω). If we could predict the precession speed for a given spin, we would largely "understand" this simple example of gyroscope motion. What can we use to attempt a prediction?

We've already used the Momentum Principle to conclude something about the forces F_N and f. What about energy? If the gyroscope precesses in the horizontal plane, there's no change in gravitational energy, because the center of mass remains always at the same height. There's no change in kinetic energy, either. So energy considerations don't seem to tell us anything interesting about precession (though they would be useful in analyzing the more complex rise and fall of nutation). The remaining principle to apply is the Angular Momentum Principle (no wonder we left gyroscopes until this chapter).

? What is the magnitude of the rotational angular momentum of the gyroscope? What is the magnitude of the translational angular momentum of the gyroscope around the support?

We said that the rotating disk has moment of inertia I and angular speed ω, so the rotational angular momentum is simply $L_{rot} = I\omega$. The magnitude of the translational angular momentum is this:

$$L_{support} = \left|\vec{R}\times\vec{P}\right| = RP$$

Evidently the magnitudes of both the rotational and translational angular momenta are constant, not changing with time. But what about the directions of these angular momenta? Do they change with time?

? How does the direction of the translational angular momentum change with time? How does the direction of the rotational angular momentum change with time?

Neither the magnitude nor the direction of the translational angular momentum changes (neglecting friction): it has constant magnitude and points vertically upward at all times. But the direction of the rotational angular momentum is constantly changing as the gyroscope precesses. We need to calculate the rate at which the rotational angular momentum vector is changing, $d\vec{L}_{rot}/dt$. Since the rotational angular momentum vector is a rotating vector, which rotates with the precession angular speed Ω, we have this:

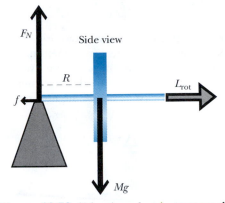

Figure 10.50 Side view showing external forces acting on the gyroscope, and the rotational angular momentum vector.

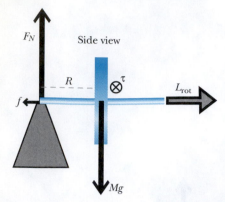

Figure 10.51 Side view showing external forces acting on the gyroscope, and the rotational angular momentum vector. There is a torque around the center of mass, into the page, due to the F_N force.

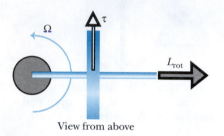

Figure 10.52 View from above, showing the torque vector and the rotational angular momentum vector.

Figure 10.53 What is the precession rate when the spin axis is at an angle θ to the vertical?

$$\left| \frac{d\vec{L}_{rot}}{dt} \right| = \Omega L_{rot}$$

? Therefore, what is the Angular Momentum Principle for the gyroscope?

The remaining element we need for the Angular Momentum Principle is the torque that acts around the center of mass, at the center of the spinning disk. Look again at the force diagram (Figure 10.51), and you see that the magnitude of the torque about the center of mass is $RF_N = RMg$, and the direction of the torque is into the page. Seen from above, the torque points at a right angle to the rotational angular momentum (Figure 10.52). Therefore the Angular Momentum Principle yields

$$\left| \frac{d\vec{L}_{rot}}{dt} \right| = \Omega L_{rot} = \tau_{cm}, \text{ where } L_{rot} = I\omega, \text{ and } \tau_{cm} = RMg$$

Solving for the precession angular speed Ω, we have

$$\Omega = \frac{\tau_{cm}}{L_{rot}} = \frac{RMg}{I\omega}$$

This is a surprisingly simple result for such a complicated system, and it agrees at least qualitatively with observations. A smaller spin ω is associated with a larger precession rate Ω. Conversely, a gyroscope which spins very fast precesses very slowly. If possible, your instructor will provide an opportunity to test the theory quantitatively by observing an actual gyroscope whose spin and precession rates can be measured.

Reflection

Reflect on how we arrived at this result. Look over the steps involved. The key point is that the rotational angular momentum varies in direction without varying in magnitude, and the magnitude of the rate of change of the rotational angular momentum is simply Ω times L_{rot}. This rate of change of the rotational angular momentum is equal to the net torque acting on the system (around the center of mass).

10.X.29 To complete this reflection, determine the relationship between ω and Ω for the case of pure precession, but with the spin axis at an arbitrary angle θ to the vertical (Figure 10.53; $\theta = 90°$ is the case we treated of horizontal precession). If you have the opportunity, see whether this relationship holds for a real gyroscope.

Uses of gyroscopes

The precession we calculated is the result of a nonzero torque acting around the center of mass. Through clever mounting of the gyroscope it is possible to make this torque vanishingly small, in which case the axis does not precess but always maintains the same direction, which has been useful in navigation. The satellite-based Global Positioning System (GPS) can tell you where your airplane is; gyroscopically based instruments can tell you the orientation and attitude of your airplane at this known location.

Large gyroscopes have been used to stabilize ships against rolling in the sea. A gyroscope with very high rotational angular momentum (large moment of inertia and large angular speed) is hard to turn quickly, and this can provide mechanical stability. Gyroscopes are also used to stabilize the orientation of spacecraft and satellites.

Friction

We deliberately neglected friction in the calculation. Friction has two main effects on a gyroscope. Friction in the spin axis makes ω decrease with time, which leads to an increased precession rate Ω. Friction on the top of the support slows down the precession, which means that Ω is no longer equal to RMg/L_{rot} as it should be for pure precession. As a result of these effects, what you observe with a gyroscope that starts out in pure precession is that the spin axis eventually starts to tip lower, accompanied by faster precession. The gyroscope starts out with a stately, dignified, slow precession, but toward the end gives the impression of motion that is more and more frantic.

*Magnetic Resonance Imaging (MRI)

Electrically charged atomic and subatomic objects that have angular momentum act like little bar magnets with north and south poles along the angular momentum axis (we say that they have "magnetic moments"). It is as though the spinning charged particle constituted little loops of current, and current loops produce a magnetic field. Even some electrically neutral particles such as the neutron have magnetic moments, because they are built out of electrically charged quarks that have magnetic moments.

In the presence of an applied magnetic field (typically produced by other current-carrying coils of wire), a bar magnet tends to twist to line up with the applied magnetic field. That is, a magnetic field exerts a torque on a bar magnet. A well-known example is that a compass needle aligns with the Earth's magnetic field, which is useful for navigation. (The compass needle is magnetized and is itself a little bar magnet.)

Since a magnetic field applies a torque to a bar magnet, or to an atomic or subatomic particles that has angular momentum, such particles precess in the presence of an applied magnetic field (Figure 10.54). This phenomenon is exploited in a technique important in physics, chemistry, and biology, called NMR—nuclear magnetic resonance. A particularly useful application of NMR is in magnetic resonance imaging (MRI).

In MRI, a large, steady, uniform magnetic field created by large direct-current-carrying coils of wire surrounding the patient makes protons in the person's body precess. Many of the most common nuclei in the body are even-even nuclei (in particular, the most common isotopes of carbon and oxygen) and have zero angular momentum and therefore no magnetic moment, and these do not precess. But hydrogen nuclei (protons) are very common in the body, have rotational angular momentum, and do precess. The stronger the magnetic field, the larger the torque acting on the proton nuclear magnets, and the faster their precession. (Note that for a gyroscope, $\Omega = \tau / L_{rot}$; the precession rate is proportional to the applied torque τ.)

With the protons precessing in the presence of the large steady magnetic field, a small high-frequency (time-varying) magnetic field tuned to match exactly the precession frequency of the protons will flip the spins upside down. This is called a "resonance" phenomenon. The high-frequency signal is turned off, and the protons revert back to being aligned with the steady magnetic field. The act of flipping back emits radiation (at the precession frequency) that can be detected by coils connected to a receiver, and the strength of the signal indicates how many protons were affected.

That's the basic physical mechanism, but this by itself would not yield spatial detail about the interior of the body. The trick is to superimpose on the large steady magnetic field a small nonuniform magnetic field, which has the effect of establishing slightly different torques on protons at different locations in the body. As a result, when the protons flip back into alignment they radiate signals whose frequencies indicate their locations. A computer algorithm calculates how much of each frequency is present in the signal, and this indicates how many protons are at each location. In this way a very detailed image is built up of the interior of the body.

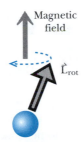

Figure 10.54 Precession of proton spin in an external magnetic field.

*Precession of spin axes in astronomy

As was mentioned earlier, the Earth is subject to small torques due to gravitational forces of the Sun and Moon acting on its nonspherical, "oblate" shape. There is a torque on the Earth's equatorial bulge, perpendicular to the spin rotation axis, and this causes the Earth's axis to precess very slowly, once around about every 26000 years, so that the "North Star" hasn't always been the star we call Polaris. This effect is called the "precession of the equinoxes" because it leads to a change in what month the spring and fall equinoxes occur. To see why there is such an effect, consider Figure 10.55.

Figure 10.55 Forces on the Earth due to the Sun. The diagram is not to scale; the asymmetry of the Earth is greatly exaggerated. The net torque about the center of mass of the Earth is out of the page.

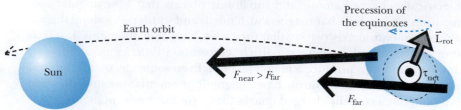

The Sun (or Moon) exerts a slightly larger force on the closer bulge, due to the $1/r^2$ dependence of the force, so there is a nonzero net torque around the center of mass of the Earth, pointing toward you, out of the page. This makes the rotational angular momentum $\vec{L}_{rot}$ of the Earth precess in a "retrograde" way—that is, in the opposite direction to most rotations in our Solar System. Slowly the direction of the axis of the Earth changes due to the torque. (The size of the equatorial bulge, and the differences in F_{near} and F_{far}, have been greatly exaggerated in Figure 10.55).

Since the Earth's axis is tipped 23° away from perpendicular to the Earth-Sun orbit, the precession of the axis makes a big change in the location of the "North Star." 13000 years from now the "North Star" will be a star that is 46° away from Polaris in the night sky.

Figure 10.56 Forces on the Earth-Moon system due to the Sun. The diagram is not to scale; the Earth and Moon are shown as if their masses were equal. The net torque on the Earth-Moon system about its center of mass is out of the page.

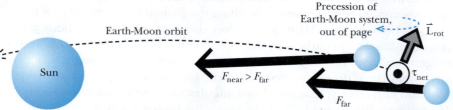

The Moon going around the Earth is a kind of gyroscope. The Earth-Moon orbit is inclined a few degrees to the plane of the Earth-Sun orbit, and the Sun exerts a nonzero torque perpendicular to the angular momentum of the Earth-Moon system. To see why, consider a simpler case where the Earth and Moon have equal mass, and you see the same effect as with the equatorial bulge of the Earth (Figure 10.56).

This torque varies in magnitude during a month, but the averaged effect is that the Earth-Moon system precesses once around in about 18 years (this precession is also "retrograde"). This has an effect on the timing of eclipses, which can occur only when the Moon in its orbit is passing through the plane of the Earth-Sun orbit.

Tidal torques

If the Earth did not rotate, the Moon would create tides in the oceans that would pile up in line with the Moon. However, the rotating Earth drags these tidal bulges so that they are no longer in line with the Moon, and the Moon exerts a small torque (Figure 10.57). This torque is directed along the axis and acts to slow down the spin rate, so that the day is getting longer than 24 hours. This effect is called "tidal friction." This interaction with the Moon has the effect that as the rotational angular momentum of the Earth decreases, the translational angular momentum of the Earth-Moon system in-

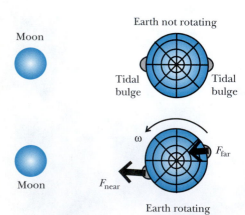

Figure 10.57 Tidal torques due to the Moon, viewed from a point above the North Pole. In the top image the Earth is not rotating; in the bottom image it is rotating. Not to scale; tidal effects greatly exaggerated.

creases (conservation of total angular momentum), with the result that the Earth and Moon are getting farther apart. (These tidal forces have a small net tangential component that acts on both Earth and Moon to increase their translational angular momenta.)

Something like this presumably happened in the past to the Moon, when it was molten. The rotational angular momentum of the Moon decreased due to tidal torques exerted by the Earth, which are much larger than the tidal torques that the low-mass Moon exerts on the Earth. When the Moon's spin angular speed had decreased to be the same as its translational angular speed (currently about 2π radians or 360° per month), the process terminated. So now the Moon always displays nearly the same face to the Earth.

10.11 *MORE COMPLEX ROTATIONAL SITUATIONS

In more advanced courses you may study nonsymmetric rotational situations like that in Figure 10.58. The angular velocity points along the axis of rotation, but the (translational) angular momentum of each mass is $\hat{r}_A \times \vec{p}$, so the total angular momentum $\vec{L}_A$ does *not* point along the axis! Moreover, the angular momentum vector continually changes direction. This change of angular momentum requires a nonzero torque, which is applied to the axle by the bearings in which the axle rotates. The effect of this "dynamic imbalance" is to cause severe wear on the axle and bearings. Car tires must be carefully balanced to prevent this.

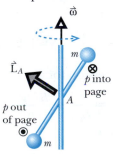

Figure 10.58 If the object lacks axial symmetry, the angular momentum and angular velocity may have different directions!

A symmetric object like that shown in Figure 10.59 has an angular momentum that does point along the axis, because the perpendicular components of the angular momenta cancel each other. This is the simpler kind of situation we have dealt with in this chapter. In this symmetric case it is true that $I = m_1 r_{\perp 1}^2 + m_2 r_{\perp 2}^2 + m_3 r_{\perp 3}^2 + m_4 r_{\perp 4}^2 + \ldots$ about the axis of rotation, and the simple formulas $\vec{L}_{\text{rot}} = I\vec{\omega}$ and $K_{\text{rot}} = \frac{1}{2}I\omega^2$ are valid.

We calculated moment of inertia as a sum (integral) over $r_\perp^2$, the perpendicular distance to an axis. To deal with rotational motion in general requires expressing

the moment of inertia as a "tensor," a 3 by 3 array of numbers representing integrals over x^2, xy, xz, y^2, yx, yz, z^2, zx, and zy. Matrix multiplication of this tensor times the angular velocity vector $\vec{\omega}$ yields an angular momentum vector that need not point in the same direction as $\vec{\omega}$. This is beyond the scope of this introductory textbook.

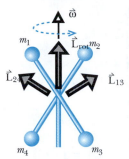

Figure 10.59 If the object is symmetric, the angular momentum is in the direction of the angular velocity.

10.12 *RATE OF CHANGE OF A ROTATING VECTOR

A minor comment: In Chapter 4 we derived the formula for the time rate of change of the unit momentum vector $\hat{p}$, which had the magnitude (v/r), which is the same as ω. In general, the magnitude of the time rate of change of any rotating unit vector $\hat{X}$ is $|d\hat{X}/dt| = \omega$, and the magnitude of the rate of change of a rotating vector $\vec{X} = |\vec{X}|\hat{X}$ whose magnitude isn't changing is $\omega|\vec{X}|$.

If you look at a diagram involving the angular velocity vector $\vec{\omega}$ you can see that the direction of $d\vec{X}/dt$ is given by the cross product $\vec{\omega} \times \vec{X}$. It can be shown rather easily that

$$\frac{d\vec{X}}{dt} = \vec{\omega} \times \vec{X}$$

gives both the magnitude and direction of a rotating vector (Figure 10.60). This is another illustration of the usefulness of cross products.

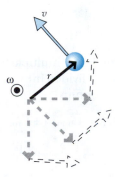

Figure 10.60 In circular motion, both $\hat{r}$ and $\vec{v}$ are rotating vectors. $\vec{\omega}$ is out of the page. At any instant, $\vec{\omega} \times \hat{r}$ is in the direction of $\hat{v}$, while $\vec{\omega} \times \vec{v}$ is toward the center, in the direction of $\vec{a}$, as predicted by the relation $d\vec{X}/dt = \vec{\omega} \times \vec{X}$.

10.13 SUMMARY

The Angular Momentum Principle for a system:

$$\frac{d\vec{L}_{tot,A}}{dt} = \vec{\tau}_{net,\,ext,\,A} \quad \text{or} \quad \Delta\vec{L}_{tot,A} = \vec{\tau}_{net,\,ext,\,A}\,\Delta t$$

Angular momentum of a particle about location A:

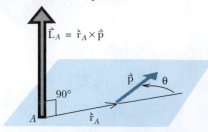

$$\vec{L}_A = \vec{r}_A \times \vec{p}$$

$$L_A = r_\perp p = r_A p \sin\theta$$

direction given by right-hand rule

$$\vec{L}_A = \langle (yp_z - zp_y), (zp_x - xp_z), (xp_y - yp_x)\rangle$$

Torque is a measure of the twist imparted by a force about location A:

$$\vec{\tau}_A = \vec{r}_A \times \vec{F}$$

Moment of inertia about an axis of rotation:

$$I = m_1 r_{\perp 1}^2 + m_2 r_{\perp 2}^2 + m_3 r_{\perp 3}^2 + \dots$$

Moments of inertia:

$$I_{disk} = I_{cylinder} = \tfrac{1}{2}MR^2$$

about center of disk, rotating around axis of cylinder

$$I_{sphere} = \tfrac{2}{5}MR^2$$

for axis passing through center of sphere.

Uniform solid cylinder of length L, radius R, about axis perpendicular to cylinder, through center of cylinder:

$$I = \tfrac{1}{12}ML^2 + \tfrac{1}{4}MR^2$$

Angular momentum for a multiparticle system about some location A can be divided into translational and rotational angular momentum

$$\vec{L}_A = \vec{L}_{trans,A} + \vec{L}_{rot}$$

$$\vec{L}_{trans,A} = \vec{r}_{cm,A} \times \vec{P}_{tot}$$

$$\vec{L}_{rot} = \vec{r}_1 \times \vec{p}_1 + \vec{r}_2 \times \vec{p}_2 + \vec{r}_3 \times \vec{p}_3$$

Rotational angular momentum in terms of moment of inertia:

$$\vec{L}_{rot} = I\vec{\omega}$$

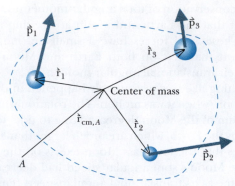

Rate of change of rotational angular momentum:

$$\frac{d\vec{L}_{rot}}{dt} = \vec{\tau}_{net,ext,cm}$$

Kinetic energy relative to center of mass in terms of moment of inertia and rotational angular momentum:

$$K_{rot} = \tfrac{1}{2}I\omega^2 = \frac{1}{2}\frac{(I\omega)^2}{I} = \frac{L_{rot}^2}{2I}$$

Angular momentum is quantized in microscopic systems:

A component of angular momentum is an integer or half-integer multiple of $\hbar$.

The square magnitude of angular momentum has quantized values

$$L^2 = l(l+1)\hbar^2$$

where l has integer or half-integer values.

Bohr model of hydrogen:

$$E = -\frac{13.6\text{ eV}}{N^2}, N = 1, 2, 3,\dots$$

Angular speed of precession for a gyroscope:

$$\Omega = \frac{\tau}{L_{rot}} = \frac{RMg}{I\omega}$$

10.14 REVIEW QUESTIONS

In addition to these review questions, note that the exercises on page 340 and page 348 give you important practice in calculating angular momentum and torque.

Angular momentum

10.RQ.30 Give an example of a situation where an object is traveling in a straight line, yet has nonzero angular momentum.

10.RQ.31 Under what circumstances is angular momentum constant? Give an example of a situation where the x component of angular momentum is constant, but the y component isn't.

Translational and rotational angular momentum

10.RQ.32 Give examples of translational angular momentum and rotational angular momentum in our Solar System.

10.RQ.33 Give an example of a physical situation in which the angular momentum is zero yet the translational and rotational angular momenta are both nonzero.

The Bohr model of hydrogen

10.RQ.34 What features of the Bohr model of hydrogen are consistent with the later, full quantum mechanical analysis? What features of the Bohr model had to be abandoned?

Particle spin

10.RQ.35 Give some examples of atomic or nuclear phenomena associated with particle spin.

Torque

10.RQ.36 Under what conditions is the torque about some location equal to zero?

10.RQ.37 Make a sketch showing a situation where the torque due to a single force about some location is 20 N·m in the positive z direction, whereas about another location the torque is 10 N·m in the negative z direction.

Equilibrium

10.RQ.38 What must be true for a system to be in equilibrium (that is, not moving)?

Gyroscopes

10.RQ.39 Two gyroscopes are made exactly alike except that the spinning disk in one is made of low-density aluminum, whereas the disk in the other is made of high-density lead. If they have the same spin angular speeds and the same torque is applied to both, which gyroscope precesses faster?

10.15 PROBLEMS

10.P.40 Angular momentum in an elliptical orbit

In Chapter 3 you may have written a program to model the motion of a planet going around a fixed star. In this problem you will build on that program.

(a) Use initial conditions that produce an elliptical orbit. At each step calculate the translational angular momentum $\vec{L}_{trans,A}$ of the planet with respect to a location A chosen to be in the orbital plane but outside the orbit. Display this in two ways (i and ii below), and briefly describe in words what you observe:

(i) Display $\vec{L}_{trans,A}$ as an arrow with its tail at location A, throughout the orbit. Since the magnitude of $\vec{L}_{trans,A}$ is quite different from the magnitudes of the distances involved, you will need to scale the arrow by some factor to fit it on the screen.

(ii) Graph the component of $\vec{L}_{trans,A}$ perpendicular to the orbital plane as a function of time.

(b) Repeat part (a), but this time choose a different location B at the center of the fixed star, and calculate and display $\vec{L}_{trans,B}$ relative to that location B. As in part (a), display $\vec{L}_{trans,B}$

as an arrow (scaled appropriately), and also graph the component of $\vec{L}_{trans,B}$ perpendicular to the orbital plane, as a function of time, and briefly describe in words what you observe.

(c) Choose a location C which is not in the orbital plane, and calculate and display $\vec{L}_{trans,C}$ as an arrow throughout the orbit. (You do not need to make a graph.) Briefly describe in words what you observe.

10.P.41 Playground ride

A playground ride consists of a disk of mass M and radius R mounted on a low-friction axle. A child of mass m runs at speed v on a line tangential to the disk and jumps onto the outer edge of the disk.

A child runs and jumps on a playground ride.

(a) If the disk was initially at rest, now how fast is it rotating? (The moment of inertia of a uniform disk is $\frac{1}{2}MR^2$.)

(b) If you were to do a lot of algebra to calculate the kinetic energies in part (a), you would find that $K_f < K_i$. Where has this energy gone?

(c) Calculate the change in *linear* momentum of the system consisting of the child plus the disk (but not including the axle), from just before to just after impact. What caused this change in the linear momentum?

(d) The child on the disk walks inward on the disk and ends up standing at a new location a distance $R/2$ from the axle. Now what is the angular speed?

(e) If you were to do a lot of algebra to calculate the kinetic energies in part (d), you would find that $K_f > K_i$. Where has this energy come from?

(f) In part (a), estimate numerical values for all of the quantities and determine a value for the spin rate ω. Is your result reasonable?

10.P.42 Rotating disk

A disk of radius 0.2 m and moment of inertia 1.5 kg·m^2 is mounted on a nearly frictionless axle. A string is wrapped tightly around the disk, and you pull on the string with a constant force of 25 N. After a while the disk has reached an angular speed of 2 radians/s. What is its angular speed 0.1 seconds later? Explain briefly.

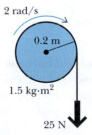

A disk rotates on a nearly frictionless axle.

10.P.43 Moment of inertia of a disk

Show that the moment of inertia of a disk of mass M and radius R is $\frac{1}{2}MR^2$. Divide the disk into narrow rings each of radius r and width dr. The contribution to I by one of these rings is simply $r^2 \, dm$, where dm is the amount of mass contained in that particular ring. The mass of any ring is the mass per unit area times the area of the ring. The area of this ring is approximately $2\pi r \, dr$. Use integral calculus to add up all the contributions.

10.P.44 Momentum and angular momentum

Two small objects each of mass m are connected by a lightweight rod of length d. At a particular instant they have velocities as shown and are subjected to external forces as shown. The system is moving in outer space.

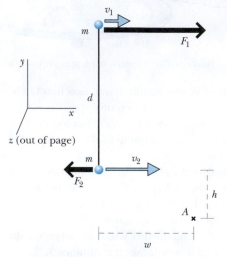

Two masses connected by a lightweight rod.

In the following questions involving vectors, give components along the axes shown, and state which axis you're using for each component.

(a) What is the total (linear) momentum $\vec{P}_{\text{total}}$ of this system?

(b) What is the velocity $\vec{v}_{\text{cm}}$ of the center of mass?

(c) What is the total angular momentum $\vec{L}_A$ of the system relative to point A?

(d) What is the rotational angular momentum $\vec{L}_{\text{rot}}$ of the system?

(e) What is the translational angular momentum $\vec{L}_{\text{trans}}$ of the system relative to point A?

After a short time interval Δt,

(f) What is the total (linear) momentum $\vec{P}_{\text{total}}$ of the system?

(g) What is the rotational angular momentum $\vec{L}_{\text{rot}}$ of the system?

10.P.45 Rotating stool and barbells

You sit on a rotating stool and hold barbells in both hands with your arms fully extended horizontally. You make one complete turn in 2 seconds. You then pull the barbells in close to your body.

(a) Estimate how long it now takes you to make one complete turn. Be clear and explicit about the principles you apply and about your assumptions and approximations.

(b) About how much energy did you expend?

10.P.46 Ice skater

An ice skater whirls with her arms and one leg stuck out as shown, making one complete turn in one second. Then she quickly moves her arms up above her head and pulls her leg in as shown.

(a) Estimate how long it now takes for her to make one complete turn. Explain your calculations, and state clearly what approximations and estimates you make.

(b) Estimate the minimum amount of chemical energy she must expend to change her configuration.

10.P.47 A collision with a rod with masses on the ends

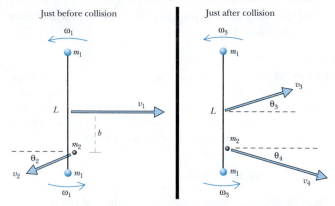

A spinning rod is struck by a small object.

Two small objects each of mass m_1 are connected by a lightweight rod of length L. At a particular instant the center of mass speed is v_1 as shown, and the object is rotating counterclockwise with angular speed ω_1. A small object of mass m_2 traveling with speed v_2 collides with the rod at an angle θ_2 as shown, at a distance b from the center of the rod. After being struck, the mass m_2 is observed to move with speed v_4, at angle θ_4. All the quantities are positive magnitudes. This all takes place in outer space.

For the object consisting of the rod with the two masses, write equations that, in principle, could be solved for the center of mass speed v_3, direction θ_3, and angular speed ω_3 in terms of the given quantities. State clearly what physical principles you use to obtain your equations.

Don't attempt to solve the equations; just set them up.

10.P.48 An asteroid collision

A spherical non-spinning asteroid of mass M and radius R moving with speed v_1 to the right collides with a similar non-spinning asteroid moving with speed v_2, to the left, and they stick together. The impact parameter is d. Note that $I_{\text{sphere}} = \frac{2}{5}MR^2$.

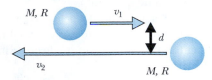

Two asteroids collide and stick together.

After the collision, what is the velocity v_{cm} of the center of mass and the angular velocity ω about the center of mass? (Note that each asteroid rotates about its own center with this same ω.)

10.P.49 Space junk

A tiny piece of space junk of mass m strikes a glancing blow to a spherical satellite After the collision the space junk is traveling in a new direction and moving more slowly. The space junk had negligible rotation both before and after the collision. The velocities of the space junk before and after the collision are shown in the diagram. The center of mass of the satellite is at its geometrical center. The satellite has mass M, radius R, and moment of inertia I about its center. Before the collision the satellite was moving and rotating as shown in the diagram.

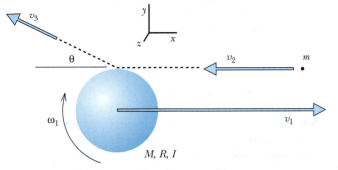

Space junk strikes a satellite.

(a) Just after the collision, what are the components of the center-of-mass velocity of the satellite (v_x and v_y) and its rotational speed ω?

(b) Calculate the rise in the thermal energy of the satellite and space junk combined. You do *not* need to substitute in the values for quantities you already calculated in part (a).

10.P.50 Rotational spectrum of a diatomic molecule

(a) Calculate the energies of the quantized rotational energy levels for O_2. The most common oxygen nucleus contains 8 protons and 8 neutrons. Estimate any quantities you need. See discussion of diatomic molecules on page 357; the parameter l has values 0, 1, 2, 3...

(b) Describe the emission spectrum for electromagnetic radiation emitted in transitions among the rotational O_2 energy levels. Include a calculation of the lowest-energy emission in electron volts (1.6×10^{-19} J).

(c) It is transitions among "electronic" states of atoms that produce visible light, with photon energies on the order of a couple of electron-volts. Each electronic energy level has quantized rotational and vibrational (harmonic oscillator) energy sublevels. Explain why this leads to a visible spectrum that contains "bands" rather than individual energies.

10.P.51 Space station

A space station has the form of a hoop of radius R, with mass M. Initially its center of mass is not moving, but it is spinning with angular speed ω_0. Then a small package of mass m is thrown by a spring-loaded gun toward a nearby spacecraft as shown; the package has a speed v after launch. Calculate the center-of-mass velocity of the space station (v_x and v_y) and its rotational speed ω after the launch.

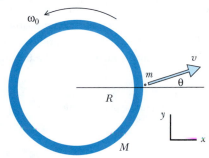

Launch a package from a space station.

10.P.52 The Bohr model

The Bohr model correctly predicts the main energy levels not only for atomic hydrogen but also for other "one-electron" atoms where all but one of the atomic electrons has been removed, such as in He$^+$ (one electron removed) or Li^{++} (two electrons removed).

(a) Predict the energy levels in eV for a system consisting of a nucleus containing Z protons and just one electron. You need not recapitulate the entire derivation for the Bohr model, but do explain the changes you have to make to take into account the factor Z.

(b) The negative muon (μ^-) behaves like a heavy electron, with the same charge as the electron but with a mass 207 times as large as the electron mass. As a moving μ^- comes to rest in matter, it tends to knock electrons out of atoms and settle down onto a nucleus to form a "one-muon" atom. For a system consisting of a lead nucleus (Pb^{208} has 82 protons and 126 neutrons) and just one negative muon, predict the energy in eV of a photon emitted in a transition from the first excited state to the ground state. The high-energy photons emitted by transitions between energy levels in such "muonic atoms" are easily observed in experiments with muons.

(c) Calculate the radius of the smallest Bohr orbit for a μ^- bound to a lead nucleus (Pb^{208} has 82 protons and 126 neutrons). Compare with the approximate radius of the lead nucleus (remember that the radius of a proton or neutron is about 10^{-15} m, and the nucleons are packed closely together in the nucleus).

Comments: This analysis in terms of the simple Bohr model hints at the result of a full quantum-mechanical analysis, which shows that in the ground state of the lead-muon system there is a rather high probability for finding the muon *inside* the lead nucleus. Nothing in quantum mechanics forbids this penetration, especially since the muon does not participate in the strong interaction.

The eventual fate of the μ^- in a muonic atom is that it either decays into an electron, neutrino, and antineutrino, or it reacts through the weak interaction with a proton in the nucleus to produce a neutron and a neutrino. This "muon capture" reaction is more likely if the probability is high for the muon to be found inside the nucleus, as is the case with heavy nuclei such as lead.

10.P.53 A comet

A certain comet of mass m at its closest approach to the Sun is observed to be at a distance r_1 from the center of the Sun, moving with speed v_1. At a later time the comet is observed to be at a distance r_2 from the center of the Sun, and the angle between $\hat{r}_2$ and the velocity vector is measured to be θ. What is v_2? Explain briefly.

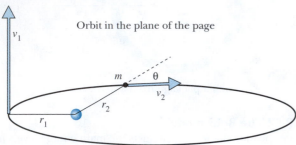

10.P.54 Yo-yo

A yo-yo is constructed of three disks: two outer disks of mass M, radius R, and thickness d, and an inner disk (around which the string is wrapped) of mass m, radius r, and thickness d. The yo-yo is suspended from the ceiling and then released with the string vertical.

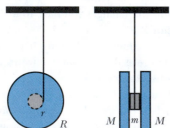

A yo-yo is released and moves downward.

Calculate the tension in the string as the yo-yo falls. Note that when the center of the yo-yo moves down a distance y, the yo-yo turns through an angle y/r, which in turn means that the angular speed ω is equal to v_{cm}/r. The moment of inertia of a uniform disk is $\frac{1}{2}MR^2$.

10.P.55 Nuclear gamma ray

The nucleus dysprosium-160 (containing 160 nucleons) acts like a spinning object with quantized angular momentum, $L^2 = l(l+1)\hbar^2$, and for this nucleus it turns out that l must be an even integer (0, 2, 4...). When a Dy-160 nucleus drops from the $l = 2$ state to the $l = 0$ state, it emits an 87 keV photon (87×10^3 eV).

(a) What is the moment of inertia of the Dy-160 nucleus?

(b) Given your result from part (a), find the approximate radius of the Dy-160 nucleus, assuming it is spherical. (In fact, these and similar experimental observations have shown that some nuclei are not quite spherical.)

(c) The radius of a (spherical) nucleus is given approximately by $(1.3 \times 10^{-15} \text{ m})A^{1/3}$, where A is the total number of protons and neutrons. Compare this prediction with your result in part (b).

10.P.56 A hovering yo-yo

String is wrapped around an object of mass M and moment of inertia I. You pull the string with your hand straight up with some constant force F such that the center of the object does not move up or down, but the object spins faster and faster. This is like a yo-yo; nothing but the vertical string touches the object.

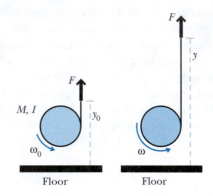

The center of this yo-yo doesn't move up or down.

When your hand is a height y_0 above the floor, the object has an angular speed ω_0. When your hand has risen to a height y above the floor, what is the angular speed ω of the object? Your result should not contain F nor the (unknown) radius of the object. Explain what physics principles you are using.

10.P.57 Bullet and stick

A stick of length L and mass M hangs from a low-friction axle. A bullet of mass m traveling at a high speed v strikes near the bottom of the stick and quickly buries itself in the stick.

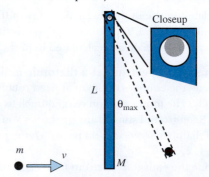

A bullet strikes a stick that is suspended from an axle.

(a) During the brief impact, is the linear momentum of the stick+bullet system constant? Explain why or why not. Include in your explanation a sketch of how the stick shifts on the axle during the impact.

(b) During the brief impact, around what point does the angular momentum of the stick+bullet system remain constant?

(c) Just after the impact, what is the angular speed ω of the stick (with the bullet embedded in it)? (Note that the center of mass of the stick has a speed $\omega L/2$. The moment of inertia of a uniform rod about its center of mass is $\frac{1}{12}ML^2$.)

(d) Calculate the change in kinetic energy from just before to just after the impact. Where has this energy gone?

(e) The stick (with the bullet embedded in it) swings through a maximum angle θ_{max} after the impact, then swings back. Calculate θ_{max}.

10.P.58 Pulling a rotating device

A string is wrapped around a uniform disk of mass M and radius R. Attached to the disk are four low-mass rods of radius b, each with a small mass m at the end.

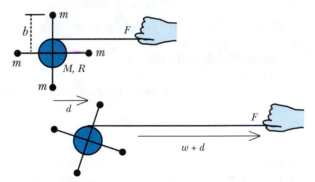

A rotating disk with four masses attached.

The apparatus is initially at rest on a nearly frictionless surface. Then you pull the string with a constant force F. At the instant when the center of the disk has moved a distance d, a length w of string has unwound off the disk.

(a) At this instant, what is the speed of the center of the apparatus? Explain your approach.

(b) At this instant, what is the angular speed of the apparatus? Explain your approach.

(c) You keep pulling with constant force F for an additional time Δt. By how much ($\Delta\omega$) does the angular speed of the apparatus increase in this time interval Δt?

10.P.59 An object rolls down a ramp

A solid object of uniform density with mass M, radius R, and moment of inertia I rolls without slipping down a ramp at an angle θ to the horizontal. The object could be a hoop, a disk, a sphere, etc.

(a) Carefully follow the complete analysis procedure explained in earlier chapters, but with the addition of the Angular Momentum Principle about the center of mass. Note that in your force diagram you must include a small frictional force f that points *up* the ramp. Without that force the object will slip. Also note that the condition of nonslipping implies that the instantaneous velocity of the atoms of the object that are momentarily in contact with the ramp is zero, so $f < \mu F_N$ (no slipping). This zero-velocity condition also implies that $v_{cm} = \omega R$, where ω

is the angular speed of the object, since the instantaneous speed of the contact point is $v_{cm} - \omega R$.

(b) The moment of inertia about the center of mass of a uniform hoop is MR^2, for a uniform disk it is $(1/2)MR^2$, and for a uniform sphere it is $(2/5)MR^2$. Calculate the acceleration dv_{cm}/dt for each of these objects.

(c) If two hoops of different mass are started from rest at the same time and the same height on a ramp, which will reach the bottom first? If a hoop, a disk, and a sphere of the same mass are started from rest at the same time and the same height on a ramp, which will reach the bottom first?

(d) Write the energy equation for the object rolling down the ramp, and for the point-particle system. Show that the time derivatives of these equations are compatible with the force and torque analyses.

10.P.60 A rotating disk with masses sliding on a rod

A rod of length L and negligible mass is attached to a uniform disk of mass M and radius R. A string is wrapped around the disk, and you pull on the string with a constant force F. Two small balls each of mass m slide along the rod with negligible friction. The apparatus starts from rest, and when the center of the disk has moved a distance d, a length of string s has come off the disk, and the balls have collided with the ends of the rod and stuck there. The apparatus slides on a nearly frictionless table. Here is a view from above:

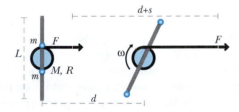

The masses slide on the rod.

(a) At this instant, what is the speed v of the center of the disk?

(b) At this instant the angular speed of the disk is ω. How much thermal energy has been produced?

(c) In the next short amount of time Δt, by how much will the angular speed change ($\Delta\omega$)?

10.P.61 Angular momentum of the Earth

(a) Calculate the magnitude of the translational angular momentum of the Earth relative to the center of the Sun. See data on inside back cover.

(b) Calculate the magnitude of the rotational angular momentum of the Earth. How does this compare to your result in part (a)?

(The angular momentum of the Earth relative to the center of the Sun is the sum of the translational and rotational angular momenta. The rotational axis of the Earth is tipped 23.5° away from a perpendicular to the plane of its orbit.)

10.P.62 Asteroid collision

Suppose an asteroid of mass 2×10^{21} kg is nearly at rest outside the solar system, far beyond Pluto. It falls toward the Sun and crashes into the Earth at the equator, coming in at an angle of

30 degrees to the vertical as shown, against the direction of rotation of the Earth. It is so large that its motion is barely affected by the atmosphere.

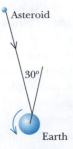

An asteroid crashes into the Earth (not to scale).

(a) Calculate the impact speed.

(b) Calculate the change in the length of a day due to the impact.

10.P.63 Diver

A diver dives from a high platform. When he leaves the platform, he tucks tightly and performs three complete revolutions in the air, then straightens out with his body fully extended before entering the water. He is in the air for a total time of 1.4 seconds. What is his angular speed ω just as he enters the water? Give a numerical answer.

Tucked position (side view).

Entering water (front view).

Be explicit about details of your model, and include (brief) explanations. You will need to estimate some quantities.

10.P.64 Stick on ice

A stick of mass M and length L is lying on ice. A small mass m traveling at high speed v_i strikes the stick a distance d from the center and bounces off with speed v_f as shown in the diagram, which is a top view of the situation. The magnitudes of the initial and final angles to the x axis of the small mass's velocity are

θ_i and θ_f. All of the symbols in the diagram represent positive numbers.

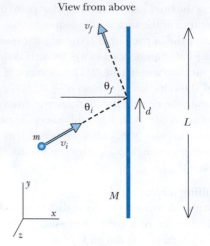

A stick lying on ice is struck by a small mass.

(a) Afterwards, what are the velocity components v_x and v_y of the center of the stick? Explain briefly.

(b) Afterwards, what is the magnitude and direction of the angular velocity $\vec{\omega}$ of the stick?

(c) What is the increase in thermal energy of the objects? You can leave your expression in terms of the initial quantities and v_x, v_y, and ω.

10.P.65 Gyroscope experiment—qualitative

This problem requires that you have a toy gyroscope available. The purpose of this problem is to make as concrete as possible the unusual motions of a gyroscope and their analysis in terms of fundamental principles. In all of the following studies, the effects are most dramatic if you give the gyroscope as large a spin angular speed as possible.

(a) Hold the spinning gyroscope firmly in your hand, and try to rotate the spin axis quickly to point in a new direction. Explain qualitatively why this feels "funny." Also explain why you *don't* feel anything odd when you move the spinning gyroscope in any direction *without* changing the direction of the spin axis.

(b) Support one end of the spinning gyroscope (on a pedestal or in an open loop of the string) so that the gyroscope precesses *counterclockwise* as seen from above. Explain this counterclockwise precession direction; include sketches of top and side views of the gyroscope.

(c) Again support one end of the spinning gyroscope so that the gyroscope precesses *clockwise* as seen from above. Explain this clockwise precession direction; include sketches of top and side views of the gyroscope.

10.P.66 Gyroscope experiment—quantitative

This problem requires that you have a toy gyroscope available. The purpose of this problem is to make as concrete as possible the unusual motions of a gyroscope and their analysis in terms of fundamental principles. In all of the following studies, the effects are most dramatic if you give the gyroscope as large a spin angular speed as possible.

(a) If you knew the spin angular speed of your gyroscope, you could predict the precession rate. Invent an appropriate experimental technique and determine the spin angular speed approximately. Explain your experimental method and your calculations. Then predict the corresponding precession rate, and compare with your measurement of the precession rate. You will have to measure and estimate some properties of the gyroscope and how it is constructed.

(b) Make a *quick* measurement of the precession rate with the spin axis horizontal, then make another quick measurement of the precession rate with the spin axis nearly vertical. (It you make quick measurements, friction on the spin axis doesn't have much time to change the spin angular speed.) Repeat, this time with the spin axis initially nearly vertical, then horizontal. Making all four of these measurements gives you some indication of how much the spin unavoidably changes due to friction while you are quickly changing the angle. What do you conclude about the dependence of the precession rate on the angle, assuming the same spin rate at these different angles? What is the theoretical prediction for the dependence of the precession rate on angle (for the same spin rate)?

10.P.67 Bicycle wheel

A bicycle wheel with a heavy rim is mounted on a lightweight axle, and one end of the axle rests on top of a post. The wheel is observed to precess in the horizontal plane. With the spin direction shown, does the wheel precess clockwise or counterclockwise? Explain in detail, including appropriate diagrams.

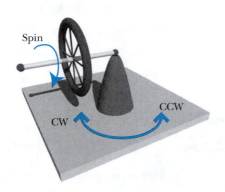

A bicycle wheel on a pivot.

10.P.68 Wood or steel top

(a) A solid wood top spins at high speed on the floor, with a spin direction shown in the diagram. Using appropriately labeled diagrams, explain the direction of motion of the top (you do not need to explain the magnitude).

A wood or steel top.

(b) How would the motion change if the top had a higher spin rate? Explain briefly.

(c) If the top were made of solid steel instead of wood, explain how this would affect the motion (for the same spin rate).

10.P.69 Gyroscope

The axis of a gyroscope is tilted at an angle of 30° to the vertical. The rotor has a radius of 15 cm, mass 3 kg, moment of inertia 0.06 kg·m^2, and spins on its axis at 30 radians/s. It is supported in a cage (not shown) in such a way that without an added weight it does not precess. Then a mass of 0.2 kg is hung from the axis at a distance of 18 cm from the center of the rotor.

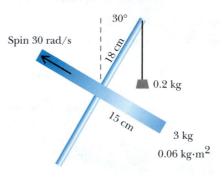

A gyroscope.

(a) Viewed from above, does the gyroscope precess in a 1) clockwise or a 2) counterclockwise direction? That is, does the top end of the axis move 1) out of the page or 2) into the page in the next instant? Explain your reasoning.

(b) How long does it take for the gyroscope to make one complete precession?

10.16 ANSWERS TO EXERCISES

10.X.1 (page 336)   A: 30 kg·m²/s, into page ($\bigotimes$)

B: 30 kg·m²/s, into page ($\bigotimes$)

C: 0

D: 50 kg·m²/s, out of page ($\bigodot$)

E: 50 kg·m²/s, out of page ($\bigodot$)

F: 50 kg·m²/s, out of page ($\bigodot$)

G: 0

H: 30 kg·m²/s, into page ($\bigotimes$)

10.X.2 (page 336)   0.38 kg·m2/s, into page ($\bigotimes$)

10.X.4 (page 338)   Nonzero; into the page; zero; (no direction)

10.X.5 (page 338)   No; nonzero; into the page; nonzero; into the page

10.X.6 (page 338)   Nonzero; up (toward the sky); nonzero; up

10.X.7 (page 339) $2M\left(\frac{d}{2}\right)^2 = \frac{1}{2}Md^2$

10.X.8 (page 339) $MR^2\omega$

10.X.9 (page 340)   395 J

10.X.10 (page 340)   (a) up; (b) $d_1 p_1 \sin\alpha$; (c) up; (d) $d_2 p_2$

10.X.11 (page 340)   (a) $(d/2)p = (d/2)[m(d/2)\omega_0]$ into page

(b) $2m(d/2)^2\omega_0$ into page ($\bigotimes$)

(c) $2m(d/2)^2$

(d) into page

(e) $2m(d/2)^2\omega_0$ into page ($\bigotimes$)

(f) they are equal

(g) $m(d/2)^2\omega_0^2$

10.X.12 (page 341)   (a) 0

(b) $b(2m)(b\omega_1)$ into page ($\bigotimes$)

(c) $b(2m)(b\omega_1)$ into page ($\bigotimes$)

10.X.13 (page 341)   (a) $2m(d/2)^2\omega_2$ into page ($\bigotimes$)

(b) $b(2m)(b\omega_1)$ into page ($\bigotimes$)

(c) $2m[b^2\omega_1 + (d/2)^2\omega_2]$ into page ($\bigotimes$)

10.X.14 (page 343)   6 N·m

10.X.15 (page 343)   2 N

10.X.16 (page 343)   (4,3.8,0) kg·m²/s

10.X.17 (page 344)   Force directed toward center of Sun, so zero torque about the location of the Sun.
Angular momentum about center of Sun *constant*.

10.X.18 (page 344) $r_1 m v_1$, into page; $r_2 m v_2$, into page; $r_1 v_1 = r_2 v_2$

10.X.19 (page 346)   Negligible external forces means negligible external torques.
Therefore magnitude and direction of angular momentum constant.

10.X.20 (page 346)   No torques around center of mass means no change in rotational angular momentum, so rotational angular momentum stays constant in magnitude (which determines length of day) and direction (which determines what "North Star" the axis points at). Doesn't matter that Earth is going around Sun; rotational angular momentum affected solely by torque around center of mass.

10.X.21 (page 348)   (a) $R\cos(45°)mv$ out of page ($\bigodot$)

(b) $R\cos(45°)mv$ out of page ($\bigodot$)

(c) $\dfrac{R\cos(45°)mv}{(M+m)R^2}$ out of page ($\bigodot$)

(d) There is a change of linear momentum of the clay (but not of the wheel), caused by a force applied by the axle to the combined system.

10.X.22 (page 348)   0.8 N·m, into the page, or <0,0,–0.8> N·m

10.X.23 (page 348)   (a) 0.0256 kg·m²/s, into page ($\bigotimes$), or

$\langle 0, 0, -0.0256\rangle$ kg·m²/s

(b) 0.1856 kg·m2/s, into page ($\bigotimes$), or

$\langle 0, 0, -0.1856\rangle$ kg·m²/s

(c) 145 rad/s, into page ($\bigotimes$), or

$\langle 0, 0, -145\rangle$ rad/s

10.X.24 (page 348)   8.33 N

10.X.25 (page 352) $\dfrac{dL_{\text{string},z}}{dt} = \dfrac{d}{dt}\left[RMv_{cm} - \dfrac{1}{2}MR^2\omega\right] = 0$,

since no torque here. Differentiating, we get $\dfrac{d\omega}{dt} = \dfrac{2}{R}\dfrac{dv_{cm}}{dt} = \dfrac{2F_T}{MR}$, as before.

10.X.26 (page 353) $\dfrac{dP_z}{dt} = F_N - M_1 g - M_2 g = 0$

$\dfrac{dL_{\text{support},z}}{dt} = d_1 M_1 g - d_2 M_2 g = 0$ (+z out of page)

10.X.27 (page 353) $\dfrac{dL_{\text{left person},z}}{dt} = d_1 F_N - (d_1 + d_2)M_2 g = 0$

Substitute $F_N = (M_1 + M_2)g$ and get

$d_1 M_1 g - d_2 M_2 g = 0$

10.X.29 (page 360)   Same as in horizontal case: $\Omega = \dfrac{RMg}{I\omega}$

CHAPTER 11

ENTROPY:
LIMITS ON THE POSSIBLE

Key concepts
- Thermal energy in a macroscopic object can be distributed among the particles of the system in many different ways, and each specific arrangement is called a "microstate."
- When two objects are placed in thermal contact, energy becomes distributed between the two objects in such a way as to maximize the number of possible ways of arranging the available energy (number of microstates).
- Entropy is determined by the number of ways of arranging the energy (number of microstates), and the entropy of a system containing two objects is the sum of the two entropies. Like the number of microstates, the entropy tends to be maximized.
- Temperature is related to the rate at which entropy increases as energy increases.
- With a simple ball-and-spring model of a solid, for a particular thermal energy one can calculate the number of microstates, the entropy, temperature, and specific heat capacity.
- The probability of finding a microscopic system in an excited state increases with temperature.

11.1 STATISTICAL ISSUES

A pen lying on a table doesn't suddenly jump upwards, despite the fact that there is plenty of energy in the table in the form of microscopic kinetic and potential energy. Why doesn't this happen? It wouldn't be a violation of the Energy Principle. In fact, if you place a single atom on the table, it will typically get so much energy from the table that it can jump thousands of meters upward (assuming there was no air or other obstacles in the way). This chapter deals with a statistical analysis of microscopic energy that puts limits on what is possible.

Much of this chapter is based on an article by Thomas A. Moore and Daniel V. Schroeder, "A different approach to introducing statistical mechanics," *American Journal of Physics* vol. 65, pp. 26-36 (January 1997).

Temperature
We have repeatedly encountered a connection between the temperature of an object and the average motion of the atoms that make up the object, but our understanding of the meaning and role of temperature is incomplete. What is the quantitative relationship between temperature and microscopic energy?

Direction of thermal energy flow
When two blocks of different temperatures are placed in contact with each other, we observe that energy flows from the hotter block to the colder block (Figure 11.1). We can understand this in a rough way. The average energy of atoms in the hotter block is greater than the average energy of atoms in the colder block. In the interface region, where atoms of the two blocks are in contact with each other, it seems more likely that an atom in

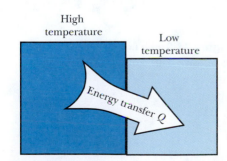

Figure 11.1 We observe that energy flows from a hotter object to a cooler one.

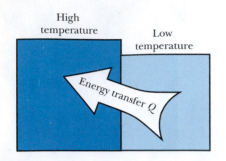

Figure 11.2 Could energy be transferred from the cooler object to the hotter one?

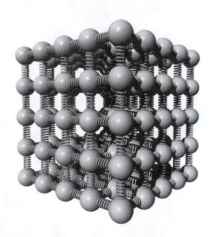

Figure 11.3 The familiar model of a solid as massive balls connected by springs.

the hotter block will lose energy rather than gain energy when it collides with an atom in the colder block.

But couldn't energy flow the other way? After all, sometimes an individual atom in the hotter block might happen to have a lot less than its average energy when it collides with an individual atom in the colder block that happens to have a lot more than its average energy. In that case energy would be transferred from the colder block to the hotter block (Figure 11.2).

This "uphill" movement of energy seems unlikely, but how unlikely is it? Should we occasionally observe a hot block get hotter when we put it in contact with a cold block? If you put an ice cube in your drink, could your drink get warmer and the ice cube get colder? This would not violate conservation of energy, because the total energy of the two objects would still be the same. So what physical principle would be violated?

Reversibility

A related question, and a very deep one, has to do with the "reversibility" of processes. The fundamental physical interactions seem to be completely "reversible" in the following technical, physics sense: Make a movie of an alpha particle scattering off a gold nucleus, then run the movie forward or reversed. Both views look entirely reasonable. In contrast, make a movie of a ball bouncing on the floor with decreasing height on each bounce, then run the movie forward or reversed. The reversed movie looks silly: you see a ball bouncing higher and higher with each bounce! Would such a motion violate conservation of energy? Not necessarily. It could be that the atoms in the floor happen on average to give energy to the ball, and the floor is getting colder as a result of this continuing loss of energy. So what physical principle is violated by a process represented by the reversed movie?

Statistical models

This chapter addresses these questions by applying probabilistic and statistical ideas to the behavior of systems. This subject is called "statistical mechanics." We will first apply statistical mechanics to a simple model of a solid, and we'll find that we can go surprisingly far with just a few new concepts. The main concepts that we will develop for the specific case of a solid apply in general to a wide variety of systems.

How will we know whether our statistical model explains anything? Our criterion for understanding is whether the predictions of our microscopic model agree with measurements of macroscopic systems, such as measurements of heat capacity (the amount of energy required to increase the temperature of an object by one degree).

11.2 A STATISTICAL MODEL OF SOLIDS

We have noted in previous chapters that the interactions between atoms in a solid are electric in nature, but that a detailed model of them involves quantum mechanics. The interatomic potential energy function encountered in Chapter 6 provides a reasonably accurate description of these interactions. Because of the similarity of this function to the potential energy curve for a harmonic oscillator (a mass on a spring), in previous chapters we have modeled a solid as a large number of tiny masses (the atoms) connected to their neighbors by springs (the interatomic bonds), as shown in Figure 11.3. This model has allowed us to understand qualitatively how solids interact with other objects. We would now like to use this model to ask detailed quantitative questions about the distribution of energy in a solid.

In this chapter we focus on calculating how probable a particular distribution of speeds or energies in a solid would be. We do this for a solid because it happens to be easier to calculate probabilities for a simple model of a solid than it is for a gas or a liquid. The atoms in a solid are nearly fixed in position, so we don't have to consider how likely different spatial arrangements might be (unlike the situation for a gas). Reasoning about the energy distribution in a solid will also enable us to draw conclusions about the transfer of thermal energy from one solid object to another.

The Einstein model of a solid

Because our goal is to calculate the probability of particular distributions of energy among atoms, we can simplify our model of a solid even further. We can think of each atom as moving independently of its neighbors, as though it were connected to rigid walls rather than to other atoms (Figure 11.4). Of course in a real solid energy is exchanged with neighboring atoms. However, to address our current questions we do not need to worry about the mechanism of energy exchange between atoms, since we will focus on calculating probabilities for various distributions of energy among the atoms of the solid.

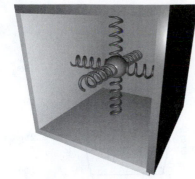

Figure 11.4 A single atom can oscillate in three dimensions. We simplify our model by assuming that each atom moves independently of the surrounding atoms—in effect, that it is connected to rigid walls instead of moving atoms.

This is a different kind of model from those we have so far constructed, because this very simple model would not allow us to predict in detail the dynamic motion of each atom in the solid, nor to ask how long it would take for energy to flow from one end of a solid to the other. In fact, there could not be propagation of sound in such a solid, with its "rigid walls." The model will not shed light on the details of a process, but it does allow us to ask questions about initial and final states. Since at the moment the questions we want to address involve initial and final states, this model is useful, and it is mathematically much simpler than the connected-atoms model.

Einstein proposed this simple model in 1907, and he found that some basic properties of solids such as heat capacity could in fact be understood using this model. This model also allows us to understand in detail the statistical nature of energy transfer between a hot object and a cold object, and why two objects come to "thermal equilibrium" (the same final temperature). This simple model of a solid will help us gain a more sophisticated and powerful understanding of the meaning of temperature.

A three-dimensional oscillator—three one-dimensional oscillators

We will consider each atom in a solid to be connected by springs to immovable walls. Each isolated atom is a three-dimensional spring-mass system, with $\vec{s}$ representing the three-dimensional vector displacement away from a fixed equilibrium position (because we are ignoring collective motions of groups of atoms). Since $p^2 = p_x^2 + p_y^2 + p_z^2$ and $s^2 = s_x^2 + s_y^2 + s_z^2$, we can write the energy of a three-dimensional classical oscillator as consisting of three parts, corresponding to the x, y, and z oscillations:

$$K_{\text{vib}} + U_s = \left(\frac{p_x^2}{2m} + \frac{1}{2}k_s s_x^2\right) + \left(\frac{p_y^2}{2m} + \frac{1}{2}k_s s_y^2\right) + \left(\frac{p_z^2}{2m} + \frac{1}{2}k_s s_z^2\right)$$

Recall from Chapter 7 that a quantized oscillator (a quantum mechanical "ball and spring") has evenly spaced energy levels. A complete quantum mechanical analysis of an oscillator that is free to oscillate in three dimensions rather than just one dimension leads to the conclusion that the motion of a 3-D oscillator can be separated into x, y, and z components, each of which has the same energy level structure as the familiar one-dimensional oscillator. This is mathematically equivalent to replacing each three-dimensional oscillator (an atom) with three ordinary one-dimensional oscillators, which we will do to simplify our model. We will think of a block as containing N one-dimensional oscillators, corresponding to $N/3$ atoms (Figure 11.5).

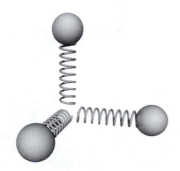

Figure 11.5 Since we are considering the atoms to be independent, in our model we can replace a single three-dimensional oscillator (one atom) by three independent one-dimensional oscillators.

In Chapter 7 we discussed energy quantization in atomic spring-mass systems. As predicted by quantum mechanics and abundantly confirmed by experiments, energy can be added to a one-dimensional atomic oscillator only

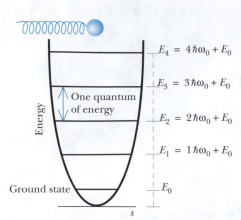

Figure 11.6 The quantized energy levels of a one-dimensional atomic oscillator.

in multiples of one "quantum" of energy $\hbar\omega_0$, where $\omega_0 = \sqrt{k_{s,i}/m_a}$ (Figure 11.6):

Added energy can be 0, $1\hbar\omega_0$, $2\hbar\omega_0$, $3\hbar\omega_0$, etc.

Here $\hbar = h/(2\pi) = 1.05 \times 10^{-34}$ joule · second, $k_{s,i}$ is the interatomic spring stiffness, and m_a is the mass of the atom. The "ground state" (lowest energy level) of the quantum oscillator is not at the bottom of the parabola but $\frac{1}{2}\hbar\omega_0$ above the bottom. This offset doesn't matter for what we are going to do; it just sets the baseline energy level, and we will measure energies starting from the ground state.

Distributing energy among objects

When two identical blocks are brought into contact, their total energy is shared among all the oscillators (atoms) in both blocks. It seems plausible that the most likely outcome would be that the total thermal energy be shared equally between identical blocks. However, might there be some probability that the first block would have more energy than the other, or even have all of it? With a large number of atoms the probability of even a small deviation from the most probable distribution turns out to be hugely unlikely. In order to look at this question in detail, we need to find a way to calculate the probability of various possible distributions of the energy. If we can find the most probable energy distribution, we can predict the eventual equilibrium distribution of energy between two objects brought into thermal contact with each other.

Distributing energy: A single atom inside a solid

Consider a single atom inside a solid, whose energy we model in terms of the energy of three one-dimensional oscillators (x, y, z), neglecting interactions with the neighbors. Each of these three oscillators can have 0, 1, 2, etc. number of "quanta" of vibrational energy ($\hbar\omega_0$) added to its ground state.

Suppose the total vibrational energy added to the atom is 4 quanta, and we ask how we might distribute this energy among the three oscillators. We will soon see that enumerating all the possible ways of distributing the energy among the oscillators leads to a deeper understanding of the statistical behavior of a solid. We could give all the energy to the first one-dimensional oscillator and none to the others, or all to the second, or all to the third, as shown in Figure 11.7.

Or we could give three quanta to one oscillator, and give the remaining quantum to one of the others, as shown in Figure 11.8.

Or we could give two quanta to one oscillator, and distribute the other two to the others, as shown in Figure 11.9.

That's it—there aren't any more ways to distribute the energy. (Check to make sure that you can't think of any other arrangements.) By explicitly listing all the possible arrangements, we see that there are 15 different ways that the four energy quanta could be distributed among the three one-dimensional oscillators.

In each of these 15 cases the total energy of the three-oscillator atom is exactly the same. When we make macroscopic measurements of the energy of a block, we don't know and we usually don't care exactly how the energy is distributed among the atoms that make up the block, because the internal energy of the block is the same for all of these distributions. However, the number of different arrangements influences interactions with other objects, as we will see.

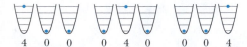

Figure 11.7 Three ways of distributing four quanta of vibrational energy among three one-dimensional oscillators.

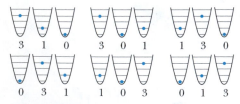

Figure 11.8 Six more ways of distributing four quanta of vibrational energy among three one-dimensional oscillators.

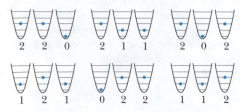

Figure 11.9 Still more ways of distributing four quanta of vibrational energy among three one-dimensional oscillators.

"Microstate" and "macrostate"

Some terminology: We say that there are 15 "microstates," such as the microstate with 3 quanta in the first oscillator, 0 in the second, and 1 in the third (the second microstate in Figure 11.8). The 15 microstates correspond to one "macrostate," which is characterized by the total energy being equal to 4 quanta of energy, no matter how distributed.

THE FUNDAMENTAL ASSUMPTION OF STATISTICAL MECHANICS

The fundamental assumption of statistical mechanics is that, over time, an isolated system in a given macrostate is equally likely to be found in any of its possible microstates.

For example, we isolate three oscillators (corresponding to one atom), containing a total energy of 4 quanta, which is a macrostate (we haven't said which oscillators have how many quanta). The fundamental assumption of statistical mechanics implies that if we repeatedly observe the detailed arrangement of the 4 quanta among the three oscillators (the microstate), over time we would find on the average that each of these microstates would occur 1/15th of the times that we looked. This fundamental assumption is plausible, but ultimately it is justified by the fact that deductions based on this assumption do agree with experimental observations.

Distributing energy: Two interacting atoms

We can now consider the case of two blocks in thermal contact (Figure 11.10). The importance of counting arrangements ("microstates") can be seen when we have systems of oscillators interacting with each other. Consider the smallest possible "blocks," two neighboring atoms (a total of 6 one-dimensional oscillators), and suppose 4 quanta of vibrational energy, $4\hbar\omega_0$, are distributed among these 6 oscillators. We already know that there are 15 ways to distribute the 4 quanta among the three one-dimensional oscillators of the first atom, and in that case there is only one way to distribute zero quanta to the second atom. Similarly, in Figure 11.11 we see that there are 15 ways to distribute the 4 quanta among the three one-dimensional oscillators of the second atom, and in that case there is only one way to distribute zero quanta to the first atom.

But since the two atoms are in contact with each other, the 4 quanta could also be shared between the atoms. For example, the first atom might have only 3 quanta, with the second atom having the other quantum.

? How many ways are there to distribute 3 quanta among 3 one-dimensional oscillators? Try to list all the possibilities, as we did for the case of 4 quanta.

You should have found ten ways, which we could list as 300 (that is, 3 quanta on the first oscillator, 0 on the second, and 0 on the third), 030, 003, 201, 210, 021, 120, 012, 102, and 111.

? For each of these arrangements, how many ways are there to distribute the one remaining quantum among the other 3 one-dimensional oscillators?

There are just three ways to arrange one quantum among the other 3 oscillators: 100, 010, and 001.

The product of these two numbers is the total number of ways of distributing the 4 quanta in such a way that there are 3 quanta on the first atom and 1 on the other. This product is $10 \cdot 3 = 30$ ways. These results are summarized in Figure 11.12.

Now let's give just 2 quanta to the first atom (leaving 2 quanta for the other atom).

? How many ways are there to distribute 2 quanta among 3 one-dimensional oscillators?

In principle you could prepare a system in one particular microstate and hope it would stay that way, but if there is the slightest bit of interaction among the pieces of the system, or with the surroundings, over time the fundamental assumption of statistical mechanics claims that all microstates will occur with equal probability. The Einstein model is simplified in that it doesn't provide a mechanism for energy to transfer from one atom to another (a change of microstate), nor even from one oscillator to another within the same atom. The Einstein model is useful for counting microstates, but for other purposes it needs to be supplemented by the notion that some slight interaction does occur.

Figure 11.10 Two small blocks in thermal contact.

atom 1	atom 2	# ways
4	0	15·1
0	4	1·15

Figure 11.11 If we give all four quanta to one atom or the other, there are 30 ways of distributing four quanta of energy between the two atoms (i.e. among six independent oscillators).

atom 1	atom 2	# ways
3	1	10·3
1	3	3·10

Figure 11.12 We find 60 more ways to distribute four quanta among two atoms (six oscillators) if we give three quanta to one atom and one quantum to the other.

atom 1	atom 2	# ways
2	2	6·6

Figure 11.13 We find 36 ways of distributing four quanta of energy by giving two quanta to each atom.

q_1	q_2	# Ways1	# Ways2	# Ways1 · # Ways2
0	4	1	15	15
1	3	3	10	30
2	2	6	6	36
3	1	10	3	30
4	0	15	1	15

Figure 11.14 Summary of the 126 different ways to distribute four quanta of vibrational energy between two atoms (each consisting of three independent oscillators).

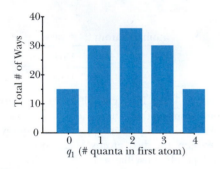

Figure 11.15 Bar graph showing the total number of ways of distributing four quanta of vibrational energy between two atoms (six one-dimensional oscillators).

You should find that there are six ways: 200, 020, 002, 101, 110, and 011.

? For each of these arrangements, how many ways are there to distribute the 2 remaining quanta among the other 3 one-dimensional oscillators?

Clearly there are also six ways to arrange the remaining 2 quanta among the other 3 oscillators. The total number of ways of distributing the 4 quanta in such a way that there are 2 quanta on the first atom and 2 on the other is $6 \cdot 6 = 36$ (Figure 11.13)

Now we have carried out enough calculations to be able to make a table of all the ways of distributing 4 quanta between two atoms (that is, among 6 one-dimensional oscillators). In this table, q_1 is the number of quanta (0 to 4) assigned to the first atom, and q_2 is the number of quanta assigned to the second (with q_1+q_2 necessarily equal to 4). In Figure 11.14 we show the number of ways to arrange the quanta among the first three one-dimensional oscillators, among the second three oscillators, and the product, which is the total number of ways of distributing q_1 and q_2 quanta.

Figure 11.15 is a graph of the total number of ways of distributing the quanta vs. q_1. Remember that the basic assumption of statistical mechanics is that all of these 126 different arrangements ($15 + 30 + 36 + 30 + 15 = 126$ microstates) with a total energy of 4 quanta are equally probable.

? Therefore, if two atoms share 4 quanta, what is the most likely division of the energy between the two? What is the probability at some instant that the energy is equally divided? What is the probability at some instant that the first atom has all the energy?

Evidently the most probable division is 2 and 2, since there are 36 ways for this to be done, whereas all other divisions have fewer ways for this to happen. The probability for a 2-2 division is 36/126, or 29%, so even though this is the most probable division, it happens less than one-third of the time. The probability that the first atom has all the energy is the number of ways for this to happen (15) divided by 126: 15/126 = 12%.

This is a microcosm of what happens when two identical blocks are brought into contact: the most probable division of the thermal energy is that it is shared equally. In order to understand in detail how this works in a real macroscopic system, we need to consider large numbers of atoms and large numbers of quanta. For numbers of atoms or quanta even slightly larger than the few we've been considering, it becomes practically impossible to figure out the number of arrangements of quanta by explicitly listing all the possibilities as we did up to now. We need a formula for calculating the number of arrangements.

11.X.1 For practice in counting microstates, determine how many ways there are to arrange 3 quanta among 4 one-dimensional oscillators. (This would represent one-and-a-third atoms, so this doesn't make physical sense.)

One system or many?

The results for distributing 4 quanta among 2 atoms (6 oscillators) can be thought of in two complementary ways. You can say, "I will make frequent observations of my isolated two-atom system, and I expect that in 29% of these observations I'll find the energy split evenly (2-2)." Alternatively, you can set up 100 of these isolated two-atom systems and say, "Whenever I look at my 100 systems, I expect that 29 of them will have the energy split evenly (2-2)." Sometimes the "one system, many observations" view is particularly helpful, and sometimes the "many systems, one observation" view is the more useful way to think about the statistical nature of a phenomenon.

A formula for the number of arrangements of quanta

Clearly it would be not only tedious but impractical to keep on counting states as we did above for systems involving very large numbers of atoms. We need a formula for calculating how many different ways a set of objects can be arranged. We can develop this formula in a general way, using concepts of probability, and then apply it to solids made up of atomic oscillators.

Suppose you have five billiard balls in a bag, numbered 1 through 5. You draw them out of the bag, one at a time, and record the sequence of numbers, such as 34152 (Figure 11.16). Then you put them back in the bag and repeat. How many different number sequences are possible?

- The first ball might be any one of the five balls.
- For each of these 5 possibilities, there are four possibilities for the second ball, or $5 \cdot 4 = 20$ choices so far.
- For each of the $5 \cdot 4 = 20$ choices of the first two balls, there are three different possible choices for the third ball, or $5 \cdot 4 \cdot 3$ choices so far.
- For each of the $5 \cdot 4 \cdot 3 = 60$ choices of the first three balls, there are 2 possible choices for the fourth ball, for a total of $5 \cdot 4 \cdot 3 \cdot 2 \cdot 1 = 120$ possibilities, since there is only one remaining ball to choose.

Evidently there are 120 different "permutations" of the five integers. It would be exceedingly tedious to list all these different arrangements, but we have a simple formula to calculate how many there are: 5!, which is the standard notation for "5 factorial," meaning $5 \cdot 4 \cdot 3 \cdot 2 \cdot 1$. Fortunately, for analyzing a solid all we care about is how many arrangements (microstates) there are for a particular total energy, not the details about how much of the energy is assigned to which oscillators. So all we need from a formula is the number of arrangements.

? We can easily check this factorial formula for the case of three balls. Explicitly list all possible permutations of the numbers 1, 2, and 3, and verify that there are indeed $3! = 3 \cdot 2 \cdot 1 = 6$ possible arrangements.

Making fewer distinctions

Now suppose that of the five balls, three are black (B) and two are colored (C), and we ask how many different arrangements of the colors are possible, such as BCBBC shown in Figure 11.18. We know that there are $5! = 120$ different arrangements of the numbered balls, but if we're only interested in the color sequence, the numerical order of the black balls is irrelevant, as is the numerical order of the colored balls.

There are $3! = 6$ permutations of the black balls among each other, and $2! = 2$ permutations of the colored balls among each other. Therefore there are many fewer than 120 distinctively different color sequences, and we need to correct for this by dividing by the extra permutations:

$$\text{\# of color sequences} = \frac{5!}{3!2!} = \frac{120}{(6)(2)} = 10$$

? Check this result by listing all the different ways of ordering 3 B's and 2 C's.

Arranging quanta among oscillators

By an appropriate choice of visual representation, we can convert our problem of calculating the number of arrangements of q quanta among N one-dimensional oscillators ($N/3$ atoms) into the problem we just solved, for which we have a formula. Consider again the specific case of $q = 4$ quanta distributed among $N = 3$ one-dimensional oscillators. We'll represent a quantum of energy by the symbol •, and a (fictitious) boundary between oscillators by a vertical bar **|**. Here is a picture of a particular situation where the first oscillator has 2 quanta, the second oscillator has 1 quantum, and

Figure 11.16 Five numbered billiard balls are taken one at a time from a bag. How many different number sequences are possible?

Figure 11.17 Five billiard balls are taken one at a time from a bag. Three are black and two are colored. How many different color sequences are possible?

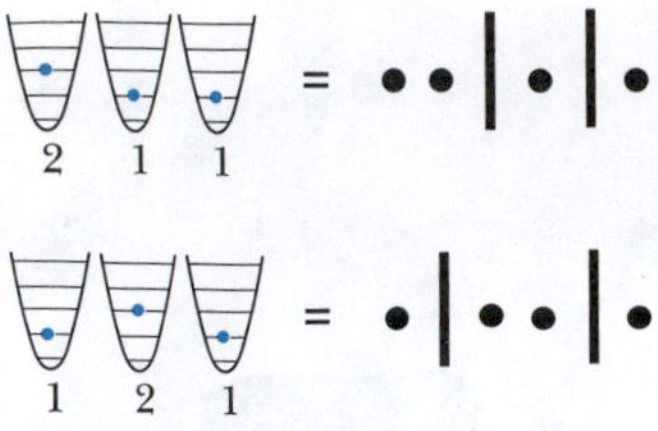

Figure 11.18 Two different representations of number of quanta in three oscillators.

the third oscillator has 1 quantum; together with the boundaries between oscillators we have 6 objects arranged in a particular sequence:

$$\cdot \cdot | \cdot | \cdot$$

This sequence represents a total energy of 4 quanta and 2 boundaries (a total of 6 things, if we consider a boundary to be a "thing"). Figure 11.18 illustrates two such sequences. Notice that we need $N - 1 = 2$ vertical bars to be able to indicate which oscillators have how much energy. How many such sequences like this are there? Rearranging quanta and boundaries is equivalent to moving quanta between oscillators. In this pictorial form, the problem is like having 4 red balls and 2 green ones. Therefore we can write a formula for the number of different arrangements (number of different microstates):

$$\frac{6!}{4!2!} = 15 \text{ ways of arranging 4 quanta among 3 oscillators}$$

This agrees with our earlier calculations. Generalizing to arbitrary numbers of quanta distributed among arbitrary numbers of oscillators, we have this important result for the number of microstates, written as Ω (Greek capital omega):

WAYS Ω TO ARRANGE q QUANTA AMONG N ONE-DIMENSIONAL OSCILLATORS

$$\Omega = \frac{(q + N - 1)!}{q!(N - 1)!}$$

11.X.2 Verify that this formula gives the correct number of ways to arrange 0, 1, 2, 3, or 4 quanta among 3 one-dimensional oscillators, given in the table on page 378.

Very big numbers

As you increase q and/or N, this formula gets very big very fast. For example, the number of ways to distribute 100 quanta among 300 oscillators (100 atoms) is about 1.7×10^{96}, which is 17 followed by 95 zeros!

In striking contrast, there is only one way to arrange to have all 100 energy quanta be placed on just one particular oscillator out of all the 300 oscillators. Although this arrangement would satisfy the requirement that the total energy of the system be 100 quanta, this is extremely improbable. The fundamental assumption of statistical mechanics is that in an isolated system, over time, all microstates are equally probable, so the odds that the actual microstate is the one with all the energy given to just one particular oscillator, of your choice, is 1 in 1.7×10^{96}, which makes this unlikely event essentially impossible.

A typical macroscopic object such as a block of ordinary size contains 10^{23} or more atoms, not a mere hundred, and the number of quanta is even larger. How likely is it that all the energy will be found concentrated on one particular atom? The mind boggles at the astronomically huge odds against this ever happening. Could it happen? Yes. Is any human ever likely to observe it? No.

11.X.3 Suppose you look once every second at a system with 300 oscillators and 100 energy quanta, to see whether your favorite oscillator happens to have all the energy (all 100 quanta) at the instant when you look. You expect that just once out of 1.7×10^{96} times you will find all of the energy concentrated on your favorite oscillator. On the average, about how many years will you have to wait? Compare this to the age of the Universe, which is thought to be about 10^{10} years. (1 year $\approx \pi \times 10^{7}$ seconds.)

11.3 THERMAL EQUILIBRIUM OF BLOCKS IN CONTACT

We now have the tools necessary for analyzing in some detail what will happen when two blocks are brought into contact and approach thermal equilibrium. We'll choose two blocks made of the same material, so a quantum of energy is the same for the atomic oscillators in both blocks (same atomic mass m; same interatomic forces, so same "spring stiffness" k_s). We'll make the analysis somewhat general by choosing blocks of different sizes. One block contains N_1 one-dimensional oscillators ($N_1/3$ atoms) and initially contains q_1 quanta of energy, and the other block contains N_2 one-dimensional oscillators ($N_2/3$ atoms) and initially contains q_2 quanta of energy.

? We want to treat the simple case where the total energy $q_1 + q_2$ of the two-block system remains fixed at all times. What simplifying assumption should we make about the situation?

During the entire process we assume that there is little energy transferred into or out of the surrounding air or the supports for the blocks, so that the total energy of the two blocks doesn't change. However, the number of quanta in each block, q_1 or q_2, need not stay fixed, since energy can flow back and forth between the two blocks.

Consider a very concrete example. Suppose $N_1 = 300$ (100 atoms), $N_2 = 200$ (about 67 atoms; or choose 201 oscillators if you wish to be exact), and there is a total energy distributed throughout the two blocks corresponding to $q_1 + q_2 = 100$ quanta (Figure 11.19). We use a computer to calculate the number of ways that q_1 quanta can be distributed among the $N_1 = 300$ oscillators of the first block (using the formula we developed earlier), and we multiply this number times the number of ways that $q_2 = (100 - q_1)$ quanta can be distributed among the $N_2 = 200$ oscillators of the second block.

The product of these two calculations is the number of ways that we can arrange the 100 quanta so that the first block has q_1 quanta and the second block has $q_2 = (100 - q_1)$ quanta, for a total energy shared between the two blocks of 100 quanta. We have the computer do this calculation for $q_1 = 0$, 1, 2, 3,...99, 100 quanta, which corresponds to $q_2 = 100$, 99, 98,...1, 0 quanta. The following table shows the first few results, where the number of microstates is denoted by Ω (Greek uppercase omega):

q_1	$q_2 = (100 - q_1)$	$\Omega_1 = \dfrac{(q_1 + 300 - 1)!}{q_1!(300 - 1)!}$	$\Omega_2 = \dfrac{(q_2 + 200 - 1)!}{q_2!(200 - 1)!}$	Total # of Ways $\Omega_1\Omega_2$
0	100	1	2.772 E+81	2.772 E+81
1	99	300	9.271 E+80	2.781 E+83
2	98	4.515 E+04	3.080 E+80	1.391 E+85
3	97	4.545 E+06	1.016 E+80	4.619 E+86
4	96	3.443 E+08	3.331 E+79	1.147 E+88
...	...	...	...	...

We see in the last column that there is an enormous number of ways of arranging 100 quanta among 500 oscillators, and that this number is growing rapidly in the table. In Figure 11.20 the possible number of arrangements of quanta is plotted on the y axis, and q_1, the number of quanta assigned to the first block, is plotted on the x axis.

The most probable arrangement (indicated by the highest point of the peak on the graph) is that 60 of the 100 quanta will be found in the first block, which contains 300 oscillators out of the total of 500 oscillators. This

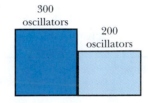

Figure 11.19 A total of 100 quanta of vibrational energy are available to be distributed between two systems, one consisting of 300 oscillators (100 atoms), the other consisting of 200 oscillators (about 67 atoms).

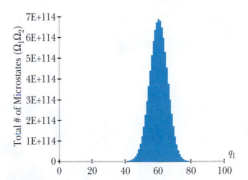

Figure 11.20 The number of ways of distributing 100 quanta of vibrational energy between two blocks having 300 and 200 oscillators, respectively. q_1 is the number of quanta in the first (larger) block.

seems reasonable, since it does seem most probable that the energy would be distributed uniformly throughout the two blocks (since they're made of the same material), and that would give 3/5 (60%) of the energy to the first block. It is gratifying that our statistical analysis leads to this plausible result.

It appears from the width of the peak shown in the graph that we shouldn't be too surprised if occasionally we would find that the first block contains anywhere between 40 to 80 of the 100 quanta, but it appears that it is very unlikely to find fewer than 40 or more than 80 of the quanta in the first block.

Relatively speaking, how likely is it for *none* of the energy to be in the first block? In that case, $q_1 = 0$, and from the table we see that the number of ways to arrange the 100 quanta this way is 2.772 E+81 (2.772×10^{81}). That's an awfully big number, but how big is it compared to the most probable arrangement, where $q_1 = 60$, for which there are about 7×10^{114} ways according to the graph? Evidently it is less likely by about a factor of 10^{33} (!) to find the energy split 0-100 rather than 60-40. Is it possible according to the laws of physics for none of the energy to be in the first block? Yes. Is it likely that we would ever observe such an unusual distribution? Most emphatically not!

Note that the number of ways to arrange the quanta 0-100 (2.772×10^{81}) isn't actually visible on the graph, because it is 10^{33} times smaller than the peak of the graph. Values of q_1 outside the 40 to 80 range are invisible on the graph, because they are relatively so very small compared to the peak.

Width of the distribution

Compared to the graph on page 378 of our earliest calculation (6 oscillators and 4 quanta), the peak shown in Figure 11.20 occupies a much narrower range of the graph, reflecting the sharply decreased relative probability of observing extreme distributions of the energy.

For further comparison, Figure 11.21 is a graph of the number of ways of distributing 1000 quanta among two blocks containing 3000 and 2000 oscillators. This peak is much narrower than the peak in the previous graph (where there were only 100 quanta distributed among only 300 and 200 oscillators). This is a general trend. The larger the number of quanta and oscillators, the narrower is the peak around the most probable distribution.

The fractional width of the peak is the width of the peak at half height divided by the value of q_1 that gives the maximum probability (in the present case, the width divided by 600). It can be shown that the fractional width of the peak is proportional to $1/\sqrt{q}$ or to $1/\sqrt{N}$, whichever is larger. That is, if you quadruple the number of quanta or the number of oscillators, the fractional width of the peak decreases by a factor of 2.

> At this point it is very useful to do problem 11.P.38, in which these ideas are made very concrete.

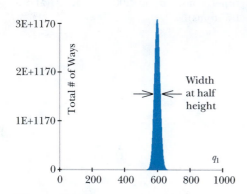

Figure 11.21 The number of ways of distributing 1000 quanta between two blocks containing 3000 and 2000 oscillators, respectively.

? Consider two blocks that are of ordinary macroscopic size, containing 3×10^{23} atoms and 2×10^{23} atoms, and many quanta per atom. Qualitatively, what would you expect about the width of the peak, if you could calculate it? How likely is it that you would ever observe a significant fluctuation away from the most probable 60-40 split?

You would expect the peak to be extremely narrow, and hence the probability of significant fluctuations would be very low.

The most probable is the only real possibility

These considerations show that in the world of macroscopic objects such as ordinary-sized blocks, the most probable arrangement is essentially the *only* arrangement that is ever observed. That is why you do not see a block suddenly leap up from the table when all of the thermal energy of the table floods into the block. On the other hand, at the microscopic level we should not be surprised if the energy of one of the atoms in a block varies a lot, since the most probable distribution includes many different arrangements of the quanta, with varying numbers of quanta on one particular atom.

Entropy and equilibrium

We now have a good statistical description of the thermal equilibrium of two blocks in contact. At equilibrium, energy is distributed between the two blocks in the most probable manner, based on having the largest number of ways of achieving this distribution (largest number of microstates for the macrostate of given total energy). We have been studying very small systems consisting of a hundred or more atoms, and we found that in small systems there is some significant probability of finding the energy distributed somewhat differently than the most probable 60-40 division. (Objects consisting of only a few hundred atoms are called "nanoparticles" and are currently the subject of intense research, because many of their properties and behaviors are intermediate between those of atoms and those of large-scale objects.) For large macroscopic objects containing 10^{20} atoms or more, the most probable distribution of the energy is essentially the only energy distribution we will ever observe, because the probability of distributions that are only slightly different is very small.

Next we will study more deeply the details of why a particular thermal equilibrium becomes established. Suppose for example that the first block (a nanoparticle with 300 oscillators, or 100 atoms) starts out with 90 quanta and the second (200 oscillators, or about 67 atoms) with only 10 quanta (Figure 11.22). When we put them together, we expect this to shift toward a 60-40 distribution (Figure 11.23). Studying the details of why this shift occurs will lead us to a deep understanding of the concepts of temperature and something called "entropy."

Since we start from a non-equilibrium energy distribution, it would be nice if we could make some kind of graph where we could see the total number of ways to arrange the quanta even when we are far from equilibrium. As we have seen, the graph of the total number of microstates is so strongly peaked that we don't see anything outside the peak. For example, on the graph in Figure 11.23 where we can see 10^{114} we can't see the relatively much smaller value of 10^{81}. A way to get around this problem is to plot the logarithm of the data, which makes the data visible across the whole range of energy distributions.

In statistical mechanics it is standard practice to use the base-e or "natural" logarithm ("ln"). For the case of distributing 100 quanta among 300 and 200 one-dimensional oscillators, which we studied before, Figure 11.24 shows a plot of the natural logarithms of the number of ways Ω_1 to arrange q_1 quanta in block 1, the number of ways Ω_2 to arrange $q_2 = (100 - q_1)$ quanta in block 2, and the total number of ways $\Omega_1\Omega_2$ to arrange the 100 quanta among the 500 oscillators. These quantities are plotted against q_1, of course running left to right, 0 to 100. You can think of q_2 as running right to left.

Because we are considering discrete microstates, each of these curves is really a set of 100 dots, but we connect the dots and make continuous curves. This is particularly appropriate when we deal with macroscopic objects, where an increment of one quantum along the x axis is practically an infinitesimal fraction of the axis.

By taking (natural) logarithms, we have converted a highly-peaked graph into a slowly varying one. To make sure you understand how this graph is related to the peaked graph, do the following:

? For this situation, we calculated on page 381 that there is just 1 way to arrange 0 quanta in block 1, and 2.772×10^{81} ways to arrange 100 quanta in block 2. Take the natural logarithm of this number and verify that all three curves on the graph make sense to you at $q_1 = 0$ (q_2

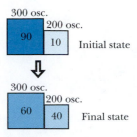

Figure 11.22 If the initial energy distribution between two systems in thermal contact is not the most probable energy distribution, energy will be exchanged until the most probable distribution is reached.

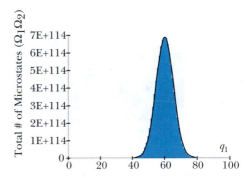

Figure 11.23 Ways of distributing 100 quanta of vibrational energy between a system of 300 oscillators and a system of 200 oscillators.

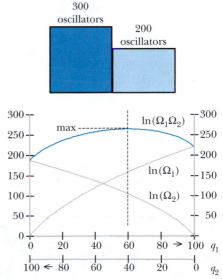

Figure 11.24 A logarithmic plot of the ways of distributing 100 quanta of vibrational energy between two systems, plotted against q_1, the number of quanta in object 1.

= 100). Note that at the right end of the graph, at q_1 = 100, the curve for $\ln(\Omega_1)$ goes higher than $\ln(2.772{\times}10^{81})$, because block 1 has a larger number of oscillators among which to distribute the 100 quanta (300 oscillators compared with 200).

Definition of entropy

We will deal repeatedly with the natural logarithm of the number of ways to arrange energy among a group of atoms (the number of microstates corresponding to a particular macrostate of specified energy). This quantity, $\ln(\Omega)$, when multiplied by the Boltzmann constant k, is called the "entropy" of the object and is denoted by the letter "S" (the triple equal sign means "is defined as"):

DEFINITION OF ENTROPY S

$$S \equiv k\ln\Omega$$

The Boltzmann constant is k = $1.38{\times}10^{-23}$ J/K, so entropy has units of joules per kelvin. The Boltzmann constant is included in the definition for consistency with an older, macroscopically based definition of entropy.

Since $k\ln(\Omega_1\Omega_2) = k\ln(\Omega_1) + k\ln(\Omega_2)$ (this is a property of logarithms), the entropy S of the two-block system is equal to the entropy S_1 of the first block plus the entropy S_2 of the second block. A consequence of defining entropy in terms of a logarithm is that we can consider entropy as describing a property of a system, and when there is more than one object in a system we get the total entropy simply by adding up the individual entropies, $S_{tot} = S_1 + S_2$, as is the case with energy ($E_{tot} = E_1 + E_2$).

Figure 11.25 is a plot of the entropies of each block, and their sum. Since both $k\ln(\Omega_1\Omega_2)$ and $\Omega_1\Omega_2$ go through a maximum at the same value of q_1, we can state the condition for thermal equilibrium of the two blocks, in terms of entropy:

CONDITION FOR THERMAL EQUILIBRIUM

In equilibrium the most probable energy distribution is that which maximizes the total entropy $S_{tot} = S_1 + S_2$ of two objects in thermal contact.

? In Figure 11.25, what is the physical significance of the fact that the top curve goes through a maximum for q_1 = 60?

The most probable energy distribution is a 60/40 sharing, corresponding to the 300/200 sizes of the two objects.

? In Figure 11.25, what is the physical significance of the point where the two lower curves cross each other?

Nothing! This point isn't anything special. It is the point where the entropy in one object is equal to the entropy in the other object (and equal to half the entropy of the combined system). It turns out that this doesn't have any real physical significance.

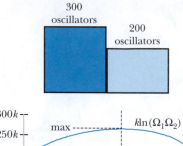

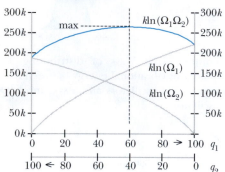

Figure 11.25 A plot of the entropies S_1, S_2, and $S_1 + S_2$ plotted against q_1, the number of quanta in object 1. Entropy is $k\ln\Omega$.

At this point it is very useful to do problem 11.P.39, to see in detail how to produce Figure 11.24.

11.4 THE SECOND LAW OF THERMODYNAMICS

The "first law of thermodynamics" is another name for the Energy Principle $\Delta E = W + Q$, which by now should be very familiar to you. The "second law of thermodynamics" is however something new. It can be stated in a number of equivalent forms, but there is a particularly useful formulation in terms of entropy:

THE SECOND LAW OF THERMODYNAMICS

If a closed system is not in equilibrium, the most probable consequence is that the entropy of the system will increase.

In other words, a closed system will tend toward maximum entropy.

As a specific example of the second law of thermodynamics, consider our two blocks. The most probable energy distribution is the one for which the total entropy is a maximum, and if initially the energy distribution is something else, it is highly likely that the entropy will increase. For nanoparticles there can be significant fluctuations away from this state of maximum entropy (with accompanying decrease in total entropy), but for ordinary-sized systems these fluctuations are extremely small, and for practical purposes the entropy of a closed macroscopic system never decreases.

Loosely, one can restate the second law of thermodynamics like this: "A closed system tends toward increasing disorder." This is overly vague without precise definitions of "order" and "disorder," and it is the definition of entropy as $k\ln\Omega$ that provides the needed precision. A closed system tends toward increasing entropy, which is a measure of the number of microstates corresponding to a particular macrostate of given energy.

Irreversibility

Any process in which the entropy of the Universe increases is "irreversible" in the technical, physics sense: a reversed movie of the process looks odd. But any process in which the entropy of the Universe doesn't change is in principle "reversible," and there do exist processes that are approximately or nearly reversible. For example, a steel ball bearing that bounces vertically on a steel plate rebounds almost to the same height from which it was dropped, and this represents a (nearly) reversible process, because the system returns (nearly) to its original state.

However, if we watch for a while, the ball bearing returns to lower and lower heights and eventually settles down to rest on the plate. If the ball bearing kept returning to its original height, this would not be a violation of energy conservation. It is the second law of thermodynamics that says that we cannot expect the process to be completely reversible. There is a larger number of ways to share the energy between the ball and the plate than the number of ways for the ball to keep all the energy.

Suppose we make a movie of the ball bearing bouncing lower and lower and coming to rest on the plate. Then we run it backwards, and our friends see the ball starting to bounce, and bouncing higher and higher (Figure 11.26). This needn't violate energy conservation: energy in the plate could be flowing into the ball. But this reversed movie certainly looks odd, presumably because we have an instinctive sense, based on lots of experience, that the entropy of the Universe increases rather than decreases.

Time running forward

Time running in reverse

REVERSIBLE AND IRREVERSIBLE PROCESSES

Reversible process: $\Delta S_{sys} + \Delta S_{surroundings} = 0$

Irreversible process: $\Delta S_{sys} + \Delta S_{surroundings} > 0$

11.X.4 You see a movie in which a shallow puddle of water coalesces into a perfectly cubical ice cube. How do you know the movie is being played backwards? Otherwise, what physical principle would be violated?

Figure 11.26 If the height of a bouncing ball increased with time instead of decreasing with time, this would not necessarily violate conservation of energy.

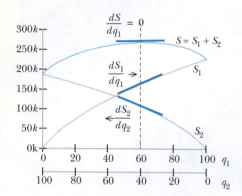

Figure 11.27 Entropy *vs.* number of quanta of energy in system 1.

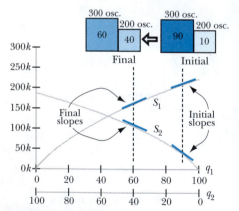

Figure 11.28 Location of the initial and final states of the two-block system on a plot of entropy *vs.* number of energy quanta in system 1.

11.5 WHAT IS TEMPERATURE?

Until now, we have associated temperature with the average energy of a molecule. The concept of entropy makes possible a deeper connection between our macroscopic measurements of temperature and a fundamental, atomic, statistical view of matter and energy. We will develop a statistically based definition of temperature.

Consider again our two blocks. On the graph in Figure 11.27 we plot entropy $S = k \ln \Omega$. Remember that we are approximating a set of dots by a smooth curve, since the dots are very close together (tiny energy spacing). At the maximum of the total entropy $S = S_1 + S_2$, the total entropy curve is horizontal (slope = 0), and the following relationship is true at equilibrium:

$$\frac{dS}{dq_1} = \frac{dS_1}{dq_1} + \frac{dS_2}{dq_1} = 0$$

Since $q_2 = (100 - q_1)$, this leads to

$$\frac{dS_1}{dq_1} - \frac{dS_2}{dq_2} = 0 \text{ (at equilibrium)}$$

dS_2/dq_2 is the slope moving from right to left and is positive, since q_2 increases from right to left. Therefore we that the slopes are the same:

$$\frac{dS_1}{dq_1} = \frac{dS_2}{dq_2} \text{ (at equilibrium; maximum total entropy)}$$

dS_1/dq_1 is a measure of the state of the first block, and dS_2/dq_2 is a measure of the state of the second block.

? What physical property can these derivatives represent? What physical property of the blocks is the same when the two blocks reach thermal equilibrium?

The blocks reach the same temperature. It must be that the derivative of the entropy with respect to the energy is somehow related to temperature.

To see what this relationship is, suppose that when the two blocks were initially brought into contact, the first block contained $q_1 = 90$ quanta and the second block only $q_2 = 100 - 90 = 10$ quanta, as illustrated in Figure 11.28. Note that in Figure 11.28 the initial slope of the entropy curve for block 2 is steeper than the initial slope of the entropy curve for block 1. (Remember: dS_2/dq_2 is the slope moving from right to left and is positive, since q_2 increases from right to left.) Therefore, if we remove one quantum of energy from block 1 and give it to block 2, we'll increase the entropy in block 2 more than we'll decrease the entropy in block 1.

? Will this result in a net increase or a net decrease in the total entropy S? (See Figure 11.28.)

Since the entropy of block 2 increases more than the entropy of block 1 decreases, there is a net increase of the entropy of the two blocks together. We already established that the state of the two blocks will evolve toward greater total entropy, not less total entropy.

? Consider which block gives up energy, and which block takes it up. Which block is initially at a higher temperature?

Evidently block 1 must be at the higher temperature, because on average it will give energy to block 2 in order that the total entropy increase.

? Does a steeper or a less steep slope of entropy *vs.* energy correspond to higher temperature?

Initially, block 1 has the higher temperature, and S_1 vs. q_1 has the smaller slope. Initially, block 2 has the lower temperature, and S_2 vs. q_2 has the larger slope.

> **?** Therefore, which of the following looks like a better guess for relating temperature to entropy: $T_1 \propto dS_1/dq_1$? Or $1/T_1 \propto dS_1/dq_1$?

Since block 1 has the higher temperature and the smaller slope (smaller dS/dq), it seems possible that $1/T_1 \propto dS_1/dq_1$.

Definition of temperature

For these reasons, we *define* temperature in terms of dS/dq. But the magnitude of one quantum of energy varies for different systems, so we can't compare different systems if we use q. So instead of q we use E_{int}, the internal energy above the ground state of a system measured in joules. For a group of oscillators, E_{int} is the number of quanta q times the energy per quantum: $E_{int} = q\hbar\sqrt{k_s/m}$ for a system of harmonic oscillators, where k_s is the spring stiffness.

DEFINITION OF TEMPERATURE T

$$\frac{1}{T} \equiv \frac{dS}{dE_{int}}$$

The Boltzmann constant k in the definition $S \equiv k\ln\Omega$ makes the units come out right. With the internal energy E_{int} of a system measured in joules (J), and entropy S measured in joules per kelvin (J/K), temperature T is measured in kelvins (K). (Remember that we're approximating a set of closely-spaced dots by a curve, so taking a derivative makes sense.)

This is a highly sophisticated and abstract way of defining temperature. How does this relate to the temperature that is measured by an ordinary thermometer? We will see in the next chapter that the temperature defined by $1/T = dS/dE_{int}$ is the same as the "absolute" temperature, which appears for example in the ideal gas law $PV = N_{moles}RT$, which you have probably encountered in previous chemistry or physics studies. On this temperature scale, ice melts at a temperature of +273.15 K (0° C), and water boils at +373.15 K (100° C). To put it another way, absolute zero Kelvin is at −273.15° Celsius. For our purposes, it will be adequate to say ice melts at 273 K. Room temperature is about 20° C or about 293 K.

To recapitulate, the greater the dependence of the entropy on the energy for an object (the steeper the slope of S vs. E_{int}), the more "eager" the object is to take in energy to contribute to increasing the total entropy of the total system. It will do this if it can take energy from another object that has a smaller dependence of entropy on energy (smaller slope), since the second object's entropy will decrease less than the first object's entropy will increase. An object with a large dS/dE_{int} has a low temperature, since it is low-temperature objects that take in energy from high-temperature objects.

Conversely, the smaller the dependence of the entropy on the energy, the less reluctant the object is to give up energy and decrease its own entropy, if it can give the energy to an object with a greater dependence of entropy on energy. An object with a small dS/dE_{int} has a high temperature, since it is high-temperature objects that donate energy to low-temperature objects.

> At this point it is useful to do problem 11.P.40, in which you calculate the temperature of an object as a function of the amount of energy in the object.

W, Q, and entropy

More formally, we should write $1/T = \partial S/\partial E_{int}$, involving a "partial derivative," which means that we hold everything but S and E constant when we take the derivative. In particular, we hold the volume constant, which means that we do no work on the system, so we should write

$1/T = (\partial S/\partial E_{int})_V$, where the subscript V means "take the (partial) derivative holding the volume constant."

One might ask whether there is any entropy change of the surroundings associated with energy exchange in the form of work. The answer turns out to be no for processes that are reversible (there do exist irreversible processes in which work can be associated with an increase in entropy). Energy transfer caused by a temperature difference between objects in contact, Q, is "disorganized" energy transfer and is associated with entropy change. Work is "organized" energy transfer and in many situations doesn't affect the entropy. It is beyond the scope of this textbook to prove this rigorously, but we can illustrate the basic issues in terms of an Einstein solid.

If you mechanically squeeze or compress a block, doing work on it to raise its energy, it can be shown that you shift the energy levels upward without changing which state an oscillator is in, as shown in Figure 11.29. The "spring" stiffnesses don't change, but there is more energy stored in the "springs." On the other hand, if there is transfer of energy into the system due to a temperature difference, Q, there is no change in the energy level, but an oscillator will jump to a higher level. This jump from one level to another affects the probability calculations and the entropy, which is the natural log of the number of ways to arrange the energy quanta.

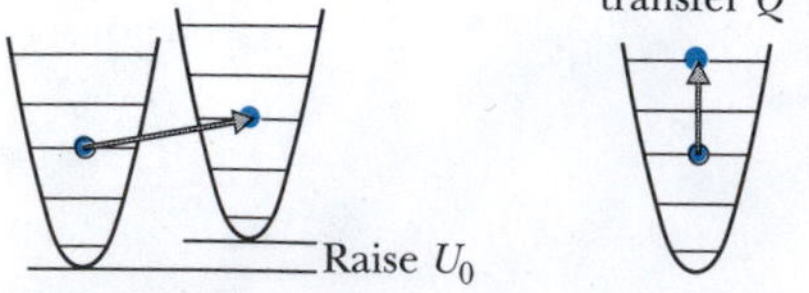

Roughly speaking, transfer of energy Q due to a temperature difference alters which state the system is in and affects the entropy. Work W alters the energy levels without changing which level the system is in, and doesn't affect the entropy. A small amount of transfer of energy into a system due to a temperature difference of amount Q, with $\Delta E_{int} = Q$, leads to an entropy change of that system.

Since $\dfrac{1}{T} = \dfrac{\Delta S}{\Delta E_{int}} = \dfrac{\Delta S}{Q}$, we conclude the following:

Figure 11.29 Mechanical work (compression or stretching of an oscillator system) compared to energy transfer due to a temperature difference to the same system.

ENTROPY CHANGE ASSOCIATED WITH SMALL Q

$$\Delta S = \frac{Q}{T} \text{ (small } Q \text{, associated with nearly constant } T)$$

If there is a great deal of energy transfer due to a temperature difference, causing a large temperature change in a system, we must add up (integrate) the contributions to the entropy change.

> *11.X.5* There was transfer of energy of 5000 J into a system due to a temperature difference, and the entropy increased by 10 J/K. What was the approximate temperature of the system?

> *11.X.6* It takes about 335 joules to melt one gram of ice. During the melting, the temperature stays constant. Which has higher entropy, a gram of liquid water at $0°$ C or a gram of ice at $0°$ C? Does this make sense? How large is the entropy difference?

Can the entropy of an object decrease?

? We analyzed two blocks that initially had a 90-10 energy distribution. In the process of changing to a 60-40 distribution, was it possible for the entropy of *one* of the blocks to decrease significantly? Is this a violation of the second law of thermodynamics?

The entropy of block 1 decreased, but the entropy of block 2 increased even more. We believe that the Universe is a closed system, and that the total entropy of the Universe continually increases and never decreases. However, some portions of the Universe may experience a decrease of entropy, as long as there is at least as much increase elsewhere.

Irreversibility again

We started with a 90-10 distribution of the energy but found the most probable final distribution to be 60-40. This change is effectively irreversible. The 60-40 distribution is so enormously more probable than the 90-10 arrangement that an observer will essentially *never* see a 90-10 distribution spontaneously recur. Closed macroscopic systems (such as two blocks inside an insulating box) always evolve toward the most probable arrangement, and stay there (or so close that you can hardly tell the difference).

11.6 SPECIFIC HEAT CAPACITY OF A SOLID

In Problem 11.P.40 you calculated the relationship between the temperature of a block and its energy. How can we compare these of calculations with experiment, when we don't have a good way to measure the total energy in a solid block? It is, however, possible to measure *changes* in energy and temperature. Suppose that we add a known amount of energy ΔE to a system, and we measure the resulting rise in temperature ΔT. The ratio of the energy input to the temperature rise of an object is called the heat capacity of that object. If we express this on a per-gram or per-mole or per-atom basis, it is called "specific heat capacity." We will work with specific heat capacity on a per-atom basis:

$$C = \frac{\Delta E_{atom}}{\Delta T}, \text{ where } \Delta E_{atom} = \frac{\Delta E_{system}}{N_{atoms}}$$

Different materials have different specific heat capacities. For example, at room temperature the specific heat capacity of water (on a per-gram basis) is 4.2 J/K/gram, while the specific heat capacity of iron at room temperature is 0.84 J/K/gram. Specific heat capacity is an important property of a material for two reasons. It has practical consequences in science and engineering because it, along with the thermal conductivity, determines the thermal interactions that the material has with other objects. Also, specific heat capacity is an experimentally measurable quantity that we can compare with theoretical calculations, to test the validity of our models.

How would we measure the heat capacity of a block? One way is to enclose the block inside an insulating box and use an electric heater inside the box to raise the temperature of the block (Figure 11.29). The electric power (in watts) times the amount of time the current runs (in seconds) gives us the input energy ΔE (in joules), and we can use a thermometer to measure the temperature rise ΔT of the block. The specific heat capacity (on a per-atom basis) is $\Delta E_{atom}/\Delta T = (\Delta E_{block}/\Delta T)/N_{atoms}$. The heater itself should have a small mass compared to the sample of material whose heat capacity we're measuring, so that the heater isn't a significant part of the material that is being warmed up.

The following exercise illustrates another way to measure specific heat capacity, on a per-gram basis:

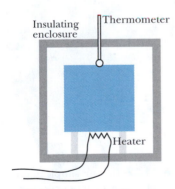

Figure 11.30 Schematic diagram of an apparatus for measuring the heat capacity of a solid.

11.X.7 A 100 gram block of metal at a temperature of 20° C is placed into an insulated container with 400 grams of water at a temperature of 0° C. The temperature of the metal and water ends up at 2° C. What is the specific heat capacity of this metal, per gram? Start from the Energy Principle. The specific heat capacity of water is 4.2 joules/K/gram.

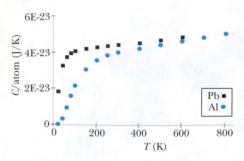

Figure 11.31 Measured heat capacities of lead and aluminum, shown on a per-atom basis.

Experimental results for specific heat capacity

Measurements of the specific heat capacity of aluminum (circles) and lead (squares) are displayed in Figure 11.31 (the numerical data are shown in a table below). They are plotted as heat capacity (J/K) per atom to facilitate comparison. Note that at low temperatures the specific heat capacity of both substances changes dramatically with temperature. It is not initially obvious why this should be the case, and in fact this temperature dependence was not predicted by the classical theory, developed in the 1800's. Note also that at room temperature and above, the specific heat capacity (on a per-atom basis) of both materials is about the same.

Will our simple statistical model of a solid be able to predict the temperature dependence of the specific heat capacity of a solid? In problem 11.P.40 you determined the relationship between energy and temperature for a block (in fact, you did the calculation for two different blocks). This calculation can easily be extended to predict the heat capacity per oscillator, which is $\Delta E_{osc}/\Delta T$, where E_{osc} is the average energy per oscillator. We can compare this prediction with experimental data by noting that the heat capacity per atom is 3 times the heat capacity per oscillator.

Effective spring stiffness

In Chapter 4 we determined the interatomic spring stiffness $k_{s, i}$ of aluminum and lead by relating it to the macroscopic stress-strain relationship (Young's modulus). In problem 11.P.41) you will use the Einstein independent-oscillator model of a solid to predict the specific heat capacity of aluminum and lead. We need to use a value for $k_{s, i}$ in this model, in order to convert the number of quanta to a value for energy in joules. Presumably the spring stiffness $k_{s, i}$ that we use in $\hbar\sqrt{k_{s, i}/m_a}$ should be related to the spring stiffness $k_{s, i}$ determined from Young's modulus. However, we shouldn't expect our predictions to agree exactly with experimental data, because our model ignores the fact that the atoms are not actually isolated from each other.

Moreover, the effective spring stiffness for oscillations in the x, y, or z directions can be expected to be larger than the value of $k_{s, i}$ estimated in Chapter 4 for two reasons.

First, each atom is attached to two "springs" (Figure 11.32). When an atomic core is displaced a distance s_x from its equilibrium position, the spring to the left and the spring to the right each exert a force of $k_{s, i}s_x$, so the combined force is $2k_{s, i}s_x$. This implies that the "effective spring stiffness" for oscillations would be $2k_{s, i}$.

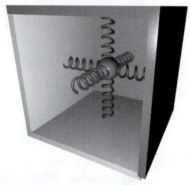

Figure 11.32 In our simple model of a solid, each atom experiences forces from two interatomic "springs." Each of these springs is half as long as an interatomic bond.

Second, in this simple model we divide the solid into "cubes" surrounding each independent atomic oscillator (Figure 11.32), so each of the two springs to the left and right of an atomic core is effectively half the length of a full interatomic spring, and each half-spring would have a spring stiffness of $2k_{s, i}$.

These considerations suggest that the effective spring stiffness for thermal oscillations in the context of the Einstein model might be a factor of 4 larger than the spring stiffness we determined in Chapter 4 from stretching a wire.

We emphasize that the Einstein model is clearly deficient, since an atom is not actually enclosed in rigid walls. However, having chosen this model we follow through in a consistent way, including using a spring stiffness k_s appropriate to the model. There are more sophisticated models for determining heat capacity that fit experimental data even better than the Einstein model does, but the Einstein model captures the main features.

At this point it is important to do problem 11.P.41, in which you compute the specific heat capacity of aluminum and lead, and compare with actual experimental data.

Specific heat capacity as a function of temperature

Figure 11.33 displays the result of a calculation of the specific heat capacity of copper on a per atom basis. The colored dots represent actual experimental data, and the solid line represents the computed values. An outline of the calculation that produced this curve is this:

- Use the mass of a copper atom and the effective interatomic spring stiffness for copper, calculated in Chapter 4 and modified by the factor discussed above, to find the value of one quantum of vibrational energy for this system.
- For a small particle of solid copper containing a given number of atoms, calculate the entropy of the system for increasing values of energy added to the system, in amounts of one quantum (as in problem 11.P.39).
- Calculate the change in entropy due to the addition of each quantum of energy, and use the definition of temperature ($1/T \approx \Delta S/\Delta E$) to find the temperature of the particle for each value of energy (as in problem 11.P.40)
- Calculate the change in temperature of the particle due to the addition of each quantum of energy, to find the specific heat capacity ($C \approx \Delta E/\Delta T$), and plot values of C (per atom) versus T (K) (as in problem 11.P.41).

An example of an iterative calculation done by hand is given on page 392.

Our simple model of a solid as a collection of independent harmonic oscillators does a surprisingly good job of fitting experimental data over a wide range of temperatures. If we were able to include more oscillators in our calculations (by using double precision arithmetic to handle larger numbers, or by using mathematical techniques such as Stirling's approximation for factorials), we could extend our predictions to lower temperatures.

The deviation of the experimental data from the prediction at very high temperatures suggests that at these temperatures a simple harmonic oscillator is not a good model of the atoms in this solid. At high quantum levels, the harmonic oscillator potential energy curve is a poor approximation to the actual potential energy curve describing interatomic interactions (see Chapter 6).

Note an interesting aspect of the graph: at room temperatures, the predicted specific heat capacity on a per-atom basis approaches a constant value of $3k$ (three times the Boltzmann constant, 1.38×10^{-23}). This value agrees quite well with the measured specific heat capacity of a variety of substances at ordinary temperatures.

Energy quantization and specific heat capacity

The key difference between our model of solids (the Einstein model) and earlier classical models that did not predict a temperature dependence of specific heat capacity lies in the quantization of energy in atomic oscillators. Statistical mechanics was originally developed in the 1800's, before the beginning of quantum theory. The classical theory did predict that at high temperatures the energy of each atom in a solid would be $3kT$ (kT per non-quantized one-dimensional oscillator), so that the heat capacity per atom would be $3k$.

If you examine your calculations, you'll find that the high-temperature limit (heat capacity per atom = $3k$) corresponds to temperatures high enough that kT is significantly larger than one quantum of energy. In a situation where the average energy per one-dimensional oscillator is about kT and is large compared to one quantum of energy, the quantization of the energy doesn't make much difference in the analysis. That is, if energy quanta are small compared to the energies of interest, mathematical analysis can be carried out adequately by considering energy to be continuous rather than discrete. In such cases pre-quantum and quantum calculations will give the same results, as is the case here at high temperatures.

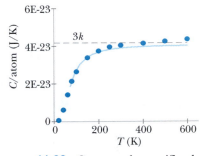

Figure 11.33 Computed specific heat capacity of copper, compared with experimental values.

On the other hand, at low temperatures the average energy per one-dimensional oscillator is comparable to or smaller than one quantum of energy, and the continuous, nonquantum calculations are not valid. The classical theory provided no explanation for the discovery that as materials were cooled down to very low temperatures, the specific heat capacity decreased with decreasing temperature.

In 1907 Einstein carried through the analysis we have just done and predicted the curve we have just plotted. The good agreement with both low-temperature *and* high-temperature measurements of specific heat capacity was strong additional evidence for the hypothesis that the energy of oscillators is indeed quantized. (At extremely low temperatures the model of a solid must be refined, taking into account the electrons in the metal, among other things, to achieve full agreement between theory and experiment.)

The fact that the specific heat capacity for all materials decreases at low temperatures has practical consequences. For example, it makes it difficult to cool a sample to a very low temperature. Cooling a sample depends essentially on putting the sample in contact with a "sink," a large object that is already at a lower temperature, so that there is transfer of energy Q due to the temperature difference out of the sample into the sink, lowering the temperature of the sample and not raising the temperature of the sink very much. But at very low temperatures the sink has a low specific heat capacity, so this is difficult to achieve.

11.X.8 In an insulated container a 100-watt electric heating element of small mass warms up a 300 gram sample of copper for 6 seconds. The initial temperature of the copper was 20° C (room temperature). Predict the final temperature of the copper, using the $3k$ specific heat capacity per atom.

11.X.9 Since $\Delta T = \Delta E / C$, what will happen at low temperatures to the temperature of the sink when some energy ΔE is transferred to it from the sample? Why is this unfortunate?

Which of our results are general?

We have analyzed simple models of solid matter. Nevertheless the basic conclusions are quite general. For example, if our two model blocks were made of different materials, so that the energy quanta were of different size in the two blocks, this would complicate the procedures for evaluating the number of ways Ω to arrange the energy, but the basic conclusion would remain, that the entropy will increase to a maximum.

Example: A lead nanoparticle

In Chapter 4, from Young's modulus for lead we found that the effective interatomic "spring" stiffness was about 5 N/m.

(a) For a nanoparticle consisting of 3 lead atoms, what is the approximate temperature when there are 5 quanta of energy in the nanoparticle?

(b) What is the approximate specific heat capacity (per atom) at this temperature? Compare with the approximate high-temperature specific heat capacity (per atom) for lead.

We'll use the Einstein model of a solid, though with only three atoms our conclusions will be very approximate.

(a) We model the three atoms as 9 independent quantized oscillators. One quantum of energy in one of these oscillators is this many joules:

$$\hbar\omega_0 = \hbar\sqrt{k_s/m} = (1.05\times10^{-34}\ \text{J·s})\sqrt{\frac{4(5\ \text{N/m})}{(207\times1.7\times10^{-27}\ \text{kg})}} = 7.92\times10^{-22}\ \text{J}$$

q	Ω	$\ln\Omega$
4	$\dfrac{12!}{4!8!} = 495$	6.20
5	$\dfrac{13!}{5!8!} = 1287$	7.16
6	$\dfrac{14!}{6!8!} = 3003$	8.01

Figure 11.34 Calculation of entropy ($S = k\ln\Omega$).

We need to calculate the number of ways to arrange q quanta in the neighborhood of $q = 5$, and the associated entropy (Figure 11.34).

Since $1/T = \partial S/\partial E$, $T \approx \Delta E/\Delta S$. Take differences from $q = 4$ to $q = 6$, in order to approximate the slope at $q = 5$; energy increase is 2 quanta:

$$T \approx \frac{N\hbar\omega_0}{k\Delta(\ln\Omega)} = \frac{2(7.92\times10^{-22}\,\text{J})}{(1.38\times10^{-23}\,\text{J/K})(8.01 - 6.20)} \approx 63.7\,\text{K}$$

(b) Find T_1 at midpoint of 4 to 5 interval, T_2 at midpoint of 5 to 6 interval, energy increase from T_1 to T_2 is one quantum:

$$T_1 \approx \frac{(7.92\times10^{-22}\,\text{J})}{(1.38\times10^{-23}\,\text{J/K})(7.16 - 6.20)} \approx 60.1\,\text{K}$$

$$T_2 \approx \frac{(7.92\times10^{-22}\,\text{J})}{(1.38\times10^{-23}\,\text{J/K})(8.01 - 7.16)} \approx 67.7\,\text{K}$$

As expected, these two temperatures bracket the temperature of 63.7 K at $q = 5$. We want the specific heat capacity on a per-atom basis, and there are 3 atoms:

$$C_{\text{per atom}} = \frac{1}{3}\frac{\Delta E}{\Delta T} \approx \frac{1}{3}\frac{(7.92\times10^{-22}\,\text{J})}{(67.7 - 60.1)\,\text{K}} \approx 3.44\times10^{-23}\,\text{J/K}$$

The approximate high-temperature specific heat capacity for a solid is $3k$ per atom, which is $3(1.38\times10^{-23}\,\text{J/K}) = 4.14\times10^{-23}\,\text{J/K}$. This suggests that at a temperature of about 63.7 K, lead is not quite at the high-temperature limit.

The actual experimental value for lead at 63.7 K (interpolation data on page 407) is 22.6 J/K/mole, which on a per-atom basis is

$$(22.6\,\text{J/K/mole})\left(\frac{1\,\text{mole}}{6\times10^{23}\,\text{atoms}}\right) = 3.8\times10^{-23}\,\text{J/K}$$

Our 3-atom calculation is not a very accurate model of a macroscopic block of lead, in part because with only three atoms it is not a very good model to say that each atom is connected by spring-like bonds to six neighboring atoms. Despite the failings of our simple model, our result is rather close to the actual value.

11.7 THE BOLTZMANN DISTRIBUTION

How does the density of Earth's atmosphere change as altitude increases? How does the distribution of the speeds of molecules in a gas change as the temperature of the gas increases? So far we have mainly been concerned with the thermal equilibrium of two blocks, and how thermal equilibrium arises as that particular distribution of energy between the two blocks that has (by far) the largest number of ways to arrange the quanta. What can we say about the probability of observing a particular amount of energy associated with one particular atom or oscillator? Addressing this question will lead us to the "Boltzmann distribution," which provides insight into the behavior of a very wide variety of physical, chemical, and biological phenomena.

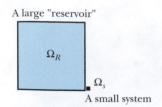

Figure 11.35 A small system in contact with a very large system (a "reservoir").

A constant-temperature reservoir

Consider a large system in contact with a small system, as shown in Figure 11.35. The two systems are isolated from their surroundings and share a fixed amount of energy E_{tot}. The large system's energy and temperature cannot change very much, so we call the large system a nearly constant-temperature "reservoir," which keeps the small system always at nearly the same temperature.

The entropy of the total system (reservoir plus small system) will increase rapidly if energy is transferred from the small system to the reservoir, because in the large system that energy can be distributed among a very much larger number of microstates than were available in the small system. Since we expect the total entropy to increase, we expect that most of the time, most of the total energy E_{tot} will be found in the reservoir, and little energy in the small system. We will show how to make this idea quantitative.

Let $\Omega_{res}(E_{res})$ be the number of microstates in the reservoir when the amount of energy in the reservoir is E_{res}. Similarly, let $\Omega(E)$ be the number of microstates in the small system when it has an amount of energy E, where E is much less than E_{res}, because the reservoir is so big. The total energy of the combined system is $E_{tot} = E_{res} + E$, which is a fixed number because the two systems are isolated from their surroundings. The total number of microstates for the combined system corresponding to E_{tot} is $\Omega_{tot}(E_{tot})$, which is also a fixed number.

Probability of a particular division of energy

The number of ways of arranging E_{res} in the reservoir and E in the small system is $\Omega_{res}(E_{res})\Omega(E)$, while the total number of ways of arranging E_{tot} in the combined system is $\Omega_{tot}(E_{tot})$. Therefore the probability $P(E)$ of finding the energy split between the reservoir and the small system so that there is energy E in the small system is this:

$$P(E) = \frac{\Omega_{res}(E_{res})\Omega(E)}{\Omega_{tot}(E_{tot})}$$

The most probable value of the energy E to be found in the small system is zero, because the more energy E that is taken away from the big system and put into the small system, the fewer the microstates in the big system (without a comparable increase in the small system), which would mean a decrease in the total entropy.

We're interested in how fast $P(E)$ decreases as we move more and more of the total energy into the small system. We take logarithms and multiply by k to express everything in terms of entropy, then see how the expression varies with E:

$$k\ln P(E) = k\ln(\Omega_{res}(E_{res})) + k\ln(\Omega(E)) - k\ln(\Omega_{tot}(E_{tot}))$$

Consider the term $k\ln(\Omega_{res}(E_{res}))$, which is the entropy $S_{res}(E_{res})$ of the reservoir when the energy in the reservoir is $E_{res} = E_{tot} - E$ due to some energy E having gone into the small system. In Figure 11.36 we show the familiar calculation of the entropy of the reservoir as a function of the energy in the reservoir. On an expanded scale we show a portion of a plot of the entropy vs. E_{res} near the maximum possible value of E_{res}, which is the total energy E_{tot} of the combined system.

We need to evaluate the entropy of the reservoir for small values of $E = (E_{tot} - E_{res})$. We can obtain this by the following argument. The nearly straight line on the graph has a slope dS_{res}/dE_{res} and goes through $S_{res}(E_{tot})$ where the energy in the reservoir is E_{tot}. Therefore in this region $S_{res}(E_{res})$ can be represented by

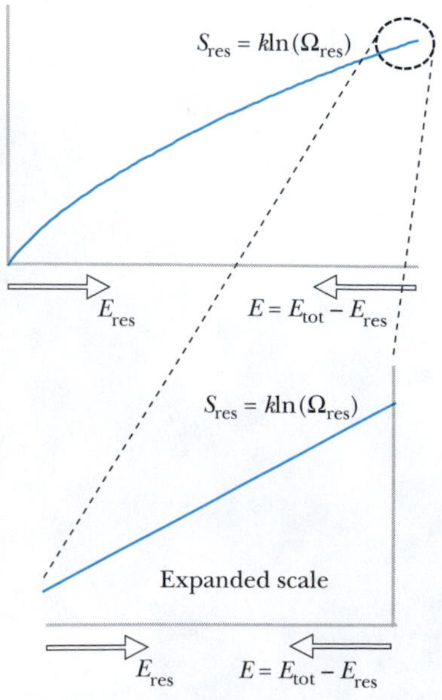

$S_{res} = k\ln(\Omega_{res})$

E_{res} $E = E_{tot} - E_{res}$

$S_{res} = k\ln(\Omega_{res})$

Expanded scale

E_{res} $E = E_{tot} - E_{res}$

Figure 11.36 Entropy (in units of k) as a function of E_{res}, for the larger object.

$$S_{\text{res}}(E_{\text{res}}) = S_{\text{res}}(E_{\text{tot}}) - \frac{dS_{\text{res}}}{dE_{\text{res}}} E$$

The entropy of the reservoir decreases as more energy E is shifted into the small system.

But $dS_{\text{res}}/dE_{\text{res}} = 1/T$, where T is the temperature of the reservoir, and the fact that the slope is nearly constant in this region reflects the fact that removing a small amount of energy from this large object hardly changes its temperature. The large object is a nearly constant-temperature reservoir, which keeps the small object at that temperature.

Making the substitution $dS_{\text{res}}/dE_{\text{res}} = 1/T$, we have

$$S_{\text{res}}(E_{\text{res}}) = k\ln(\Omega_{\text{res}}(E_{\text{res}})) = S_{\text{res}}(E_{\text{tot}}) - \frac{E}{T}$$

Substitute this into the equation for $k\ln P(E)$ and then divide by k:

$$\ln P(E) = \frac{S_{\text{res}}(E_{\text{tot}})}{k} - \frac{E}{kT} + \ln(\Omega(E)) - \ln(\Omega_{\text{tot}}(E_{\text{tot}}))$$

Note that the terms that don't involve E are all constants because E_{tot} is constant. Take the exponential of this equation—that is, raise e to both sides of the equation:

$$e^{\ln P(E)} = e^{\text{constant}} e^{\ln(\Omega(E))} e^{-\frac{E}{kT}}$$

Since e raised to the natural logarithm of some quantity is that quantity, we have the following, where A is some constant:

$$P(E) = A\Omega(E)e^{-\frac{E}{kT}}$$

We have calculated the probability of finding a small amount of energy E in a small system that is in contact with a large reservoir. This is called the Boltzmann distribution in honor of the Austrian physicist who developed statistical mechanics in the 19th century:

THE BOLTZMANN DISTRIBUTION

The probability of finding energy E in a small system in contact with a large reservoir is proportional to

$$\Omega(E)e^{-\frac{E}{kT}}$$

The exponential part, $e^{-\frac{E}{kT}}$ is called the "Boltzmann factor."

$\Omega(E)$ is the number of microstates corresponding to energy E.

This result is very general and applies to a very wide variety of phenomena. In many situations the number of microstates $\Omega(E)$ changes much more slowly with energy than does the Boltzmann factor, in which case the qualitative behavior of the system can be determined just from the exponential.

? What is the most likely value of E, the energy to be found in a microscopic system that is in thermal equilibrium with a large system, if the Boltzmann factor is the important factor? Are you likely to find an energy that is much larger than kT?

Since the exponent is negative, the exponential factor is largest for $E = 0$, so you're most likely to find the system in its ground state. This agrees with our expectation that taking energy out of the reservoir and putting it in the small system will reduce the number of total microstates and represent a decrease in entropy for the combined system, which is unlikely.

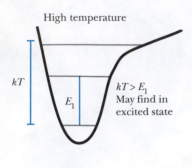

High temperature

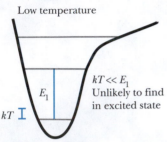

Low temperature

Figure 11.37 At a low temperature it is unlikely to find a system in an excited state, but at a high temperature the likelihood increases.

Since e^{-1} is 0.37, the probability of the energy being much larger than kT is rather small.

If the temperature is so low that kT is small compared to the energy E_1 of the first excited state, you are unlikely ever to find the system in one of its excited states. The system is thermally inert, because the probability of taking in *any* amount of energy is very small. In contrast, if the temperature is high enough that $kT > E_1$, then you will sometimes find the system in one of its excited states, although it is still true that the system is most likely to be found in its ground state (Figure 11.37).

The Boltzmann distribution has far-reaching consequences. For example, chemical and biochemical reaction rates typically depend strongly on temperature because with higher temperature the reactants are moving faster and may be found in excited states. Physical reaction rates are also affected. At the very high temperatures found in the interior of our Sun, kinetic energies are high enough to overcome the electric repulsion between nuclei and to allow the nuclei to come in contact, so that they can undergo thermonuclear fusion reactions. A gas becomes a plasma if the temperature is so high that kT is comparable to the ionization energy.

11.X.10 At room temperature, show that $kT \approx \frac{1}{40}$ eV. It is useful to memorize this result, because it tells a lot about what phenomena are likely to occur at room temperature.

11.X.11 A microscopic oscillator has its first and second excited states 0.05 eV and 0.10 eV above the ground state energy. Calculate the Boltzmann factor for the ground state, first excited state, and second excited state, at room temperature.

Many observations vs. many systems

If a microscopic system (such as a single oscillator) is in contact with a large system, the probability of finding a particular amount of energy in the microscopic system is governed by the Boltzmann distribution.

One can think about this in two ways. The first is to imagine measuring the energy in this particular system repeatedly, over time, and recording the results. The fraction of the results that indicate a particular energy is predicted by the Boltzmann distribution.

A second way to think about this is to imagine assembling a large number of identical systems, each in contact with a large system, and taking one "snapshot" in which the energy of each system is recorded simultaneously. The fraction of systems with a particular energy is predicted by the Boltzmann distribution.

The second approach also gives us a way to predict the distribution of energy in a single large system. For example, consider a system consisting of all the air molecules in a room. Imagine that each molecule is identified by a letter: A, B, C... One could consider molecule A to be a microscopic system in contact with the large system consisting of all other molecules (B, C, ...). However, one could also consider B to be the microscopic system, in contact with all other molecules (A, C, ...). We can use the Boltzmann distribution to predict the fraction of all molecules in the room that have a particular energy.

These two views of the Boltzmann distribution, one microscopic system observed repeatedly, or a large number of microscopic systems observed once, complement each other. Sometimes one view is more helpful, sometimes the other.

11.8 THE BOLTZMANN DISTRIBUTION IN A GAS

The Boltzmann distribution applies to any kind of system—not just a solid. As a major application of the Boltzmann distribution, we will study a gas consisting of molecules which don't interact much with each other. Examples are the so-called "ideal gas" (with no interactions at all), and any real gas at sufficiently low density that the molecules seldom come near each other.

In order to apply the Boltzmann distribution, we need a formula for the energy of a molecule in the gas. We will omit rest energy, and we will also omit nuclear energy and electronic energy from our total, because at ordinary temperatures there is not enough energy available in the surroundings to raise the molecule to an excited nuclear state or an excited electronic state. Also, for simplicity, instead of writing ΔE_{vib} to represent an amount of vibrational energy above the ground vibrational state, we will simply write E_{vib}.

The energy of a single gas molecule in the gravitational field near the Earth's surface (excluding rest energy, nuclear energy, and electronic energy) is this:

$$K_{trans} + E_{vib} + E_{rot} + Mgy_{cm}$$

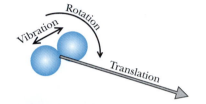

Figure 11.38 Energy of an oxygen molecule.

where E_{vib} and E_{rot} are the vibrational and rotational energies relative to the center of mass (if the molecule contains more than one atom, as in Figure 11.38) and y_{cm} is the height of the molecule's center of mass above the Earth's surface. The mass of the molecule is M. For brevity we will speak of "the energy of the gas molecule" but of course the gravitational potential energy really applies to the system of gas molecule plus Earth.

? For this expression for the energy to be accurate, why must the gas be an "ideal" gas (or a real gas at low density)?

In a real gas at somewhat high density we may not be able to neglect the potential energy associated with the intermolecular forces, in which case this expression for the energy would not be adequate.

If the temperature T is the same everywhere in the (ideal) gas, the probability that a particular molecule will have a certain amount of energy is proportional to

$$\Omega(E)e^{-\frac{[K_{trans} + E_{vib} + E_{rot} + Mgy_{cm}]}{kT}}$$

We mentally divide the gas into two systems: one particular molecule of interest, and all the rest of the molecules. These two systems are in thermal equilibrium with each other, because the gas molecules are continually colliding with each other and can share energy. The energy for the one particular molecule is then expected to follow the Boltzmann distribution.

To avoid excess subscripts, in the following discussion we will simply write v for the center-of-mass speed v_{cm}, and y for the center-of-mass height y_{cm}.

Separating the various factors

It is useful to group the terms of the Boltzmann factor like this:

$$\left[e^{-\frac{K_{trans}}{kT}}\right]\left[e^{-\frac{Mgy}{kT}}\right]\left[e^{-\frac{E_{vib}}{kT}}\right]\left[e^{-\frac{E_{rot}}{kT}}\right]$$

The first bracket is associated with the distribution of velocities, the second with the distribution of positions, the third with the distribution of vibrational energy, and the fourth with the distribution of rotational energy. We will discuss each of these individually.

Height distribution in a gas

A striking property of Earth's atmosphere is that in high mountains the air density and pressure are significantly lower than at sea level. The air density is so low on top of Mount Everest that most climbers carry oxygen tanks. Can we explain this? In order to get at the main issue, the variation of density with height, we make the rough approximation that the temperature is constant—the same in the mountains as at sea level.

? How bad an approximation is it to consider the temperature to be the same at all altitudes?

Even in high mountains the temperature is typically above –29° C (–20° F), which is 244 K, and this is only 17% lower than room temperature of 293 K (20° C or 68° F). So maybe this approximation isn't too bad.

We speak of the probability that the x component of the gas molecule's position lies between some x and $x+dx$, and similarly for y and z. Here dx is considered to be a very short distance, small compared to the size of the container, but large compared to the size of a molecule.

Focus just on that part of the Boltzmann distribution that deals with position, where the distribution is proportional to the probability of finding one particular molecule between x and $x+dx$, y and $y+dy$, and z and $z+dz$:

$$e^{-\frac{Mgy}{kT}}\, dx\, dy\, dz$$

Evidently there's nothing very interesting about x and z (directions parallel to the ground). But in the vertical direction there is an exponential fall-off with increasing height for the probability of finding a particular molecule at height y (Figure 11.39).

Looked at another way, our exponential formula tells us how the number density of the atmosphere depends on height, because in telling us about the behavior of one representative molecule, the formula also tells us something about all the molecules.

We shifted gears from thinking about one molecule to thinking about many. Think again about one single air molecule. Suppose you place it on a table, which is at room temperature. It is not a very unusual event for a thermally agitated atom in the table to hit our air molecule hard enough to send it kilometers high into the air! Actually, our particular molecule will very soon run into another air molecule and not make it very high in one great leap, but on average we do find lots of air molecules very high above sea level rather than finding them all lying on the ground.

Notice again that kT is the important factor in understanding the statistical behavior of matter. Here it sets the scale for the variation with height of the atmosphere's number density.

At a height where the gravitational potential energy $U_g = mgy$ is equal to kT, the number density has dropped by a factor of $e^{-1} = 1/e = 0.37$. Dry air at sea level is 78% nitrogen (one mole N_2 = 28 grams), 21% oxygen (one mole O_2 = 32 grams), about 1% argon, and 0.03% CO_2. An average mass of 29 grams per mole is good enough for most calculational purposes.

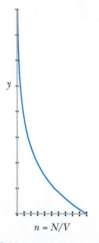

Figure 11.39 Number density vs. height in a constant-temperature atmosphere.

11.X.12 Suppose you put one air molecule on your desk, so that it is in thermal equilibrium with the desk at room temperature. Suppose that there is no atmosphere to get in the way of this one molecule bouncing up and down on the desk. Calculate the typical height that the air molecule will be above your desk, so that $Mgy \approx kT$.

11.X.13 Marbles of mass M = 10 grams are lying on the floor. They are of course in thermal equilibrium with their surroundings. What is a typical height above the floor for one of these marbles? That is, for what value of y is $Mgy \approx kT$?

11.X.14 Approximately what fraction of the sea-level air density is found at the top of Mount Everest, a height of 8848 meters above sea level?

Distribution of velocities in a gas

Next we look at the distribution of molecular speeds. When the gas is confined inside a finite container the momentum (or velocity) and height are quantized, but under almost all conditions the size of the energy quantum associated with momentum (or velocity) and position is so small compared to kT that it is appropriate to take a nonquantum approach for these variables.

We speak of the probability that the gas molecule has an x component of velocity within the range between some v_x and $v_x + dv_x$, and similarly for v_y and v_z. Here dv_x is considered to be a very small amount, small compared to the average speed of the molecules.

Since we're explicitly interested in v_{cm}, it will be useful to express translational kinetic energy as $\frac{1}{2}Mv_{cm}^2$ rather than $p^2/(2M)$ in this discussion. Since $v^2 = v_x^2 + v_y^2 + v_z^2$, we can write the formula for the distribution of velocity in a gas as follows (remember that we are simply writing v for the center-of-mass speed v_{cm}):

$$\left[e^{-\frac{\frac{1}{2}Mv_x^2}{kT}}\, dv_x \right]\left[e^{-\frac{\frac{1}{2}Mv_y^2}{kT}}\, dv_y \right]\left[e^{-\frac{\frac{1}{2}Mv_z^2}{kT}}\, dv_z \right]$$

The distribution for each velocity component is a bell-shaped curve, called a "Gaussian." In Figure 11.40 we show the distribution of the x component of velocity for helium at room temperature.

? Judging from this graph, what is the average value for v_x? What are the average values of v_y and v_z? Why are these results plausible?

Evidently the average value for the x component of the velocity is zero. This is reassuring, because a molecule is as likely to be headed to the right as it is to be headed to the left. Similarly, the average values for the y and z components of velocity are zero.

It would not be particularly surprising to find a gas molecule with an x component of velocity such that $\frac{1}{2}Mv_x^2 \approx kT$. On the other hand, it would be very surprising to find a gas molecule with a value of v_x many times larger. As usual, kT sets the scale for thermal phenomena.

Distribution of speeds in a gas

By converting from rectangular coordinates to "spherical" coordinates, the velocity-component distribution can be converted into a speed distribution. We won't go into the details, but the main idea is that the volume element in rectangular coordinates in "velocity space" $dv_x dv_y dv_z$ turns into the volume of a spherical shell with surface area $4\pi v^2$, and thickness dv, as shown in Figure 11.41.

Transferring to spherical coordinates, we have the following:

$$e^{-\frac{\frac{1}{2}Mv^2}{kT}}\, dv_x dv_y dv_z = e^{-\frac{\frac{1}{2}Mv^2}{kT}}\, 4\pi v^2\, dv$$

The only thing missing is a "normalization" factor in front to make the integral over all speeds from 0 to infinity be equal to 1.0 (since our one molecule must have a speed somewhere in that range). Here is the "Maxwell-Boltzmann" distribution for a low-density gas:

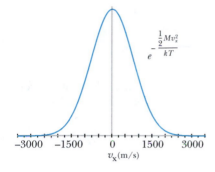

Figure 11.40 Distribution of the x component of velocity for helium at room temperature.

Figure 11.41 The volume of a shell with radius v and thickness dv is $4\pi v^2 dv$.

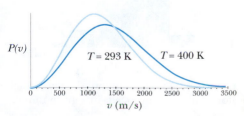

Figure 11.42 Distribution of speeds of helium atoms at two temperatures.

MAXWELL-BOLTZMANN SPEED DISTRIBUTION (LOW-DENSITY GAS)

$$P(v) = 4\pi\left(\frac{M}{2\pi kT}\right)^{\frac{3}{2}} v^2 \, e^{-\frac{\frac{1}{2}Mv^2}{kT}}$$

The probability that a molecule of a gas has a center-of-mass speed within the range v to $v+dv$ is given by $P(v)dv$.

The distribution function for helium is shown for two different temperatures in Figure 11.42. You can see that the average speed of a helium atom is predicted to be about 1200 m/s at room temperature (20° Celsius, which is 293 Kelvin). The prediction was first made by Maxwell, in the mid-1800's, long before the development of quantum mechanics. In retrospect, this worked because the quantum levels of the translational kinetic energy are so close together as to be almost a continuum, so the classical model is a good approximation.

The average number of helium atoms in a container that have speeds in the range of 415.4 m/s to 415.7 m/s can be calculated by evaluating Maxwell's formula with $v = 415.4$ m/s and multiplying by $dv = 0.3$ m/s, which gives the probability of finding one molecule in this speed range. If there are N atoms in the container, multiplying by N gives the total number of molecules likely to be in that speed range at any given instant. The area of the vertical slice shown in Figure 11.43 represents the fraction of helium atoms that are likely to have speeds between 500 m/s and 600 m/s. At higher temperatures the distribution shifts to higher speeds (Figure 11.42).

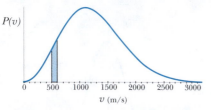

Figure 11.43 Fraction of helium atoms at 293 K with speeds between 500 and 600 m/s.

Measuring the distribution of speeds in a gas

The actual distribution of speeds of molecules in any gas can be measured by an ingenious experiment. Make a tiny hole in a container of gas and let the molecules leak out into a vacuum (that is, a region from which the air is continually pumped out). The moving molecules run through collimating holes and then through a slot in a rapidly rotating drum and strike a row of devices at the other side of the drum that can detect gas molecules (Figure 11.44). Such measurements confirm the Maxwell prediction.

Fast molecules strike soonest, followed by medium-speed molecules, and finally slow molecules, so molecules with different speeds strike at different locations along the rotating drum. The number of molecules striking various locations along the drum is a direct indication of the distribution of speeds of molecules in the gas. (A correction has to be applied to these data to obtain the speed distribution inside the gas, because high-speed molecules emerge through the hole more frequently than low-speed molecules do, even if their numbers are the same in the interior of the gas. Also, the apparatus measures the distribution of v_x rather than v.)

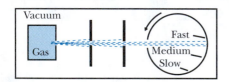

Figure 11.44 Apparatus for measuring the distribution of speeds of gas molecules.

How might we display the results of this experiment? We could present the data as a graph (Figure 11.45) of the fraction $\Delta N/N$ of the N helium atoms that have speeds between 0 and 500 m/s, between 500 and 1000 m/s, between 1000 and 1500 m/s, etc.

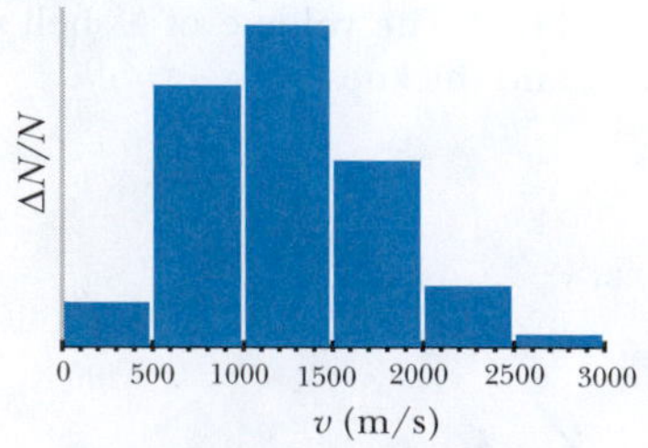

Average translational kinetic energy in a gas

For a gas confined inside a container, quantum mechanics predicts that the kinetic energy $\frac{1}{2}Mv^2$ is quantized, but if the container is of ordinary macroscopic size, the energy quanta are extremely small compared to the average kinetic energy, even at very low temperatures. Most gases liquefy before the temperature drops so low as to invalidate the pre-quantum analysis for the velocity distribution.

Figure 11.45 Distribution showing fraction of atoms with speeds in a given range.

Whenever the energy has a term containing a square of a position or momentum component, such as x^2 or p_x^2, pre-quantum theory predicts an associated average energy of $\frac{1}{2}kT$. This result is valid in quantum mechanics for high temperatures, where kT is large compared to the quantum energy spacing. This result follows from using an integral to take the average value of a quadratic term when the distribution is a Gaussian involving that quadratic term. That is, using integral tables (or integrating by parts), you can show that the average value of a quadratic term is $\frac{1}{2}kT$:

$$\overline{w^2} = \frac{\displaystyle\int_0^\infty w^2 e^{-\frac{w^2}{kT}}dw}{\displaystyle\int_0^\infty e^{-\frac{w^2}{kT}}dw} = \frac{1}{2}kT$$

The average value is obtained by weighting the values of w^2 by the probability of finding that value; the denominator takes care of normalizing the distribution properly.

As an example, when we did quantum-based calculations for the Einstein solid, we found that at high temperatures the specific heat capacity on a per-atom basis was $3k$, which implies an average energy of $3kT$. We modeled an atomic oscillator as three one-dimensional oscillators, each having an average energy of $\frac{1}{2}kT$ corresponding to kinetic energy $p^2/(2M)$ and another $\frac{1}{2}kT$ for the potential energy $\frac{1}{2}k_s x^2$. This is a total of six quadratic terms, implying an average energy per atom of $6(\frac{1}{2}kT)$, or $3kT$. We therefore expect the heat capacity per atom in a solid at high temperature to be $3k$, which is indeed what is observed.

HIGH-TEMPERATURE AVERAGE ENERGY

If $kT \gg$ the energy quantum, the average energy

associated with a quadratic energy term is $\frac{1}{2}kT$.

The number of quadratic terms in the expression for the energy is often called the "degrees of freedom."

? What is the average translational kinetic energy $\frac{1}{2}M\overline{v^2}$ in terms of kT? (A bar over a quantity means "average value.") What is the associated contribution to the specific heat capacity of the gas?

Since there are three quadratic terms in the translational kinetic energy, $K_{\text{trans}} = \frac{1}{2}mv_x^2 + \frac{1}{2}mv_y^2 + \frac{1}{2}mv_z^2$, the average value is three times $\frac{1}{2}kT$. We have the following important result:

AVERAGE K_{trans} FOR AN IDEAL GAS

$$\overline{K}_{\text{trans}} = \frac{1}{2}M\overline{v^2} = \frac{3}{2}kT$$

The contribution to the specific heat capacity for a gas is $\frac{3}{2}k$.

Root-mean-square speed

The square root of $\overline{v^2}$ is called the "root-mean-square" or "rms" speed:

$$v_{\text{rms}} \equiv \sqrt{\overline{v^2}}$$

With this definition of v_{rms}, we write $\frac{1}{2}m\overline{v^2} = \frac{1}{2}mv_{\text{rms}}^2 = \frac{3}{2}kT$.

11.X.15 Calculate v_{rms} for a helium atom in the room you're in (whose temperature is probably about 293 K).

11.X.16 Calculate v_{rms} for a nitrogen molecule (N_2; molecular mass 28) in the room you're in (whose temperature is probably about 293 K). Air is about 80% nitrogen.

Average speed vs. rms speed

The way you find the average speed of molecules in a gas is to weight the speed by the number of molecules that have that speed, and divide by the total number of molecules:

$$\bar{v} = \frac{N_1 v_1 + N_2 v_2 + \text{etc.}}{N} = \frac{N_1}{N} v_1 + \frac{N_2}{N} v_2 + \text{etc.}$$

For a continuous distribution of speeds this turns into an integral, where the weighting factors are given by the Maxwell speed distribution, which gives the probability that a molecule have a center-of-mass speed in the range v to $v+dv$. Using integral tables one finds this:

$$\bar{v} = \int_0^{\infty} 4\pi \left(\frac{m}{2\pi kT}\right)^{\frac{3}{2}} e^{-\frac{\frac{1}{2}mv^2}{kT}} (v) v^2 \, dv = \sqrt{\frac{8}{3\pi}} \sqrt{\frac{3kT}{m}} = 0.92 \, v_{rms}$$

This shows that the average speed is smaller than the rms speed ($0.92 v_{rms}$). This same calculational scheme can be used to determine any average. For example, if you calculate the average value of v^2 by this method, you find $\overline{v^2} = 3kT/m$, as expected.

On page 400 we commented that the average speed $\bar{v}$ of helium atoms at room temperature is about 1200 m/s. In 11.X.15 on page 401 you found that v_{rms} at room temperature is 1360 m/s. This is an example of the fact that the rms speed is higher than the average speed.

11.X.17 The rms speed is somewhat higher than the average speed due to the averaging of squared speeds. Calculate the average of the numbers 1, 2, 3, and 4, then calculate the rms average (the square root of the average of their squares), and show that the rms average is larger than the simple average. Squaring gives extra weight to larger contributions.

Application: Retaining a gas in the atmosphere

Since $v_{rms} = \sqrt{3k(T/m)}$, helium atoms typically travel much faster than nitrogen molecules in our atmosphere, due to the small mass of the helium atoms. Some few helium atoms will be going much faster than the average and may attain a high enough speed to escape from the Earth entirely (escape velocity from the Earth is about 1.1×10^4 m/s). This leads to a continuous leakage of high-speed helium atoms and other low mass species such as hydrogen molecules (Figure 11.46). Other processes may also contribute to the flow of helium away from the Earth.

? Why doesn't the Moon have any atmosphere at all?

The Moon's gravitational field is so weak that escape speed is quite small, and all common gases can escape, not just hydrogen and helium.

11.X.18 Calculate the escape speed from the Moon and compare with typical speeds of gas molecules. The mass of the Moon is 7×10^{22} kg, and its radius is 1.75×10^6 m.

Application: Speed of sound

Sound waves in a gas consist of propagation of variations in density, and the fundamental mechanism for this kind of wave propagation involves collisions between neighboring molecules, whose speeds are proportional to v_{rms} (and roughly comparable to the speed of sound). For example, compare the v_{rms} for nitrogen, 496 m/s, which you calculated in Ex. 11.X.16

Figure 11.46 Low mass atoms or molecules in our atmosphere may have speeds high enough to escape from the Earth entirely.

There's almost no helium in our atmosphere. Where do we obtain helium for party balloons and low-temperature refrigeration and scientific experiments? It is extracted from natural gas when the gas is pumped out of the ground. Heavy radioactive elements such as uranium in the Earth's crust emit alpha particles (helium nuclei) which capture electrons and become helium atoms. These atoms are trapped with natural gas in underground cavities. When the helium-bearing natural gas is pumped to the surface, the helium is extracted (at some cost).

(page 402): with the speed of sound in air (which is mostly nitrogen) at 293 K, which is measured to be 344 m/s. (You may know the approximate rule that a 1-second delay between lightning and thunder indicates a distance of about 1000 feet, which is about 300 m.)

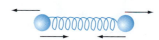

High gas density   Low gas density

Figure 11.47 Periodic variations in gas density make a sound wave, which travels through a gas with speed v (the speed of sound).

11.X.19 Should the speed of sound in air increase or decrease with increasing temperature? What percentage change would result from doubling the absolute temperature? (This effect is readily observed by measuring the speed of sound in a gas as a function of absolute temperature. The excellent agreement between theory and experiment provides additional evidence for our understanding of gases.)

Vibrational energy in a diatomic gas molecule

We have treated the distribution of velocity and position. Next we discuss the distribution of vibrational energy, with the Boltzmann factor

$$e^{-\frac{E_{vib}}{kT}}$$

For a monatomic gas such as helium, there is no vibrational energy term. But for a diatomic molecule such as N_2 or HCl, the vibrational energy is

$$E_{vib} = \frac{p_1^2}{2m_1} + \frac{p_2^2}{2m_2} + \frac{1}{2}k_s s^2$$

where k_s is the effective spring stiffness corresponding to the interatomic electric force (not to be confused with k, the Boltzmann constant), and s is the stretch of the interatomic bond. This formula for the vibrational energy is essentially the same as the formula for a one-dimensional spring-mass oscillator, which we have studied in detail (Figure 11.48).

Since $p_1 = p_2$ for momenta p_1 and p_2 relative to the center of mass, we can write

$$\frac{p_2^2}{2m_2} = \frac{p_1^2}{2m_2} = \left(\frac{m_1}{m_2}\right)\left(\frac{p_1^2}{2m_1}\right)$$

$$\frac{p_1^2}{2m_1} + \left(\frac{m_1}{m_2}\right)\frac{p_1^2}{2m_1} + \frac{1}{2}k_s s^2 = \left(1 + \frac{m_1}{m_2}\right)\frac{p_1^2}{2m_1} + \frac{1}{2}k_s s^2$$

This equation has a form exactly like that for a spring-mass oscillator, with a different mass.

Earlier we modeled a solid as a large number of isolated three-dimensional atomic oscillators (each corresponding to three one-dimensional oscillators, because there are spring-like interatomic forces on an atom from neighboring atoms in all directions). This is an overly simplified model of a solid, because in a solid the atoms interact with each other. For example, if an atom moves to the left this affects the atoms to the right and to the left.

In a low-density gas however the vibrational oscillators really are nearly independent of each other, because the gas molecules aren't even in contact with each other except when they happen to collide. So the analysis we carried out for the Einstein model of a solid applies even better to the vibrational portion of the energy in a real gas than it does to a real solid.

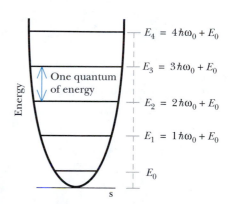

$E_4 = 4\hbar\omega_0 + E_0$

$E_3 = 3\hbar\omega_0 + E_0$

One quantum of energy

$E_2 = 2\hbar\omega_0 + E_0$

$E_1 = 1\hbar\omega_0 + E_0$

E_0

Figure 11.48 The quantized vibrational energy levels of a diatomic oscillator.

Among the results that apply immediately are that the specific heat capacity associated with the vibrational energy of one oscillator is k at high temperatures. The specific heat capacity decreases at very low temperatures where kT is small compared to the energy spacing of the quantized oscillator energies. Just as it was a surprise when the specific heat capacity of metals was found to decrease at low temperatures, there was a similar surprise in the measurements of the specific heat capacity of diatomic gases at low temperatures, because the contribution of the vibrational motion vanished.

? There are two quadratic terms in the vibrational energy (kinetic and spring). Therefore, at high temperatures what is the average vibrational energy in terms of kT? What is the associated contribution to the specific heat capacity of the gas?

At high temperatures the vibrational motion of a diatomic molecule (a one-dimensional oscillator) is expected to have an average energy that is approximately $2(\frac{1}{2}kT) = kT$, and contributes k to the specific heat capacity.

Rotational energy in a diatomic gas molecule

Finally we consider the rotational-energy portion of the distribution, for which the Boltzmann factor is

$$e^{-\frac{E_{rot}}{kT}}$$

For a monatomic gas there is no rotational term. But for a diatomic gas there is rotational kinetic energy associated with rotation of the "dumbbell" consisting of the two nuclei. The rotational kinetic energy is this (Figure 11.49):

$$E_{rot} = \frac{1}{2}I\omega_x^2 + \frac{1}{2}I\omega_y^2 = \frac{L_{rot,x}^2}{2I} + \frac{L_{rot,y}^2}{2I}$$

This corresponds to rotational angular momenta $L_{rot,x}$ and $L_{rot,y}$ about the x axis and about the y axis. Note that only the nuclei contribute significantly to the rotational kinetic energy, because the electrons have much less mass.

Rotation about the z axis connecting the two nuclei is irrelevant, for a somewhat subtle reason. The angular momentum is quantized, which leads to energy quantization. For rotations around the z axis (Figure 11.50) the energy is $L_{rot,z}^2/(2I)$, where $L_{rot,z}$ is the z component of the rotational angular momentum. Since I about the z axis for the tiny nuclei is extremely small compared with the moment of inertia about the x and y axes of the diatomic molecule, the rotational energy of the associated first excited state is enormous, and this state is not excited at ordinary temperatures.

Since there are two quadratic energy terms associated with rotation, we conclude that at high temperatures the rotational motion of a diatomic molecule has an average energy that is approximately $2(\frac{1}{2}kT) = kT$, and contributes k to the specific heat capacity.

The spacing between quantized rotational energies for a diatomic molecule is even smaller than the vibrational energy quantum. As a result, the gas must be cooled to a very low temperature before the pre-quantum results become invalid. Many gases liquefy before this low-temperature regime is reached.

Specific heat capacity of a gas

We are now in a position to discuss the specific heat capacity of a gas as a function of temperature. This property is important in calculating the thermal interactions of a gas. Historically, measurements of the specific heat capacity of gases were also important in testing theories of statistical mechanics. We'll concentrate on calculating c_v, the specific heat capacity at constant volume (meaning no mechanical work is done on the gas). The as-

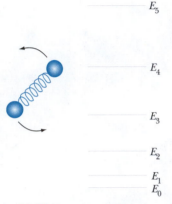

E_5

E_4

E_3

E_2

E_1
E_0

Figure 11.49 Rotational kinetic energy levels of a diatomic molecule.

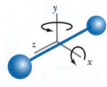

Figure 11.50 A diatomic molecule can rotate around the x or the y axis. We say there are two rotational degrees of freedom.

sociated experiment would be to add a known amount of energy to a gas in a rigid container and measure the temperature rise of the gas.

Consider the average energy of a diatomic molecule such as N_2 or HCl, consisting of the translational kinetic energy associated with the motion of the center of mass, plus the vibrational energy, plus the rotational energy, plus the gravitational energy of the molecule plus the Earth:

$$\left\{\frac{1}{2}m\overline{v_x^2} + \frac{1}{2}m\overline{v_y^2} + \frac{1}{2}m\overline{v_z^2}\right\} + \left\{\left(1 + \frac{m_1}{m_2}\right)\frac{\overline{p_1^2}}{2m_1} + \frac{1}{2}k_s\overline{s^2}\right\} + \left\{\frac{\overline{L_{rot,x}^2}}{2I} + \frac{\overline{L_{rot,y}^2}}{2I}\right\} + Mg\overline{y}_{cm}$$

$$\quad\quad\text{Translation} \quad\quad\quad\quad\quad\quad \text{Vibration} \quad\quad\quad\quad\quad \text{Rotation} \quad\quad \text{Gravitation}$$

? At high temperature, what should the specific heat capacity at constant volume be?

There are seven quadratic terms in the expression for the energy, so if the temperature is very high, we expect an average energy of $7(\frac{1}{2}kT)$ and a specific heat capacity at constant volume $c_v = \frac{7}{2}k$.

The discrete energy levels for a diatomic molecule include electronic, vibrational, and rotational energy levels (Figure 11.51). The electronic energy levels correspond to particular configurations of the electron clouds, and the spacing between these levels is typically 1 eV or more. Since at room temperature kT is about $\frac{1}{40}$ eV, the electronic levels do not contribute to the specific heat capacity at ordinary temperatures. The vibrational energy spacing is much smaller, and the rotational energy spacing is smaller still. A diatomic gas at high temperature has a band-like spectrum, since transitions between electronic levels may be accompanied by one or more vibrational or rotational quanta.

What about Mgy?

What about the Mgy_{cm} term in the energy? Remember that we are trying to predict the specific heat capacity of a gas at constant volume (so that no mechanical work is done on the system). We could measure c_v by having a known amount of energy transfer Q due to a temperature difference into a closed container of gas and observing the temperature rise of the gas. During this process the height y_{cm} of the center of mass of the gas in the container hardly changes, so the Mgy term doesn't contribute to the specific heat capacity.

Actually there is an extremely small increase in y_{cm} of the gas in the container due to the temperature dependence of the height distribution (page 398), but the associated contribution to the specific heat capacity is negligible.

Specific heat capacity *vs.* temperature

Figure 11.52 is a graph of the specific heat capacity (at constant volume) of a diatomic gas as a function of temperature. In the following exercises, see whether you can explain these values of the specific heat capacity in terms of our analyses of the various contributions to the energy of a diatomic gas molecule.

Remember that kT is about $\frac{1}{40}$ eV at room temperature.

11.X.20 What is the specific heat capacity on a per-molecule basis (at constant volume) of a monatomic gas such as helium or neon? Why doesn't the specific heat capacity depend on temperature?

11.X.21 At high temperatures, what is the specific heat capacity (at constant volume) on a per-molecule basis of a diatomic gas such as oxygen or nitrogen?

11.X.22 Suppose we lower the temperature of a diatomic gas to a point where kT is small compared to the energy of the first excited vibrational

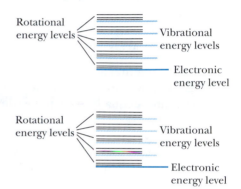

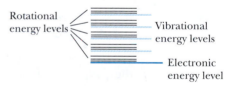

Figure 11.51 Electronic, vibrational, and rotational energy levels for a diatomic molecule.

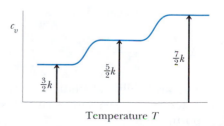

Figure 11.52 Specific heat capacity c_v of a diatomic gas *vs.* temperature.

state, but still large compared to the energy of the first excited rotational state. Now what is the specific heat capacity (at constant volume)?

11.X.23 Lower the temperature even more, so that kT is small compared to the energy of the first excited rotational state (some gases liquefy before this low temperature is reached). What is the specific heat capacity (at constant volume) at this low temperature?

11.X.24 As the temperature is decreased, the specific heat capacity (at constant volume) for H_2 eventually decreases to $\frac{3}{2}k$, before the gas liquefies. Does this transition to $c_v = \frac{3}{2}k$ happen at a higher or lower temperature for D_2 (deuterium, whose nuclei each contain a proton plus a neutron)? Why? (Hint: Remember that it is the rotational angular momentum that is quantized.)

11.9 SUMMARY

The fundamental assumption of statistical mechanics is that, over time, an isolated system in a given macrostate is equally likely to be found in any of its possible microstates.

Two blocks in thermal contact evolve to that division of energy that has associated with it the largest number of ways $\Omega = \Omega_1\Omega_2$ of arranging the total energy among the atoms (largest number of microstates for a macrostate of given energy).

Second law of thermodynamics: The entropy of a closed system never decreases. Only in a reversible process does the entropy of a closed system stay constant.

ways to arrange q quanta of energy among N one-dimensional oscillators

$$\Omega = \frac{(q+N-1)!}{q!(N-1)!} \quad \text{(number of microstates)}$$

Entropy: $S \equiv k\ln\Omega$, where $k = 1.38\times10^{-23}$ J/K

Temperature: $\dfrac{1}{T} = \dfrac{\partial S}{\partial E}$

A small flow of energy Q into a system raises entropy by $\Delta S = \dfrac{Q}{T}$

Specific heat capacity per atom: $C = \dfrac{\Delta E_{atom}}{\Delta T}$, where

$$\Delta E_{atom} = \frac{\Delta E_{system}}{N_{atoms}}$$

The probability of finding energy E in a small system in contact with a large reservoir is proportional to

$$\Omega(E)e^{-\frac{E}{kT}}$$

$\Omega(E)$ is the number of microstates corresponding to energy E

For each energy term involving a quadratic term such as x^2 or p_x^2 ("degree of freedom"), if the average energy is large compared to kT the average energy is $\frac{1}{2}kT$ and the contribution to the specific heat capacity is $\frac{1}{2}k$. This contribution decreases at low temperatures. At high temperatures, specific heat capacity on a per-atom basis in a solid $\approx 3k$.

Speed distribution for a gas:

$$P(v) = 4\pi\left(\frac{M}{2\pi kT}\right)^{\frac{3}{2}} e^{-\frac{\frac{1}{2}Mv^2}{kT}} v^2$$

11.10 REVIEW QUESTIONS

Arranging things

11.RQ.25 List explicitly all the ways of arranging 2 quanta among 4 one-dimensional oscillators.

A formula for the number of arrangements

11.RQ.26 How many different ways are there to get 5 heads in 10 throws of a true coin? How many different ways are there to get no heads in 10 throws of a true coin?

11.RQ.27 How many different ways are there to arrange 4 quanta among 3 atoms in a solid?

Thermal equilibrium

11.RQ.28 Energy conservation for two blocks in contact with each other is satisfied if all the energy is in one block and none in the other. Would you expect to observe this distribution in practice? Why or why not?

11.RQ.29 What is the advantage of plotting the (natural) logarithm of the number of ways of arranging the energy among the many atoms (natural logarithm of the number of microstates)?

11.RQ.30 Which has a higher temperature, a system whose entropy changes rapidly with increasing energy, or one whose entropy changes little with increasing energy?

Specific heat capacity

11.RQ.31 Explain why it is a disadvantage for some purposes that the specific heat capacity of all materials decreases at low temperatures.

11.RQ.32 Sketch and label graphs of specific heat capacity *vs.* temperature for hydrogen gas (H_2) and oxygen gas (O_2), using the same temperature scale. Explain briefly.

11.RQ.33 Which has more internal energy at room temperature—a mole of helium or a mole of air?

Second law of thermodynamics

11.RQ.34 Two blocks with different temperatures had entropies of 10 J/K and 35 J/K before they were brought in contact.

What can you say about the entropy of the combined system after the two came in contact with each other?

The Boltzmann distribution

11.RQ.35 Explain qualitatively the basis for the Boltzmann distribution. Never mind the details of the math for the moment. Focus on the trade-offs involved with giving energy to a single oscillator *vs.* giving that energy to a large object.

11.RQ.36 Many chemical reactions proceed at rates that depend on the temperature. Discuss this from the point of view of the Boltzmann distribution.

Distributions

11.RQ.37 Here is the distribution of speeds of atoms in a particular gas at a particular temperature. Approximately what is the average speed? Is the rms (root-mean-square) speed bigger or smaller than this? Approximately what fraction of the molecules have speeds greater than 1000 m/s?

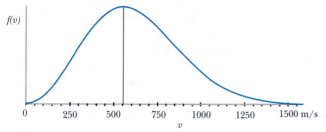

Distribution of speeds of atoms in a gas at a particular temperature.

11.11 PROBLEMS

> *It is important to work through the first four problems in detail in order to make the ideas concrete.*

11.P.38 Probability distribution

(a) Model a system consisting of two atoms (three oscillators each), among which 4 quanta of energy are to be distributed. Write a program to display a histogram showing the total number of possible microstates of the two-atom system *vs.* the number of quanta assigned to atom 1. Compare your calculations and histogram to the one on page 378 (you should get the same distribution).

(b) Model a system consisting of two solid blocks, block 1 containing 300 oscillators and block 2 containing 200 oscillators. Find the possible distributions of 100 quanta among these blocks, and plot number of microstates *vs.* number of quanta assigned to block 1. Compare your calculations and histogram to those on page 381. Determine the distribution of quanta for which the probability is half as large as the most probable 60-40 distribution.

(c) Do a series of calculations distributing 100 quanta between two blocks whose total number of oscillators is 500, but whose relative number of atoms varies. For example, consider equal numbers of oscillators, and ratios of 2:1, 5:1, etc. Describe your observations.

11.P.39 Natural logarithm (ln) of ways to arrange energy

Start with your solution to Problem 11.P.38 (b). For the same system of two blocks, with $N_1 = 300$ oscillators and $N_2 = 200$ oscillators, plot $\ln(\Omega_1)$, $\ln(\Omega_2)$, and $\ln(\Omega_1\Omega_2)$, for q_1 running from 0 to 100 quanta. Your graph should look like the one in Figure 11.24. Determine the maximum value of $\ln(\Omega_1\Omega_2)$ and the value of q_1 where this maximum occurs. What is the significance of this value of q_1?

11.P.40 Temperature

Modify your calculations from Problem 11.P.39 to plot the temperature in kelvins of block 1 as a function of the number of quanta q_1 present in the first block. On the same graph plot the temperature in kelvins of block 2 as a function of q_1 (of course, $q_2 = q_{tot} - q_1$).

In order to plot the temperature in kelvins, you must determine the values of ΔE and ΔS that correspond to a one-quantum change in energy. Consider the model we are using. The energy of one quantum, in joules, is $\Delta E = \hbar\sqrt{k_s/m}$. The increment in entropy corresponding to this increment in energy is $\Delta S = k\Delta(\ln\Omega)$. Assume that the blocks are made of aluminum. In Chapter 4 you made an estimate of the interatomic spring constant k_s for aluminum. We show later, on page 390, that the effective k_s for oscillations in the Einstein solid is expected to be about 4 times the value obtained from measuring Young's modulus.

What is the significance of the value of q_1 (and of q_2) where the temperature curves for the two blocks cross (the temperatures are equal)?

11.P.41 Specific heat capacity

Modify your analysis of Problem 11.P.40 to determine the specific heat capacity as a function of temperature for a single block of metal. In order to see all of the important effects, consider a single block of 35 atoms (105 oscillators) with up to 300 quanta of energy. Note that in this analysis you are calculating quantities for a *single* block, not two blocks in contact.

T (K)	C, Al (J/K/mole)	C, Pb (J/K/mole)
20	0.23	11.01
40	2.09	19.57
60	5.77	22.43
80	9.65	23.69
100	13.04	24.43
150	18.52	25.27
200	21.58	25.87
250	23.25	26.36
300	24.32	26.82
400	25.61	27.45

To make specific comparison with experimental data, consider the cases of aluminum (Al) and lead (Pb). For each metal, plot the theoretical specific heat capacity C per atom *vs.* T (K), with dots showing the actual experimental data (converted to the same basis, J/K per atom).

Adjust the interatomic spring stiffness k_s until your calculations approximately fit the experimental data given in the adjoining table. (You will need to convert to a per-atom value.) In trying to fit the data, ignore the lowest-temperature data, where electrons in the metal make a significant contribution that is not included in the Einstein model.

What value of k_s gives a good fit? Compare with the estimated values of k_s obtained in Chapter 4, but remember that those estimated values are probably 4 times smaller than the effective spring constant for oscillations.

Show from your graph that the high-temperature limit of the specific heat capacity is about $3k$ per atom. The rise above this limit at high temperatures is probably due to the fact that the assumed uniform spacing of the quantized oscillator energy levels isn't a good approximation for highly excited states. See Chapter 6.

11.P.42 Square root of *N*

For some examples of your choice, demonstrate by carrying out actual computer calculations that the "square-root" rule holds true for the fractional width of the peak representing the most probable arrangements of the energy. Warning: Check to see what is the largest number you can use in your computations; some programs or programming environments won't handle numbers bigger than about 1×10^{307}, for example, and larger numbers are treated as "infinite."

11.P.43 Copper and aluminum blocks

A 50-gram block of copper (one mole has a mass of 63.5 grams) at a temperature of 35° C is put in contact with a 100-gram block of aluminum (molar mass 27 grams) at a temperature of 20° C. The blocks are inside an insulated enclosure, with little contact with the walls. At these temperatures, the high-temperature limit is valid for the specific heat capacity. Calculate the final temperature of the two blocks. Do NOT look up the specific heat capacities of aluminum and copper; you should be able to figure them out on your own.

11.P.44 Low-temperature specific heat capacity of copper

Young's modulus for copper is measured by stretching a copper wire to be about 1.2×10^{11} N/m². The density of copper is about 9 grams/cm³, and the mass of a mole is 63.5 grams. Starting from a very low temperature, use these data to estimate roughly the temperature T at which we expect the specific heat capacity for copper to approach $3k$. Compare your estimate with the data shown on page 391.

11.P.45 Experiment: Measurement of the specific heat capacity of water

The goal of this experiment is to understand, in a concrete way, what specific heat capacity is and how it can be measured.

You will need a microwave oven, a styrofoam coffee cup, and a clock or watch.

In the range of temperature where water is a liquid (0 C to 100 C), it is approximately true that it takes 4.2 J of energy (1 calorie) to raise the temperature of 1 gram of water through 1 degree Kelvin. To measure this specific heat capacity of water, we need some way to raise a known mass of water from a known initial temperature to a final temperature which can also be measured, while we keep track of the energy supplied to the water. One way to do this, as discussed in the text, is to warm up water in a well-insulated container within which a heater, whose power output is known, warms up the water. In this experiment, instead of a well-insulated box with a heater, we will use microwave power which preferentially warms up water by exciting rotational modes of the water molecules, as opposed to burners or heaters which warm up water in a pan by first warming up the pan.

Use a styrofoam coffee cup of known volume in which water can be warmed up. The density of water is 1 gram/cm³.

> It is a good idea not to fill the cup completely full, because this makes it more likely to spill.

One method of recording the initial temperature of the water is to get water from the faucet and wait for it to equilibrate with room temperature (which can either be read off a thermostat or estimated based on past experience). After waiting about a half hour for this to happen, place the cup in the microwave oven and turn on the oven at maximum power. The cup needs to be watched as it warms up, so that when the water starts to boil, the elapsed time can be noted accurately.

> BE CAREFUL! A styrofoam cup full of hot liquid can buckle if you hold it near the rim. Hold the cup near the bottom. If the cup is full, do not attempt to move the cup while the water is hot. A spill can cause a painful burn.

On the back of the microwave oven (or inside the front door), there is usually a sticker with specifications which says "Output Power = ...Watts" which can be used to calculate the energy supplied. If there is no indication, use a typical value of 600 watts for a standard microwave oven. Using all the quantities measured above and knowing the temperature interval over which you have warmed up the water, you can calculate the specific heat capacity of water.

(a) Show and explain all your data and calculations, and compare with the accepted value for water (4.2 J/K per gram).

(b) Discuss why your result might be expected to differ from the accepted value. For each effect that you consider, state whether this effect would lead to a result that is larger or smaller than the accepted value.

11.P.46 Atmosphere at very high altitudes

In studying a voyage to the Moon in Chapter 3 we somewhat arbitrarily started at a height of 50 kilometers above the surface of the Earth.

(a) At this altitude, what is the density of the air as a fraction of the density at sea-level?

(b) Approximately how many air molecules are there in one cubic centimeter at this altitude?

(c) At what altitude is air density one-millionth (1×10^{-6}) that at sea level?

11.P.47 Two nanoparticles of iron

A nanoparticle consisting of four iron atoms (object 1) initially has 1 quantum of energy. It is brought into contact with a nanoparticle consisting of two iron atoms (object 2) which initially has 2 quanta of energy. The mass of one mole of iron is 56 grams. (Continued on next page.)

(a) Using the Einstein model of a solid, calculate and plot ln Ω_1 *vs.* q_1 (the number of quanta in object 1), ln Ω_2 *vs.* q_1, and ln Ω_{total} *vs.* q_1 (put all three plots on the same graph). Show your work and explain briefly.

(b) Calculate the approximate temperature of the objects at equilibrium. State what assumptions or approximations you made.

11.P.48 Creating a plasma

At sufficiently high temperatures, the thermal speeds of gas molecules may be high enough that collisions may ionize a molecule (that is, remove an outer electron). An ionized gas in which each molecule has lost an electron is called a "plasma." Determine approximately the temperature at which air becomes a plasma.

11.P.49 Specific heat capacity

100 joules of energy transfer due to a temperature difference are given to air in a 50-liter rigid container and to helium in a 50-liter rigid container, both initially at STP (standard temperature and pressure).

(a) Which gas experiences a greater temperature rise?

(b) What is the temperature rise of the helium gas?

11.P.50 Specific heat capacity of hydrogen

(a) Below about 80 K the specific heat capacity at constant volume for hydrogen gas (H_2) is $\frac{3}{2}k$ per molecule, but at higher temperatures the specific heat capacity increases to $\frac{5}{2}k$ per molecule due to contributions from rotational energy states. Use these observations to estimate the distance between the hydrogen nuclei in an H_2 molecule.

(b) At about 2000 K the specific heat capacity at constant volume for hydrogen gas (H_2) increases to $\frac{7}{2}k$ per molecule due to contributions from vibrational energy states. Use these observations to estimate the stiffness of the "spring" that approximately represents the interatomic force.

11.P.51 Five microscopic objects

Here is a one-dimensional row of 5 microscopic objects each of mass 4×10^{-26} kg, connected by forces that can be modeled by springs of stiffness 15 N/m. These objects can move only along the x axis.

A one-dimensional row of microscopic masses and springs.

(a) Using the Einstein model, calculate the approximate entropy of this system for total energy of 0, 1, 2, 3, 4, and 5 quanta. Think carefully about what the Einstein model is, and apply those concepts to this one-dimensional situation.

(b) Calculate the approximate temperature of the system when the total energy is 4 quanta.

(c) Calculate the approximate specific heat capacity on a per-object basis when the total energy is 4 quanta.

(d) If the temperature is raised very high, what is the approximate specific heat capacity on a per-object basis? Give a numerical value and compare with your result in part (c).

11.P.52 Pluto's atmosphere

In 1988 telescopes viewed Pluto as it crossed in front of a distant star. As the star emerged from behind the planet, light from the star was slightly dimmed as it went through Pluto's atmosphere. The observations indicated that the atmospheric density at a height of 50 km above the surface of Pluto is about one-third the density at the surface. The mass of Pluto is known to be about 1.5×10^{22} kg, and its radius is about 1200 km. Spectroscopic data indicate that the atmosphere is mostly nitrogen (N_2). Estimate the temperature of Pluto's atmosphere. State what approximations and/or simplifying assumptions you made.

11.P.53 Buckminsterfullerene

Buckminsterfullerene, C_{60}, is a large molecule consisting of sixty carbon atoms connected to form a hollow sphere. The diameter of a C_{60} molecule is about 7×10^{-10} m. It has been hypothesized that C_{60} molecules might be found in clouds of interstellar dust, which often contain interesting chemical compounds. The temperature of an interstellar dust cloud may be very low, around 3 K. Suppose you are planning to try to detect the presence of C_{60} in such a cold dust cloud by detecting photons emitted when molecules undergo transitions from one rotational energy state to another.

Approximately, what is the highest-numbered rotational level from which you would expect to observe emissions? Rotational levels are $l = 0, 1, 2, 3, ...$

11.P.54 Coin tosses

The reasoning developed for counting microstates applies to many other situations involving probability. For example, if you flip a coin 5 times, how many different sequences of 3 heads and 2 tails are possible? Answer: 10 different sequences, such as HTHHT or TTHHH. In contrast, how many different sequences of 5 heads and 0 tails are possible? Obviously only one, HHHHH, and our formula gives $5!/[5!0!] = 1$, using the standard definition that 0! is defined to equal 1.

If the coin is equally likely on a single throw to come up heads or tails, any specific sequence like HTHHT or HHHHH is equally likely. But there is only one way to get HHHHH, while there are 10 ways to get 3 heads and 2 tails, so this is 10 times more probable than getting all heads.

Use the formula $5!/[N!(5-N)!]$ to calculate the number of ways to get 0 heads, 1 head, 2 heads, 3 heads, 4 heads, or 5 heads in a sequence of 5 coin tosses. Make a graph of the number of ways *vs.* the number of heads.

11.P.55 A disk and brake

A box contains a uniform disk of mass M and radius R which is pivoted on a low-friction axle through its center. A block of mass m is pressed against the disk by a spring, so that the block acts like a brake, making the disk hard to turn. The box and the spring have negligible mass. A string is wrapped around the disk (out of the way of the brake) and passes through a hole in the box. A force of constant magnitude F acts on the end of the string. The motion takes place in outer space. At time t_i the speed of the box is v_i, and the rotational speed of the disk is ω_i. At time t_f the box has moved a distance x, and the end of the string has moved a longer distance d, as shown.

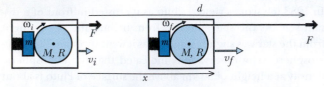

The box in shown both at its initial position, and after it has moved a distance x.

(a) At time t_f, what is the speed v_f of the box?

(b) During this process, the brake exerts a tangential friction force of magnitude f. At time t_f what is the angular speed ω_f of the disk?

(c) At time t_f assume you know (from part b) the rotational speed ω_f of the disk. From time t_i to time t_f what is the increase in thermal energy of the apparatus?

(d) Suppose the increase in thermal energy in part c is 8×10^4 J. The disk and brake are made of iron, and their total mass is 1.2 kg. At time t_i their temperature was 350 K. At time t_f what is their approximate temperature?

11.12 Answers to Exercises

11.X.1 (page 378) 20

11.X.3 (page 380) 5.4×10^{88} years; 5.4×10^{78} times the lifetime of the Universe!

11.X.4 (page 385) Looks *very* improbable; second law of thermodynamics being violated.

11.X.5 (page 388) 500 K

Liquid; yes, liquid more disordered; 1.23 J/K (note that $0°$ C is 273 K)

11.X.7 (page 389) 1.9 J/K/gram

11.X.8 (page 392) $25°$ C

11.X.9 (page 392) Large rise in temperature of sink; no longer as cold as before

11.X.11 (page 396) 1; 0.14; 0.02

11.X.12 (page 398) About 8.4 kilometers! (about 5 miles)

11.X.13 (page 398) About 4×10^{-20} m !

11.X.14 (page 399) About 1/3

11.X.15 (page 401) 1360 m/s

11.X.16 (page 402) 496 m/s

11.X.17 (page 402) 2.50; 2.74

2300 m/s; not very large compared with 500 m/s

11.X.19 (page 403) Increase; factor of $\sqrt{2}$, so increase about 40%

11.X.20 (page 405) $\frac{3}{2}k$; translational energy quanta are very small.

11.X.21 (page 405) $\frac{7}{2}k$

11.X.22 (page 405) $\frac{5}{2}k$

11.X.23 (page 406) $\frac{3}{2}k$

11.X.24 (page 406) Lower temperature. Rotational energy is $\frac{S_x^2}{2I}$, so energy of rotational first excited state for D_2 is smaller (larger nuclear mass means larger moment of inertia, since internuclear distance about the same).

CHAPTER 12

GASES AND ENGINES

Key concepts

- We can model a gas as tiny balls in rapid random motion.
- Applying statistical ideas to this microscopic model allows us to predict macroscopic relationships among pressure, volume, and temperature, including the ideal gas law.
- Agreement between the ideal gas law and the Boltzmann distribution in the previous chapter shows that the definition of temperature in terms of entropy is consistent with the everyday temperature scale.
- Energy transfer due to a temperature difference, and work, can change the temperature and/or volume of a gas in a chamber with a movable piston.
- The attainable efficiency of engines that convert thermal energy into useful work is constrained by the second law of thermodynamics.

12.1 GASES, SOLIDS, AND LIQUIDS

Remember that a gas has no fixed structure, unlike a solid. The molecules are not bound to each other but move around very freely, which is why a gas does not have a well-defined shape of its own; it fills whatever container you put it in (Figure 12.1). Think for example of the constantly shifting shape of a cloud, or the deformability of a balloon, in contrast with the rigidity of a block of aluminum. On average, gas molecules are sufficiently far apart that most of the time they hardly interact with each other. This low level of interaction is what makes it feasible to model a gas in some detail, using relatively simple concepts.

The molecular motion must be sufficiently violent that molecules can't stay stuck together. At high enough temperatures, any molecules that do manage to bind to each other temporarily soon get knocked apart again by high-speed collisions with other molecules. At a low enough temperature however, molecules move sufficiently slowly that collisions are no longer violent enough to break intermolecular bonds. Rather, more and more molecules stick to each other in a growing mass as the gas turns into a liquid or, at still lower temperatures, a solid.

Why we don't study liquids

For completeness, we should say why we don't discuss liquids in this textbook. A liquid is intermediate between a solid and a gas. The molecules are sufficiently attracted to each other that the liquid doesn't fly apart like a gas (Figure 12.2), yet the attraction is not strong enough to keep each molecule near a fixed equilibrium position. The molecules can slide past each other, giving liquids their special property of fluid flow (unlike solids) with fixed volume (unlike gases).

The analysis of liquids in terms of atomic, microscopic models is quite difficult compared with gases, where the molecules only rarely come in contact with other atoms, or compared with solids, where the atoms never move very far away from their equilibrium positions. For this reason, in this introductory textbook with its emphasis on atomic-level description and analysis we concentrate mostly on understanding gases and solids.

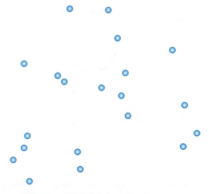

Figure 12.1 Molecules in a gas.

Figure 12.2 Molecules in a liquid.

12.2 GAS LEAKS THROUGH A HOLE

In the previous chapter we used statistical mechanics to determine the average speed of a gas molecule. Here we'll see some interesting phenomena where the average speed plays a role. One example is the leakage rate of a gas through a small hole in a container filled with the gas. First we'll consider a simplified one-directional example, in order to understand the basic issues before stating the results for a real three-dimensional gas.

One-directional gas

The chain of reasoning that we follow is basically geometric. Consider a situation where many gas molecules are all traveling to the right inside a tube. For the moment, temporarily assume that they all have the same speed v (Figure 12.3). The cross-sectional area of the tube is A (Figure 12.4). There are N gas molecules in the tube of volume V, and we will frequently use the symbol $n = N/V$ to stand for the number of molecules per unit volume, which we will express in SI units as number per cubic meter:

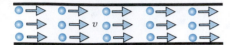

Figure 12.3 Side view of molecules all traveling to the right with speed v inside a tube. There are n molecules per cubic meter inside the tube.

NUMBER DENSITY: NUMBER PER CUBIC METER

$$\text{Definition: } n \equiv \frac{N}{V} \quad \frac{\text{number of molecules}}{\text{m}^3}$$

Figure 12.4 End view of molecules all traveling the same direction in a tube of cross-sectional area A.

Because we eventually want to be able to calculate how fast a gas will leak through a hole in a container, we will calculate how many molecules leave this tube in a short time interval Δt. For a molecule that is traveling at speed v to be able to reach the right end of the tube in this time interval Δt, it must be within a distance $v\Delta t$ of the end (Figure 12.5).

Since the cross-sectional area of the tube is A, the volume V of the tube that contains just those molecules that will leave the tube during the time interval Δt is simply $A(v\Delta t)$; see Figure 12.6.

There are n molecules per m^3 inside the tube, and $N = nV$ molecules inside the volume V. So the number N of molecules that will leave in the time interval Δt is

$$N = nV = n(Av\Delta t)$$

Dividing by the time interval Δt, we find the following result:

> \# of molecules crossing area a per second $= nAv$
>
> (one-directional case; all molecules have same speed)

? Does this formula make sense? What would you expect if you increased the number of molecules per cubic meter, or the cross-sectional area, or the speed?

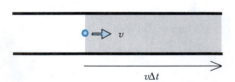

Figure 12.5 A molecule that will leave the tube in a time Δt must be within a distance $v\Delta t$ from the end.

The formula does make sense. The more molecules per unit volume ($n = N/V$), the more molecules will reach the right end of the tube per unit time. The bigger the cross-sectional area (A), the more molecules that will pass through that area per unit time. The faster the molecules are moving (v), the more molecules from farther away that can reach the end of the tube in a given time interval. The units are right: $(\#/\text{m}^3)(\text{m}^2)(\text{m/s}) = \#/\text{s}$.

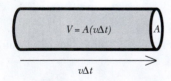

Figure 12.6 Volume containing the molecules that will leave the tube in time Δt.

Effect of different speeds

We need to account for the fact that the gas molecules don't all have the same speed v. Suppose n_1 molecules per unit volume have speeds of approximately v_1, n_2 molecules per unit volume have speeds of approximately v_2, etc. The number of molecules crossing an area A per second is

$$n_1 A v_1 + n_2 A v_2 + \text{etc.}$$

The average speed of all the molecules is by definition the following, where we weight each different speed by the number of molecules per cubic meter that have that approximate speed:

$$\bar{v} = \frac{n_1 v_1 + n_2 v_2 + \text{etc.}}{n}$$

Therefore,

$$n_1 A v_1 + n_2 A v_2 + \text{etc.} = nA\bar{v}$$

A horizontal bar over a symbol is a standard notation for "average," and $\bar{v}$ means average speed.

Finally we have a valid formula for the number of molecules in a one-directional flow leaving the right end of the tube, even in the situation where they have a distribution of different speeds:

OF MOLECULES CROSSING AREA *A* PER SECOND (1-D)

$nA\bar{v}$ (one-directional case; molecules have various speeds)

A "one-directional" gas may sound a bit silly, but this formula does apply to real one-directional flows such as the flow of water or gas through a pipe or the flow of electrons through a copper wire in an electric circuit.

A three-dimensional gas

A more realistic model of a gas would have approximately equal numbers of molecules heading to the left as well as to the right in the tube, in which case the number of molecules leaving the right end of the tube would be only $\frac{1}{2}nA\bar{v}$. In a real three-dimensional gas, molecules are moving in all directions. Only those molecules that are headed in the $+x$ direction can pass through a hole located to the right, not those moving in the $-x$ direction nor in the $\pm y$ or $\pm z$ directions. The molecules are headed randomly in all six directions, so we might expect our formula would have a factor of $1/6$.

However, the actual factor is $1/4$, which comes from detailed averaging over all directions and is related to our use of the average speed (magnitude of velocity), rather than averages of velocity components v_x or v_y or v_z. We don't want to get bogged down in the rather heavyweight mathematics required to prove this, so we just state that the factor is $1/4$ rather than $1/6$.

OF MOLECULES CROSSING AREA *A* PER SECOND (3-D)

$\frac{1}{4}nA\bar{v}$ (three-dimensional case; molecules have various speeds)

$$\text{Remember that } n = \frac{N}{V} \frac{\text{\# of molecules}}{\text{m}^3}$$

Let's see what this formula predicts for the rate at which helium will escape from a balloon through a hole. You probably learned in chemistry that at "standard temperature and pressure" (STP, which means a temperature of 0° Celsius or 273 Kelvin, and a pressure of 1 atmosphere), one mole of a gas occupies 22.4 liters (a liter is 1000 cubic centimeters). As we saw in the previous chapter, the average speed $\bar{v}$ of helium atoms at ordinary temperatures is about 1200 m/s.

12.X.1 Suppose we make a circular hole 1 millimeter in diameter in a balloon. Calculate the initial rate at which helium escapes through the hole (at 0° C), in number of helium atoms leaving the balloon per second.

Cooling of the gas

There is an interesting and important effect of gas escaping through a hole. If you look back over the derivations of the formulas, you can see that faster molecules escape disproportionately to their numbers. Faster molecules can be farther away from the hole than is true for slower molecules and still escape through the hole in the next short time interval. As a result, the distribution of speeds of molecules inside the container becomes somewhat depleted of high speeds.

12.X.2 What can you say about the temperature of the gas inside the container as the gas escapes? Why?

Example: Leakage from a balloon

A party balloon is about one foot in diameter (about 30 centimeters). The volume of a sphere is $\frac{4}{3}\pi r^3$, where r is the radius of the sphere. Refer to the calculation you made in Exercise 12.X.1 on page 413 for the initial leak rate through a circular hole 1 millimeter in diameter.

(a) If the helium were to escape at a constant rate equal to the initial rate, about how long would it take for all the helium to leak out?

(b) As the helium escapes, the balloon shrinks, and the number of helium atoms per cubic meter n will stay roughly constant, in which case our analysis is pretty good. However, remember that the temperature of the gas drops due to a preferential loss of high-speed atoms. Would this effect make the amount of time to empty be more or less than the value you calculated in part (a)?

(a) The leak rate was 6.3×10^{21} atoms per second. Calculate the number of helium atoms in the balloon originally:

$$\frac{4}{3}\pi(0.3 \text{ m})^3\left(\frac{6\times10^{23} \text{ atoms}}{22.4\times10^3 \text{ cm}^3}\right)\left(\frac{10^6 \text{ cm}^3}{\text{m}^3}\right) = 3\times10^{24} \text{ atoms}$$

Assuming constant rate:

$$\frac{3\times10^{24} \text{ atoms}}{6.3\times10^{21} \text{ atoms/s}} = 500 \text{ s}$$

(b) With the preferential loss of high-speed atoms, the speed distribution inside the balloon shifts to lower speeds (corresponding to a lower temperature). If the average speed is lower, the leak rate is lower, and it should take longer for the balloon to empty than we calculated in part (a), where we assumed a constant leak rate.

We've implicitly done the analysis in vacuum. If the balloon is in air, air molecules enter the balloon through the same hole.

When you blow up an ordinary rubber balloon the pressure is higher than one atmosphere, which means higher number density n but also a correspondingly higher leak rate, which is proportional to n. To a first approximation the density doesn't matter in this estimate of the time to empty. Also note that often when you puncture a balloon the balloon rips, creating a large opening; this is not the case we are considering.

12.3 MEAN FREE PATH

There is another important property of a gas that we can understand at this point in terms of our atomic model. Suppose we place a special molecule somewhere in the helium gas, one whose movements we can trace. For example, it might be a molecule of perfume. On average, how far does this molecule go before it runs into a molecule of the gas? The average distance between collisions is called the "mean free path." It plays an important role in many phenomena, including the creation of electric sparks in air (discussed in Volume II of this textbook).

The calculation of the mean free path depends on a simple geometrical argument. Draw a cylinder along the direction of motion of the special molecule, with length d and radius $R+r$, where R is the radius of the molecule, and r is the radius of a gas molecule, as shown in Figure 12.7 and Figure 12.8.

The geometrical significance of this cylinder is that if the path of the special molecule comes within one molecular radius of a gas molecule, there will be a collision, so any gas molecule whose center is inside the cylinder will be hit. How long should the cylinder be for there to be a collision?

We define d to be the average distance the special molecule will travel before colliding with another molecule, so the cylinder drawn in Figure 12.7 should contain on average about one gas molecule. The cross-sectional area of the cylinder is $A \approx \pi(R+r)^2$ and the volume of the cylinder is Ad. If n stands for the number of gas molecules per cubic meter, we can write a formula involving the mean free path:

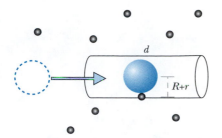

Figure 12.7 If a "special molecule" (colored) enters a cylinder of length d and radius $R+r$, it will collide with a gas molecule (black).

MEAN FREE PATH

$$nAd \approx 1 \text{ where } A \approx \pi(R+r)^2 \text{ (and } n = N/V)$$

We are ignoring some subtle effects in this calculation, but this analysis gives us the main picture: the mean free path is shorter for higher density or larger molecular size.

To get an idea of the order of magnitude of a typical mean free path, assume that an N_2 molecule in the air has a radius of approximately 2×10^{-10} m (the radius of one of the N atoms being about 10^{-10} m), and let's calculate approximately the mean free path d of an N_2 molecule moving through air. At "standard temperature and pressure" (STP, 0° Celsius, 1 atmosphere pressure) one mole of air (6×10^{23} molecules) has a volume of 22.4 liters (22.4×10^3 cm^3).

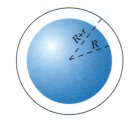

Figure 12.8 End view of cylinder.

$$n = \frac{6\times10^{23} \text{ molecules}}{(22.4\times10^3 \text{ cm}^3)(10^{-6} \text{ m}^3/\text{cm}^3)} = 2.7\times10^{25} \frac{\text{molecules}}{\text{m}^3}$$

$$A \approx \pi(2 \times 2\times10^{-10} \text{ m})^2 = 5\times10^{-19} \text{ m}^2$$

$$d \approx \frac{1}{nA} \approx \frac{1}{\left(2.7\times10^{25} \dfrac{\text{molecules}}{\text{m}^3}\right)(5\times10^{-19} \text{ m}^2)} = 7\times10^{-8} \text{ m}$$

12.X.3 It is interesting to compare the mean free path of about 7×10^{-8} m to the average spacing L between air molecules, which is the cube root of the volume occupied on average by one molecule (Figure 12.9). Calculate L. You may be surprised to find that the mean free path d is much larger than the average molecular spacing L. The molecules represent rather small targets.

12.4 PRESSURE AND TEMPERATURE

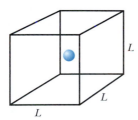

Figure 12.9 The volume of air occupied on average by one molecule.

Next we will use statistical ideas to relate what we know about the motion of molecules in a gas to the pressure that the gas exerts on its container.

In a party balloon, helium atoms are continually hitting the rubber walls of the balloon. In 12.X.1 on page 413 you found that on average, at STP (standard temperature and pressure) about 6×10^{21} helium atoms hit a 1-mm-diameter section of the balloon every second. The time-and space-averaged effect of this bombardment is an average force exerted on every square millimeter of the balloon. This average force per unit area is called the "pressure" P and is measured in N/m^2, also called a "pascal." We are going to calculate how big the pressure is in terms of the average speed of the helium atoms, thus building a link between the microscopic behavior of the helium atoms and the macroscopic time- and space-averaged pressure.

When the velocity distribution is stable, on average a helium atom bounces off the wall with no change of kinetic energy. We emphasize that this is the average behavior. Any individual atom may happen to gain or lose energy in the collision with a vibrating atom in a rubber molecule, but if the balloon is not being warmed up or cooled down, the velocity distribution in the gas is stable, and on average the helium atoms rebound from the wall with the same kinetic energy they had just before hitting the wall (Figure 12.10).

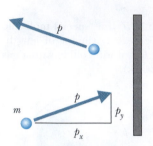

Figure 12.10 On average an atom bounces off the wall without changing speed.

? If the kinetic energy doesn't change when an atom bounces off the wall, is there any change in the atom's momentum?

Unlike speed or kinetic energy, momentum is a vector quantity, and there is a large change in the x component of momentum, from $+p_x$ to $-p_x$. (There is no change in the y component.) Therefore the momentum change of the helium atom is $\Delta p_{x,\,\text{helium}} = -2p_x$.

? What caused this change in the momentum of the helium atom?

A force is required to change the momentum of an object. In this case the force was applied by the wall of the container (or in more detail, by an atom in a rubber molecule in the balloon). By the principle of reciprocity (Newton's third law), the helium atom must have applied an equal and opposite force to the wall (or more precisely, to that atom in the rubber molecule). Therefore the wall must have acquired an amount of momentum $\Delta p_{x,\,\text{wall}} = 2p_x$.

If we could calculate the average time Δt between collisions of helium atoms with a small area A of the wall, we could express the pressure (force per unit area) as follows, since $d\vec{\mathbf{p}}/dt = \vec{\mathbf{F}}$:

$$P = \frac{F}{A} = \frac{1}{A}\frac{\Delta p_x}{\Delta t}$$

Assuming one direction and one speed

We already know that if there are n_{right} helium atoms per unit volume that have an x component of velocity equal to $+v_x$ (that is, moving to the right toward the wall), the number of such atoms that hit an area A to their right in a time Δt is $nAv_x\Delta t$. Therefore the average time between collisions is

$$\Delta t = 1/(n_{\text{right}}Av_x)$$

and we find

$$P = \frac{F}{A} = \frac{1}{A}\frac{\Delta p_x}{\left(\dfrac{1}{n_{\text{right}}Av_x}\right)} = (n_{\text{right}}v_x)(2p_x) = 2n_{\text{right}}\left(\frac{p_x}{m}\right)p_x = 2n_{\text{right}}\frac{p_x^2}{m}$$

An alternative argument

Alternatively, we can say that $n_{\text{right}}Av_x$ helium atoms with this x component of velocity hit the area A every second, and each of them delivers $2p_x$ of momentum to the wall, so that the total momentum transfer to the wall per second is

$$(\text{\# hits per second})(\text{momentum transfer per hit})$$

$$(n_{\text{right}}Av_x)(2p_x)$$

Divide by A to get the force per unit area, and you again get

$$P = (n_{\text{right}}v_x)(2p_x) = 2n_{\text{right}}\frac{p_x^2}{m}$$

Taking the speed distribution into account

Actually, this is just the contribution to the pressure made by those atoms that happened to have this particular value of v_x to the right. On average only half the atoms are headed to the right, so we replace n_{right} by $n/2$, where n is the number of atoms per cubic meter (going in either direction).

$$P = 2 n_{\text{right}} \frac{p_x^2}{m} = n \frac{p_x^2}{m}$$

We also need to average over the slow and fast atoms:

$$P = n \frac{\overline{p_x^2}}{m}$$

We'll explain what we mean by this average. Suppose n_1 atoms per unit volume have x components of momentum of approximately p_{x1}, n_2 atoms per unit volume have x components of momentum of approximately p_{x2}, etc. The pressure is

$$P = n_1 \left(\frac{p_{x1}^2}{m} \right) + n_2 \left(\frac{p_{x2}^2}{m} \right) + \dots$$

The average value of p_x^2 for all the atoms is by definition the following, where we weight each different value of p_x^2 by the number of atoms per cubic meter that have that approximate value:

$$\overline{p_x^2} = \frac{n_1 p_{x1}^2 + n_2 p_{x2}^2 + \text{etc.}}{n} , \text{ so we have } n_1 p_{x1}^2 + n_2 p_{x2}^2 + \text{etc.} = n \overline{p_x^2}$$

Therefore we can write this:

$$P = n_1 \left(\frac{p_{x1}^2}{m} \right) + n_2 \left(\frac{p_{x2}^2}{m} \right) + \dots = n \frac{\overline{p_x^2}}{m}$$

Taking direction into account

We can re-express the pressure in terms of momentum p rather than p_x by the following steps:

First, note that $p = \sqrt{p_x^2 + p_y^2 + p_z^2}$, or $p^2 = p_x^2 + p_y^2 + p_z^2$.

Second, since the atoms are flying around in random directions, there should be no difference in averaging in the x, y, or z direction, so $\overline{p_x^2} = \overline{p_y^2} = \overline{p_z^2}$, which implies that $\overline{p^2} = 3 \overline{p_x^2}$.

Taking these factors into account, instead of $P = n \dfrac{\overline{p_x^2}}{m}$ we can write the following important result:

GAS PRESSURE IN TERMS OF ATOMIC QUANTITIES

$$P = \tfrac{1}{3} n \frac{\overline{p^2}}{m} , \text{ where } n = N/V$$

? Look back over the line of reasoning that led us to this result, and reflect on the nature of the argument. Stripped of the details, try to summarize the major steps leading to this result.

We reached this result by combining two effects: the number of molecules hitting an area per second is proportional to v, and the momentum transfer is also proportional to v. Hence the force per unit area is proportional to v^2.

The ideal gas law

It is useful to rewrite our result for the pressure, factoring out the term $\overline{p^2}/(2m)$, which is the average translational kinetic energy K_{trans} of a molecule of mass m:

$$P = \tfrac{2}{3} n\left(\frac{\overline{p^2}}{2m}\right) = \tfrac{2}{3} n \overline{K}_{\text{trans}}$$

The pressure of a gas is proportional to the number density (number of molecules per cubic meter, N/V, and proportional to the average translational kinetic energy of the gas molecules. This gives us a connection to temperature, because we found in the previous chapter that $\overline{K}_{\text{trans}} = \tfrac{3}{2} kT$.

Substituting into the pressure formula, $P = \tfrac{2}{3} n \overline{K}_{\text{trans}} = \tfrac{2}{3} n(\tfrac{3}{2} kT) = nkT$, which relates pressure to temperature and is called the "ideal gas law":

MOLECULAR VERSION OF THE IDEAL GAS LAW

$$P = nkT$$

$n = N/V$ = number of molecules per m^3

$k = R/6\times10^{23}$ is the "Boltzmann constant" (1.4×10^{-23} J/K)

12.X.4 A mole of a gas (6×10^{23} molecules) occupies 22.4 liters under standard conditions (temperature of 0° Celsius and one atmosphere pressure). Calculate the value of one atmosphere pressure in units of N/m^2.

The macroscopic ideal gas law

We can compare our molecular version of the gas law with experiments that measure the macroscopic properties of gases. For gases with fairly low densities, measurements of the pressure for all gases (helium, oxygen, nitrogen, carbon dioxide, etc.) are well summarized by the macroscopic "ideal gas law" which is probably familiar to you from chemistry, and which describes the observed behavior of any low-density gas:

$$P = \frac{(\# \text{ of moles})RT}{V} \quad \text{(mole version of the gas law)}$$

Here R is the "gas constant" (8.3 J/K/mole), T is the absolute temperature in Kelvins, and V is the volume in cubic meters. To compare with our microscopic prediction, we can convert this macroscopic version of the ideal gas law to a version involving microscopic quantities. Note the following:

$$\# \text{ of moles} = \frac{(\# \text{ of molecules})}{(6\times10^{23} \text{ molecules/mole})} = \frac{N}{6\times10^{23}}$$

where N is the number of molecules in the gas. Therefore we have

$$P = \left(\frac{N}{6\times10^{23}}\right)\frac{RT}{V} = \left(\frac{N}{V}\right)\left(\frac{R}{6\times10^{23}}\right)T$$

But this is simply $P = nkT$, where $n = N/V$ and $R/6\times10^{23}$ is the Boltzmann constant k:

$$k = \frac{R}{6\times10^{23}} = \frac{8.3 \text{ J/K/mole}}{6\times10^{23} \text{ molecules/mole}} = 1.4\times10^{-23} \text{ J/K}$$

Warning about the meaning of "*n*"

You may be familiar with the gas law written in the form $PV = nRT$, where n is the number of moles, whereas we mean something else by "n"—the number of molecules per cubic meter, N/V. On the rare occasions when we need to refer to the number of moles, we'll write it out as "# of moles."

Temperature from entropy or from the ideal gas law

In the previous chapter we found the relationship $\overline{K}_{trans} = \frac{3}{2}kT$, based on the Boltzmann distribution as applied to a low-density gas. The Boltzmann distribution in turn was based on the statistical mechanics definition of temperature in terms of entropy as $1/T = \partial E/\partial S$.

When we inserted $\overline{K}_{trans} = \frac{3}{2}kT$ into the kinetic theory result for pressure, $P = \frac{1}{3}n(\overline{p^2}/m)$, we obtained the molecular version of the ideal gas law, $P = nkT$, which we showed was equivalent to the macroscopic version of the ideal gas law, $P = (\text{\# of moles})RT/V$.

Low-density gases are described well by the ideal gas law, and for that reason gases are used to make accurate thermometers. You measure the pressure P and volume V of a known number of moles of a low-density gas and determine the temperature from the macroscopic gas law:

$$T = \frac{PV}{(\text{\# of moles})R} \quad \text{(gas thermometer)}$$

The fact that we could start from $1/T = \partial E/\partial S$ and derive the ideal gas law proves that the temperature measured by a gas thermometer is exactly the same as the "thermodynamic temperature" defined in terms of entropy.

Real gases

Our analysis works well for low-density gases. For high-density gases, there are two major complications. First, the molecules themselves take up a considerable fraction of the space, so the effective volume is less than the geometrical volume, and V in the gas law must be replaced by a smaller value.

Second, at high densities the short-range electric forces between molecules have some effect. In this context these intermolecular forces are called "van der Waals" forces, which are the gradient of the interatomic potential energy discussed in Chapter 6. In a low-density gas almost all of the energy is kinetic energy, but in a high-density gas some of the energy goes into configurational energy associated with the interatomic potential energy. The effect is that in the gas law P must be replaced by a smaller value.

When these two effects are taken into effect, the resulting "van der Waals" equation fits the experimental data quite well for all densities of a gas, although the corrections are different for different gases due to differences in molecular sizes and intermolecular forces.

Energy of a diatomic gas

The formula for the average translational kinetic energy $\overline{K}_{trans} = \frac{3}{2}kT$ is valid even for a gas with multiatom molecules such as nitrogen (N_2) or oxygen (O_2). But, as you will recall from Chapter 8, this translational kinetic energy is only a part of the total energy of the molecule. In addition to translational energy, a diatomic molecule can have rotational and vibrational energy relative to the center of mass (Figure 12.11), so the energy contains additional terms. We write the energy of a diatomic molecule in the following way:

$$K_{trans} + E_{vib} + E_{rot} + Mgy_{cm}$$

The key point, as we discussed in the previous chapter, is that the average energy of a diatomic molecule is greater than the average energy of an atom in a monatomic gas at the same temperature. For a monatomic gas such as helium the average total energy per molecule is just $\overline{K}_{trans} = \frac{3}{2}kT$.

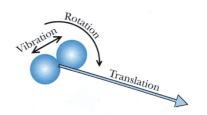

Figure 12.11 Energy of an oxygen molecule.

Application: Weight of a gas in a box

The exponential dependence of number density ($n \propto e^{-\frac{mgy}{kT}}$) and therefore of pressure on height makes it possible to understand something that otherwise might be rather odd about weighing a box that contains a gas (Figure 12.12). Let's do the weighing in a vacuum so we don't have to worry about

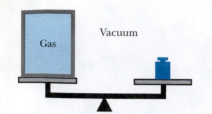

Figure 12.12 Weighing a box containing a gas.

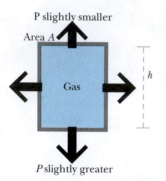

Figure 12.13 The force due to gas pressure on the top of the box differs slightly from the force on the bottom of the box.

buoyancy forces due to surrounding air. The box definitely weighs more on the scales if there is gas in it than if there is a vacuum in it. In fact, if the mass of the gas is M, the additional weight is Mg, because the momentum principle for multiparticle systems refers to the net force on a system, and the gravitational contribution to the net force is the sum of the gravitational forces on each individual gas molecule in the box, Mg.

That seems reasonable until you think about the details of what is going on inside the box. At any given instant, the vast majority of the gas molecules are not touching the box! And some of the gas molecules are colliding with the top of the box, exerting an upward force on the box. How can these molecules possibly contribute to the weight measured by the scales? What is the mechanism for the momentum principle working out correctly for this multiparticle system?

The number density n at the bottom of the box is slightly larger than the number density at the top of the box, if all the gas is at the same temperature, which is a good assumption in most situations. Therefore the pressure, $P = nkT$, is larger at the bottom than at the top. Let's calculate the y components of the forces associated with these slightly different pressures, where A is the area of the top (and bottom) of the box, and h is the height of the box (Figure 12.13).

$$F_{net,y} = P_{top}A - P_{bottom}A = A\Delta P$$

where ΔP is the pressure difference from bottom to top inside the box. We can calculate ΔP directly, by starting from the fact that for a small height change Δy, we have this:

$$\frac{\Delta P}{\Delta y} \approx \frac{dP}{dy}, \text{ so } \Delta P \approx \frac{dP}{dy}\Delta y$$

If we can evaluate dP/dy, we can determine the small pressure difference ΔP. Since $P = nkT$, the height dependence of the pressure is the same as for the number density (for constant temperature), and we can write this:

$$\frac{dP}{dy} = \frac{d}{dy}\left[P_{bottom}e^{-\frac{mgy}{kT}}\right]$$

$$\frac{dP}{dy} = P_{bottom}e^{-\frac{mgy}{kT}}\left[-\frac{mg}{kT}\right] \approx -P_{bottom}\left[\frac{mg}{kT}\right]$$

because the factor $e^{-\frac{mgy}{kT}}$ is very close to 1. The result is negative because the force the gas exerts on the top of the box is smaller than the force on the bottom. Choosing $\Delta y = h$, and writing $P_{bottom} = nkT$, we have this result for ΔP:

$$\Delta P = \frac{dP}{dy}\Delta y \approx -nkT\left[\frac{mg}{kT}\right]h = -nmgh$$

Now that we know the pressure difference, we can calculate the y component of the net force:

$$F_{net,y} = A\Delta P = -nmg(Ah)$$

? But Ah is the volume V of the box, and $n = N/V$ is the number of molecules per m^3 in the box, so what does this formula reduce to?

We have the striking result that the air inside the box pushes down on the box with a force equal to the combined weight Mg of all N molecules in the box:

$$F_y = -Nmg = -Mg \text{ !!}$$

We have shown that the difference in the time- and space-averaged momentum transfers by molecular collisions to the top and bottom of the box is equal to the weight of the gas in the box, as predicted by the momentum principle for a multiparticle system. The pressure difference is very slight, but then the weight of the gas is very small, for that matter. What is surprising is that any particular instant, relatively few of the molecules are actually in contact with the box, yet the effects of these relatively few molecules is the same as though they were all sitting on the bottom of the box.

You could think of a box full of water in the same way. The water pressure is larger at the bottom than at the top, but with water this difference is quite large, corresponding to the much higher density of water. In fact, a column of water only 10 meters high makes a pressure equal to that produced by the many kilometers of atmosphere (atmospheric pressure at sea level is about 10^5 N/m^2).

> *12.X.5* In the preceding discussion the box was in vacuum, but under normal conditions the box we were weighing would have been surrounded by air. Suppose the box has thin walls and is initially open to the air. Then we close the lid, trapping air inside it. Consider all forces including buoyancy forces, and determine what the scales will read: the weight of the box alone or the box plus air?

Application: Weight of a bouncing molecule

There is a related calculation that is amusing. If you did a related homework problem in Chapter 9 on collisions, you already went through a similar calculation. Consider a single molecule (otherwise in vacuum) that bounces up and down on a scale without losing significant energy. Figure 12.14 shows a snapshot at the instant that the molecule is just bouncing up after hitting the scale.

What do the scales read? If the scales can respond quickly, we will see brief spikes each time the molecule strikes it (Figure 12.15). Let's determine what the time-averaged force is.

? How long does it take for the molecule to reach the top of its trajectory, h, starting with a speed v? What then is the time Δt between impacts? How much momentum transfer is there to the scale on each impact?

It takes a time interval v/g to go up (for the speed to decrease from v to 0 with acceleration $-g$) and another time interval v/g to come down, so the time between impacts is $\Delta t = 2v/g$. Each impact transfers an amount of momentum $\Delta P = 2mv$. Therefore the time-averaged force is

$$\frac{\Delta P}{\Delta t} = \frac{2mv}{2v/g} = mg \ !!$$

If the scale is sluggish, and can't respond in a time as short as $\Delta t = 2v/g$, the scale will simply register the value mg, just as though the molecule were sitting quietly on the scale.

12.5 ENERGY TRANSFERS WITH THE SURROUNDINGS

In this section we offer an analysis of energy transfers between a gas and its surroundings. We will address such questions as these: How much energy transfer Q due to a temperature difference is required to raise the temperature of a gas by one degree (the heat capacity of a gas)? How does the temperature of a gas change when you compress it quickly? Do the answers to these questions depend on what kind of gas is involved?

We need a device that lets us control the flow of energy into and out of a gas, in the form of work W or energy transfer Q. We will use a system consisting of a cylindrical container containing gas that is enclosed by a piston

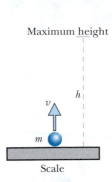

Figure 12.14 The ball has just hit the scale and has rebounded upward.

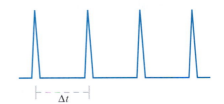

Figure 12.15 Impacts of the ball on the scale.

Figure 12.16 A cylinder with a piston that can move vertically.

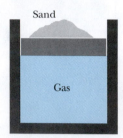

Figure 12.17 The piston can be loaded with varying amounts of sand.

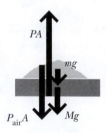

Figure 12.18 Forces on the piston.

that can move in and out of the cylinder with little friction but which fits tightly enough to keep the gas from leaking out (Figure 12.16). This is similar to a cylinder in an automobile engines, into which is sprayed a mixture of gasoline vapor and air. The mixture is ignited by a spark, and the chemical reactions raise the temperature and pressure very high very quickly. The piston is pushed outward, which turns a shaft which ultimately drives the wheels.

Force and pressure

Figure 12.17 shows a cylinder with a vertical-running piston on which we can load varying amounts of sand, in order to be able to control the pressure of the contained gas, and to be able to do controlled amounts of work on the gas. We will also put the cylinder in contact with hot or cold objects and allow energy transfer into or out of the cylinder.

> **?** Consider the piston plus sand as the system of interest for a moment and think about what forces act on this system.

A physics diagram for the piston+sand system (Figure 12.18) includes the downward gravitational forces on the piston (Mg) and on the sand (mg), an upward time- and space-averaged force due to the pressure of the enclosed gas on the lower surface of the piston, and a downward time- and space-averaged force due to the pressure of the outside atmosphere on the upper surface of the system.

Since pressure is force per unit area, the upward force is PA, where A is the surface area of the bottom of the piston. Similarly, the downward force contributed by the outside atmosphere is $P_{air}A$. These pressure-related forces are not actually continuous, since they are the result of collisions of individual gas molecules with the piston and sand. However, the rate of collisions is so extremely high over the area of the piston that the force seems essentially constant. For example, in 12.X.1 on page 413 you found that on average, at STP (standard temperature and pressure) about 6×10^{21} helium atoms strike a tiny 1-mm-diameter section every second.

> **?** In mechanical equilibrium (velocity of piston not changing), solve for the gas pressure inside the cylinder.

The momentum principle tells us that in equilibrium the net force on the piston+sand system must be zero, from which we are able to deduce that the pressure of the gas inside the cylinder is

$$P = P_{air} + \frac{Mg + mg}{A}$$

A sudden change

By varying the amount of sand (m) we can vary the pressure of the gas under study.

> **?** If you suddenly add a lot of sand, what happens?

If you suddenly add a lot of sand to the piston, there is suddenly a sizable nonzero net downward force on the piston+sand system: $F_{net} = AP_{air} + Mg + mg - AP \neq 0$. The piston starts to pick up speed downwards. As it does so, it runs into gas molecules and tends to increase their speeds.

> **?** What happens to the temperature of the gas in the cylinder?

Since higher average speed means higher temperature, the gas temperature starts to increase, at the same time that the volume of the gas is decreasing. This is a double whammy: both increased temperature and decreased volume contribute to increased pressure, since $P = (N/V)kT$, where N is the number of gas molecules in the cylinder. Therefore the pressure in the gas quickly rises, and eventually there will be a new equilibrium with a lower piston (supporting more sand) and a higher gas pressure in the cylinder.

However, getting to that new equilibrium is pretty complicated. If there is no friction or other energy dissipation, the piston will oscillate down and up, with the gas pressure going up and down. It is even possible to determine an effective "spring constant" for the gas and calculate the frequency of the oscillation. However, in any real system there will be some friction, so we know that the system will eventually settle down to a new equilibrium configuration.

Quasistatic processes

To avoid these complicated (though interesting) transient effects, we will study what happens when we add sand very carefully and very slowly, one grain at a time, and we assume that the new equilibrium is established almost immediately. This is called a "quasistatic compression" because the system is at all times very nearly in equilibrium (Figure 12.19). Similarly, if we slowly remove one grain at a time, we can carry out a "quasistatic expansion." Note in particular that at no time does the piston have any significant amount of kinetic energy, and macroscopic kinetic energy is essentially zero at all times. Of course there is plenty of microscopic kinetic energy in the gas molecules and the outside air molecules, and in the thermal motion of atoms in the cylinder walls, piston, and sand.

Suppose we carry out a lengthy, time-consuming quasistatic compression by adding lots of sand, one grain at a time, very slowly. The piston goes down quite a ways, the pressure in the gas is a lot higher, and the volume of the gas is a lot smaller. If we know the pressure and volume, we can calculate the temperature by using the ideal gas law $P = (N/V)kT$ (assuming the gas density isn't too high to make this invalid).

But can we predict how low the piston will go for a given amount of added sand? Oddly enough, no—not without knowing something more about this device. There are two extreme cases that are both important in practice and calculable for an ideal gas. If the apparatus is made of metal (a very good thermal conductor) and is in good thermal contact with a large object at temperature T, the process will proceed at nearly constant temperature, with energy transfer from the gas into the surroundings. If the apparatus is made of glass (a very poor thermal conductor), the temperature of the gas inside the cylinder will rise in a predictable way, with negligible energy transfer to the surroundings in the form of energy transfer Q. We will analyze both kinds of processes—constant temperature processes, and no-Q processes.

Constant-temperature (isothermal) compression

Suppose the cylinder is made of metal (which is a very good thermal conductor) and is sitting in a very big tub of water whose temperature is T, as shown in Figure 12.20.

As we compress the gas, the temperature in the gas starts to increase. However, this will lead to energy flowing out of the gas into the water, because whenever the temperatures differ in two objects that are in thermal contact with each other, we have seen that there is a transfer of energy from the hotter object into the colder object. In fact, for many materials the rate of energy transfer is proportional to the temperature difference—double the temperature difference, double the rate at which energy transfers from the hotter object into the colder one.

The mechanism for energy transfer due to a temperature difference is that atoms in the hotter object are on average moving faster than atoms in

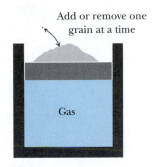

Figure 12.19 By adding or removing one grain of sand at a time we effect a "quasistatic" compression or expansion.

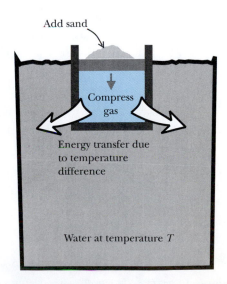

Figure 12.20 A cylinder of gas immersed in a large volume of water.

the colder object, so in collisions between atoms at the boundary between the two systems it is likely that energy will be gained by the colder object and lost by the hotter object.

Energy transfer out of the gas will lower the temperature of the gas, since the total energy of the gas is proportional to the temperature. Quickly the temperature of the gas will fall back to the temperature of the water. The temperature of the big tub of water on the other hand will hardly change as a result of the energy added to it from the gas, because there is such a large mass of water to warm up.

Therefore the entire quasistatic compression takes place essentially at the temperature of the water, and the final temperature of the gas is nearly the same as the initial temperature of the gas. This is a constant-temperature compression (also called an "isothermal" compression, which just means constant temperature). Similarly, we can slowly remove sand from the piston and carry out a quasistatic constant-temperature expansion.

? Suppose the original gas pressure and volume were P_1 and V_1, and as the result of a constant-temperature process the final volume is V_2. What is the final pressure P_2?

Since the temperature hasn't changed, from the ideal gas law $P = (N/V)kT$ we can deduce that $P_1 V_1 = P_2 V_2$, and therefore $P_2 = P_1(V_1/V_2)$.

A constant-temperature compression

A more difficult question we can ask (and answer!) is this: How much energy was added to the water in the constant-temperature compression?

? What energy inputs and outputs were made to the gas? What energy change occurred in the gas?

The piston did work on the gas, and there was energy transfer Q out of the gas (and into the water). The Earth's gravitational force did work on the gas (since the center of mass of the gas went down), but this is negligibly small compared to the work done by the piston (the lowering of the heavy piston involves much more gravitational energy than the lowering of the low-mass gas).

? Did the total energy of the gas change (ignoring the small gravitational energy change)?

Since the total energy of an ideal gas is proportional to temperature (including rotational and vibrational energy if the gas is not monatomic), and we made sure that we kept the temperature constant, the total energy of the gas did not change. Therefore we have this energy equation for the open system that is the gas:

? Δ(energy of gas) = (W by piston) − ($|Q|$ that flowed into water) = 0

Symbolically, $\Delta E_{gas} = W + Q = 0$, where Q is negative (transfer out of system). In thermal processes of this kind, the energy equation is called "the first law of thermodynamics."

If we can calculate the work done by the piston, we can equate that result to the amount of energy transfer Q into the water. The piston exerts a (time- and space-averaged) force PA on the gas, where P is the pressure in the gas and A is the cross-sectional area of the piston (Figure 12.21).

? If the piston drops a distance h, is the work it does PAh?

As the piston drops, the pressure in the gas increases, so the force is not constant. We have to integrate the variable force through the distance h in order to determine the work. If we measure x downward from the initial piston position, at each step dx of the way the increment of work done is $PA\,dx$. But

Figure 12.21 Force on the gas by the piston.

$F_{\text{on gas}} = PA$

Adx is an increment of volume (base area *A* time altitude *dx*), and the change in the volume is negative. Putting this all together, we have this:

WORK DONE BY A PISTON ON A GAS

$$W = -\int_{V_1}^{V_2} P \, dV$$

Check the sign: If the volume decreases, the integral will be negative, and the minus sign in front of the integral makes the work done on the gas be positive, which is correct. Conversely, in an expansion of a gas, the integral is positive, and the work is negative, because the gas is doing work on the piston rather than the other way round.

? In a constant-temperature (isothermal) compression, work is done on the gas. Where does this energy flow go? How does the temperature stay constant?

As the piston moves down, it increases the average energy of the gas molecules that run into it. Therefore the input work starts to raise the temperature of the gas (higher speeds), but this higher temperature causes energy transfer *Q* out of the cylinder, into the (very slightly) lower-temperature water. The net effect is for energy to flow into the gas in the form of mechanical work, and out of the gas into the water in the form of energy transfer *Q*. There is no change in the total energy of the gas, because the temperature of the gas didn't change. The outward energy transfer brings the gas temperature back to what it was before the falling piston tried to increase the temperature.

Now we can calculate quantitatively the work done by the piston in the constant-temperature compression, which is equal to the energy transfer from the gas to the surrounding water. We add lots of sand, one grain at a time (to maintain quasi-static equilibrium), and we compress the gas from an initial volume V_1 to a final volume V_2 (Figure 12.22).

Replace *P* by *NkT/V*, where *T* is a constant:

$$W = -\int_{V_1}^{V_2} P \, dV = -\int_{V_1}^{V_2} \frac{NkT}{V} \, dV$$

$$W = -NkT \int_{V_1}^{V_2} \frac{dV}{V} = -NkT [\ln V]_{V_1}^{V_2} = -NkT \ln\left(\frac{V_2}{V_1}\right)$$

Since the process was a compression, V_2 is smaller than V_1. Since $-\ln(V_2/V_1) = \ln(V_1/V_2)$, we have

$$W_{\text{by piston}} = |Q_{\text{into water}}| = NkT \ln\left(\frac{V_1}{V_2}\right)$$

Just as the symbol *W* is normally used for "work," so the symbol *Q* is normally used for "energy transfer due to a temperature difference." Here *Q* is negative (energy output).

Graphical representation

There is a useful graphical representation of this constant-temperature process. We plot the pressure *P* vs. the volume *V*, as shown in Figure 12.23, and the process is represented by a curve of constant *T* (which is also a curve of constant *PV*, since *PV = NkT*). The area under the process curve represents the work *W*, which is also the energy transfer *Q* that flows to the surroundings.

$$W = -\int_{V_1}^{V_2} P \, dV$$

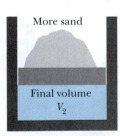

Figure 12.22 The gas is compressed at a constant temperature *T* (the cylinder is immersed in a large tub of water at temperature *T*).

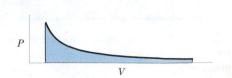

Figure 12.23 If *T* is kept constant, then the product *PV* is also constant.

The first law of thermodynamics

For historical reasons, in thermal processes the energy principle is called "the first law of thermodynamics."

THE FIRST LAW OF THERMODYNAMICS

$$\Delta E_{sys} = W + Q$$

In the particular case of constant-temperature (isothermal) compression of an ideal gas, we have

$$\Delta E_{int} = W + Q = 0$$

where E_{int} is the "internal" energy of the gas (the sum of the translational, rotational, vibrational, and other energy terms of all the molecules).

12.X.6 A mole of nitrogen is compressed (by piling lots of sand on the piston) to a volume of 12 liters at room temperature (293 K). The cylinder is placed on an electric heating element whose temperature is maintained at 293.001 K. A quasistatic expansion is carried out at constant temperature by very slowly removing grains of sand from the top of the piston, with the temperature of the gas staying constant at 293 K. When the volume is 18 liters, how much energy transfer Q has gone from the heating element into the gas? How much work W has been done on the piston by the gas? How much has the energy of the gas changed?

(You must assume that there is no energy transfer from the gas to the surrounding air, and no friction in the motion of the piston, all of which is pretty unrealistic in the real world! Nevertheless there are processes that can be approximated by a constant-temperature expansion. This problem is an idealization of a real process.)

Heat capacity

We were able to calculate the important properties of a constant-temperature compression, where the apparatus was in good thermal contact with its surroundings. Before analyzing the opposite extreme, where the apparatus is thermally insulated from its surroundings and no energy transfer due to a temperature difference is involved, we will review the concept of heat capacity in this context.

Lock the piston in position so that the volume of the gas won't change (Figure 12.24), and so no work is done ($W = 0$). Put the apparatus in a big tub of water whose temperature is higher than the gas temperature, so that energy transfer will go from the water to the gas. Allow an amount of energy transfer Q due to the temperature difference to enter the gas and observe how much temperature rise ΔT has occurred in the gas. We define the "specific heat capacity C_V at constant volume" on a per-molecule basis in the following way, where N as usual is the total number of molecules in the gas:

$$\Delta E_{thermal} = NC_V\Delta T \text{ (constant volume)}$$

Since $W = 0$, this is also equal to Q. The larger the specific heat capacity, the smaller the temperature rise ΔT for a given energy input Q.

? What is C_V for a monatomic gas such as helium?

Since the total energy in a monatomic gas is $N\overline{K}_{trans} = \frac{3}{2}NkT$, the energy transfer Q will increase the total energy of the gas by an amount $\frac{3}{2}Nk\Delta T$, so $C_V = \frac{3}{2}k$. Experimental measurements of the specific heat capacity of monatomic gases agree with this prediction.

For gases that are not monatomic, such as nitrogen (N_2), c_V can be larger than $\frac{3}{2}k$ because the total energy can be larger than $\frac{3}{2}NkT$ due to rotational and vibrational energy relative to the center of mass of the molecule. Before

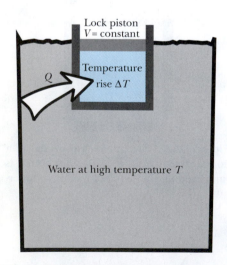

Figure 12.24 If the piston is locked so the volume of the gas cannot change, we can measure c_v of the gas.

quantum mechanics, theoretical predictions for the contribution of the rotational and vibrational energies to the specific heat capacity of gases did not agree with experimental measurements. This was one of the puzzles that was eventually resolved by quantum mechanics, as we found in the previous chapter.

SPECIFIC HEAT CAPACITY AT CONSTANT VOLUME ON A PER-MOLECULE BASIS

$$C_V = \tfrac{3}{2}k \text{ for a monatomic gas (He, etc.)}$$

$$C_V \geq \tfrac{3}{2}k \text{ for other gases (N}_2\text{, etc.)}$$

Heat capacity at constant pressure

If we don't lock the piston but let the gas expand at constant pressure, the incoming energy transfer Q due to the temperature difference not only raises the energy of the gas by an amount $\Delta E_{\text{thermal}} = NC_V\Delta T$ but also raises the piston, which involves an amount of work W (Figure 12.25). We define the "specific heat capacity C_P at constant pressure" on a per-molecule basis as follows:

$$Q = NC_P\Delta T \text{ (constant pressure)}$$

Evidently C_P is bigger than C_V, because $NC_P\Delta T = NC_V\Delta T + W$. We can calculate the work done on the piston by the gas as the gas expands, which is the negative of the work done by the piston on the gas:

$$W = \int_{V_1}^{V_2} P\,dV = P\int_{V_1}^{V_2} dV = PV_2 - PV_1 = NkT_2 - NkT_1 = Nk\Delta T$$

Therefore for an ideal gas there is a simple relationship between C_V and C_P:

$$NC_P\Delta T = NC_V\Delta T + Nk\Delta T$$

$$C_P = C_V + k$$

12.X.7 What is the specific heat capacity at constant pressure on a per-molecule basis C_P for helium?

Molar specific heat capacity

Specific heat capacity is often given on a per-mole basis rather than a per-molecule basis. The molar specific heat capacity at constant volume for an ideal gas is $C_V = \tfrac{3}{2}R$, where R is the molar gas constant (8.3 J/K), which is $6\times10^{23}k$. The molar specific heat capacity at constant pressure is $C_P = C_V + R$ for an ideal gas.

No-Q (adiabatic) compression

With the apparatus made of metal and sitting in a big tub of water, the temperature during the compression didn't change. Now we analyze the opposite extreme. We make the cylinder and piston out of glass or ceramic (which are poor conductors for energy transfer due to a temperature difference) to minimize energy transfer out of the gas as the gas gets hotter during the compression. In fact, we make the approximation that there is no energy transfer at all (Figure 12.26). Such a no-Q process is also called "adiabatic" (which means "no energy transfer due to a temperature difference").

How realistic is such a no-Q approximation? For many real situations this can be a rather good approximation if the compression or expansion is fast, so that there isn't enough time for significant energy transfer due to a tem-

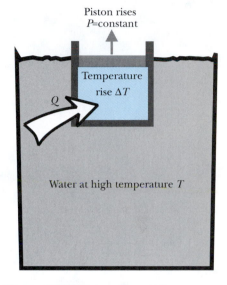

Figure 12.25 If the piston is free to move, the pressure inside the cylinder remains the same, and we can measure c_p of the gas.

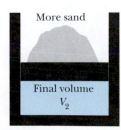

Figure 12.26 If the cylinder and piston are made of insulating material, no energy transfer Q can occur between the gas and its surroundings. This is called an adiabatic compression.

perature difference. Such a flow of energy from one object to another is a rather slow process. For example, a cup of coffee sitting on a table may stay quite hot for ten minutes or more.

But didn't we say that we were going to carry out compressions and expansions very slowly, "quasistatically"? Yes, so there is a contradiction. However, it is often the case that motion may be slow enough to be a good approximation to a quasistatic process and nevertheless may also be fast compared to the time required for significant energy transfer due to a temperature difference. Again, think of how long it takes a cup of coffee to cool off.

We again consider the gas in the cylinder as the system of interest, and the work done by the piston is equal to the increase in energy of the gas:

$$W = -\int_{V_1}^{V_2} P dV = N C_V \Delta T$$

? Why did we use C_V in this equation? The volume isn't constant in this compression!

An ideal gas is unique among materials in that its total energy is entirely determined by the temperature, independent of volume or pressure. So the change of energy of the gas itself is just $N C_V \Delta T$, even when the volume is not constant. As a practical matter, real gases behave this way as long as the density is not too high. At high densities the energy of the gas includes a significant term associated with the interatomic forces.

We can work through the integration, starting from a "differential" form of the equation shown above:

$$-P dV = N C_V dT$$

Use the gas law $P = (N/V)kT$ to substitute for the pressure:

$$-NkT \frac{dV}{V} = N C_V dT$$

Divide through by kT and rearrange:

$$\left(\frac{C_V}{k}\right)\frac{dT}{T} + \frac{dV}{V} = 0$$

$$\left(\frac{C_V}{k}\right)\int\frac{dT}{T} + \int\frac{dV}{V} = 0$$

Now integrate:

$$\left(\frac{C_V}{k}\right)\ln T + \ln V = \text{constant}$$

Rewrite using the properties of logarithms:

$$\ln T^{\left(\frac{C_V}{k}\right)} + \ln V = \ln\left[T^{\left(\frac{C_V}{k}\right)}V\right] = \text{constant}$$

And therefore we have

$$T^{\left(\frac{C_V}{k}\right)}V = \text{constant}$$

12.X.8 Use the ideal gas law to eliminate T from this expression, and show that $PV^{\gamma} = $ constant in a no-Q process, where the parameter γ is defined as the ratio of the constant-pressure specific heat capacity to the constant-volume specific heat capacity, $\gamma \equiv C_P/C_V$. (Since C_p is greater than C_v, γ is greater than 1.)

In particular, if the initial pressure and volume are P_1 and V_1, and the final pressure and volume are P_2 and V_2, then $P_2 V_2^{\gamma} = P_1 V_1^{\gamma}$. Alternatively, we also have $T_2^{(C_V/k)} V_2 = T_1^{(C_V/k)} V_1$.

Graphical representation

Again, there is a useful graphical representation of this no-Q process. In Figure 12.27 we plot the pressure P vs. the volume V, and the process is represented by a curve along which $PV^\gamma = $ constant. The area under the process curve represents the work W. For comparison we also show the curve for a constant-temperature process (T and PV constant). The no-Q curve is much steeper than the constant-temperature curve.

Work vs. energy transfer due to a temperature difference

We have studied the response of a gas to energy inputs and outputs, both mechanical W (work) and thermal Q (due to a temperature difference). Work represents organized, orderly, macroscopic energy input. Energy transfer due to a temperature difference represents disorganized, disorderly, microscopic energy input. There is randomness at the atomic level in the collisions of atoms, which is the basic mechanism for energy transfer due to a temperature difference. In the previous chapter we saw deep consequences of the distinction between work W and energy transfer Q.

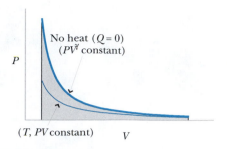

Figure 12.27 *PV* curves for a no-Q (adiabatic) process and a constant temperature (isothermal) process.

12.6 *APPLICATION: A RANDOM WALK

If you would find it useful to study one more application of the basic statistical concepts underlying our analysis of a gas, here is an interesting one. If we could watch one special molecule wandering around in the air, colliding frequently with air molecules, its path would look something like Figure 12.28. This kind of motion is called a "random walk." It may surprise you to find that despite the random nature of this motion we can calculate something significant about the motion, using simple statistical reasoning.

One dimension

For simplicity, first we'll consider just the x component of the motion. Pick the origin of the x axis to be at the original position of the special molecule. The first thing that happens is that the special molecule moves until it collides with an air molecule. We call this first x component of the displacement Δx_1. This component of the displacement may be to the right ($+x$ direction) or to the left ($-x$ direction). As a result of the collision, the special molecule may change direction, and change speed. The next displacement before another collision we call Δx_2, etc. After N collisions, the net displacement Δx away from the origin is

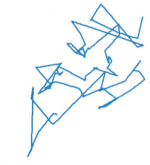

Figure 12.28 A random walk, generated by a computer program, using random numbers.

$$\Delta x = \Delta x_1 + \Delta x_2 + \Delta x_3 + \Delta x_4 + ... + \Delta x_N$$

After N collisions, what is the average (most probable) position of the special molecule? On average, it is just as likely that it has moved to the right as moved to the left, so the average x component of displacement ought to be zero. It is easy to see that this will be the case, by taking the average value:

$$\overline{\Delta x} = \overline{\Delta x_1} + \overline{\Delta x_2} + \overline{\Delta x_3} + \overline{\Delta x_4} + ... + \overline{\Delta x_N}$$

Each of the individual x displacements are equally likely to be to the left or the right, so the average value of the nth x displacement (for $n = 1, 2, 3, ...N$) is zero. Therefore the average value of the net x component of displacement is also zero.

On average, after N collisions the special molecule ends up to the left of the origin as often as it ends up to the right. But we can ask the question, "On average, how far away from the origin (along the x axis) does the special molecule get, no matter whether it ends up to the left or the right?" A good indicator of this distance is the average value of the square of the net x-displacement, $(\Delta x)^2$, because that's always a positive number. We can calculate this average.

Since $\Delta x = \Delta x_1 + \Delta x_2 + \Delta x_3 + \Delta x_4 + ... + \Delta x_N$, we have

$$(\Delta x)^2 = (\Delta x_1)^2 + (\Delta x_2)^2 + ... + (\Delta x_N)^2 + 2\Delta x_1 \Delta x_2 + 2\Delta x_1 \Delta x_3 + 2\Delta x_1 \Delta x_4 + ...$$

In this square of the net x displacement, there are two kinds of terms: squares of individual x displacements such as $(\Delta x_2)^2$ and "cross terms" like $2\Delta x_1 \Delta x_3$. Given the random nature of the process, we expect that one fourth of the time both Δx_1 and Δx_3 are positive (product is positive), one-fourth of the time they're both negative (product is positive), one-fourth of the time Δx_1 is positive with Δx_3 negative (product is negative), and one-fourth of the time Δx_1 is negative with Δx_3 positive (product is negative). Therefore, the average value of each cross term $\overline{2\Delta x_i \Delta x_j}$ $(i \neq j)$ is zero.

As for the other terms, the squares of individual x-displacements such as $(\Delta x_2)^2$, the average value of each of these terms is some number d_x^2, related to d^2, the square of the mean free path that we discussed in Section on page 414.

? Count up how many of these square terms there are for N collisions and calculate the average value of the square of the net displacement:

$$\overline{(\Delta x)^2} = ?$$

There are N of these terms, and the square root of this quantity is the "root-mean-square" or "rms" value of the net displacement:

$$\Delta x_{rms} = \sqrt{\overline{(\Delta x)^2}} = (\sqrt{N})d_x$$

We can also write this formula in terms of time. If we let v be the average speed of the special molecule between collisions, and we let T be the average time between collisions, we have $v = d/T$. Also, the total time t for the N collisions is $t = NT$.

? Rewrite the formula for the rms displacement in terms just of d_x, v, and t (that is, eliminate N and T): $\Delta x_{rms} = ?$

We find that $\Delta x_{rms} = \sqrt{Nd_x d_x} = \sqrt{N(vT)d_x} = \sqrt{vd_x}\sqrt{t}$. This is a somewhat curious result. For ordinary motion at constant speed, displacement increases proportional to time: double the time, double the displacement. In contrast, the rms displacement in a random walk grows with the square root of the time: on average, the rms displacement doubles when the time quadruples.

This calculation was done for the x component of the motion, but we can generalize our result to real three-dimensional motion in a gas. In Figure 12.29 is a three-dimensional displacement involving Δx, Δy, and Δz.

You can see in Figure 12.29 a three-dimensional version of the Pythagorean theorem for triangles. We can assume that the motion in each dimension is independent of the motions in the other two dimensions, and we have the following result:

$$\Delta r_{rms} = \sqrt{\overline{(\Delta x)^2} + \overline{(\Delta y)^2} + \overline{(\Delta z)^2}} = \sqrt{Nd_x^2 + Nd_y^2 + Nd_z^2} = (\sqrt{N})d$$

(Here, d_x is the component in the x direction of the three-dimensional mean free path d.)

Writing the rms displacement in terms of the average speed v, the time t, and the mean-free path d, we have

$$\Delta r_{rms} = \sqrt{vd}\sqrt{t}$$

This result is somewhat unusual, because it predicts (correctly, it turns out) that for this random process the displacement from the starting location is proportional to the square root of the time rather than to the time.

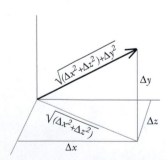

Figure 12.29 Pythagorean theorem in three dimensions.

12.X.9 As you have calculated, the average speed of an air molecule at room temperature is about 500 m/s. What is Δr_{rms} for an air molecule after one second?

12.7 FUNDAMENTAL LIMITATIONS ON EFFICIENCY

An important technology involves the conversion of energy transfer Q due to a temperature difference into useful work W (Figure 12.30). For example, in a steam-powered electricity generating plant, coal is burned to warm up water that drives a steam engine to turn a generator, which converts the work done by the steam engine into electric energy. Energy conservation of course puts limits on how much useful work you can get from the burning of the coal. But there is a further limitation due to the second law of thermodynamics. It turns out that in a practical generating plant only about one-third of the energy of the coal can be turned into useful work! The fundamental problem is that energy transfer due to a temperature difference is a disorderly energy transfer, and the second law of thermodynamics takes that into account.

We will find that the most efficient processes are "reversible" processes, that is, processes that produce no change in the entropy of the Universe (a reversed movie of such a process looks possible). In the next sections we discuss two processes, mechanical friction and finite-temperature-difference energy transfer, which are major contributors to entropy production and whose effects must be minimized in order to obtain the most efficient performance in a mechanical system. Where we are headed is to establish limits on how efficiently thermal energy can be turned into mechanical energy in an engine, as a consequence of the second law of thermodynamics. This is a revealing example of the power of the second law of thermodynamics to set limits on possible phenomena.

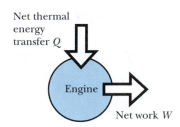

Figure 12.30 An engine converts net energy transfer Q into net work W.

Friction and entropy production

The second law of thermodynamics says that the entropy of the Universe never decreases. Portions of the Universe may experience a decrease of entropy, but only if the entropy of other portions increases at least as much (and typically more). An example of a process that increases the total entropy of the Universe is sliding friction. A block sliding across a table slows down as kinetic energy associated with the overall motion of the block turns into thermal energy inside the block, with an increase in entropy.

This friction process is irreversible. We would be astonished if after coming to rest the block suddenly started picking up speed back toward its starting position, although this would not violate conservation of energy. But the probability is vanishingly small that the thermal energy in the block could convert back into an orderly motion of the block.

Energy transfer and entropy production

Mechanical friction contributes to increasing the entropy of the Universe. We will show that energy transfer between two objects that have different temperatures also makes the total entropy of the Universe increase. This is then a process to avoid if possible in an efficient engine.

Connect a metal bar between a large block at a high temperature T_H and another large block at a low temperature T_L (Figure 12.31). We write the rate of energy transfer from the high-temperature block (the "source") to the lower-temperature block (the "sink") as $\dot{Q}$. The dot over the letter Q means "rate" or "amount per second." This energy transfer rate is

- proportional to the "thermal conductivity" σ of the bar (metals have higher thermal conductivity than glass or plastic),

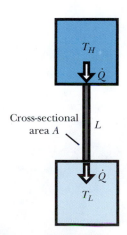

Figure 12.31 Energy flows at a rate $\dot{Q}$ from high temperature to low.

- proportional to the temperature difference (twice the temperature difference, twice the rate of energy transfer),
- proportional to the cross-sectional area A of the bar (twice the cross-sectional area is like having two bars), and
- inversely proportional to the length L of the bar (twice the length of the bar, half the energy transfer rate):

RATE OF ENERGY TRANSFER DUE TO A TEMPERATURE DIFFERENCE

$$\dot{Q} = \sigma A \frac{(T_H - T_L)}{L} \quad (\text{J/s})$$

The quantity $(T_H - T_L)/L$ is called the "temperature gradient." The larger the gradient (the more rapidly the temperature changes with distance along the bar), the larger the number of joules per second of energy transfer.

We want to show you that energy transfer between different temperatures increases the total entropy of the Universe. Suppose the source and sink are so large that for a while the energy transfer doesn't change the temperatures much. Since $1/T = (dS)/(dE)$, it follows that a small energy input (dE) into a system, of amount Q, leads to an entropy change of that system:

$$\Delta S = \frac{Q}{T}$$

Suppose the energy transfer *rate* is $\dot{Q}$. In a short time interval Δt, the high-temperature source at temperature T_H loses an amount of energy $\dot{Q}\Delta t$.

? In this short time interval Δt, does the entropy of the source increase or decrease? By how much?

Since $\Delta S_H = (\Delta E)/T_H$, and the high-temperature source loses energy, we have $\Delta S_H = -(\dot{Q}\Delta t)/T_H$.

In this same time interval, the low-temperature sink at temperature T_L gains the same amount of energy, $\dot{Q}\Delta t$.

? In this short time interval Δt, does the entropy of the sink increase or decrease? By how much?

Since the low-temperature sink gains energy, we have $\Delta S_L = +(\dot{Q}\Delta t)/T_L$

Here is a crucial and perhaps not entirely obvious point: Does the entropy of the bar change during this time interval? No. The blocks are so large that their temperatures don't change much in a short time interval, so the temperatures at the ends of the bar, and all along the bar, are not changing. Every second, energy enters the bar, but the same amount of energy leaves the bar, so the energy of the bar isn't changing. Nothing about the bar is changing. It is merely a conduit for the energy transfer, but its state is not changing.

Therefore the change in the total entropy of the Universe in a time interval Δt is

$$\Delta S_{\text{source}} + \Delta S_{\text{sink}} + \Delta S_{\text{bar}} = -\frac{\dot{Q}\Delta t}{T_H} + \frac{\dot{Q}\Delta t}{T_L} + 0 = \dot{Q}\Delta t\left(\frac{1}{T_L} - \frac{1}{T_H}\right)$$

? Is this quantity positive or negative? Is this consistent with the second law of thermodynamics?

Since $T_L < T_H$, $1/T_L > 1/T_H$, and the entropy change of the Universe in this process is positive. This is consistent with the second law of thermodynamics, which states that the entropy of the Universe should never decrease.

Reversible and irreversible processes

An increase in the total entropy of the Universe is associated with an irreversible process, because to return to the earlier state would require that the total entropy of the Universe actually decrease, which won't happen with macroscopic systems. We would be astonished if the energy transfer suddenly reversed and ran from the colder block thermally back "uphill" to the hotter block!

There is only one way to carry out energy transfer due to a temperature difference reversibly or nearly so (that is, with little or no change in the total entropy of the Universe)—do it between systems whose temperatures are very nearly equal to each other ($T_H \approx T_L$). But there's a practical problem.

? If the two temperatures are nearly equal (to avoid increasing the entropy of the Universe), at what rate is energy transferred from the (slightly) hotter source to the (slightly) cooler sink?

Energy transfer will flow extremely slowly, because $\dot{Q}$ is proportional to the temperature difference (or more specifically to the temperature gradient). If there is hardly any temperature difference, the energy transfer rate will be very small. So we can carry out such energy transfers nearly reversibly, hardly changing the entropy of the Universe, but only if we do it so slowly as to be of little practical use in driving some kind of mechanical engine that converts thermal energy into work.

12.X.10 The thermal conductivity of copper (a good thermal conductor) is 400 watt/K/m (for comparison, the thermal conductivity of iron is about 70 watt/K/m, and that of glass is only about 1 watt/K/m). One end of a copper bar 1 cm on a side and 30 cm long is immersed in a large pot of boiling water (100° C), and the other end is embedded in a large block of ice (0° C). It takes about 335 joules to melt one gram of ice. How long does it take to melt one gram of ice? Does the entropy of the Universe increase, decrease, or stay the same?

12.8 A MAXIMALLY EFFICIENT PROCESS

Despite the serious practical problem that reversible energy transfer due to a nearly zero difference in temperature proceeds infinitesimally slowly, we will analyze reversible "engines" in which (nearly) reversible energy input with tiny temperature differences is used (very slowly!) to do something mechanically useful, such as lift a weight or turn an electric generator. We will also assume that we can nearly eliminate sliding friction. The idea that we will pursue is to see how efficiently we can convert disorderly thermal energy into orderly mechanical energy (a lifted weight). We expect that reversible processes, which don't increase the total entropy of the Universe, should be the most efficient processes, though we recognize that these most efficient processes must proceed excruciatingly slowly.

Since we will use nearly reversible processes to do work, we could run the engine backwards, and such a backwards-running engine turns out to be a refrigerator—an engine in which work input to the engine can lead to extracting thermal energy from something to make it colder or to keep it cold.

The conception of a theoretically most efficient possible engine, and the recognition that this ideal engine would have to be reversible and not increase what we now call the total entropy of the Universe, was due to a young French engineer, Sadi Carnot, in 1824. His achievement is all the more remarkable because it came before the principle of energy conservation was established!

After using the second law of thermodynamics to determine the maximum possible efficiency of infinitely slow engines, in the last part of the chapter we will analyze engines that run at useful speeds. We will find that they are even less efficient than the infinitely slow engines.

A cyclic process of a reversible engine

Some of the first practical engines drove pumps to pump water out of deep mines in England, repeatedly lifting large amounts of water large distances. The energy came from burning coal, which boiled water to make steam in a cylinder which pushed up on a piston, and operated the pump. People started wondering how efficient such an engine could be. What limits the amount of useful work you can get from burning a ton of coal?

We describe a scheme for running an engine in a reversible way, which ought to be as efficient as possible. For concreteness, our engine is a device consisting of a cylinder containing an ideal gas, with a piston and some sand on the piston to adjust the pressure on the gas (Figure 12.32). This is a simple and familiar device which permits energy exchanges in the form of work and energy transfer due to a temperature difference. As we'll see, the actual construction details of the engine don't matter for the theoretical question regarding the maximum possible efficiency, though they matter very much in the actual construction of a useful engine.

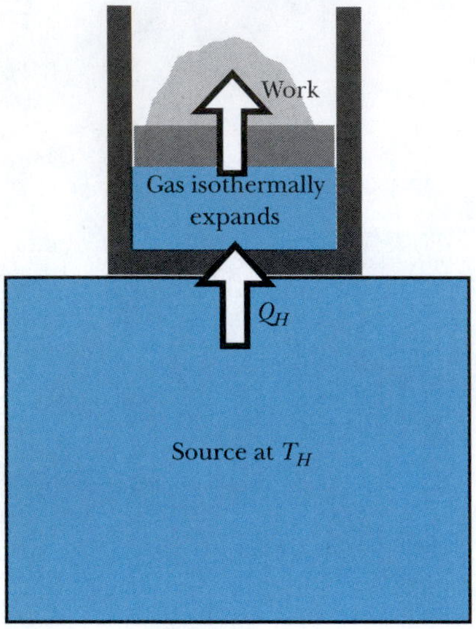

A constant-temperature (isothermal) expansion

In addition to the engine itself, we need a large high-temperature source, large enough that extracting some energy from it will cause only a negligible decrease in its temperature. Start by arranging that the engine (the gas cylinder with its piston) has a temperature very slightly lower than the temperature T_H of the high-temperature source. We connect the engine to the high-temperature source and perform a reversible constant-temperature (isothermal) expansion of the engine, doing some work on the surroundings in the process (Figure 12.32). We lift the piston by slowly shifting some sand sideways from the piston onto nearby shelves. This requires almost no work on our part but results in the lifting of some weight. We make sure that the temperature of the gas doesn't change during this process. There is energy transfer into the engine because it is very slightly cooler, but the temperature difference is so slight that there is negligible change in the entropy of the Universe.

Figure 12.32 Constant-temperature expansion; gas temperature just slightly lower than T_H.

Entropy and energy in lifting the piston

Remember that the entropy change of a system when there is energy transfer Q is $\Delta S = Q/T$. In lifting the piston the entropy of the high-temperature source has changed by an amount $\Delta S_H = -Q_H/T_H$, where Q_H is a positive quantity. The entropy of the gas has changed by an amount $\Delta S_{gas} = +Q_H/T_H$. The entropy of the Universe hasn't changed at all. (The increased entropy of the gas is associated with the fact that there are more ways to arrange the molecules in a large volume than in a small volume.)

Because the temperature hasn't changed, there is no change in the energy of the ideal gas. Recall that the energy of an ideal gas, or a low-density real gas, is a function solely of the temperature, not the volume. (A dense real gas is more complicated, because the electric potential energy for pairs of molecules depends on distance, and therefore the energy of the gas depends on volume as well as temperature.)

We have succeeded in converting all of the energy transfer from the high-temperature source into useful work (lifting the piston), because none of the input energy went into changing the energy of the gas. This is 100% efficiency in converting thermal energy into useful work on the surroundings. What's the problem?

The need for a cycle

The problem is this: We need to be able to do this again, and again, and again. For example, we want to keep pumping water out of the deep mine, over and over. But to repeat the lifting process, we have to return to the original state, with the gas compressed. We could simply reverse the process, doing work on the gas (constant-temperature compression), with energy transfer from the gas into the high-temperature source. But the net effect would be that there was zero net energy transfer to the gas, and zero net work done on the surroundings.

We need to run our engine in a non-trivial repetitive "cycle," where we can lift the piston repeatedly. Each time we lift the piston and do useful work, we need to get the piston back down again without having to do the same amount of work to push it down. If we can get the gas back to its original state with net work done on the surroundings, we will have a useful device.

One possibility for bringing the piston down would be to let the gas cool down. But to lower the temperature we would have to place the gas cylinder in contact with a large object at some low temperature T_L, called the "sink" because, as we'll see, we will dump some energy into it. We can't place the gas cylinder in contact with the sink immediately, because the gas is at a high temperature T_H, and placing the hot gas in contact with the cold sink would mean that there would be a large temperature difference, and there would be a large increase in the entropy of the Universe associated with the energy transfer from the hot gas to the cold sink.

A no-*Q* (adiabatic) expansion

To avoid this large production of entropy, we need to bring the temperature of the gas down almost to T_L *before* making contact with the sink. To do this, we disconnect the gas cylinder from the high-temperature source and perform a reversible no-Q (adiabatic) expansion, which does some more work on the surroundings and lowers the temperature of the engine (Figure 12.33). This is accomplished by slowly removing some weight from the piston, allowing the gas to expand and the temperature to fall. We stop the expansion when the temperature of the gas is just slightly higher than the low temperature T_L of the sink.

A constant-temperature (isothermal) compression

We can now safely place the gas cylinder in contact with the low-temperature sink. This is okay as far as entropy production is concerned, because we made sure that the temperature of the gas is only very slightly higher than the sink temperature T_L. The piston is now quite high, and we need to bring it down. We do this by slowly adding some weight to the piston, performing a constant-temperature (isothermal) compression (Figure 12.34). There is energy transfer into the sink. Because the temperature of the ideal gas is nearly T_L at all times, there is no energy change in the gas. Therefore the work that we do is equal to the energy transfer Q_L into the sink.

There is an increase in the entropy of the sink $\Delta S_L = +Q_L / T_L$, and there is an entropy decrease in the entropy of the gas $\Delta S_{gas} = -Q_L / T_L$. The entropy of the Universe doesn't change. (The decreased entropy of the gas is associated with the fact that there are fewer ways to arrange the molecules in a small volume than in a large volume.

A no-*Q* (adiabatic) compression

We stop the compression at a particular state of the gas chosen with care so that we can do the following: We disconnect the engine from the low-temperature sink and perform a no-Q compression that raises the temperature of the gas to a temperature just barely less than T_H, the temperature of the

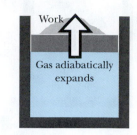

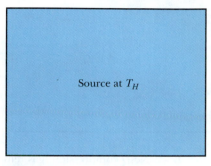

Figure 12.33 No-Q expansion; gas temperature drops to T_L.

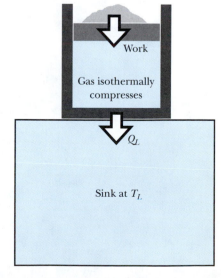

Figure 12.34 Constant-temperature compression; gas temperature just slightly higher than T_L.

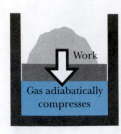

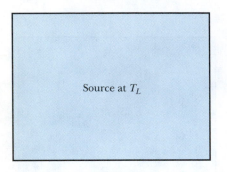

Figure 12.35 No-Q compression; gas temperature rises to T_H. Engine is now back to its original state, ready to begin another cycle.

high-temperature source. We do some work to perform this compression (Figure 12.35).

This compression brings the system back to its original state (density, pressure, temperature), so we can repeat the cycle of four processes all over again. The possible usefulness of this engine is that we can run it repeatedly, over and over. What remains to be analyzed is how much work is done on the surroundings in one cycle, and how much energy input we need to supply. Of course if we find that the net work is zero, the engine won't be useful. However, explicit calculations for an ideal gas show that at least in that case there is net work done on the surroundings. We will find that this is true for any reversible engine run in such a cycle.

This reversible cycle of two constant-temperature (isothermal) processes and two no-Q (adiabatic) processes running between a high-temperature source and a low-temperature sink is called a "Carnot cycle."

Summary of the cycle
Here is a summary of the four processes of the cycle:
- Isothermal expansion in contact with T_H
- Adiabatic expansion (no contact); temperature falls to T_L
- Isothermal compression in contact with T_L
- Adiabatic compression (no contact); temperature rises to T_H

The four processes return the gas to its original temperature and volume. The question is, did we get any net work out of this cycle?

Entropy in a cycle of a reversible engine
Let Q_H be the (absolute value of the) energy extracted from the source at temperature T_H during the constant-temperature expansion, and let Q_L be the (absolute value of the) energy dumped into the sink at temperature T_L during the constant-temperature compression. Remember that the entropy change of a system when there is energy transfer Q is $\Delta S = Q/T$.

The entropy change of the high-temperature source in one cycle $= -\dfrac{Q_H}{T_H}$

The entropy change of the low-temperature sink in one cycle $= +\dfrac{Q_L}{T_L}$

? Is there any entropy change *in the engine* in one cycle?

The entropy change of the engine in one cycle is zero, because the state of the gas is brought back to its original state.

? Therefore, in one cycle, how much entropy change is there in the surroundings as a result of the work done on and by the surroundings?

Only the energy transfer due to a temperature difference affects the entropy during one cycle (work doesn't contribute), and we calculated these through the formula $\Delta S = Q/T$. We were careful to run the engine in a reversible way, avoiding making any increase in the total entropy of the Universe, so in one cycle we have

$$-\frac{Q_H}{T_H} + \frac{Q_L}{T_L} + 0 = 0$$

Therefore we can write this important relationship for one cycle of the engine:

ONE CYCLE OF A REVERSIBLE ENGINE

$$\frac{Q_H}{T_H} = \frac{Q_L}{T_L}$$

This is a surprisingly simple result for such a complicated process. What does this simple result depend on? Solely on the second law of thermodynamics, that statistically the total entropy of the Universe will essentially never decrease.

We illustrated the processes in the cycle with a cylinder filled with an ideal gas, but the result above doesn't actually depend at all on the details of how the engine is constructed. In particular, it doesn't matter whether the engine contains helium gas, or a solid block of rubber, or a liter of liquid alcohol. There would be an easily visible difference between using a solid or liquid rather than a gas in the engine, because the distance through which the piston would move would be much smaller than with a gas. But the simple result shown above would remain the same.

Energy in a cycle of a reversible engine

Let's review the energy changes in the engine in one cycle. The source inputs an amount of energy Q_H. The sink removes an amount of energy Q_L. The working substance returns to its original state, so it undergoes no change of energy. Is there anything else? Yes, it may be that there has been net work W done on the surroundings. How can we tell? We can use conservation of energy of the engine for one cycle:

$$\Delta E_{engine} = Q_H - Q_L + W = 0$$

? Use the result that $Q_H / T_H = Q_L / T_L$ (because the total entropy of the Universe doesn't change), to determine whether W is positive or negative.

Because $Q_H = (T_H / T_L) Q_L > Q_L$, we have $W = Q_L - Q_H < 0$. The fact that W is negative means that our engine does net work on the surroundings in one cycle. Figure 12.36 outlines the basic scheme. In one cycle the net effect is that the engine takes in energy Q_H from a hot source, does some work on something (for example, turns an electric generator), and exhausts the remaining energy Q_L into a cold sink. The magnitude of the work done is $|W| = Q_H - Q_L$.

The crucial issue is that the engine will not run in repeatable cycles without exhausting some energy to the low-temperature sink. You *cannot* convert all of the energy Q_H into useful work. We need the low-temperature block to allow us to begin the constant-temperature compression phase of the cycle in a way that avoids any energy transfer with significant temperature difference, which would increase the total entropy of the Universe.

We have to pay for the high-temperature energy Q_H which we use, and the loss of some of the input energy in the form of Q_L is an unfortunate fact of life. We pay for coal or fuel oil or electricity to warm something up to the high temperature T_H, from which we can draw energy to do work for us.

For example, in an old railroad steam engine a fire maintained water at a high temperature to constitute the high-temperature source. In an automobile engine we burn gasoline to create a high temperature and push the pistons. In an electricity generating station we burn coal or fuel oil, or use nuclear fission reactions, to create a high-temperature source from which to drive the generators. We'd like to get our full money's worth (Q_H), but we don't. We only get $|W| = Q_H - Q_L$, having "lost" some of the energy to a

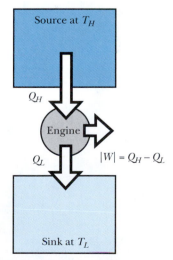

Figure 12.36 Conservation of energy lets us determine the net work output.

low-temperature sink, which typically is our ordinary surroundings at around 20° C (293 K).

Efficiency

This leads to the question, what fraction of the energy Q_H we pay for turns into useful work W? This fraction is called the "efficiency" of the engine:

$$\text{efficiency} \equiv \frac{|W|}{Q_H}$$

? Given this definition of efficiency, prove the following result:

THE EFFICIENCY OF A REVERSIBLE ENGINE

$$\text{Efficiency} = 1 - \frac{T_L}{T_H}$$

We have $|W|/Q_H = (Q_H - Q_L)/Q_H = 1 - Q_L/Q_H = 1 - T_L/T_H$. This may seem surprising. The efficiency of a reversible engine depends solely on the ratio of the absolute temperatures of the source and sink. We emphasize that it doesn't matter how the engine is designed, or what kind of materials it is made of. This is the highest efficiency we can ever achieve for converting thermal energy into useful work.

> *12.X.11* What is the efficiency of a reversible engine if the source is a large container of boiling water and the sink is a large block of melting ice?
>
> *12.X.12* An actual electricity generating plant is powered by a nuclear reactor, with a high temperature of 300° C and low temperature of 25° C (near room temperature). What is the efficiency, assuming that we can treat the processes as being reversible?

No other engine can be more efficient

We can show that no engine running between temperatures T_H and T_L can be more efficient than a reversible engine.

The proof is a "proof by contradiction." Suppose that an inventor claims to have invented some cleverly designed new kind of (cyclic) engine, with a higher efficiency than $(1 - T_L/T_H)$ when run between these same two temperatures. In that case for a given Q_H the new engine would exhaust less Q_L than is exhausted from a reversible engine, and the entropy change of the Universe with this new engine would be negative instead of zero:

$$-\frac{Q_H}{T_H} + \frac{Q_L}{T_L} < 0 \quad \text{(Impossible!)}$$

This would be a violation of the second law of thermodynamics, so it is impossible. We can be quite sure that the inventor's claims are not valid. Note carefully that the inventor's claims don't violate energy conservation. Nothing about energy conservation forbids converting 100% of the input energy Q_H into work W in a cycle. The impossibility lies rather in the massive improbability of seeing the total entropy of the Universe decrease.

We took great pains to make a reversible engine, for which the total entropy change of the Universe was zero. No other engine can be more efficient. In fact, any real engine will have some friction and some energy transfer across differing temperatures, in which case the entropy change of the Universe will be positive. So the second law of thermodynamics leads to the following, where "= 0" applies only to ideal, reversible engines, and "> 0" applies to real engines:

ENTROPY CHANGE OF THE UNIVERSE FOR REAL ENGINES

$$\Delta S_{\text{Universe}} = -\frac{Q_H}{T_H} + \frac{Q_L}{T_L} \geq 0$$

We should not underestimate the need for the ingenuity of inventors, however. The reversible engine made with an ideal gas cylinder is pretty useless for practical purposes. The challenge of ingenuity is to design engines that are practical, and this involves many kinds of engineering design decisions and trade-offs. But no matter how ingenious the design, the second law of thermodynamics puts a rigid limitation on how efficient *any* engine can be.

12.X.13 If $\Delta S_{\text{Universe}} > 0$ in one cycle of an engine, show that the efficiency is less than the ideal efficiency $(1 - T_L / T_H)$ obtained with a reversible engine.

Running the engine backwards: A refrigerator or heat pump

We could run our engine backwards, since we took care to make all aspects of the cycle reversible. Starting from our original starting point, we would disconnect from the hot block, perform a no-Q expansion to lower the temperature to that of the cold block, connect to the cold block, do a constant-temperature expansion, then disconnect from the cold block. Next we do a no-Q compression to raise the temperature to that of the hot block, connect to the hot block, and do a constant-temperature compression back to the original state.

The net effect in one cycle is that the low-temperature block gives some energy Q_L to the engine, the high-temperature block absorbs some energy Q_H from the engine, and we do some (positive) work on the system (instead of the engine doing work on the surroundings). Here is the formula for conservation of energy of the engine when we run this reversed engine:

$$\Delta E_{\text{engine}} = Q_L - Q_H + W = 0$$

? Use the result that $Q_H/T_H = Q_L/T_L$ (which is still valid for our reversed engine, as you can convince yourself) to determine whether W is positive or negative.

Since $Q_H = (T_H/T_L)Q_L > Q_L$, we have $W = Q_H - Q_L > 0$. This seems an odd kind of "engine": it doesn't do any work for us—we have to do work on it. Is that of any use? Yes! At the cost of doing some work, we extract energy from a low-temperature block and exhaust it into a high-temperature block. This is a refrigerator (Figure 12.37).

Consider how we keep food cold in an ordinary home refrigerator. A reversed engine maintains the inside of the refrigerator at a low temperature T_L, while the room is at a higher temperature T_H. Although the door and walls are heavily insulated, some energy does leak through from the room, and this energy, Q_L, must be removed to keep the food at a constant low temperature T_L. We exhaust an amount of energy Q_H into the room at temperature T_H. It takes an amount of work W to achieve this effect of moving energy out of the cold region and into the warm region.

In the most favorable case (no increase in total entropy) we have the following two conditions stemming from entropy and energy considerations (the high-temperature exhaust energy must equal the sum of the low-temperature energy plus the work we put in to drive the cycle):

$$\frac{Q_H}{T_H} = \frac{Q_L}{T_L} \text{ and } Q_H = Q_L + W$$

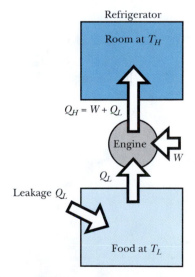

Figure 12.37 Run the engine backwards and you have a refrigerator.

Here our view of "efficiency" shifts a bit. What we care about is how much work W we have to do to remove an amount of leakage energy Q_L from the inside of the refrigerator.

? Show that the energy removal we get per amount of work done is as follows:

$$\frac{Q_L}{W} = \frac{1}{(T_H/T_L) - 1} > 1$$

Isn't this result somewhat surprising? Are we getting something for nothing when Q_L is bigger than the work W that we do? No, there's no violation of energy conservation. It's just that Q_L plus W equals the output energy Q_H, and we only have to add a modest amount of work to drive the energy "uphill."

Heat pump

A related device is the "heat pump" used in some areas to warm homes by driving low-temperature energy "uphill" into a higher-temperature house. Figure 12.38 shows a house whose interior is maintained at a temperature T_H despite continual leakage Q_H into the outside air. Energy Q_L at low temperature T_L is pumped out of the ground into the house, by the addition of some work W that we do in an engine called a heat pump. This is somewhat similar to the refrigerator situation, except that now what we care about is how much high-temperature energy Q_H we get per amount of work W done by the heat pump.

? Show that this ratio is given by the following formula for the heat pump:

$$\frac{Q_H}{W} = \frac{1}{1 - (T_L/T_H)} > 1$$

Again, this feels like we're getting something for nothing, since Q_H is greater than the work W that we do. After all, if you warm the house directly with gas or oil or electric energy, you have to pay for every joule of leakage, not some small fraction of the leakage. So why aren't heat pumps much more commonly used than they are? Basically because we've been calculating the best deal we can possibly obtain (the case of the reversible engine—that is, one whose operation leads to no increase in the total entropy of the Universe). Real engines running either forward or reversed don't attain the theoretical maximum performance, as we will see in the next sections.

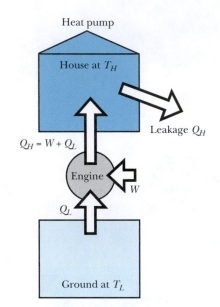

Figure 12.38 Run the engine backwards and you have a heat pump.

(Labels in figure: Heat pump; House at T_H; Leakage Q_H; $Q_H = W + Q_L$; Engine; W; Q_L; Ground at T_L)

12.X.14 Suppose there is a leakage rate of 50 watts through the insulation into a refrigerator, which we maintain at 3° C. What is the minimum electric power required to continually remove this leakage energy, to maintain the low temperature? In that case, what is the rate at which energy is exhausted into the room? Room temperature is about 20° C.

12.X.15 Suppose the temperature underground from where we draw low-temperature energy for a heat pump is about 5° C, and we want to keep the house at a temperature of 20° C. How many joules of work must our heat pump supply for every joule of leakage of energy there is out of the house?

12.9 *WHY DON'T WE ATTAIN THE THEORETICAL EFFICIENCY?

In 12.X.12 on page 438 you calculated the reversible-engine efficiency of an actual nuclear-powered electricity generating plant to be 48%, but the observed efficiency of this plant is only 30%. Real engines do not attain the efficiency predicted by the second law of thermodynamics for ideal reversible engines. In fact, the real efficiency is often only about half of the theoretical efficiency. The reason for less than optimum performance in real engines is partly due to mechanical friction, but this effect can be minimized by good design and proper lubrication. The main limitation on performance comes from the practical necessity of incorporating energy transfers between parts of the system that are at significantly different temperatures, which leads to sizable increases in the total entropy of the Universe, and to much reduced efficiency.

The problem is speed. As we saw earlier in the chapter, the rate of energy transfer $\dot{Q}$ in joules per second between two objects (such as the hot or cold block and the engine) at temperatures T_H and T_L connected by some conducting material of length L and cross-sectional area A, with thermal conductivity σ is this:

$$\dot{Q} = \sigma A \frac{(T_H - T_L)}{L}$$

? So what is the rate of energy transfer in a reversible engine when in contact with the hot block or the cold block?

Alas, a reversible engine is totally impractical when it comes to getting anything done in a finite amount of time, because the rate of energy transfer is zero. If an engine is perfectly reversible, a cycle takes an infinite amount of time!

Consider the portion of an engine's cycle where the working substance is in thermal contact with the high-temperature source. In order to carry out the expansion quickly, there must be a high rate of energy transfer from the source into the working substance. That means we need a large contact area A, a short distance L, and a high thermal conductivity σ. Most significant of all, the temperature of the source (T_H) must be considerably higher than the temperature of the engine, leading to irreversibility and increase of the entropy of the Universe.

The efficiency of a nonreversible engine

It is actually possible to calculate the effect of such nonreversible energy transfer, as was pointed out in an article titled "Efficiency of a Carnot engine at maximum power output," by F. L. Curzon and B. Ahlborn, *American Journal of Physics* volume **43**, pages 22-24 (January 1975).

Suppose that when the engine is in contact with the high-temperature block at temperature T_H, the engine is at a lower temperature T_1, so that there is a finite energy transfer rate into the engine of $\dot{Q}_H = b(T_H - T_1)$ joules per second, where b is a constant that lumps together the factors of thermal conductivity, cross-sectional area, and length ($\sigma A/L$). Similarly, assume that when the engine is in contact with the low-temperature block at temperature T_L, the engine is at a higher temperature T_2, so that there is a finite energy transfer rate into the working substance of $\dot{Q}_L = b(T_2 - T_L)$ joules per second. We're making the factor b be the same for both the high-temperature and the low-temperature contacts. This simplifies the algebra, and it turns out that using different b factors doesn't affect the final result anyway.

Now the energy flow diagram looks like Figure 12.39, with an ideal reversible engine running between the temperatures T_1 and T_2, but with irreversible finite-rate energy transfer between this reversible engine and the source and sink at temperatures T_H and T_L.

Energy enters the engine from the high-temperature source at a rate $\dot{Q}_H$, work is done at a rate $\dot{W}$, and energy is dumped into the low-temperature

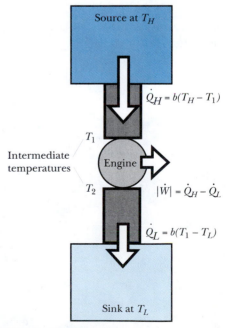

Figure 12.39 An engine that runs at a nonzero rate.

sink at a rate $\dot{Q}_L$. All of these quantities are measured in watts (joules per second). From energy conservation we have

$$\dot{Q}_H = \dot{W} + \dot{Q}_L$$

Curzon and Ahlborn pointed out that although this engine will necessarily be less efficient than a reversible engine (due to the irreversible energy transfers), it might be useful to ask the question, "What is the *maximum* output power $\dot{W}$ per input power $\dot{Q}_H$?" This question comes down to the question of what value of b will maximize the output power. This is essentially an engineering design question, yet there is an unusually simple answer to this question about the maximum output power.

There must be a maximum power output

By considering two extreme designs (Figure 12.40) it is easy to see that there must be a maximum possible power output (per power input). One extreme set of operating conditions is the reversible engine we've already considered, in which all the energy flow rates are zero, including the output power $\dot{W}$.

Another extreme is to arrange conditions in such a way that $T_1 \approx T_2$, so that we have essentially connected the source and sink by a conducting bar. In that case the input power $\dot{Q}_H$ is equal to $\dot{Q}_L$, and the useful output power $\dot{W}$ is again zero.

Somewhere between these two extremes, for both of which there is zero output power, we expect to find a maximum output power. We now look for the conditions that maximize the output power. The search will be carried out by expressing the output power in terms of T_1, the higher of the two temperatures to which the engine is directly exposed in Figure 12.39. Then we will vary T_1 (by adjusting b in Figure 12.39) until we maximize the power output. The details are given in a derivation at the end of the chapter. We obtain the following surprisingly simple result:

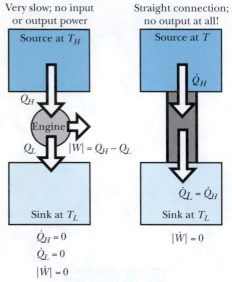

Very slow; no input or output power

Source at T_H

Q_H

Engine

Q_L $|W| = Q_H - Q_L$

Sink at T_L

$\dot{Q}_H = 0$
$\dot{Q}_L = 0$
$|\dot{W}| = 0$

Straight connection; no output at all!

Source at T

$\dot{Q}_H$

$\dot{Q}_L = \dot{Q}_H$

Sink at T_L

$|\dot{W}| = 0$

Figure 12.40 Two extreme designs where there is zero power output.

THE EFFICIENCY OF A MAXIMUM-POWER ENGINE

$$\frac{\dot{W}}{\dot{Q}_H} \le 1 - \sqrt{\frac{T_L}{T_H}}$$

This gives us the upper bound on the fraction of the thermal input power that can end up as output mechanical power if we maximize the output power. The formula looks quite similar to the reversible-engine zero-power efficiency, but the square root makes a big difference.

Comparison with the real world

Curzon and Ahlborn give an example of a plant powered by a nuclear reactor, with high temperature 300° C and low temperature 25° C. Using the formula given above, the calculated efficiency is 28%, and the actual observed efficiency is 30% (see exercise at the end of this section). This confirms the estimate that Curzon and Ahlborn made that

- the main source of irreversibility is in finite-temperature-difference energy transfer, and
- power plants will normally be run in such a way as to maximize output power.

There is an economic issue here. Suppose that you are responsible for building and operating an electricity generating plant, and you must supply one megawatt of power (one million joules per second). If the operating expense of buying fuel is extremely high but capital investment in generators is inexpensive, it makes sense to build lots of generators and run them very slowly, nearly reversibly, to make the required megawatt of electric power.

If the operating expense of buying fuel is relatively low but capital investment in generators is expensive, it makes sense to build few generators and go for maximum power, even though the fuel is not used very efficiently in producing the megawatt of electric power.

Similar issues apply to refrigerators and heat pumps. Energy leaks into a refrigerator at some rate (joules per second) and must be removed by "pushing it thermally uphill" to a higher temperature, the temperature of the room. But in order to have a nonzero rate of energy flow from the food to the engine, the engine must reach an even lower temperature than the food, with entropy-producing energy transfer from the food into the engine. Also, in order for there to be a nonzero rate of energy flow from the engine to the room air, the engine must reach an even higher temperature than the air, with entropy-producing energy transfer from the engine into the air. Schematically the situation looks like Figure 12.41.

Often a refrigerator or freezer has exposed coils (a "heat exchanger") where the energy transfer occurs between the working substance circulating in the coils and the air. If you touch the coils, you find that they are indeed much hotter than room temperature, in order to drive a sufficiently high rate of energy transfer into the air. Notice not only that the heat exchanger must be hotter than the air, but that it is a rather large and costly device because it has to have a large surface area to get a large rate of energy transfer.

In the case of heat pumps, where the heat pump picks up low-temperature energy underground, the heat pump must reach an even lower temperature in order to get a nonzero rate of energy flow from the ground into the engine. Also, in order for there to be a nonzero rate of energy flow into the house, the engine must reach an even higher temperature than the air, with entropy-producing energy transfer from the engine into the air. The radiators (heat exchangers) in the house must be considerably hotter than the air. There is considerable expense in all the metal in the ground and in the house that enables adequate energy transfer rate. Schematically the situation looks like Figure 12.42.

So although a heat pump does have a theoretical advantage in warming a house in part from energy in the cold ground, the expense of the heat exchangers and the problems of going to rather low temperatures in the ground are practical limitations, especially in very cold climates. Heat pumps are more useful in climates where the winters are not too severe.

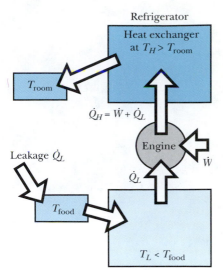

Figure 12.41 A refrigerator with nonzero cooling rate.

12.X.16 Curzon and Ahlborn give an example of a plant powered by a nuclear reactor, with high temperature 300° C and low temperature 25° C. In 12.X.12 on page 438 you calculated the reversible-engine zero-power efficiency to be 48%. Now calculate the non-reversible-engine maximum-power efficiency for this power plant and compare with the observed efficiency, which is 30%.

12.10 *DERIVATION: MAXIMUM-POWER EFFICIENCY

In this appendix we find the conditions under which the power output is maximized for an engine that connects to high- and low-temperature reservoirs through finite-temperature-difference connections.

Expressing the power in terms of T_1

We will express the power output in terms of T_1, the higher temperature to which the reversible engine is subjected. We assume that the engine, the part that runs between the temperatures T_1 and T_2, can be considered to be a reversible engine because there are no different-temperature energy transfers within it, and we're assuming that we can make mechanical friction negligibly small.

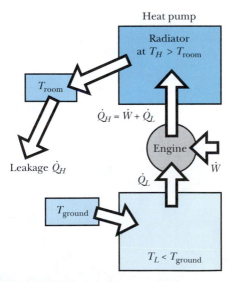

Figure 12.42 A heat pump with nonzero energy transfer rate due to a temperature difference.

? In terms of the temperatures T_1 and T_2, what does the second law of thermodynamics say about the relationship between $\dot{Q}_H$ and $\dot{Q}_L$?

Since we have a reversible engine, we know that $\dot{Q}_H / T_1 = \dot{Q}_L / T_2$. Using this relationship, we can write a formula for the energy transfer into the low-temperature sink in two different ways:

$$\dot{Q}_L = b(T_2 - T_L), \text{ and also } \dot{Q}_L = \frac{T_2}{T_1}\dot{Q}_H = \frac{T_2}{T_1}b(T_H - T_1)$$

Therefore we have

$$b(T_2 - T_L) = \frac{T_2}{T_1}b(T_H - T_1)$$

With a bit of manipulation we can solve this equation for T_2 in terms of T_1:

$$T_2 = \frac{T_L T_1}{2T_1 - T_H}$$

? For the inner, reversible engine, the input power is of course $\dot{Q}_H = b(T_H - T_1)$, and using the result for the efficiency of a reversible engine, show that

$$\dot{W} = b(T_H - T_1)\left(1 - \frac{T_2}{T_1}\right)$$

Using our solution for T_2 in terms of T_1, this becomes

$$\dot{W} = b(T_H - T_1)\left(1 - \frac{T_L}{2T_1 - T_H}\right)$$

Varying the thermal conduction to maximize the power

We consider the source and sink temperatures T_H and T_L to be fixed, so the only temperature that is unspecified in this equation is T_1, which we will vary until we maximize the output power $\dot{W}$. Physically, this corresponds to varying the parameter b that determines the energy transfer rate. As usual, we can maximize $\dot{W}$ by differentiating it with respect to T_1, then set this derivative equal to zero (corresponding to a maximum, where the slope is zero):

$$\frac{d\dot{W}}{dT_1} = b\left[-1 + \frac{T_L}{2T_1 - T_H} + (T_H - T_1)\frac{2T_L}{(2T_1 - T_H)^2}\right] = 0$$

Solving for T_1 and simplifying, we get a quadratic equation:

$$4T_1^2 - (4T_H)T_1 + (T_H^2 - T_L T_H) = 0$$

There are two solutions to the quadratic equation:

$$T_1 = \frac{T_H \pm \sqrt{T_L T_H}}{2}$$

Only the "+" sign makes physical sense in this situation, because if T_L were nearly as large as T_H, the solution with the "–" sign would make T_1 nearly 0 instead of lying between T_L and T_H.

We conclude that if $T_1 = (T_H + \sqrt{T_L T_H})/2$, the output power $\dot{W}$ will be maximized. Since $\dot{Q}_H = b(T_H - T_1)$, our value for T_1 determines the value of b ($= \sigma A/L$) that will make T_1 come out right to maximize the power output for a given $\dot{Q}_H$.

Finally we can plug our value for T_1 back into the expression for $\dot{W}$ to find out what the efficiency at maximum power turns out to be:

$$\frac{\dot{W}}{\dot{Q}_H} = \frac{\dot{W}}{b(T_H - T_1)} = 1 - \frac{T_L}{2T_1 - T_H} = 1 - \frac{T_L}{\sqrt{T_L T_H}}$$

Here is the final result, where we write "≤" because we have calculated an upper bound on the efficiency (having taken into account issues of energy transfer rate due to a temperature difference but not mechanical friction):

THE EFFICIENCY OF A MAXIMUM-POWER ENGINE

$$\frac{\dot{W}}{\dot{Q}_H} \leq 1 - \sqrt{\frac{T_L}{T_H}}$$

This gives us the fraction of the thermal input power that ends up as output mechanical power if we maximize the output power. The formula looks quite similar to the reversible-engine zero-power efficiency, but the square root makes a big difference.

12.11 SUMMARY

Mean free path d: $n[\pi(R + r)^2(d)] \approx 1$

Root-mean-square speed: $v_{rms} \equiv \sqrt{\overline{v^2}}$

Work done on a gas: $W = -\int_{V_1}^{V_2} P\,dV$

First law of thermodynamics: $\Delta E_{sys} = W + Q$

Molecular specific heat capacity at constant volume C_v: $Q = NC_V \Delta T$

Molecular specific heat capacity at constant pressure C_p: $Q = NC_p \Delta T$

Random walk root-mean-square displacement:
$$\Delta r_{rms} = (\sqrt{N})d = \sqrt{v}d\sqrt{t}$$

# of gas molecules hitting an area A per second =
$\frac{1}{4}nA\bar{v}$ (3-D; various speeds)

$$P = \frac{2}{3}n\left(\frac{\overline{p^2}}{2m}\right) = nkT$$

where $n = N/V$ (# of molecules per cubic meter)
For a multiatom gas molecule,

$$\Delta \bar{E}_{tot} = \Delta(\tfrac{3}{2}kT + \bar{E}_{rot} + \bar{E}_{vib})$$

$C_V = \frac{3}{2}k$ for a monatomic gas (He, etc.)

$C_V \geq \frac{3}{2}k$ for other gases (N$_2$, etc.)

$C_P = C_V + k$ (molecular specific heat capacity);

$C_P = C_V + R$ (molar specific heat capacity)

In a constant-temperature (isothermal) compression,

$$W_{by\ piston} = Q_{into\ surroundings} = NkT\ln\left(\frac{V_1}{V_2}\right)$$

In a no-Q (adiabatic) compression,

$PV^\gamma = $ constant, where $\gamma = C_p / C_v$, and also

$$T^{\left(\frac{C_V}{k}\right)}V = \text{constant}$$

$k = 1.38 \times 10^{-23}$ J/K $R = (6 \times 10^{23})k = 8.3$ J/K

At "standard temperature and pressure" (STP, which means a temperature of 0° Celsius or 273 Kelvin, and a pressure of 1 atmosphere), one mole of a gas occupies 22.4 liters (a liter is 1000 cubic centimeters).

Sea-level pressure (1 atm) is 10^5 N/m^2

Rate of energy transfer due to
a temperature difference:

$$\dot{Q} = \sigma A \frac{(T_H - T_L)}{L} \quad \text{(J/s)}$$

For a cyclic engine running between high temperature T_H and low temperature T_L we have

Entropy change of the universe for real engines

$$\Delta S_{\text{Universe}} = -\frac{Q_H}{T_H} + \frac{Q_L}{T_L} \geq 0$$

The efficiency of a Real engine

$$\text{Efficiency} = \frac{W}{Q_H} \leq 1 - \frac{T_L}{T_H}$$

The "=" sign applies only if the engine is reversible (extremely slow processes, no friction). A heat engine run in reverse is a refrigerator (or a heat pump).

If we maximize power output, we have

The efficiency of a maximum-power engine

$$\frac{\dot{W}}{\dot{Q}_H} \leq 1 - \sqrt{\frac{T_L}{T_H}}$$

12.12 REVIEW QUESTIONS

Gas leaks

12.RQ.17 Gas leaks at a rate L (in molecules per second) through a small circular hole. If the density of the gas is doubled, and the Kelvin temperature is doubled, and the radius of the hole is doubled, what is the new leak rate?

Mean free path

12.RQ.18 How does the mean free path of an atom in a gas change if the temperature is increased, with the volume kept constant?

12.RQ.19 What is the approximate time between collisions for one particular air molecule?

Speed of sound

12.RQ.20 How does the speed of sound in a gas change when you raise the temperature from 0° C to 20° C? Explain briefly.

Compressions and expansions

12.RQ.21 If you expand the volume of a gas containing N molecules to twice its original volume, while maintaining a constant temperature, how much energy transfer due to a temperature difference is there from the surroundings?

12.RQ.22 If you compress a volume of helium containing N atoms to half the original volume in a well-insulated cylinder, what is the ratio of the final pressure to the original pressure?

Thermal conduction

12.RQ.23 An aluminum bar 30 cm long and 3 cm by 4 cm on its sides is connected between two large metal blocks at temperatures of 135° C and 20° C, and energy is transferred from the hotter block to the cooler block at a rate $\dot{Q}_1$. If instead the two blocks were connected by an aluminum bar that is 20 cm long and 3 cm by 2 cm on its sides, what would be the energy transfer rate?

Engines

12.RQ.24 A (nearly) reversible engine is used to melt ice as well as do some useful work. If the engine does 1000 joules of work and dumps 400 joules into the ice, what is the temperature of the high-temperature source?

12.RQ.25 For engines where the high and low temperatures are not very different from each other, show that the efficiency of a maximum-power engine is about half of the efficiency of a reversible engine.

12.13 PROBLEMS

12.P.26 Leakage from a tank

In the example problem (page 414) we considered leakage from a flexible party balloon. If the leakage is from a rigid container (a metal storage tank, for example), the number of atoms per cubic meter, n, will decrease with time t. If the total volume of the tank is V, there are $N = nV$ atoms in the tank at any instant, so we can write the following "differential" equation (that is, an equation that involves derivatives):

$$\frac{d}{dt}(nV) = -\frac{1}{4}nA\bar{v}$$

This says "the rate of change of the number of atoms in the tank is equal to the (negative) of the rate at which atoms are leaving the tank."

(a) Show that the differential equation is satisfied if $n = n_{initial}e^{-\frac{A\bar{v}}{4V}t}$, where t is the time elapsed since the hole was made. Just plug this function of n, and its derivative, into the

equation and show that the two sides of the equation are equal for all values of the time t. Also show that the initial particle density n is equal to $n_{initial}$.

(b) Despite the fancy math, this solution is really only approximate, because the average speed isn't a constant but is decreasing. Suppose however that we use a heater to keep the container and the gas at a nearly constant temperature, so that the average speed does remain nearly constant. Suppose the (rigid) container is again a sphere 30 cm in diameter, with a circular hole 1 millimeter in diameter. About how long would it take for most of the helium to leak out? Explain your choice of what you mean by "most."

12.P.27 A helium leak

A rigid, thermally insulated container with a volume of 22.4 liters is filled with one mole of helium gas (4 grams per mole) at a temperature of 0 Celsius (273 K). The container is sitting in a room, surrounded by air at STP.

(a) Calculate the pressure inside the container in N/m^2.

(b) Calculate the root-mean-square average speed of the helium atoms.

(c) Now open a tiny square hole in the container, with area 10^{-8} m^2 (the hole is 0.1 mm on a side). After 5 seconds, how many helium atoms have left the container?

(d) Air molecules from the room enter the container through the hole during these 5 seconds. Which is greater, the number of air molecules that enter the container or the number of helium atoms that leave the container? Explain briefly.

(e) Does the pressure inside the container increase slightly, stay the same, or decrease slightly during these 5 seconds? Explain briefly. If you have to make any simplifying assumptions, state them clearly.

12.P.28 Gaseous diffusion

Natural uranium ore consists mostly of the isotope U-238 (92 protons and 146 neutrons), but 0.7% of the ore consists of the isotope U-235 (92 protons and 143 neutrons). Because only U-235 fissions in a reactor, industrial processes are used to enrich the uranium by enhancing the U-235 content.

One of the enrichment methods is "gaseous diffusion." The gas UF_6, uranium hexafluoride, is manufactured from supplies of natural uranium and fluorine (each of the six fluorine atoms has 9 protons and 10 neutrons). A container is filled with UF_6 gas. There are tiny holes in the container, and gas molecules leak through these holes into an adjoining container, where pumps sweep out the leaked gas.

(a) Explain why the gas that initially leaks into the second container has a slightly higher fraction of U-235 than is found in natural uranium.

(b) Estimate roughly the practical change in the concentration of U-235 that can be achieved in this single-stage separation process. Explain what approximations or simplifying assumptions you have made to obtain your estimate.

(c) A typical nuclear reactor requires uranium that has been enriched to the point where about 3% of the uranium is U-235. Estimate roughly the number of stages of gaseous diffusion required (that is, the number of times the gas must be allowed to leak from one container into another). Note that the effects are multiplicative.

This is why a practical gaseous diffusion plant has a large number of stages, each operating at high pressure, which makes this an expensive process. The first large gaseous diffusion plant was constructed during World War II at Oak Ridge, Tennessee, and used cheap Tennessee Valley Authority electricity.

12.P.29 Spacecraft emergency

You are on a spacecraft measuring 8 m by 3 m by 3 m when it is struck by a piece of space junk, leaving a circular hole of radius 4 mm, unfortunately in a place that can be reached only by making a time-consuming spacewalk. About how much time do you have to patch the leak? Explain what approximations you make in assessing the seriousness of the situation.

12.P.30 Meteor in air

A roughly spherical meteor made mainly of iron (density about 8 grams per cubic centimeter) is hurtling downward through the air at low altitude. At an instant when its speed is 10^4 m/s, calculate the approximate rate of change of the meteor's speed. Do the analysis for two different meteors—one with a radius of 10 meters and one with a radius of 100 meters.

Start from fundamental principles. Do not try to use some existing formula that applies to a very different situation. Follow the kind of *reasoning* used in this chapter, applied to the new situation, rather than trying to use the *results* of this chapter.

A major difference from our earlier analyses is that the meteor is traveling much faster than the average thermal speed of the air molecules, so it is a good approximation to consider the air molecules below the meteor to be essentially at rest, and to assume that no air molecules manage to catch up with the meteor and hit it from behind. The meteor drills a temporary hole in the atmosphere, a vacuum, that gets filled explosively by air rushing in after the meteor has passed.

There is good evidence that a very large meteor, perhaps 10 kilometers in diameter, hit the Earth near the Yucatan Peninsula in southern Mexico 65 million years ago and caused so much damage that the dinosaurs became extinct. See the excellent account in *T. rex and the Crater of Doom*, by Walter Alvarez (Princeton University Press, 1997). Alvarez is the geologist who made the first discoveries leading to our current understanding of this cataclysmic event.

12.P.31 Manipulating a gas

A horizontal cylinder 10 cm in diameter contains helium gas at room temperature and atmospheric pressure. A piston keeps the gas inside a region of the cylinder 20 cm long.

(a) If you *quickly* pull the piston outward a distance of 12 cm, what is the approximate temperature of the helium immediately afterwards? What approximations did you make?

(b) How much work did you do, including sign? (Note that you need to consider the outside air as well as the inside helium in calculating the amount of work *you* do.)

(c) Immediately after the pull, what force must you exert on the piston to hold it in position (with the helium enclosed in a volume that is 20+12 = 32 cm long)?

(d) You wait while the helium slowly returns to room temperature, maintaining the piston at its current location. After this wait, what force must you exert to hold the piston in position?

(e) Next, you very *slowly* allow the piston to move back into the cylinder, stopping when the region of helium gas is 25 cm long. What force must you exert to hold the piston at this position? What approximations did you make?

12.P.32 Pumping up a bicycle tire

Atmospheric pressure at sea level is about 10^5 N/m^2, which is about 15 pounds per square inch (psi). A bicycle tire typically is pumped up to 50 psi above atmospheric pressure (psi "gauge"), for an actual pressure of about 65 psi. In rapidly pumping up a bicycle tire, starting from atmospheric pressure, about how high does the temperature of the air rise? Explain what approximations and simplifying assumptions you make.

12.P.33 An engine containing an ideal gas

In one cycle of a reversible engine running between a high-temperature source and a low-temperature sink, a consequence of the second law of thermodynamics is that $Q_H/T_H = Q_L/T_L$. This result is independent of what kind of material the engine contains. It is instructive to check this general result for a specific model where we can calculate everything explicitly. Consider a reversible cycle of an ideal gas of N molecules, starting at high temperature T_H and volume V_1.

(1) Perform a constant-temperature (isothermal) expansion to volume V_2, and calculate the associated energy transfer Q_H into the gas.

(2) Next perform a no-Q (adiabatic) expansion to volume V_3 and temperature T_L.

(3) Next perform a constant-temperature (isothermal) compression to volume V_4, and calculate the associated energy transfer Q_L out of the gas.

(4) Finally perform a no-Q (adiabatic) compression back to the starting state, temperature T_H and volume V_1.

12.14 ANSWERS TO EXERCISES

12.X.1 (page 413) 6.3×10^{21} atoms per second

12.X.2 (page 414) T decreases because average v decreases.

12.X.3 (page 415) $L \approx 0.3 \times 10^{-10}$ m

12.X.4 (page 418) 1×10^5 N/m^2

12.X.5 (page 421) Box alone; buoyancy force is Mg upward.

12.X.6 (page 426) 1000 J; 1000 J; 0

12.X.7 (page 427) $\frac{3}{2}k + k = \frac{5}{2}k$

12.X.8 (page 428) $\left(\dfrac{PV}{Nk}\right)^{c_v/k} V = $ constant

$P^{c_v/k} V^{c_v/k+1} = $ new constant

Take (c_v/k) root:

$PV^{1+k/c_v} = PV^{(c_v+k)/c_v} = PV^{c_p/c_v}$

12.X.9 (page 431) 6 millimeters

12.X.10 (page 433) 25 seconds; entropy of Universe increases

12.X.11 (page 438) 0.27

12.X.12 (page 438) 0.48

12.X.13 (page 439) $\Delta S_{\text{Universe}} > 0$ implies $\dfrac{Q_L}{T_L} > \dfrac{Q_H}{T_H}$, which implies

$\dfrac{Q_L}{Q_H} > \dfrac{T_L}{T_H}$, **so**

$$W = Q_H - Q_L = Q_H\left(1 - \frac{Q_L}{Q_H}\right) < Q_H\left(1 - \frac{T_L}{T_H}\right)$$

12.X.14 (page 440) 3 watts; 53 watts

12.X.15 (page 440) 0.05 J

12.X.16 (page 443) 28%, which is close to but less than the observed 30%. This implies that the power plant is not run at maximum power.

APPENDIX A

BASIC DERIVATIVES

A.1 DERIVATIVES

In case you have not already studied calculus, or it has been a while since you last used calculus, we offer a calculation of some simple derivatives that are particularly useful in physics. You are probably familiar with the rate of change of velocity as the ratio of the velocity change $\Delta \vec{v}$ to the time interval Δt, $\Delta \vec{v}/\Delta t$. If we let the time interval Δt get very small, this ratio approaches what we call the "derivative" of the velocity with respect to the time:

$$\text{As } \Delta t \to 0, \ \frac{\Delta \vec{v}}{\Delta t} \to \frac{d\vec{v}}{dt}$$

More generally, we can approximate the rate of change of any physical quantity Q, as some other physical quantity R is varied, by the ratio $\Delta Q/\Delta R$, and if we consider a very small change ΔR, we approach the derivative of Q with respect to R:

$$\frac{\Delta Q}{\Delta R} \to \frac{dQ}{dR}$$

For concreteness, we'll take derivatives with respect to the time t, but the same principle applies if you take a derivative with respect to some other varying quantity, such as position x or y.

Derivative of a constant with respect to t

$$\frac{d(\text{constant})}{dt} = 0 \ \text{ because the change } \Delta(\text{constant}) = 0.$$

Derivative of t with respect to t

$$\frac{\Delta(t)}{\Delta t} = 1, \text{ no matter whether } \Delta t \text{ is large or small, so } \frac{d(t)}{dt} = 1$$

Derivative of t^2 with respect to t

The rate of change of t^2 with respect to t can be obtained starting from the fact that when t increases by an amount Δt, the quantity t^2 increases to $(t + \Delta t)^2$:

$$\frac{\Delta(t^2)}{\Delta t} = \frac{(t + \Delta t)^2 - t^2}{\Delta t}$$

$$= \frac{t^2 + 2t\Delta t + (\Delta t)^2 - t^2}{\Delta t}$$

$$= \frac{2t\Delta t + (\Delta t)^2}{\Delta t}$$

$$= 2t + \Delta t$$

This is nearly equal to $2t$ if Δt is very small compared to $2t$. Therefore we conclude this:

$$\frac{d(t^2)}{dt} = 2t$$

In a similar fashion it is possible to show the following:

$$\frac{d(t^n)}{dt} = nt^{n-1} \text{ for any constant value of } n \text{ (positive } or \text{ negative)}$$

Derivative with a constant multiplier

The derivative of a quantity that contains a constant multiplier C is simply C times the basic derivative:

$$\frac{\Delta(CQ)}{\Delta t} = \frac{C\Delta Q}{\Delta t} = C\frac{\Delta Q}{\Delta t}$$

so $\frac{d(CQ)}{dt} = C\frac{dQ}{dt}$

Derivative of sine and cosine

We can calculate the rate of change (the derivative) of the sine and cosine functions by using the trigonometric identities for the sines and cosines of sums of angles:

$$\sin(A + B) = \sin A \cos B + \cos A \sin B$$

$$\cos(A + B) = \cos A \cos B - \sin A \sin B$$

Here is the derivative of the sine:

$$\frac{\Delta(\sin t)}{\Delta t} = \frac{\sin(t + \Delta t) - \sin t}{\Delta t}$$

$$\frac{\Delta(\sin t)}{\Delta t} = \frac{\sin t \cos \Delta t + \cos t \sin \Delta t - \sin t}{\Delta t}$$

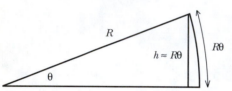

Figure 1.1 The sine of a small angle is approximately equal to the angle, measured in radians.

The cosine of a very small angle is approximately equal to 1, and the sine of a very small angle is approximately equal to the angle, measured in radians. You can see from Figure B1.1 that if the angle θ is small, the height h is approximately equal to the arc length $R\theta$ (with θ measured in radians), in which case $\sin(\theta) = h/R \approx \theta$. As an example, $30° = \pi/6$ radians = 0.524 radians, which is very close to $\sin(30°) = 0.5$. The approximation gets even better for smaller angles.

Therefore if Δt is very small, we have approximately

$$\frac{\Delta(\sin t)}{\Delta t} \approx \frac{\sin t + (\cos t)\Delta t - \sin t}{\Delta t} = \cos t$$

Therefore the derivative of the sine is a cosine. You can go through the same kind of reasoning to find that the derivative of the cosine is a negative sine. In summary:

$$\frac{d(\sin t)}{dt} = \cos t$$

$$\frac{d(\cos t)}{dt} = -\sin t$$

Do these derivatives make sense? Note that at $t = 0$, the function $\sin(t)$ is growing, and its derivative $\cos(0) = +1$, which makes sense. At $t = \pi/2$ (90°), the sine function reaches a maximum and is momentarily neither increasing nor decreasing, and its derivative $\cos(\pi/2) = 0$, which makes sense.

The chain rule

Often we need to calculate the rate of change of a function of some variable with respect to some *other* variable, such as differentiating a function of x with respect to t:

$$\frac{d[f(x)]}{dt}$$

In this example, suppose x changes by an amount Δx when t changes by an amount Δt. Then we have the following:

$$\frac{\Delta[f(x)]}{\Delta t} = \left\{\frac{\Delta[f(x)]}{\Delta x}\right\}\left\{\frac{\Delta x}{\Delta t}\right\}$$

$$\text{But } \frac{\Delta[f(x)]}{\Delta x} \rightarrow \frac{d[f(x)]}{dx}$$

$$\text{and } \frac{\Delta x}{\Delta t} \rightarrow \frac{dx}{dt}$$

As we let Δt get very small, we obtain the chain rule:

$$\frac{d[f(x)]}{dt} = \left\{\frac{d[f(x)]}{dx}\right\}\left\{\frac{dx}{dt}\right\}$$

Here is an example of the use of the chain rule in physics (ω is a constant):

$$\frac{d[\cos(\omega t)]}{dt} = \left\{\frac{d[\cos(\omega t)]}{d(\omega t)}\right\}\left\{\frac{d(\omega t)}{dt}\right\} = \{-\sin(\omega t)\}\{\omega\}$$

A.2 SELF-TEST

Here are some short questions on derivatives that you can use as a self-test. Try not to look back through this appendix as you work on these questions. If you find that you do need to look up something in order to answer a question, or if you find that your answer disagrees with the answers given at the end of this appendix, be sure to make a note of which question this was, to guide your further study and review.

2.X.1 What is the derivative of $4t^2$ with respect to t?

2.X.2 What is the rate of change of $(5x^{-1} + 6)$ as x varies?

2.X.3 Consider the behavior of the derivative of $\cos(t)$ at $t = 0$ and $t = \pi/2$. Do you get values that make sense, given how the cosine is changing near those times?

2.X.4 Calculate $\dfrac{d[\sin(3t)]}{dt}$.

A.3 ANSWERS TO SELF-TEST

A.2.X.1 (page 3) $8t$

A.2.X.2 (page 3) $-5x^{-2}$

A.2.X.3 (page 3) At $t = 0$, the cosine is at a maximum, so its derivative should be zero; indeed, $-\sin(0) = 0$.

At $t = \pi/2$, the cosine is decreasing, so its rate of change should be negative; indeed, $-\sin(\pi/2) = -1$.

A.2.X.4 (page 3) $3\cos(3t)$

Greek alphabet, and its uses in this course

Alpha	A		α	alpha particle (helium-4 nucleus)
Beta	B		β	
Gamma	Γ		γ	relativistic factor, photon
Delta	Δ	change of; small quantity of	δ	
Epsilon	E		ε	a small quantity
Zeta	Z		ζ	
Eta	H		η	
Theta	Θ		θ	angle
Iota	I		ι	
Kappa	K		κ	
Lambda	Λ		λ	wavelength
Mu	M	mega (10^6)	μ	micro (10^{-6}), muon, coefficient of friction
Nu	N		ν	neutrino; frequency ($= f$)
Xi	Ξ		ξ	
Omicron	O		o	
Pi	Π		π	circle: circumference/diameter
Rho	P		ρ	density, resistivity
Sigma	Σ	sum	σ	conductivity
Tau	T		τ	torque
Upsilon	Y		υ	
Phi	Φ	flux	ϕ	phase angle
Chi	X		χ	
Psi	Ψ		ψ	
Omega	Ω	# of microstates of a system, ohm, precession angular frequency	ω	angular frequency, angular speed

Units of measurement

kg = kilogram	m = meter	s = second	C = coulomb
N = newton ($kg \cdot m/s^2$)	J = joule ($N \cdot m$)	W = watt (J/s)	
A = ampere (C/s)	V = volt (J/C)	T = tesla (N/C)/(m/s)	eV = electron volt = 1.6×10^{-19} J

Important prefixes

pico (p) = 10^{-12}	nano (n) = 10^{-9}	micro (μ) = 10^{-6}	milli (m) = 10^{-3}
centi (c) = 10^{-2}	kilo (k) = 10^3	mega (M) = 10^6	giga (G) = 10^9

Unit conversions

1 kg = 2.2 pounds	1 inch = 2.54 cm	1 foot = 30.5 cm	1 liter = 1.1 quart
1 cal = 4.2 J	1 kcal (1 food calorie, 1 "large calorie") = 4.2×10^3 J		$T_{Kelvin} = T_{Celsius} + 273.15$